W9-DFQ-224

CONVERSION FACTORS

Length

1 m = 39.37 in. = 3.281 ft
1 in. = 2.54 cm
1 km = 0.621 mi
1 mi = 5280 ft = 1.609 km
1 light year (ly) = 9.461×10^{15} m
1 angstrom (Å) = 10^{-10} m

Mass

1 kg = 10^3 g = 6.85×10^{-2} slug
1 slug = 14.59 kg
1 u = 1.66×10^{-27} kg = 931.5 MeV/c^2

Time

1 min = 60 s
1 h = 3600 s
1 day = 8.64×10^4 s
1 yr = 365.242 days = 3.156×10^7 s

Volume

1 L = 1000 cm^3 = 3.531×10^{-2} ft^3
1 ft^3 = 2.832×10^{-2} m^3
1 gal = 3.786 L = 231 $in.^3$

Angle

$180° = \pi$ rad
1 rad = 57.30°
1° = 60 min = 1.745×10^{-2} rad

Speed

1 km/h = 0.278 m/s = 0.621 mi/h
1 m/s = 2.237 mi/h = 3.281 ft/s
1 mi/h = 1.61 km/h = 0.447 m/s = 1.47 ft/s

Force

1 N = 0.2248 lb = 10^5 dynes
1 lb = 4.448 N
1 dyne = 10^{-5} N = 2.248×10^{-6} lb

Work and energy

1 J = 10^7 erg = 0.738 ft · lb = 0.239 cal
1 cal = 4.186 J
1 ft · lb = 1.356 J
1 Btu = 1.054×10^3 J = 252 cal
1 J = 6.24×10^{18} eV
1 eV = 1.602×10^{-19} J
1 kWh = 3.60×10^6 J

Pressure

1 atm = 1.013×10^5 N/m^2 (or Pa) = 14.70 lb/$in.^2$
1 Pa = 1 N/m^2 = 1.45×10^{-4} lb/$in.^2$
1 lb/$in.^2$ = 6.895×10^3 N/m^2

Power

1 hp = 550 ft · lb/s = 0.746 kW
1 W = 1 J/s = 0.738 ft · lb/s
1 Btu/h = 0.293 W

College Physics

seventh edition

VOLUME 1 / CHAPTERS 1–14

Raymond A. Serway
Emeritus, James Madison University

Jerry S. Faughn
Emeritus, Eastern Kentucky University

Chris Vuille
Embry-Riddle Aeronautical University

Charles A. Bennett
The University of North Carolina, Asheville
Contributing Author, Media

THOMSON
BROOKS/COLE

Australia · Canada · Mexico · Singapore · Spain · United Kingdom · United States

THOMSON
BROOKS/COLE

Physics Acquisitions Editor: Chris Hall
Publisher: David Harris
Vice President, Editor-in-Chief, Sciences: Michelle Julet
Development Editor: Ed Dodd
Assistant Editor: Annie Mac
Editorial Assistant: Kyra Engelberg
Technology Project Manager: Sam Subity
Marketing Manager: Erik Evans, Julie Conover
Marketing Assistant: Leyla Jowza
Advertising Project Manager: Stacey Purviance
Project Manager, Editorial Production: Teri Hyde
Print/Media Buyer: Barbara Britton
Permissions Editor: Audrey Pettengill
Production Service: Progressive Publishing Alternatives
Text Designer: Patrick Devine
Photo Researcher: Jane Sanders
Copy Editor: Progressive Publishing Alternatives
Illustrator: Rolin Graphics, Progressive Information Technologies
Cover Designer: Patrick Devine
Cover Image: ML Sinibaldi/CORBIS
Cover Printer: Phoenix Color Corp., MD
Compositor: Progressive Information Technologies
Printer: R.R. Donnelley

COPYRIGHT © 2006 by Raymond A. Serway.

ALL RIGHTS RESERVED. No part of this work covered by the copyright hereon may be reproduced or used in any form or by any means—graphic, electronic, or mechanical, including but not limited to photocopying, recording, taping, Web distribution, information networks, or information storage and retrieval systems without the written permission of the publisher.

Printed in the United States of America
1 2 3 4 5 6 7 11 10 09 08 07

For more information about our products, contact us at:
Thomson Learning Academic Resource Center
1-800-423-0563

For permission to use material from this text or product, submit a request online at **http://www.thomsonrights.com**.
Any additional questions about permissions can be submitted by email to **thomsonrights@thomson.com.**

Thomson Brooks/Cole
10 Davis Drive
Belmont, CA 94002
USA

Library of Congress Control Number: 2004113839

Student Edition: ISBN 0-534-99918-2
Instructor's Edition: ISBN 0-534-99724-4

Asia
Thomson Learning
5 Shenton Way #01-01
UIC Building
Singapore 068808

Australia/New Zealand
Thomson Learning
102 Dodds Street
Southbank, Victoria 3006
Australia

Canada
Nelson
1120 Birchmount Road
Toronto, Ontario M1K 5G4
Canada

Europe/Middle East/Africa
Thomson Learning
High Holborn House
50/51 Bedford Row
London WC1R 4LR
United Kingdom

Latin America
Thomson Learning
Seneca, 53
Colonia Polanco
11560 Mexico D.F.
Mexico

Spain/Portugal
Paraninfo
Calle Magallanes, 25
28015 Madrid, Spain

Your quick start guide to PhysicsNow™

Welcome to PhysicsNow, your fully integrated system for physics tutorials and self-assessment on the web. To get started, just follow these simple instructions.

Your first visit to PhysicsNow

1. Go to **http://www.cp7e.com** and click the **Register** button.
2. The first time you visit, you will be asked to select your school. Choose your state from the drop-down menu, then type in your school's name in the box provided and click **Search**. A list of schools with names similar to what you entered will show on the right. Find your school and click on it.
3. On the next screen, enter the access code from the card that came with your textbook in the "Content or Course Access Code" box*. Enter your email address in the next box and click **Submit.**

   * PhysicsNow access codes may be purchased separately. Should you need to purchase an access code, go back to **http://www.cp7e.com** and click the **Buy** button.
4. On the next screen, choose a password and click **Submit.**

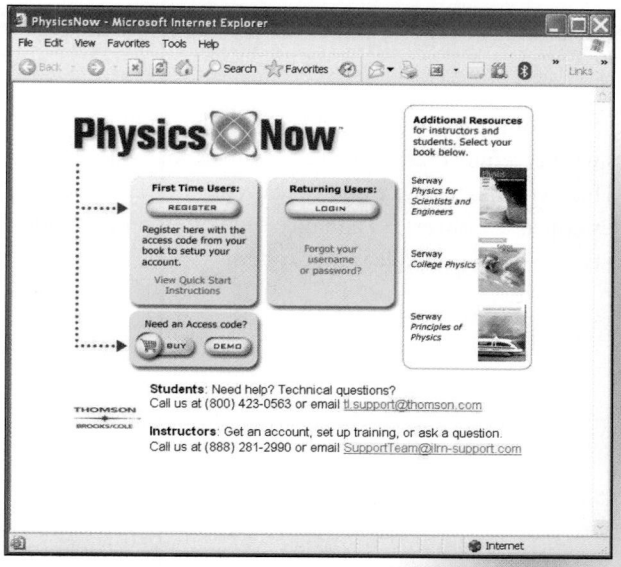

5. Lastly, fill out the registration form and click **Register and Enter iLrn.** This information will only be used to contact you if there is a problem with your account.
6. You should now see the **PhysicsNow** homepage. Select a chapter and begin!

   Note: Your account information will be sent to the email address that you entered in Step 3, so be sure to enter a valid email address. You will use your email address as your username the next time you login.

Second and later visits

1. Go to **http://www.cp7e.com** and click the **Login** button.
2. Enter your user name (the email address you entered when you registered) and your password and then click **Login.**

SYSTEM REQUIREMENTS:
(Please see the System Requirements link at **www.ilrn.com** *for complete list.)*
PC: Windows 98 or higher, Internet Explorer 5.5 or higher
Mac: OS X or higher, Mozilla browser 1.2.1 or higher

TECHNICAL SUPPORT:
For online help, click on **Technical Support** in the upper right corner of the screen, or contact us at:

1-800-423-0563 Monday–Friday • 8:30 A.M. to 6:00 P.M. EST

tl.support@thomson.com

Turn the page to learn more about **PhysicsNow** and how it can help you achieve success in your course!

THOMSON BROOKS/COLE™

What do you need to learn now?

Take charge of your learning with **PhysicsNow**™, a powerful student-learning tool for physics! This interactive resource helps you gauge your unique study needs, then gives you a *Personalized Learning Plan* that will help you focus in on the concepts and problems that will most enhance your understanding. With **PhysicsNow**, you have the resources you need to take charge of your learning!

The access code card included with this new copy of ***College Physics*** is your ticket to all of the resources in **PhysicsNow.** (See the previous page for login instructions.)

Interact at every turn with the POWER and SIMPLICITY of PhysicsNow!

PhysicsNow combines Serway and Faughn's best-selling *College Physics* with carefully crafted media resources that will help you learn. This dynamic resource and the Seventh Edition of the text were developed in concert, to enhance each other and provide you with a seamless, integrated learning system.

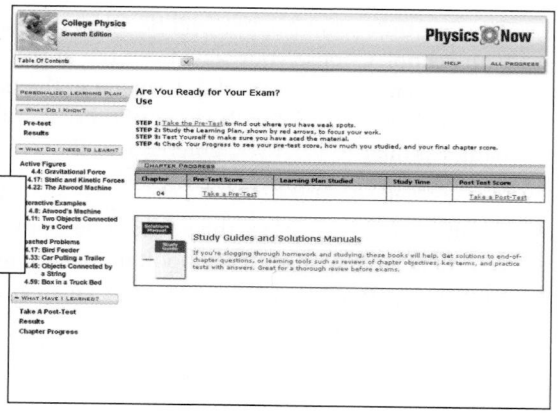

Physics Now™

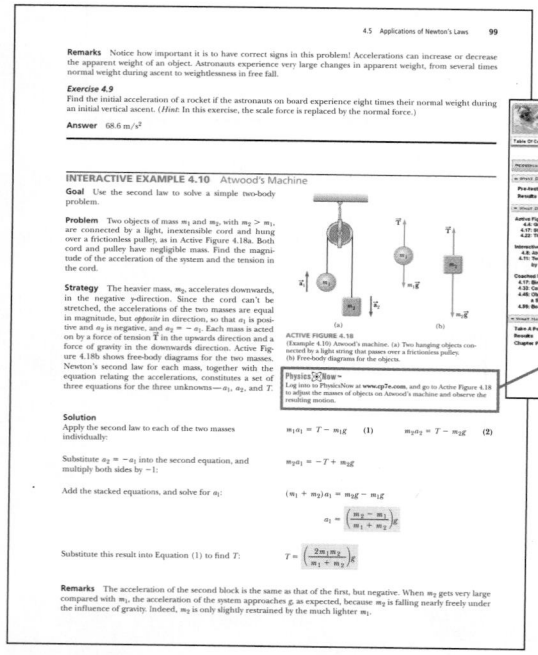

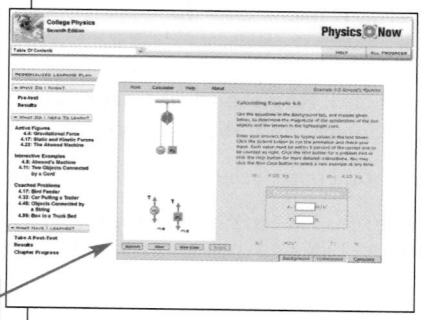

As you work through the text, you will see notes that direct you to the media-enhanced activities in **PhysicsNow.** This precise page-by-page integration means you'll spend less time flipping through pages or navigating websites looking for useful exercises. These multimedia exercises will make all the difference when you're studying and taking exams . . . after all, it's far easier to understand physics if it's seen in action, and PhysicsNow enables you to become a part of the action!

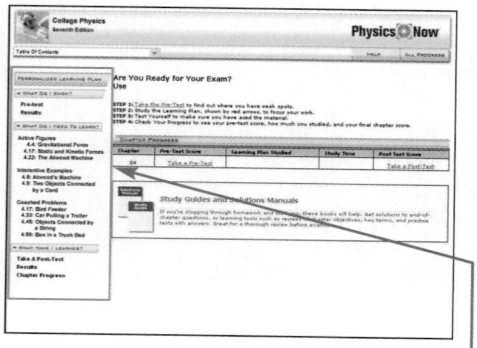

Begin at http://www.cp7e.com and build your own Personalized Learning Plan now!

Log into PhysicsNow at **http://www.cp7e.com** by using the free access code packaged with the text. You'll immediately notice the system's simple, browser-based format. You can build a complete *Personalized Learning Plan* for yourself by taking advantage of all three powerful components found on PhysicsNow:

- **What I Know**
- **What I Need to Learn**
- **What I've Learned**

The best way to maximize the system and optimize your time is to start by taking the *Pre-Test* ▶▶▶

What I Know

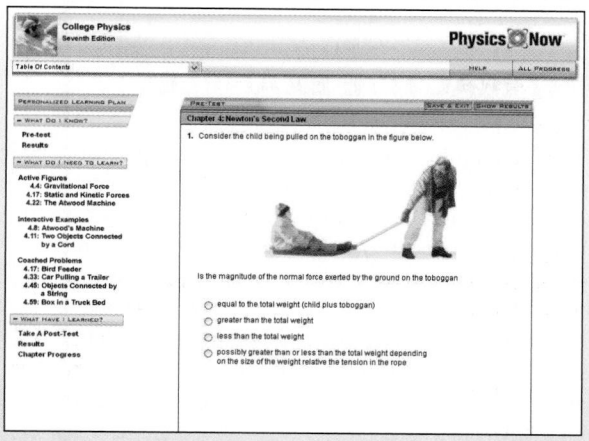

▲ You take a *Pre-Test* to measure your level of comprehension after reading a chapter. Each *Pre-Test* includes approximately 15 questions. The *Pre-Test* is your first step in creating your custom-tailored *Personalized Learning Plan.*

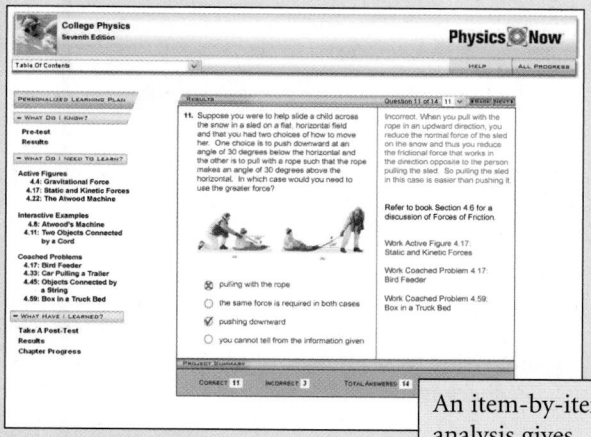

An item-by-item analysis gives you feedback on each of your answers.

▲ Once you've completed the "What I Know" *Pre-Test,* you are presented with a detailed *Personalized Learning Plan,* with text references that outline the elements you need to review in order to master the chapter's most essential concepts. This roadmap to concept mastery guides you to exercises designed to improve skills and to increase your understanding of the basic concepts.

At each stage, the *Personalized Learning Plan* refers to ***College Physics*** to reinforce the connection between text and technology as a powerful learning tool.

What I Need to Learn

Once you've completed the *Pre-Test,* you're ready to work through tutorials and exercises that will help you master the concepts that are essential to your success in the course.

ACTIVE FIGURES

A remarkable bank of more than 200 animated figures helps you visualize physics in action. Taken straight from illustrations in the text, these *Active Figures* help you master key concepts from the book. By interacting with the animations and accompanying quiz questions, you come to an even greater understanding of the concepts you need to learn from each chapter. ▼

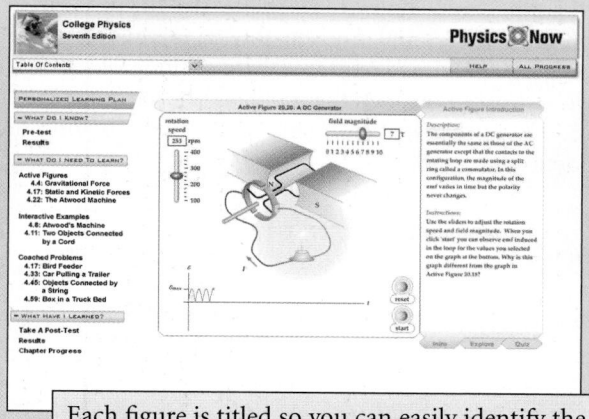

Each figure is titled so you can easily identify the concept you are seeing. The final tab features a *Quiz*. The *Explore* tab guides you through the animation so you understand what you should be seeing and learning.

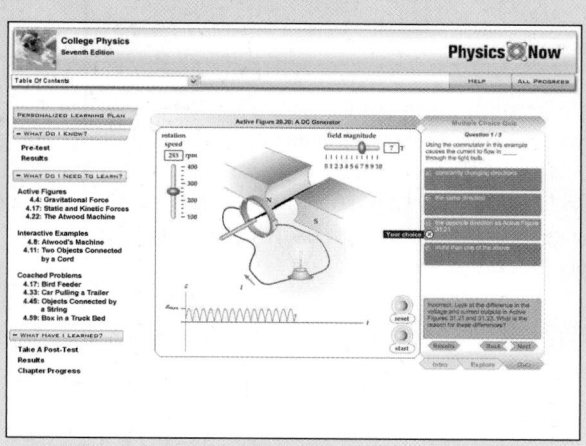

▲ The brief *Quiz* ensures that you mastered the concept played out in the animation—and gives you feedback on each response.

Continued on the next page

COACHED PROBLEMS

Engaging *Coached Problems* reinforce the lessons in the text by taking a step-by-step approach to problem-solving methodology. Each *Coached Problem* gives you the option of breaking down a problem from the text into steps with feedback to 'coach' you toward the solution. There are approximately five *Coached Problems* per chapter.

You can choose to work through the *Coached Problems* by inputting an answer directly or working in steps with the program. If you choose to work in steps, the problem is solved with the same problem-solving methodology used in ***College Physics*** to reinforce these critical skills. Once you've worked through the problem, you can click **Try Another** to change the variables in the problem for more practice.

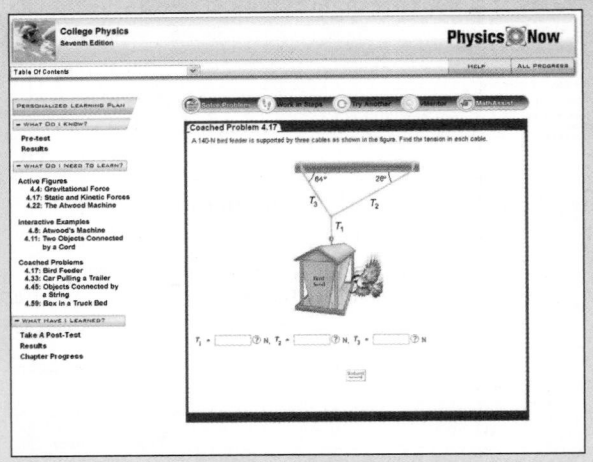

▲ Also built into each *Coached Problem* is a link to Brooks/Cole's exclusive **vMentor**™ web-based tutoring service site that lets you interact directly with a live physics tutor. If you're stuck on math, a *MathAssist* link on each *Coached Problem* launches tutorials on math specific to that problem.

INTERACTIVE EXAMPLES

You'll strengthen your problem-solving and visualization skills with *Interactive Examples*. Extending selected examples from the text, *Interactive Examples* utilize the proven and trusted problem-solving methodology presented in ***College Physics.*** These animated learning modules give you all the tools you need to solve a problem type—you're then asked to apply what you have learned to different scenarios. You will find approximately two *Interactive Examples* for each chapter of the text.▼

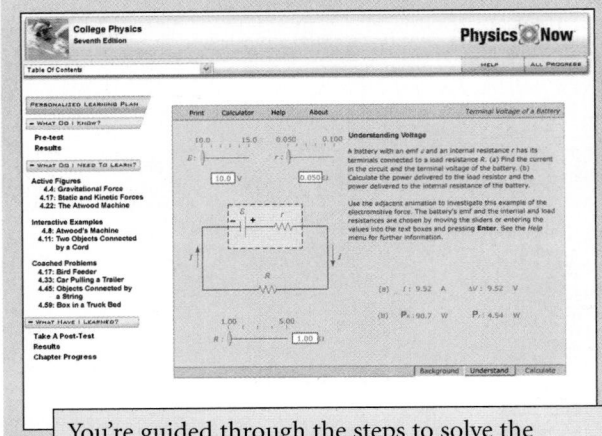

You're guided through the steps to solve the problem and then asked to input an answer in a simulation to see if your result is correct. Feedback is instantaneous.

What I've Learned

► After working through the problems highlighted in your *Personalized Learning Plan,* you move on to a *Post Test,* about 15 questions per chapter

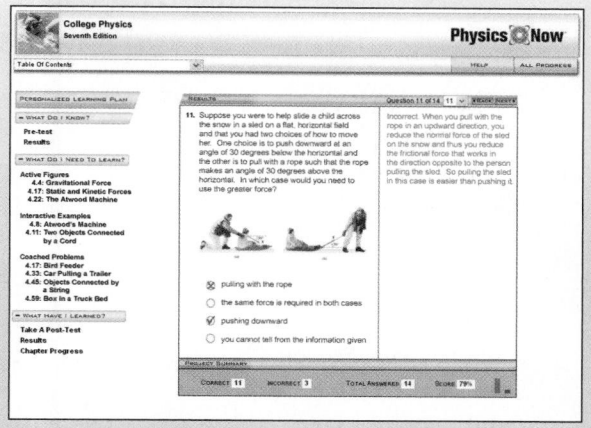

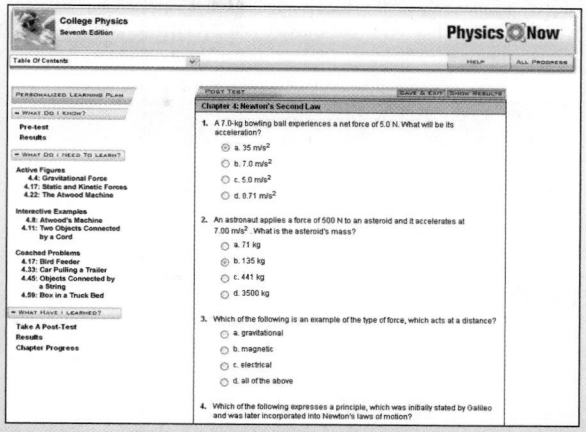

◄ Once you've completed the *Post Test*, you receive your percentage score and specific feedback on each answer. The *Post Tests* give you a new set of questions with each attempt, so you can take them over and over as you continue to build your knowledge and skills and master concepts.

Also available to help you succeed in your course

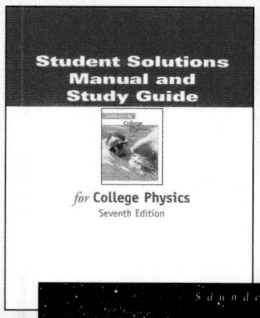

Student Solutions Manual and Study Guide
Volume I (Ch. 1–14) ISBN: 0-534-99920-4
Volume II (Ch. 15–30) ISBN: 0-534-99930-1
These manuals contain detailed solutions to approximately 12 problems per chapter. These problems are indicated in the textbook with boxed problem numbers. Each manual also features a skills section, important notes from key sections of the text, and a list of important equations and concepts.

Core Concepts in College Physics CD-ROM, Version 2.0
ISBN: 0-03-033701-1
Explore the core of physics with this powerful CD-ROM/workbook program! Content screens provide in-depth coverage of abstract and often difficult principles, building connections between physical concepts and mathematics. The presentation contains more than 350 movies—both animated and live video—including laboratory demonstrations, real-world examples, graphic models, and step-by-step explanations of essential mathematics. An accompanying workbook contains practical physics problems directly related to the presentation, along with worked solutions. Package includes three discs and a workbook.

Welcome to your MCAT Test Preparation Guide

The **MCAT Test Preparation Guide** makes your copy of *College Physics,* **Seventh Edition,** the most comprehensive MCAT study tool and classroom resource in introductory physics. The grid, which begins below and continues on the next two pages, outlines twelve concept-based **study courses** for the physics part of your MCAT exam. Use it to prepare for the MCAT, class tests, and your homework assignments.

Vectors

Skill Objectives: To calculate distance, angles between vectors, and magnitudes.

Review Plan:

Distance and Angles:
- Chapter 1, Sections 1.7, 1.8
- Active Figure 1.6
- Chapter Problems 35, 39, 42

Using Vectors:
- Chapter 3, Sections 3.1, 3.2
- Quick Quizzes 3.1–3.3
- Examples 3.1–3.3
- Active Figure 3.3
- Chapter Problems 4, 11, 19

Motion

Skill Objectives: To understand motion in two dimensions, to calculate speed and velocity, centripetal acceleration, and acceleration in free fall problems.

Review Plan:

Motion in 1 Dimension:
- Chapter 2, Sections 2.1–2.6
- Quick Quizzes 2.1–2.8
- Examples 2.1–2.10
- Active Figure 2.15
- Chapter Problems 1, 8, 17, 21, 29, 33, 38, 47, 51

Motion in 2 Dimensions:
- Chapter 3, Sections 3.3, 3.4
- Quick Quizzes 3.4–3.7
- Examples 3.4–3.8
- Active Figures 3.14, 3.15
- Chapter Problems 27, 33

Centripetal Acceleration:
- Chapter 7, Section 7.4
- Quick Quizzes 7.6, 7.7
- Example 7.6

Force

Skill Objectives: To know and understand Newton's Laws, to calculate resultant forces, and weight.

Review Plan:

Newton's Laws:
- Chapter 4, Sections 4.1–4.4
- Quick Quizzes 4.1–4.5
- Examples 4.1–4.4
- Active Figure 4.6
- Chapter Problems 5, 7, 11

Resultant Forces:
- Chapter 4, Section 4.5
- Quick Quizzes 4.6, 4.7
- Examples 4.7, 4.9, 4.10
- Chapter Problems 17, 25, 33

Equilibrium

Skill Objectives: To calculate momentum and impulse, center of gravity, and torque.

Review Plan:

Momentum:
- Chapter 6, Sections 6.1–6.3
- Quick Quizzes 6.2–6.6
- Examples 6.1–6.4, 6.6
- Active Figures 6.7, 6.10, 6.13
- Chapter Problems 7, 16, 21

Torque:
- Chapter 8, Sections 8.1–8.4
- Examples 8.1–8.7
- Chapter Problems 5, 9

Work

Skill Objectives: To calculate friction, work, kinetic energy, potential energy, and power.

Review Plan:

Friction:
- Chapter 4, Section 4.6
- Quick Quizzes 4.8–4.10
- Active Figure 4.19

Work:
- Chapter 5, Section 5.1
- Quick Quiz 5.1
- Example 5.1
- Active Figure 5.5
- Chapter Problems 11, 17

Energy:
- Chapter 5, Sections 5.2, 5.3
- Examples 5.4, 5.5
- Quick Quizzes 5.2, 5.3

Power:
- Chapter 5, Section 5.6
- Examples 5.12, 5.13

Matter

Skill Objectives: To calculate pressure, density, specific gravity, and flow rates.

Review Plan:

Properties:
- Chapter 9, Sections 9.1–9.3
- Quick Quiz 9.1
- Examples 9.1, 9.3, 9.4
- Active Figure 9.3
- Chapter Problem 7

Pressure:
- Chapter 9, Sections 9.3–9.6
- Quick Quizzes 9.2–9.6
- Examples 9.4–9.9
- Active Figures 9.19, 9.20
- Chapter Problems 13, 23, 39

Flow rates:
- Chapter 9, Sections 9.7, 9.8
- Quick Quiz 9.7
- Examples 9.11–9.14
- Chapter Problem 42

Waves

Skill Objectives: To understand interference of waves, to calculate basic properties of waves, properties of springs, and properties of pendulums.

Review Plan:

Wave Properties:
- Chapters 13, Sections 13.1–13.4, 13.7–13.11
- Quick Quizzes 13.1–13.6
- Examples 13.1, 13.6, 13.8–13.10
- Active Figures 13.1, 13.8, 13.12, 13.13, 13.24, 13.26, 13.32, 13.33, 13.34, 13.35
- Chapter Problems 9, 15, 23, 29, 41, 47, 53

Pendulum:
- Chapter 13, Sections 13.5
- Quick Quizzes 13.7–13.9
- Example 13.7
- Active Figures 13.15, 13.16
- Chapter Problem 35

Sound

Skill Objectives: To understand interference of waves, to calculate properties of waves, the speed of sound, Doppler shifts, and intensity.

Review Plan:

Sound Properties:
- Chapter 14, Sections 14.1–14.4, 14.6
- Quick Quizzes 14.1, 14.2
- Examples 14.1, 14.2, 14.4, 14.5
- Active Figures 14.6, 14.11
- Chapter Problems 7, 13,25

Interference/Beats:
- Chapter 14, Sections 14.7, 14.8, 14.11
- Quick Quiz 14.7
- Examples 14.6, 14.11
- Active Figures 14.18, 14.25
- Chapter Problems 33, 37, 51

Light

Skill Objectives: To understand mirrors and lenses, to calculate the angles of reflection, to use the index of refraction, and to find focal lengths.

Review Plan:

Reflection and Refraction:
- Chapter 22, Sections 22.1–22.4
- Quick Quizzes 22.2–22.4
- Examples 22.1–22.4
- Active Figures 22.4, 22.6, 22.7
- Chapter Problems 9, 17, 21, 25

Mirrors and Lenses:
- Chapter 23, Sections 23.1–23.6
- Quick Quizzes 23.1, 23.2, 23.4–23.6
- Examples 23.7, 23.8, 23.9
- Active Figures 23.2, 23.16, 23.25
- Chapter Problems 25, 29, 33, 39

Electrostatics

Skill Objectives: To understand and calculate the electric field, the electrostatic force, and the electric potential.

Review Plan:

Coulomb's Law:
- Chapter 15, Sections 15.1–15.3
- Quick Quiz 15.2
- Examples 15.1–15.3
- Active Figure 15.6
- Chapter Problems 11

Electric Field:
- Chapter 15, Sections 15.4, 15.5
- Quick Quizzes 15.3–15.6
- Examples 15.4, 15.5
- Active Figures 15.11, 15.16
- Chapter Problems 19, 23, 27

Potential:
- Chapter 16, Sections 16.1–16.3
- Quick Quizzes 16.1–16.5
- Examples 16.1, 16.4
- Active Figure 16.7
- Chapter Problems 1, 7, 15

Circuits

Skill Objectives: To understand and calculate current, resistance, voltage, power, and energy, and to use circuit analysis.

Review Plan:

Ohm's Law:
- Chapter 17, Sections 17.1–17.4
- Quick Quizzes 17.1, 17.3
- Examples 17.1, 17.3
- Chapter Problem 15

Power and energy:
- Chapter 17, Section 17.8
- Quick Quizzes 17.6–17.8
- Example 17.6
- Active Figure 17.11
- Chapter Problem 36

Circuits:
- Chapter 18, Sections 18.2, 18.3
- Quick Quizzes 18.1–18.5
- Examples 18.1–18.3
- Active Figures 18.2, 18.6
- Chapter Problems 4, 13

Atoms

Skill Objectives: To calculate half-life, and to understand decay processes and nuclear reactions.

Review Plan:

Atoms:
- Chapter 29, Sections 29.1, 29.2

Radioactive Decay:
- Chapter 29, Sections 29.3–29.5
- Examples 29.3, 29.4, 29.7
- Active Figures 29.6, 29.7
- Chapter Problems 15, 21, 29, 35

Nuclear reactions:
- Chapter 29, Sections 29.6
- Quick Quiz 29.4
- Examples 29.8, 29.9
- Chapter Problems 41, 45

WE DEDICATE THIS BOOK

To our wives, children, grandchildren, and relatives who have provided so much love, support, and understanding through the years, and to the students for whom this book was written.

Contents Overview

VOLUME 1

PART 1 **Mechanics**

1 Introduction 1
2 Motion in One Dimension 23
3 Vectors and Two-Dimensional Motion 53
4 The Laws of Motion 81
5 Energy 118
6 Momentum and Collisions 160
7 Rotational Motion and the Law of Gravity 189
8 Rotational Equilibrium and Rotational Dynamics 226
9 Solids and Fluids 266

PART 2 **Thermodynamics**

10 Thermal Physics 321
11 Energy in Thermal Processes 352
12 The Laws of Thermodynamics 386

PART 3 **Vibrations and Waves**

13 Vibrations and Waves 424
14 Sound 458

APPENDIX A Mathematical Review A.1
APPENDIX B An Abbreviated Table of Isotopes A.9
APPENDIX C Some Useful Tables A.14
APPENDIX D Si Units A.16

Answers to Quick Quizzes and Odd-numbered Conceptual Questions and Problems A.17

Index I.1

Contents

Volume 1

Part 1: Mechanics

Chapter 1
Introduction 1
1.1 Standards of Length, Mass, and Time 1
1.2 The Building Blocks of Matter 4
1.3 Dimensional Analysis 5
1.4 Uncertainty in Measurement and Significant Figures 7
1.5 Conversion of Units 9
1.6 Estimates and Order-of-Magnitude Calculations 11
1.7 Coordinate Systems 13
1.8 Trigonometry 14
1.9 Problem-Solving Strategy 16
Summary 17

Chapter 2
Motion in One Dimension 23
2.1 Displacement 24
2.2 Velocity 25
2.3 Acceleration 30
2.4 Motion Diagrams 33
2.5 One-Dimensional Motion with Constant Acceleration 35
2.6 Freely Falling Objects 41
Summary 45

Chapter 3
Vectors and Two-Dimensional Motion 53
3.1 Vectors and Their Properties 53
3.2 Components of a Vector 56
3.3 Displacement, Velocity, and Acceleration in Two Dimensions 59
3.4 Motion in Two Dimensions 60
3.5 Relative Velocity 68
Summary 72

Chapter 4
The Laws of Motion 81
4.1 Forces 81
4.2 Newton's First Law 83
4.3 Newton's Second Law 84
4.4 Newton's Third Law 90
4.5 Applications of Newton's Laws 92
4.6 Forces of Friction 100
Summary 107

Chapter 5
Energy 118
5.1 Work 118
5.2 Kinetic Energy and the Work–Energy Theorem 123
5.3 Gravitational Potential Energy 126
5.4 Spring Potential Energy 134
5.5 Systems and Energy Conservation 140
5.6 Power 142
5.7 Work Done by a Varying Force 146
Summary 148

Chapter 6
Momentum and Collisions 160
6.1 Momentum and Impulse 160
6.2 Conservation of Momentum 166
6.3 Collisions 168
6.4 Glancing Collisions 175
6.5 Rocket Propulsion 177
Summary 180

Chapter 7
Rotational Motion and the Law of Gravity 189
7.1 Angular Speed and Angular Acceleration 189
7.2 Rotational Motion Under Constant Angular Acceleration 193
7.3 Relations Between Angular and Linear Quantities 194
7.4 Centripetal Acceleration 198
7.5 Newtonian Gravitation 206
7.6 Kepler's Laws 214
Summary 217

Chapter 8
Rotational Equilibrium and Rotational Dynamics 226
8.1 Torque 227
8.2 Torque and the Two Conditions for Equilibrium 230
8.3 The Center of Gravity 232
8.4 Examples of Objects in Equilibrium 234
8.5 Relationship Between Torque and Angular Acceleration 237
8.6 Rotational Kinetic Energy 244
8.7 Angular Momentum 247
Summary 251

Chapter 9
Solids and Fluids 266
9.1 States of Matter 266
9.2 The Deformation of Solids 268
9.3 Density and Pressure 274
9.4 Variation of Pressure with Depth 277
9.5 Pressure Measurements 281
9.6 Buoyant Forces and Archimedes's Principle 282
9.7 Fluids in Motion 288
9.8 Other Applications of Fluid Dynamics 294
9.9 Surface Tension, Capillary Action, and Viscous Fluid Flow 298
9.10 Transport Phenomena 305
Summary 309

Part 2: Thermodynamics

Chapter 10
Thermal Physics 321
10.1 Temperature and the Zeroth Law of Thermodynamics 322
10.2 Thermometers and Temperature Scales 323
10.3 Thermal Expansion of Solids and Liquids 327
10.4 Macroscopic Description of an Ideal Gas 334
10.5 The Kinetic Theory of Gases 339
Summary 344

Chapter 11
Energy in Thermal Processes 352
11.1 Heat and Internal Energy 352
11.2 Specific Heat 355
11.3 Calorimetry 357
11.4 Latent Heat and Phase Change 359
11.5 Energy Transfer 365
11.6 Global Warming and Greenhouse Gases 375
Summary 377

Chapter 12
The Laws of Thermodynamics 386
12.1 Work in Thermodynamic Processes 386
12.2 The First Law of Thermodynamics 389
12.3 Heat Engines and the Second Law of Thermodynamics 399
12.4 Entropy 408
12.5 Human Metabolism 413
Summary 416

Part 3: Vibrations and Waves

Chapter 13
Vibrations and Waves 424
13.1 Hooke's Law 424
13.2 Elastic Potential Energy 427
13.3 Comparing Simple Harmonic Motion with Uniform Circular Motion 432
13.4 Position, Velocity, and Acceleration as a Function of Time 435
13.5 Motion of a Pendulum 438
13.6 Damped Oscillations 441
13.7 Waves 441
13.8 Frequency, Amplitude, and Wavelength 444
13.9 The Speed of Waves on Strings 446
13.10 Interference of Waves 447
13.11 Reflection of Waves 448
Summary 449

Chapter 14
Sound 458
14.1 Producing a Sound Wave 458
14.2 Characteristics of Sound Waves 459
14.3 The Speed of Sound 461
14.4 Energy and Intensity of Sound Waves 463
14.5 Spherical and Plane Waves 465

14.6 The Doppler Effect **467**
14.7 Interference of Sound Waves **473**
14.8 Standing Waves **475**
14.9 Forced Vibrations and Resonance **479**
14.10 Standing Waves in Air Columns **480**
14.11 Beats **484**
14.12 Quality of Sound **486**
14.13 The Ear **487**
Summary **489**

Appendix A
Mathematical Review A.1

Appendix B
An Abbreviated Table of Isotopes A.9

Appendix C
Some Useful Tables A.14

Appendix D
SI Units A.16

Answers to Quick Quizzes and Odd-numbered Conceptual Questions and Problems A.17

Index I.1

About the Authors

Raymond A. Serway received his doctorate at Illinois Institute of Technology and is Professor Emeritus at James Madison University. In 1990, he received the Madison Scholar Award at James Madison University, where he taught for 17 years. Dr. Serway began his teaching career at Clarkson University, where he conducted research and taught from 1967 to 1980. He was the recipient of the Distinguished Teaching Award at Clarkson University in 1977 and of the Alumni Achievement Award from Utica College in 1985. As Guest Scientist at the IBM Research Laboratory in Zurich, Switzerland, he worked with K. Alex Müller, 1987 Nobel Prize recipient. Dr. Serway also was a visiting scientist at Argonne National Laboratory, where he collaborated with his mentor and friend, Sam Marshall. In addition to earlier editions of this textbook, Dr. Serway is the co-author of *Physics for Scientists and Engineers,* Sixth Edition; *Principles of Physics,* Fourth Edition; and *Modern Physics,* Third Edition. He also is the author of the high-school textbook *Physics,* published by Holt, Rinehart, & Winston. In addition, Dr. Serway has published more than 40 research papers in the field of condensed matter physics and has given more than 70 presentations at professional meetings. Dr. Serway and his wife Elizabeth enjoy traveling, golfing, and spending quality time with their four children and six grandchildren.

Jerry S. Faughn earned his doctorate at the University of Mississippi. He is Professor Emeritus and former Chair of the Department of Physics and Astronomy at Eastern Kentucky University. Dr. Faughn has also written a microprocessor interfacing text for upper-division physics students. He is co-author of a non-mathematical physics text and a physical science text for general education students, and (with Dr. Serway) the high-school textbook *Physics,* published by Holt, Reinhart, & Winston. He has taught courses ranging from the lower division to the graduate level, but his primary interest is in students just beginning to learn physics. Dr. Faughn has a wide variety of hobbies, among which are reading, travel, geneology, and old-time radio. His wife Mary Ann is an avid gardener, and he contributes to her efforts by staying out of the way. His daughter Laura is in family practice and his son David is an attorney.

Chris Vuille is an associate professor of physics at Embry-Riddle Aeronautical University, Daytona Beach, Florida, the world's premier institution for aviation higher education. He received his doctorate in physics at the University of Florida in 1989, moving to Daytona after a year at ERAU's Prescott, Arizona campus. While he has taught courses at all levels, including post-graduate, his primary interest has been the delivery of introductory physics. He has received several awards for teaching excellence, including the Senior Class Appreciation Award (three times). He conducts research in general relativity and quantum theory, and was a participant in the JOVE program, a special three-year NASA grant program during which he studied neutron stars. His work has appeared in a number of scientific journals, and he has been a featured science writer in Analog Science Fiction/Science Fact magazine. Dr. Vuille enjoys tennis, lap swimming, guitar and classical piano, and is a former chess champion of St. Petersburg and Atlanta. In his spare time he writes science fiction and goes to the beach. His wife, Dianne Kowing, is an optometrist for a local VA clinic. Teen daughter Kira Vuille-Kowing is an accomplished swimmer, violinist, and mall shopper. He has two sons, twelve-year-old Christopher, a cellist and electronic game expert, and three-year-old James, master of tinkertoys.

Charles A. Bennett received his doctorate at North Carolina State University, and is Professor of Physics at the University of North Carolina at Asheville. His research interests include quantum and physical optics, and laser applications in environmental and fusion energy research. He has collaborated with Oak Ridge National Laboratory since 1983, where he is currently an adjunct research and development associate of the Advanced Laser and Optical Technology and Development group. In addition to his work in optics, Dr. Bennett has a long record of innovation in educational technology, particularly in the integration of active media into on-line homework. He is a past director of the UNCA Center for Teaching and Learning, and has received UNCA's most prestigious recognition for scholarship: the Ruth and Leon Feldman Professorship for 1996–1997. He is an avid hiker, and enjoys spending time with his wife Karen and four children.

Preface

College Physics is written for a one-year course in introductory physics usually taken by students majoring in biology, the health professions, and other disciplines including environmental, earth, and social sciences, and technical fields such as architecture. The mathematical techniques used in this book include algebra, geometry, and trigonometry, but not calculus.

The main objectives of this introductory textbook are twofold: to provide the student with a clear and logical presentation of the basic concepts and principles of physics, and to strengthen an understanding of the concepts and principles through a broad range of interesting applications to the real world. To meet these objectives, we have emphasized sound physical arguments and problem-solving methodology. At the same time, we have attempted to motivate the student through practical examples that demonstrate the role of physics in other disciplines.

This textbook, which covers the standard topics in classical physics and 20th-century physics, is divided into six parts. Part I (Chapters 1–9) deals with Newtonian mechanics and the physics of fluids; Part II (Chapters 10–12) is concerned with heat and thermodynamics; Part III (Chapters 13–14) covers wave motion and sound; Part IV (Chapters 15–21) develops the concepts of electricity and magnetism; Part V (Chapters 22–25) treats the properties of light and the field of geometric and wave optics; and Part VI (Chapters 26–30) provides an introduction to special relativity, quantum physics, atomic physics, and nuclear physics.

CHANGES TO THE SEVENTH EDITION

A number of new features, changes, and improvements have been added to this edition. Based on comments from users of the sixth edition and reviewers' suggestions, a major effort was made to improve clarity of presentation, precision of language, and accuracy throughout. The new pedagogical features added to this edition are based on current trends in science education. The following represent the major changes in the seventh edition.

Pedagogical Changes

- **Examples** All in-text worked examples have been reconstituted for the seventh edition, and now are presented in a two-column format that better aids student learning and helps to better reinforce physical concepts. Special care has been taken in this new edition to present a range of levels within each chapter's collection of worked examples, so that students are better prepared to solve the end-of-chapter problems. The examples are set off from the text for ease of location and are given titles to describe their content. All worked examples now include the following parts:

  (a) **Goal,** which describes the physical concepts being explored within the Worked Example.
  (b) **Problem,** which presents the problem itself.
  (c) **Strategy,** which helps students analyze the problem and create a framework for working out the solution.
  (d) **Solution,** which uses a two-column format that gives the explanation for each step of the solution in the left-hand column, while giving each accompanying mathematical step in the right-hand column. This layout facilitates matching the idea with its execution, and helps students learn how to organize their work. Another benefit: students can easily use this format as a training tool, covering up the solution on the right and solving the problem using the comments on the left as a guide.

(e) **Remarks,** which follow each Solution and highlight some of the underlying concepts and methodology used in arriving at a correct solution. In addition, the remarks are often used to put the problem into a larger, real world context.

(f) **Exercise/Answer** By popular demand of users and reviewers, every worked example in the new edition is now followed immediately by an exercise with an answer. These exercises allow the students to reinforce their understanding by working a similar or related problem, with the answers giving them instant feedback. At the option of the instructor, the exercises can also be assigned as homework. Students who work through these exercises on a regular basis will find the end-of-chapter problems less intimidating.

- ***PhysicsNow*™** Directly links with *College Physics, Seventh Edition. PhysicsNow*™ and *College Physics, Seventh Edition* were built in concert to enhance each other and provide a seamless, integrated learning system. Throughout the text, readers will see **Physics Now**™ icons that direct them to media-enhanced activities at the *PhysicsNow*™ Web site **(www.cp7e.com)**. The precise page-by-page integration means professors and students will spend less time flipping through pages and navigating Web sites for useful exercises.
- **Active Figures** Many diagrams from the text have been animated to form Active Figures, part of the new *PhysicsNow*™ integrated Web-based learning system. There are over 50 Active Figures in Volume 1, available at **www.cp7e.com.** By visualizing phenomena and processes that cannot be fully represented on a static page, students greatly increase their conceptual understanding. Active Figures are easily identified by their red figure legend and **Physics Now**™ icon, and the caption after the icon describes briefly the nature and contents of the animation. In addition to viewing animations of the figures, students can change variables to see the effects, conduct suggested explorations of the principles involved in the figure, and take and receive feedback on quizzes related to the figure.
- **Interactive Examples** Eighteen of the worked examples in Volume 1 have been identified as interactive. As part of the *PhysicsNow*™ Web-based learning system, students can engage in an extension of the problem solved in the example. This often includes elements of both visualization and calculation, and may also involve prediction and intuition-building. Interactive Examples are available at **www.cp7e.com.**
- **Quick Quizzes** All of the Quick Quizzes in this edition have been cast in an objective format, including multiple choice, true-false, and ranking. Quick Quizzes provide students with opportunities to test their understanding of the physical concepts presented. The questions require students to make decisions on the basis of sound reasoning, and some of them have been written to help students overcome common misconceptions. Answers to all Quick Quiz questions are found at the end of the textbook, while answers with detailed explanations are provided in the *Instructor's Manual.* Many instructors choose to use Quick Quiz questions in a "peer instruction" teaching style.
- **Summary** The Summary is now organized by each individual chapter heading, for ease of reference.
- **Problems and Conceptual Questions** All problems have been carefully worded and have been checked for clarity and accuracy, and approximately 15% of the conceptual questions and problems are new to this edition. Solutions to approximately 20% of the end-of-chapter problems are included in the *Student Solutions Manual and Study Guide.* These problems are identified with a box around the problem number. A smaller subset of problems will be available with coached solutions as part of the *PhysicsNow*™ Web-based learning system and will be accessible to students and instructors using *College Physics.* The **Physics Now**™ icon identifies these problems.
- **Line-by-Line Revision** The text has been carefully edited to improve clarity of presentation and precision of language. We hope that the result is a book both accurate and enjoyable to read.

CONTENT CHANGES

Although the overall content and organization of the textbook is similar to that of the sixth edition, several changes were implemented.

- Vectors are now denoted in bold face with arrows over them (for example, $\vec{\mathbf{v}}$), making them easier to recognize.
- In many chapters, one or two examples have been modified so as to combine new concepts with previously studied concepts. This not only reviews prior material, but also integrates different concepts, showing how they work together to enhance our understanding of the physical world.
- In Chapter 4, on Newton's laws, we have doubled the number of examples, from seven in the sixth edition to fourteen in the seventh, to help aid student understanding.
- Chapter 5, on work and energy, has been extensively reorganized, with gravitational potential energy and spring potential energy handled in separate sections. The connection between work done by conservative forces and the definition of the potential energy of a system has been emphasized and clarified.
- The number of examples in Chapter 9, Solids and Fluids, has been increased from thirteen to nineteen.
- Chapter 12, The Laws of Thermodynamics, now covers the important concepts of degrees of freedom and the equipartition of energy, and relates them to determining molar specific heats. The different kinds of thermodynamic processes have been organized into their own subsections, and the work done during an isothermal process is now presented and used in examples. Generic processes are also briefly addressed. The methods for calculating the quantities appearing in the first law are organized into a table that makes clear the similarities and differences between the processes.
- The concepts of electric potential and electric potential energy have been clarified in Chapter 16.
- In Chapter 23, the tables for sign conventions for mirrors and lenses have been reorganized for greater clarity.
- The methods of handling interference in thin films have been organized into a table in Chapter 24, facilitating student understanding and problem-solving in that area.
- Additional detail on prescribing lenses has been added to the example problems of Chapter 25, Optical Instruments.
- In Chapter 26, the Twin Paradox and concepts of general relativity were clarified and enhanced.

TEXTBOOK FEATURES

Most instructors would agree that the textbook assigned in a course should be the student's primary guide for understanding and learning the subject matter. Furthermore, the textbook should be easily accessible and written in a style that facilitates instruction and learning. With this in mind, we have included many pedagogical features that are intended to enhance the textbook's usefulness to both students and instructors. These features are as follows:

Style To facilitate rapid comprehension, we have attempted to write the book in a style that is clear, logical, relaxed, and engaging. The somewhat informal and relaxed writing style is designed to connect better with students and enhance their reading enjoyment. New terms are carefully defined, and we have tried to avoid the use of jargon.

Previews All chapters begin with a brief preview that includes a discussion of the chapter's objectives and content.

Units The international system of units (SI) is used throughout the text. The U.S. customary system of units is used only to a limited extent in the chapters on mechanics and thermodynamics.

Marginal Notes Comments and notes appearing in the margin can be used to locate important statements, equations, and concepts in the text.

Problem-Solving Strategies A general strategy to be followed by the student is outlined at the end of Chapter 1 and provides students with a structured process for solving problems. In most chapters, more specific strategies and suggestions are included for solving the types of problems featured in both the worked examples and end-of-chapter problems. This feature, highlighted by a surrounding box, is intended to help students identify the essential steps in solving problems and increases their skills as problem solvers.

Biomedical Applications For biology and pre-med students, icons point the way to various practical and interesting applications of physical principles to biology and medicine. Where possible, an effort was made to include more problems that would be relevant to these disciplines.

Applying Physics provide students with an additional means of reviewing concepts presented in that section. Some Applying Physics examples demonstrate the connection between the concepts presented in that chapter and other scientific disciplines. These examples also serve as models for students when assigned the task of responding to the Conceptual Questions presented at the end of each chapter.

Tips are placed in the margins of the text and address common student misconceptions and situations in which students often follow unproductive paths. Approximately 60 Tips are provided in Volume 1 to help students avoid common mistakes and misunderstandings.

Important Statements and Equations Most important statements and definitions are set in **boldface** type or are highlighted with a background screen for added emphasis and ease of review. Similarly, important equations are highlighted with a tan background screen to facilitate location.

Illustrations and Tables The readability and effectiveness of the text material, worked examples, and end-of-chapter conceptual questions and problems are enhanced by the large number of figures, diagrams, photographs, and tables. Full color adds clarity to the artwork and makes illustrations as realistic as possible. Three-dimensional effects are rendered with the use of shaded and lightened areas where appropriate. Vectors are color coded, and curves in graphs are drawn in color. Color photographs have been carefully selected, and their accompanying captions have been written to serve as an added instructional tool. A complete description of the pedagogical use of color appears on the inside front cover.

Significant Figures Significant figures in both worked examples and end-of-chapter problems have been handled with care. Most numerical examples and problems are worked out to either two or three significant figures, depending on the accuracy of the data provided. Intermediate results presented in the examples are rounded to the proper number of significant figures, and only those digits are carried forward.

Conceptual Questions are provided at the end of each chapter (over 250 questions are included in Volume 1). The **Applying Physics** examples presented in the text serve as models for students when conceptual questions are assigned, and show how the concepts can be applied to understanding the physical world. The conceptual questions provide the student with a means of self-testing the concepts presented in the chapter. Some conceptual questions are appropriate for initiating classroom discussions. Answers to all odd-numbered conceptual questions are located in the answer section at the end of the book.

End-of-Chapter Problems An extensive set of problems is included at the end of each chapter (in all, over 1,000 problems are given throughout Volume 1). Answers to odd-numbered problems are given at the end of the book. For the convenience of both the student and instructor, about two thirds of the problems are keyed to specific sections of the chapter. The remaining problems, labeled "Additional Problems," are not keyed to specific sections. There are three levels of problems that are graded according to their difficulty. Straightforward problems are numbered in black, intermediate level problems are numbered in blue, and the most challenging problems are numbered in magenta. The icon identifies problems dealing with applications to the life sciences and medicine.

Solutions to approximately 20% of the problems in each chapter are in the *Student Solutions Manual and Study Guide.* Among these, selected problems are identified with *PhysicsNow*™ icons and have coached solutions available at **www.cp7e.com.**

Activities have been included at the end of each chapter's problem set to encourage students to engage in scientific activities outside of the classroom.

Appendices and Endpapers Several appendices are provided at the end of the textbook. Most of the appendix material represents a review of mathematical concepts and techniques used in the text, including scientific notation, algebra, geometry, trigonometry, differential calculus, and integral calculus. Reference to these appendices is made as needed throughout the text. Most of the mathematical review sections include worked examples and exercises with answers. In addition to the mathematical review, some appendices contain useful tables that supplement textual information. For easy reference, the front endpapers contain a chart explaining the use of color throughout the book and a list of frequently used conversion factors.

TEACHING OPTIONS

This book contains more than enough material for a one-year course in introductory physics. This serves two purposes. First, it gives the instructor more flexibility in choosing topics for a specific course. Second, the book becomes more useful as a resource for students. On the average, it should be possible to cover about one chapter each week for a class that meets three hours per week. Those sections, examples, and end-of-chapter problems dealing with applications of physics to life sciences are identified with the DNA icon . We offer the following suggestions for shorter courses for those instructors who choose to move at a slower pace through the year.

Option A: If you choose to place more emphasis on contemporary topics in physics, you should consider omitting all of parts of Chapter 8 (Rotational Equilibrium and Rotational Dynamics), Chapter 21 (Alternating Current Circuits and Electromagnetic Waves), and Chapter 25 (Optical Instruments).

Option B: If you choose to place more emphasis on classical physics, you could omit all or parts of Part VI of the textbook, which deals with special relativity and other topics in 20th-century physics.

The *Instructor's Manual* offers additional suggestions for specific sections and topics that may be omitted without loss of continuity if time presses.

ANCILLARIES

The ancillary package has been updated substantially and streamlined in response to suggestions from users of the sixth edition. The most essential parts of the student package are the two-volume *Student Solutions Manual and Study Guide* with a tight focus on problem-solving and the Web-based *PhysicsNow*™ learning system. Instructors will find increased support for their teaching efforts with new electronic materials.

STUDENT ANCILLARIES

Thomson • Brooks/Cole offers several items to supplement and enhance the classroom experience. These ancillaries will allow instructors to customize the textbook to their students' needs and to their own style of instruction. One or more of these ancillaries may be shrink-wrapped with the text at a reduced price:

Student Solutions Manual and Study Guide by John R. Gordon, Charles Teague, and Raymond A. Serway. Now offered in two volumes, this manual features detailed solutions to approximately 12 problems per chapter. Boxed numbers identify those problems in the textbook for which complete solutions are found in the manual. The manual also features a skills section, important notes from key sections of the text, and a list of important equations and concepts. Volume 1 contains Chapters 1–14 and Volume 2 contains Chapters 15–30.

PhysicsNow™ Students log into *PhysicsNow*™ at **www.cp7e.com** by using the free access code packaged with this text.* The *PhysicsNow*™ system is made up of three interrelated parts:

- How much do you know?
- What do you need to learn?
- What have you learned?

Students maximize their success by starting with the Pre-Test for the relevant chapter. Each Pre-Test is a mix of conceptual and numerical questions. After completing the Pre-Test, each student is presented with a detailed Learning Plan. The Learning Plan outlines elements to review in the text and Web-based media (Active Figures, Interactive Examples, and Coached Problems) in order to master the chapter's most essential concepts. After working through these materials, students move on to a multiple-choice Post-Test presenting them with questions similar to those that might appear on an exam. Results can be e-mailed to instructors.

Physics Laboratory Manual, 2nd edition by David Loyd. This manual supplements the learning of basic physical principles while introducing laboratory procedures and equipment. Each chapter of the manual includes a pre-laboratory assignment, objectives, an equipment list, the theory behind the experiment, experimental procedures, graphs, and questions. A laboratory report is provided for each experiment so the student can record data, calculations, and experimental results. Students are encouraged to apply statistical analysis to their data in order to develop their ability to judge the validity of their results.

WebTutor™ on WebCT and Blackboard **WebTutor™** offers students real-time access to a full array of study tools, including PhysicsNow's Active Figures, chapter outlines, summaries, and learning objectives. New to this edition is a bank of end-of-chapter problems from each chapter of the book coded in multiple choice format and ready to assign as quizzes or homework.

The Brooks/Cole Physics Resource Center You'll find additional online quizzes, Web links, and animations at **http://physics.brookscole.com.**

INSTRUCTOR ANCILLARIES

The following ancillaries are available to qualified adopters. Please contact your local Thomson • Brooks/Cole sales representative for details.
Ancillaries offered in two volumes are split as follows: Volume 1 contains Chapters 1–14 and Volume 2 contains Chapters 15–30.

Instructor's Solutions Manual by Jerry Faughn and Charles Teague. Available in two volumes, this manual consists of complete solutions to all the problems in the text,

*Free pincodes are only available with new copies of *College Physics,* 7th edition.

answers to the even-numbered problems and conceptual questions, and full answers with explanations to the Quick Quizzes.

Test Bank by Ed Oberhofer. Contains approximately 1 750 multiple-choice problems and questions. Answers are provided in a separate key. It is provided in print form (in two volumes) for the instructor who does not have access to a computer, and instructors may duplicate pages for distribution to students. The questions in the ***Test Bank*** are also available in electronic format with complete answers and solutions in iLrn's Computerized Testing.

Overhead Transparency Acetates The collection of transparencies in two volumes consists of approximately 200 full-color figures and photographs from the text to enhance lectures. These transparencies feature large print for easy viewing in the classroom.

Instructor's Manual for Physics Laboratory Manual 2nd edition by David Loyd. Each chapter contains a discussion of the experiment, teaching hints, answers to selected questions from the student laboratory manual, and a post-laboratory quiz with short answers and essay questions. The author has also included a list of the suppliers of scientific equipment and a summary of the equipment needed for all the experiments in the manual.

Personal Response System Content To support your use of personal response systems in the classroom, we've coded the questions from the Quick Quiz questions for use with the system of your choice. Contact your Thomson representative for more details.

Multimedia Manager Instructor's Resource CD This easy-to-use lecture tool allows you to quickly assemble art, photos, and multimedia with notes to create fluid lectures. The CD-ROM set (Volume 1, Chapters 1–14; Volume 2, Chapters 15–30) includes a database of animations, video clips and digital art from the text as well as PowerPoint lectures and electronic files of the Instructor's Solutions Manual and **Test Bank.** You'll also find additional content from the textbook like the Quick Quizzes and Conceptual Questions.

Physics Now™ ***PhysicsNow*™ Course Management Tools powered by iLrn** This extension to the student tutorial environment of PhysicsNow™ allows instructors to deliver online assignments in an environment that is familiar to students. This powerful system is your gateway to managing online homework, testing, and course administration all in one shell with the proven content to make your course a success. To see a demonstration of this powerful system, contact your Thomson representative or go to **www.cp7e.com.**

Physics Now™ ***PhysicsNow*™ Homework Management powered by iLrn** To further enhance PhysicsNow, contributing author Chuck Bennett has created ready-to-use homework assignments which are based on the in-chapter examples. These assignments feature step-by-step guides through algorithmic versions of the in-text examples paired with more challenging examples to extend your students' problem-solving abilities. The problems include hints and feedback to help students pinpoint their errors. The goal of this set of exercises is to bring students' problem solving abilities up to the level they will need to work the end-of-chapter problems.

PhysicsNow™ gives you a rich array of problem types and grading options. Its library of assignable questions includes all of the end-of-chapter problems from the text so that you can select the problems you want to include in your online homework assignments. These well-crafted problems are algorithmically generated so that you can assign the same problem with different variables for each student. A flexible grading tolerance feature allows you to specify a percentage range of correct answers so that your students are not penalized for rounding errors. You can give students the option to work an assignment multiple times and record the highest score or limit the times they are able to attempt it. In addition, you can create your own problems to complement the problems from the text. Results flow automatically to an exportable grade book so that instructors are better able to assess student understanding of the material, even prior to class or to an actual test.

PhysicsNow™ iLrn Computerized Testing Extend the student experience with iLrn into a testing or quizzing environment. The test item file from the text is included to give you a bank of well-crafted questions that you can deliver online or print out. As with the homework problems, you can use the program's friendly interface to craft your own questions to complement the Serway/Faughn questions. You have complete control over grading, deadlines, and availability and can create multiple tests based on the same material.

WebTutor™ on WebCT and Blackboard With **WebTutor**™'s text-specific, preformatted content and total flexibility, instructors can easily create and manage their own personal Web site. WebTutor™'s course management tool gives instructors the ability to provide virtual office hours, post syllabi, set up threaded discussions, track student progress with the quizzing material, and much more. WebTutor™ also provides robust communication tools, such as a course calendar, asynchronous discussion, real-time chat, a whiteboard, and an integrated e-mail system.

Additional Options for Online Homework

WebAssign: A Web-Based Homework System WebAssign is the most utilized homework system in physics. Designed by physicists for physicists, this system is a trusted companion to your teaching. An enhanced version of WebAssign is available for College Physics. In addition to selected end-of-chapter problems, this enhanced version includes animations with conceptual questions. We've also added tutorial problems with feedback and hints to guide students' content mastery. Take a look at this new innovation from the most trusted name in physics homework. **www.webassign.net**

LON-CAPA This Web-based course management system developed at Michigan State University includes selected end of chapter problems from *College Physics*. For more information, visit the LON-CAPA Web site at **www.lon-capa.org.**

UT Austin Homework Service Details about and a demonstration of this service are available at **http://hw.ph.utexas.edu/hw.html.**

ACKNOWLEDGEMENTS

In preparing the seventh edition of this textbook, we have been guided by the expertise of many people who have reviewed manuscript and/or provided pre-revision suggestions. We wish to acknowledge the following reviewers and express our sincere appreciation for their helpful suggestions, criticism, and encouragement.

Seventh edition reviewers:

Gary B. Adams, *Arizona State University*
Ricardo Alarcon, *Arizona State University*
Natalie Batalha, *San Jose State University*
Kapila Clara Castoldi, *Oakland University*
Andrew Cornelius, *University of Nevada, Las Vegas*
Yesim Darici, *Florida International University*
N. John DiNardo, *Drexel University*
Hasan Fakhruddin, *Ball State University/ The Indiana Academy*
Lewis Ford, *Texas A & M University*
James R. Goff, *Pima Community College*
Yadin Y. Goldschmidt, *University of Pittsburgh*
Raymond Hall, *California State University, Fresno*
Patrick Hamill, *San Jose State University*
Grant W. Hart, *Brigham Young University*
James E. Heath, *Austin Community College*
Grady Hendricks, *Blinn College*
Rhett Herman, *Radford University*

Aleksey Holloway, *University of Nebraska at Omaha*
Joey Huston, *Michigan State University*
Mark James, *Northern Arizona University*
Joseph Keane, *St. Thomas Aquinas College*
Dorina Kosztin, *University of Missouri-Columbia*
Martha Lietz, *Niles West High School*
Rafael Lopez-Mobilia, *University of Texas at San Antonio*
John A. Milsom, *University of Arizona*
Monty Mola, *Humboldt State University*
Charles W. Myles, *Texas Tech University*
Ed Oberhofer, *Lake Sumter Community College*
Chris Pearson, *University of Michigan-Flint*
Alexey A. Petrov, *Wayne State University*
Dubravka Rupnik, *Louisiana State University*
Surajit Sen, *State University of New York at Buffalo*
George Strobel, *University of Georgia*
James Wanliss, *Embry-Riddle Aeronautical University*

Volume 1 of *College Physics,* seventh edition was carefully checked for accuracy by Ken Bolland (Ohio State University), Steve Ellis (University of Kentucky), Steve Hagen (University of Florida), Mark James (Northern Arizona University), Marllin L. Simon (Auburn University), George Strobel (University of Georgia), M. Anthony Reynolds (Embry-Riddle Aeronautical University), Emily Flynn, Joel Handley and Matthew Sirocky. Though responsibility for any remaining errors rests with us, we thank them for their dedication and vigilance.

We thank the following people for their suggestions and assistance during the preparation of earlier editions of this textbook:

Marilyn Akins, *Broome Community College*
Albert Altman, *University of Lowell*
John Anderson, *University of Pittsburgh*
Lawrence Anderson-Huang, *University of Toledo*
Subhash Antani, *Edgewood College*
Neil W. Ashcroft, *Cornell University*
Charles R. Bacon, *Ferris State University*
Dilip Balamore, *Nassau Community College*
Ralph Barnett, *Florissant Valley Community College*
Lois Barrett, *Western Washington University*
Paul D. Beale, *University of Colorado at Boulder*
Paul Bender, *Washington State University*
David H. Bennum, *University of Nevada at Reno*
Jeffery Braun, *University of Evansville*
John Brennan, *University of Central Florida*
Michael Bretz, *University of Michigan, Ann Arbor*
Michael E. Browne, *University of Idaho*
Joseph Cantazarite, *Cypress College*
Ronald W. Canterna, *University of Wyoming*
Clinton M. Case, *Western Nevada Community College*
Neal M. Cason, *University of Notre Dame*
Roger W. Clapp, *University of South Florida*
Giuseppe Colaccico, *University of South Florida*
Lattie F. Collins, *East Tennessee State University*
Lawrence B. Colman, *University of California, Davis*
Jorge Cossio, *Miami Dade Community College*
Terry T. Crow, *Mississippi State College*
Stephen D. Davis, *University of Arkansas at Little Rock*
John DeFord, *University of Utah*
Chris J. DeMarco, *Jackson Community College*
Michael Dennin, *University of California, Irvine*
N. John DiNardo, *Drexel University*
Robert J. Endorf, *University of Cincinnati*
Paul Feldker, *Florissant Valley Community College*
Leonard X. Finegold, *Drexel University*
Tom French, *Montgomery County Community College*
Albert Thomas Frommhold, Jr., *Auburn University*
Lothar Frommhold, *University of Texas at Austin*
Eric Ganz, *University of Minnesota*
Teymoor Gedayloo, *California Polytechnic State University*
Simon George, *California State University, Long Beach*
John R. Gordon, *James Madison University*
George W. Greenlees, *University of Minnesota*
Wlodzimierz Guryn, *Brookhaven National Laboratory*
James Harmon, *Oklahoma State University*
Grant W. Hart, *Brigham Young University*
Christopher Herbert, *New Jersey City University*
John Ho, *State University of New York at Buffalo*
Murshed Hossain, *Rowan University*
Robert C. Hudson, *Roanoke College*
Fred Inman, *Mankato State University*
Ronald E. Jodoin, *Rochester Institute of Technology*
Drasko Jovanovic, *Fermilab*
George W. Kattawar, *Texas A & M University*
Frank Kolp, *Trenton State University*
Joan P.S. Kowalski, *George Mason University*
Ivan Kramer, *University of Maryland, Baltimore County*
Sol Krasner, *University of Chicago*
Karl F. Kuhn, *Eastern Kentucky University*
David Lamp, *Texas Tech University*
Harvey S. Leff, *California State Polytechnic University*
Joel Levine, *Orange Coast College*
Michael Lieber, *University of Arkansas*
James Linbald, *Saddleback Community College*
Bill Lochslet, *Pennsylvania State University*
Michael LoPresto, *Henry Ford Community College*
Bo Lou, *Ferris State University*
Jeffrey V. Mallow, *Loyola University of Chicago*
David Markowitz, *University of Connecticut*
Steven McCauley, *California State Polytechnic University, Pomona*
Joe McCauley, Jr., *University of Houston*
Ralph V. McGrew, *Broome Community College*
Bill F. Melton, *University of North Carolina at Charlotte*
H. Kent Moore, *James Madison University*
John Morack, *University of Alaska, Fairbanks*
Steven Morris, *Los Angeles Harbor College*

Carl R. Nave, *Georgia State University*
Martin Nikolo, *Saint Louis University*
Blaine Norum, *University of Virginia*
M. E. Oakes, *University of Texas at Austin*
Lewis J. Oakland, *University of Minnesota*
Ed Oberhofer, *Lake Sumter Community College*
Lewis O'Kelly, *Memphis State University*
David G. Onn, *University of Delaware*
J. Scott Payson, *Wayne State University*
T. A. K. Pillai, *University of Wisconsin, La Crosse*
Lawrence S. Pinsky, *University of Houston*
William D. Ploughe, *Ohio State University*
Patrick Polley, *Beloit College*
Brooke M. Pridmore, *Clayton State University*
Joseph Priest, *Miami University*
James Purcell, *Georgia State University*
W. Steve Quon, *Ventura College*
Michael Ram, *State University of New York at Buffalo*
Kurt Reibel, *Ohio State University*
Barry Robertson, *Queen's University*
Virginia Roundy, *California State University, Fullerton*
Larry Rowan, *University of North Carolina, Chapel Hill*
William R. Savage, *The University of Iowa*
Reinhard A. Schumacher, *Carnegie Mellon University*
John Simon, *University of Toledo*
Donald D. Snyder, *Indiana University at Southbend*
Carey E. Stronach, *Virginia State University*
Thomas W. Taylor, *Cleveland State University*
Perry A. Tompkins, *Samford University*
L. L. Van Zandt, *Purdue University*
Howard G. Voss, *Arizona State University*
Larry Weaver, *Kansas State University*
Donald H. White, *Western Oregon State College*
Bernard Whiting, *University of Florida*
George A. Williams, *The University of Utah*
Jerry H. Wilson, *Metropolitan State College*
Robert M. Wood, *University of Georgia*
Clyde A. Zaidins, *University of Colorado at Denver*

Randall Jones generously contributed many of the new end-of-chapter problems, especially those of interest to the life sciences. Edward F. Redish of the University of Maryland graciously allowed us to list some of his problems from the Activity Based Physics Project as both problems and activities. Some of the remaining activities were written by Robert J. Beichner of North Carolina State University.

We are extremely grateful to the publishing team at the Brooks/Cole Publishing Company for their expertise and outstanding work in all aspects of this project. In particular, we'd like to thank Ed Dodd, who tirelessly coordinated and directed our efforts in preparing the manuscript in its various stages, and Annie Mac, who worked behind the scenes securing text and accuracy reviews, and coordinated and transmitted all the print ancillaries. Jane Sanders, the photo researcher, did a great job finding photos of physical phenomena, Sam Subity coordinated the building of the ***PhysicsNow***™ Web site, and Rob Hugel helped translate our rough sketches into accurate, compelling art. Donna King of Progressive Publishing Alternatives managed the difficult task of keeping production moving and on schedule. Chris Hall and Teri Hyde also made numerous valuable contributions. Chris provided just the right amount of guidance and vision throughout the project. We would like to thank David Harris, a great team builder and motivator with loads of enthusiasm and an infectious sense of humor.

Finally, we are deeply indebted to our wives and children for their love, support, and long-term sacrifices.

Raymond A. Serway
St. Petersburg, Florida

Jerry S. Faughn
Richmond, Kentucky

Chris Vuille
Daytona Beach, Florida

Charles A. Bennett
Asheville, North Carolina

Applications

APPLICATION

Directs you to sections discussing applied principles of physics

BIOMEDICAL APPLICATION

Although physics is relevant to so much in our modern lives, this may not be obvious to students in an introductory course. In this seventh edition of *College Physics*, we continue a design feature begun in the previous edition. This feature makes the relevance of physics to everyday life more obvious by pointing out specific applications in the form of a marginal note. Some of these applications pertain to the life sciences and are marked with the DNA icon. The list below is not intended to be a complete listing of all the applications of the principles of physics found in this textbook. Many other applications are to be found within the text and especially in the worked examples, conceptual questions, and end-of-chapter problems.

Chapter 3
Long jumping, p. 65

Chapter 4
Seat belts, p. 84
Helicopter flight, p. 91
Colliding vehicles, p. 92
Skydiving, p. 107

Chapter 5
Flagellar movement; bioluminescence, p. 141
Asteroid impact, p. 141
Diet versus exercise in weight-loss programs, p. 146

Chapter 6
Boxing and brain injury, p. 162
Injury to passengers in car collisions, p. 165
Glaucoma testing, p. 169
Professor Goddard was right all along—rockets work in space! p. 177
Multistage rockets, p. 178

Chapter 7
ESA launch sites, p. 196
Phonograph records and compact discs, p. 197
Artificial gravity, p. 202
Banked roadways, pp. 204–205
Why is the Sun hot?, p. 212
Geosynchronous orbit and telecommunications satellites, p. 216

Chapter 8
A weighted forearm, p. 234
Bicycle gears, p. 239
Warming up, p. 242
Figure skating, p. 248
Aerial somersaults, p. 248
Rotating neutron stars, pp. 248–249

Chapter 9
Arch structures in buildings, p. 273
Snowshoes, p. 275
Bed of nails trick, p. 276
Hydraulic lifts, p. 279
Building the pyramids, p. 281
Measuring blood pressure, p. 282
Ballpoint pens, p. 282
Swim bladders in fish, p. 284
Cerebrospinal fluid, p. 284
Testing your car's antifreeze, p. 284
Checking the battery charge, p. 285
Flight of a golf ball, pp. 294–295
"Atomizers" in perfume bottles and paint sprayers, p. 295
Vascular flutter and aneurysms, p. 295
Lift on aircraft wings, p. 295
Sailing upwind, p. 296
Home plumbing, p. 297
Rocket engines, p. 297
Air sac surface tension, p. 299
Detergents and waterproofing agents, p. 301
Effect of osmosis on living cells, p. 306
Kidney function and dialysis, p. 306

Chapter 10
Skin temperature, p. 326
Thermal expansion joints, p. 328
Pyrex glass, p. 328
Bimetallic strips and thermostats, p. 329–330
Rising sea levels, p. 332
Bursting pipes in winter, p. 333
Expansion and temperature, p. 343

Chapter 11
Working off breakfast, p. 354
Physiology of exercise, p. 354
Sea breezes and thermals, p. 355
Home insulation, p. 368
Staying warm in the arctic, p. 369
Cooling automobile engines, p. 371
Algal blooms in ponds and lakes, p. 371
Body temperature, p. 372
Light-colored summer clothing, p. 373
Thermography, p. 373
Radiation thermometers for measuring body temperature, p. 373
Thermal radiation and night vision, p. 374
Thermos bottles, p. 375
Global warming and greenhouse gases, pp. 375–377

Chapter 12
Refrigerators and heat pumps, p. 403
"Perpetual motion" machines, p. 409
The direction of time, p. 412
Human metabolism, pp. 413–416
Fighting fat, p. 415

Chapter 13
Archery, p. 429
Pistons and drive wheels, p. 432
Bungee jumping, p. 434
Pendulum clocks, p. 439
Use of pendulum in prospecting, p. 439
Shock absorbers, p. 441
Bass guitar strings, p. 446

Chapter 14
Medical uses of ultrasound, p. 459
Ultrasonic ranging unit for cameras, p. 460
The sounds heard during a storm, p. 462
OSHA noise level regulations, p. 465
Sonic booms, p. 472
Connecting your stereo speakers, p. 474
Tuning a musical instrument, p. 477
Guitar fundamentals, p. 477
Shattering goblets with the voice, p. 480
Structural resonance in bridges and buildings, p. 480
Oscillations in a harbor, p. 482
Why are instruments warmed up? p. 482
How do bugles work? p. 482
Using beats to tune a musical instrument, p. 484
Why does the professor sound like Donald Duck? p. 487
The ear, pp. 487–489

To the Student

As a student, it's important that you understand how to use the book most effectively, and how best to go about learning physics. Scanning through the Preface will acquaint you with the various features available, both in the book and online. Awareness of your educational resources and how to use them is essential. Though physics is challenging, it can be mastered with the correct approach.

HOW TO STUDY

Students often ask how best to study physics and prepare for examinations. There's no simple answer to this question, but we'd like to offer some suggestions based on our own experiences in learning and teaching over the years.

First and foremost, maintain a positive attitude toward the subject matter. Like learning a language, physics takes time. Those who keep applying themselves on a daily basis can expect to reach understanding. Keep in mind that physics is the most fundamental of all natural sciences. Other science courses that follow will use the same physical principles, so it is important that you understand and are able to apply the various concepts and theories discussed in the text.

CONCEPTS AND PRINCIPLES

Students often try to do their homework without first studying the basic concepts. It is essential that you understand the basic concepts and principles *before* attempting to solve assigned problems. You can best accomplish this goal by carefully reading the textbook *before* you attend your lecture on the covered material. When reading the text, you should jot down those points that are not clear to you. Also be sure to make a diligent attempt at answering the questions in the Quick Quizzes as you come to them in your reading. We have worked hard to prepare questions that help you judge for yourself how well you understand the material. Pay careful attention to the many Tips throughout the text. These will help you avoid misconceptions, mistakes, and misunderstandings as well as maximize the efficiency of your time by minimizing adventures along fruitless paths. During class, take careful notes and ask questions about those ideas that are unclear to you. Keep in mind that few people are able to absorb the full meaning of scientific material after only one reading.

Be sure to take advantage of the features available in the ***PhysicsNow***™ learning system, such as the Active Figures, Interactive Examples, and Coached Problems. Your lectures and laboratory work supplement your textbook and should clarify some of the more difficult material. You should minimize rote memorization of material. Successful memorization of passages from the text, equations, and derivations does not necessarily indicate that you understand the fundamental principles.

Your understanding will be enhanced through a combination of efficient study habits, discussions with other students and with instructors, and your ability to solve the problems presented in the textbook. Ask questions whenever you feel clarification of a concept is necessary.

STUDY SCHEDULE

It is important for you to set up a regular study schedule, preferably a daily one. Make sure you read the syllabus for the course and adhere to the schedule set by your instructor. As a general rule, you should devote about two hours of study time for every hour you are in class. If you are having trouble with the course, seek the advice of the instructor or other students who have taken the course. You may find it necessary to seek further instruction from experienced students. Very often, instructors offer review sessions in addition to regular class periods. It is important that you avoid the practice of delaying study until a day or two before an exam. One hour of study a day for fourteen days is far more effective than fourteen hours the day before the exam. "Cramming" usually produces disastrous results, especially in science. Rather than undertake an all-night study session just before an exam, briefly review the basic concepts and equations and get a good night's rest. If you feel you need additional help in understanding the concepts, in preparing for exams, or in problem-solving, we suggest that you acquire a copy of the *Student Solutions Manual and Study Guide* that accompanies this textbook; this manual should be available at your college bookstore.

USE THE FEATURES

You should make full use of the various features of the text discussed in the preface. For example, marginal notes are useful for locating and describing important equations and concepts, and **boldfaced** type indicates important statements and definitions. Many useful tables are contained in the Appendices, but most tables are incorporated in the text where they are most often referenced. Appendix A is a convenient review of mathematical techniques.

Answers to all Quick Quizzes, as well as odd-numbered conceptual questions and problems, are given at the end of the textbook. Answers to selected end-of-chapter problems are provided in the *Student Solutions Manual and Study Guide.* Problem-Solving Strategies included in selected chapters throughout the text give you additional information about how you should solve problems. The Table of Contents provides an overview of the entire text, while the Index enables you to locate specific material quickly. Footnotes sometimes are used to supplement the text or to cite other references on the subject discussed.

After reading a chapter, you should be able to define any new quantities introduced in that chapter and to discuss the principles and assumptions used to arrive at certain key relations. The chapter summaries and the review sections of the *Student Solutions Manual and Study Guide* should help you in this regard. In some cases, it may be necessary for you to refer to the index of the text to locate certain topics. You should be able to correctly associate with each physical quantity the symbol used to represent that quantity and the unit in which the quantity is specified. Furthermore, you should be able to express each important relation in a concise and accurate prose statement.

PROBLEM-SOLVING

R. P. Feynman, Nobel laureate in physics, once said, "You do not know anything until you have practiced." In keeping with this statement, we strongly advise that you develop the skills necessary to solve a wide range of problems. Your ability to solve problems will be one of the main tests of your knowledge of physics; therefore, you should try to solve as many problems as possible. It is essential that you understand basic concepts and principles before attempting to solve problems. It is good practice to try to find alternate solutions to the same problem. For example, you can solve problems in mechanics using Newton's laws, but very often an alternate method that draws on energy considerations is more direct. You should not deceive yourself into thinking you understand a problem merely because you have seen it solved in class. You must be able to solve the problem and similar problems on your own. We have recast the examples in this book to help you in this regard. After studying an example, see if you can cover up the right-hand side and do it yourself, using only the written descriptions on the left as hints. Once you succeed at that, try solving the example completely on your own. Finally, solve the exercise. Once you have accomplished all this, you will have a good mastery of the problem, its concepts, and mathematical technique.

The approach to solving problems should be carefully planned. A systematic plan is especially important when a problem involves several concepts. First, read the problem several times until you are confident you understand what is being asked. Look for any key words that will help you interpret the problem and perhaps allow you to make certain assumptions. Your ability to interpret a question properly is an integral part of problem-solving. Second, you should acquire the habit of writing down the information given in a problem and those quantities that need to be found; for example, you might construct a table listing both the quantities given and the quantities to be found. This procedure is sometimes used in the worked examples of the textbook. After you have decided on the method you feel is appropriate for a given problem, proceed with your solution. Finally, check your results to see if they are reasonable and consistent with your initial understanding of the problem. General problem-solving strategies of this type are included in the text and are highlighted with a surrounding box. If you follow the steps of this procedure, you will find it easier to come up with a solution and also gain more from your efforts.

Often, students fail to recognize the limitations of certain equations or physical laws in a particular situation. It is very important that you understand and remember the assumptions underlying a particular theory or formalism. For example, certain equations in kinematics apply only to a particle moving with constant acceleration. These equations are not valid for describing motion whose acceleration is not constant, such as the motion of an object connected to a spring or the motion of an object through a fluid.

EXPERIMENTS

Physics is a science based on experimental observations. In view of this fact, we recommend that you try to supplement the text by performing various types of "hands-on" experiments, either at home or in the laboratory. For example, the common Slinky™ toy is excellent for studying traveling waves; a ball swinging on the end of a long string can be used to investigate pendulum motion; various masses attached to the end of a vertical spring or rubber band can be used to determine their elastic nature; an old pair of Polaroid sunglasses and some discarded lenses and a magnifying glass are the components of various experiments in optics; and the approximate measure of the free-fall acceleration can be determined simply by measuring with a stopwatch the time it takes for a ball to drop from a known height. The list of such experiments is endless. When physical models are not available, be imaginative and try to develop models of your own.

Some of the Activities at the end of each chapter can be performed on your own or with the help of a friend and will give you specific guidelines for acquiring "hands-on" experience.

PhysicsNow™ with Active Figures and Interactive Examples

We strongly encourage you to use the ***PhysicsNow***™ Web-based learning system that accompanies this textbook. It is far easier to understand physics if you see it in action, and these new materials will enable you to become a part of that action. ***PhysicsNow***™ media described in the Preface are accessed at the URL **www.cp7e.com**, and feature a three-step learning process consisting of a Pre-Test, a personalized learning plan, and a Post-Test.

In addition to the Coached Problems identified with icons, ***PhysicsNow***™ includes the following Active Figures and Interactive Examples from Volume 1:

Chapter 1

Active Figures 1.6 and 1.7

Chapter 2

Active Figures 2.2, 2.12, 2.13, and 2.15

Interactive Examples 2.5 and 2.9

Chapter 3

Active Figures 3.3, 3.14, and 3.15
Interactive Examples 3.3 and 3.7

Chapter 4

Active Figures 4.6, 4.18, and 4.19
Interactive Example 4.10

Chapter 5

Active Figures 5.5, 5.15, 5.20, and 5.29
Interactive Example 5.5

Chapter 6

Active Figure 6.7, 6.10, 6.13, and 6.15
Interactive Examples 6.3 and 6.7

Chapter 7

Active Figures 7.5, 7.17, and 7.21
Interactive Example 7.7

Chapter 8

Active Figure 8.25
Interactive Examples 8.8 and 8.11

Chapter 9

Active Figures 9.3, 9.5, 9.6, and 9.19
Interactive Examples 9.7 and 9.13

Chapter 10

Active Figures 10.10, 10.12, and 10.15

Chapter 11

Interactive Example 11.10

Chapter 12

Active Figures 12.1, 12.9, 12.12, 12.15, and 12.16

Chapter 13

Active Figures 13.1, 13.8, 13.12, 13.13, 13.15, 13.16, 13.19, 13.24, 13.26, 13.32, 13.33, 13.34, and 13.35
Interactive Example 13.10

Chapter 14

Active Figures 14.8, 14.10, 14.18, and 14.25
Interactive Examples 14.1, 14.5, and 14.7

An Invitation to Physics

It is our hope that you too will find physics an exciting and enjoyable experience and that you will profit from this experience, regardless of your chosen profession. Welcome to the exciting world of physics!

To see the World in a Grain of Sand
And a Heaven in a Wild Flower,
Hold infinity in the palm of your hand
And Eternity in an hour.

William Blake, "Auguries of Innocence"

Stone/Getty Images

Thousands of years ago, people in southern England built Stonehenge, which was used as a calendar. The position of the sun and stars relative to the stones determined seasons for planting or harvesting.

CHAPTER

1

Introduction

OUTLINE

1.1 Standards of Length, Mass, and Time
1.2 The Building Blocks of Matter
1.3 Dimensional Analysis
1.4 Uncertainty in Measurement and Significant Figures
1.5 Conversion of Units
1.6 Estimates and Order-of-Magnitude Calculations
1.7 Coordinate Systems
1.8 Trigonometry
1.9 Problem-Solving Strategy

The goal of physics is to provide an understanding of the physical world by developing theories based on experiments. A physical theory is essentially a guess, usually expressed mathematically, about how a given physical system works. The theory makes certain predictions about the physical system which can then be checked by observations and experiments. If the predictions turn out to correspond closely to what is actually observed, then the theory stands, although it remains provisional. No theory to date has given a complete description of all physical phenomena, even within a given subdiscipline of physics. Every theory is a work in progress.

The basic laws of physics involve such physical quantities as force, velocity, volume, and acceleration, all of which can be described in terms of more fundamental quantities. In mechanics, the three most fundamental quantities are **length** (L), **mass** (M), and **time** (T); all other physical quantities can be constructed from these three.

1.1 STANDARDS OF LENGTH, MASS, AND TIME

To communicate the result of a measurement of a certain physical quantity, a *unit* for the quantity must be defined. For example, if our fundamental unit of length is defined to be 1.0 meter, and someone familiar with our system of measurement reports that a wall is 2.0 meters high, we know that the height of the wall is twice the fundamental unit of length. Likewise, if our fundamental unit of mass is defined as 1.0 kilogram, and we are told that a person has a mass of 75 kilograms, then that person has a mass 75 times as great as the fundamental unit of mass.

In 1960, an international committee agreed on a standard system of units for the fundamental quantities of science, called **SI** (Système International). Its units of length, mass, and time are the meter, kilogram, and second, respectively.

PhysicsNow™ Throughout the text, the PhysicsNow icon indicates an opportunity for you to test yourself on key concepts and to explore animations and interactions on the PhysicsNow website at **www.cp7e.com.**

Length

In 1799, the legal standard of length in France became the meter, defined as one ten-millionth of the distance from the equator to the North Pole. Until 1960, the official length of the meter was the distance between two lines on a specific bar of platinum-iridium alloy stored under controlled conditions. This standard was abandoned for several reasons, the principal one being that measurements of the separation between the lines are not precise enough. In 1960, the meter was defined as 1 650 763.73 wavelengths of orange-red light emitted from a krypton-86 lamp. In October 1983, this definition was abandoned also, and **the meter was redefined as the distance traveled by light in vacuum during a time interval of 1/299 792 458 second**. This latest definition establishes the speed of light at 299 792 458 meters per second.

Definition of the meter ▶

Mass

Definition of the kilogram ▶

The SI unit of mass, the kilogram, is defined as the mass of a specific platinum-iridium alloy cylinder kept at the International Bureau of Weights and Measures at Sèvres, France (similar to that shown in Figure 1.1a). As we'll see in Chapter 4, mass is a quantity used to measure the resistance to a change in the motion of an object. It's more difficult to cause a change in the motion of an object with a large mass than an object with a small mass.

Time

Before 1960, the time standard was defined in terms of the average length of a solar day in the year 1900. (A solar day is the time between successive appearances of the Sun at the highest point it reaches in the sky each day.) The basic unit of time, the second, was defined to be $(1/60)(1/60)(1/24) = 1/86\,400$ of the average solar day. In 1967, the second was redefined to take advantage of the high precision attainable with an atomic clock, which uses the characteristic frequency of the light emitted from the cesium-133 atom as its "reference clock." **The second is now defined as 9 192 631 700 times the period of oscillation of radiation from the cesium atom**. The newest type of cesium atomic clock is shown in Figure 1.1b.

Definition of the second ▶

(a)

(b)

Courtesy of National Institute of Standards and Technology, U.S. Dept. of Commerce

Figure 1.1 (a) The National Standard Kilogram No. 20, an accurate copy of the International Standard Kilogram kept at Sèvres, France, is housed under a double bell jar in a vault at the National Institute of Standards and Technology. (b) The nation's primary time standard is a cesium fountain atomic clock developed at the National Institute of Standards and Technology laboratories in Boulder, Colorado. This clock will neither gain nor lose a second in 20 million years.

TABLE 1.1

Approximate Values of Some Measured Lengths

	Length (m)
Distance from Earth to most remote known quasar	1×10^{26}
Distance from Earth to most remote known normal galaxies	4×10^{25}
Distance from Earth to nearest large galaxy (M31, the Andromeda galaxy)	2×10^{22}
Distance from Earth to nearest star (Proxima Centauri)	4×10^{16}
One light year	9×10^{15}
Mean orbit radius of Earth about Sun	2×10^{11}
Mean distance from Earth to Moon	4×10^{8}
Mean radius of Earth	6×10^{6}
Typical altitude of satellite orbiting Earth	2×10^{5}
Length of football field	9×10^{1}
Length of housefly	5×10^{-3}
Size of smallest dust particles	1×10^{-4}
Size of cells in most living organisms	1×10^{-5}
Diameter of hydrogen atom	1×10^{-10}
Diameter of atomic nucleus	1×10^{-14}
Diameter of proton	1×10^{-15}

Approximate Values for Length, Mass, and Time Intervals

Approximate values of some lengths, masses, and time intervals are presented in Tables 1.1, 1.2, and 1.3, respectively. Note the wide ranges of values. Study these tables to get a feel for a kilogram of mass (this book has a mass of about 2 kilograms), a time interval of 10^{10} seconds (one century is about 3×10^{9} seconds), or two meters of length (the approximate height of a forward on a basketball team). Appendix A reviews the notation for powers of 10, such as the expression of the number 50 000 in the form 5×10^{4}.

Systems of units commonly used in physics are the Système International, in which the units of length, mass, and time are the meter (m), kilogram (kg), and second (s); the cgs, or Gaussian, system, in which the units of length, mass, and time are the centimeter (cm), gram (g), and second; and the U.S. customary system, in which the units of length, mass, and time are the foot (ft), slug, and second. SI units are almost universally accepted in science and industry, and will be used throughout the book. Limited use will be made of Gaussian and U.S. customary units.

TABLE 1.2

Approximate Values of Some Masses

	Mass (kg)
Observable Universe	1×10^{52}
Milky Way galaxy	7×10^{41}
Sun	2×10^{30}
Earth	6×10^{24}
Moon	7×10^{22}
Shark	1×10^{2}
Human	7×10^{1}
Frog	1×10^{-1}
Mosquito	1×10^{-5}
Bacterium	1×10^{-15}
Hydrogen atom	2×10^{-27}
Electron	9×10^{-31}

TABLE 1.3

Approximate Values of Some Time Intervals

	Time Interval (s)
Age of Universe	5×10^{17}
Age of Earth	1×10^{17}
Average age of college student	6×10^{8}
One year	3×10^{7}
One day (time required for one revolution of Earth about its axis)	9×10^{4}
Time between normal heartbeats	8×10^{-1}
Period[a] of audible sound waves	1×10^{-3}
Period[a] of typical radio waves	1×10^{-6}
Period[a] of vibration of atom in solid	1×10^{-13}
Period[a] of visible light waves	2×10^{-15}
Duration of nuclear collision	1×10^{-22}
Time required for light to travel across a proton	3×10^{-24}

[a]A *period* is defined as the time required for one complete vibration.

TABLE 1.4

Some Prefixes for Powers of Ten Used with "Metric" (SI and cgs) Units

Power	Prefix	Abbreviation
10^{-18}	atto-	a
10^{-15}	femto-	f
10^{-12}	pico-	p
10^{-9}	nano-	n
10^{-6}	micro-	μ
10^{-3}	milli-	m
10^{-2}	centi-	c
10^{-1}	deci-	d
10^{1}	deka-	da
10^{3}	kilo-	k
10^{6}	mega-	M
10^{9}	giga-	G
10^{12}	tera-	T
10^{15}	peta-	P
10^{18}	exa-	E

Some of the most frequently used "metric" (SI and cgs) prefixes representing powers of 10 and their abbreviations are listed in Table 1.4. For example, 10^{-3} m is equivalent to 1 millimeter (mm), and 10^{3} m is 1 kilometer (km). Likewise, 1 kg is equal to 10^{3} g, and 1 megavolt (MV) is 10^{6} volts (V).

1.2 THE BUILDING BLOCKS OF MATTER

A 1-kg ($\approx$ 2-lb) cube of solid gold has a length of about 3.73 cm ($\approx$ 1.5 in.) on a side. Is this cube nothing but wall-to-wall gold, with no empty space? If the cube is cut in half, the two resulting pieces retain their chemical identity as solid gold. But what if the pieces of the cube are cut again and again, indefinitely? Will the smaller and smaller pieces always be the same substance, gold? Questions such as these can be traced back to early Greek philosophers. Two of them—Leucippus and Democritus—couldn't accept the idea that such cutting could go on forever. They speculated that the process ultimately would end when it produced a particle that could no longer be cut. In Greek, *atomos* means "not sliceable." From this term comes our English word *atom*, once believed to be the smallest, ultimate particle of matter, but since found to be a composite of more elementary particles.

The atom can be visualized as a miniature Solar System, with a dense, positively charged nucleus occupying the position of the Sun, with negatively charged electrons orbiting like planets. This model of the atom, first developed by the great Danish physicist Niels Bohr nearly a century ago, led to the understanding of certain properties of the simpler atoms such as hydrogen, but failed to explain many fine details of atomic structure.

Notice the size of a hydrogen atom, listed in Table 1.1, and the size of a proton—the nucleus of a hydrogen atom—one hundred thousand times smaller. If the proton were the size of a Ping Pong ball, the electron would be a tiny speck about the size of a bacterium, orbiting the proton a kilometer away! Other atoms are similarly constructed. So there is a surprising amount of empty space in ordinary matter.

After the discovery of the nucleus in the early 1900s, questions arose concerning its structure. Is the nucleus a single particle or a collection of particles? The exact composition of the nucleus hasn't been defined completely even today, but by the early 1930s a model evolved that helped us understand how the nucleus behaves. Scientists determined that two basic entities—protons and neutrons—occupy the nucleus. The *proton* is nature's fundamental carrier of positive charge (equal in magnitude but opposite in sign to the charge on the electron), and the number of protons in a nucleus determines what the element is. For instance, a nucleus containing only one proton is the nucleus of an atom of hydrogen, regardless of how many neutrons may be present. Extra neutrons correspond to different isotopes of hydrogen—deuterium and tritium—which react chemically in exactly the same way as hydrogen, but are more massive. An atom having two protons in its nucleus, similarly, is always helium, although again, differing numbers of neutrons are possible.

The existence of *neutrons* was verified conclusively in 1932. A neutron has no charge and has a mass about equal to that of a proton. One of its primary purposes is to act as a "glue" to hold the nucleus together. If neutrons were not present, the repulsive electrical force between the positively charged protons would cause the nucleus to fly apart.

The division doesn't stop here; it turns out that protons, neutrons, and a zoo of other exotic particles are now thought to be composed of six particles called **quarks** (rhymes with "forks," though some rhyme it with "sharks"). These particles have been given the names *up*, *down*, *strange*, *charm*, *bottom*, and *top*. The up, charm, and top quarks each carry a charge equal to $+\frac{2}{3}$ that of the proton, whereas the down, strange, and bottom quarks each carry a charge equal to $-\frac{1}{3}$ the proton charge. The proton consists of two up quarks and one down quark (see Fig. 1.2), giving the correct charge for the proton, $+1$. The neutron is composed of two down quarks and one up quark and has a net charge of zero.

The up and down quarks are sufficient to describe all normal matter, so the existence of the other four quarks indirectly observed in high-energy experiments,

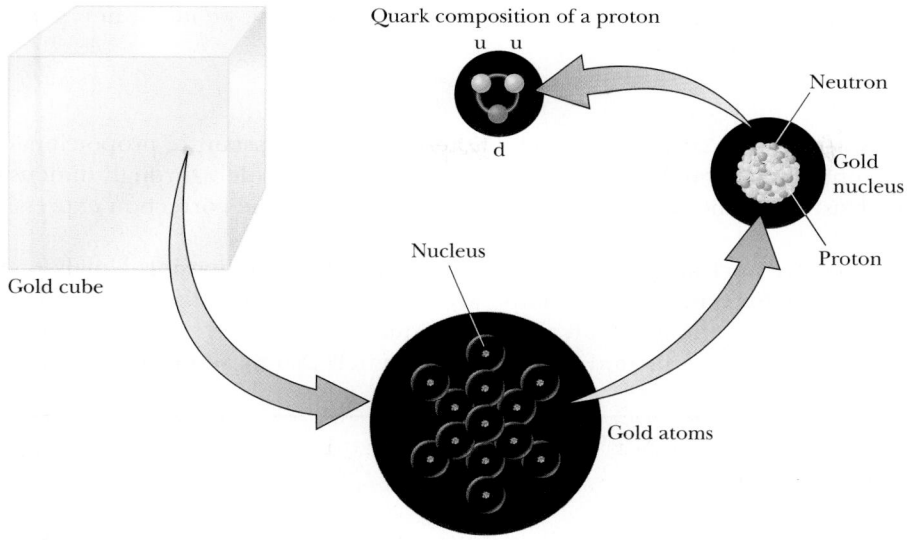

Figure 1.2 Levels of organization in matter. Ordinary matter consists of atoms, and at the center of each atom is a compact nucleus consisting of protons and neutrons. Protons and neutrons are composed of quarks. The quark composition of a proton is shown.

is something of a mystery. It's also possible that quarks themselves have internal structure. Many physicists believe that the most fundamental particles may be tiny loops of vibrating string.

1.3 DIMENSIONAL ANALYSIS

In physics, the word *dimension* denotes the physical nature of a quantity. The distance between two points, for example, can be measured in feet, meters, or furlongs, which are different ways of expressing the dimension of *length*.

The symbols that we use in this section to specify the dimensions of length, mass, and time are L, M, and T, respectively. Brackets [ ] will often be used to denote the dimensions of a physical quantity. For example, in this notation the dimensions of velocity v are written $[v] = \mathrm{L/T}$, and the dimensions of area A are $[A] = \mathrm{L}^2$. The dimensions of area, volume, velocity, and acceleration are listed in Table 1.5, along with their units in the three common systems. The dimensions of other quantities, such as force and energy, will be described later as they are introduced.

In physics, it's often necessary either to derive a mathematical expression or equation or to check its correctness. A useful procedure for doing this is called **dimensional analysis**, which makes use of the fact that **dimensions can be treated as algebraic quantities**. Such quantities can be added or subtracted only if they have the same dimensions. It follows that the terms on the opposite sides of an equation must have the same dimensions. If they don't, the equation is wrong. If they do, the equation is probably correct, except for a possible constant factor.

To illustrate this procedure, suppose we wish to derive a formula for the distance x traveled by a car in a time t if the car starts from rest and moves with constant acceleration a. The quantity x has the dimension length: $[x] = \mathrm{L}$. Time t, of course, has dimension $[t] = \mathrm{T}$. Acceleration is the change in velocity v with time. Since v has dimensions of length per unit time, or $[v] = \mathrm{L/T}$, acceleration must have dimensions $[a] = \mathrm{L/T^2}$. We organize this information in the form of an equation:

$$[a] = \frac{[v]}{[t]} = \frac{\mathrm{L/T}}{\mathrm{T}} = \frac{\mathrm{L}}{\mathrm{T}^2} = \frac{[x]}{[t]^2}$$

TABLE 1.5

Dimensions and Some Units of Area, Volume, Velocity, and Acceleration

System	Area (L^2)	Volume (L^3)	Velocity (L/T)	Acceleration ($\mathrm{L/T^2}$)
SI	m^2	m^3	m/s	$\mathrm{m/s^2}$
cgs	cm^2	cm^3	cm/s	$\mathrm{cm/s^2}$
U.S. customary	ft^2	ft^3	ft/s	$\mathrm{ft/s^2}$

Looking at the left- and right-hand sides of this equation, we might now guess that

$$a = \frac{x}{t^2} \quad \rightarrow \quad x = at^2$$

This is not quite correct, however, because there's a constant of proportionality—a simple numerical factor—that can't be determined solely through dimensional analysis. As will be seen in Chapter 2, it turns out that the correction expression is $x = \frac{1}{2}at^2$.

When we work algebraically with physical quantities, dimensional analysis allows us to check for errors in calculation, which often show up as discrepancies in units. If, for example, the left-hand side of an equation is in meters and the right-hand side is in meters per second, we know immediately that we've made an error.

EXAMPLE 1.1 Analysis of an Equation

Goal Check an equation using dimensional analysis.

Problem Show that the expression $v = v_0 + at$, is dimensionally correct, where v and v_0 represent velocities, a is acceleration, and t is a time interval.

Strategy Analyze each term, finding its dimensions, and then check to see if all the terms agree with each other.

Solution

Find dimensions for v and v_0.

$$[v] = [v_0] = \frac{\mathrm{L}}{\mathrm{T}}$$

Find the dimensions of at.

$$[at] = \frac{\mathrm{L}}{\mathrm{T}^2}(\mathrm{T}) = \frac{\mathrm{L}}{\mathrm{T}}$$

Remarks All the terms agree, so the equation is dimensionally correct.

Exercise 1.1

Determine whether the equation $x = vt^2$ is dimensionally correct. If not, provide a correct expression, up to an overall constant of proportionality.

Answer Incorrect. The expression $x = vt$ is dimensionally correct.

EXAMPLE 1.2 Find an Equation

Goal Derive an equation by using dimensional analysis.

Problem Find a relationship between a constant acceleration a, speed v, and distance r from the origin for a particle traveling in a circle.

Strategy Start with the term having the most dimensionality, a. Find its dimensions, and then rewrite those dimensions in terms of the dimensions of v and r. The dimensions of time will have to be eliminated with v, since that's the only quantity in which the dimension of time appears.

Solution

Write down the dimensions of a:

$$[a] = \frac{\mathrm{L}}{\mathrm{T}^2}$$

Solve the dimensions of speed for T:

$$[v] = \frac{\mathrm{L}}{\mathrm{T}} \quad \rightarrow \quad \mathrm{T} = \frac{\mathrm{L}}{[v]}$$

Substitute the expression for T into the equation for $[a]$:

$$[a] = \frac{\mathrm{L}}{\mathrm{T}^2} = \frac{\mathrm{L}}{(\mathrm{L}/[v])^2} = \frac{[v]^2}{\mathrm{L}}$$

Substitute L = $[r]$, and guess at the equation: $$[a] = \frac{[v]^2}{[r]} \rightarrow a = \frac{v^2}{r}$$

Remarks This is the correct equation for centripetal acceleration—acceleration towards the center of motion—to be discussed in Chapter 7. There isn't any need in this case, to introduce a numerical factor. Such a factor is often displayed explicitly as a constant k in front of the right-hand side—for example, $a = kv^2/r$. As it turns out, $k = 1$ gives the correct expression.

Exercise 1.2

In physics, energy E carries dimensions of mass times length squared, divided by time squared. Use dimensional analysis to derive a relationship for energy in terms of mass m and speed v, up to a constant of proportionality. Set the speed equal to c, the speed of light, and the constant of proportionality equal to 1 to get the most famous equation in physics.

Answer $E = kmv^2 \rightarrow E = mc^2$ when $k = 1$ and $v = c$.

1.4 UNCERTAINTY IN MEASUREMENT AND SIGNIFICANT FIGURES

Physics is a science in which mathematical laws are tested by experiment. No physical quantity can be determined with complete accuracy, because our senses are physically limited, even when extended with microscopes, cyclotrons, and other gadgets.

Knowing the experimental uncertainties in any measurement is very important. Without this information, little can be said about the final measurement. Using a crude scale, for example, we might find that a gold nugget has a mass of 3 kilograms. A prospective client interested in purchasing the nugget would naturally want to know about the accuracy of the measurement, to ensure paying a fair price. He wouldn't be happy to find that the measurement was good only to within a kilogram, because he might pay for three kilograms and get only two. Of course, he might get four kilograms for the price of three, but most people would be hesitant to gamble that an error would turn out in their favor.

Accuracy of measurement depends on the sensitivity of the apparatus, the skill of the person carrying out the measurement, and the number of times the measurement is repeated. There are many ways of handling uncertainties, and here we'll develop a basic and reliable method of keeping track of them in the measurement itself and in subsequent calculations.

Suppose that in a laboratory experiment we measure the area of a rectangular plate with a meter stick. Let's assume that the accuracy to which we can measure a particular dimension of the plate is ± 0.1 cm. If the length of the plate is measured to be 16.3 cm, we can claim only that it lies somewhere between 16.2 cm and 16.4 cm. In this case, we say that the measured value has three significant figures. Likewise, if the plate's width is measured to be 4.5 cm, the actual value lies between 4.4 cm and 4.6 cm. This measured value has only two significant figures. We could write the measured values as 16.3 ± 0.1 cm and 4.5 ± 0.1 cm. In general, **a significant figure is a reliably known digit** (other than a zero used to locate a decimal point).

Suppose we would like to find the area of the plate by multiplying the two measured values together. The final value can range between $(16.3 - 0.1\text{ cm})(4.5 - 0.1\text{ cm}) = (16.2\text{ cm})(4.4\text{ cm}) = 71.28\text{ cm}^2$ and $(16.3 + 0.1\text{ cm})(4.5 + 0.1\text{ cm}) = (16.4\text{ cm})(4.6\text{ cm}) = 75.44\text{ cm}^2$. Claiming to know anything about the hundredths place, or even the tenths place, doesn't make any sense, because it's clear we can't even be certain of the units place, whether it's the 1 in 71, the 5 in 75, or somewhere in between. The tenths and the hundredths places are clearly not significant. We have some information about the units place, so that number is significant. Multiplying the numbers at the middle of the uncertainty ranges gives $(16.3\text{ cm})(4.5\text{ cm}) = 73.35\text{ cm}^2$, which is also in the middle of the area's uncertainty range. Since the hundredths and tenths are not significant, we drop them and take the answer to be 73 cm^2, with an uncertainty of $\pm 2\text{ cm}^2$. Note that the answer has two significant figures, the same number of figures as the least accurately known quantity being multiplied, the 4.5-cm width.

There are two useful rules of thumb for determining the number of significant figures. The first, concerning multiplication and division, is as follows: **In multiplying (dividing) two or more quantities, the number of significant figures in the final product (quotient) is the same as the number of significant figures in the *least accurate* of the factors being combined, where *least accurate* means *having the lowest number of significant figures.***

To get the final number of significant figures, it's usually necessary to do some rounding. If the last digit dropped is less than 5, simply drop the digit. If the last digit dropped is greater than or equal to 5, raise the last retained digit by one.

EXAMPLE 1.3 Installing a Carpet

Goal Apply the multiplication rule for significant figures.

Problem A carpet is to be installed in a room of length 12.71 m and width 3.46 m. Find the area of the room, retaining the proper number of significant figures.

Strategy Count the significant figures in each number. The smaller result is the number of significant figures in the answer.

Solution

Count significant figures: 12.71 m → 4 significant figures

3.46 m → 3 significant figures

Multiply the numbers, keeping only three digits: $12.71\text{ m} \times 3.46\text{ m} = 43.9766\text{ m}^2 \rightarrow 44.0\text{ m}^2$

Remarks In reducing 43.976 6 to three significant figures, we used our rounding rule, adding 1 to the 9, which made 10 and resulted in carrying 1 to the unit's place.

Exercise 1.3

Repeat this problem, but with a room measuring 9.72 m long by 5.3 m wide.

Answer 52 m^2

TIP 1.1 Using Calculators

Calculators were designed by engineers to yield as many digits as the memory of the calculator chip permitted, so be sure to round the final answer down to the correct number of significant figures.

Zeros may or may not be significant figures. Zeros used to position the decimal point in such numbers as 0.03 and 0.007 5 are not significant (but are useful in avoiding errors). Hence, 0.03 has one significant figure, and 0.007 5 has two.

When zeros are placed after other digits in a whole number, there is a possibility of misinterpretation. For example, suppose the mass of an object is given as 1 500 g. This value is ambiguous, because we don't know whether the last two zeros are being used to locate the decimal point or whether they represent significant figures in the measurement.

Using scientific notation to indicate the number of significant figures removes this ambiguity. In this case, we express the mass as 1.5×10^3 g if there are two significant figures in the measured value, 1.50×10^3 g if there are three significant figures, and 1.500×10^3 g if there are four. Likewise, 0.000 15 is expressed in scientific notation as 1.5×10^{-4} if it has two significant figures or as 1.50×10^{-4} if it has three significant figures. The three zeros between the decimal point and the digit 1 in the number 0.000 15 are not counted as significant figures because they only locate the decimal point. In this book, **most of the numerical examples and end-of-chapter problems will yield answers having two or three significant figures.**

For addition and subtraction, it's best to focus on the number of decimal places in the quantities involved rather than on the number of significant figures. **When numbers are added (subtracted), the number of decimal places in the result should equal the smallest number of decimal places of any term in the sum (difference).** For example, if we wish to compute 123 (zero decimal places) + 5.35 (two decimal places), the answer is 128 (zero decimal places) and not 128.35. If we compute the sum 1.000 1 (four decimal places) + 0.000 3 (four decimal places) = 1.000 4, the result has the correct number of decimal places, namely four. Observe that the rules for multiplying significant figures don't work here because the answer has five significant figures even though one of the terms in the sum, 0.000 3, has only one significant figure. Likewise, if we perform the subtraction 1.002 − 0.998 = 0.004, the result has three decimal places because each term in the subtraction has three decimal places.

TIP 1.2 No Commas in Numbers with Many Digits

In science, numbers with more than three digits are written in groups of three digits separated by spaces rather than commas; so that 10 000 is the same as the common American notation 10,000. Similarly, $\pi = 3.14159265$ is written as 3.141 592 65.

To show why this rule should hold, we return to the first example in which we added 123 and 5.35, and rewrite these numbers as 123.*xxx* and 5.35*x*. Digits written with an *x* are completely unknown and can be any digit from 0 to 9. Now we line up 123.*xxx* and 5.35*x* relative to the decimal point and perform the addition, using the rule that an unknown digit added to a known or unknown digit yields an unknown:

$$\begin{array}{r} 123.xxx \\ +\quad 5.35x \\ \hline 128.xxx \end{array}$$

The answer of 128.*xxx* means that we are justified only in keeping the number 128 because everything after the decimal point in the sum is actually unknown. The example shows that the controlling uncertainty is introduced into an addition or subtraction by the term with the smallest number of decimal places.

In performing any calculation, especially one involving a number of steps, there will always be slight discrepancies introduced by both the rounding process and the algebraic order in which steps are carried out. For example, consider $2.35 \times 5.89/1.57$. This computation can be performed in three different orders. First, we have $2.35 \times 5.89 = 13.842$, which rounds to 13.8, followed by $13.8/1.57 = 8.789\,8$, rounding to 8.79. Second, $5.89/1.57 = 3.751\,6$, which rounds to 3.75, resulting in $2.35 \times 3.75 = 8.812\,5$, rounding to 8.81. Finally, $2.35/1.57 = 1.496\,8$ rounds to 1.50, and $1.50 \times 5.89 = 8.835$ rounds to 8.84. So three different algebraic orders, following the rules of rounding, lead to answers of 8.79, 8.81, and 8.84, respectively. Such minor discrepancies are to be expected, because the last significant digit is only one representative from a range of possible values, depending on experimental uncertainty. The discrepancies can be reduced by carrying one or more extra digits during the calculation. In our examples, however, intermediate results will be rounded off to the proper number of significant figures, and only those digits will be carried forward. In experimental work, more sophisticated techniques are used to determine the accuracy of an experimental result.

1.5 CONVERSION OF UNITS

Sometimes it's necessary to convert units from one system to another. Conversion factors between the SI and U.S. customary systems for units of length are as follows:

$$1 \text{ mile} = 1\,609 \text{ m} = 1.609 \text{ km} \qquad 1 \text{ ft} = 0.304\,8 \text{ m} = 30.48 \text{ cm}$$

$$1 \text{ m} = 39.37 \text{ in.} = 3.281 \text{ ft} \qquad 1 \text{ in.} = 0.025\,4 \text{ m} = 2.54 \text{ cm}$$

A more extensive list of conversion factors can be found on the inside front cover of this book.

Billy E. Barnes/Stock Boston

This road sign near Raleigh, North Carolina, shows distances in miles and kilometers. How accurate are the conversions?

Units can be treated as algebraic quantities that can "cancel" each other. We can make a fraction with the conversion that will cancel the units we don't want, and multiply that fraction by the quantity in question. For example, suppose we want to convert 15.0 in. to centimeters. Because 1 in. = 2.54 cm, we find that

$$15.0 \text{ in.} = 15.0 \cancel{\text{in.}} \times \left(\frac{2.54 \text{ cm}}{1.00 \cancel{\text{in.}}}\right) = 38.1 \text{ cm}$$

The next two examples show how to deal with problems involving more than one conversion and with powers.

EXAMPLE 1.4 Pull Over, Buddy!

Goal Convert units using several conversion factors.

Problem If a car is traveling at a speed of 28.0 m/s, is it exceeding the speed limit of 55.0 mi/h?

Strategy Meters must be converted to miles and seconds to hours, using the conversion factors listed on the inside front cover of the book. This requires two or three conversion ratios.

Solution

Convert meters to miles:

$$28.0 \text{ m/s} = \left(28.0 \frac{\cancel{\text{m}}}{\text{s}}\right)\left(\frac{1.00 \text{ mi}}{1\,609 \cancel{\text{m}}}\right) = 1.74 \times 10^{-2} \text{ mi/s}$$

Convert seconds to hours:

$$\begin{aligned} &1.74 \times 10^{-2} \text{ mi/s} \\ &= \left(1.74 \times 10^{-2} \frac{\text{mi}}{\cancel{\text{s}}}\right)\left(60.0 \frac{\cancel{\text{s}}}{\cancel{\text{min}}}\right)\left(60.0 \frac{\cancel{\text{min}}}{\text{h}}\right) \\ &= 62.6 \text{ mi/h} \end{aligned}$$

Remarks The driver should slow down because he's exceeding the speed limit. An alternate approach is to use the single conversion relationship 1.00 m/s = 2.24 mi/h:

$$28.0 \text{ m/s} = \left(28.0 \frac{\cancel{\text{m}}}{\cancel{\text{s}}}\right)\left(\frac{2.24 \text{ mi/h}}{1.00 \cancel{\text{m}}/\cancel{\text{s}}}\right) = 62.7 \text{ mi/h}$$

Answers to conversion problems may differ slightly, as here, due to rounding during intermediate steps.

Exercise 1.4

Convert 152 mi/h to m/s.

Answer 68.0 m/s

EXAMPLE 1.5 Press the Pedal to the Metal

Goal Convert a quantity featuring powers of a unit.

Problem The traffic light turns green, and the driver of a high-performance car slams the accelerator to the floor. The accelerometer registers 22.0 m/s^2. Convert this reading to km/min^2.

Strategy Here we need one factor to convert meters to kilometers and another two factors to convert seconds squared to minutes squared.

Solution

Insert the necessary factors:

$$\frac{22.0 \cancel{\text{m}}}{1.00 \cancel{\text{s}}^2}\left(\frac{1.00 \text{ km}}{1.00 \times 10^3 \cancel{\text{m}}}\right)\left(\frac{60.0 \cancel{\text{s}}}{1.00 \text{ min}}\right)^2 = 79.2 \frac{\text{km}}{\text{min}^2}$$

Remarks Notice that in each conversion factor the numerator equals the denominator when units are taken into account. A common error in dealing with squares is to square the units inside the parentheses while forgetting to square the numbers!

Exercise 1.5
Convert 4.50×10^3 kg/m^3 to g/cm^3.

Answer 4.50 g/cm^3

1.6 ESTIMATES AND ORDER-OF-MAGNITUDE CALCULATIONS

Getting an exact answer to a calculation may often be difficult or impossible, either for mathematical reasons or because limited information is available. In these cases, estimates can yield useful approximate answers that can determine whether a more precise calculation is necessary. Estimates also serve as a partial check if the exact calculations are actually carried out. If a large answer is expected but a small exact answer is obtained, there's an error somewhere.

For many problems, knowing the approximate value of a quantity—within a factor of 10 or so—is sufficient. This approximate value is called an **order-of-magnitude** estimate, and requires finding the power of 10 that is closest to the actual value of the quantity. For example, 75 kg $\sim 10^2$ kg, where the symbol $\sim$ means "is on the order of" or "is approximately." Increasing a quantity by three orders of magnitude means that its value increases by a factor of $10^3 = 1\,000$.

Occasionally, the process of making such estimates results in fairly crude answers, but answers ten times or more too large or small are still useful. For example, suppose you're interested in how many people have contracted a certain disease. Any estimates under ten thousand are small compared with Earth's total population, but a million or more would be alarming. So even relatively imprecise information can provide valuable guidance.

In developing these estimates, you can take considerable liberties with the numbers. For example, $\pi \sim 1$, $27 \sim 10$, and $65 \sim 100$. To get a less crude estimate, it's permissible to use slightly more accurate numbers (e.g., $\pi \sim 3$, $27 \sim 30$, $65 \sim 70$). Better accuracy can also be obtained by systematically underestimating as many numbers as you overestimate. Some quantities may be completely unknown, but it's standard to make reasonable guesses, as the examples show.

EXAMPLE 1.6 How Much Gasoline Do We Use?

Goal Develop a complex estimate.

Problem Estimate the number of gallons of gasoline used by all cars in the United States each year.

Strategy Estimate the number of people in the United States, and then estimate the number of cars per person. Multiply to get the number of cars. Guess at the number of miles per gallon obtained by a typical car and the number of miles driven per year, and from that get the number of gallons each car uses every year. Multiply by the estimated number of cars to get a final answer, the number of gallons of gas used.

Solution

The number of cars equals the number of people times the number of cars per person:

$$\text{number of cars} = (3.00 \times 10^8 \text{ people}) \times (0.5 \text{ cars/person}) \sim 10^8 \text{ cars}$$

The number of gallons used by one car in a year is the number of miles driven divided by the miles per gallon.

$$\frac{\#\text{ gal/yr}}{\text{car}} \approx \frac{\left(\dfrac{10^4 \text{ mi/yr}}{\text{car}}\right)}{10\,\dfrac{\text{mi}}{\text{gal}}} = 10^3\,\frac{\text{gal/yr}}{\text{car}}$$

Multiply these two results together to get an estimate of the number of gallons of gas used per year.

$$\#\text{ gal} \sim (10^8 \text{ cars}) \times \left(10^3\,\frac{\text{gal/yr}}{\text{car}}\right) = 10^{11} \text{ gal/yr}$$

Remarks Notice the inexact, and somewhat high, figure for the number of people in the United States, the estimate on the number of cars per person (figuring that every other person has a car of one kind or another), and the truncation of 1.5×10^8 cars to 10^8 cars. A similar estimate was used on the number of miles driven per year by a typical vehicle, and the average fuel economy, 10 mi/gal, looks low. None of this is important, because we are interested only in an order-of-magnitude answer. Few people owning a car would drive just 1 000 miles in a year, and very few would drive 100 000 miles, so 10 000 miles is a good estimate. Similarly, most cars get between 10 and 30 mi/gal, so using 10 is a reasonable estimate, while very few cars would get 100 mi/gal or 1 mi/gal.

In making estimates, it's okay to be cavalier! Feel free to take liberties ordinarily denied.

Exercise 1.6

How many new car tires are purchased in the United States each year? (Use the fact that tires wear out after about 50 000 miles.)

Answer $\sim 10^8$ tires (Individual answers may vary.)

Example 1.7 Stack One-Dollar Bills to the Moon

Goal Estimate the number of stacked objects required to reach a given height.

Problem How many one-dollar bills, stacked one on top of the other, would reach the Moon?

Strategy The distance to the Moon is about 400 000 km. Guess at the number of dollar bills in a millimeter, and multiply the distance by this number, after converting to consistent units.

Solution

We estimate that ten stacked bills form a layer of 1 mm. Convert mm to km:

$$\frac{10 \text{ bills}}{1 \cancel{\text{mm}}}\left(\frac{10^3 \cancel{\text{mm}}}{1 \cancel{\text{m}}}\right)\left(\frac{10^3 \cancel{\text{m}}}{1 \text{ km}}\right) = \frac{10^7 \text{ bills}}{1 \text{ km}}$$

Multiply this value by the approximate lunar distance:

$$\#\text{ of dollar bills} \sim (4 \times 10^5 \text{ km})\left(\frac{10^7 \text{ bills}}{1 \text{ km}}\right)$$

$$= 4 \times 10^{12} \text{ bills}$$

Remarks That's the same order of magnitude as the U.S. national debt!

Exercise 1.7

How many pieces of cardboard, typically found at the back of a bound pad of paper, would you have to stack up to match the height of the Washington monument, about 170 m tall?

Answer $\sim 10^5$ (Answers may vary.)

EXAMPLE 1.8 Number of Galaxies in the Universe

Goal Estimate a volume and a number density, and combine.

Problem Given that astronomers can see about 10 billion light years into space and that there are 14 galaxies in our local group, 2 million light years from the next local group, estimate the number of galaxies in the observable universe. (*Note*: One light year is the distance traveled by light in one year, about 9.5×10^{15} m.) (See Fig. 1.3.)

Strategy From the known information, we can estimate the number of galaxies per unit volume. The local group of 14 galaxies is contained in a sphere a million light years in radius, with the Andromeda group in a similar sphere, so there are about 10 galaxies within a volume of radius 1 million light years. Multiply that number density by the volume of the observable universe.

R. Williams (STScI), the HDF-S team, and NASA

Figure 1.3 In this deep-space photograph, there are few stars—just galaxies without end.

Solution

Compute the approximate volume V_{lg} of the local group of galaxies:

$$V_{lg} = \tfrac{4}{3}\pi r^3 \sim (10^6\ \text{ly})^3 = 10^{18}\ \text{ly}^3$$

Compute the number of galaxies per cubic light year:

$$\frac{\#\text{ of galaxies}}{\text{ly}^3} = \frac{\#\text{ of galaxies}}{V_{lg}} \sim \frac{10\ \text{galaxies}}{10^{18}\ \text{ly}^3} = 10^{-17}\ \frac{\text{galaxies}}{\text{ly}^3}$$

Compute the approximate volume of the observable universe:

$$V_u = \tfrac{4}{3}\pi r^3 \sim (10^{10}\ \text{ly})^3 = 10^{30}\ \text{ly}^3$$

Multiply the density of galaxies by V_u:

$$\begin{aligned}\#\text{ of galaxies} &\sim \left(\frac{\#\text{ of galaxies}}{\text{ly}^3}\right)V_u \\ &= \left(10^{-17}\ \frac{\text{galaxies}}{\text{ly}^3}\right)(10^{30}\ \text{ly}^3) \\ &= \boxed{10^{13}\ \text{galaxies}}\end{aligned}$$

Remarks Notice the approximate nature of the computation, which uses $4\pi/3 \sim 1$ on two occasions and $14 \sim 10$ for the number of galaxies in the local group. This is completely justified: Using the actual numbers would be pointless, because the other assumptions in the problem—the size of the observable universe and the idea that the local galaxy density is representative of the density everywhere—are also very rough approximations. Further, there was nothing in the problem that required using volumes of spheres rather than volumes of cubes. Despite all these arbitrary choices, the answer still gives useful information, because it rules out a lot of reasonable possible answers. Before doing the calculation, a guess of a billion galaxies might have seemed plausible.

Exercise 1.8

Given that the nearest star is about 4 light years away and that the galaxy is roughly a disk 100 000 light years across and a thousand light years thick, estimate the number of stars in the Milky Way galaxy.

Answer $\sim 10^{12}$ stars (Estimates will vary. The actual answer is probably close to 4×10^{11} stars.)

1.7 COORDINATE SYSTEMS

Many aspects of physics deal with locations in space, which require the definition of a coordinate system. A point on a line can be located with one coordinate, a point in a plane with two coordinates, and a point in space with three.

A coordinate system used to specify locations in space consists of the following:

- A fixed reference point O, called the *origin*
- A set of specified axes, or directions, with an appropriate scale and labels on the axes
- Instructions on labeling a point in space relative to the origin and axes

One convenient and commonly used coordinate system is the **Cartesian coordinate system**, sometimes called the **rectangular coordinate system**. Such a system in two dimensions is illustrated in Figure 1.4. An arbitrary point in this system is labeled with the coordinates (x, y). For example, the point P in the figure has coordinates $(5, 3)$. If we start at the origin O, we can reach P by moving 5 meters horizontally to the right and then 3 meters vertically upwards. In the same way, the point Q has coordinates $(-3, 4)$, which corresponds to going 3 meters horizontally to the left of the origin and 4 meters vertically upwards from there.

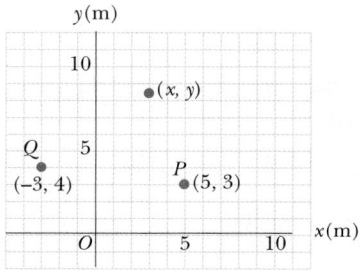

Figure 1.4 Designation of points in a two-dimensional Cartesian coordinate system. Every point is labeled with coordinates (x, y).

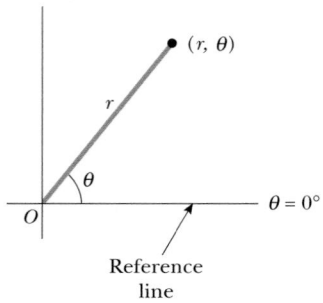

Figure 1.5 A polar coordinate system.

Positive x is usually selected as right of the origin and positive y upward from the origin, but in two dimensions this choice is largely a matter of taste. (In three dimensions, however, there are "right-handed" and "left-handed" coordinates, which lead to minus sign differences in certain operations. These will be addressed as needed.)

Sometimes it's more convenient to locate a point in space by its **plane polar coordinates** (r, θ), as in Figure 1.5. In this coordinate system, an origin O and a reference line are selected as shown. A point is then specified by the distance r from the origin to the point and by the angle θ between the reference line and a line drawn from the origin to the point. The standard reference line is usually selected to be the positive x-axis of a Cartesian coordinate system. The angle θ is considered positive when measured counterclockwise from the reference line and negative when measured clockwise. For example, if a point is specified by the polar coordinates 3 m and 60°, we locate this point by moving out 3 m from the origin at an angle of 60° above (counterclockwise from) the reference line. A point specified by polar coordinates 3 m and $-60°$ is located 3 m out from the origin and 60° below (clockwise from) the reference line.

1.8 TRIGONOMETRY

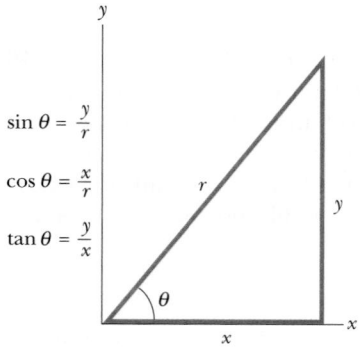

ACTIVE FIGURE 1.6
Certain trigonometric functions of a right triangle.

Physics Now™ Log into PhysicsNow at **www.cp7e.com**, and go to Active Figure 1.6 to move the point and see the changes to the rectangular and polar coordinates and to the sine, cosine, and tangent of angle θ.

Consider the right triangle shown in Active Figure 1.6, where side y is opposite the angle θ, side x is adjacent to the angle θ, and side r is the hypotenuse of the triangle. The basic trigonometric functions defined by such a triangle are the ratios of the lengths of the sides of the triangle. These relationships are called the sine (sin), cosine (cos), and tangent (tan) functions. In terms of θ, the basic trigonometric functions are as follows:[1]

$$\sin\theta = \frac{\text{side opposite } \theta}{\text{hypotenuse}} = \frac{y}{r}$$

$$\cos\theta = \frac{\text{side adjacent to } \theta}{\text{hypotenuse}} = \frac{x}{r} \qquad [1.1]$$

$$\tan\theta = \frac{\text{side opposite } \theta}{\text{side adjacent to } \theta} = \frac{y}{x}$$

For example, if the angle θ is equal to 30°, then the ratio of y to r is always 0.50; that is, $\sin 30° = 0.50$. Note that the sine, cosine, and tangent functions are quantities without units because each represents the ratio of two lengths.

Another important relationship, called the **Pythagorean theorem**, exists between the lengths of the sides of a right triangle:

$$r^2 = x^2 + y^2 \qquad [1.2]$$

Finally, it will often be necessary to find the values of inverse relationships. For example, suppose you know that the sine of an angle is 0.866, but you need to know the value of the angle itself. The inverse sine function may be expressed as $\sin^{-1}(0.866)$, which is a shorthand way of asking the question "What angle has a sine of 0.866?" Punching a couple of buttons on your calculator reveals that this angle is 60.0°. Try it for yourself and show that $\tan^{-1}(0.400) = 21.8°$. Be sure that your calculator is set for degrees and not radians. In addition, the inverse tangent function can return only values between $-90°$ and $+90°$, so when an angle is in the second or third quadrant, it's necessary to add 180° to the answer in the calculator window.

TIP 1.3 Degrees vs. Radians
When calculating trigonometric functions, make sure your calculator setting—degrees or radians—is consistent with the degree measure you're using in a given problem.

The definitions of the trigonometric functions and the inverse trigonometric functions, as well as the Pythagorean theorem, can be applied to *any* right triangle, regardless of whether its sides correspond to x- and y-coordinates.

These results from trigonometry are useful in converting from rectangular coordinates to polar coordinates, or vice versa, as the next example shows.

[1]Many people use the mnemonic *SOHCAHTOA* to remember the basic trigonometric formulas: *S*ine = *O*pposite/*H*ypotenuse, *C*osine = *A*djacent/*H*ypotenuse, and *T*angent = *O*pposite/*A*djacent. (Thanks go to Professor Don Chodrow for pointing this out.)

EXAMPLE 1.9 Cartesian and Polar Coordinates

Goal Understand how to convert from plane rectangular coordinates to plane polar coordinates and vice versa.

Problem **(a)** The Cartesian coordinates of a point in the xy-plane are $(x, y) = (-3.50, -2.50)$ m, as shown in Active Figure 1.7. Find the polar coordinates of this point. **(b)** Convert $(r, \theta) = (5.00 \text{ m}, 37.0°)$ to rectangular coordinates.

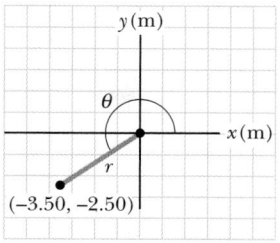

ACTIVE FIGURE 1.7 (Example 1.9) Converting from Cartesian coordinates to polar coordinates.

PhysicsNow™ Log into PhysicsNow at **www.cp7e.com**, and go to Active Figure 1.7 to move the point in the xy-plane and see how its Cartesian and polar coordinates change.

Strategy Apply the trigonometric functions and their inverses, together with the Pythagorean theorem.

Solution

(a) Cartesian to Polar

Take the square root of both sides of Equation 1.2 to find the radial coordinate:

$$r = \sqrt{x^2 + y^2} = \sqrt{(-3.50 \text{ m})^2 + (-2.50 \text{ m})^2} = \boxed{4.30 \text{ m}}$$

Use Equation 1.1 for the tangent function to find the angle with the inverse tangent, adding 180° because the angle is actually in third quadrant:

$$\tan\theta = \frac{y}{x} = \frac{-2.50 \text{ m}}{-3.50 \text{ m}} = 0.714$$

$$\theta = \tan^{-1}(0.714) = 35.5° + 180° = \boxed{216°}$$

(b) Polar to Cartesian

Use the trigonometric definitions, Equation 1.1.

$$x = r\cos\theta = (5.00 \text{ m})\cos 37.0° = \boxed{3.99 \text{ m}}$$

$$y = r\sin\theta = (5.00 \text{ m})\sin 37.0° = \boxed{3.01 \text{ m}}$$

Remarks When we take up vectors in two dimensions in Chapter 3, we will routinely use a similar process to find the direction and magnitude of a given vector from its components, or, conversely, to find the components from the vector's magnitude and direction.

Exercise 1.9

(a) Find the polar coordinates corresponding to $(x, y) = (-3.25, 1.50)$ m. (b) Find the Cartesian coordinates corresponding to $(r, \theta) = (4.00 \text{ m}, 53.0°)$

Answers **(a)** $(r, \theta) = (3.58 \text{ m}, 155°)$ **(b)** $(x, y) = (2.41 \text{ m}, 3.19 \text{ m})$

EXAMPLE 1.10 How High Is the Building?

Goal Apply basic results of trigonometry.

Problem A person measures the height of a building by walking out a distance of 46.0 m from its base and shining a flashlight beam toward the top. When the beam is elevated at an angle of 39.0° with respect to the horizontal, as shown in Figure 1.8, the beam just strikes the top of the building. Find the height of the building and the distance the flashlight beam has to travel before it strikes the top of the building.

Strategy Refer to the right triangle shown in the figure. We know the angle, 39.0°, and the length of the side adjacent to it. Since the height of the building is the side opposite the angle, we can use the tangent function. With the adjacent and opposite sides known, we can then find the hypotenuse with the Pythagorean theorem.

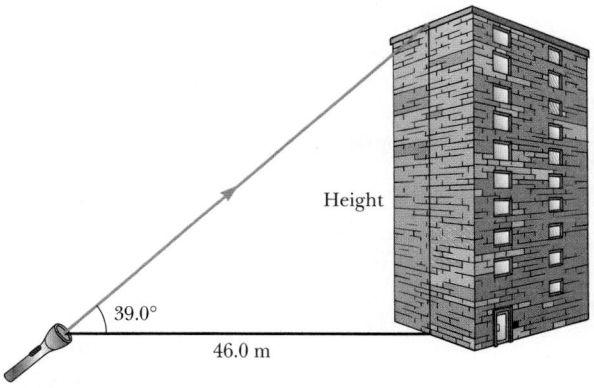

Figure 1.8 (Example 1.10)

Solution

Use the tangent of the given angle:

$$\tan 39.0^\circ = \frac{\text{height}}{46.0 \text{ m}}$$

Solve for the height:

$$\text{Height} = (\tan 39.0^\circ)(46.0 \text{ m}) = (0.810)(46.0 \text{ m}) = 37.3 \text{ m}$$

Find the hypotenuse of the triangle:

$$r = \sqrt{x^2 + y^2} = \sqrt{(37.3 \text{ m})^2 + (46.0 \text{ m})^2} = 59.2 \text{ m}$$

Remarks In a later chapter, right-triangle trigonometry is often used when working with vectors.

Exercise 1.10

High atop a building 50.0 m tall, you spot a friend standing on a street corner. Using a protractor and dangling a plumb bob, you find that the angle between the horizontal and the direction of your friend is 25.0°. Your eyes are located 1.75 m above the top of the building. How far away from the foot of the building is your friend?

Answer 111 m

1.9 PROBLEM-SOLVING STRATEGY

Most courses in general physics require the student to learn the skills used in solving problems, and examinations usually include problems that test such skills. This brief section presents some useful suggestions that will help increase your success in solving problems. An organized approach to problem solving will also enhance your understanding of physical concepts and reduce exam stress. Throughout the book, there will be a number of sections labeled "Problem-Solving Strategy," many of them just a specializing of the list given below (and illustrated in Figure 1.9).

General Problem-Solving Strategy

1. **Read** the problem carefully at least twice. Be sure you understand the nature of the problem before proceeding further.
2. **Draw** a diagram while rereading the problem.
3. **Label** all physical quantities in the diagram, using letters that remind you what the quantity is (e.g., m for mass). Choose a coordinate system and label it.
4. **Identify** physical principles, the knowns and unknowns, and list them. Put circles around the unknowns.
5. **Equations**, the relationships between the labeled physical quantities, should be written down next. Naturally, the selected equations should be consistent with the physical principles identified in the previous step.
6. **Solve** the set of equations for the unknown quantities in terms of the known. Do this algebraically, without substituting values until the next step, except where terms are zero.
7. **Substitute** the known values, together with their units. Obtain a numerical value with units for each unknown.
8. **Check** your answer. Do the units match? Is the answer reasonable? Does the plus or minus sign make sense? Is your answer consistent with an order of magnitude estimate?

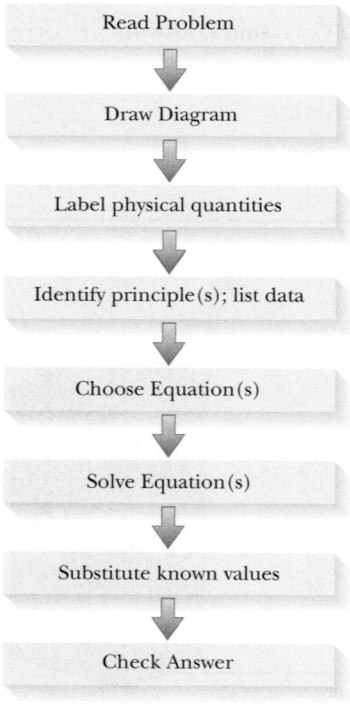

Figure 1.9 A guide to problem solving.

This same procedure, with minor variations, should be followed throughout the course. The first three steps are extremely important, because they get you mentally oriented. Identifying the proper concepts and physical principles assists you in choosing the correct equations. The equations themselves are essential, because when you understand them, you also understand the relationships between the physical quantities. This understanding comes through a lot of daily practice.

Equations are the tools of physics: To solve problems, you have to have them at hand, like a plumber and his wrenches. Know the equations, and understand what they mean and how to use them. Just as you can't have a conversation without knowing the local language, you can't solve physics problems without knowing and understanding the equations. This understanding grows as you study and apply the concepts and the equations relating them.

Carrying through the algebra for as long as possible, substituting numbers only at the end, is also important, because it helps you think in terms of the physical quantities involved, not merely the numbers that represent them. Many beginning physics students are eager to substitute, but once numbers are substituted, it's harder to understand relationships and easier to make mistakes.

The physical layout and organization of your work will make the final product more understandable and easier to follow. Although physics is a challenging discipline, your chances of success are excellent if you maintain a positive attitude and keep trying.

EXAMPLE 1.11 A Round Trip by Air

Goal Illustrate the Problem-Solving Strategy.

Problem An airplane travels 4.50×10^2 km due east and then travels an unknown distance due north. Finally, it returns to its starting point by traveling a distance of 525 km. How far did the airplane travel in the northerly direction?

Strategy We've finished reading the problem (step 1), and have drawn a diagram (step 2) in Figure 1.10 and labeled it (step 3). From the diagram, we recognize a right triangle and identify (step 4) the principle involved: the Pythagorean theorem. Side y is the unknown quantity, and the other sides are known.

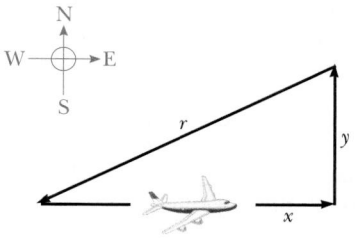

$x = 450$ km
$r = 525$ km
$y = ?$

Figure 1.10 (Example 1.11)

Solution

Write the Pythagorean theorem (step 5): $r^2 = x^2 + y^2$

Solve symbolically for y (step 6): $y^2 = r^2 - x^2 \rightarrow y = +\sqrt{r^2 - x^2}$

Substitute the numbers, with units (step 7): $y = \sqrt{(525\text{ km})^2 - (4.50 \times 10^2\text{ km})^2} = 270\text{ km}$

Remarks Note that the negative solution has been disregarded, because it's not physically meaningful. In checking (step 8), note that the units are correct and that an approximate answer can be obtained by using the easier quantities, 500 km and 400 km. Doing so gives an answer of 300 km, which is approximately the same as our calculated answer of 270 km.

Exercise 1.11

A plane flies 345 km due south, then turns and flies northeast 615 km, until it's due east of its starting point. If the plane now turns and heads for home, how far will it have to go?

Answer 509 km

SUMMARY

PhysicsNow™ Take a practice test by logging into PhysicsNow at **www.cp7e.com** and clicking on the Pre-Test link for this chapter.

1.1 Standards of Length, Mass, and Time

The physical quantities in the study of mechanics can be expressed in terms of three fundamental quantities: length, mass, and time, which have the SI units meters (m), kilograms (kg), and seconds (s), respectively.

1.2 The Building Blocks of Matter

Matter is made of atoms, which in turn are made up of a relatively small nucleus of protons and neutrons within a cloud of electrons. Protons and neutrons are composed of still smaller particles, called quarks.

1.3 Dimensional Analysis

Dimensional analysis can be used to check equations and to assist in deriving them. When the dimensions on both sides of the equation agree, the equation is often correct up to a numerical factor. When the dimensions don't agree, the equation must be wrong.

1.4 Uncertainty in Measurement and Significant Figures

No physical quantity can be determined with complete accuracy. The concept of significant figures affords a basic method of handling these uncertainties. A significant figure is a reliably known digit, other than a zero, used to locate the decimal point. The two rules of significant figures are as follows:

1. When multiplying or dividing using two or more quantities, the result should have the same number of significant figures as the quantity having the fewest significant figures.
2. When quantities are added or subtracted, the number of decimal places in the result should be the same as in the quantity with the fewest decimal places.

Use of scientific notation can avoid ambiguity in significant figures. In rounding, if the last digit dropped is less than 5, simply drop the digit, otherwise raise the last retained digit by one.

1.5 Conversion of Units

Units in physics equations must always be consistent. In solving a physics problem, it's best to start with consistent units, using the table of conversion factors on the inside front cover as necessary.

Converting units is a matter of multiplying the given quantity by a fraction, with one unit in the numerator and its equivalent in the other units in the denominator, arranged so the unwanted units in the given quantity are cancelled out in favor of the desired units.

1.6 Estimates and Order-of-Magnitude Calculations

Sometimes it's useful to find an approximate answer to a question, either because the math is difficult or because information is incomplete. A quick estimate can also be used to check a more detailed calculation. In an order-of-magnitude calculation, each value is replaced by the closest power of ten, which sometimes must be guessed or estimated when the value is unknown. The computation is then carried out. For quick estimates involving known values, each value can first be rounded to one significant figure.

1.7 Coordinate Systems

The Cartesian coordinate system consists of two perpendicular axes, usually called the x-axis and y-axis, with each axis labeled with all numbers from negative infinity to positive infinity. Points are located by specifying the x- and y-values. Polar coordinates consist of a radial coordinate r which is the distance from the origin, and an angular coordinate θ, which is the angular displacement from the positive x-axis.

1.8 Trigonometry

The three most basic trigonometric functions of a right triangle are the sine, cosine, and tangent, defined as follows:

$$\sin\theta = \frac{\text{side opposite } \theta}{\text{hypotenuse}} = \frac{y}{r}$$

$$\cos\theta = \frac{\text{side adjacent to } \theta}{\text{hypotenuse}} = \frac{x}{r} \qquad [1.1]$$

$$\tan\theta = \frac{\text{side opposite } \theta}{\text{side adjacent to } \theta} = \frac{y}{x}$$

The **Pythagorean theorem** is an important relationship between the lengths of the sides of a right triangle:

$$r^2 = x^2 + y^2 \qquad [1.2]$$

where r is the hypotenuse of the triangle and x and y are the other two sides.

CONCEPTUAL QUESTIONS

1. Estimate the order of magnitude of the length, in meters, of each of the following: (a) a mouse, (b) a pool cue, (c) a basketball court, (d) an elephant, (e) a city block.
2. What types of natural phenomena could serve as time standards?
3. (a) Estimate the number of times your heart beats in a month. (b) Estimate the number of human heartbeats in an average lifetime.
4. An object with a mass of 1 kg weighs approximately 2 lb. Use this information to estimate the mass of the following objects: (a) a baseball; (b) your physics textbook; (c) a pickup truck.
5. Find the order of magnitude of your age in seconds.
6. Estimate the number of atoms in 1 cm^3 of a solid. (Note that the diameter of an atom is about 10^{-10} m.)
7. The height of a horse is sometimes given in units of "hands." Why is this a poor standard of length?
8. How many of the lengths or time intervals given in Tables 1.2 and 1.3 could you verify, using only equipment found in a typical dormitory room?
9. An ancient unit of length called the *cubit* was equal to six palms, where a palm was the width of the four fingers of an open hand. Noah's ark was 300 cubits long, 50 cubits wide, and 30 cubits high. Estimate the volume of the ark in cubic meters. Also, estimate the volume of a typical home in cubic meters, and compare it with the volume of the ark.
10. Do an order-of-magnitude calculation for an everyday situation you encounter. For example, how far do you walk or drive each day?
11. If an equation is dimensionally correct, does this mean that the equation must be true? If an equation is not dimensionally correct, does this mean that the equation can't be true?

12. Figure Q1.12 is a photograph showing unit conversions on the labels of some grocery-store items. Check the accuracy of these conversions. Are the manufacturers using significant figures correctly?

Courtesy of George Semple

Figure Q1.12

PROBLEMS

1, 2, 3 = straightforward, intermediate, challenging □ = full solution available in *Student Solutions Manual/Study Guide*
PhysicsNow™ = coached solution with hints available at **www.cp7e.com** = biomedical application

Section 1.3 Dimensional Analysis

1. A shape that covers an area A and has a uniform height h has a volume $V = Ah$. (a) Show that $V = Ah$ is dimensionally correct. (b) Show that the volumes of a cylinder and of a rectangular box can be written in the form $V = Ah$, identifying A in each case. (Note that A, sometimes called the "footprint" of the object, can have any shape and that the height can, in general, be replaced by the average thickness of the object.)

2. (a) Suppose that the displacement of an object is related to time according to the expression $x = Bt^2$. What are the dimensions of B? (b) A displacement is related to time as $x = A\sin(2\pi ft)$, where A and f are constants. Find the dimensions of A. (*Hint*: A trigonometric function appearing in an equation must be dimensionless.)

3. The period of a simple pendulum, defined as the time necessary for one complete oscillation, is measured in time units and is given by

$$T = 2\pi\sqrt{\frac{\ell}{g}}$$

where ℓ is the length of the pendulum and g is the acceleration due to gravity, in units of length divided by time squared. Show that this equation is dimensionally consistent. (You might want to check the formula using your keys at the end of a string and a stopwatch.)

4. Each of the following equations was given by a student during an examination:

$$\tfrac{1}{2}mv^2 = \tfrac{1}{2}mv_0^2 + \sqrt{mgh} \qquad v = v_0 + at^2 \qquad ma = v^2$$

Do a dimensional analysis of each equation and explain why the equation can't be correct.

5. Newton's law of universal gravitation is represented by

$$F = G\frac{Mm}{r^2}$$

where F is the gravitational force, M and m are masses, and r is a length. Force has the SI units $\text{kg}\cdot\text{m/s}^2$. What are the SI units of the proportionality constant G?

6. (a) One of the fundamental laws of motion states that the acceleration of an object is directly proportional to the resultant force on it and inversely proportional to its mass. If the proportionality constant is defined to have no dimensions, determine the dimensions of force. (b) The newton is the SI unit of force. According to the results for (a), how can you express a force having units of newtons by using the fundamental units of mass, length, and time?

Section 1.4 Uncertainty in Measurement and Significant Figures

7. How many significant figures are there in (a) 78.9 ± 0.2, (b) 3.788×10^9, (c) 2.46×10^{-6}, (d) 0.0032?

8. A rectangular plate has a length of (21.3 ± 0.2) cm and a width of (9.8 ± 0.1) cm. Calculate the area of the plate, including its uncertainty.

9. PhysicsNow™ Carry out the following arithmetic operations: (a) the sum of the measured values 756, 37.2, 0.83, and 2.5; (b) the product 0.0032×356.3; (c) the product $5.620 \times \pi$.

10. The speed of light is now defined to be $2.99\ 7924\ 58 \times 10^8$ m/s. Express the speed of light to (a) three significant figures, (b) five significant figures, and (c) seven significant figures.

11. A farmer measures the perimeter of a rectangular field. The length of each long side of the rectangle is found to be 38.44 m, and the length of each short side is found to be 19.5 m. What is the perimeter of the field?

12. The radius of a circle is measured to be (10.5 ± 0.2) m. Calculate (a) the area and (b) the circumference of the circle, and give the uncertainty in each value.

13. A fisherman catches two striped bass. The smaller of the two has a measured length of 93.46 cm (two decimal places, four significant figures), and the larger fish has a measured length of 135.3 cm (one decimal place, four significant figures). What is the total length of fish caught for the day?

14. (a) Using your calculator, find, in scientific notation with appropriate rounding, (a) the value of $(2.437 \times 10^4)(6.5211 \times 10^9)/(5.37 \times 10^4)$ and (b) the value of $(3.14159 \times 10^2)(27.01 \times 10^4)/(1\,234 \times 10^6)$.

Section 1.5 Conversion of Units

15. A fathom is a unit of length, usually reserved for measuring the depth of water. A fathom is approximately 6 ft in length. Take the distance from Earth to the Moon to be 250 000 miles, and use the given approximation to find the distance in fathoms.

16. Find the height or length of these natural wonders in kilometers, meters, and centimeters: (a) The longest cave system in the world is the Mammoth Cave system in Central Kentucky, with a mapped length of 348 miles. (b) In the United States, the waterfall with the greatest single drop is Ribbon Falls in California, which drops 1 612 ft. (c) At 20 320 feet, Mount McKinley in Alaska is America's highest mountain. (d) The deepest canyon in the United States is King's Canyon in California, with a depth of 8 200 ft.

17. A rectangular building lot measures 100 ft by 150 ft. Determine the area of this lot in square meters (m^2).

18. Suppose your hair grows at the rate of 1/32 inch per day. Find the rate at which it grows in nanometers per second. Since the distance between atoms in a molecule is on the order of 0.1 nm, your answer suggests how rapidly atoms are assembled in this protein synthesis.

19. Using the data in Table 1.1 and the appropriate conversion factors, find the distance to the nearest star, in feet.

20. Using the data in Table 1.3 and the appropriate conversion factors, find the age of Earth in years.

21. The speed of light is about 3.00×10^8 m/s. Convert this figure to miles per hour.

22. A house is 50.0 ft long and 26 ft wide and has 8.0-ft-high ceilings. What is the volume of the interior of the house in cubic meters and in cubic centimeters?

23. The amount of water in reservoirs is often measured in acre-ft. One acre-ft is a volume that covers an area of one acre to a depth of one foot. An acre is 43 560 ft^2. Find the volume in SI units of a reservoir containing 25.0 acre-ft of water.

24. The base of a pyramid covers an area of 13.0 acres (1 acre = 43 560 ft^2) and has a height of 481 ft (Fig. P1.24). If the volume of a pyramid is given by the expression $V = bh/3$, where b is the area of the base and h is the height, find the volume of this pyramid in cubic meters.

© Sylvain Grandadam/Photo Researchers, Inc.

Figure P1.24

25. Physics Now™ A quart container of ice cream is to be made in the form of a cube. What should be the length of a side, in centimeters? (Use the conversion 1 gallon = 3.786 liter.)

26. (a) Find a conversion factor to convert from miles per hour to kilometers per hour. (b) For a while, federal law mandated that the maximum highway speed would be 55 mi/h. Use the conversion factor from part (a) to find the speed in kilometers per hour. (c) The maximum highway speed has been raised to 65 mi/h in some places. In kilometers per hour, how much of an increase is this over the 55-mi/h limit?

27. One cubic centimeter ($1.0\ cm^3$) of water has a mass of 1.0×10^{-3} kg. (a) Determine the mass of $1.0\ m^3$ of water. (b) Assuming that biological substances are 98% water, estimate the masses of a cell with a diameter of 1.0 μm, a human kidney, and a fly. Take a kidney to be roughly a sphere with a radius of 4.0 cm and a fly to be roughly a cylinder 4.0 mm long and 2.0 mm in diameter.

28. A billionaire offers to give you $1 billion if you can count out that sum with only $1 bills. Should you accept her offer? Assume that you can count at an average rate of one bill every second, and be sure to allow for the fact that you need about 8 hours a day for sleeping and eating.

Section 1.6 Estimates and Order-of-Magnitude Calculations

Note: In developing answers to the problems in this section, you should state your important assumptions, including the numerical values assigned to parameters used in the solution.

29. Imagine that you are the equipment manager of a professional baseball team. One of your jobs is to keep baseballs on hand for games. Balls are sometimes lost when players hit them into the stands as either home runs or foul balls. Estimate how many baseballs you have to buy per season in order to make up for such losses. Assume that your team plays an 81-game home schedule in a season.

30. A hamburger chain advertises that it has sold more than 50 billion hamburgers. Estimate how many pounds of hamburger meat must have been used by the chain and how many head of cattle were required to furnish the meat.

31. An automobile tire is rated to last for 50 000 miles. Estimate the number of revolutions the tire will make in its lifetime.

32. Grass grows densely everywhere on a quarter-acre plot of land. What is the order of magnitude of the number of blades of grass? Explain your reasoning. Note that 1 acre = 43 560 ft^2.

33. Estimate the number of Ping-Pong balls that would fit into a typical-size room (without being crushed). In your solution, state the quantities you measure or estimate and the values you take for them.

34. Soft drinks are commonly sold in aluminum containers. To an order of magnitude, how many such containers are thrown away or recycled each year by U.S. consumers? How many tons of aluminum does this represent? In your solution, state the quantities you measure or estimate and the values you take for them.

Section 1.7 Coordinate Systems

35. A point is located in a polar coordinate system by the coordinates $r = 2.5$ m and $\theta = 35°$. Find the x- and

y-coordinates of this point, assuming that the two coordinate systems have the same origin.

36. A certain corner of a room is selected as the origin of a rectangular coordinate system. If a fly is crawling on an adjacent wall at a point having coordinates (2.0, 1.0), where the units are meters, what is the distance of the fly from the corner of the room?

37. Express the location of the fly in Problem 36 in polar coordinates.

38. Two points in a rectangular coordinate system have the coordinates (5.0, 3.0) and (− 3.0, 4.0), where the units are centimeters. Determine the distance between these points.

Section 1.8 Trigonometry

39. Physics Now™ For the triangle shown in Figure P1.39, what are (a) the length of the unknown side, (b) the tangent of θ, and (c) the sine of ϕ?

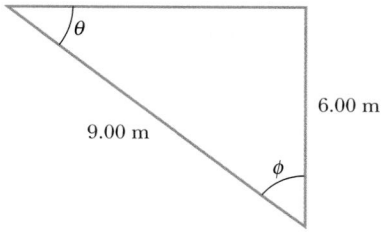

Figure P1.39

40. A ladder 9.00 m long leans against the side of a building. If the ladder is inclined at an angle of 75.0° to the horizontal, what is the horizontal distance from the bottom of the ladder to the building?

41. A high fountain of water is located at the center of a circular pool as shown in Figure P1.41. Not wishing to get his feet wet, a student walks around the pool and measures its circumference to be 15.0 m. Next, the student stands at the edge of the pool and uses a protractor to gauge the angle of elevation at the bottom of the fountain to be 55.0°. How high is the fountain?

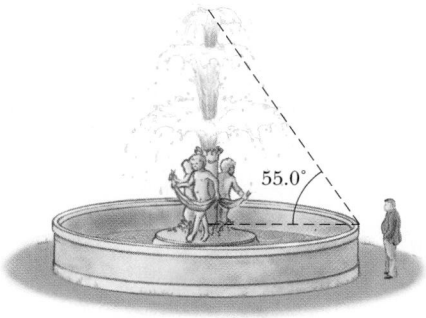

Figure P1.41

42. A right triangle has a hypotenuse of length 3.00 m, and one of its angles is 30.0°. What are the lengths of (a) the side opposite the 30.0° angle and (b) the side adjacent to the 30.0° angle?

43. In Figure P1.43, find (a) the side opposite θ, (b) the side adjacent to ϕ, (c) $\cos\theta$, (d) $\sin\phi$, and (e) $\tan\phi$.

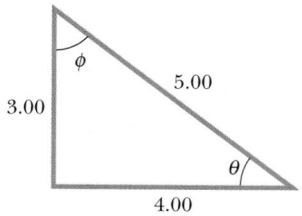

Figure P1.43

44. In a certain right triangle, the two sides that are perpendicular to each other are 5.00 m and 7.00 m long. What is the length of the third side of the triangle?

45. In Problem 44, what is the tangent of the angle for which 5.00 m is the opposite side?

46. A surveyor measures the distance across a straight river by the following method: Starting directly across from a tree on the opposite bank, he walks 100 m along the riverbank to establish a baseline. Then he sights across to the tree. The angle from his baseline to the tree is 35.0°. How wide is the river?

ADDITIONAL PROBLEMS

47. A restaurant offers pizzas in two sizes: small, with a radius of six inches; and large, with a radius of nine inches. A customer argues that if the small one sells for six dollars, the large should sell for nine dollars. Without doing any calculations, is the customer correct? Defend your answer. Calculate the area of each pizza to find out how much pie you are getting in each case. If the small one costs six dollars how much should the large cost?

48. The radius of the planet Saturn is 5.85×10^7 m, and its mass is 5.68×10^{26} kg (Fig. P1.48). (a) Find the density of Saturn (its mass divided by its volume) in grams per cubic centimeter. (The volume of a sphere is given by $(4/3)\pi r^3$.) (b) Find the area of Saturn in square feet. (The surface area of a sphere is given by $4\pi r^2$.)

NASA

Figure P1.48 A view of Saturn.

49. The displacement of an object moving under uniform acceleration is some function of time and the acceleration. Suppose we write this displacement as $s = ka^m t^n$, where k is a dimensionless constant. Show by dimensional analysis that this expression is satisfied if $m = 1$ and $n = 2$. Can the analysis give the value of k?

50. Compute the order of magnitude of the mass of (a) a bathtub filled with water and (b) a bathtub filled with pennies. In your solution, list the quantities you estimate and the value you estimate for each.

51. **PhysicsNow™** You can obtain a rough estimate of the size of a molecule by the following simple experiment: Let a droplet of oil spread out on a smooth surface of water. The resulting oil slick will be approximately one molecule thick. Given an oil droplet of mass 9.00×10^{-7} kg and density 918 kg/m^3 that spreads out into a circle of radius 41.8 cm on the water surface, what is the order of magnitude of the diameter of an oil molecule?

52. In 2003, the U.S. national debt was about \$7 trillion. (a) If payments were made at the rate of \$1 000 per second, how many years would it take to pay off the debt, assuming that no interest were charged? (b) A dollar bill is about 15.5 cm long. If seven trillion dollar bills were laid end to end around the Earth's equator, how many times would they encircle the planet? Take the radius of the Earth at the equator to be 6 378 km. (*Note:* Before doing any of these calculations, try to guess at the answers. You may be very surprised.)

53. Estimate the number of piano tuners living in New York City. This question was raised by the physicist Enrico Fermi, who was well known for making order-of-magnitude calculations.

54. Sphere 1 has surface area A_1 and volume V_1, and sphere 2 has surface area A_2 and volume V_2. If the radius of sphere 2 is double the radius of sphere 1, what is the ratio of (a) the areas, A_2/A_1 and (b) the volumes, V_2/V_1?

55. (a) How many seconds are there in a year? (b) If one micrometeorite (a sphere with a diameter on the order of 10^{-6} m) struck each square meter of the Moon each second, estimate the number of years it would take to cover the Moon with micrometeorites to a depth of one meter. (*Hint:* Consider a cubic box, 1 m on a side, on the Moon, and find how long it would take to fill the box.)

ACTIVITIES

A.1. Choose a variety of objects that range in length from a few centimeters to a few meters. Try guessing the lengths in a unit appropriate to their size, and then use a meter stick supplied by your instructor to check your guesses. Keep trying until you can estimate consistently to within 20% of an object's actual length.

A.2. Choose a variety of objects that range in mass from a few grams to a few kilograms. Estimate the masses by hefting the objects, then use a balance supplied by your instructor to check your guesses. Keep trying until you can estimate consistently to within 30% of an object's actual mass.

A.3. You know that the measurements of a typical sheet of paper are 8.50 in. by 11.0 in. Convert these measurements to millimeters. Use your results to calculate the length of the diagonal of the sheet of paper by using the Pythagorean theorem. Measure the diagonal with a ruler to see how well you have done. Finally, use a suitable trig function to calculate the angle that the diagonal line makes with the horizontal. Use a protractor to verify your calculation.

Caron/Corbis Sygma

Gravity propels a ski jumper down a straight, snow-covered slope at an acceleration that is approximately constant. The equations of kinematics, studied in this chapter, can give his position and velocity along the slope at any time.

CHAPTER

2

Motion In One Dimension

OUTLINE

2.1 Displacement
2.2 Velocity
2.3 Acceleration
2.4 Motion Diagrams
2.5 One-Dimensional Motion with Constant Acceleration
2.6 Freely Falling Objects

Life is motion. Our muscles coordinate motion microscopically to enable us to walk and jog. Our hearts pump tirelessly for decades, moving blood through our bodies. Cell wall mechanisms move select atoms and molecules in and out of cells. From the prehistoric chase of antelopes across the savanna to the pursuit of satellites in space, mastery of motion has been critical to our survival and success as a species.

The study of motion and of physical concepts such as force and mass is called **dynamics**. The part of dynamics that describes motion without regard to its causes is called **kinematics**. In this chapter, the focus is on kinematics in one dimension: motion along a straight line. This kind of motion—and, indeed, *any* motion—involves the concepts of displacement, velocity, and acceleration. Here, we use these concepts to study the motion of objects undergoing constant acceleration. In Chapter 3 we will repeat this discussion for objects moving in two dimensions.

The first recorded evidence of the study of mechanics can be traced to the people of ancient Sumeria and Egypt, who were interested primarily in understanding the motions of heavenly bodies. The most systematic and detailed early studies of the heavens were conducted by the Greeks from about 300 B.C. to A.D. 300. Ancient scientists and laypeople regarded the Earth as the center of the Universe. This **geocentric model** was accepted by such notables as Aristotle (384–322 B.C.) and Claudius Ptolemy (about A.D. 140). Largely because of the authority of Aristotle, the geocentric model became the accepted theory of the Universe until the 17th century.

About 250 B.C., the Greek philosopher Aristarchus worked out the details of a model of the Solar System based on a spherical Earth that rotated on its axis and revolved around the Sun. He proposed that the sky appeared to turn westward because the Earth was turning eastward. This model wasn't given much consideration, because it was believed that if the Earth turned, it would set up a great wind as it moved through the air. We know now that the Earth carries the air and everything else with it as it rotates.

© 2003 Cartoonists & Writers Syndicate

Figure 2.1

The Polish astronomer Nicolaus Copernicus (1473–1543) is credited with initiating the revolution that finally replaced the geocentric model. In his system, called the **heliocentric model**, Earth and the other planets revolve in circular orbits around the Sun.

This early knowledge formed the foundation for the work of Galileo Galilei (1564–1642), who stands out as the dominant facilitator of the entrance of physics into the modern era. In 1609, he became one of the first to make astronomical observations with a telescope. He observed mountains on the Moon, the larger satellites of Jupiter, spots on the Sun, and the phases of Venus. Galileo's observations convinced him of the correctness of the Copernican theory. His quantitative study of motion formed the foundation of Newton's revolutionary work in the next century.

2.1 DISPLACEMENT

Motion involves the displacement of an object from one place in space and time to another. Describing the motion requires some convenient coordinate system and a specified origin. A **frame of reference** is a choice of coordinate axes that defines the starting point for measuring any quantity, an essential first step in solving virtually any problem in mechanics (Fig. 2.1). In Active Figure 2.2a, for example, a car moves along the x-axis. The coordinates of the car at any time describe its position in space and, more importantly, its *displacement* at some given time of interest.

Definition of displacement ▶

The **displacement** Δx of an object is defined as its *change in position*, and is given by

$$\Delta x \equiv x_f - x_i \qquad [2.1]$$

where the initial position of the car is labeled x_i and the final position is x_f. (The indices i and f stand for initial and final, respectively.)

SI unit: meter (m)

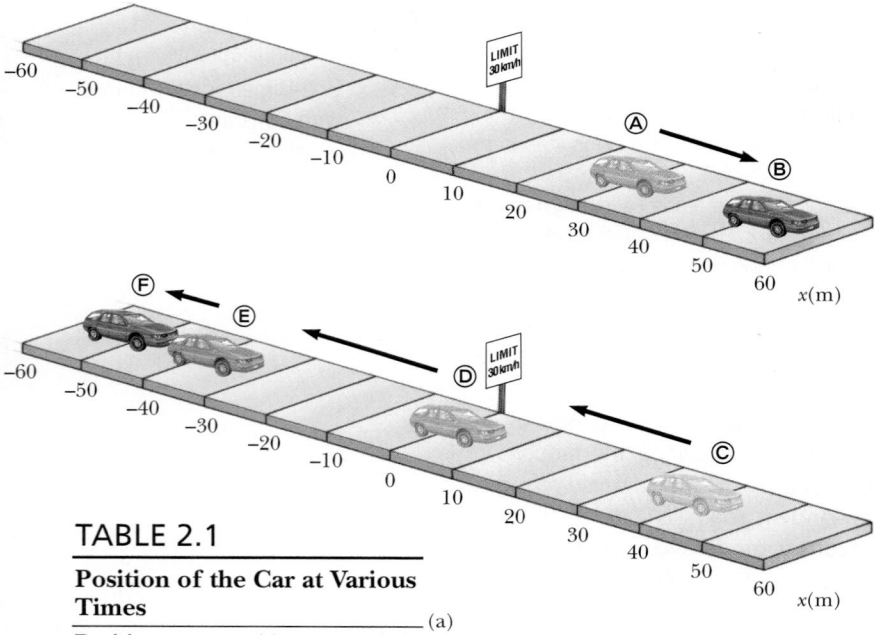

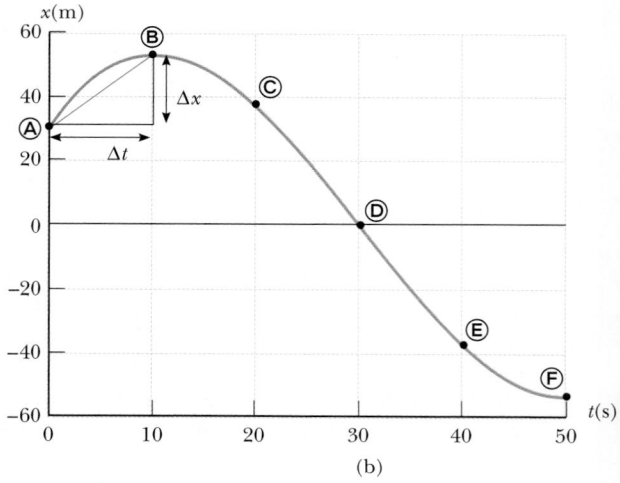

TABLE 2.1

Position of the Car at Various Times

Position	t (s)	x (m)
Ⓐ	0	30
Ⓑ	10	52
Ⓒ	20	38
Ⓓ	30	0
Ⓔ	40	−37
Ⓕ	50	−53

ACTIVE FIGURE 2.2

(a) A car moves back and forth along a straight line taken to be the x-axis. Because we are interested only in the car's translational motion, we can model it as a particle. (b) Graph of position vs. time for the motion of the "particle."

PhysicsNow™

Log into PhysicsNow at **www.cp7e.com**, and go to Active Figure 2.2 to move each of the six points Ⓐ through Ⓕ and observe the motion of the car pictorially and graphically as it follows a smooth path through the points.

We will use the Greek letter delta, Δ, to denote a change in any physical quantity. From the definition of displacement, we see that Δx (read "delta ex") is positive if x_f is greater than x_i and negative if x_f is less than x_i. For example, if the car moves from point Ⓐ to point Ⓑ, so that the initial position is $x_i = 30$ m and the final position is $x_f = 52$ m, the displacement is $\Delta x = x_f - x_i = 52\text{ m} - 30\text{ m} = +22$ m. However, if the car moves from point Ⓒ to point Ⓕ, then the initial position is $x_i = 38$ m and the final position is $x_f = -53$ m, the displacement is $\Delta x = x_f - x_i = -53\text{ m} - 38\text{ m} = -91$ m. A positive answer indicates a displacement in the positive x-direction, whereas a negative answer indicates a displacement in the negative x-direction. Active Figure 2.2b displays the graph of the car's position as a function of time.

TIP 2.1 A Displacement Isn't a Distance!

The displacement of an object is *not* the same as the distance it travels. Toss a tennis ball up and catch it. The ball travels a *distance* equal to twice the maximum height reached, but its *displacement* is zero.

Because displacement has both a magnitude (size) and a direction, it's a vector quantity, as are velocity and acceleration. In general, **a vector quantity is characterized by having both a magnitude and a direction**. By contrast, **a scalar quantity has magnitude, but no direction**. Scalar quantities such as mass and temperature are completely specified by a numeric value with appropriate units; no direction is involved.

Vector quantities will be usually denoted in boldface type with an arrow over the top of the letter. For example, $\vec{\mathbf{v}}$ represents velocity and $\vec{\mathbf{a}}$ denotes an acceleration, both vector quantities. In this chapter, however, it won't be necessary to use that notation, because in one-dimensional motion an object can only move in one of two directions, and these directions are easily specified by plus and minus signs.

TIP 2.2 Vectors Have Both a Magnitude and a Direction.

Scalars have size. Vectors, too, have size, and they also point in a direction.

2.2 VELOCITY

In day-to-day usage, the terms *speed* and *velocity* are interchangeable. In physics, however, there's a clear distinction between them: Speed is a scalar quantity, having only magnitude, while velocity is a vector, having both magnitude and direction.

Why must velocity be a vector? If you want to get to a town 70 km away in an hour's time, it's not enough to drive at a speed of 70 km/h; you must travel in the correct direction as well. This is obvious, but shows that velocity gives considerably more information than speed, as will be made more precise in the formal definitions.

◀ Definition of average speed

The **average speed** of an object over a given time interval is defined as the total distance traveled divided by the total time elapsed:

$$\text{Average speed} \equiv \frac{\text{total distance}}{\text{total time}}$$

SI unit: meter per second (m/s)

In symbols, this equation might be written $v = d/t$, with the letter v understood in context to be the average speed, and not a velocity. Because total distance and total time are always positive, the average speed will be positive, also. The definition of average speed completely ignores what may happen between the beginning and the end of the motion. For example, you might drive from Atlanta, Georgia, to St. Petersburg, Florida, a distance of about 500 miles, in 10 hours. Your average speed is 500 mi/10 h = 50 mi/h. It doesn't matter if you spent two hours in a traffic jam traveling only 5 mi/h and another hour at a rest stop. For average speed, only the total distance traveled and total elapsed time are important.

EXAMPLE 2.1 The Tortoise and The Hare

Goal Apply the concept of average speed.

Problem A turtle and a rabbit engage in a footrace over a distance of 4.00 km. The rabbit runs 0.500 km and then stops for a 90.0-min nap. Upon awakening, he remembers the race and runs twice as fast. Finishing the course in a total time of 1.75 h, the rabbit wins the race. **(a)** Calculate the average speed of the rabbit. **(b)** What was his average speed before he stopped for a nap?

Strategy Finding the overall average speed in part (a) is just a matter of dividing the total distance by the total time. Part (b) requires two equations and two unknowns, the latter turning out to be the two different average speeds: v_1 before the nap and v_2 after the nap. One equation is given in the statement of the problem ($v_2 = 2v_1$), while the other comes from the fact the rabbit ran for only fifteen minutes because he napped for ninety minutes.

Solution

(a) Find the rabbit's overall average speed.

Apply the equation for average speed:

$$\text{Average speed} \equiv \frac{\text{total distance}}{\text{total time}} = \frac{4.00\ \text{km}}{1.75\ \text{h}} = 2.29\ \text{km/h}$$

(b) Find the rabbit's average speed before his nap.

Sum the running times, and set the sum equal to 0.25 h:

$$t_1 + t_2 = 0.250\ \text{h}$$

Substitute $t_1 = d_1/v_1$ and $t_2 = d_2/v_2$:

$$\frac{d_1}{v_1} + \frac{d_2}{v_2} = 0.250\ \text{h} \qquad \textbf{(1)}$$

Equation (1) and $v_2 = 2v_1$ are the two equations needed, and d_1 and d_2 are known. Solve for v_1 by substitution:

$$\frac{d_1}{v_1} + \frac{d_2}{v_2} = \frac{0.500\ \text{km}}{v_1} + \frac{3.50\ \text{km}}{2v_1} = 0.250\ \text{h}$$

$$v_1 = 9.00\ \text{km/h}$$

Remark As seen in this example, average speed can be calculated regardless of any variation in speed over the given time interval.

Exercise 2.1

Estimate the average speed of the Apollo spacecraft in m/s, given that the craft took five days to reach the Moon from Earth. (The Moon is 3.8×10^8 m from Earth.)

Answer ~900 m/s

Unlike average speed, **average velocity** is a vector quantity, having both a magnitude and a direction. Consider again the car of Figure 2.2, moving along the road (the x-axis). Let the car's position be x_i at some time t_i and x_f at a later time t_f. In the time interval $\Delta t = t_f - t_i$, the displacement of the car is $\Delta x = x_f - x_i$.

Definition of average velocity ▶

The average velocity $\bar{v}$ during a time interval Δt is the displacement Δx divided by Δt:

$$\bar{v} = \frac{\Delta x}{\Delta t} = \frac{x_f - x_i}{t_f - t_i} \qquad \textbf{[2.2]}$$

SI Unit: meter per second (m/s)

Unlike the average speed, which is always positive, the average velocity of an object in one dimension can be either positive or negative, depending on the sign of the displacement. (The time interval Δt is always positive.) For example, in Figure 2.2a, the average velocity of the car is positive in the upper illustration, a positive sign indicating motion to the right along the x-axis. Similarly, a negative average velocity for the car in the lower illustration of the figure indicates that it moves to the left along the x-axis.

As an example, we can use the data in Table 2.1 to find the average velocity in the time interval from point Ⓐ to point Ⓑ (assume two digits are significant):

$$\bar{v} = \frac{\Delta x}{\Delta t} = \frac{52\text{ m} - 30\text{ m}}{10\text{ s} - 0\text{ s}} = 2.2\text{ m/s}$$

Aside from meters per second, other common units for average velocity are feet per second (ft/s) in the U.S. customary system and centimeters per second (cm/s) in the cgs system.

To further illustrate the distinction between speed and velocity, suppose we're watching a drag race from the Goodyear blimp. In one run we see a car follow the straight-line path from Ⓟ to Ⓠ shown in Figure 2.3 during the time interval Δt, and in a second run a car follows the curved path during the same interval. From the definition in Equation 2.2, the two cars had the same average velocity, because they had the same displacement $\Delta x = x_f - x_i$ during the same time interval Δt. The car taking the curved route, however, traveled a greater distance and had the higher average speed.

Figure 2.3 A drag race viewed from a blimp. One car follows the red straight-line path from Ⓟ to Ⓠ, and a second car follows the blue curved path.

Quick Quiz 2.1

Figure 2.4 shows the unusual path of a confused football player. After receiving a kickoff at his own goal, he runs downfield to within inches of a touchdown, then reverses direction and races back until he's tackled at the exact location where he first caught the ball. During this run, what is (a) the total distance he travels, (b) his displacement, and (c) his average velocity in the x-direction?

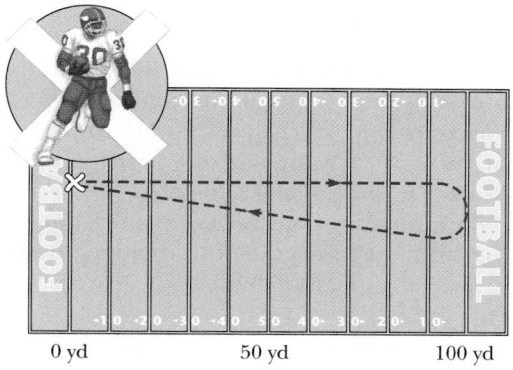

Figure 2.4 (Quick Quiz 2.1) The path followed by a confused football player.

Graphical Interpretation of Velocity

If a car moves along the x-axis from Ⓐ to Ⓑ to Ⓒ, and so forth, we can plot the positions of these points as a function of the time elapsed since the start of the motion. The result is a **position vs. time graph** like those of Figure 2.5. In Figure 2.5a, the graph is a straight line, because the car is moving at constant velocity. The same displacement Δx occurs in each time interval Δt. In this case, the average velocity is always the same and is equal to $\Delta x/\Delta t$. Figure 2.5b is a graph of the data in Table 2.1. Here, the position vs. time graph is not a straight line, because the velocity of the car is changing. Between any two points, however, we can draw a straight line just as in Figure 2.5a, and the slope of that line is the average velocity $\Delta x/\Delta t$ in that time interval. In general, **the average velocity of an object during the time interval Δt is equal to the slope of the straight line joining the initial and final points on a graph of the object's position versus time**.

From the data in Table 2.1 and the graph in Figure 2.5b, we see that the car first moves in the positive x-direction as it travels from Ⓐ to Ⓑ, reaches a position of 52 m at time $t = 10$ s, then reverses direction and heads backwards. In the first 10 s of its motion, as the car travels from Ⓐ to Ⓑ, its average velocity is 2.2 m/s, as previously calculated. In the first 40 seconds, as the car goes from Ⓐ to Ⓔ, its displacement is $\Delta x = -37\text{ m} - (30\text{ m}) = -67\text{ m}$. So the average velocity in this interval, which equals the slope of the blue line in Figure 2.5b from Ⓐ to Ⓔ, is $\bar{v} = \Delta x/\Delta t = (-67\text{ m})/(40\text{ s}) = -1.7\text{ m/s}$. In general, there will be a different average velocity between any distinct pair of points.

TIP 2.3 Slopes of Graphs

The word *slope* is often used in reference to the graphs of physical data. Regardless of the type of data, the *slope* is given by

$$\text{Slope} = \frac{\text{change in vertical axis}}{\text{change in horizontal axis}}$$

Slope carries units.

TIP 2.4 Average Velocity Versus Average Speed

Average velocity is *not* the same as average speed. If you run from $x = 0$ m to $x = 25$ m and back to your starting point in a time interval of 5 s, the average velocity is zero, while the average speed is 10 m/s.

Instantaneous Velocity

Average velocity doesn't take into account the details of what happens *during* an interval of time. On a car trip, for example, you may speed up or slow down a

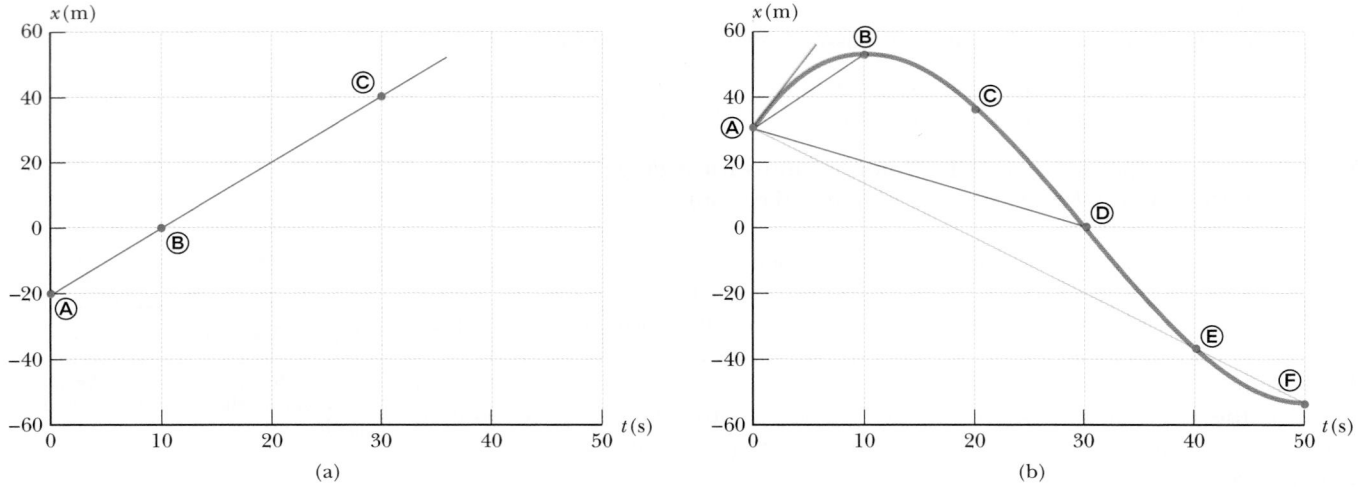

Figure 2.5 (a) Position vs. time graph for the motion of a car moving along the x-axis at constant velocity. (b) Position vs. time graph for the motion of a car with changing velocity, using the data in Table 2.1. The average velocity in the time interval Δt is the slope of the blue straight line connecting Ⓐ and Ⓓ.

number of times in response to the traffic and the condition of the road, and on rare occasions even pull over to chat with a police officer about your speed. What is most important to the police (and to your own safety) is the speed of your car and the direction it was going at a particular instant in time, which together determine the car's **instantaneous velocity**.

So in driving a car between two points, the average velocity must be computed over an interval of time, but the magnitude of instantaneous velocity can be read on the car's speedometer.

Definition of instantaneous velocity ▶

The instantaneous velocity v is the limit of the average velocity as the time interval Δt becomes infinitesimally small:

$$v \equiv \lim_{\Delta t \to 0} \frac{\Delta x}{\Delta t} \qquad [2.3]$$

SI unit: meter per second (m/s)

The notation $\lim_{\Delta t \to 0}$ means that the ratio $\Delta x/\Delta t$ is repeatedly evaluated for smaller and smaller time intervals Δt. As Δt gets extremely close to zero, the ratio $\Delta x/\Delta t$ gets closer and closer to a fixed number, which is defined as the instantaneous velocity.

To better understand the formal definition, consider data obtained on our vehicle via radar (Table 2.2). At $t = 1.00$ s, the car is at $x = 5.00$ m, and at $t = 3.00$ s, it's at $x = 52.5$ m. The average velocity computed for this interval is $\Delta x/\Delta t = (52.5 \text{ m} - 5.00 \text{ m})/(3.00 \text{ s} - 1.00 \text{ s}) = 23.8$ m/s. This result could be used as an estimate for the velocity at $t = 1.00$ s, but it wouldn't be very accurate, because the

TABLE 2.2

Positions of a Car at Specific Instants of Time

t (s)	x (m)
1.00	5.00
1.01	5.47
1.10	9.67
1.20	14.3
1.50	26.3
2.00	34.7
3.00	52.5

TABLE 2.3

Calculated Values of the Time Intervals, Displacements, and Average Velocities for the Car of Table 2.2

Time Interval (s)	Δt (s)	Δx (m)	$\bar{v}$ (m/s)
1.00 to 3.00	2.00	47.5	23.8
1.00 to 2.00	1.00	29.7	29.7
1.00 to 1.50	0.50	21.3	42.6
1.00 to 1.20	0.20	9.30	46.5
1.00 to 1.10	0.10	4.67	46.7
1.00 to 1.01	0.01	0.470	47.0

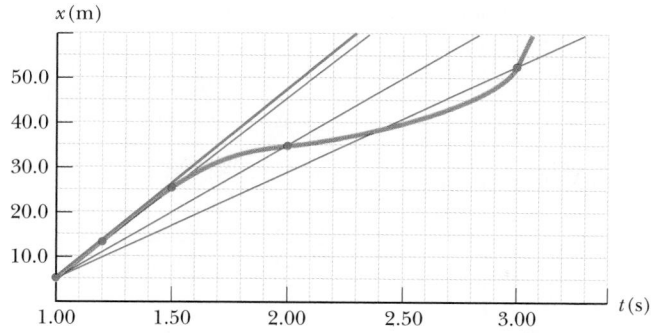

Figure 2.6 Graph representing the motion of the car from the data in Table 2.2. The slope of the blue line represents the average velocity for smaller and smaller time intervals and approaches the slope of the green tangent line.

speed changes considerably in the two-second time interval. Using the rest of the data, we can construct Table 2.3. As the time interval gets smaller, the average velocity more closely approaches the instantaneous velocity. Using the final interval of only 0.010 0 s, we find that the average velocity is $\bar{v} = \Delta x/\Delta t = 0.470\ \text{m}/0.010\ 0\ \text{s} = 47.0\ \text{m/s}$. Since 0.010 0 s is a very short time interval, the actual instantaneous velocity is likely to be very close to this latter average velocity. Finally using the conversion factor on the inside front cover of the book, we see that this is 105 mi/h, a likely violation of the speed limit.

As can be seen in Figure 2.6, the chord formed by the line gradually approaches a tangent line as the time interval becomes smaller. **The slope of the line tangent to the position vs. time curve at *A* is defined to be the instantaneous velocity at that time.**

The instantaneous speed of an object, which is a scalar quantity, is defined as the magnitude of the instantaneous velocity. Like average speed, instantaneous speed (which we will usually call, simply, "speed") has no direction associated with it and hence carries no algebraic sign. For example, if one object has an instantaneous velocity of + 15 m/s along a given line and another object has an instantaneous velocity of − 15 m/s along the same line, both have an instantaneous speed of 15 m/s.

EXAMPLE 2.2 Slowly Moving Train

Goal Obtain average and instantaneous velocities from a graph.

Problem A train moves slowly along a straight portion of track according to the graph of position versus time in Figure 2.7a. Find **(a)** the average velocity for the total trip, **(b)** the average velocity during the first 4.00 s of motion, **(c)** the average velocity during the next 4.00 s of motion, **(d)** the instantaneous velocity at $t = 2.00$ s, and **(e)** the instantaneous velocity at $t = 9.00$ s.

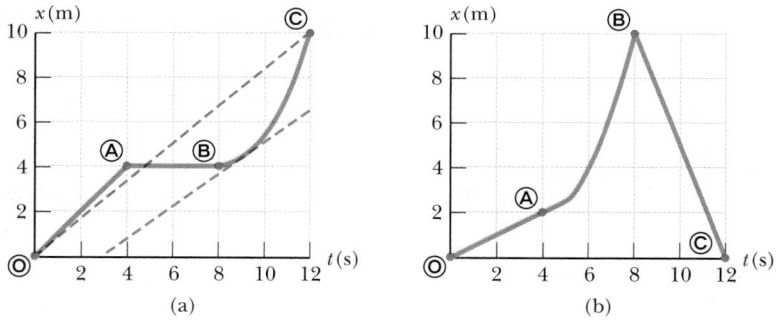

Figure 2.7 (a) (Example 2.2) (b) (Exercise 2.2).

Strategy The average velocities can be obtained by substituting the data into the definition. The instantaneous velocity at $t = 2.00$ s is the same as the average velocity at that point, because the position vs. time graph is a straight line, indicating constant velocity. Finding the instantaneous velocity when $t = 9.00$ s requires sketching a line tangent to the curve at that point and finding its slope.

Solution

(a) Find the average velocity from Ⓞ to Ⓒ.

Calculate the slope of the dashed blue line:

$$\bar{v} = \frac{\Delta x}{\Delta t} = \frac{10.0\ \text{m}}{12.0\ \text{s}} = +0.833\ \text{m/s}$$

(b) Find the average velocity during the first 4 seconds of the train's motion.

Again, find the slope:

$$\bar{v} = \frac{\Delta x}{\Delta t} = \frac{4.00 \text{ m}}{4.00 \text{ s}} = \boxed{+1.00 \text{ m/s}}$$

(c) Find the average velocity during the next four seconds.

Here, there is no change in position, so the displacement Δx is zero:

$$\bar{v} = \frac{\Delta x}{\Delta t} = \frac{0 \text{ m}}{4.00 \text{ s}} = \boxed{0 \text{ m/s}}$$

(d) Find the instantaneous velocity at $t = 2.00$ s.

This is the same as the average velocity found in **(b)**, because the graph is a straight line:

$$v = \boxed{1.00 \text{ m/s}}$$

(e) Find the instantaneous velocity at $t = 9.00$ s.

The tangent line appears to intercept the x-axis at (3.0 s, 0 m) and graze the curve at (9.0 s, 4.5 m). The instantaneous velocity at $t = 9.00$ s equals the slope of the tangent line through these points.

$$v = \frac{\Delta x}{\Delta t} = \frac{4.5 \text{ m} - 0 \text{ m}}{9.0 \text{ s} - 3.0 \text{ s}} = \boxed{0.75 \text{ m/s}}$$

Remarks From the origin to Ⓐ, the train moves at constant speed in the positive x-direction for the first 4.00 s, because the position vs. time curve is rising steadily toward positive values. From Ⓐ to Ⓑ, the train stops at $x = 4.00$ m for 4.00 s. From Ⓑ to Ⓒ, the train travels at increasing speed in the positive x-direction.

Exercise 2.2

Figure 2.7b graphs another run of the train. Find (a) the average velocity from Ⓞ to Ⓒ; (b) the average and instantaneous velocities from Ⓞ to Ⓐ; (c) the approximate instantaneous velocity at $t = 6.0$ s; and (d) the average and instantaneous velocity at $t = 9.0$ s.

Answers (a) 0 m/s (b) both are + 0.5 m/s (c) 2 m/s (d) both are − 2.5 m/s

2.3 ACCELERATION

Going from place to place in your car, you rarely travel long distances at constant velocity. The velocity of the car increases when you step harder on the gas pedal and decreases when you apply the brakes. The velocity also changes when you round a curve, altering your direction of motion. The changing of an object's velocity with time is called **acceleration**.

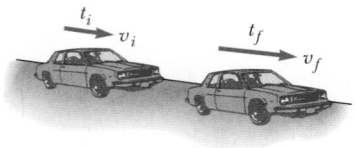

Figure 2.8 A car moving to the right accelerates from a velocity of v_i to a velocity of v_f in the time interval $\Delta t = t_f - t_i$.

Average Acceleration

A car moves along a straight highway as in Figure 2.8. At time t_i it has a velocity of v_i, and at time t_f its velocity is v_f, with $\Delta v = v_f - v_i$ and $\Delta t = t_f - t_i$.

Definition of average acceleration ▶

The average acceleration $\bar{a}$ during the time interval Δt is the change in velocity Δv divided by Δt:

$$\bar{a} \equiv \frac{\Delta v}{\Delta t} = \frac{v_f - v_i}{t_f - t_i} \qquad [2.4]$$

SI unit: meter per second per second (m/s^2)

For example, suppose the car shown in Figure 2.8 accelerates from an initial velocity of $v_i = +10$ m/s to a final velocity of $v_f = +20$ m/s in a time interval of 2 s.

(Both velocities are toward the right, selected as the positive direction.) These values can be inserted into Equation 2.4 to find the average acceleration:

$$\bar{a} = \frac{\Delta v}{\Delta t} = \frac{20 \text{ m/s} - 10 \text{ m/s}}{2 \text{ s}} = +5 \text{ m/s}^2$$

Acceleration is a vector quantity having dimensions of length divided by the time squared. Common units of acceleration are meters per second per second ((m/s)/s, which is usually written m/s^2) and feet per second per second (ft/s^2). An average acceleration of $+5 \text{ m/s}^2$ means that, on average, the car increases its velocity by 5 m/s every second in the positive x-direction.

For the case of motion in a straight line, the direction of the velocity of an object and the direction of its acceleration are related as follows: **When the object's velocity and acceleration are in the same direction, the speed of the object increases with time. When the object's velocity and acceleration are in opposite directions, the speed of the object decreases with time.**

TIP 2.5 Negative Acceleration

Negative acceleration doesn't necessarily mean an object is slowing down. If the acceleration is negative and the velocity is also negative, the object is speeding up!

To clarify this point, suppose the velocity of a car changes from -10 m/s to -20 m/s in a time interval of 2 s. The minus signs indicate that the velocities of the car are in the negative x-direction; they do *not* mean that the car is slowing down! The average acceleration of the car in this time interval is

$$\bar{a} = \frac{\Delta v}{\Delta t} = \frac{-20 \text{ m/s} - (-10 \text{ m/s})}{2 \text{ s}} = -5 \text{ m/s}^2$$

The minus sign indicates that the acceleration vector is also in the negative x-direction. Because the velocity and acceleration vectors are in the same direction, the speed of the car must increase as the car moves to the left. Positive and negative accelerations specify directions relative to chosen axes, not "speeding up" or "slowing down." The terms "speeding up" or "slowing down" refer to an increase and a decrease in speed, respectively.

TIP 2.6 Deceleration

The word *deceleration* means a reduction in speed, a slowing down. Some confuse it with a negative acceleration, which can speed something up. (See Tip 2.5.)

Quick Quiz 2.2

True or False? Define east as the negative direction and west as the positive direction. **(a)** If a car is traveling east, its acceleration must be eastward. **(b)** If a car is slowing down, its acceleration may be positive. **(c)** An object with constant nonzero acceleration can never stop and stay stopped.

Instantaneous Acceleration

The value of the average acceleration often differs in different time intervals, so it's useful to define the **instantaneous acceleration**, which is analogous to the instantaneous velocity discussed in Section 2.2.

◀ Definition of instantaneous acceleration

The instantaneous acceleration a is the limit of the average acceleration as the time interval Δt goes to zero:

$$a \equiv \lim_{\Delta t \to 0} \frac{\Delta v}{\Delta t} \qquad [2.5]$$

SI unit: meter per second per second (m/s^2)

Here again, the notation $\lim_{\Delta t \to 0}$ means that the ratio $\Delta v/\Delta t$ is evaluated for smaller and smaller values of Δt. The closer Δt gets to zero, the closer the ratio gets to a fixed number, which is the instantaneous acceleration.

Figure 2.9, a **velocity vs. time graph**, plots the velocity of an object against time. The graph could represent, for example, the motion of a car along a busy street. The average acceleration of the car between times t_i and t_f can be found by determining the slope of the line joining points Ⓟ and Ⓠ. If we imagine that point Ⓠ is brought closer and closer to point Ⓟ, the line comes closer and closer to becoming tangent at Ⓟ. **The instantaneous acceleration of an object at a given time**

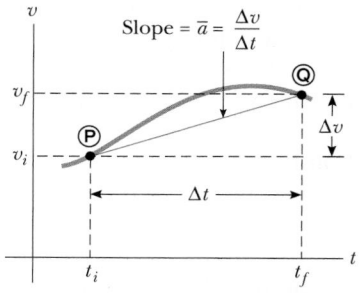

Figure 2.9 Velocity vs. time graph for an object moving in a straight line. The slope of the blue line connecting points Ⓟ and Ⓠ is defined as the average acceleration in the time interval $\Delta t = t_f - t_i$.

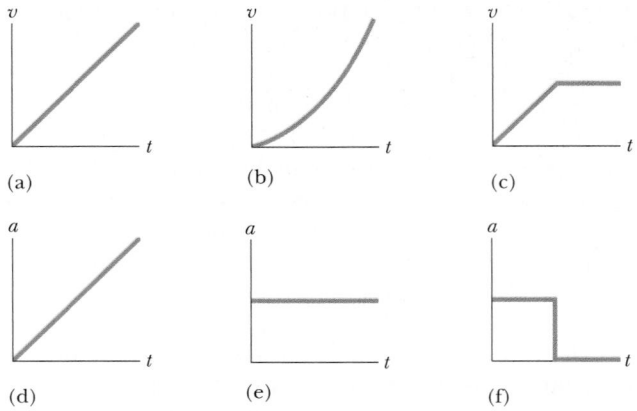

Figure 2.10 (Quick Quiz 2.3) Match each velocity vs. time graph to its corresponding acceleration vs. time graph.

equals the slope of the tangent to the velocity vs. time graph at that time. From now on, we will use the term *acceleration* to mean "instantaneous acceleration."

In the special case where the velocity vs. time graph of an object's motion is a straight line, the instantaneous acceleration of the object at any point is equal to its average acceleration. This also means that the tangent line to the graph overlaps the graph itself. In that case, the object's acceleration is said to be *uniform*, which means that it has a constant value. Constant acceleration problems are important in kinematics and will be studied extensively in this and the next chapter.

Quick Quiz 2.3

Parts (a), (b), and (c) of Figure 2.10 represent three graphs of the velocities of different objects moving in straight-line paths as functions of time. The possible accelerations of each object as functions of time are shown in parts (d), (e), and (f). Match each velocity vs. time graph with the acceleration vs. time graph that best describes the motion.

EXAMPLE 2.3 Catching a Fly Ball

Goal Apply the definition of instantaneous acceleration.

Problem A baseball player moves in a straight-line path in order to catch a fly ball hit to the outfield. His velocity as a function of time is shown in Figure 2.11a. Find his instantaneous acceleration at points Ⓐ, Ⓑ, and Ⓒ.

Strategy At each point, the velocity vs. time graph is a straight line segment, so the instantaneous acceleration will be the slope of that segment. Select two points on each segment and use them to calculate the slope.

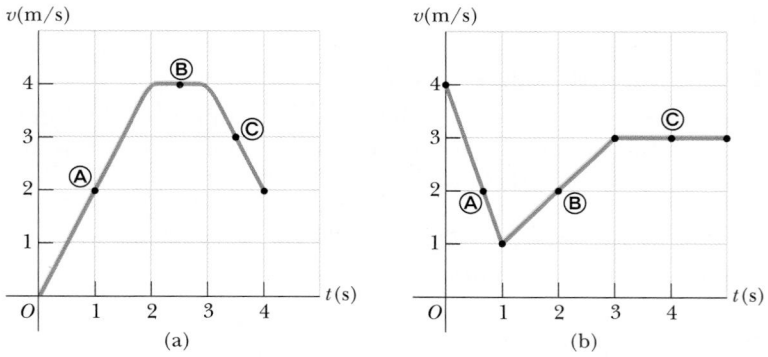

Figure 2.11 (a) (Example 2.3) (b) (Exercise 2.3)

Solution

Acceleration at Ⓐ.

The acceleration equals the slope of the line connecting the points (0 s, 0 m/s) and (2.0 s, 4.0 m/s):

$$a = \frac{\Delta v}{\Delta t} = \frac{4.0 \text{ m/s} - 0}{2.0 \text{ s} - 0} = \boxed{+2.0 \text{ m/s}^2}$$

Acceleration at Ⓑ.

$\Delta v = 0$, because the segment is horizontal:

$$a = \frac{\Delta v}{\Delta t} = \frac{4.0 \text{ m/s} - 4.0 \text{ m/s}}{3.0 \text{ s} - 2.0 \text{ s}} = \boxed{0 \text{ m/s}^2}$$

Acceleration at Ⓒ.

The acceleration equals the slope of the line connecting the points (3.0 s, 4.0 m/s) and (4.0 s, 2.0 m/s):

$$a = \frac{\Delta v}{\Delta t} = \frac{2.0\ \text{m/s} - 4.0\ \text{m/s}}{4.0\ \text{s} - 3.0\ \text{s}} = -2.0\ \text{m/s}^2$$

Remarks For the first 2.0 s, the ballplayer moves in the positive x-direction (the velocity is positive) and steadily accelerates (the curve is steadily rising) to a maximum speed of 4.0 m/s. He moves for 1.0 s at a steady speed of 4.0 m/s and then slows down in the last second (the v vs. t curve is falling), still moving in the positive x-direction (v is always positive).

Exercise 2.3

Repeat the problem, using Figure 2.11b.

Answer The accelerations at Ⓐ, Ⓑ, and Ⓒ are $-3.0\ \text{m/s}^2$, $1.0\ \text{m/s}^2$, and $0\ \text{m/s}^2$, respectively.

2.4 MOTION DIAGRAMS

Velocity and acceleration are sometimes confused with each other, but they're very different concepts, as can be illustrated with the help of motion diagrams. A **motion diagram** is a representation of a moving object at successive time intervals, with velocity and acceleration vectors sketched at each position, red for velocity vectors and violet for acceleration vectors, as in Active Figure 2.12. The time intervals between adjacent positions in the motion diagram are assumed equal.

A motion diagram is analogous to images resulting from a stroboscopic photograph of a moving object. Each image is made as the strobe light flashes. Active Figure 2.12 represents three sets of strobe photographs of cars moving along a straight roadway from left to right. The time intervals between flashes of the stroboscope are equal in each diagram.

In Active Figure 2.12a, the images of the car are equally spaced: The car moves the same distance in each time interval. This means that the car moves with *constant positive velocity* and has *zero acceleration.* The red arrows are all the same length (constant velocity) and there are no violet arrows (zero acceleration).

In Active Figure 2.12b, the images of the car become farther apart as time progresses and the velocity vector increases with time, because the car's displacement between adjacent positions increases as time progresses. The car is moving with a *positive velocity* and a constant *positive acceleration.* The red arrows are successively longer in each image, and the violet arrows point to the right.

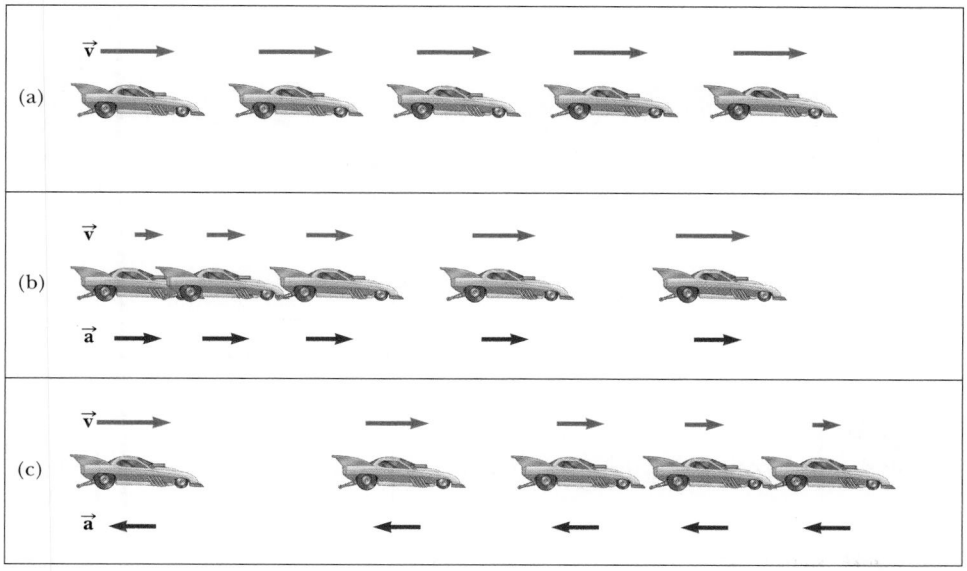

ACTIVE FIGURE 2.12
(a) Motion diagram for a car moving at constant velocity (zero acceleration). (b) Motion diagram for a car undergoing constant acceleration in the direction of its velocity. The velocity vector at each instant is indicated by a red arrow, and the constant acceleration vector by a violet arrow. (c) Motion diagram for a car undergoing constant acceleration in the direction *opposite* the velocity at each instant.

PhysicsNow™
Log into PhysicsNow at **www.cp7e.com**, and go to Active Figure 2.12, where you can select the constant acceleration and initial velocity of the car and observe pictorial and graphical representations of its motion.

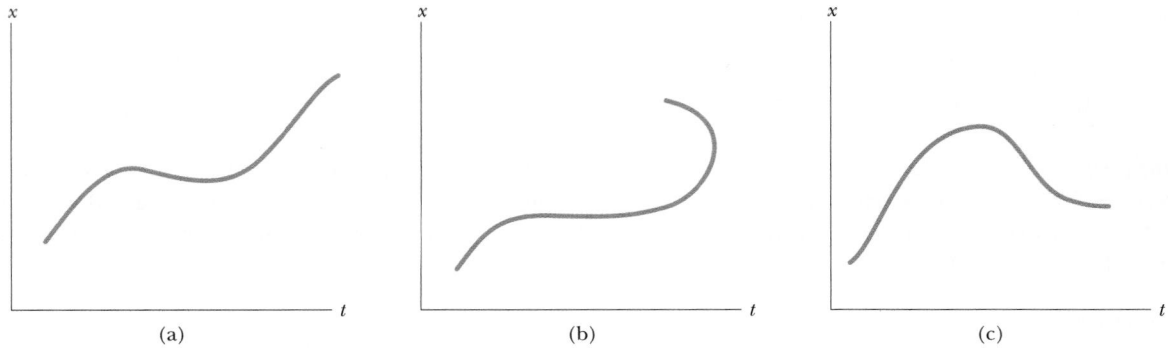

ACTIVE FIGURE 2.13
(Quick Quiz 2.4) Which position vs. time curve is impossible?

PhysicsNow™
Log into PhysicsNow at **www.cp7e.com**, and go to Active Figure 2.13, where you can practice matching appropriate velocity vs. time graphs and acceleration vs. time graphs.

In Active Figure 2.12c, the car slows as it moves to the right because its displacement between adjacent positions decreases with time. In this case, the car moves initially to the right with a constant negative acceleration. The velocity vector decreases in time (the red arrows get shorter) and eventually reaches zero, as would happen when the brakes are applied. Note that the acceleration and velocity vectors are *not* in the same direction. The car is moving with a *positive velocity,* but with a *negative acceleration.*

Try constructing your own diagrams for various problems involving kinematics.

Quick Quiz 2.4

The three graphs in Active Figure 2.13 represent the position vs. time for objects moving along the x-axis. Which, if any, of these graphs is not physically possible?

Quick Quiz 2.5

Figure 2.14a is a diagram of a multiflash image of an air puck moving to the right on a horizontal surface. The images sketched are separated by equal time intervals, and the first and last images show the puck at rest. (a) In Figure 2.14b, which color graph best shows the puck's position as a function of time? (b) In Figure 2.14c, which color graph best shows the puck's velocity as a function of time? (c) In Figure 2.14d, which color graph best shows the puck's acceleration as a function of time?

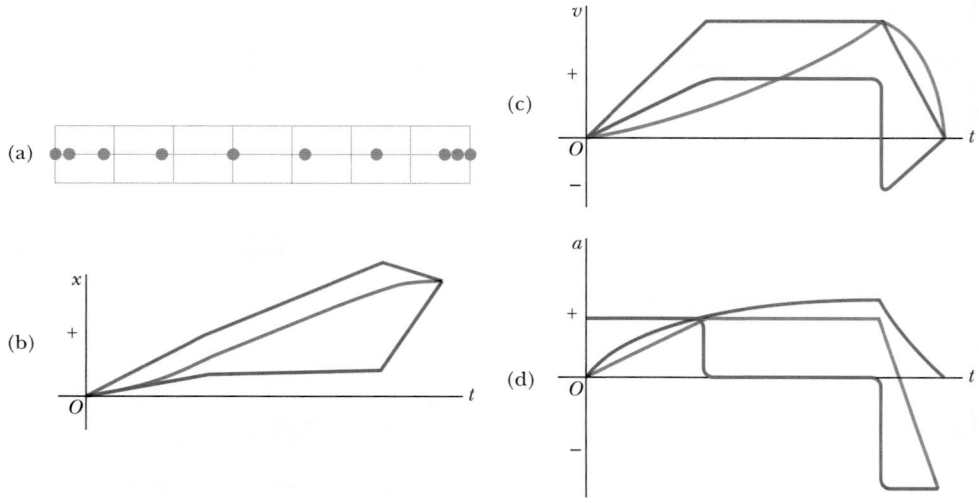

Figure 2.14 (Quick Quiz 2.5) Choose the correct graphs.

2.5 ONE-DIMENSIONAL MOTION WITH CONSTANT ACCELERATION

Many applications of mechanics involve objects moving with *constant acceleration.* This type of motion is important because it applies to numerous objects in nature, such as an object in free fall near Earth's surface (assuming that air resistance can be neglected). A graph of acceleration versus time for motion with constant acceleration is shown in Active Figure 2.15a. **When an object moves with constant acceleration, the instantaneous acceleration at any point in a time interval is equal to the value of the average acceleration over the entire time interval.** Consequently, the velocity increases or decreases at the same rate throughout the motion, and a plot of v versus t gives a straight line with either positive, zero, or negative slope.

Because the average acceleration equals the instantaneous acceleration when a is constant, we can eliminate the bar used to denote average values from our defining equation for acceleration, writing $\bar{a} = a$, so that Equation 2.4 becomes

$$a = \frac{v_f - v_i}{t_f - t_i}$$

The observer timing the motion is always at liberty to choose the initial time, so for convenience, let $t_i = 0$ and t_f be any arbitrary time t. Also, let $v_i = v_0$ (the initial velocity at $t = 0$) and $v_f = v$ (the velocity at any arbitrary time t). With this notation, we can express the acceleration as

$$a = \frac{v - v_0}{t}$$

or

$$v = v_0 + at \qquad \text{(for constant } a\text{)} \qquad \textbf{[2.6]}$$

Equation 2.6 states that the acceleration a steadily changes the initial velocity v_0 by an amount at. For example, if a car starts with a velocity of $+2.0$ m/s to the right and accelerates to the right with $a = +6.0\ \text{m/s}^2$, it will have a velocity of $+14$ m/s after 2.0 s have elapsed:

$$v = v_0 + at = +2.0\ \text{m/s} + (6.0\ \text{m/s}^2)(2.0\ \text{s}) = +14\ \text{m/s}$$

The graphical interpretation of v is shown in Active Figure 2.15b. The velocity varies linearly with time according to Equation 2.6, as it should for constant acceleration.

Because the velocity is increasing or decreasing *uniformly* with time, we can express the average velocity in any time interval as the arithmetic average of the initial velocity v_0 and the final velocity v:

$$\bar{v} = \frac{v_0 + v}{2} \qquad \text{(for constant } a\text{)} \qquad \textbf{[2.7]}$$

Remember that this expression is valid only when the acceleration is constant, in which case the velocity increases uniformly.

We can now use this result along with the defining equation for average velocity, Equation 2.2, to obtain an expression for the displacement of an object as a function of time. Again, we choose $t_i = 0$ and $t_f = t$, and for convenience, we write $\Delta x = x_f - x_i = x - x_0$. This results in

$$\Delta x = \bar{v}t = \left(\frac{v_0 + v}{2}\right)t$$

$$\Delta x = \tfrac{1}{2}(v_0 + v)t \qquad \text{(for constant } a\text{)} \qquad \textbf{[2.8]}$$

We can obtain another useful expression for displacement by substituting the equation for v (Eq. 2.6) into Equation 2.8:

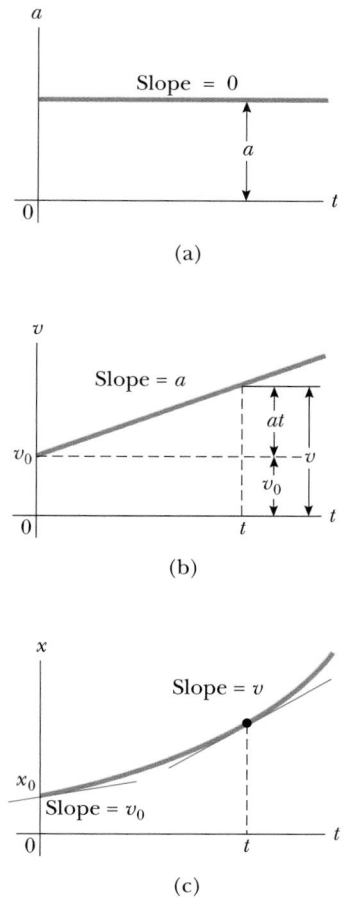

ACTIVE FIGURE 2.15
A particle moving along the x-axis with constant acceleration a.
(a) the acceleration vs. time graph,
(b) the velocity vs. time graph, and
(c) the position vs. time graph.

PhysicsNow™
Log into PhysicsNow at **www.cp7e.com**, and go to Active Figure 2.15, where you can adjust the constant acceleration and observe the effect on the position and velocity graphs.

TABLE 2.4

Equations for Motion in a Straight Line Under Constant Acceleration

Equation	Information Given by Equation
$v = v_0 + at$	Velocity as a function of time
$\Delta x = v_0 t + \frac{1}{2}at^2$	Displacement as a function of time
$v^2 = v_0^2 + 2a\Delta x$	Velocity as a function of displacement

Note: Motion is along the x-axis. At $t = 0$, the velocity of the particle is v_0.

$$\Delta x = \tfrac{1}{2}(v_0 + v_0 + at)\,t$$

$$\Delta x = v_0 t + \tfrac{1}{2}at^2 \qquad \text{(for constant } a\text{)} \qquad \textbf{[2.9]}$$

This equation can also be written in terms of the position x, since $\Delta x = x - x_0$. Active Figure 2.15c shows a plot of x versus t for Equation 2.9, which is related to the graph of velocity vs. time: The area under the curve in Active Figure 2.15b is equal to $v_0 t + \frac{1}{2}at^2$, which is equal to the displacement Δx. In fact, **the area under the graph of v versus t for any object is equal to the displacement Δx of the object.**

Finally, we can obtain an expression that doesn't contain time by solving Equation 2.6 for t and substituting into Equation 2.8, resulting in

$$\Delta x = \tfrac{1}{2}(v + v_0)\left(\frac{v - v_0}{a}\right) = \frac{v^2 - v_0^2}{2a}$$

$$v^2 = v_0^2 + 2a\Delta x \qquad \text{(for constant } a\text{)} \qquad \textbf{[2.10]}$$

Equations 2.6 and 2.9 together can solve any problem in one-dimensional motion with constant acceleration, but Equations 2.7, 2.8, and, especially, 2.10 are sometimes convenient. The three most useful equations—Equations 2.6, 2.9, and 2.10—are listed in Table 2.4.

The best way to gain confidence in the use of these equations is to work a number of problems. There is usually more than one way to solve a given problem, depending on which equations are selected and what quantities are given. The difference lies mainly in the algebra.

Problem-Solving Strategy Accelerated Motion

The following procedure is recommended for solving problems involving accelerated motion.

1. **Read** the problem.
2. **Draw** a diagram, choosing a coordinate system, labeling initial and final points, and indicating directions of velocities and accelerations with arrows.
3. **Label** all quantities, circling the unknowns. Convert units as needed.
4. **Equations** from Table 2.4 should be selected next. All kinematics problems in this chapter can be solved with the first two equations, and the third is often convenient.
5. **Solve** for the unknowns. Doing so often involves solving two equations for two unknowns. It's usually more convenient to substitute all known values before solving.
6. **Check** your answer, using common sense and estimates.

TIP 2.7 Pigs Don't Fly

After solving a problem, you should think about your answer and decide whether it seems reasonable. If it isn't, look for your mistake!

Most of these problems reduce to writing the kinematic equations from Table 2.4 and then substituting the correct values into the constants a, v_0, and x_0 from the given information. Doing this produces two equations—one linear and one quadratic—for two unknown quantities.

EXAMPLE 2.4 The Daytona 500

Goal Apply the basic kinematic equations.

Problem A race car starting from rest accelerates at a constant rate of 5.00 m/s^2. What is the velocity of the car after it has traveled 1.00×10^2 ft?

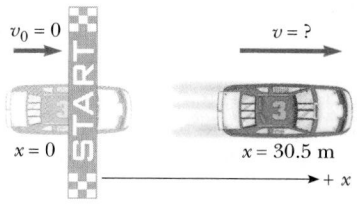

Figure 2.16 (Example 2.4)

Strategy We've read the problem, drawn the diagram in Figure 2.16, and chosen a coordinate system (steps 1 and 2). We'd like to find the velocity v after a certain known displacement Δx. The acceleration a is also known, as is the initial velocity v_0 (step 3, labeling, is complete), so the third equation in Table 2.4 looks most useful. The rest is simple substitution.

Solution

Convert units of Δx to SI, using the information in the inside front cover.

$$1.00 \times 10^2 \text{ ft} = (1.00 \times 10^2 \text{ ft})\left(\frac{1 \text{ m}}{3.28 \text{ ft}}\right) = 30.5 \text{ m}$$

Write the kinematics equation for v^2 (step 4):

$$v^2 = v_0^2 + 2a\,\Delta x$$

Solve for v, taking the positive square root because the car moves to the right (step 5):

$$v = \sqrt{v_0^2 + 2a\,\Delta x}$$

Substitute $v_0 = 0$, $a = 5.00$ m/s^2, and $\Delta x = 30.5$ m:

$$v = \sqrt{v_0^2 + 2a\,\Delta x} = \sqrt{(0)^2 + 2(5.00 \text{ m/s}^2)(30.5 \text{ m})}$$
$$= 17.5 \text{ m/s}$$

Remarks The answer is easy to check. An alternate technique is to use $\Delta x = v_0 t + \frac{1}{2}at^2$ to find t and then use the equation $v = v_0 + at$ to find v.

Exercise 2.4

Suppose the driver in this example now slams on the brakes, stopping the car in 4.00 s. Find (a) the acceleration and (b) the distance the car travels, assuming the acceleration is constant.

Answers (a) $a = -4.38$ m/s^2 (b) $d = 35.0$ m

INTERACTIVE EXAMPLE 2.5 Car Chase

Goal Solve a problem involving two objects, one moving at constant acceleration and the other at constant velocity.

Problem A car traveling at a constant speed of 24.0 m/s passes a trooper hidden behind a billboard, as in Figure 2.17. One second after the speeding car passes the billboard, the trooper sets off in chase with a constant acceleration of 3.00 m/s^2. **(a)** How long does it take the trooper to overtake the speeding car? **(b)** How fast is the trooper going at that time?

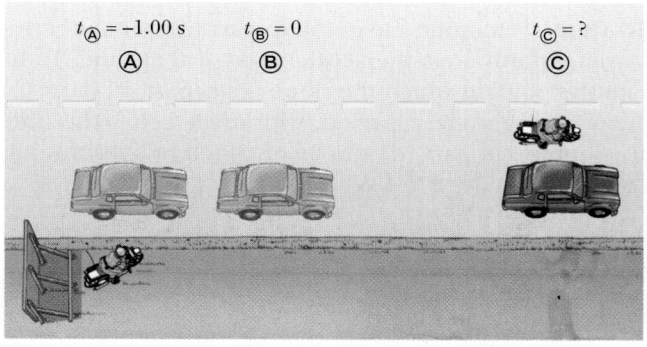

Figure 2.17 (Example 2.5) A speeding car passes a hidden trooper. When does the trooper catch up to the car?

Strategy Solving this problem involves two simultaneous kinematics equations of position, one for the police motorcycle and the other for the car. Choose $t = 0$ to correspond to the time the trooper takes up the chase, when the car is at $x_{car} = 24.0$ m because of its head start (24.0 m/s $\times$ 1.00 s). The trooper catches up with the car when their positions are the same, which suggests setting $x_{trooper} = x_{car}$ and solving for time, which can then be used to find the trooper's speed in part (b).

Solution

(a) How long does it take the trooper to overtake the car?

Write the equation for the car's displacement:

$$\Delta x_{car} = x_{car} - x_0 = v_0 t + \tfrac{1}{2}a_{car}t^2$$

Take $x_0 = 24.0$ m, $v_0 = 24.0$ m/s and $a_{car} = 0$. Solve for x_{car}:

$$x_{car} = x_0 + vt = 24.0 \text{ m} + (24.0 \text{ m/s})t$$

Write the equation for the trooper's position, taking $x_0 = 0$, $v_0 = 0$, and $a_{trooper} = 3.00$ m/s^2:

$$x_{trooper} = \tfrac{1}{2}a_{trooper}t^2 = \tfrac{1}{2}(3.00 \text{ m/s}^2)t^2 = (1.50 \text{ m/s}^2)t^2$$

Set $x_{trooper} = x_{car}$, and solve the quadratic equation. (The quadratic formula appears in Appendix A, Equation A.8.) Only the positive root is meaningful.

$$(1.50 \text{ m/s}^2)t^2 = 24.0 \text{ m} + (24.0 \text{ m/s})t$$

$$(1.50 \text{ m/s}^2)t^2 - (24.0 \text{ m/s})t - 24.0 \text{ m} = 0$$

$$t = 16.9 \text{ s}$$

(b) Find the trooper's speed at this time.

Substitute the time into the trooper's velocity equation:

$$v_{trooper} = v_0 + a_{trooper}t = 0 + (3.00 \text{ m/s}^2)(16.9 \text{ s})$$

$$= 50.7 \text{ m/s}$$

Remarks The trooper, traveling about twice as fast as the car, must swerve or apply his brakes strongly to avoid a collision! This problem can also be solved graphically, by plotting position versus time for each vehicle on the same graph. The intersection of the two graphs corresponds to the time and position at which the trooper overtakes the car.

Exercise 2.5

A motorist with an expired license tag is traveling at 10.0 m/s down a street, and a policeman on a motorcycle, taking another 5.00 s to finish his donut, gives chase at an acceleration of 2.00 m/s^2. Find (a) the time required to catch the car and (b) the distance the trooper travels while overtaking the motorist.

Answers (a) 13.7 s (b) 188 m

PhysicsNow™ You can study the motion of the car and the trooper for various velocities of the car by logging into PhysicsNow at **www.cp7e.com** and going to Interactive Example 2.5.

EXAMPLE 2.6 The Acela: The Porsche of American Trains

Problem The sleek high-speed electric train known as the Acela (pronounced ahh-sell-ah) is currently in service on the Washington-New York-Boston run and is shown in Figure 2.18a. The Acela consists of two power cars and six coaches and can carry 304 passengers at speeds up to 170 mi/h. In order to negotiate curves comfortably at high speeds, the train carriages tilt as much as 6° from the vertical, to prevent passengers from being pushed to the side. A velocity vs. time graph for the Acela is shown in Figure 2.18b. **(a)** Describe the motion of the Acela. **(b)** Find the peak acceleration of the Acela in miles per hour per second ((mi/h)/s) as the train speeds up from 45 mi/h to 170 mi/h. **(c)** Find the train's displacement in miles between $t = 0$ and $t = 200$ s. **(d)** Find the average acceleration of the Acela and its displacement in miles in the interval from 200 s to 300 s. (The train has regenerative braking, which means that it feeds energy back into the utility lines each time it stops!) **(e)** Find the total displacement in the interval from 0 to 400 s.

Strategy Examine the graph in part (a), remembering that the slope of the tangent line at any point of the velocity vs. time graph gives the acceleration at that time. To find the peak acceleration in part (b), study the graph and locate the point at which the slope is steepest. In parts (c)–(e), estimating the area under the curve gives the displacement during a given period, with areas below the time axis, as in part (e), subtracted from the total. The average acceleration in part (d) can be obtained by substituting numbers taken from the graph into the definition of average acceleration, $\bar{a} = \Delta v/\Delta t$.

Solution

(a) Describe the motion.

From about −50 s to 50 s, the Acela cruises at a constant velocity in the $+x$-direction. Then the train accelerates in the $+x$-direction from 50 s to 200 s, reaching a top speed of about 170 mi/h, whereupon it brakes to rest at 350 s and reverses, steadily gaining speed in the $-x$-direction.

(b) Find the peak acceleration.

Calculate the slope of the steepest tangent line, which connects the points (50 s, 50 mi/h) and (100 s, 150 mi/h) (the light blue line in Figure 2.18c):

$$a = \text{slope} = \frac{\Delta v}{\Delta t} = \frac{(1.5 \times 10^2 - 5.0 \times 10^1) \text{ mi/h}}{(1.0 \times 10^2 - 5.0 \times 10^1) \text{ s}}$$

$$= 2.0 \text{ (mi/h)/s}$$

(c) Find the displacement between 0 s and 200 s.

Using triangles and rectangles, approximate the area in Figure 2.18d:

$$\begin{aligned}\Delta x_{0\to 200\text{ s}} &= \text{area}_1 + \text{area}_2 + \text{area}_3 + \text{area}_4 + \text{area}_5 \\ &\approx (5.0\times 10^1\text{ mi/h})(5.0\times 10^1\text{ s}) \\ &\quad + (5.0\times 10^1\text{ mi/h})(5.0\times 10^1\text{ s}) \\ &\quad + (1.6\times 10^2\text{ mi/h})(1.0\times 10^2\text{ s}) \\ &\quad + \tfrac{1}{2}(5.0\times 10^1\text{ s})(1.0\times 10^2\text{ mi/h}) \\ &\quad + \tfrac{1}{2}(1.0\times 10^2\text{ s})(1.7\times 10^2\text{ mi/h} - 1.6\times 10^2) \\ &= 2.4\times 10^4(\text{mi/h})\text{s}\end{aligned}$$

Courtesy Amtrak NEC Media Relations

(a)

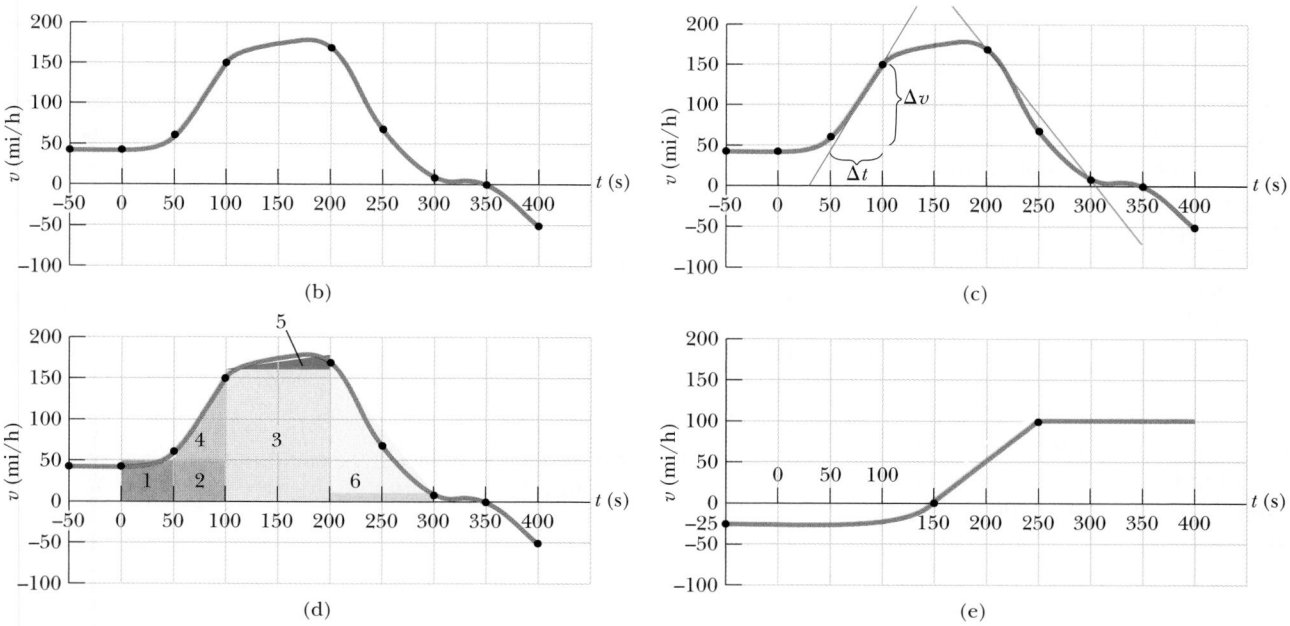

(b) (c) (d) (e)

Figure 2.18 (Example 2.6) (a) The Acela, 1 250 000 lb of cold steel thundering along at 170 mi/h. (b) Velocity vs. time graph for the Acela. (c) The slope of the steepest tangent blue line gives the peak acceleration, while the slope of the green line is the average acceleration between 200 s and 300 s. (d) The area under the velocity vs. time graph in some time interval gives the displacement of the Acela in that time interval. (e) (Exercise 2.6).

Convert units to miles by converting hours to seconds:

$$\Delta x_{0\to 200\text{ s}} \approx 2.4\times 10^4\,\frac{\text{mi}\cdot\text{s}}{\text{h}}\left(\frac{1\text{ h}}{3\,600\text{ s}}\right) = \boxed{6.7\text{ mi}}$$

(d) Find the average acceleration from 200 s to 300 s, and find the displacement.

The slope of the green line is the average acceleration from 200 s to 300 s (Fig. 2.18c):

$$\begin{aligned}\bar{a} = \text{slope} = \frac{\Delta v}{\Delta t} &= \frac{(1.0\times 10^1 - 1.7\times 10^2)\text{ mi/h}}{1.0\times 10^2\text{ s}} \\ &= \boxed{-1.6(\text{mi/h})/\text{s}}\end{aligned}$$

The displacement from 200 s to 300 s is equal to area$_6$, which is the area of a triangle plus the area of a very narrow rectangle beneath the triangle:

$$\Delta x_{200\to300\text{ s}} \approx \tfrac{1}{2}(1.0\times10^2\text{ s})(1.7\times10^2 - 1.0\times10^1)\text{ mi/h} + (1.0\times10^1\text{ mi/h})(1.0\times10^2\text{ s}) = 9.0\times10^3(\text{mi/h})(\text{s}) = \boxed{2.5\text{ mi}}$$

(e) Find the total displacement from 0 s to 400 s.

The total displacement is the sum of all the individual displacements. We still need to calculate the displacements for the time intervals from 300 s to 350 s and from 350 s to 400 s. The latter is negative, because it's below the time axis.

$$\Delta x_{300\to350\text{ s}} \approx \tfrac{1}{2}(5.0\times10^1\text{ s})(1.0\times10^1\text{ mi/h}) = 2.5\times10^2(\text{mi/h})(\text{s})$$

$$\Delta x_{350\to400\text{ s}} \approx \tfrac{1}{2}(5.0\times10^1\text{ s})(-5.0\times10^1\text{ mi/h}) = -1.3\times10^3(\text{mi/h})(\text{s})$$

Find the total displacement by summing the parts:

$$\Delta x_{0\to400\text{ s}} \approx (2.4\times10^4 + 9.0\times10^3 + 2.5\times10^2 - 1.3\times10^3)(\text{mi/h})(\text{s}) = \boxed{8.9\text{ mi}}$$

Remarks There are a number of ways of finding the approximate area under a graph. Choice of technique is a personal preference.

Exercise 2.6

Suppose the velocity vs. time graph of another train is given in Figure 2.18e. Find (a) the maximum instantaneous acceleration and (b) the total displacement in the interval from 0 s to 4.00×10^2 s.

Answers (a) 1.0(mi/h)/s (b) 4.7 mi

EXAMPLE 2.7 Runway Length

Goal Apply kinematics to horizontal motion with two phases.

Problem A typical jetliner lands at a speed of 160 mi/h and decelerates at the rate of (10 mi/h)/s. If the plane travels at a constant speed of 160 mi/h for 1.0 s after landing before applying the brakes, what is the total displacement of the aircraft between touchdown on the runway and coming to rest?

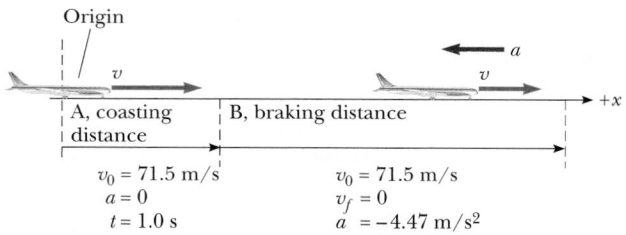

Figure 2.19 (Example 2.7) Coasting and braking distances for a landing jetliner.

Strategy See Figure 2.19. First, convert all quantities to SI units. The problem must be solved in two parts, or phases, corresponding to the initial coast after touchdown, followed by braking. Using the kinematic equations, find the displacement during each part and add the two displacements.

Solution

Convert units to SI:

$$v_0 = (160\text{ mi/h})\left(\frac{0.447\text{ m/s}}{1.00\text{ mi/h}}\right) = 71.5\text{ m/s}$$

$$a = (-10.0\text{ (mi/h)/s})\left(\frac{0.447\text{ m/s}}{1.00\text{ mi/h}}\right) = -4.47\text{ m/s}^2$$

Taking $a = 0$, $v_0 = 71.5$ m/s, and $t = 1.00$ s, find the displacement while the plane is coasting:

$$\Delta x_{\text{coasting}} = v_0 t + \tfrac{1}{2}at^2 = (71.5\text{ m/s})(1.00\text{ s}) + 0 = 71.5\text{ m}$$

Use the time-independent kinematic equation to find the displacement while the plane is braking.

$$v^2 = v_0^2 + 2a\,\Delta x_{\text{braking}}$$

Take $a = -4.47$ m/s^2 and $v_0 = 71.5$ m/s. The negative sign on a means that the plane is slowing down.

$$\Delta x_{\text{braking}} = \frac{v^2 - v_0^2}{2a} = \frac{0 - (71.5\text{ m/s})^2}{2.00(-4.47\text{ m/s}^2)} = 572\text{ m}$$

Sum the two results to find the total displacement: $\Delta x_{\text{coasting}} + \Delta x_{\text{braking}} = 72\text{ m} + 572\text{ m} = 644\text{ m}$

Remarks To find the displacement while braking, we could have used the two kinematics equations involving time, namely, $\Delta x = v_0 t + \frac{1}{2}at^2$ and $v = v_0 + at$, but because we weren't interested in time, the time-independent equation was easier to use.

Exercise 2.7

A jet lands at 80.0 m/s, applying the brakes 2.00 s after landing. Find the acceleration needed to stop the jet within 5.00×10^2 m.

Answer $a = -9.41\text{ m/s}^2$

2.6 FREELY FALLING OBJECTS

When air resistance is negligible, all objects dropped under the influence of gravity near Earth's surface fall toward Earth with the same constant acceleration. This idea may seem obvious today, but it wasn't until about 1600 that it was accepted. Prior to that time, the teachings of the great philosopher Aristotle (384–322 B.C.) had held that heavier objects fell faster than lighter ones.

According to legend, Galileo discovered the law of falling objects by observing that two different weights dropped simultaneously from the Leaning Tower of Pisa hit the ground at approximately the same time. Although it's unlikely that this particular experiment was carried out, we know that Galileo performed many systematic experiments with objects moving on inclined planes. In his experiments, he rolled balls down a slight incline and measured the distances they covered in successive time intervals. The purpose of the incline was to reduce the acceleration and enable Galileo to make accurate measurements of the intervals. (Some people refer to this experiment as "diluting gravity.") By gradually increasing the slope of the incline, he was finally able to draw mathematical conclusions about freely falling objects, because a falling ball is equivalent to a ball going down a vertical incline. Galileo's achievements in the science of mechanics paved the way for Newton in his development of the laws of motion, which we will study in Chapter 4.

Try the following experiment: Drop a hammer and a feather simultaneously from the same height. The hammer hits the floor first, because air drag has a greater effect on the much lighter feather. On August 2, 1971, this same experiment was conducted on the Moon by astronaut David Scott, and the hammer and feather fell with exactly the same acceleration, as expected, hitting the lunar surface at the same time. In the idealized case where air resistance is negligible, such motion is called *free fall.*

The expression *freely falling object* doesn't necessarily refer to an object dropped from rest. **A freely falling object is any object moving freely under the influence of gravity alone, regardless of its initial motion.** Objects thrown upward or downward and those released from rest are all considered freely falling.

We denote the magnitude of the **free-fall acceleration** by the symbol g. The value of g decreases with increasing altitude, and varies slightly with latitude, as well. At Earth's surface, the value of g is approximately 9.80 m/s^2. Unless stated otherwise, we will use this value for g in doing calculations. For quick estimates, use $g \approx 10\text{ m/s}^2$.

If we neglect air resistance and assume that the free-fall acceleration doesn't vary with altitude over short vertical distances, then the motion of a freely falling object is the same as motion in one dimension under constant acceleration. This means that the kinematics equations developed in Section 2.6 can be applied. It's conventional to define "up" as the $+y$-direction and to use y as the position variable. In that case, the acceleration is $a = -g = -9.80\text{ m/s}^2$. In Chapter 7, we study how to deal with variations in g with altitude.

North Wind Archive

GALILEO GALILEI Italian Physicist and Astronomer (1564–1642)

Galileo formulated the laws that govern the motion of objects in free fall. He also investigated the motion of an object on an inclined plane, established the concept of relative motion, invented the thermometer, and discovered that the motion of a swinging pendulum could be used to measure time intervals. After designing and constructing his own telescope, he discovered four of Jupiter's moons, found that our own Moon's surface is rough, discovered sunspots and the phases of Venus, and showed that the Milky Way consists of an enormous number of stars. Galileo publicly defended Nicolaus Copernicus's assertion that the Sun is at the center of the Universe (the heliocentric system). He published *Dialogue Concerning Two New World Systems* to support the Copernican model, a view the Church declared to be heretical. After being taken to Rome in 1633 on a charge of heresy, he was sentenced to life imprisonment and later was confined to his villa at Arcetri, near Florence, where he died in 1642.

Quick Quiz 2.6

A tennis player on serve tosses a ball straight up. While the ball is in free fall, does its acceleration (a) increase, (b) decrease, (c) increase and then decrease, (d) decrease and then increase, or (e) remain constant?

Quick Quiz 2.7

As the tennis ball of Quick Quiz 2.6 travels through the air, its speed (a) increases, (b) decreases, (c) decreases and then increases, (d) increases and then decreases, or (e) remains the same.

Quick Quiz 2.8

A skydiver jumps out of a hovering helicopter. A few seconds later, another skydiver jumps out, so they both fall along the same vertical line relative to the helicopter. Both sky divers fall with the same acceleration. Does the vertical distance between them (a) increase, (b) decrease, or (c) stay the same? Does the difference in their velocities (d) increase, (e) decrease, or (f) stay the same? (Assume that g is constant.)

EXAMPLE 2.8 Look Out Below!

Goal Apply the basic kinematics equations to an object falling from rest under the influence of gravity.

Problem A golf ball is released from rest at the top of a very tall building. Neglecting air resistance, calculate the position and velocity of the ball after 1.00 s, 2.00 s, and 3.00 s.

Strategy Make a simple sketch. Because the height of the building isn't given, it's convenient to choose coordinates so that $y = 0$ at the top of the building. Use the velocity and position kinematic equations, substituting known and given values.

Solution

Write the kinematics Equations 2.6 and 2.9:

$$v = at + v_0$$

$$\Delta y = y - y_0 = v_0 t + \tfrac{1}{2}at^2$$

Substitute $y_0 = 0$, $v_0 = 0$, and $a = -g = -9.80\ \text{m/s}^2$ into the preceding two equations:

$$v = at = (-9.80\ \text{m/s}^2)\,t$$

$$y = \tfrac{1}{2}(-9.80\ \text{m/s}^2)\,t^2 = -(4.90\ \text{m/s}^2)\,t^2$$

Substitute in the different times, and create a table.

t (s)	v (m/s)	y (m)
1.00	−9.8	−4.9
2.00	−19.6	−19.6
3.00	−29.4	−44.1

Remarks The minus signs on v mean that the velocity vectors are directed downward, while the minus signs on y indicates positions below the origin. The velocity of a falling object is directly proportional to the time, and the position is proportional to the time squared, results first proven by Galileo.

Exercise 2.8

Calculate the position and velocity of the ball after 4.00 s has elapsed.

Answer −78.4 m, −39.2 m/s

INTERACTIVE EXAMPLE 2.9
Not a Bad Throw for a Rookie!

Goal Apply the kinematic equations to a freely falling object with a nonzero initial velocity.

Problem A stone is thrown from the top of a building with an initial velocity of 20.0 m/s straight upward, at an initial height of 50.0 m above the ground. The stone just misses the edge of the roof on its way down, as shown in Figure 2.20. Determine **(a)** the time needed for the stone to reach its maximum height, **(b)** the maximum height, **(c)** the time needed for the stone to return to the height from which it was thrown and the velocity of the stone at that instant; **(d)** the time needed for the stone to reach the ground, and **(e)** the velocity and position of the stone at $t = 5.00$ s.

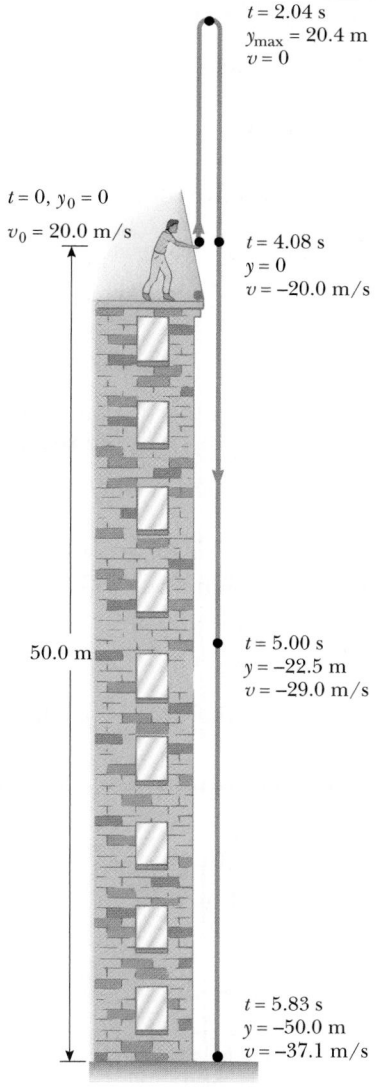

Figure 2.20 (Example 2.9) A freely falling object is thrown upward with an initial velocity of $v_0 = +20.0$ m/s. Positions and velocities are given for several times.

Strategy The diagram in Figure 2.20 establishes a coordinate system with $y_0 = 0$ at the level at which the stone is released from the thrower's hand, with y positive upward. Write the velocity and position kinematic equations for the stone, and substitute the given information. All the answers come from these two equations by using simple algebra or by just substituting the time. In part (a), for example, the stone comes to rest for an instant at its maximum height, so set $v = 0$ at this point and solve for time. Then substitute the time into the displacement equation, obtaining the maximum height.

Solution

(a) Find the time when the stone reaches its maximum height.

Write the velocity and position kinematic equations:

$$v = at + v_0$$

$$\Delta y = y - y_0 = v_0 t + \tfrac{1}{2}at^2$$

Substitute $a = -9.80$ m/s^2, $v_0 = 20.0$ m/s, and $y_0 = 0$ into the preceding two equations:

$$v = (-9.80 \text{ m/s}^2)t + 20.0 \text{ m/s} \quad \textbf{(1)}$$

$$y = (20.0 \text{ m/s})t - (4.90 \text{ m/s}^2)t^2 \quad \textbf{(2)}$$

Substitute $v = 0$, the velocity at maximum height, into Equation (1) and solve for time:

$$0 = (-9.80 \text{ m/s}^2)t + 20.0 \text{ m/s}$$

$$t = \frac{-20.0 \text{ m/s}}{-9.80 \text{ m/s}^2} = \boxed{2.04 \text{ s}}$$

(b) Determine the stone's maximum height.

Substitute the time $t = 2.04$ s into Equation (2):

$$y_{max} = (20.0 \text{ m/s})(2.04) - (4.90 \text{ m/s}^2)(2.04)^2 = \boxed{20.4 \text{ m}}$$

(c) Find the time the stone takes to return to its initial position, and find the velocity of the stone at that time.

Set $y = 0$ in Equation (2) and solve t:

$$0 = (20.0 \text{ m/s})t - (4.90 \text{ m/s}^2)t^2$$

$$= t(20.0 \text{ m/s} - 4.90 \text{ m/s}^2\, t)$$

$$t = \boxed{4.08 \text{ s}}$$

Substitute the time into Equation (1) to get the velocity: $v = 20.0\ \text{m/s} + (-9.80\ \text{m/s}^2)(4.08\ \text{s}) = -20.0\ \text{m/s}$

(d) Find the time required for the stone to reach the ground.

In Equation 2, set $y = -50.0$ m: $-50.0\ \text{m} = (20.0\ \text{m/s})t - (4.90\ \text{m/s}^2)t^2$

Apply the quadratic formula and take the positive root: $t = 5.83\ \text{s}$

(e) Find the velocity and position of the stone at $t = 5.00$ s. Substitute values into Equations (1) and (2):

$$v = (-9.80\ \text{m/s}^2)(5.00\ \text{s}) + 20.0\ \text{m/s} = -29.0\ \text{m/s}$$

$$y = (20.0\ \text{m/s})(5.00\ \text{s}) - (4.90\ \text{m/s}^2)(5.00\ \text{s})^2 = -22.5\ \text{m}$$

Remarks Notice how everything follows from the two kinematic equations. Once they are written down, and constants correctly identified as in Equations (1) and (2), the rest is relatively easy. If the stone were thrown downward, the initial velocity would have been negative.

Exercise 2.9

A projectile is launched straight up at 60.0 m/s from a height of 80.0 m, at the edge of a sheer cliff. The projectile falls, just missing the cliff and hitting the ground below. Find (a) the maximum height of the projectile above the point of firing, (b) the time it takes to hit the ground at the base of the cliff, and (c) its velocity at impact.

Answers (a) 184 m (b) 13.5 s (c) -72.3 m/s

PhysicsNow™ You can study the motion of the thrown ball by logging into PhysicsNow at **www.cp7e.com** and going to Interactive Example 2.9.

EXAMPLE 2.10 A Rocket Goes Ballistic

Goal Solve a problem involving a powered ascent followed by free fall motion.

Problem A rocket moves straight upward, starting from rest with an acceleration of $+29.4\ \text{m/s}^2$. It runs out of fuel at the end of 4.00 s and continues to coast upward, reaching a maximum height before falling back to Earth. **(a)** Find the rocket's velocity and position at the end of 4.00 s. **(b)** Find the maximum height the rocket reaches. **(c)** Find the velocity the instant before the rocket crashes on the ground.

Strategy Take $y = 0$ at the launch point and y positive upward, as in Figure 2.21. The problem consists of two phases. In phase 1, the rocket has a net *upward* acceleration of $29.4\ \text{m/s}^2$, and we can use the kinematic equations with constant acceleration a to find the height and velocity of the rocket at the end of phase 1, when the fuel is burned up. In phase 2, the rocket is in free fall and has an acceleration of $-9.80\ \text{m/s}^2$, with initial velocity and position given by the results of phase 1. Apply the kinematic equations for free fall.

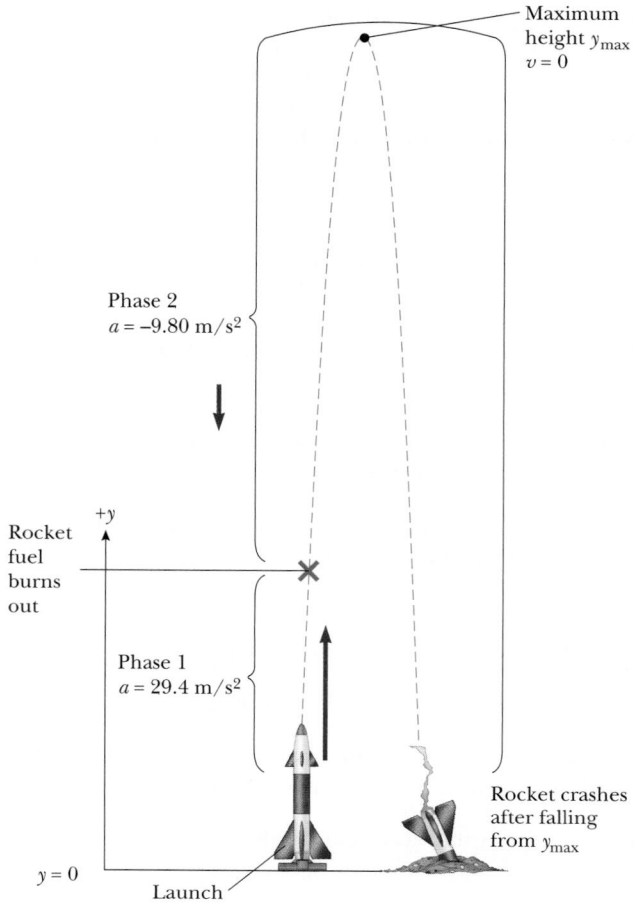

Figure 2.21 (Example 2.10) Two linked phases of motion for a rocket that is launched, uses up its fuel, and crashes.

Solution

(a) Phase 1: Find the rocket's velocity and position after 4.00 s.

Write the velocity and position kinematic equations:

$$v = v_0 + at \quad \textbf{(1)}$$

$$\Delta y = y - y_0 = v_0 t + \tfrac{1}{2}at^2 \quad \textbf{(2)}$$

Adapt these equations to phase 1, substituting $a = 29.4\ \text{m/s}^2$, $v_0 = 0$, and $y_0 = 0$:

$$v = (29.4\ \text{m/s}^2)t \quad \textbf{(3)}$$

$$y = \tfrac{1}{2}(29.4\ \text{m/s}^2)t^2 = (14.7\ \text{m/s}^2)t^2 \quad \textbf{(4)}$$

Substitute $t = 4.00$ s into Equations (3) and (4) to find the rocket's velocity v and position y at the time of burnout. These will be called v_b and y_b, respectively.

$$v_b = 118\ \text{m/s} \text{ and } y_b = 235\ \text{m}$$

(b) Phase 2: Find the maximum height the rocket attains.

Adapt Equations (1) and (2) to phase 2, substituting $a = -9.8\ \text{m/s}^2$, $v_0 = v_b = 118$ m/s, and $y_0 = y_b = 235$ m:

$$v = (-9.8\ \text{m/s}^2)t + 118\ \text{m/s} \quad \textbf{(5)}$$

$$y = 235\ \text{m} + (118\ \text{m/s})t - (4.90\ \text{m/s}^2)t^2 \quad \textbf{(6)}$$

Substitute $v = 0$ (the rocket's velocity at maximum height) in Equation 5 to get the time it takes the rocket to reach its maximum height:

$$0 = (-9.8\ \text{m/s}^2)t + 118\ \text{m/s} \rightarrow t = \frac{118\ \text{m/s}}{9.80\ \text{m/s}^2} = 12.0\ \text{s}$$

Substitute $t = 12.0$ s into Equation (6) to find the rocket's maximum height:

$$y_{\text{max}} = 235\ \text{m} + (118\ \text{m/s})(12.0\ \text{s}) - (4.90\ \text{m/s}^2)(12.0\ \text{s})^2 = 945\ \text{m}$$

(c) Phase 2: Find the velocity of the rocket just prior to impact.

Find the time to impact by setting $y = 0$ in Equation (6) and using the quadratic formula:

$$0 = 235\ \text{m} + (118\ \text{m/s})t - (4.90\ \text{m/s}^2)t^2$$

$$t = 25.9\ \text{s}$$

Substitute this value of t into Equation (5):

$$v = (-9.80\ \text{m/s}^2)(25.9\ \text{s}) + 118\ \text{m/s} = -136\ \text{m/s}$$

Remarks You may think that it is more natural to break this problem into three phases, with the second phase ending at the maximum height and the third phase a free fall from maximum height to the ground. Although this approach gives the correct answer, it's an unnecessary complication. Two phases are sufficient, one for each different acceleration.

Exercise 2.10

An experimental rocket designed to land upright falls freely from a height of 2.00×10^2 m, starting at rest. At a height of 80.0 m, the rocket's engines start and provide constant upward acceleration until the rocket lands. What acceleration is required if the speed on touchdown is to be zero? (Neglect air resistance.)

Answer $14.7\ \text{m/s}^2$

SUMMARY

PhysicsNow™ Take a practice test by logging into PhysicsNow at **www.cp7e.com** and clicking on the Pre-Test link for this chapter.

2.1 Displacement

The **displacement** of an object moving along the x-axis is defined as the change in position of the object,

$$\Delta x \equiv x_f - x_i \quad \textbf{[2.1]}$$

where x_i is the initial position of the object and x_f is its final position.

A **vector** quantity is characterized by both a magnitude and a direction. A **scalar** quantity has a magnitude only.

2.2 Velocity

The **average speed** of an object is given by

$$\text{Average speed} \equiv \frac{\text{total distance}}{\text{total time}}$$

The **average velocity** $\bar{v}$ during a time interval Δt is the displacement Δx divided by Δt.

$$\bar{v} \equiv \frac{\Delta x}{\Delta t} = \frac{x_f - x_i}{t_f - t_i} \qquad [2.2]$$

The average velocity is equal to the slope of the straight line joining the initial and final points on a graph of the position of the object versus time.

The slope of the line tangent to the position vs. time curve at some point is equal to the **instantaneous velocity** at that time. The **instantaneous speed** of an object is defined as the magnitude of the instantaneous velocity.

2.3 Acceleration

The **average acceleration** $\bar{a}$ of an object undergoing a change in velocity Δv during a time interval Δt is

$$\bar{a} \equiv \frac{\Delta v}{\Delta t} = \frac{v_f - v_i}{t_f - t_i} \qquad [2.4]$$

The **instantaneous acceleration** of an object at a certain time equals the slope of a velocity vs. time graph at that instant.

2.5 One-Dimensional Motion with Constant Acceleration

The most useful equations that describe the motion of an object moving with constant acceleration along the x axis are as follows:

$$v = v_0 + at \qquad [2.6]$$

$$\Delta x = v_0 t + \tfrac{1}{2}at^2 \qquad [2.9]$$

$$v^2 = v_0^2 + 2a\Delta x \qquad [2.10]$$

All problems can be solved with the first two equations alone, the last being convenient when time doesn't explicitly enter the problem. After the constants are properly identified, most problems reduce to one or two equations in as many unknowns.

2.6 Freely Falling Objects

An object falling in the presence of Earth's gravity exhibits a free-fall acceleration directed toward Earth's center. If air friction is neglected and if the altitude of the falling object is small compared with Earth's radius, then we can assume that the free-fall acceleration $g = 9.8\ \text{m/s}^2$ is constant over the range of motion. Equations 2.6, 2.9, and 2.10 apply, with $a = -g$.

CONCEPTUAL QUESTIONS

1. If the velocity of a particle is nonzero, can the particle's acceleration be zero? Explain.
2. If the velocity of a particle is zero, can the particle's acceleration be zero? Explain.
3. If a car is traveling eastward, can its acceleration be westward? Explain.
4. The speed of sound in air is 331 m/s. During the next thunderstorm, try to estimate your distance from a lightning bolt by measuring the time lag between the flash and the thunderclap. You can ignore the time it takes for the light flash to reach you. Why?
5. Can the equations of kinematics be used in a situation where the acceleration varies with time? Can they be used when the acceleration is zero?
6. If the average velocity of an object is zero in some time interval, what can you say about the displacement of the object during that interval?
7. A child throws a marble into the air with an initial speed v_0. Another child drops a ball at the same instant. Compare the accelerations of the two objects while they are in flight.
8. Figure Q2.8 shows strobe photographs taken of a disk moving from left to right under different conditions. The time interval between images is constant. Taking the direction to the right to be positive, describe the motion of the disk in each case. For which case is (a) the acceleration positive? (b) the acceleration negative? (c) the velocity constant?
9. Can the instantaneous velocity of an object at an instant of time ever be greater in magnitude than the average velocity over a time interval containing that instant? Can it ever be less?

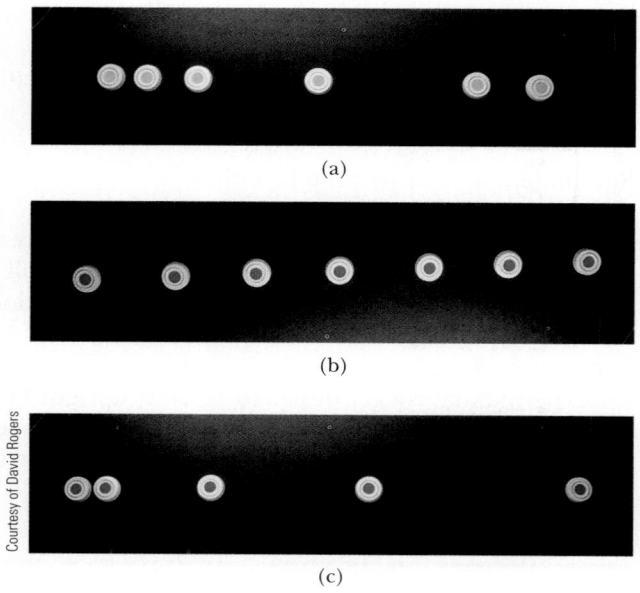

Figure Q2.8

10. Car A, traveling from New York to Miami, has a speed of 25 m/s. Car B, traveling from New York to Chicago, also has a speed of 25 m/s. Are their velocities equal? Explain.
11. A ball is thrown vertically upward. (a) What are its velocity and acceleration when it reaches its maximum altitude? (b) What is the acceleration of the ball just before it hits the ground?

12. A rule of thumb for driving is that a separation of one car length for each 10 mi/h of speed should be maintained between moving vehicles. Assuming a constant reaction time, discuss the relevance of this rule for (a) motion with constant velocity and (b) motion with constant acceleration.

13. Two cars are moving in the same direction in parallel lanes along a highway. At some instant, the velocity of car A exceeds the velocity of car B. Does this mean that the acceleration of A is greater than that of B? Explain.

14. Consider the following combinations of signs and values for the velocity and acceleration of a particle with respect to a one-dimensional x-axis:

	Velocity	Acceleration
a.	Positive	Positive
b.	Positive	Negative
c.	Positive	Zero
d.	Negative	Positive
e.	Negative	Negative
f.	Negative	Zero
g.	Zero	Positive
h.	Zero	Negative

Describe what the particle is doing in each case, and give a real-life example for an automobile on an east-west one-dimensional axis, with east considered the positive direction.

15. A student at the top of a building of height h throws one ball upward with a speed of v_0 and then throws a second ball downward with the same initial speed, v_0. How do the final velocities compare when the balls reach the ground?

16. A ball is thrown straight upward and moves in free fall. Choose a coordinate system with its origin at the release point of the ball and the positive direction upward. (a) What is the sign of the velocity of the ball just before the ball reaches its maximum height, just after it reaches its maximum height, and at its maximum height. (b) What is the sign of the acceleration of the ball just before the ball reaches its maximum height, just after it reaches its maximum height, and at its maximum height. (c) If the ball takes time t_1 to reach its maximum height, how long will it take to return to ground level? (d) If the ball is thrown upward with a velocity of $+v_0$, what will be the ball's velocity upon returning to ground level?

17. A pebble is dropped into a water well, and the splash is heard 16 s later, as illustrated in the cartoon strip shown in Figure Q2.17. Estimate the distance from the rim of the well to the water's surface.

18. A ball rolls in a straight line along the horizontal direction. Using motion diagrams (or multiflash photographs), describe the velocity and acceleration of the ball for each of the following situations: (a) The ball moves to the right at a constant speed. (b) The ball moves from right to left and continually slows down. (c) The ball moves from right to left and continually speeds up. (d) The ball moves to the right, first speeding up at a constant rate and then slowing down at a constant rate.

19. You drop a ball from a window on an upper floor of a building. The ball strikes the ground with speed v. You now repeat the drop, but you have a friend down on the street who throws another ball upward at speed v. Your friend throws the ball upward at exactly the same time that you drop yours from the window. At some location, the balls pass each other. Is this location *at* the halfway point between window and ground, *above* that point, or *below* that point?

By permission of John Hart and Field Enterprises, Inc./Creator's Syndicate

Figure Q2.17

PROBLEMS

1, 2, 3 = straightforward, intermediate, challenging □ = full solution available in *Student Solutions Manual/Study Guide*
PhysicsNow™ = coached problem with hints available at **www.cp7e.com** = biomedical application

Section 2.1 Displacement
Section 2.2 Velocity

1. A person travels by car from one city to another with different constant speeds between pairs of cities. She drives for 30.0 min at 80.0 km/h, 12.0 min at 100 km/h, and 45.0 min at 40.0 km/h and spends 15.0 min eating lunch and buying gas. (a) Determine the average speed for the trip. (b) Determine the distance between the initial and final cities along the route.

2. (a) Sand dunes on a desert island move as sand is swept up the windward side to settle in the leeward side. Such "walking" dunes have been known to travel 20 feet in a year and can travel as much as 100 feet per year in particularly windy times. Calculate the average speed in each case in m/s. (b) Fingernails grow at the rate of drifting continents, about 10 mm/yr. Approximately how long did it take for North America to separate from Europe, a distance of about 3 000 mi?

3. Two boats start together and race across a 60-km-wide lake and back. Boat A goes across at 60 km/h and returns at 60 km/h. Boat B goes across at 30 km/h, and its crew, realizing how far behind it is getting, returns at 90 km/h. Turnaround times are negligible, and the boat that completes the round trip first wins. (a) Which boat wins and by how much? (Or is it a tie?) (b) What is the average velocity of the winning boat?

4. The Olympic record for the marathon is 2 h, 9 min, 21 s. The marathon distance is 26 mi, 385 yd. Determine the average speed (in miles per hour) of the record.

5. A motorist drives north for 35.0 minutes at 85.0 km/h and then stops for 15.0 minutes. He then continues north, traveling 130 km in 2.00 h. (a) What is his total displacement? (b) What is his average velocity?

6. A graph of position versus time for a certain particle moving along the x-axis is shown in Figure P2.6. Find the average velocity in the time intervals from (a) 0 to 2.00 s, (b) 0 to 4.00 s, (c) 2.00 s to 4.00 s, (d) 4.00 s to 7.00 s, and (e) 0 to 8.00 s.

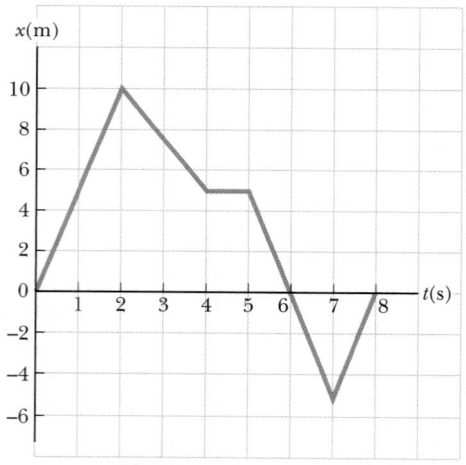

Figure P2.6 (Problems 6 and 15)

7. A tennis player moves in a straight-line path as shown in Figure P2.7. Find her average velocity in the time intervals from (a) 0 to 1.0 s, (b) 0 to 4.0 s, (c) 1.0 s to 5.0 s, and (d) 0 to 5.0 s.

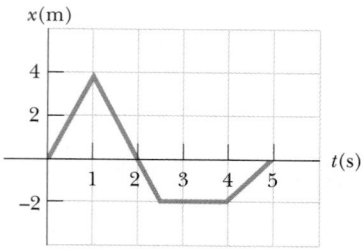

Figure P2.7 (Problems 7 and 17)

8. Two cars travel in the same direction along a straight highway, one at a constant speed of 55 mi/h and the other at 70 mi/h. (a) Assuming that they start at the same point, how much sooner does the faster car arrive at a destination 10 mi away? (b) How far must the faster car travel before it has a 15-min lead on the slower car?

9. An athlete swims the length of a 50.0-m pool in 20.0 s and makes the return trip to the starting position in 22.0 s. Determine her average velocities in (a) the first half of the swim, (b) the second half of the swim, and (c) the round trip.

10. If the average speed of an orbiting space shuttle is 19 800 mi/h, determine the time required for it to circle Earth. Make sure you consider the fact that the shuttle is orbiting about 200 mi above Earth's surface, and assume that Earth's radius is 3 963 miles.

11. **Physics Now™** A person takes a trip, driving with a constant speed of 89.5 km/h, except for a 22.0-min rest stop. If the person's average speed is 77.8 km/h, how much time is spent on the trip and how far does the person travel?

12. A tortoise can run with a speed of 0.10 m/s, and a hare can run 20 times as fast. In a race, they both start at the same time, but the hare stops to rest for 2.0 minutes. The tortoise wins by a shell (20 cm). (a) How long does the race take? (b) What is the length of the race?

13. In order to qualify for the finals in a racing event, a race car must achieve an average speed of 250 km/h on a track with a total length of 1 600 m. If a particular car covers the first half of the track at an average speed of 230 km/h, what minimum average speed must it have in the second half of the event in order to qualify?

14. Runner A is initially 4.0 mi west of a flagpole and is running with a constant velocity of 6.0 mi/h due east. Runner B is initially 3.0 mi east of the flagpole and is running with a constant velocity of 5.0 mi/h due west. How far are the runners from the flagpole when they meet?

15. A graph of position versus time for a certain particle moving along the x-axis is shown in Figure P2.6. Find the instantaneous velocity at the instants (a) $t = 1.00$ s, (b) $t = 3.00$ s, (c) $t = 4.50$ s, and (d) $t = 7.50$ s.

16. A race car moves such that its position fits the relationship

$$x = (5.0 \text{ m/s})t + (0.75 \text{ m/s}^3)t^3$$

where x is measured in meters and t in seconds. (a) Plot a graph of the car's position versus time. (b) Determine the instantaneous velocity of the car at $t = 4.0$ s, using time intervals of 0.40 s, 0.20 s, and 0.10 s. (c) Compare the average velocity during the first 4.0 s with the results of (b).

17. Find the instantaneous velocities of the tennis player of Figure P2.7 at (a) 0.50 s, (b) 2.0 s, (c) 3.0 s, and (d) 4.5 s.

Section 2.3 Acceleration

18. Secretariat ran the Kentucky Derby with times of 25.2 s, 24.0 s, 23.8 s, and 23.0 s for the quarter mile. (a) Find his average speed during each quarter-mile segment. (b) Assuming that Secretariat's instantaneous speed at the finish line was the same as his average speed during the final quarter mile, find his average acceleration for the entire race. (*Hint:* Recall that horses in the Derby start from rest.)

19. A steam catapult launches a jet aircraft from the aircraft carrier *John C. Stennis*, giving it a speed of 175 mi/h in 2.50 s. (a) Find the average acceleration of the plane. (b) Assuming that the acceleration is constant, find the distance the plane moves.

20. A car traveling in a straight line has a velocity of +5.0 m/s at some instant. After 4.0 s, its velocity is +8.0 m/s. What is the car's average acceleration during the 4.0-s time interval?

21. Physics Now™ A certain car is capable of accelerating at a rate of $+0.60\ \text{m/s}^2$. How long does it take for this car to go from a speed of 55 mi/h to a speed of 60 mi/h?

22. The velocity vs. time graph for an object moving along a straight path is shown in Figure P2.22. (a) Find the average acceleration of the object during the time intervals 0 to 5.0 s, 5.0 s to 15 s, and 0 to 20 s. (b) Find the instantaneous acceleration at 2.0 s, 10 s, and 18 s.

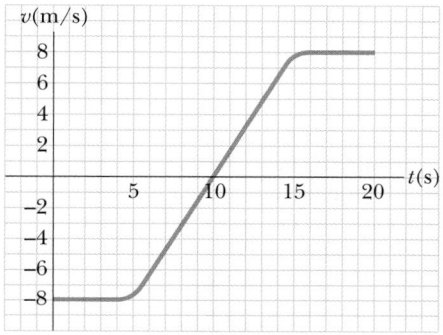

Figure P2.22

23. The engine of a model rocket accelerates the rocket vertically upward for 2.0 s as follows: At $t = 0$, the rocket's speed is zero; at $t = 1.0$ s, its speed is 5.0 m/s; and at $t = 2.0$ s, its speed is 16 m/s. Plot a velocity vs. time graph for this motion, and use the graph to determine (a) the rocket's average acceleration during the 2.0-s interval and (b) the instantaneous acceleration of the rocket at $t = 1.5$ s.

Section 2.5 One-Dimensional Motion with Constant Acceleration

24. A car traveling in a straight-line path has a velocity of +10.0 m/s at some instant. After 3.00 s, its velocity is +6.00 m/s. What is the average acceleration of the car during this time interval?

25. In 1865, Jules Verne proposed sending men to the Moon by firing a space capsule from a 220-m-long cannon with final speed of 10.97 km/s. What would have been the unrealistically large acceleration experienced by the space travelers during their launch? (A human can stand an acceleration of $15g$ for a short time.) Compare your answer with the free-fall acceleration, $9.80\ \text{m/s}^2$.

26. A truck covers 40.0 m in 8.50 s while smoothly slowing down to a final speed of 2.80 m/s. (a) Find the truck's original speed. (b) Find its acceleration.

27. A speedboat increases its speed uniformly from 20 m/s to 30 m/s in a distance of 200 m. Find (a) the magnitude of its acceleration and (b) the time it takes the boat to travel the 200-m distance.

28. Two cars are traveling along a straight line in the same direction, the lead car at 25.0 m/s and the other car at 30.0 m/s. At the moment the cars are 40.0 m apart, the lead driver applies the brakes, causing his car to have an acceleration of $-2.00\ \text{m/s}^2$. (a) How long does it take for the lead car to stop? (b) Assuming that the chasing car brakes at the same time as the lead car, what must be the chasing car's minimum negative acceleration so as not to hit the lead car? (c) How long does it take for the chasing car to stop?

29. A Cessna aircraft has a lift-off speed of 120 km/h. (a) What minimum constant acceleration does the aircraft require if it is to be airborne after a takeoff run of 240 m? (b) How long does it take the aircraft to become airborne?

30. A truck on a straight road starts from rest and accelerates at $2.0\ \text{m/s}^2$ until it reaches a speed of 20 m/s. Then the truck travels for 20 s at constant speed until the brakes are applied, stopping the truck in a uniform manner in an additional 5.0 s. (a) How long is the truck in motion? (b) What is the average velocity of the truck during the motion described?

31. A drag racer starts her car from rest and accelerates at $10.0\ \text{m/s}^2$ for a distance of 400 m ($\frac{1}{4}$ mile). (a) How long did it take the race car to travel this distance? (b) What is the speed of the race car at the end of the run?

32. A jet plane lands with a speed of 100 m/s and can accelerate at a maximum rate of $-5.00\ \text{m/s}^2$ as it comes to rest. (a) From the instant the plane touches the runway, what is the minimum time needed before it can come to rest? (b) Can this plane land on a small tropical island airport where the runway is 0.800 km long?

33. A driver in a car traveling at a speed of 60 mi/h sees a deer 100 m away on the road. Calculate the minimum constant acceleration that is necessary for the car to stop without hitting the deer (assuming that the deer does not move in the meantime).

34. A record of travel along a straight path is as follows:

1. Start from rest with a constant acceleration of $2.77\ \text{m/s}^2$ for 15.0 s.
2. Maintain a constant velocity for the next 2.05 min.
3. Apply a constant negative acceleration of $-9.47\ \text{m/s}^2$ for 4.39 s.

(a) What was the total displacement for the trip? (b) What were the average speeds for legs 1, 2, and 3 of the trip, as well as for the complete trip?

35. A train is traveling down a straight track at 20 m/s when the engineer applies the brakes, resulting in an acceleration of $-1.0\ \text{m/s}^2$ as long as the train is in motion. How far does the train move during a 40-s time interval starting at the instant the brakes are applied?

36. A car accelerates uniformly from rest to a speed of 40.0 mi/h in 12.0 s. Find (a) the distance the car travels during this time and (b) the constant acceleration of the car.

37. A car starts from rest and travels for 5.0 s with a uniform acceleration of $+1.5\ \text{m/s}^2$. The driver then applies the brakes, causing a uniform acceleration of $-2.0\ \text{m/s}^2$. If the brakes are applied for 3.0 s, (a) how fast is the car going at the end of the braking period, and (b) how far has the car gone?

38. A train 400 m long is moving on a straight track with a speed of 82.4 km/h. The engineer applies the brakes at a crossing, and later the last car passes the crossing with a speed of 16.4 km/h. Assuming constant acceleration, determine how long the train blocked the crossing. Disregard the width of the crossing.

39. A hockey player is standing on his skates on a frozen pond when an opposing player, moving with a uniform speed of 12 m/s, skates by with the puck. After 3.0 s, the first player makes up his mind to chase his opponent. If he accelerates uniformly at 4.0 m/s^2, (a) how long does it take him to catch his opponent, and (b) how far has he traveled in that time? (Assume that the player with the puck remains in motion at constant speed.)

40. A glider on an air track carries a flag of length ℓ through a stationary photogate that measures the time interval Δt_d during which the flag blocks a beam of infrared light passing across the gate. The ratio $v_d = \ell/\Delta t_d$ is the average velocity of the glider over this part of its motion. Suppose the glider moves with constant acceleration. (a) Argue for or against the idea that v_d is equal to the instantaneous velocity of the glider when it is halfway through the photogate in terms of distance. (b) Argue for or against the idea that v_d is equal to the instantaneous velocity of the glider when it is halfway through the photogate in terms of time.

41. In the Daytona 500 auto race, a Ford Thunderbird and a Mercedes Benz are moving side by side down a straightaway at 71.5 m/s. The driver of the Thunderbird realizes that she must make a pit stop, and she smoothly slows to a stop over a distance of 250 m. She spends 5.00 s in the pit and then accelerates out, reaching her previous speed of 71.5 m/s after a distance of 350 m. At this point, how far has the Thunderbird fallen behind the Mercedes Benz, which has continued at a constant speed?

42. A certain cable car in San Francisco can stop in 10 s when traveling at maximum speed. On one occasion, the driver sees a dog a distance d m in front of the car and slams on the brakes instantly. The car reaches the dog 8.0 s later, and the dog jumps off the track just in time. If the car travels 4.0 m beyond the position of the dog before coming to a stop, how far was the car from the dog? (*Hint*: You will need three equations.)

Section 2.6 Freely Falling Objects

43. A ball is thrown vertically upward with a speed of 25.0 m/s. (a) How high does it rise? (b) How long does it take to reach its highest point? (c) How long does the ball take to hit the ground after it reaches its highest point? (d) What is its velocity when it returns to the level from which it started?

44. It is possible to shoot an arrow at a speed as high as 100 m/s. (a) If friction is neglected, how high would an arrow launched at this speed rise if shot straight up? (b) How long would the arrow be in the air?

45. A certain freely falling object requires 1.50 s to travel the last 30.0 m before it hits the ground. From what height above the ground did it fall?

46. Traumatic brain injury such as concussion results when the head undergoes a very large acceleration. Generally, an acceleration less than 800 m/s^2 lasting for any length of time will not cause injury, whereas an acceleration greater than 1 000 m/s^2 lasting for at least 1 ms will cause injury. Suppose a small child rolls off a bed that is 0.40 m above the floor. If the floor is hardwood, the child's head is brought to rest in approximately 2.0 mm. If the floor is carpeted, this stopping distance is increased to about 1.0 cm. Calculate the magnitude and duration of the deceleration in both cases, to determine the risk of injury. Assume that the child remains horizontal during the fall to the floor. Note that a more complicated fall could result in a head velocity greater or less than the speed you calculate.

47. Physics Now™ A small mailbag is released from a helicopter that is descending steadily at 1.50 m/s. After 2.00 s, (a) what is the speed of the mailbag, and (b) how far is it below the helicopter? (c) What are your answers to parts (a) and (b) if the helicopter is *rising* steadily at 1.50 m/s?

48. A ball thrown vertically upward is caught by the thrower after 2.00 s. Find (a) the initial velocity of the ball and (b) the maximum height the ball reaches.

49. A model rocket is launched straight upward with an initial speed of 50.0 m/s. It accelerates with a constant upward acceleration of 2.00 m/s^2 until its engines stop at an altitude of 150 m. (a) What is the maximum height reached by the rocket? (b) How long after lift-off does the rocket reach its maximum height? (c) How long is the rocket in the air?

50. A parachutist with a camera descends in free fall at a speed of 10 m/s. The parachutist releases the camera at an altitude of 50 m. (a) How long does it take the camera to reach the ground? (b) What is the velocity of the camera just before it hits the ground?

51. A student throws a set of keys vertically upward to his fraternity brother, who is in a window 4.00 m above. The brother's outstretched hand catches the keys 1.50 s later. (a) With what initial velocity were the keys thrown? (b) What was the velocity of the keys just before they were caught?

52. It has been claimed that an insect called the froghopper (*Philaenus spumarius*) is the best jumper in the animal kingdom. This insect can accelerate at 4 000 m/s^2 over a distance of 2.0 mm as it straightens its specially designed "jumping legs." (a) Assuming a uniform acceleration, what is the velocity of the insect after it has accelerated through this short distance, and how long did it take to reach that velocity? (b) How high would the insect jump if air resistance could be ignored? Note that the actual height obtained is about 0.7 m, so air resistance is important here.

ADDITIONAL PROBLEMS

53. A truck tractor pulls two trailers, one behind the other, at a constant speed of 100 km/h. It takes 0.600 s for the big rig to completely pass onto a bridge 400 m long. For what duration of time is all or part of the truck-trailer combination on the bridge?

54. A speedboat moving at 30.0 m/s approaches a no-wake buoy marker 100 m ahead. The pilot slows the boat with a constant acceleration of -3.50 m/s^2 by reducing the throttle. (a) How long does it take the boat to reach the buoy? (b) What is the velocity of the boat when it reaches the buoy?

55. A bullet is fired through a board 10.0 cm thick in such a way that the bullet's line of motion is perpendicular to the face of the board. If the initial speed of the bullet is 400 m/s and it emerges from the other side of the board

with a speed of 300 m/s, find (a) the acceleration of the bullet as it passes through the board and (b) the total time the bullet is in contact with the board.

56. An indestructible bullet 2.00 cm long is fired straight through a board that is 10.0 cm thick. The bullet strikes the board with a speed of 420 m/s and emerges with a speed of 280 m/s. (a) What is the average acceleration of the bullet through the board? (b) What is the total time that the bullet is in contact with the board? (c) What thickness of board (calculated to 0.1 cm) would it take to stop the bullet, assuming that the acceleration through all boards is the same?

57. A ball is thrown upward from the ground with an initial speed of 25 m/s; at the same instant, another ball is dropped from a building 15 m high. After how long will the balls be at the same height?

58. A ranger in a national park is driving at 35.0 mi/h when a deer jumps into the road 200 ft ahead of the vehicle. After a reaction time t, the ranger applies the brakes to produce an acceleration $a = -9.00\ \text{ft/s}^2$. What is the maximum reaction time allowed if she is to avoid hitting the deer?

59. Two students are on a balcony 19.6 m above the street. One student throws a ball vertically downward at 14.7 m/s; at the same instant, the other student throws a ball vertically upward at the same speed. The second ball just misses the balcony on the way down. (a) What is the difference in the two balls' time in the air? (b) What is the velocity of each ball as it strikes the ground? (c) How far apart are the balls 0.800 s after they are thrown?

60. The driver of a truck slams on the brakes when he sees a tree blocking the road. The truck slows down uniformly with an acceleration of $-5.60\ \text{m/s}^2$ for 4.20 s, making skid marks 62.4 m long that end at the tree. With what speed does the truck then strike the tree?

61. A young woman named Kathy Kool buys a sports car that can accelerate at the rate of $4.90\ \text{m/s}^2$. She decides to test the car by drag racing with another speedster, Stan Speedy. Both start from rest, but experienced Stan leaves the starting line 1.00 s before Kathy. If Stan moves with a constant acceleration of $3.50\ \text{m/s}^2$ and Kathy maintains an acceleration of $4.90\ \text{m/s}^2$, find (a) the time it takes Kathy to overtake Stan, (b) the distance she travels before she catches him, and (c) the speeds of both cars at the instant she overtakes him.

62. A mountain climber stands at the top of a 50.0-m cliff that overhangs a calm pool of water. She throws two stones vertically downward 1.00 s apart and observes that they cause a single splash. The first stone had an initial velocity of -2.00 m/s. (a) How long after release of the first stone did the two stones hit the water? (b) What initial velocity must the second stone have had, given that they hit the water simultaneously? (c) What was the velocity of each stone at the instant it hit the water?

63. An ice sled powered by a rocket engine starts from rest on a large frozen lake and accelerates at $+40\ \text{ft/s}^2$. After some time t_1, the rocket engine is shut down and the sled moves with constant velocity v for a time t_2. If the total distance traveled by the sled is 17 500 ft and the total time is 90 s, find (a) the times t_1 and t_2 and (b) the velocity v. At the 17 500-ft mark, the sled begins to accelerate at $-20\ \text{ft/s}^2$. (c) What is the final position of the sled when it comes to rest? (d) How long does it take to come to rest?

64. In Bosnia, the ultimate test of a young man's courage used to be to jump off a 400-year-old bridge (now destroyed) into the River Neretva, 23 m below the bridge. (a) How long did the jump last? (b) How fast was the jumper traveling upon impact with the river? (c) If the speed of sound in air is 340 m/s, how long after the jumper took off did a spectator on the bridge hear the splash?

65. A person sees a lightning bolt pass close to an airplane that is flying in the distance. The person hears thunder 5.0 s after seeing the bolt and sees the airplane overhead 10 s after hearing the thunder. The speed of sound in air is 1 100 ft/s. (a) Find the distance of the airplane from the person at the instant of the bolt. (Neglect the time it takes the light to travel from the bolt to the eye.) (b) Assuming that the plane travels with a constant speed toward the person, find the velocity of the airplane. (c) Look up the speed of light in air, and defend the approximation used in (a).

66. Another scheme to catch the roadrunner has failed! Now a safe falls from rest from the top of a 25.0-m-high cliff toward Wile E. Coyote, who is standing at the base. Wile first notices the safe after it has fallen 15.0 m. How long does he have to get out of the way?

67. **Physics Now™** A stunt man sitting on a tree limb wishes to drop vertically onto a horse galloping under the tree. The constant speed of the horse is 10.0 m/s, and the man is initially 3.00 m above the level of the saddle. (a) What must be the horizontal distance between the saddle and the limb when the man makes his move? (b) How long is he in the air?

68. A hard rubber ball, released at chest height, falls to the pavement and bounces back to nearly the same height. When the ball is in contact with the pavement, its lower side is temporarily flattened. Before the dent in the ball pops out, suppose that its maximum depth is on the order of 1 cm. Compute an order-of-magnitude estimate for the maximum acceleration of the ball. State your assumptions, the quantities you estimate, and the values you estimate for them.

69. *Vroom—vroom!* As soon as a traffic light turns green, a car speeds up from rest to 50.0 mi/h with a constant acceleration of 9.00 mi/h · s. In the adjoining bike lane, a cyclist speeds up from rest to 20.0 mi/h with a constant acceleration of 13.0 mi/h · s. Each vehicle maintains a constant velocity after reaching its cruising speed. (a) For how long is the bicycle ahead of the car? (b) By what maximum distance does the bicycle lead the car?

70. In order to pass a physical education class at a university, a student must run 1.0 mi in 12 min. After running for 10 min, she still has 500 yd to go. If her maximum acceleration is $0.15\ \text{m/s}^2$, can she make it? If the answer is no, determine what acceleration she would need to be successful.

71. One swimmer in a relay race has a 0.50-s lead and is swimming at a constant speed of 4.0 m/s. He has 50 m to swim before reaching the end of the pool. A second swimmer moves in the same direction as the leader. What constant speed must the second swimmer have in order to catch up to the leader at the end of the pool?

ACTIVITIES

A.1. Estimate a few speeds in metric units, using a stopwatch or a wristwatch. For example, roll a ball across a table and estimate the number of centimeters it moves each second to find its speed. Other speeds you might try are for someone walking across the room, a jogger running, a car moving through some distance, and so forth. To see how well you did, make some actual measurements for those situations in which it is feasible to do so.

A.2. Use what you know about falling objects to measure your reaction time. Hold the index finger and thumb of your dominant hand about 2.5 cm apart, and then have your co-worker hold a ruler vertically in the space between your finger and thumb, as shown in Figure A2.2. Note the position of the ruler relative to your index finger. Your co-worker must release the ruler, and you must catch it (without moving your hand downward) as quickly as you can. The ruler (a freely falling object) falls through a distance $d = \frac{1}{2}gt^2$, where t is the reaction time and $g = 9.80\ \text{m/s}^2$. Repeat this measurement of d five times, average your results, and calculate an average value of t. Now measure your co-worker's reaction time, using the same procedure. Compare your results. For most people, the reaction time is at best about 0.2 s. As an extension to this experiment, replace the ruler with a crisp dollar bill. Hold the bill such that your thumb and index finger are just at the level of Washington's face. Unless you are anticipating the time of release, you will not be able to catch the bill when it is released, because the time required for the top to pass out of your hand is less than the typical 0.2-s reaction time.

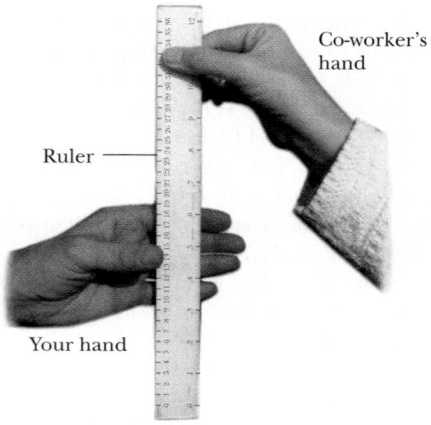

Figure A2.2

A.3. Galileo studied accelerated motion by allowing objects to roll down inclined planes so that their motion would be slow enough to make reasonable observations. Try a similar procedure. Make a mark at the top of an inclined plane as the starting point for the motion, and use a metal barrier at the end as a sound cue for stopping a stopwatch. Measure the length of the plane. Record the average time for several trials of a ball rolling down the plane at a measured angle. From the information you obtain, calculate the acceleration. Repeat the experiment for a larger angle of inclination. Do this for several trials, until you can plot a graph of acceleration versus angle. From your graph, can you guess what the acceleration would be if the inclined plane were vertical? Would the results of your experiment be different if you had used a significantly more massive ball? If you are unsure, repeat the experiment to see if there is a difference.

A.4. Perform the activities that follow to verify that all objects fall with the same acceleration. First, try dropping a coffee filter oriented horizontally and also dropping a pencil. Then repeat the experiment with the filter in a loose ball, a tight ball, and, finally, in a compacted wad. You should note that compacting the filter tends to reduce the effects of air resistance and makes the two objects fall more nearly at the same rate.

© Bettmann/Corbis

Legendary motorcycle stuntman Evel Knievel blasts off in his custom rocket-powered Harley-Davidson Skycycle in an attempt to jump the Snake River Canyon in 1974. A parachute prematurely deployed and caused the craft to fall into the canyon, just short of the other side. Knievel survived.

CHAPTER

3

Vectors and Two-Dimensional Motion

OUTLINE

3.1 Vectors and Their Properties

3.2 Components of a Vector

3.3 Displacement, Velocity, and Acceleration in Two Dimensions

3.4 Motion in Two Dimensions

3.5 Relative Velocity

In our discussion of one-dimensional motion in Chapter 2, we used the concept of vectors only to a limited extent. In our further study of motion, manipulating vector quantities will become increasingly important, so much of this chapter is devoted to vector techniques. We'll then apply these mathematical tools to two-dimensional motion, especially that of projectiles, and to the understanding of relative motion.

3.1 VECTORS AND THEIR PROPERTIES

Each of the physical quantities we will encounter in this book can be categorized as either a *vector quantity* or a *scalar quantity*. As noted in Chapter 2, a vector has both direction and magnitude (size). A scalar can be completely specified by its magnitude with appropriate units; it has no direction. An example of each kind of quantity is shown in Figure 3.1 (page 54).

As described in Chapter 2, displacement, velocity, and acceleration are vector quantities. Temperature is an example of a scalar quantity. If the temperature of an object is $-5°C$, that information completely specifies the temperature of the object; no direction is required. Masses, time intervals, and volumes are scalars as well. Scalar quantities can be manipulated with the rules of ordinary arithmetic. Vectors can also be added and subtracted from each other, and multiplied, but there are a number of important differences, as will be seen in the following sections.

When a vector quantity is handwritten, it is often represented with an arrow over the letter ($\vec{A}$). As mentioned in Section 2.1, a vector quantity in this book

Mack Henley/Visuals Unlimited

George Semple

Figure 3.1 (a) The number of grapes in this bunch ripe for picking is one example of a scalar quantity. Can you think of other examples? (b) This helpful person pointing in the right direction tells us to travel five blocks north to reach the courthouse. A vector is a physical quantity that must be specified by both magnitude and direction.

will be represented by boldface type with an arrow on top (for example, $\vec{\mathbf{A}}$). The magnitude of the vector $\vec{\mathbf{A}}$ will be represented by italic type, as A. Italic type will also be used to represent scalars.

Equality of Two Vectors. Two vectors $\vec{\mathbf{A}}$ and $\vec{\mathbf{B}}$ are equal if they have the same magnitude and the same direction. This property allows us to translate a vector parallel to itself in a diagram without affecting the vector. In fact, for most purposes, any vector can be moved parallel to itself without being affected. (See Fig. 3.2.)

Adding Vectors. When two or more vectors are added, they must all have the same units. For example, it doesn't make sense to add a velocity vector, carrying units of meters per second, to a displacement vector, carrying units of meters. Scalars obey the same rule: It would be similarly meaningless to add temperatures to volumes or masses to time intervals.

Vectors can be added geometrically or algebraically. (The latter is discussed at the end of the next section.) To add vector $\vec{\mathbf{B}}$ to vector $\vec{\mathbf{A}}$ geometrically, first draw $\vec{\mathbf{A}}$ on a piece of graph paper to some scale, such as 1 cm = 1 m, and so that its direction is specified relative a coordinate system. Then draw vector $\vec{\mathbf{B}}$ to the same scale with the tail of $\vec{\mathbf{B}}$ starting at the tip of $\vec{\mathbf{A}}$, as in Active Figure 3.3a. Vector $\vec{\mathbf{B}}$ must be drawn along the direction that makes the proper angle relative vector $\vec{\mathbf{A}}$. The **resultant vector** $\vec{\mathbf{R}} = \vec{\mathbf{A}} + \vec{\mathbf{B}}$ is the vector drawn from the tail of $\vec{\mathbf{A}}$ to the tip of $\vec{\mathbf{B}}$. This procedure is known as the **triangle method of addition**.

When two vectors are added, their sum is independent of the order of the addition: $\vec{\mathbf{A}} + \vec{\mathbf{B}} = \vec{\mathbf{B}} + \vec{\mathbf{A}}$. This relationship can be seen from the geometric construction in Active Figure 3.3b, and is called the **commutative law of addition**.

This same general approach can also be used to add more than two vectors, as is done in Figure 3.4 for four vectors. The resultant vector sum $\vec{\mathbf{R}} = \vec{\mathbf{A}} + \vec{\mathbf{B}} + \vec{\mathbf{C}} + \vec{\mathbf{D}}$ is the vector drawn from the tail of the first vector to the tip of the last. Again, the order in which the vectors are added is unimportant.

Negative of a Vector. The negative of the vector $\vec{\mathbf{A}}$ is defined as the vector that gives zero when added to $\vec{\mathbf{A}}$. This means that $\vec{\mathbf{A}}$ and $-\vec{\mathbf{A}}$ have the same magnitude but opposite directions.

TIP 3.1 Vector Addition Versus Scalar Addition

$\vec{\mathbf{A}} + \vec{\mathbf{B}} = \vec{\mathbf{C}}$ is very different from $A + B = C$. The first is a vector sum, which must be handled graphically or with components, while the second is a simple arithmetic sum of numbers.

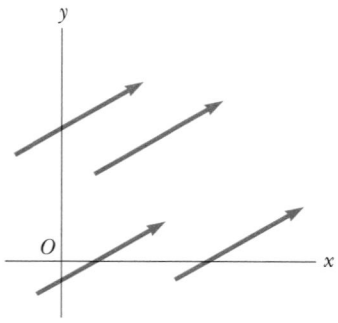

Figure 3.2 These four vectors are equal because they have equal lengths and point in the same direction.

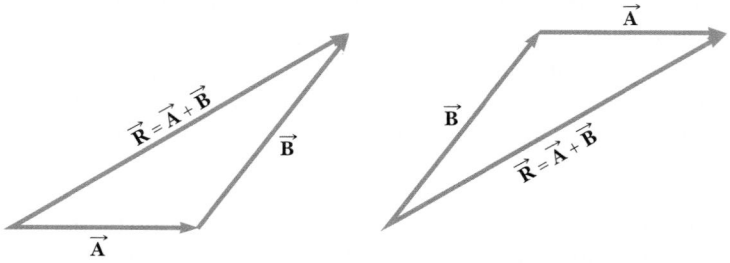

(a) (b)

ACTIVE FIGURE 3.3

(a) When vector $\vec{\mathbf{B}}$ is added to vector $\vec{\mathbf{A}}$, the vector sum $\vec{\mathbf{R}}$ is the vector that runs from the tail of $\vec{\mathbf{A}}$ to the tip of $\vec{\mathbf{B}}$. (b) Here the resultant runs from the tail of $\vec{\mathbf{B}}$ to the tip of $\vec{\mathbf{A}}$. These constructions prove that $\vec{\mathbf{A}} + \vec{\mathbf{B}} = \vec{\mathbf{B}} + \vec{\mathbf{A}}$.

PhysicsNow™

Log into PhysicsNow at **www.cp7e.com**, and go to Active Figure 3.3 to vary $\vec{\mathbf{A}}$ and $\vec{\mathbf{B}}$ and see the effect on the resultant.

Subtracting Vectors. Vector subtraction makes use of the definition of the negative of a vector. We define the operation $\vec{\mathbf{A}} - \vec{\mathbf{B}}$ as the vector $-\vec{\mathbf{B}}$ added to the vector $\vec{\mathbf{A}}$:

$$\vec{\mathbf{A}} - \vec{\mathbf{B}} = \vec{\mathbf{A}} + (-\vec{\mathbf{B}}) \qquad [3.1]$$

Vector subtraction is really a special case of vector addition. The geometric construction for subtracting two vectors is shown in Figure 3.5.

Multiplying or Dividing a Vector by a Scalar. Multiplying or dividing a vector by a scalar gives a vector. For example, if vector $\vec{\mathbf{A}}$ is multiplied by the scalar number 3, the result, written $3\vec{\mathbf{A}}$, is a vector with a magnitude three times that of $\vec{\mathbf{A}}$ and pointing in the same direction. If we multiply vector $\vec{\mathbf{A}}$ by the scalar -3, the result is $-3\vec{\mathbf{A}}$, a vector with a magnitude three times that of $\vec{\mathbf{A}}$ and pointing in the opposite direction (because of the negative sign).

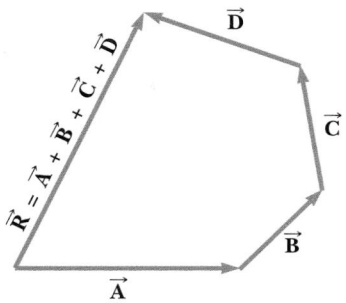

Figure 3.4 A geometric construction for summing four vectors. The resultant vector $\vec{\mathbf{R}}$ is the vector that completes the polygon.

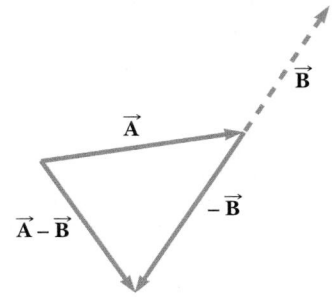

Figure 3.5 This construction shows how to subtract vector $\vec{\mathbf{B}}$ from vector $\vec{\mathbf{A}}$. The vector $-\vec{\mathbf{B}}$ has the same magnitude as the vector $\vec{\mathbf{B}}$, but points in the opposite direction.

Quick Quiz 3.1

The magnitudes of two vectors $\vec{\mathbf{A}}$ and $\vec{\mathbf{B}}$ are 12 units and 8 units, respectively. What are the largest and smallest possible values for the magnitude of the resultant vector $\vec{\mathbf{R}} = \vec{\mathbf{A}} + \vec{\mathbf{B}}$? (a) 14.4 and 4; (b) 12 and 8; (c) 20 and 4; (d) none of these.

Quick Quiz 3.2

If vector $\vec{\mathbf{B}}$ is added to vector $\vec{\mathbf{A}}$, the resultant vector $\vec{\mathbf{A}} + \vec{\mathbf{B}}$ has magnitude $A + B$ when $\vec{\mathbf{A}}$ and $\vec{\mathbf{B}}$ are (a) perpendicular to each other; (b) oriented in the same direction; (c) oriented in opposite directions; (d) none of these answers.

EXAMPLE 3.1 Taking a Trip

Goal Find the sum of two vectors by using a graph.

Problem A car travels 20.0 km due north and then 35.0 km in a direction 60° west of north, as in Figure 3.6. Using a graph, find the magnitude and direction of a single vector that gives the net effect of the car's trip. This vector is called the car's *resultant displacement.*

Strategy Draw a graph, and represent the displacement vectors as arrows. Graphically locate the vector resulting from the sum of the two displacement vectors. Measure its length and angle with respect to the vertical.

Solution
Let $\vec{\mathbf{A}}$ represent the first displacement vector, 20.0 km north, $\vec{\mathbf{B}}$ the second displacement vector, extending west of north. Carefully graph the two vectors, drawing a resultant vector $\vec{\mathbf{R}}$ with its base touching the base of $\vec{\mathbf{A}}$ and extending to the tip of $\vec{\mathbf{B}}$. Measure the length of this vector, which turns out to be about 48 km. The angle β, measured with a protractor, is about 39° west of north.

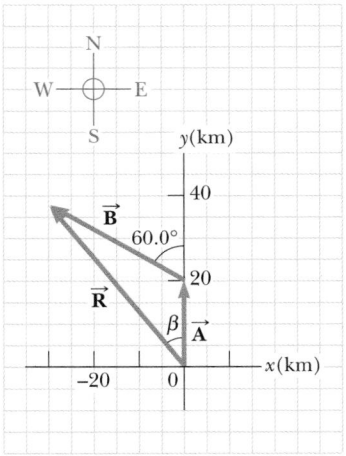

Figure 3.6 (Example 3.1) A graphical method for finding the resultant displacement vector $\vec{\mathbf{R}} = \vec{\mathbf{A}} + \vec{\mathbf{B}}$.

Remarks Notice that ordinary arithmetic doesn't work here: the correct answer of 48 km is not equal to 20.0 km + 35.0 km = 55.0 km!

Exercise 3.1
Graphically determine the magnitude and direction of the displacement if a man walks 30.0 km 45° north of east and then walks due east 20.0 km.

Answer 46 km, 27° north of east

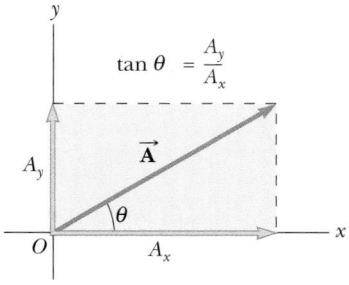

Figure 3.7 Any vector $\vec{A}$ lying in the xy-plane can be represented by its rectangular components A_x and A_y.

3.2 COMPONENTS OF A VECTOR

One method of adding vectors makes use of the projections of a vector along the axes of a rectangular coordinate system. These projections are called **components**. Any vector can be completely described by its components.

Consider a vector $\vec{A}$ in a rectangular coordinate system, as shown in Figure 3.7. $\vec{A}$ can be expressed as the sum of two vectors: $\vec{A}_x$, parallel to the x-axis; and $\vec{A}_y$, parallel to the y-axis. Mathematically,

$$\vec{A} = \vec{A}_x + \vec{A}_y$$

where $\vec{A}_x$ and $\vec{A}_y$ are the component vectors of $\vec{A}$. The projection of $\vec{A}$ along the x-axis, A_x, is called the x-component of $\vec{A}$, and the projection of $\vec{A}$ along the y-axis, A_y, is called the y-component of $\vec{A}$. These components can be either positive or negative numbers with units. From the definitions of sine and cosine, we see that $\cos\theta = A_x/A$ and $\sin\theta = A_y/A$, so the components of $\vec{A}$ are

$$A_x = A\cos\theta \qquad A_y = A\sin\theta \qquad [3.2]$$

These components form two sides of a right triangle having a hypotenuse with magnitude A. It follows that $\vec{A}$'s magnitude and direction are related to its components through the Pythagorean theorem and the definition of the tangent:

$$A = \sqrt{A_x^2 + A_y^2} \qquad [3.3]$$

$$\tan\theta = \frac{A_y}{A_x} \qquad [3.4]$$

To solve for the angle θ, which is measured from the positive x-axis by convention, we can write Equation 3.4 in the form

$$\theta = \tan^{-1}\left(\frac{A_y}{A_x}\right)$$

This formula gives the right answer only half the time! The inverse tangent function returns values only from $-90°$ to $+90°$, so the answer in your calculator window will only be correct if the vector happens to lie in first or fourth quadrant. If it lies in second or third quadrant, adding 180° to the number in the calculator window will always give the right answer. The angle in Equations 3.2 and 3.4 must be measured from the positive x-axis. Other choices of reference line are possible, but certain adjustments must then be made. (See Tip 3.2 and Figure 3.8.)

If a coordinate system other than the one shown in Figure 3.7 is chosen, the components of the vector must be modified accordingly. In many applications it's more convenient to express the components of a vector in a coordinate system having axes that are not horizontal and vertical, but are still perpendicular to each other. Suppose a vector $\vec{B}$ makes an angle θ' with the x'-axis defined in Figure 3.9. The rectangular components of $\vec{B}$ along the axes of the figure are given by $B_{x'} = B\cos\theta'$ and $B_{y'} = B\sin\theta'$, as in Equations 3.2. The magnitude and direction of $\vec{B}$ are then obtained from expressions equivalent to Equations 3.3 and 3.4.

TIP 3.2 x- and y-Components

Equation 3.2 for the x- and y-components of a vector associates cosine with the x-component and sine with the y-component, as in Figure 3.8a. This association is due *solely* to the fact that we chose to measure the angle θ with respect to the positive x-axis. If the angle were measured with respect to the y-axis, as in Figure 3.8b, the components would be given by $A_x = A\sin\theta$ and $A_y = A\cos\theta$.

TIP 3.3 Inverse Tangents on Calculators: Right Half the Time

The inverse tangent function on calculators returns an angle between $-90°$ and $+90°$. If the vector lies in the second or third quadrant, the angle, as measured from the positive x-axis, will be the angle returned by your calculator plus 180°.

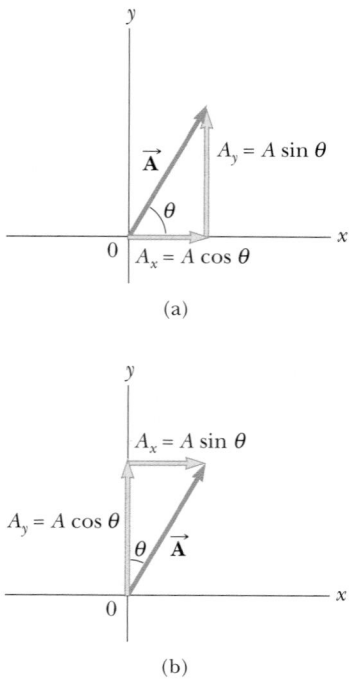

Figure 3.8 The angle θ need not always be defined from the positive x-axis.

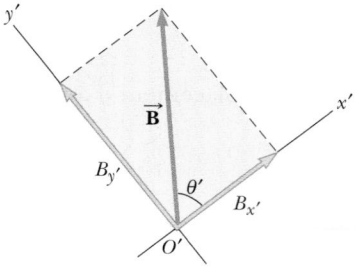

Figure 3.9 The components of vector $\vec{B}$ in a tilted coordinate system.

Quick Quiz 3.3

Figure 3.10 shows two vectors lying in the xy-plane. Determine the signs of the x- and y-components of $\vec{\mathbf{A}}$, $\vec{\mathbf{B}}$, and $\vec{\mathbf{A}} + \vec{\mathbf{B}}$, and place your answers in the following table:

Vector	x-component	y-component
$\vec{\mathbf{A}}$		
$\vec{\mathbf{B}}$		
$\vec{\mathbf{A}} + \vec{\mathbf{B}}$		

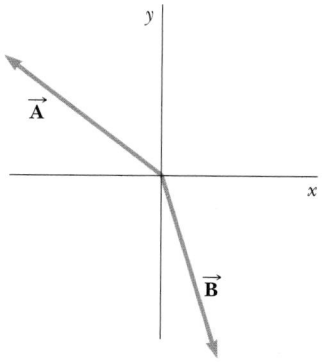

Figure 3.10 (Quick Quiz 3.3)

EXAMPLE 3.2 Help Is on the Way!

Goal Find vector components, given a magnitude and direction, and vice versa.

Problem **(a)** Find the horizontal and vertical components of the 1.00×10^2 m displacement of a superhero who flies from the top of a tall building along the path shown in Figure 3.11a. **(b)** Suppose instead the superhero leaps in the other direction along a displacement vector $\vec{\mathbf{B}}$, to the top of a flagpole where the displacement components are given by $B_x = -25.0$ m and $B_y = 10.0$ m. Find the magnitude and direction of the displacement vector.

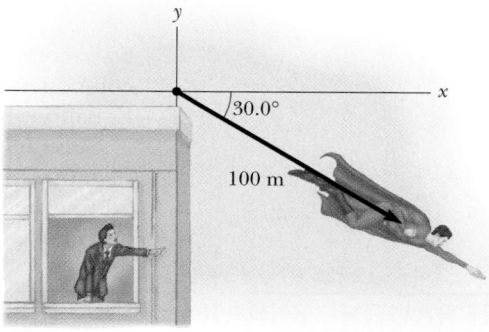

(a)

Strategy **(a)** The triangle formed by the displacement and its components is shown in Figure 3.11b. Simple trigonometry gives the components relative the standard x-y coordinate system: $A_x = A \cos \theta$ and $A_y = A \sin \theta$ (Equations 3.2). Note that $\theta = -30.0°$, negative because it's measured clockwise from the positive x-axis. **(b)** Apply Equations 3.3 and 3.4 to find the magnitude and direction of the vector.

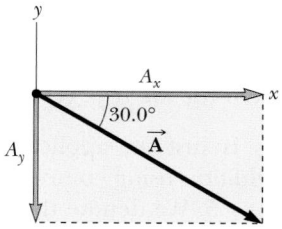

(b)

Figure 3.11 (Example 3.2)

Solution

(a) Find the vector components of $\vec{\mathbf{A}}$ from its magnitude and direction.

Use Equations 3.2 to find the components of the displacement vector $\vec{\mathbf{A}}$:

$$A_x = A \cos \theta = (1.00 \times 10^2 \text{ m}) \cos(-30.0°) = \boxed{+86.6 \text{ m}}$$

$$A_y = A \sin \theta = (1.00 \times 10^2 \text{ m}) \sin(-30.0°) = \boxed{-50.0 \text{ m}}$$

(b) Find the magnitude and direction of the displacement vector $\vec{\mathbf{B}}$ from its components.

Compute the magnitude of $\vec{\mathbf{B}}$ from the Pythagorean theorem:

$$B = \sqrt{B_x^2 + B_y^2} = \sqrt{(-25.0 \text{ m})^2 + (10.0 \text{ m})^2} = \boxed{26.9 \text{ m}}$$

Calculate the direction of $\vec{\mathbf{B}}$ using the inverse tangent, remembering to add 180° to the answer in your calculator window, because the vector lies in the second quadrant:

$$\theta = \tan^{-1}\left(\frac{B_y}{B_x}\right) = \tan^{-1}\left(\frac{10.0}{-25.0}\right) = -21.8°$$

$$\theta = \boxed{158°}$$

Remarks In part (a), note that $\cos(-\theta) = \cos \theta$; however, $\sin(-\theta) = -\sin \theta$. The negative sign of A_y reflects the fact that displacement in the y-direction is *downward*.

Exercise 3.2

(a) Suppose the superhero had flown 150 m at a 120° angle with respect to the positive x-axis. Find the components of the displacement vector. (b) Suppose instead, the superhero had leaped with a displacement having an x-component of 32.5 m and a y-component of 24.3 m. Find the magnitude and direction of the displacement vector.

Answers (a) $A_x = -75$ m, $A_y = 130$ m (b) 40.6 m, 36.8°

Adding Vectors Algebraically

The graphical method of adding vectors is valuable in understanding how vectors can be manipulated, but most of the time vectors are added algebraically in terms of their components. Suppose $\vec{\mathbf{R}} = \vec{\mathbf{A}} + \vec{\mathbf{B}}$. Then the components of the resultant vector $\vec{\mathbf{R}}$ are given by

$$R_x = A_x + B_x \qquad \textbf{[3.5a]}$$

$$R_y = A_y + B_y \qquad \textbf{[3.5b]}$$

So x-components are added only to x-components, and y-components only to y-components. The magnitude and direction of $\vec{\mathbf{R}}$ can subsequently be found with Equations 3.3 and 3.4.

Subtracting two vectors works the same way, because it's a matter of adding the negative of one vector to another vector. You should make a rough sketch when adding or subtracting vectors, in order to get an approximate geometric solution as a check.

INTERACTIVE EXAMPLE 3.3 Take a Hike

Goal Add vectors algebraically and find the resultant vector.

Problem A hiker begins a trip by first walking 25.0 km southeast from her base camp. On the second day she walks 40.0 km in a direction 60.0° north of east, at which point she discovers a forest ranger's tower. **(a)** Determine the components of the hiker's displacements in the first and second days. **(b)** Determine the components of the hiker's total displacement for the trip. **(c)** Find the magnitude and direction of the displacement from base camp.

Strategy This is just an application of vector addition using components, Equations 3.5. We denote the displacement vectors on the first and second days by $\vec{\mathbf{A}}$ and $\vec{\mathbf{B}}$, respectively. Using the camp as the origin of the coordinates, we get the vectors shown in Figure 3.12a. After finding x- and y-components for each vector, we add them "componentwise." Finally, we determine the magnitude and direction of the resultant vector $\vec{\mathbf{R}}$, using the Pythagorean theorem and the inverse tangent function.

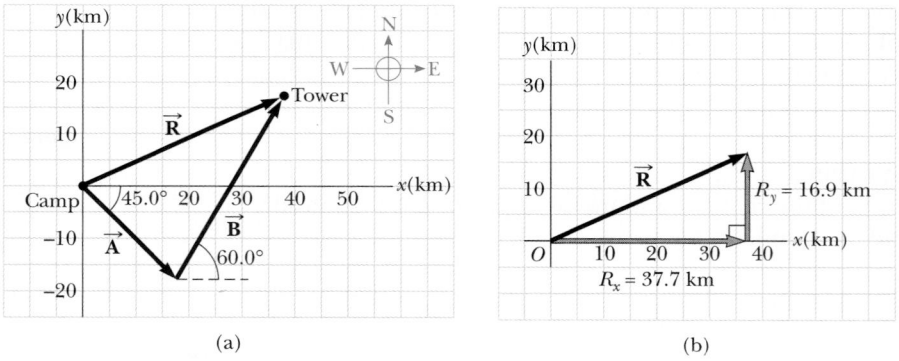

Figure 3.12 (Example 3.3) (a) Hiker's path and the resultant vector. (b) Components of the hiker's total displacement from camp.

Solution

(a) Find the components of $\vec{\mathbf{A}}$.

Use Equations 3.2 to find the components of $\vec{\mathbf{A}}$:

$$A_x = A\cos(-45.0°) = (25.0\text{ km})(0.707) = \boxed{17.7\text{ km}}$$

$$A_y = A\sin(-45.0°) = -(25.0\text{ km})(0.707) = \boxed{-17.7\text{ km}}$$

Find the components of $\vec{\mathbf{B}}$:

$$B_x = B\cos 60.0° = (40.0\text{ km})(0.500) = \boxed{20.0\text{ km}}$$

$$B_y = B\sin 60.0° = (40.0\text{ km})(0.866) = \boxed{34.6\text{ km}}$$

(b) Find the components of the resultant vector, $\vec{\mathbf{R}} = \vec{\mathbf{A}} + \vec{\mathbf{B}}$.

To find R_x, add the x-components of $\vec{\mathbf{A}}$ and $\vec{\mathbf{B}}$:

$$R_x = A_x + B_x = 17.7 \text{ km} + 20.0 \text{ km} = 37.7 \text{ km}$$

To find R_y, add the y-components of $\vec{\mathbf{A}}$ and $\vec{\mathbf{B}}$:

$$R_y = A_y + B_y = -17.7 \text{ km} + 34.6 \text{ km} = 16.9 \text{ km}$$

(c) Find the magnitude and direction of $\vec{\mathbf{R}}$.

Use the Pythagorean theorem to get the magnitude:

$$R = \sqrt{R_x^2 + R_y^2} = \sqrt{(37.7 \text{ km})^2 + (16.9 \text{ km})^2} = 41.3 \text{ km}$$

Calculate the direction of $\vec{\mathbf{R}}$ using the inverse tangent function:

$$\theta = \tan^{-1}\left(\frac{16.9 \text{ km}}{37.7 \text{ km}}\right) = 24.1°$$

Remarks Figure 3.12b shows a sketch of the components of $\vec{\mathbf{R}}$ and their directions in space. The magnitude and direction of the resultant can also be determined from such a sketch.

Exercise 3.3

A cruise ship leaving port, travels 50.0 km 45.0° north of west and then 70.0 km at a heading 30.0° north of east. Find (a) the ship's displacement vector and (b) the displacement vector's magnitude and direction.

Answer (a) $R_x = 25.3$ km, $R_y = 70.4$ km (b) 74.8 km, 70.2° north of east

PhysicsNow™ Investigate this problem further by logging into PhysicsNow at **www.cp7e.com** and going to Interactive Example 3.3.

3.3 DISPLACEMENT, VELOCITY, AND ACCELERATION IN TWO DIMENSIONS

In one-dimensional motion, as discussed in Chapter 2, the direction of a vector quantity such as a velocity or acceleration can be taken into account by specifying whether the quantity is positive or negative. The velocity of a rocket, for example, is positive if the rocket is going up and negative if it's going down. This simple solution is no longer available in two or three dimensions. Instead, we must make full use of the vector concept.

Consider an object moving through space as shown in Figure 3.13. When the object is at some point Ⓟ at time t_i, its position is described by the position vector $\vec{\mathbf{r}}_i$, drawn from the origin to Ⓟ. When the object has moved to some other point Ⓠ at time t_f, its position vector is $\vec{\mathbf{r}}_f$. From the vector diagram in Figure 3.13, the final position vector is the sum of the initial position vector and the displacement $\Delta\vec{\mathbf{r}}$: $\vec{\mathbf{r}}_f = \vec{\mathbf{r}}_i + \Delta\vec{\mathbf{r}}$. From this relationship, we obtain the following one:

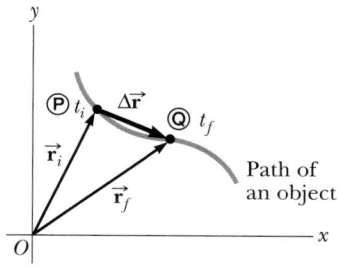

Figure 3.13 An object moving along some curved path between points Ⓟ and Ⓠ. The displacement vector $\Delta\vec{\mathbf{r}}$ is the difference in the position vectors: $\Delta\vec{\mathbf{r}} = \vec{\mathbf{r}}_f - \vec{\mathbf{r}}_i$.

An object's **displacement** is defined as the change in its position vector, or

$$\Delta\vec{\mathbf{r}} \equiv \vec{\mathbf{r}}_f - \vec{\mathbf{r}}_i \qquad [3.6]$$

SI unit: meter (m)

◀ Displacement vector

We now present several generalizations of the definitions of velocity and acceleration given in Chapter 2.

An object's **average velocity** during a time interval Δt is its displacement divided by Δt:

$$\vec{\mathbf{v}}_{\text{av}} \equiv \frac{\Delta\vec{\mathbf{r}}}{\Delta t} \qquad [3.7]$$

SI unit: meter per second (m/s)

◀ Average velocity

Because the displacement is a vector quantity and the time interval is a scalar quantity, we conclude that the average velocity is a *vector* quantity directed along $\Delta\vec{\mathbf{r}}$.

Instantaneous velocity ▶

An object's **instantaneous velocity** $\vec{\mathbf{v}}$ is the limit of its average velocity as Δt goes to zero:

$$\vec{\mathbf{v}} \equiv \lim_{\Delta t \to 0} \frac{\Delta\vec{\mathbf{r}}}{\Delta t} \qquad [3.8]$$

SI unit: meter per second (m/s)

The direction of the instantaneous velocity vector is along a line that is tangent to the object's path and in the direction of its motion.

Average acceleration ▶

An object's **average acceleration** during a time interval Δt is the change in its velocity $\Delta\vec{\mathbf{v}}$ divided by Δt, or

$$\vec{\mathbf{a}}_{\text{av}} \equiv \frac{\Delta\vec{\mathbf{v}}}{\Delta t} \qquad [3.9]$$

SI unit: meter per second squared ($\mathbf{m/s^2}$)

Instantaneous acceleration ▶

An object's **instantaneous acceleration** vector $\vec{\mathbf{a}}$ is the limit of its average acceleration vector as Δt goes to zero:

$$\vec{\mathbf{a}} \equiv \lim_{\Delta t \to 0} \frac{\Delta\vec{\mathbf{v}}}{\Delta t} \qquad [3.10]$$

SI unit: meter per second squared ($\mathbf{m/s^2}$)

It's important to recognize that an object can accelerate in several ways. First, the magnitude of the velocity vector (the speed) may change with time. Second, the direction of the velocity vector may change with time, even though the speed is constant, as can happen along a curved path. Third, both the magnitude and the direction of the velocity vector may change at the same time.

Quick Quiz 3.4

Which of the following objects can't be accelerating? (a) An object moving with a constant speed; (b) an object moving with a constant velocity; (c) an object moving along a curve.

Quick Quiz 3.5

Consider the following controls in an automobile: gas pedal, brake, steering wheel. The controls in this list that cause an acceleration of the car are (a) all three controls, (b) the gas pedal and the brake, (c) only the brake, or (d) only the gas pedal.

3.4 MOTION IN TWO DIMENSIONS

In Chapter 2, we studied objects moving along straight-line paths, such as the x-axis. In this chapter, we look at objects that move in both the x- and y-directions simultaneously under constant acceleration. An important special case of this two-dimensional motion is called **projectile motion**.

Projectile motion ▶

Anyone who has tossed any kind of object into the air has observed projectile motion. If the effects of air resistance and the rotation of Earth are neglected, the

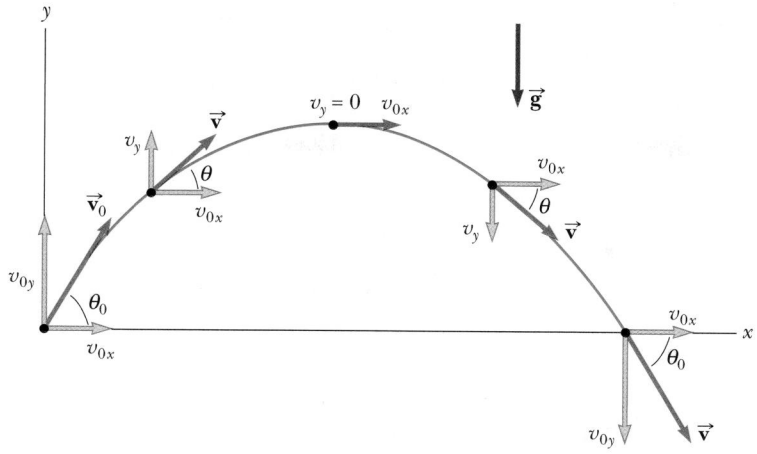

ACTIVE FIGURE 3.14
The parabolic trajectory of a particle that leaves the origin with a velocity of $\vec{v}_0$. Note that $\vec{v}$ changes with time. However, the x-component of the velocity, v_x, remains constant in time. Also, $v_y = 0$ at the peak of the trajectory, but the acceleration is always equal to the free-fall acceleration and acts vertically downward.

PhysicsNow™
Log into PhysicsNow at **www.cp7e.com**, and go to Active Figure 3.14, where you can change the particle's launch angle and initial speed. You can also observe the changing components of velocity along the trajectory of the projectile.

path of a projectile in Earth's gravity field is curved in the shape of a parabola, as shown in Active Figure 3.14.

The positive x-direction is horizontal and to the right, and the y-direction is vertical and positive upward. The most important experimental fact about projectile motion in two dimensions is that **the horizontal and vertical motions are completely independent of each other**. This means that motion in one direction has no effect on motion in the other direction. If a baseball is tossed in a parabolic path, as in Active Figure 3.14, the motion in the y-direction will look just like a ball tossed straight up under the influence of gravity. Active Figure 3.15 shows the effect of various initial angles; note that complementary angles give the same horizontal range.

In general, the equations of constant acceleration developed in Chapter 2 follow separately for both the x-direction and the y-direction. An important difference is that the initial velocity now has two components, not just one as in that chapter. We assume that at $t = 0$, the projectile leaves the origin with an initial velocity $\vec{v}_0$. If the velocity vector makes an angle θ_0 with the horizontal, where θ_0 is called the *projection angle*, then from the definitions of the cosine and sine functions and Active Figure 3.14, we have

$$v_{0x} = v_0 \cos \theta_0 \qquad \text{and} \qquad v_{0y} = v_0 \sin \theta_0$$

where v_{0x} is the initial velocity (at $t = 0$) in the x-direction and v_{0y} is the initial velocity in the y-direction.

Now, Equations 2.6, 2.9, and 2.10 developed in Chapter 2 for motion with constant acceleration in one dimension carry over to the two-dimensional case; there is one set of three equations for each direction, with the initial velocities modified as just discussed. In the x-direction, with a_x constant, we have

TIP 3.4 Acceleration at the Highest Point
The acceleration in the y-direction is *not* zero at the top of a projectile's trajectory. Only the y-component of the velocity is zero there. If the acceleration were zero, too, the projectile would never come down!

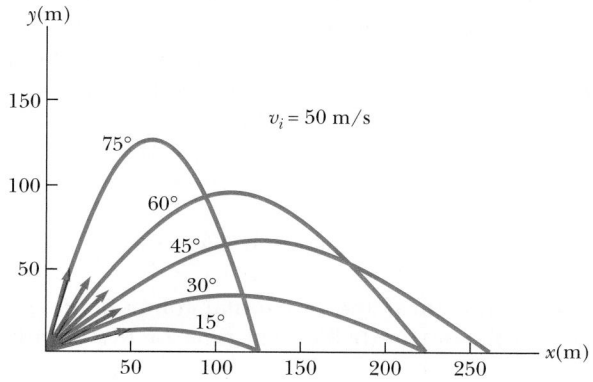

ACTIVE FIGURE 3.15
A projectile launched from the origin with an initial speed of 50 m/s at various angles of projection. Note that complementary values of the initial angle θ result in the same value of R (the range of the projectile).

PhysicsNow™
Log into PhysicsNow at **www.cp7e.com**, and go to Active Figure 3.15, where you can vary the projection angle to observe the effect on the trajectory and measure the flight time.

HIRB/Index Stock

A water fountain. The individual water streams follow parabolic trajectories. The horizontal range and maximum height of a given stream of water depend on the elevation angle of that stream's initial velocity as well as its initial speed.

$$v_x = v_{0x} + a_x t \quad \textbf{[3.11a]}$$

$$\Delta x = v_{0x} t + \tfrac{1}{2} a_x t^2 \quad \textbf{[3.11b]}$$

$$v_x^2 = v_{0x}^2 + 2a_x \Delta x \quad \textbf{[3.11c]}$$

where $v_{0x} = v_0 \cos \theta_0$. In the y-direction, we have

$$v_y = v_{0y} + a_y t \quad \textbf{[3.12a]}$$

$$\Delta y = v_{0y} t + \tfrac{1}{2} a_y t^2 \quad \textbf{[3.12b]}$$

$$v_y^2 = v_{0y}^2 + 2a_y \Delta y \quad \textbf{[3.12c]}$$

where $v_{0y} = v_0 \sin \theta_0$ and a_y is constant. The object's speed v can be calculated from the components of the velocity using the Pythagorean theorem:

$$v = \sqrt{v_x^2 + v_y^2}$$

The angle that the velocity vector makes with the x-axis is given by

$$\theta = \tan^{-1}\left(\frac{v_y}{v_x}\right)$$

This formula for θ, as previously stated, must be used with care, because the inverse tangent function returns values only between $-90°$ and $+90°$. Adding 180° is necessary for vectors lying in the second or third quadrant.

The kinematic equations are easily adapted and simplified for projectiles close to the surface of the Earth. In that case, assuming air friction is negligible, the acceleration in the x-direction is 0 (because air resistance is neglected). **This means that $a_x = 0$, and the projectile's velocity component along the x-direction remains constant.** If the initial value of the velocity component in the x-direction is $v_{0x} = v_0 \cos \theta_0$, then this is also the value of v_x at any later time, so

$$v_x = v_{0x} = v_0 \cos \theta_0 = \text{constant} \quad \textbf{[3.13a]}$$

while the horizontal displacement is simply

$$\Delta x = v_{0x} t = (v_0 \cos \theta_0) t \quad \textbf{[3.13b]}$$

For the motion in the y-direction, we make the substitution $a_y = -g$ and $v_{0y} = v_0 \sin \theta_0$ in Equations 3.12, giving

$$v_y = v_0 \sin \theta_0 - gt \quad \textbf{[3.14a]}$$

$$\Delta y = (v_0 \sin \theta_0) t - \tfrac{1}{2} g t^2 \quad \textbf{[3.14b]}$$

$$v_y^2 = (v_0 \sin \theta_0)^2 - 2g \Delta y \quad \textbf{[3.14c]}$$

The important facts of projectile motion can be summarized as follows:

1. Provided air resistance is negligible, the horizontal component of the velocity v_x remains constant because there is no horizontal component of acceleration.
2. The vertical component of the acceleration is equal to the free fall acceleration $-g$.
3. The vertical component of the velocity v_y and the displacement in the y-direction are identical to those of a freely falling body.
4. Projectile motion can be described as a superposition of two independent motions in the x- and y-directions.

EXAMPLE 3.4 Projectile Motion with Diagrams

Goal Approximate answers in projectile motion using a motion diagram.

Problem A ball is thrown so that its initial vertical and horizontal components of velocity are 40 m/s and 20 m/s, respectively. Use a motion diagram to estimate the ball's total time of flight and the distance it traverses before hitting the ground.

Strategy Use the diagram, estimating the acceleration of gravity as $-10\ \mathrm{m/s^2}$. By symmetry, the ball goes up and comes back down to the ground at the same y-velocity as when it left, except with opposite sign. With this fact and the fact that the acceleration of gravity decreases the velocity in the y-direction by 10 m/s every second, we can find the total time of flight and then the horizontal range.

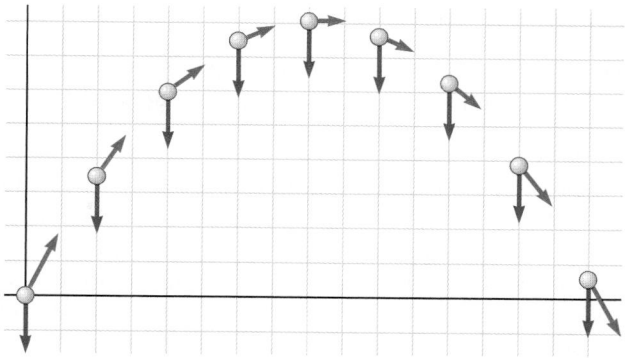

Figure 3.16 (Example 3.4) Motion diagram for a projectile.

Solution

In the motion diagram shown in Figure 3.16, the acceleration vectors are all the same, pointing downward with magnitude of nearly $10\ \mathrm{m/s^2}$. By symmetry, we know that the ball will hit the ground at the same speed in the y-direction as when it was thrown, so the velocity in the y-direction goes from 40 m/s to -40 m/s in steps of -10 m/s every second; hence, approximately 8 seconds elapse during the motion.

The velocity vector constantly changes direction, but the horizontal velocity never changes, because the acceleration in the horizontal direction is zero. Therefore, the displacement of the ball in the x-direction is given by Equation 3.13b, $\Delta x \approx v_{0x}t = (20\ \mathrm{m/s})(8\ \mathrm{s}) = 160\ \mathrm{m}$.

Remarks This example emphasizes the independence of the x- and y-components in projectile motion problems.

Exercise 3.4

Estimate the maximum height in this same problem.

Answer 80 m

Quick Quiz 3.6

Suppose you are carrying a ball and running at constant speed, and wish to throw the ball and catch it as it comes back down. Should you (a) throw the ball at an angle of about 45° above the horizontal and maintain the same speed, (b) throw the ball straight up in the air and slow down to catch it, or (c) throw the ball straight up in the air and maintain the same speed?

Quick Quiz 3.7

As a projectile moves in its parabolic path, the velocity and acceleration vectors are perpendicular to each other (a) everywhere along the projectile's path, (b) at the peak of its path, (c) nowhere along its path, or (d) not enough information is given.

Problem-Solving Strategy Projectile Motion

1. Select a coordinate system and sketch the path of the projectile, including initial and final positions, velocities, and accelerations.
2. Resolve the initial velocity vector into x- and y-components.
3. Treat the horizontal motion and the vertical motion independently.
4. Follow the techniques for solving problems with constant velocity to analyze the horizontal motion of the projectile.
5. Follow the techniques for solving problems with constant acceleration to analyze the vertical motion of the projectile.

EXAMPLE 3.5 Stranded Explorers

Goal Solve a two-dimensional projectile motion problem in which an object has an initial horizontal velocity.

Problem An Alaskan rescue plane drops a package of emergency rations to a stranded hiker, as shown in Figure 3.17. The plane is traveling horizontally at 40.0 m/s at a height of 1.00×10^2 m above the ground. **(a)** Where does the package strike the ground relative to the point at which it was released? **(b)** What are the horizontal and vertical components of the velocity of the package just before it hits the ground?

Strategy Here, we're just taking Equations 3.13 and 3.14, filling in known quantities, and solving for the remaining unknown quantities. Sketch the problem using a coordinate system as in Figure 3.17. In part (a), set the y-component of the displacement equations equal to -1.00×10^2 m—the ground level where the package lands—and solve for the time it takes the package to reach the ground. Substitute this time into the displacement equation for the x-component to find the range. In part (b), substitute the time found in part (a) into the velocity components. Notice that the initial velocity has only an x-component, which simplifies the math.

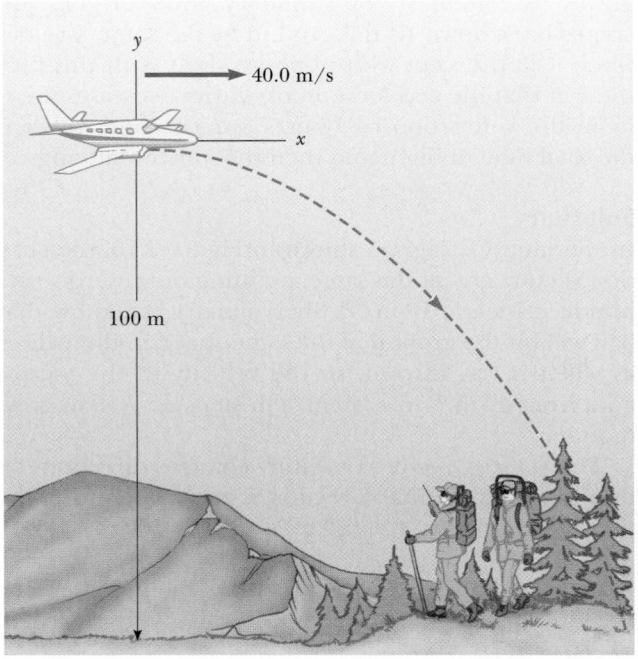

Figure 3.17 (Example 3.5) From the point of view of an observer on the ground, a package released from the rescue plane travels along the path shown.

Solution

(a) Find the range of the package.

Use Equation 3.14b to find the y-displacement:

$$\Delta y = y - y_0 = v_{0y}t - \tfrac{1}{2}gt^2$$

Substitute $y_0 = 0$ and $v_{0y} = 0$, set $y = -1.00 \times 10^2$ m—the final vertical position of the package relative the airplane—and solve for time:

$$y = -(4.90 \text{ m/s}^2)t^2 = -1.00 \times 10^2 \text{ m}$$
$$t = 4.52 \text{ s}$$

Use Equation 3.13b to find the x-displacement:

$$\Delta x = x - x_0 = v_{0x}t$$

Substitute $x_0 = 0$, $v_{0x} = 40.0$ m/s, and the time:

$$x = (40.0 \text{ m/s})(4.52 \text{ s}) = \boxed{181 \text{ m}}$$

(b) Find the components of the package's velocity at impact:

Find the x-component of the velocity at the time of impact:

$$v_x = v_0 \cos\theta = (40.0 \text{ m/s}) \cos 0° = \boxed{40.0 \text{ m/s}}$$

Find the y-component of the velocity at the time of impact:

$$v_y = v_0 \sin\theta - gt = 0 - (9.80 \text{ m/s}^2)(4.52 \text{ s})$$
$$= \boxed{-44.3 \text{ m/s}}$$

Remarks Notice how motion in the x-direction and motion in the y-direction are handled separately.

Exercise 3.5

A bartender slides a beer mug at 1.50 m/s towards a customer at the end of a frictionless bar that is 1.20 m tall. The customer makes a grab for the mug and misses, and the mug sails off the end of the bar. (a) How far away from the end of the bar does the mug hit the floor? (b) What are the speed and direction of the mug at impact?

Answers (a) 0.742 m (b) 5.08 m/s, $\theta = -72.8°$

EXAMPLE 3.6 The Long Jump

APPLICATION
Long Jumping

Goal Solve a two-dimensional projectile motion problem involving an object starting and ending at the same height.

Problem A long jumper (Fig. 3.18) leaves the ground at an angle of 20.0° to the horizontal and at a speed of 11.0 m/s. **(a)** How long does it take for him to reach maximum height? **(b)** What is the maximum height? **(c)** How far does he jump? (Assume that his motion is equivalent to that of a particle, disregarding the motion of his arms and legs.) **(d)** Find the maximum height he reaches using Equation 3.14c.

Strategy Again, we take the projectile equations, fill in the known quantities, and solve for the unknowns. At the maximum height, the velocity in the y-direction is zero, so setting Equation 3.14a equal to zero and solving gives the time it takes him to reach his maximum height. By symmetry, given that his trajectory starts and ends at the same height, doubling this time gives the total time of the jump.

Mike Powell/Allsport/Getty Images

Figure 3.18 (Example 3.6) Mike Powell, current holder of the world long-jump record of 8.95 m.

Solution

(a) Find the time t_{max} taken to reach maximum height.

Set $v_y = 0$ in Equation 3.14b and solve for t_{max}:

$$v_y = v_0 \sin\theta_0 - g t_{max} = 0$$

$$t_{max} = \frac{v_0 \sin\theta_0}{g} = \frac{(11.0\ \text{m/s})(\sin 20.0°)}{9.80\ \text{m/s}^2}$$

$$= \boxed{0.384\ \text{s}}$$

(b) Find the maximum height he reaches.

Substitute the time t_{max} into the equation for the y-displacement:

$$y_{max} = (v_0 \sin\theta_0)\, t_{max} - \tfrac{1}{2} g (t_{max})^2$$

$$y_{max} = (11.0\ \text{m/s})(\sin 20.0°)(0.384\ \text{s}) - \tfrac{1}{2}(9.80\ \text{m/s}^2)(0.384\ \text{s})^2$$

$$y_{max} = \boxed{0.722\ \text{m}}$$

(c) Find the horizontal distance he jumps.

First find the time for the jump, which is twice t_{max}:

$$t = 2t_{max} = 2(0.384\ \text{s}) = 0.768\ \text{s}.$$

Substitute this result into the equation for the x-displacement:

$$\Delta x = (v_0 \cos\theta_0)\, t = (11.0\ \text{m/s})(\cos 20.0°)(0.768\ \text{s})$$

$$= \boxed{7.94\ \text{m}}$$

(d) Use an alternate method to find the maximum height.

Use Equation 3.14c, solving for Δy:

$$v_y^2 - v_{0y}^2 = -2g\Delta y$$

$$\Delta y = \frac{v_y^2 - v_{0y}^2}{-2g}$$

Substitute $v_y = 0$ at maximum height, and the fact that $v_{0y} = (11.0\ \text{m/s}) \sin 20.0°$:

$$\Delta y = \frac{0 - ((11.0\ \text{m/s}) \sin 20.0°)^2}{-2(9.80\ \text{m/s}^2)} = \boxed{0.722\ \text{m}}$$

Remarks Although modelling the long jumper's motion as that of a projectile is an oversimplification, the values obtained are reasonable.

Exercise 3.6

A grasshopper jumps 1.00 m from rest, with an initial velocity at a 45.0° angle with respect to the horizontal. Find (a) the initial speed of the grasshopper and (b) the maximum height reached.

Answers (a) 3.13 m/s (b) 0.250 m

INTERACTIVE EXAMPLE 3.7 That's Quite an Arm

Goal Solve a two-dimensional kinematics problem with a nonhorizontal initial velocity, starting and ending at different heights.

Problem A stone is thrown upward from the top of a building at an angle of 30.0° to the horizontal and with an initial speed of 20.0 m/s, as in Figure 3.19. The point of release is 45.0 m above the ground. **(a)** How long does it take for the stone to hit the ground? **(b)** Find the stone's speed at impact. **(c)** Find the horizontal range of the stone.

Strategy Choose coordinates as in the figure, with the origin at the point of release. **(a)** Fill in the constants of Equation 3.14b for the y-displacement, and set the displacement equal to -45.0 m, the y-displacement when the stone hits the ground. Using the quadratic formula, solve for the time. To solve **(b)**, substitute the time from part (a) into the components of the velocity, and substitute the same time into the equation for the x-displacement to solve **(c)**.

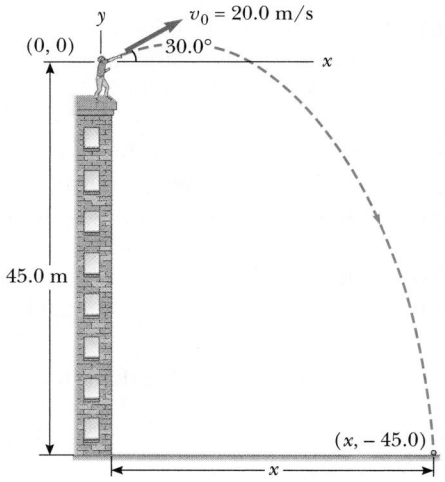

Figure 3.19 (Example 3.7)

Solution

(a) Find the time of flight.

Find the initial x- and y-components of the velocity:

$$v_{0x} = v_0 \cos \theta_0 = (20.0 \text{ m/s})(\cos 30.0°) = +17.3 \text{ m/s}$$
$$v_{0y} = v_0 \sin \theta_0 = (20.0 \text{ m/s})(\sin 30.0°) = +10.0 \text{ m/s}$$

Find the y-displacement, taking $y_0 = 0$, $y = -45.0$ m, and $v_{0y} = 10.0$ m/s:

$$\Delta y = y - y_0 = v_{0y}t - \tfrac{1}{2}gt^2$$
$$-45.0 \text{ m} = (10.0 \text{ m/s})t - (4.90 \text{ m/s}^2)t^2$$

Reorganize the equation into standard form and use the quadratic formula (see Appendix A) to find the positive root:

$$t = 4.22 \text{ s}$$

(b) Find the speed at impact.

Substitute the value of t found in part (a) into Equation 3.14a to find the y-component of the velocity at impact:

$$v_y = v_{0y} - gt = 10.0 \text{ m/s} - (9.80 \text{ m/s}^2)(4.22 \text{ s})$$
$$= -31.4 \text{ m/s}$$

Use this value of v_y, the Pythagorean theorem, and the fact that $v_x = v_{0x} = 17.3$ m/s to find the speed of the stone at impact:

$$v = \sqrt{v_x^2 + v_y^2} = \sqrt{(17.3 \text{ m/s})^2 + (-31.4 \text{ m/s})^2}$$
$$= 35.9 \text{ m/s}$$

(c) Find the horizontal range of the stone.

Substitute the time of flight into the range equation:

$$\Delta x = x - x_0 = (v_0 \cos \theta)t = (20.0 \text{ m/s})(\cos 30.0°)(4.22 \text{ s})$$
$$= 73.1 \text{ m}$$

Exercise 3.7

Suppose the stone is thrown at an angle of 30.0° degrees below the horizontal. If it strikes the ground 57.0 m away, find (a) the time of flight, (b) the initial speed, and (c) the speed and the angle of the velocity vector with respect to the horizontal at impact. (*Hint:* For part (a), use the equation for the x-displacement to eliminate $v_0 t$ from the equation for the y-displacement.)

Answers (a) 1.57 s (b) 41.9 m/s (c) 51.3 m/s, −45.0°

PhysicsNow™ Investigate this problem further by logging into PhysicsNow at **www.cp7e.com** and going to Interactive Example 3.7.

So far we have studied only problems in which an object with an initial velocity follows a trajectory determined by the acceleration of gravity alone. In the more general case, other agents, such as air drag, surface friction, or engines, can cause accelerations. These accelerations, taken together, form a vector quantity with components a_x and a_y. When both components are constant, we can use Equations 3.11 and 3.12 to study the motion, as in the next example.

EXAMPLE 3.8 The Rocket

Goal Solve a problem involving accelerations in two directions.

Problem A jet plane traveling horizontally at 1.00×10^2 m/s drops a rocket from a considerable height. (See Figure 3.20.) The rocket immediately fires its engines, accelerating at 20.0 m/s^2 in the x-direction while falling under the influence of gravity in the y-direction. When the rocket has fallen 1.00 km, find **(a)** its velocity in the y-direction, **(b)** its velocity in the x-direction, and **(c)** the magnitude and direction of its velocity. Neglect air drag and aerodynamic lift.

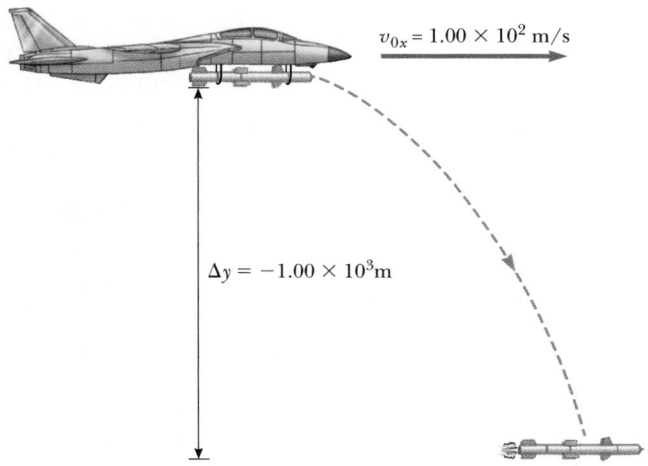

Figure 3.20 (Example 3.8)

Strategy Because the rocket maintains a horizontal orientation (say, through gyroscopes), the x- and y-components of acceleration are independent of each other. Use the time-independent equation for the velocity in the y-direction to find the y-component of the velocity after the rocket falls 1.00 km. Then calculate the time of the fall, and use that time to find the velocity in the x-direction.

Solution

(a) Find the velocity in the y-direction.

Use Equation 3.14c:

$$v_y^2 = v_{0y}^2 - 2g\Delta y$$

Substitute $v_{0y} = 0$, $g = -9.80$ m/s^2, and $\Delta y = -1.00 \times 10^3$ m, and solve:

$$v_y^2 - 0 = 2(-9.8 \text{ m/s}^2)(-1.00 \times 10^3 \text{ m})$$

$$v_y = -1.40 \times 10^2 \text{ m/s}$$

(b) Find the velocity in the x-direction.

Find the time it takes the rocket to drop 1.00×10^3 m, using the y-component of the velocity:

$$v_y = v_{0y} + a_y t$$

$$-1.40 \times 10^2 \text{ m/s} = 0 - (9.80 \text{ m/s}^2)t \quad \rightarrow \quad t = 14.3 \text{ s}$$

Substitute t, v_{0x}, and a_x into Equation 3.11a to find the velocity in the x-direction:

$$v_x = v_{0x} + a_x t = 1.00 \times 10^2 \text{ m/s} + (20.0 \text{ m/s}^2)(14.3 \text{ s})$$

$$= 386 \text{ m/s}$$

(c) Find the magnitude and direction of the velocity.

Find the magnitude using the Pythagorean theorem and the results of parts (a) and (b):

$$v = \sqrt{v_x^2 + v_y^2} = \sqrt{(-1.40 \times 10^2 \text{ m/s})^2 + (386 \text{ m/s})^2}$$
$$= 411 \text{ m/s}$$

Use the inverse tangent function to find the angle:

$$\theta = \tan^{-1}\left(\frac{v_y}{v_x}\right) \tan^{-1}\left(\frac{-1.40 \times 10^2 \text{ m/s}}{386 \text{ m/s}}\right) = -19.9°$$

Remarks Notice the symmetry: The kinematic equations for the x- and y-directions are handled in exactly the same way. Having a nonzero acceleration in the x-direction doesn't greatly increase the difficulty of the problem.

Exercise 3.8

Suppose a rocket-propelled motorcycle is fired from rest horizontally across a canyon 1.00 km wide. (a) What minimum constant acceleration in the x-direction must be provided by the engines so the cycle crosses safely if the opposite side is 0.750 km lower than the starting point? (b) At what speed does the motorcycle land if it maintains this constant horizontal component of acceleration? Neglect air drag, but remember that gravity is still acting in the negative y-direction.

Answers (a) 13.1 m/s^2 (b) 202 m/s

In a stunt similar to that described in Exercise 3.8, motorcycle daredevil Evel Knievel tried to vault across Hells Canyon, part of the Snake River system in Idaho, on his rocket-powered Harley-Davidson X-2 "Skycycle." (See the chapter-opening photo on page 53). He lost consciousness at takeoff and released a lever, prematurely deploying his parachute and falling short of the other side. He landed safely in the canyon.

3.5 RELATIVE VELOCITY

Relative velocity is all about relating the measurements of two different observers, one moving with respect to the other. The measured velocity of an object depends on the velocity of the observer with respect to the object. On highways, for example, cars moving in the same direction are often moving at high speed relative to Earth, but relative each other they hardly move at all. To an observer at rest at the side of the road, a car might be traveling at 60 mi/h, but to an observer in a truck traveling in the same direction at 50 mi/h, the car would appear to be traveling only 10 mi/h.

So measurements of velocity depend on the **reference frame** of the observer. Reference frames are just coordinate systems. Most of the time, we use a **stationary frame of reference** relative to Earth, but occasionally we use a **moving frame of reference** associated with a bus, car, or plane moving with constant velocity relative to Earth.

In two dimensions, relative velocity calculations can be confusing, so a systematic approach is important and useful. Let E be an observer, assumed stationary with respect to Earth. Let two cars be labeled A and B, and introduce the following notation (see Figure 3.21):

$\vec{\mathbf{r}}_{AE}$ = the position of Car A as measured by E (in a coordinate system fixed with respect to Earth).

$\vec{\mathbf{r}}_{BE}$ = the position of Car B as measured by E.

$\vec{\mathbf{r}}_{AB}$ = the position of Car A as measured by an observer in Car B.

According to the preceding notation, the first letter tells us what the vector is pointing at and the second letter tells us where the position vector starts. The position vectors of Car A and Car B relative to E, $\vec{\mathbf{r}}_{AE}$ and $\vec{\mathbf{r}}_{BE}$, are given in the figure. How do we find $\vec{\mathbf{r}}_{AB}$, the position of Car A as measured by an observer in Car B? We simply draw an arrow pointing from Car B to Car A, which can be obtained by subtracting $\vec{\mathbf{r}}_{BE}$ from $\vec{\mathbf{r}}_{AE}$:

$$\vec{\mathbf{r}}_{AB} = \vec{\mathbf{r}}_{AE} - \vec{\mathbf{r}}_{BE} \qquad [3.15]$$

Now, the rate of change of these quantities with time gives us the relationship

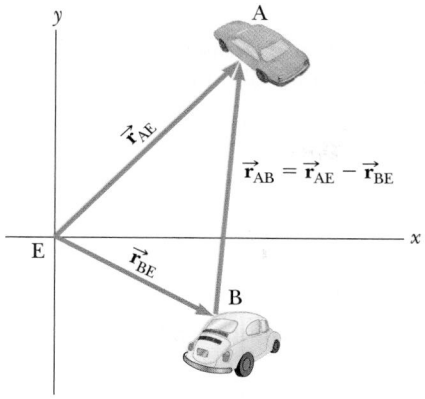

Figure 3.21 The position of Car A relative to Car B can be found by vector subtraction. The rate of change of the resultant vector with respect to time is the relative velocity equation.

between the associated velocities:

$$\vec{\mathbf{v}}_{AB} = \vec{\mathbf{v}}_{AE} - \vec{\mathbf{v}}_{BE} \qquad [3.16]$$

The coordinate system of observer E need not be fixed to Earth, although it often is. Take careful note of the pattern of subscripts; rather than memorize Equation 3.15, it's better to study the short derivation shown in Figure 3.21. Note also that the equation doesn't work for observers traveling a sizeable fraction of the speed of light, when Einstein's theory of special relativity comes into play.

PROBLEM-SOLVING STRATEGY Relative Velocity

1. Label each object involved (usually three) with a letter that reminds you of what it is (for example, E for Earth).
2. Look through the problem for phrases such as "The velocity of A relative to B," and write the velocities as '$\vec{\mathbf{v}}_{AB}$'. When a velocity is mentioned but it isn't explicitly stated as relative to something, it's almost always relative to Earth.
3. Take the three velocities you've found and assemble them into an equation just like Equation 3.16, with subscripts in an analogous order.
4. There will be two unknown components. Solve for them with the x- and y-components of the equation developed in step 3.

EXAMPLE 3.9 Pitching Practice on the Train

Goal Solve a one-dimensional relative velocity problem.

Problem A train is traveling with a speed of 15.0 m/s relative to Earth. A passenger standing at the rear of the train pitches a baseball with a speed of 15.0 m/s relative to the train off the back end, in the direction opposite the motion of the train. What is the velocity of the baseball relative to Earth?

Strategy Solving these problems involves putting the proper subscripts on the velocities and arranging them as in Equation 3.16. In the first sentence of the problem statement, we are informed that the train travels at "15.0 m/s relative to Earth." This quantity is $\vec{\mathbf{v}}_{TE}$, with T for train and E for Earth. The passenger throws the baseball at "15 m/s relative to the train," so this quantity is $\vec{\mathbf{v}}_{BT}$, where B stands for baseball. The second sentence asks for the velocity of the baseball relative to Earth, $\vec{\mathbf{v}}_{BE}$. The rest of the problem can be solved by identifying the correct components of the known quantities and solving for the unknowns, using an analog of Equation 3.16.

Solution

Write the x-components of the known quantities:

$$(\vec{\mathbf{v}}_{TE})_x = +15 \text{ m/s}$$

$$(\vec{\mathbf{v}}_{BT})_x = -15 \text{ m/s}$$

Follow Equation 3.16:

$$(\vec{\mathbf{v}}_{BT})_x = (\vec{\mathbf{v}}_{BE})_x - (\vec{\mathbf{v}}_{TE})_x$$

Insert the given values, and solve:

$$-15 \text{ m/s} = (\vec{\mathbf{v}}_{BE})_x - 15 \text{ m/s}$$

$$(\vec{\mathbf{v}}_{BE})_x = \boxed{0}$$

Exercise 3.9

A train is traveling at 27 m/s relative Earth, and a passenger standing in the train throws a ball at 15 m/s relative the train in the same direction as the train's motion. Find the speed of the ball relative to Earth.

Answer 42 m/s

EXAMPLE 3.10 Crossing a River

Goal Solve a simple two-dimensional relative motion problem.

Problem The boat in Figure 3.22 is heading due north as it crosses a wide river with a velocity of 10.0 km/h relative to the water. The river has a uniform velocity of 5.00 km/h due east. Determine the velocity of the boat with respect to an observer on the riverbank.

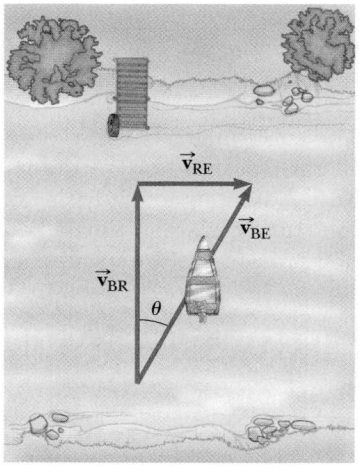

Figure 3.22 (Example 3.10)

Strategy Again, we look for key phrases. "The boat (has) . . . a velocity of 10.0 km/h relative to the water" gives $\vec{v}_{BR}$. "The river has a uniform velocity of 5.00 km/h due east" gives $\vec{v}_{RE}$, because this implies velocity with respect to Earth. The observer on the riverbank is in a reference frame at rest with respect to Earth. Because we're looking for the velocity of the boat with respect to that observer, this last velocity is designated $\vec{v}_{BE}$. Take east to be the $+x$-direction, north the $+y$-direction.

Solution

Arrange the three quantities into the proper relative velocity equation:

$$\vec{v}_{BR} = \vec{v}_{BE} - \vec{v}_{RE}$$

Write the velocity vectors in terms of their components. For convenience, these are organized in the following table:

Vector	x-component (km/h)	y-component (km/h)
$\vec{v}_{BR}$	0	10.0
$\vec{v}_{BE}$	v_x	v_y
$\vec{v}_{RE}$	5.00	0

Find the x-component of velocity:

$$0 = v_x - 5.00 \text{ km/h} \quad \rightarrow \quad v_x = 5.00 \text{ km/h}$$

Find the y-component of velocity:

$$10.0 \text{ km/h} = v_y - 0 \quad \rightarrow \quad v_y = 10.0 \text{ km/h}$$

Find the magnitude of $\vec{v}_{BE}$:

$$v_{BE} = \sqrt{v_x^2 + v_y^2}$$
$$= \sqrt{(5.00 \text{ km/h})^2 + (10.0 \text{ km/h})^2} = 11.2 \text{ km/h}$$

Find the direction of $\vec{v}_{BE}$:

$$\theta = \tan^{-1}\left(\frac{v_x}{v_y}\right) = \tan^{-1}\left(\frac{5.00 \text{ m/s}}{10.0 \text{ m/s}}\right) = 26.6°$$

The boat travels at a speed of 11.2 km/h in the direction 26.6° east of north with respect to Earth.

Exercise 3.10

Suppose the river is flowing east at 3.00 m/s and the boat is traveling south at 4.00 m/s with respect to the river. Find the speed and direction of the boat relative to Earth.

Answer 5.00 m/s, 53.1° south of east

EXAMPLE 3.11 Bucking the Current

Goal Solve a complex two-dimensional relative motion problem.

Problem If the skipper of the boat of Example 3.10 moves with the same speed of 10.0 km/h relative to the water, but now wants to travel due north, as in Figure 3.23, in what direction should he head? What is the speed of the boat, according to an observer on the shore? The river is flowing east at 5.00 km/h.

Strategy Proceed as in the previous problem. In this situation, we must find the heading of the boat and its velocity with respect to the water, using the fact that the boat travels due north.

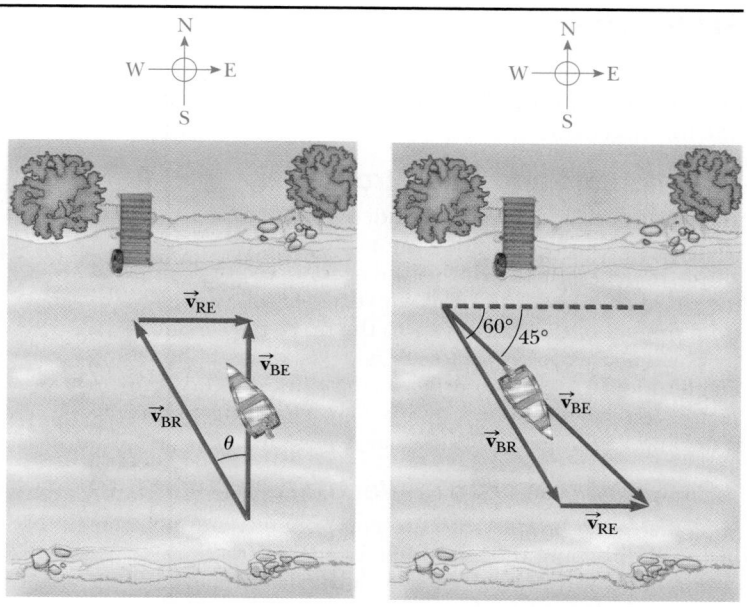

(a) (b)

Figure 3.23 (a) (Example 3.11) (b) (Exercise 3.11)

Solution

Arrange the three quantities, as before:

$$\vec{v}_{BR} = \vec{v}_{BE} - \vec{v}_{RE}$$

Organize a table of velocity components:

Vector	x-component (km/h)	y-component (km/h)
$\vec{v}_{BR}$	$-(10.0\text{ km/h})\sin\theta$	$(10.0\text{ km/h})\cos\theta$
$\vec{v}_{BE}$	0	v
$\vec{v}_{RE}$	5.00 km/h	0

The x-component of the relative velocity equation can be used to find θ:

$$-(10.0\text{ m/s})\sin\theta = 0 - 5.00\text{ km/h}$$

$$\sin\theta = \frac{5.00\text{ km/h}}{10.0\text{ km/h}} = \frac{1.00}{2.00}$$

Apply the inverse sine function and find θ, which is the boat's heading, east of north:

$$\theta = \sin^{-1}\left(\frac{1.00}{2.00}\right) = 30.0°$$

The y-component of the relative velocity equation can be used to find v:

$$(10.0\text{ km/h})\cos\theta = v \quad \rightarrow \quad v = 8.66\text{ km/h}$$

Remarks From Figure 3.23, we see that this problem can be solved with the Pythagorean theorem, because the problem involves a right triangle: The boat's x-component of velocity exactly cancels the river's velocity. When this is not the case, a more general technique is necessary, as shown in the following exercise. Notice that in the x-component of the relative velocity equation a minus sign had to be included in the term $-(10.0\text{ km/h})\sin\theta$ because the x-component of the boat's velocity with respect to the river is negative.

Exercise 3.11

Suppose the river is moving east at 5.00 km/h and the boat is traveling 45.0° south of east with respect to Earth. Find (a) the speed of the boat with respect to Earth and (b) the speed of the boat with respect to the river if the boat's heading in the water is 60.0° south of east. (See Figure 3.23b.) You will have to solve two equations and two unknowns.

Answers (a) 16.7 km/h (b) 13.7 km/h

SUMMARY

PhysicsNow™ Take a practice test by logging into PhysicsNow at **www.cp7e.com** and clicking on the Pre-Test link for this chapter.

3.1 Vectors and their Properties

Two vectors $\vec{\mathbf{A}}$ and $\vec{\mathbf{B}}$ can be added geometrically with the **triangle method**. The two vectors are drawn to scale on graph paper, with the tail of the second vector located at the tip of the first. The **resultant** vector is the vector drawn from the tail of the first vector to the tip of the second.

The negative of a vector $\vec{\mathbf{A}}$ is a vector with the same magnitude as $\vec{\mathbf{A}}$, but pointing in the opposite direction. A vector can be multiplied by a scalar, changing its magnitude, and its direction if the scalar is negative.

3.2 Components of a Vector

A vector $\vec{\mathbf{A}}$ can be split into two components, one pointing in the x-direction and the other in the y-direction. These components form two sides of a right triangle having a hypotenuse with magnitude A and are given by

$$A_x = A\cos\theta \qquad A_y = A\sin\theta \qquad [3.2]$$

The magnitude and direction of $\vec{\mathbf{A}}$ are related to its components through the Pythagorean theorem and the definition of the tangent:

$$A = \sqrt{A_x^2 + A_y^2} \qquad [3.3]$$

$$\tan\theta = \frac{A_y}{A_x} \qquad [3.4]$$

If $\vec{\mathbf{R}} = \vec{\mathbf{A}} + \vec{\mathbf{B}}$, then the components of the resultant vector $\vec{\mathbf{R}}$ are

$$R_x = A_x + B_x \qquad [3.5a]$$

$$R_y = A_y + B_y \qquad [3.5b]$$

3.3 Displacement, Velocity, and Acceleration in Two Dimensions

The displacement of an object in two dimensions is defined as the change in the object's position vector:

$$\Delta\vec{\mathbf{r}} \equiv \vec{\mathbf{r}}_f - \vec{\mathbf{r}}_i \qquad [3.6]$$

The average velocity of an object during the time interval Δt is

$$\vec{\mathbf{v}}_{\text{av}} \equiv \frac{\Delta\vec{\mathbf{r}}}{\Delta t} \qquad [3.7]$$

Taking the limit of this expression as Δt gets arbitrarily small gives the instantaneous velocity $\vec{\mathbf{v}}$:

$$\vec{\mathbf{v}} \equiv \lim_{\Delta t \to 0} \frac{\Delta\vec{\mathbf{r}}}{\Delta t} \qquad [3.8]$$

The direction of the instantaneous velocity vector is along a line that is tangent to the path of the object and in the direction of its motion.

The average acceleration of an object with a velocity changing by $\Delta\vec{\mathbf{v}}$ in the time interval Δt is

$$\vec{\mathbf{a}}_{\text{av}} \equiv \frac{\Delta\vec{\mathbf{v}}}{\Delta t} \qquad [3.9]$$

Taking the limit of this expression as Δt gets arbitrarily small gives the instantaneous acceleration vector $\vec{\mathbf{a}}$:

$$\vec{\mathbf{a}} \equiv \lim_{\Delta t \to 0} \frac{\Delta\vec{\mathbf{v}}}{\Delta t} \qquad [3.10]$$

3.4 Motion in Two Dimensions

The general kinematic equations in two dimensions for objects with constant acceleration are, for the x-direction,

$$v_x = v_{0x} + a_x t \qquad [3.11a]$$

$$\Delta x = v_{0x}t + \tfrac{1}{2}a_x t^2 \qquad [3.11b]$$

$$v_x^2 = v_{0x}^2 + 2a_x\Delta x \qquad [3.11c]$$

where $v_{0x} = v_0\cos\theta_0$, and, for the y-direction,

$$v_y = v_{0y} + a_y t \qquad [3.12a]$$

$$\Delta y = v_{0y}t + \tfrac{1}{2}a_y t^2 \qquad [3.12b]$$

$$v_y^2 = v_{0y}^2 + 2a_y\Delta y \qquad [3.12c]$$

where $v_{0y} = v_0\sin\theta_0$. The speed v of the object at any instant can be calculated from the components of velocity at that instant using the Pythagorean theorem:

$$v = \sqrt{v_x^2 + v_y^2}$$

The angle that the velocity vector makes with the x-axis is given by

$$\theta = \tan^{-1}\left(\frac{v_y}{v_x}\right)$$

The kinematic equations are easily adapted and simplified for projectiles close to the surface of the Earth. The equations for the motion in the horizontal or x-direction are

$$v_x = v_{0x} = v_0\cos\theta_0 = \text{constant} \qquad [3.13a]$$

$$\Delta x = v_{0x}t = (v_0\cos\theta_0)t \qquad [3.13b]$$

while the equations for the motion in the vertical or y-direction are

$$v_y = v_0\sin\theta_0 - gt \qquad [3.14a]$$

$$\Delta y = (v_0\sin\theta_0)t - \tfrac{1}{2}gt^2 \qquad [3.14b]$$

$$v_y^2 = (v_0\sin\theta_0)^2 - 2g\Delta y \qquad [3.14c]$$

Problems are solved by algebraically manipulating one or more of these equations, which often reduces the system to two equations and two unknowns.

3.5 Relative Velocity

Let E be an observer, and B a second observer traveling with velocity $\vec{\mathbf{v}}_{\text{BE}}$ as measured by E. If E measures the velocity of an object A as $\vec{\mathbf{v}}_{\text{AE}}$, then B will measure A's velocity as

$$\vec{\mathbf{v}}_{\text{AB}} = \vec{\mathbf{v}}_{\text{AE}} - \vec{\mathbf{v}}_{\text{BE}} \qquad [3.16]$$

Solving relative velocity problems involves identifying the velocities properly and labeling them correctly, substituting into Equation 3.16, and then solving for unknown quantities.

CONCEPTUAL QUESTIONS

1. Vector $\vec{\mathbf{A}}$ lies in the xy-plane. For what orientations of vector $\vec{\mathbf{A}}$ will both of its components be negative? When will the components have opposite signs?
2. If $\vec{\mathbf{B}}$ is added to $\vec{\mathbf{A}}$, under what conditions does the resultant vector have a magnitude equal to $A + B$? Under what conditions is the resultant vector equal to zero?
3. A wrench is dropped from the top of a 10-m mast on a sailing ship while the ship is traveling at 20 knots. Where will the wrench hit the deck? (Galileo posed this problem.)
4. Two vectors have unequal magnitudes. Can their sum be zero? Explain.
5. A projectile is fired on Earth with some initial velocity. Another projectile is fired on the Moon with the same initial velocity. If air resistance is neglected, which projectile has the greater range? Which reaches the greater altitude? (Note that the free-fall acceleration on the Moon is about $1.6\ \mathrm{m/s^2}$.)
6. Can a vector $\vec{\mathbf{A}}$ have a component greater than its magnitude A?
7. Is it possible to add a vector quantity to a scalar quantity?
8. Under what circumstances would a vector have components that are equal in magnitude?
9. As a projectile moves in its path, is there any point along the path where the velocity and acceleration vectors are (a) perpendicular to each other? (b) parallel to each other?
10. A rock is dropped at the same instant that a ball is thrown horizontally from the same initial elevation. Which will have the greater speed when it reaches ground level?
11. Explain whether the following particles do or do not have an acceleration: (a) a particle moving in a straight line with constant speed and (b) a particle moving around a curve with constant speed.
12. Correct the following statement: "The racing car rounds the turn at a constant velocity of 90 miles per hour."
13. A spacecraft drifts through space at a constant velocity. Suddenly, a gas leak in the side of the spacecraft causes it to constantly accelerate in a direction perpendicular to the initial velocity. The orientation of the spacecraft does not change, so the acceleration remains perpendicular to the original direction of the velocity. What is the shape of the path followed by the spacecraft?
14. A ball is projected horizontally from the top of a building. One second later, another ball is projected horizontally from the same point with the same velocity. At what point in the motion will the balls be closest to each other? Will the first ball always be traveling faster than the second? What will be the time difference between them when the balls hit the ground? Can the horizontal projection velocity of the second ball be changed so that the balls arrive at the ground at the same time?
15. Two projectiles are thrown with the same initial speed, one at an angle θ with respect to the level ground and the other at angle $90° - \theta$. Both projectiles strike the ground at the same distance from the projection point. Are both projectiles in the air for the same length of time?
16. A baseball is thrown such that its initial x- and y-components of velocity are known. Neglecting air resistance, describe how you would calculate the ball's (a) coordinates, (b) velocity, and (c) acceleration at the instant the ball reaches the top of its trajectory. How would these results change if air resistance were taken into account?
17. A projectile is fired at some angle to the horizontal with some initial speed v_0, and air resistance is neglected. Is the projectile a freely falling body? What is its acceleration in the vertical direction? What is its acceleration in the horizontal direction?
18. Determine which of the following moving objects obey the equations of projectile motion developed in this chapter: (a) A ball is thrown in an arbitrary direction. (b) A jet airplane crosses the sky with its engines thrusting the plane forward. (c) A rocket leaves the launch pad. (d) A rocket moves through the sky after its engines have failed. (e) A stone is thrown under water.
19. How can you throw a projectile so that it has zero speed at the top of its trajectory? So that it has nonzero speed at the top of its trajectory?
20. A ball is thrown upward in the air by a passenger on a train that is moving with constant velocity. (a) Describe the path of the ball as seen by the passenger. Describe the path as seen by a stationary observer outside the train. (b) How would these observations change if the train were accelerating along the track?

PROBLEMS

1, 2, 3 = straightforward, intermediate, challenging □ = full solution available in *Student Solutions Manual/Study Guide*
PhysicsNow™ = coached problem with hints available at **www.cp7e.com** = biomedical application

Section 3.1 Vectors and Their Properties

1. A roller coaster moves 200 ft horizontally and then rises 135 ft at an angle of 30.0° above the horizontal. Next, it travels 135 ft at an angle of 40.0° below the horizontal. Use graphical techniques to find the roller coaster's displacement from its starting point to the end of this movement.
2. An airplane flies 200 km due west from city A to city B and then 300 km in the direction of 30.0° north of west from city B to city C. (a) In straight-line distance, how far is city C from city A? (b) Relative to city A, in what direction is city C?
3. A man lost in a maze makes three consecutive displacements so that at the end of his travel he is right back where he started. The first displacement is 8.00 m westward, and the second is 13.0 m northward. Use the graphical method to find the magnitude and direction of the third displacement.

4. A jogger runs 100 m due west, then changes direction for the second leg of the run. At the end of the run, she is 175 m away from the starting point at an angle of 15.0° north of west. What were the direction and length of her second displacement? Use graphical techniques.

5. A plane flies from base camp to lake A, a distance of 280 km at a direction of 20.0° north of east. After dropping off supplies, the plane flies to lake B, which is 190 km and 30.0° west of north from lake A. Graphically determine the distance and direction from lake B to the base camp.

6. Vector $\vec{\mathbf{A}}$ has a magnitude of 8.00 units and makes an angle of 45.0° with the positive x-axis. Vector $\vec{\mathbf{B}}$ also has a magnitude of 8.00 units and is directed along the negative x-axis. Using graphical methods, find (a) the vector sum $\vec{\mathbf{A}} + \vec{\mathbf{B}}$ and (b) the vector difference $\vec{\mathbf{A}} - \vec{\mathbf{B}}$.

7. Vector $\vec{\mathbf{A}}$ is 3.00 units in length and points along the positive x-axis. Vector $\vec{\mathbf{B}}$ is 4.00 units in length and points along the negative y-axis. Use graphical methods to find the magnitude and direction of the vectors (a) $\vec{\mathbf{A}} + \vec{\mathbf{B}}$ and (b) $\vec{\mathbf{A}} - \vec{\mathbf{B}}$.

8. Each of the displacement vectors $\vec{\mathbf{A}}$ and $\vec{\mathbf{B}}$ shown in Figure P3.8 has a magnitude of 3.00 m. Graphically find (a) $\vec{\mathbf{A}} + \vec{\mathbf{B}}$, (b) $\vec{\mathbf{A}} - \vec{\mathbf{B}}$, (c) $\vec{\mathbf{B}} - \vec{\mathbf{A}}$, and (d) $\vec{\mathbf{A}} - 2\vec{\mathbf{B}}$.

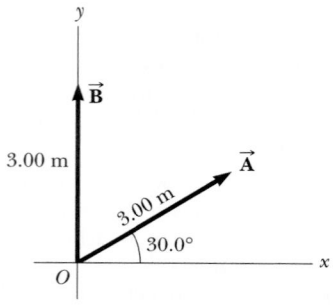

Figure P3.8

Section 3.2 Components of a Vector

9. A golfer takes two putts to get his ball into the hole once he is on the green. The first putt displaces the ball 6.00 m east, the second 5.40 m south. What displacement would have been needed to get the ball into the hole on the first putt?

10. A person walks 25.0° north of east for 3.10 km. How far would the person walk due north and due east to arrive at the same location?

11. PhysicsNow™ A girl delivering newspapers covers her route by traveling 3.00 blocks west, 4.00 blocks north, and then 6.00 blocks east. (a) What is her resultant displacement? (b) What is the total distance she travels?

12. While exploring a cave, a spelunker starts at the entrance and moves the following distances: 75.0 m north, 250 m east, 125 m at an angle 30.0° north of east, and 150 m south. Find the resultant displacement from the cave entrance.

13. A vector has an x-component of -25.0 units and a y-component of 40.0 units. Find the magnitude and direction of the vector.

14. A quarterback takes the ball from the line of scrimmage, runs backwards for 10.0 yards, then runs sideways parallel to the line of scrimmage for 15.0 yards. At this point, he throws a 50.0-yard forward pass straight downfield, perpendicular to the line of scrimmage. What is the magnitude of the football's resultant displacement?

15. The eye of a hurricane passes over Grand Bahama Island in a direction 60.0° north of west with a speed of 41.0 km/h. Three hours later, the course of the hurricane suddenly shifts due north, and its speed slows to 25.0 km/h. How far from Grand Bahama is the hurricane 4.50 h after it passes over the island?

16. A small map shows Atlanta to be 730 miles in a direction 5° north of east from Dallas. The same map shows that Chicago is 560 miles in a direction 21° west of north from Atlanta. Assume a flat Earth, and use the given information to find the displacement from Dallas to Chicago.

17. A commuter airplane starts from an airport and takes the route shown in Figure P3.17. The plane first flies to city A, located 175 km away in a direction 30.0° north of east. Next, it flies for 150 km 20.0° west of north, to city B. Finally, the plane flies 190 km due west, to city C. Find the location of city C relative to the location of the starting point.

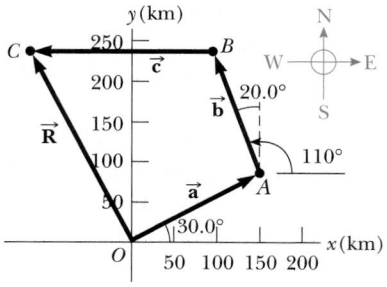

Figure P3.17

18. The helicopter view in Figure P3.18 shows two people pulling on a stubborn mule. Find (a) the single force that is equivalent to the two forces shown and (b) the force that a third person would have to exert on the mule to make the net force equal to zero. The forces are measured in units of newtons (N).

Figure P3.18

19. A man pushing a mop across a floor causes the mop to undergo two displacements. The first has a magnitude of 150 cm and makes an angle of 120° with the positive x-axis. The resultant displacement has a magnitude of 140 cm and is directed at an angle of 35.0° to the positive x-axis. Find the magnitude and direction of the second displacement.

20. An airplane starting from airport A flies 300 km east, then 350 km at 30.0° west of north, and then 150 km north to arrive finally at airport B. (a) The next day, another plane

flies directly from A to B in a straight line. In what direction should the pilot travel in this direct flight? (b) How far will the pilot travel in the flight? Assume there is no wind during either flight.

21. Long John Silver, a pirate, has buried his treasure on an island with five trees located at the following points: A (30.0 m, −20.0 m), B (60.0 m, 80.0 m), C (−10.0 m, −10.0 m), D (40.0 m, −30.0 m), and E (−70.0 m, 60.0 m). All of the points are measured relative to some origin, as in Figure P3.21. Long John's map instructs you to start at A and move toward B, but cover only one-half the distance between A and B. Then move toward C, covering one-third the distance between your current location and C. Then move toward D, covering one-fourth the distance between where you are and D. Finally, move toward E, covering one-fifth the distance between you and E, stop, and dig. (a) What are the coordinates of the point where the pirate's treasure is buried? (b) Rearrange the order of the trees—for instance, B (30 m, −20 m), A (60 m, 80 m), E (−10 m, −10 m), C (40 m, −30 m), and D (−70 m, 60 m)—and repeat the calculation to show that the answer does not depend on the order.

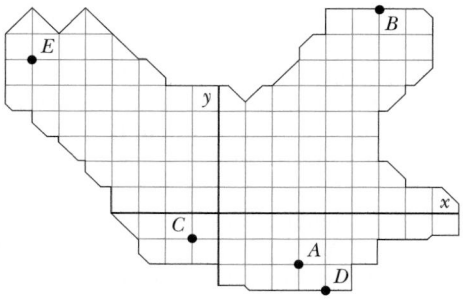

Figure P3.21

Section 3.3 Displacement, Velocity, and Acceleration in Two Dimensions

Section 3.4 Motion in Two Dimensions

22. One of the fastest recorded pitches in major-league baseball, thrown by Billy Wagner in 2003, was clocked at 101.0 mi/h (Fig. P3.22). If a pitch were thrown horizontally with this velocity, how far would the ball fall vertically by the time it reached home plate, 60.5 ft away?

AP/Wide World Photos

Figure P3.22 Billy Wagner throws a baseball.

23. A peregrine falcon is the fastest bird, flying at a speed of 200 mi/h. Nature has adapted the bird to reach such a speed by placing baffles in its nose to prevent air from rushing in and slowing it down. Also, the bird's eyes adjust their focus faster than the eyes of any other creature, so the falcon can focus quickly on its prey. Assume that a peregrine falcon is moving horizontally at its top speed at a height of 100 m above the ground when it brings its wings into its sides and begins to drop in free fall. How far will the bird fall vertically while traveling horizontally a distance of 100 m?

Joseph Kayne/Dembinsky Photo Associates

Figure P3.23 Notice the structure within the peregrine falcon's nostrils.

24. A student stands at the edge of a cliff and throws a stone horizontally over the edge with a speed of 18.0 m/s. The cliff is 50.0 m above a flat, horizontal beach, as shown in Figure P3.24. How long after being released does the stone strike the beach below the cliff? With what speed and angle of impact does the stone land?

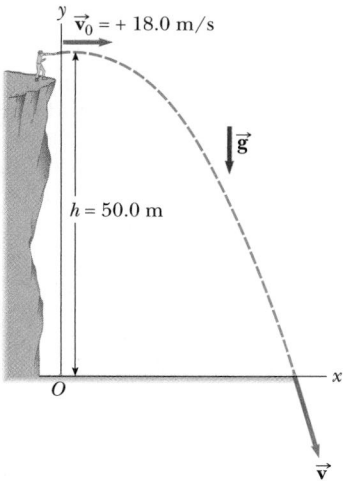

Figure P3.24

25. The best leaper in the animal kingdom is the puma, which can jump to a height of 12 ft when leaving the ground at an angle of 45°. With what speed, in SI units, must the animal leave the ground to reach that height?

26. Tom the cat is chasing Jerry the mouse across the surface of a table 1.5 m above the floor. Jerry steps out of the way at the last second, and Tom slides off the edge of the table at a speed of 5.0 m/s. Where will Tom strike the floor, and what velocity components will he have just before he hits?

27. PhysicsNow™ A tennis player standing 12.6 m from the net hits the ball at 3.00° above the horizontal. To clear the net, the ball must rise at least 0.330 m. If the ball just clears the net at the apex of its trajectory, how fast was the ball moving when it left the racquet?

28. An artillery shell is fired with an initial velocity of 300 m/s at 55.0° above the horizontal. To clear an avalanche, it explodes on a mountainside 42.0 s after firing. What are the x- and y-coordinates of the shell where it explodes, relative to its firing point?

29. A brick is thrown upward from the top of a building at an angle of 25° to the horizontal and with an initial speed of 15 m/s. If the brick is in flight for 3.0 s, how tall is the building?

30. A placekicker must kick a football from a point 36.0 m (about 39 yd) from the goal, and the ball must clear the crossbar, which is 3.05 m high. When kicked, the ball leaves the ground with a velocity of 20.0 m/s at an angle of 53° to the horizontal. (a) By how much does the ball clear or fall short of clearing the crossbar? (b) Does the ball approach the crossbar while still rising or while falling?

31. A car is parked on a cliff overlooking the ocean on an incline that makes an angle of 24.0° below the horizontal. The negligent driver leaves the car in neutral, and the emergency brakes are defective. The car rolls from rest down the incline with a constant acceleration of 4.00 m/s^2 for a distance of 50.0 m to the edge of the cliff, which is 30.0 m above the ocean. Find (a) the car's position relative to the base of the cliff when the car lands in the ocean and (b) the length of time the car is in the air.

32. A fireman 50.0 m away from a burning building directs a stream of water from a ground-level fire hose at an angle of 30.0° above the horizontal. If the speed of the stream as it leaves the hose is 40.0 m/s, at what height will the stream of water strike the building?

33. A projectile is launched with an initial speed of 60.0 m/s at an angle of 30.0° above the horizontal. The projectile lands on a hillside 4.00 s later. Neglect air friction. (a) What is the projectile's velocity at the highest point of its trajectory? (b) What is the straight-line distance from where the projectile was launched to where it hits its target?

34. A soccer player kicks a rock horizontally off a 40.0-m-high cliff into a pool of water. If the player hears the sound of the splash 3.00 s later, what was the initial speed given to the rock? Assume the speed of sound in air to be 343 m/s.

Section 3.5 Relative Velocity

35. A jet airliner moving initially at 300 mi/h due east enters a region where the wind is blowing at 100 mi/h in a direction 30.0° north of east. What is the new velocity of the aircraft relative to the ground?

36. A boat moves through the water of a river at 10 m/s relative to the water, regardless of the boat's direction. If the water in the river is flowing at 1.5 m/s, how long does it take the boat to make a round trip consisting of a 300-m displacement downstream followed by a 300-m displacement upstream?

37. A Chinook (King) salmon (Genus *Oncorynchus*) can jump out of water with a speed of 6.26 m/s. (See Problem 4.9, page 109 for an investigation of how the fish can leave the water at a higher speed than it can swim underwater.) If the salmon is in a stream with water speed equal to 1.50 m/s, how high in the air can the fish jump if it leaves the water traveling vertically upwards relative to the Earth?

38. A river flows due east at 1.50 m/s. A boat crosses the river from the south shore to the north shore by maintaining a constant velocity of 10.0 m/s due north relative to the water. (a) What is the velocity of the boat relative to the shore? (b) If the river is 300 m wide, how far downstream has the boat moved by the time it reaches the north shore?

39. A rowboat crosses a river with a velocity of 3.30 mi/h at an angle 62.5° north of west relative to the water. The river is 0.505 mi wide and carries an eastward current of 1.25 mi/h. How far upstream is the boat when it reaches the opposite shore?

40. Suppose a Chinook salmon needs to jump a waterfall that is 1.50 m high. If the fish starts from a distance 1.00 m from the base of the ledge over which the waterfall flows, find the x- and y-components of the initial velocity the salmon would need to just reach the ledge at the top of its trajectory. Can the fish make this jump? (Remember that a Chinook salmon can jump out of the water with a speed of 6.26 m/s.)

41. How long does it take an automobile traveling in the left lane of a highway at 60.0 km/h to overtake (become even with) another car that is traveling in the right lane at 40.0 km/h when the cars' front bumpers are initially 100 m apart?

42. A science student is riding on a flatcar of a train traveling along a straight horizontal track at a constant speed of 10.0 m/s. The student throws a ball along a path that she judges to make an initial angle of 60.0° with the horizontal and to be in line with the track. The student's professor, who is standing on the ground nearby, observes the ball to rise vertically. How high does the ball rise?

ADDITIONAL PROBLEMS

43. A particle undergoes two displacements. The first has a magnitude of 150 cm and makes an angle of 120.0° with the positive x-axis. The resultant of the two displacements is 140 cm, directed at an angle of 35.0° to the positive x-axis. Find the magnitude and direction of the second displacement.

44. Find the sum of these four vector forces: 12.0 N to the right at 35.0° above the horizontal, 31.0 N to the left at 55.0° above the horizontal, 8.40 N to the left at 35.0° below the horizontal, and 24.0 N to the right at 55.0° below the horizontal. (*Hint:* N stands for newton, the SI unit of force. The component method allows the addition of any vectors—forces as well as displacements and velocities. Make a drawing of this situation, and select the best axes for x and y so that you have the least number of components.)

45. A car travels due east with a speed of 50.0 km/h. Rain is falling vertically with respect to Earth. The traces of the rain on the side windows of the car make an angle of 60.0° with the vertical. Find the velocity of the rain with respect to (a) the car and (b) Earth.

46. You can use any coordinate system you like in order to solve a projectile motion problem. To demonstrate the

truth of this statement, consider a ball thrown off the top of a building with a velocity $\vec{\mathbf{v}}$ at an angle θ with respect to the horizontal. Let the building be 50.0 m tall, the initial horizontal velocity be 9.00 m/s, and the initial vertical velocity be 12.0 m/s. Choose your coordinates such that the positive y-axis is upward, the x-axis is to the right, and the origin is at the point where the ball is released. (a) With these choices, find the ball's maximum height above the ground, and the time it takes to reach the maximum height. (b) Repeat your calculations choosing the origin at the base of the building.

47. **Physics Now™** Towns A and B in Figure P3.47 are 80.0 km apart. A couple arranges to drive from town A and meet a couple driving from town B at the lake, L. The two couples leave simultaneously and drive for 2.50 h in the directions shown. Car 1 has a speed of 90.0 km/h. If the cars arrive simultaneously at the lake, what is the speed of car 2?

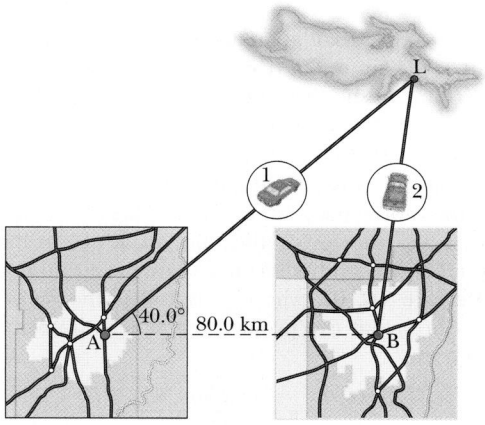

Figure P3.47

48. A Chinook salmon has a maximum underwater speed of 3.58 m/s, but it can jump out of water with a speed of 6.26 m/s. To move upstream past a waterfall, the salmon does not need to jump to the top of the fall, but only to a point in the fall where the water speed is less than 3.58 m/s; it can then swim up the fall for the remaining distance. Because the salmon must make forward progress in the water, let's assume that it can swim to the top if the water speed is 3.00 m/s. If water has a speed of 1.50 m/s as it passes over a ledge, how far below the ledge will the water be moving with a speed of 3.00 m/s? (Note that water undergoes projectile motion once it leaves the ledge.) If the salmon is able to jump vertically upward from the base of the fall, what is the maximum height of waterfall that the salmon can clear?

49. A rocket is launched at an angle of 53.0° above the horizontal with an initial speed of 100 m/s. The rocket moves for 3.00 s along its initial line of motion with an acceleration of 30.0 m/s^2. At this time, its engines fail and the rocket proceeds to move as a projectile. Find (a) the maximum altitude reached by the rocket, (b) its total time of flight, and (c) its horizontal range.

50. Two canoeists in identical canoes exert the same effort paddling and hence maintain the same speed relative to the water. One paddles directly upstream (and moves upstream), whereas the other paddles directly downstream. With downstream as the positive direction, an observer on shore determines the velocities of the two canoes to be -1.2 m/s and $+2.9$ m/s, respectively. (a) What is the speed of the water relative to the shore? (b) What is the speed of each canoe relative to the water?

51. If a person can jump a maximum horizontal distance (by using a 45° projection angle) of 3.0 m on Earth, what would be his maximum range on the Moon, where the free-fall acceleration is $g/6$ and $g = 9.80$ m/s^2? Repeat for Mars, where the acceleration due to gravity is $0.38g$.

52. A daredevil decides to jump a canyon. Its walls are equally high and 10 m apart. He takes off by driving a motorcycle up a short ramp sloped at an angle of 15°. What minimum speed must he have in order to clear the canyon?

53. (a) Vector $\vec{\mathbf{A}}$ is in the first quadrant of a Cartesian coordinate system. What is the sign of the x-component of $\vec{\mathbf{A}}$? What is the sign of the y-component? (b) Vector $\vec{\mathbf{B}}$ is in the second quadrant of a Cartesian coordinate system. What is the sign of the x-component of $\vec{\mathbf{B}}$? What is the sign of the y-component? (c) The vector sum $\vec{\mathbf{A}} + \vec{\mathbf{B}}$ _____. Choose the correct fill-in-the-blank answer from (i) must be in either the first or the second quadrant or (ii) could be in any quadrant. (d) Let $\vec{\mathbf{A}} = 30$ m at an angle of 30° from the positive x-axis and $\vec{\mathbf{B}} = 20$ m at an angle of 40° from the negative x-axis. Test your predictions of parts (a) through (c).

54. A boy and a girl are tossing an apple back and forth between them. Figure P3.54 shows one path the apple follows when watched by an observer looking on from the side. The apple is moving from left to right. Five points are marked on the path. Ignore air resistance. (a) Make a copy of this figure. At each of the marked points, draw an arrow that indicates the magnitude and direction of the apple's velocity when it passes through that point. (b) Make a second copy of the figure. This time, at each marked point, place an arrow indicating the magnitude and direction of any acceleration the apple exhibits at that point.

Figure P3.54

55. A home run is hit in such a way that the baseball just clears a wall 21 m high, located 130 m from home plate. The ball is hit at an angle of 35° to the horizontal, and air resistance is negligible. Find (a) the initial speed of the ball, (b) the time it takes the ball to reach the wall, and (c) the velocity components and the speed of the ball when it reaches the wall. (Assume that the ball is hit at a height of 1.0 m above the ground.)

56. A ball is thrown straight upward and returns to the thrower's hand after 3.00 s in the air. A second ball is thrown at an angle of 30.0° with the horizontal. At what speed must the second ball be thrown so that it reaches the same height as the one thrown vertically?

57. A quarterback throws a football toward a receiver with an initial speed of 20 m/s at an angle of 30° above the horizontal. At that instant, the receiver is 20 m from the

quarterback. In what direction and with what constant speed should the receiver run in order to catch the football at the level at which it was thrown?

58. A 2.00-m-tall basketball player is standing on the floor 10.0 m from the basket, as in Figure P3.58. If he shoots the ball at a 40.0° angle with the horizontal, at what initial speed must he throw the basketball so that it goes through the hoop without striking the backboard? The height of the basket is 3.05 m.

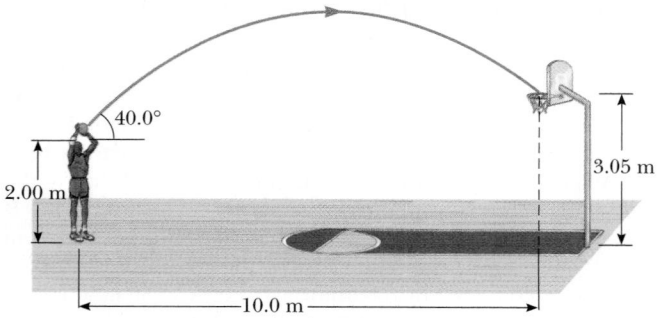

Figure P3.58

59. In a very popular lecture demonstration, a projectile is fired at a falling target as in Figure P3.59. The projectile leaves the gun at the same instant that the target is dropped from rest. Assuming that the gun is initially aimed at the target, show that the projectile will hit the target. (One restriction of this experiment is that the projectile must reach the target before the target strikes the floor.)

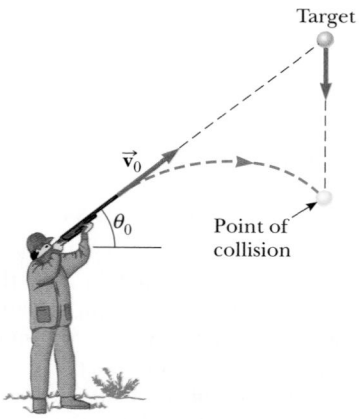

Figure P3.59

60. Figure P3.60 illustrates the difference in proportions between the male (m) and female (f) anatomies. The displacements $\vec{d}_{1m}$ and $\vec{d}_{1f}$ from the bottom of the feet to the navel have magnitudes of 104 cm and 84.0 cm, respectively. The displacements $\vec{d}_{2m}$ and $\vec{d}_{2f}$ have magnitudes of 50.0 cm and 43.0 cm, respectively. (a) Find the vector sum of the displacements $\vec{d}_1$ and $\vec{d}_2$ in each case. (b) The male figure is 180 cm tall, the female 168 cm. Normalize the displacements of each figure to a common height of 200 cm, and re-form the vector sums as in part (a). Then find the vector difference between the two sums.

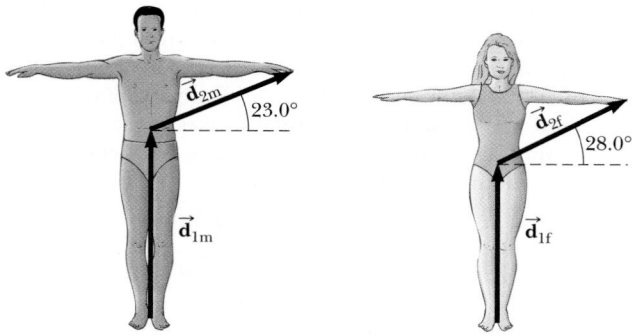

Figure P3.60

61. By throwing a ball at an angle of 45°, a girl can throw the ball a maximum horizontal distance R on a level field. How far can she throw the same ball vertically upward? Assume that her muscles give the ball the same speed in each case. (Is this assumption valid?)

62. A projectile is fired with an initial speed v_0 at an angle θ_0 to the horizontal, as in Figure 3.14. When it reaches its peak, the projectile has (x, y) coordinates given by $(R/2, h)$, and when it strikes the ground, its coordinates are $(R, 0)$, where R is called the *horizontal range.* (a) Show that the projectile reaches a maximum height given by

$$h = \frac{v_0^2 \sin^2 \theta_0}{2g}$$

(b) Show that the horizontal range of the projectile is given by

$$R = \frac{v_0^2 \sin 2\theta_0}{g}$$

63. A hunter wishes to cross a river that is 1.5 km wide and flows with a speed of 5.0 km/h parallel to its banks. The hunter uses a small powerboat that moves at a maximum speed of 12 km/h with respect to the water. What is the minimum time necessary for crossing?

64. A water insect maintains an average position on the surface of a stream by darting upstream (against the current) then drifting downstream (with the current) to its original position. The current in the stream is 0.500 m/s relative to the shore, and the insect darts upstream 0.560 m (relative to a spot on shore) in 0.800 s during the first part of its motion. Take upstream as the positive direction. (a) Determine the velocity of the insect relative to the water (i) during its dash upstream and (ii) during its drift downstream. (b) How far upstream relative to the water does the insect move during one cycle of its motion? (c) What is the average velocity of the insect relative to the water?

65. **Physics Now™** A daredevil is shot out of a cannon at 45.0° to the horizontal with an initial speed of 25.0 m/s. A net is positioned a horizontal distance of 50.0 m from the cannon. At what height above the cannon should the net be placed in order to catch the daredevil?

66. Chinook salmon are able to move upstream faster by jumping out of the water periodically; this behavior is called *porpoising*. Suppose a salmon swimming in still water jumps out of the water with a speed of 6.26 m/s at an angle of 45°, sails through the air a distance L before returning to the water, and then swims a distance L underwater at a speed of 3.58 m/s before beginning another porpoising maneuver. Determine the average speed of the fish.

67. A student decides to measure the muzzle velocity of a pellet shot from his gun. He points the gun horizontally. He places a target on a vertical wall a distance x away from the gun. The pellet hits the target a vertical distance y below the gun. (a) Show that the position of the pellet when traveling through the air is given by $y = Ax^2$, where A is a constant. (b) Express the constant A in terms of the initial (muzzle) velocity and the free-fall acceleration. (c) If $x = 3.00$ m and $y = 0.210$ m, what is the initial speed of the pellet?

68. A sailboat is heading directly north at a speed of 20 knots (1 knot = 0.514 m/s). The wind is blowing towards the east with a speed of 17 knots. Determine the magnitude and direction of the wind velocity as measured on the boat. What is the component of the wind velocity in the direction parallel to the motion of the boat? (See Problem 4.54 for an explanation of how a sailboat can move "into the wind.")

69. Instructions for finding a buried treasure include the following: Go 75 paces at 240°, turn to 135° and walk 125 paces, and then travel 100 paces at 160°. Determine the resultant displacement from the starting point.

70. When baseball outfielders throw the ball, they usually allow it to take one bounce, on the theory that the ball arrives at its target sooner that way. Suppose that, after the bounce, the ball rebounds at the same angle θ that it had when it was released (as in Fig. P3.70), but loses half its speed. (a) Assuming that the ball is always thrown with the same initial speed, at what angle θ should the ball be thrown in order to go the same d istance D with one bounce as a ball thrown upward at 45.0° with no bounce? (b) Determine the ratio of the times for the one-bounce and no-bounce throws.

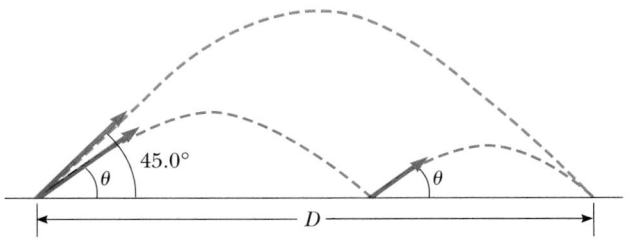

Figure P3.70

71. One strategy in a snowball fight is to throw a snowball at a high angle over level ground. Then, while your opponent is watching that snowball, you throw a second one at a low angle timed to arrive before or at the same time as the first one. Assume that both snowballs are thrown with a speed of 25.0 m/s. The first is thrown at an angle of 70.0° with respect to the horizontal. (a) At what angle should the second snowball be thrown to arrive at the same point as the first? (b) How many seconds later should the second snowball be thrown after the first in order for both to arrive at the same time?

72. A dart gun is fired while being held horizontally at a height of 1.00 m above ground level and while it is at rest relative to the ground. The dart from the gun travels a horizontal distance of 5.00 m. A college student holds the same gun in a horizontal position while sliding down a 45.0° incline at a constant speed of 2.00 m/s. How far will the dart travel if the student fires the gun when it is 1.00 m above the ground?

73. The determined Wile E. Coyote is out once more to try to capture the elusive roadrunner. The coyote wears a new pair of Acme power roller skates, which provide a constant horizontal acceleration of 15 m/s^2, as shown in Figure P3.73. The coyote starts off at rest 70 m from the edge of a cliff at the instant the roadrunner zips by in the direction of the cliff. (a) If the roadrunner moves with constant speed, find the minimum speed the roadrunner must have in order to reach the cliff before the coyote. (b) If the cliff is 100 m above the base of a canyon, find where the coyote lands in the canyon. (Assume that his skates are still in operation when he is in "flight" and that his horizontal component of acceleration remains constant at 15 m/^{2}.)

Figure P3.73

ACTIVITIES

A.1. Take three steps, turn 90°, and then walk four steps. Now count the number of steps it takes to walk in a straight line back to your starting point. Verify your result mathematically.

A.2. For this investigation, you need to be outside with a small ball such as a tennis ball and a wristwatch with a second hand. Throw the ball vertically upward as hard as you can, and find the initial speed of your throw and the approximate maximum height of the ball solely with the use of your wristwatch. What happens when you throw the ball at some angle other than 90°? Does this change the time of flight? Can you still determine the maximum height and initial speed? Give careful explanations for your answers. Your co-worker can eyeball the maximum height by standing at a distance and noting the angle.

A.3. Using as projectiles the drops of water spraying from a garden hose at ground level, test the statement that the maximum range occurs when the angle of projection is 45°. As an additional part of this experiment, hold the hose horizontally above the ground and have your co-worker position a marker at the location where the water strikes the ground. Increase the angle of inclination by about 10°, and record the strike position again. Repeat until you have reached an angle of about 75°. Note the pattern produced. Are there two angles at which the range is the same? Explain the reasoning behind your observations.

A.4. Use a chessboard as a coordinate system, with the intersection of the lines on the board as the positions of the coordinates. Select an origin for your coordinate system and, on index cards, write down several vector displacements that are at right angles to one another. For example, displacement 1 might be a movement of four units to the right, displacement 2 an upward movement of six units, and so forth. Continue this for a total of seven or eight movements until you end up at some particular location on the chessboard. Let your co-worker start at the origin and follow your vector directions to see whether he or she arrives at the expected final location. Now shuffle the cards and repeat the experiment. Does the order of the displacements make any difference as to where you eventually end up?

A.5. Roll a ball off a table. At the very instant the rolling ball leaves the table, drop a second ball from the same height above the floor. (Doing this will require a sharp eye and good reflexes!) Do the two balls hit the floor at the same time? Try varying the speed at which you roll the ball off the table. Does this change affect the time at which the balls strike the floor? Finally, roll one of the balls down an incline, and drop the other ball from the base of the incline at the instant the first ball leaves the slope. Which of these balls hits the floor first in this situation? Explain the reasoning behind your observations.

© Royalty-Free/Corbis

Forces exerted by Earth, wind, and water, properly channeled by the strength and skill of these windsurfers, combine to create a non-zero net force on their surfboards, driving them forward through the waves.

CHAPTER

4

The Laws of Motion

OUTLINE

4.1 Forces
4.2 Newton's First Law
4.3 Newton's Second Law
4.4 Newton's Third Law
4.5 Applications of Newton's Laws
4.6 Forces of Friction

Classical mechanics describes the relationship between the motion of objects found in our everyday world and the forces acting on them. As long as the system under study doesn't involve objects comparable in size to an atom or traveling close to the speed of light, classical mechanics provides an excellent description of nature.

This chapter introduces Newton's three laws of motion and his law of gravity. The three laws are simple and sensible. The first law states that a force must be applied to an object in order to change its velocity. Changing an object's velocity means accelerating it, which implies a relationship between force and acceleration. This relationship, the second law, states that the net force on an object equals the object's mass times its acceleration. Finally, the third law says that whenever we push on something, it pushes back with equal force in the opposite direction. These are the three laws in a nutshell.

Newton's three laws, together with his invention of calculus, opened avenues of inquiry and discovery that are used routinely today in virtually all areas of mathematics, science, engineering, and technology. Newton's theory of universal gravitation had a similar impact, starting a revolution in celestial mechanics and astronomy that continues to this day. With the advent of this theory, the orbits of all the planets could be calculated to high precision and the tides understood. The theory even led to the prediction of "dark stars," now called black holes, over two centuries before any evidence for their existence was observed.[1] Newton's three laws of motion, together with his law of gravitation, are considered among the greatest achievements of the human mind.

4.1 FORCES

A **force** is commonly imagined as a push or a pull on some object, perhaps rapidly, as when we hit a tennis ball with a racket. (See Figure 4.1.) We can hit the ball at different speeds and direct it into different parts of the opponent's court. This

© Lorenzo Ciniglio/Corbis

Figure 4.1 Tennis champion Andy Roddick strikes the ball with his racket, applying a force and directing the ball into the open part of the court.

[1]In 1783, John Michell combined Newton's theory of light and theory of gravitation, predicting the existence of "dark stars" from which light itself couldn't escape.

means that we can control the magnitude of the applied force and also its direction, so force is a vector quantity, just like velocity and acceleration.

If you pull on a spring (Fig. 4.2a), the spring stretches. If you pull hard enough on a wagon (Fig. 4.2b), the wagon moves. When you kick a football (Fig. 4.2c), it deforms briefly and is set in motion. These are all examples of **contact forces**, so named because they result from physical contact between two objects.

Another class of forces doesn't involve any direct physical contact. Early scientists, including Newton, were uneasy with the concept of forces that act between two disconnected objects. Nonetheless, Newton used this 'action-at-a-distance' concept in his law of gravity, whereby a mass at one location, such as the Sun, affects the motion of a distant object such as Earth despite no evident physical connection between the two objects. To overcome the conceptual difficulty associated with action at a distance, Michael Faraday (1791–1867) introduced the concept of a *field*. The corresponding forces are called **field forces**. According to this approach, an object of mass M, such as the Sun, creates an invisible influence that stretches throughout space. A second object of mass m, such as Earth, interacts with the *field* of the Sun, not directly with the Sun itself. So the force of gravitational attraction between two objects, illustrated in Figure 4.2d, is an example of a field force. The force of gravity keeps objects bound to Earth and also gives rise to what we call the *weight* of those objects.

Another common example of a field force is the electric force that one electric charge exerts on another (Fig. 4.2e). A third example is the force exerted by a bar magnet on a piece of iron (Fig. 4.2f).

The known fundamental forces in nature are all field forces. These are, in order of decreasing strength, (1) the strong nuclear force between subatomic particles; (2) the electromagnetic forces between electric charges; (3) the weak nuclear force, which arises in certain radioactive decay processes; and (4) the gravitational force between objects. The strong force keeps the nucleus of an atom from flying apart due to the repulsive electric force of the protons. The weak force

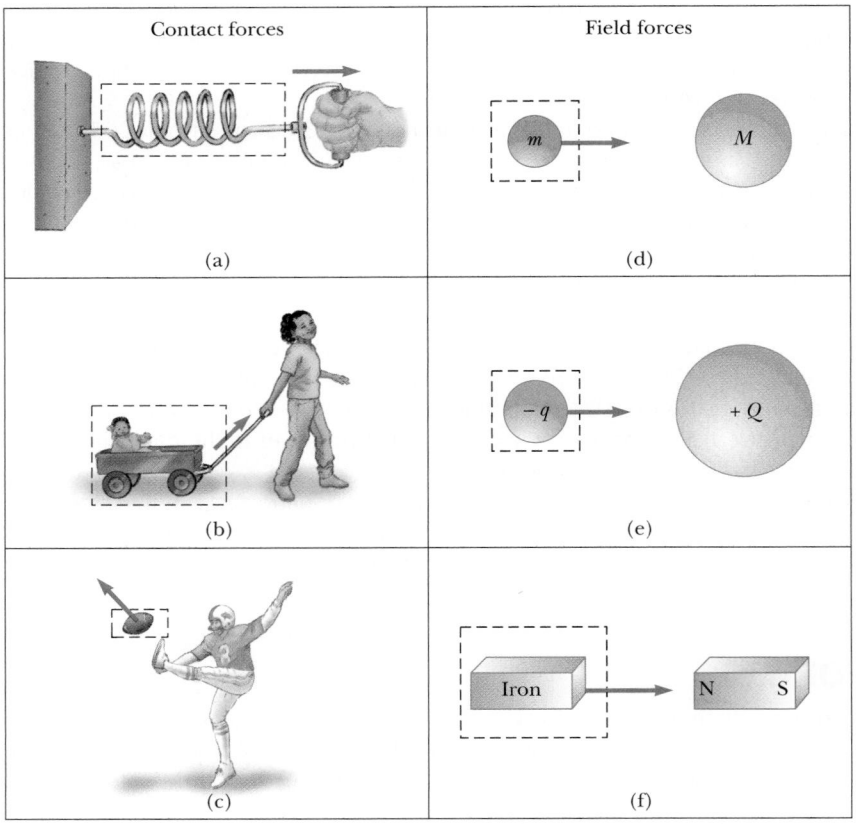

Figure 4.2 Examples of forces applied to various objects. In each case, a force acts on the object surrounded by the dashed lines. Something in the environment external to the boxed area exerts the force.

is involved in most radioactive processes and plays an important role in the nuclear reactions that generate the Sun's energy output. The strong and weak forces operate only on the nuclear scale, with a very short range on the order of 10^{-15} m. Outside this range, they have no influence. Classical physics, however, deals only with gravitational and electromagnetic forces, which have infinite range.

Forces exerted on an object can change the object's shape. For example, striking a tennis ball with a racquet, as in Figure 4.1, deforms the ball to some extent. Even objects we usually consider rigid and inflexible are deformed under the action of external forces. Often the deformations are permanent, as in the case of a collision between automobiles.

4.2 NEWTON'S FIRST LAW

Consider a book lying on a table. Obviously, the book remains at rest if left alone. Now imagine pushing the book with a horizontal force great enough to overcome the force of friction between the book and the table, setting the book in motion. Because the magnitude of the applied force exceeds the magnitude of the friction force, the book accelerates. When the applied force is withdrawn, friction soon slows the book to a stop.

Now imagine pushing the book across a smooth, waxed floor. The book again comes to rest once the force is no longer applied, but not as quickly as before. Finally, if the book is moving on a horizontal frictionless surface, it continues to move in a straight line with constant velocity until it hits a wall or some other obstruction.

Before about 1600, scientists felt that the natural state of matter was the state of rest. Galileo, however, devised thought experiments—such as an object moving on a frictionless surface, as just described—and concluded that **it's not the nature of an object to stop, once set in motion, but rather to continue in its original state of motion**. This approach was later formalized as **Newton's first law of motion**:

> An object moves with a velocity that is constant in magnitude and direction, unless acted on by a nonzero net force.

◀ Newton's first law

The net force on an object is defined as the vector sum of all external forces exerted on the object. External forces come from the object's environment. If an object's velocity isn't changing in either magnitude or direction, then its acceleration and the net force acting on it must both be zero.

Internal forces originate within the object itself and can't change the object's velocity (although they can change the object's rate of rotation, as described in Chapter 8). As a result, internal forces aren't included in Newton's second law. It's not really possible to "pull yourself up by your own bootstraps."

A consequence of the first law is the feasibility of space travel. After just a few moments of powerful thrust, the spacecraft coasts for months or years, its velocity only slowly changing with time under the relatively faint influence of the distant sun and planets.

Roger Viollet, Mill Valley, CA, University Science Books, 1982

Unless acted on by an external force, an object at rest will remain at rest and an object in motion will continue in motion with constant velocity. In this case, the wall of the building did not exert a large enough external force on the moving train to stop it.

Mass and Inertia

Imagine hitting a golf ball off a tee with a driver. If you're a good golfer, the ball will sail over two hundred yards down the fairway. Now imagine teeing up a bowling ball and striking it with the same club (an experiment we don't recommend). Your club would probably break, you might sprain your wrist, and the bowling ball, at best, would fall off the tee, take half a roll and come to rest.

From this thought experiment, we conclude that while both balls resist changes in their state of motion, the bowling ball offers much more effective resistance. The tendency of an object to continue in its original state of motion is called **inertia**.

While inertia is the tendency of an object to continue its motion in the absence of a force, **mass** is a measure of the object's resistance to changes in its motion due

APPLICATION

Seat Belts

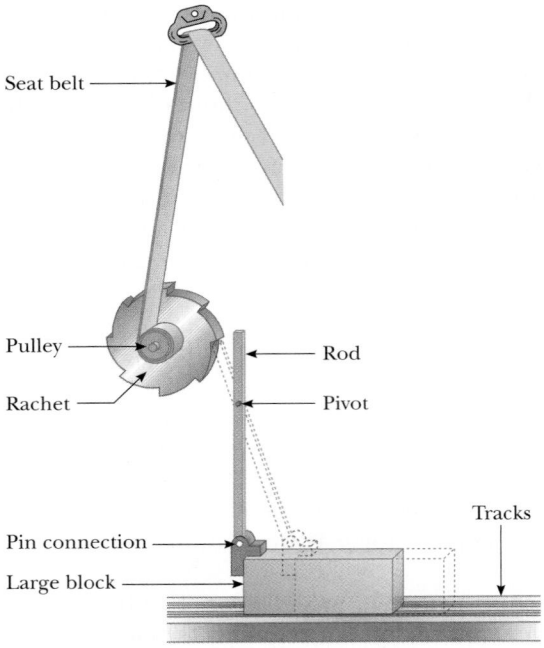

Figure 4.3 A mechanical arrangement for an automobile seat belt.

to a force. The greater the mass of a body, the less it accelerates under the action of a given applied force. The SI unit of mass is the kilogram. Mass is a scalar quantity that obeys the rules of ordinary arithmetic.

Inertia can be used to explain the operation of one type of seat belt mechanism. In the event of an accident, the purpose of the seat belt is to hold the passenger firmly in place relative to the car, to prevent serious injury. Figure 4.3 illustrates how one type of shoulder harness operates. Under normal conditions, the ratchet turns freely to allow the harness to wind on or unwind from the pulley as the passenger moves. In an accident, the car undergoes a large acceleration and rapidly comes to rest. Because of its inertia, the large block under the seat continues to slide forward along the tracks. The pin connection between the block and the rod causes the rod to pivot about its center and engage the ratchet wheel. At this point, the ratchet wheel locks in place and the harness no longer unwinds.

4.3 NEWTON'S SECOND LAW

Newton's first law explains what happens to an object that has no net force acting on it: The object either remains at rest or continues moving in a straight line with constant speed. Newton's second law answers the question of what happens to an object that *does* have a net force acting on it.

TIP 4.1 Force Causes *Changes* in Motion

Motion can occur even in the absence of forces. Force causes *changes* in motion.

Imagine pushing a block of ice across a frictionless horizontal surface. When you exert some horizontal force on the block, it moves with an acceleration of, say, 2 m/s^2. If you apply a force twice as large, the acceleration doubles to 4 m/s^2. Pushing three times as hard triples the acceleration, and so on. From such observations, we conclude that **the acceleration of an object is directly proportional to the net force acting on it**.

Mass also affects acceleration. Suppose you stack identical blocks of ice on top of each other while pushing the stack with constant force. If the force applied to one block produces an acceleration of 2 m/s^2, then the acceleration drops to half that value, 1 m/s^2, when two blocks are pushed, to one-third the initial value when three blocks are pushed, and so on. We conclude that **the acceleration of an object is inversely proportional to its mass**. These observations are summarized in **Newton's second law**:

Newton's second law ▶

> The acceleration $\vec{\mathbf{a}}$ of an object is directly proportional to the net force acting on it and inversely proportional to its mass.

The constant of proportionality is equal to one, so in mathematical terms the preceding statement can be written

$$\vec{\mathbf{a}} = \frac{\Sigma \vec{\mathbf{F}}}{m}$$

where $\vec{\mathbf{a}}$ is the acceleration of the object, m is its mass, and $\Sigma\vec{\mathbf{F}}$ is the vector sum of all forces acting on it. Multiplying through by m, we have

$$\Sigma\vec{\mathbf{F}} = m\vec{\mathbf{a}} \qquad [4.1]$$

Physicists commonly refer to this equation as '$F = ma$'. The second law is a vector equation, equivalent to the following three component equations:

$$\Sigma F_x = ma_x \qquad \Sigma F_y = ma_y \qquad \Sigma F_z = ma_z \qquad [4.2]$$

When there is no net force on an object, its acceleration is zero, which means the velocity is constant.

TIP 4.2 $m\vec{\mathbf{a}}$ Is Not a Force

Equation 4.1 does *not* say that the product $m\vec{\mathbf{a}}$ is a force. All forces exerted on an object are summed as vectors to generate the net force on the left side of the equation. This net force is then equated to the product of the mass and resulting acceleration of the object. Do *not* include an "$m\vec{\mathbf{a}}$ force" in your analysis.

Units of Force and Mass

The SI unit of force is the **newton**. When 1 newton of force acts on an object that has a mass of 1 kg, it produces an acceleration of 1 m/s^2 in the object. From this definition and Newton's second law, we see that the newton can be expressed in terms of the fundamental units of mass, length, and time as

$$1\ \text{N} \equiv 1\ \text{kg}\cdot\text{m/s}^2 \qquad [4.3]$$

◀ Definition of newton

In the U.S. customary system, the unit of force is the **pound**. The conversion from newtons to pounds is given by

$$1\ \text{N} = 0.225\ \text{lb} \qquad [4.4]$$

The units of mass, acceleration, and force in the SI and U.S. customary systems are summarized in Table 4.1.

Giraudon/Art Resource

ISAAC NEWTON English Physicist and Mathematician (1642–1727)

Newton was one of the most brilliant scientists in history. Before he was 30, he formulated the basic concepts and laws of mechanics, discovered the law of universal gravitation, and invented the mathematical methods of the calculus. As a consequence of his theories, Newton was able to explain the motions of the planets, the ebb and flow of the tides, and many special features of the motions of the Moon and Earth. He also interpreted many fundamental observations concerning the nature of light. His contributions to physical theories dominated scientific thought for two centuries and remain important today.

Quick Quiz 4.1

True or false? (a) It's possible to have motion in the absence of a force. (b) If an object isn't moving, no external force acts on it.

Quick Quiz 4.2

True or false? (a) If a single force acts on an object, the object accelerates. (b) If an object is accelerating, a force is acting on it. (c) If an object is not accelerating, no external force is acting on it.

Quick Quiz 4.3

True or false? If the net force acting on an object is in the positive x-direction, the object moves only in the positive x-direction.

TABLE 4.1

Units of Mass, Acceleration, and Force

System	Mass	Acceleration	Force
SI	kg	m/s^2	N = kg·m/s^2
U.S. customary	slug	ft/s^2	lb = slug·ft/s^2

EXAMPLE 4.1 Airboat

Goal Apply Newton's law in one dimension, together with the equations of kinematics.

Problem An airboat with mass 3.50×10^2 kg, including passengers, has an engine that produces a net horizontal force of 7.70×10^2 N, after accounting for forces of resistance. **(a)** Find the acceleration of the airboat. **(b)** Starting from rest, how long does it take the airboat to reach a speed of 12.0 m/s? **(c)** After reaching this speed, the pilot turns off the engine and drifts to a stop over a distance of 50.0 m. Find the resistance force, assuming it's constant.

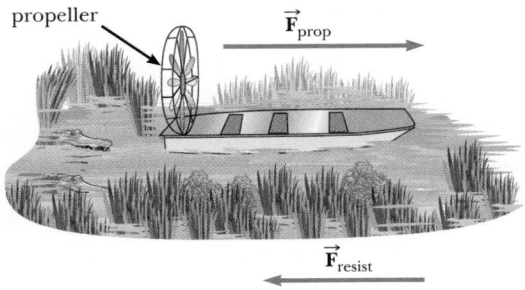

Figure 4.4 (Example 4.1)

Strategy In part **(a)**, apply Newton's second law to find the acceleration, and in part **(b)** use this acceleration in the one-dimensional kinematics equation for the velocity. When the engine is turned off in part **(c)**, only the resistance forces act on the boat, so their net acceleration can be found from $v^2 - v_0^2 = 2a\Delta x$. Then Newton's second law gives the resistance force.

Solution

(a) Find the acceleration of the airboat.

Apply Newton's second law and solve for the acceleration:

$$ma = F_{net} \rightarrow a = \frac{F_{net}}{m} = \frac{7.70 \times 10^2 \text{ N}}{3.50 \times 10^2 \text{ kg}} = 2.20 \text{ m/s}^2$$

(b) Find the time necessary to reach a speed of 12.0 m/s.

Apply the kinematics velocity equation:

$$v = at + v_0 = (2.20 \text{ m/s}^2)t = 12.0 \text{ m/s} \rightarrow t = 5.45 \text{ s}$$

(c) Find the resistance force after the engine is turned off.

Using kinematics, find the net acceleration due to resistance forces:

$$v^2 - v_0^2 = 2a\Delta x$$

$$0 - (12 \text{ m/s})^2 = 2a(50.0 \text{ m}) \rightarrow a = -1.44 \text{ m/s}^2$$

Substitute the acceleration into Newton's second law, finding the resistance force:

$$F_{resist} = ma = (3.50 \times 10^2 \text{ kg})(-1.44 \text{ m/s}^2) = -504 \text{ N}$$

Remarks The negative answer for the acceleration in part **(c)** means that the airboat is slowing down. In this problem, the *components* of the vectors $\vec{\mathbf{a}}$ and $\vec{\mathbf{F}}$ were used, so they were written in italics without arrows. It's important to bear in mind that $\vec{\mathbf{a}}$ and $\vec{\mathbf{F}}$ are vectors, not scalars.

TIP 4.3 Newton's Second Law is a *Vector* Equation

In applying Newton's second law, add all of the forces on the object as vectors and then find the resultant vector acceleration by dividing by m. Don't find the individual magnitudes of the forces and add them like scalars.

Exercise 4.1

Suppose the pilot, starting again from rest, opens the throttle partway. At a constant acceleration, the airboat then covers a distance of 60.0 m in 10.0 s. Find the net force acting on the boat.

Answer 4.20×10^2 N

EXAMPLE 4.2 Horses Pulling a Barge

Goal Apply Newton's second law in a two-dimensional problem.

Problem Two horses are pulling a barge with mass 2.00×10^3 kg along a canal, as shown in Figure 4.5. The cable connected to the first horse makes an angle of 30.0° with respect to the direction of the canal, while the cable connected to the second horse makes an angle of 45.0°. Find the initial acceleration of the barge, starting at rest, if each horse exerts a force of magnitude 6.00×10^2 N on the barge. Ignore forces of resistance on the barge.

Strategy Using trigonometry, find the vector force exerted by each horse on the barge. Add the x-components together to get the x-component of the resultant force, and then do the same with the y-components. Divide by the mass of the barge to get the accelerations in the x- and y-directions.

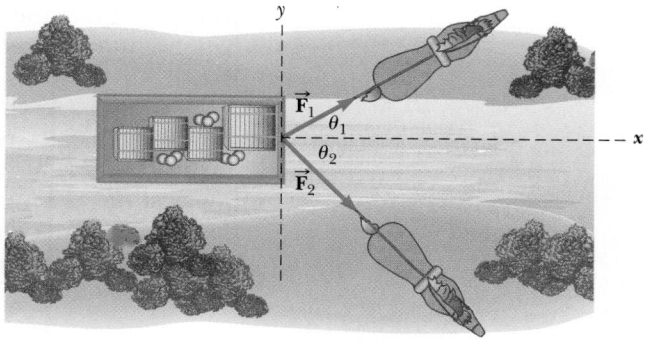

Figure 4.5 (Example 4.2)

Solution

Find the x-components of the forces exerted by the horses.

$$F_{1x} = F_1 \cos\theta_1 = (6.00 \times 10^2 \text{ N})\cos(30.0°) = 5.20 \times 10^2 \text{ N}$$
$$F_{2x} = F_2 \cos\theta_2 = (6.00 \times 10^2 \text{ N})\cos(-45.0°) = 4.24 \times 10^2 \text{ N}$$

Find the total force in the x-direction by adding the x-components:

$$F_x = F_{1x} + F_{2x} = 5.20 \times 10^2 \text{ N} + 4.24 \times 10^2 \text{ N} = 9.44 \times 10^2 \text{ N}$$

Find the y-components of the forces exerted by the horses:

$$F_{1y} = F_1 \sin\theta_1 = (6.00 \times 10^2 \text{ N})\sin 30.0° = 3.00 \times 10^2 \text{ N}$$
$$F_{2y} = F_2 \sin\theta_2 = (6.00 \times 10^2 \text{ N})\sin(-45.0°) = -4.24 \times 10^2 \text{ N}$$

Find the total force in the y-direction by adding the y-components:

$$F_y = F_{1y} + F_{2y} = 3.00 \times 10^2 \text{ N} - 4.24 \times 10^2 \text{ N} = -1.24 \times 10^2 \text{ N}$$

Find the components of the acceleration by dividing the force components by the mass:

$$a_x = \frac{F_x}{m} = \frac{9.44 \times 10^2 \text{ N}}{2.00 \times 10^3 \text{ kg}} = 0.472 \text{ m/s}^2$$
$$a_y = \frac{F_y}{m} = \frac{-1.24 \times 10^2 \text{ N}}{2.00 \times 10^3 \text{ kg}} = -0.0620 \text{ m/s}^2$$

Find the magnitude of the acceleration:

$$a = \sqrt{a_x^2 + a_y^2} = \sqrt{(0.472 \text{ m/s}^2)^2 + (-0.0620 \text{ m/s}^2)^2} = 0.476 \text{ m/s}^2$$

Find the direction of the acceleration.

$$\tan\theta = \frac{a_y}{a_x} = \frac{-0.0620}{0.472} = -0.131$$
$$\theta = \tan^{-1}(-0.131) = -7.46°$$

Exercise 4.2

Repeat Example 4.2, but assume that the upper horse pulls at a 40.0° angle, the lower horse at 20.0°.

Answer 0.520 m/s^2, 10.0°

The Gravitational Force

The **gravitational force** is the mutual force of attraction between any two objects in the Universe. Although the gravitational force can be very strong between very large objects, it's the weakest of the fundamental forces. A good demonstration of how weak it is can be carried out with a small balloon. Rubbing the balloon in your hair gives the balloon a tiny electric charge. Through electric forces, the balloon then adheres to a wall, resisting the gravitational pull of the entire Earth!

Law of universal gravitation ▶

In addition to contributing to the understanding of motion, Newton studied gravity extensively. **Newton's law of universal gravitation states that every particle in the Universe attracts every other particle with a force that is directly proportional to the product of the masses of the particles and inversely proportional to the square of the distance between them.** If the particles have masses m_1 and m_2 and are separated by a distance r, as in Active Figure 4.6, the magnitude of the gravitational force, F_g is

$$F_g = G\frac{m_1 m_2}{r^2} \qquad [4.5]$$

where $G = 6.67 \times 10^{-11}\ \mathrm{N \cdot m^2/kg^2}$ is the **universal gravitation constant**. We examine the gravitational force in more detail in Chapter 7.

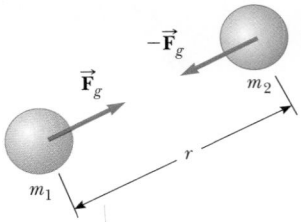

ACTIVE FIGURE 4.6
The gravitational force between two particles is attractive.

PhysicsNow™
Log into to PhysicsNow at **www.cp7e.com,** and go to Active Figure 4.6 to change the masses of the particles and the separation between the particles to see the effect on the gravitational force.

Weight

The magnitude of the gravitational force acting on an object of mass m near Earth's surface is called the *weight, w,* of the object, given by

$$w = mg \qquad [4.6]$$

where g is the acceleration of gravity.
SI unit: newton (N)

From Equation 4.5, an alternate definition of the weight of an object with mass m can be written as

$$w = G\frac{M_E m}{r^2} \qquad [4.7]$$

where M_E is the mass of Earth and r is the distance from the object to Earth's center. If the object is at rest on Earth's surface, then r is equal to Earth's radius R_E. Since r is in the denominator of Equation 4.7, the weight decreases with increasing r. So the weight of an object on a mountaintop is less than the weight of the same object at sea level.

Comparing Equations 4.6 and 4.7, we see that

$$g = G\frac{M_E}{r^2} \qquad [4.8]$$

Unlike mass, weight is not an inherent property of an object because it can take different values, depending on the value of g in a given location. If an object has a mass of 70.0 kg, for example, then its weight at a location where $g = 9.80\ \mathrm{m/s^2}$ is $mg = 686$ N. In a high-altitude balloon, where g might be $9.76\ \mathrm{m/s^2}$, the object's weight would be 683 N. The value of g also varies slightly due to the density of matter in a given locality.

Equation 4.8 is a general result that can be used to calculate the acceleration of an object falling near the surface of any massive object if the more massive object's radius and mass are known. Using the values in Table 7.3 (p. 216), you should be able to show that $g_{\mathrm{Sun}} = 274\ \mathrm{m/s^2}$ and $g_{\mathrm{Moon}} = 1.62\ \mathrm{m/s^2}$. An important fact is that for spherical bodies, distances are calculated from the centers of the objects, a consequence of Gauss's law (explained in Chapter 15), which holds for both gravitational and electric forces.

NASA

Astronaut Edwin E. "Buzz" Aldrin, Jr., walking on the Moon after the *Apollo 11* lunar landing. Aldrin's weight on the Moon is less than it is on Earth, but his mass is the same in both places.

Quick Quiz 4.4

A friend, calling from the Moon, tells you she has just won 1 newton of gold in a contest. Excitedly, you tell her that you entered Earth's version of the same contest and also won 1 newton of gold! Who won the prize of greatest value? (a) your friend (b) you (c) the values are equal.

EXAMPLE 4.3 May the Force Be with You

Goal Calculate the magnitude of a gravitational force using Newton's law of gravitation.

Problem Find the gravitational force exerted by the Sun on a 70.0-kg man located on Earth. The distance from the Sun to the Earth is about 1.50×10^{11} m, and the Sun's mass is 1.99×10^{30} kg.

Strategy Substitute numbers into Newton's law of gravitation, Equation 4.5, making sure to use the correct units.

Solution

Apply Equation 4.5, substituting values:

$$F_{\text{sun}} = G\frac{mM_S}{r^2}$$

$$= (6.67 \times 10^{-11}\ \text{kg}^{-1}\text{m}^3\text{s}^{-2})\frac{(70.0\ \text{kg})(1.99 \times 10^{30}\ \text{kg})}{(1.50 \times 10^{11}\ \text{m})^2}$$

$$= \boxed{0.413\ \text{N}}$$

Remarks The gravitational attraction between the Sun and objects on Earth is easily measurable and has been exploited in experiments to determine whether gravitational attraction depends on the composition of the object. As the exercise shows, the gravitational force on Earth due to the Moon is much weaker than the gravitational force on Earth due to the Sun. Paradoxically, the Moon's effect on the tides is over twice that of the Sun, because the tides depend on *differences* in the gravitational force across the Earth, and those differences are greater for the Moon's gravitational force because it's much closer to Earth than the Sun.

Exercise 4.3

To one significant digit, find the force exerted by the Moon on a 70-kg man on Earth. The Moon has a mass of 7.36×10^{22} kg and is 3.84×10^{8} m from Earth.

Answer $F_{\text{Moon}} = 0.002$ N

EXAMPLE 4.4 Weight on Planet X

Goal Understand the effect of a planet's mass and radius on the weight of an object on the planet's surface.

Problem An astronaut on a space mission lands on a planet with three times the mass and twice the radius of Earth. What is her weight w_x on this planet as a multiple of her Earth weight w_E?

Strategy Write M_X and r_X, the mass and radius of the planet, in terms of M_E and R_E, the mass and radius of Earth, respectively, and substitute into the law of gravitation.

Solution

From the statement of the problem, we have the following relationships:

$$M_X = 3M_E \qquad r_X = 2R_E$$

Substitute the preceding expressions into Equation 4.5 and simplify, algebraically associating the terms giving the weight on Earth:

$$w_X = G\frac{M_X m}{r_X^2} = G\frac{3M_E m}{(2R_E)^2} = \frac{3}{4}G\frac{M_E m}{R_E^2} = \boxed{\frac{3}{4}w_E}$$

Remarks This problem shows the interplay between a planet's mass and radius in determining the weight of objects on its surface. Because of Earth's much smaller radius, the weight of an object on Jupiter is only 2.64 times its weight on Earth, despite the fact that Jupiter has over 300 times as much mass.

Exercise 4.4

An astronaut lands on Ganymede, a giant moon of Jupiter that is larger than the planet Mercury. Ganymede has one-fortieth the mass of Earth and two-fifths the radius. Find the weight of the astronaut standing on Ganymede in terms of his Earth weight w_E.

Answer $w_G = (5/32)\,w_E$

4.4 NEWTON'S THIRD LAW

In Section 4.1 we found that a force is exerted on an object when it comes into contact with some other object. Consider the task of driving a nail into a block of wood, for example, as illustrated in Figure 4.7a. To accelerate the nail and drive it into the block, the hammer must exert a net force on the nail. Newton recognized, however, that a single isolated force (such as the force exerted by the hammer on the nail), couldn't exist. Instead, **forces in nature always exist in pairs**. According to Newton, as the nail is driven into the block by the force exerted by the hammer, the hammer is slowed down and stopped by the force exerted by the nail.

Newton described such paired forces with his **third law**:

Newton's third law ▶

> If object 1 and object 2 interact, the force $\vec{\mathbf{F}}_{12}$ exerted by object 1 on object 2 is equal in magnitude but opposite in direction to the force $\vec{\mathbf{F}}_{21}$ exerted by object 2 on object 1.

This law, which is illustrated in Figure 4.7b, states that **a single isolated force can't exist**. The force $\vec{\mathbf{F}}_{12}$ exerted by object 1 on object 2 is sometimes called the *action force*, and the force $\vec{\mathbf{F}}_{21}$ exerted by object 2 on object 1 is called the *reaction force*. In reality, either force can be labeled the action or reaction force. **The action force is equal in magnitude to the reaction force and opposite in direction. In all cases, the action and reaction forces act on different objects**. For example, the force acting on a freely falling projectile is the force exerted by Earth on the projectile, $\vec{\mathbf{F}}_g$, and the magnitude of this force is its weight mg. The reaction to force $\vec{\mathbf{F}}_g$ is the force exerted by the projectile on Earth, $\vec{\mathbf{F}}_g{}' = -\vec{\mathbf{F}}_g$. The reaction force $\vec{\mathbf{F}}_g{}'$ must accelerate the Earth towards the projectile, just as the action force $\vec{\mathbf{F}}_g$ accelerates the projectile towards the Earth.

TIP 4.4 Action-Reaction Pairs

In applying Newton's third law, remember that an action and its reaction force always act on *different* objects. Two external forces acting on the same object, even if they are equal in magnitude and opposite in direction, *can't* be an action-reaction pair.

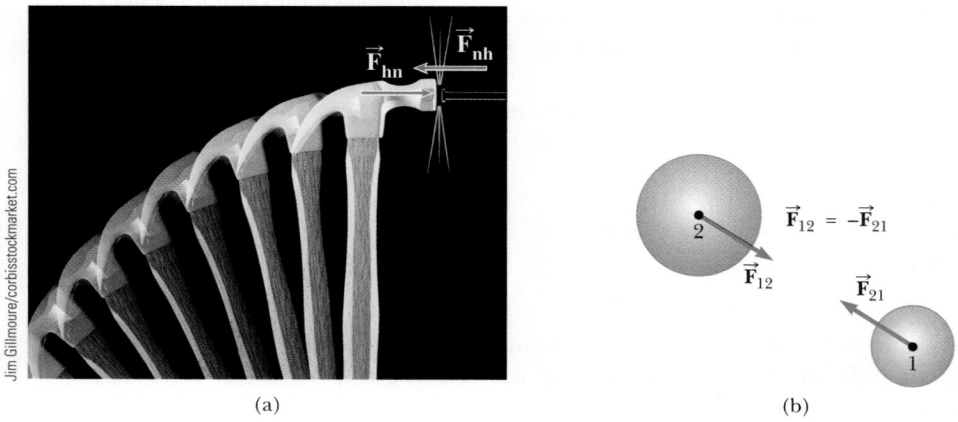

Figure 4.7 Newton's third law. (a) The force exerted by the hammer on the nail is equal in magnitude and opposite in direction to the force exerted by the nail on the hammer. (b) The force $\vec{\mathbf{F}}_{12}$ exerted by object 1 on object 2 is equal in magnitude and opposite in direction to the force $\vec{\mathbf{F}}_{21}$ exerted by object 2 on object 1.

Because the Earth has such a large mass, however, its acceleration due to this reaction force is negligibly small.

Newton's third law constantly affects our activities in everyday life. Without it, no locomotion of any kind would be possible, whether on foot, on a bicycle, or in a motorized vehicle. When walking, for example, we exert a frictional force against the ground. The reaction force of the ground against our foot propels us forward. In the same way, the tires on a bicycle exert a frictional force against the ground, and the reaction of the ground pushes the bicycle forward. As we'll see shortly, friction plays a large role in such reaction forces.

APPLICATION
Helicopter Flight

For another example of Newton's third law, consider the helicopter. Most helicopters have a large set of blades rotating in a horizontal plane above the body of the vehicle and another, smaller set rotating in a vertical plane at the back. Other helicopters have two large sets of blades above the body rotating in opposite directions. Why do helicopters always have two sets of blades? In the first type of helicopter, the engine applies a force to the blades, causing them to change their rotational motion. According to Newton's third law, however, the blades must exert a force on the engine of equal magnitude and in the opposite direction. This force would cause the body of the helicopter to rotate in the direction opposite the blades. A rotating helicopter would be impossible to control, so a second set of blades is used. The small blades in the back provide a force opposite to that tending to rotate the body of the helicopter, keeping the body oriented in a stable position. In helicopters with two sets of large counterrotating blades, engines apply forces in opposite directions, so there is no net force rotating the helicopter.

As mentioned earlier, the Earth exerts a force $\vec{\mathbf{F}}_g$ on any object. If the object is a TV at rest on a table, as in Figure 4.8a, the reaction force to $\vec{\mathbf{F}}_g$ is the force the TV exerts on the Earth, $\vec{\mathbf{F}}_g{}'$. The TV doesn't accelerate downward because it's held up by the table. The table, therefore, exerts an upward force $\vec{\mathbf{n}}$, called the **normal force**, on the TV. (*Normal*, a technical term from mathematics, means "perpendicular" in this context.) The normal force is an elastic force arising from the cohesion of matter and is electromagnetic in origin. It balances the gravitational force acting on the TV, preventing the TV from falling through the table, and can have any value needed, up to the point of breaking the table. The reaction to $\vec{\mathbf{n}}$ is the force exerted by the TV on the table, $\vec{\mathbf{n}}'$. Therefore,

$$\vec{\mathbf{F}}_g = -\vec{\mathbf{F}}_g{}' \qquad \text{and} \qquad \vec{\mathbf{n}} = -\vec{\mathbf{n}}'$$

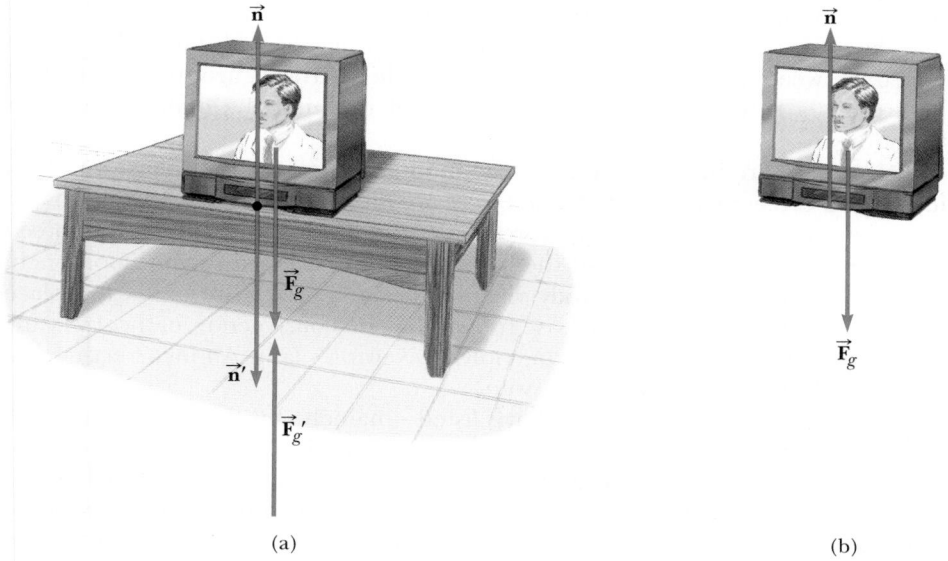

Figure 4.8 When a TV set is sitting on a table, the forces acting on the set are the normal force $\vec{\mathbf{n}}$ exerted by the table and the force of gravity, $\vec{\mathbf{F}}_g$, as illustrated in (b). The reaction to $\vec{\mathbf{n}}$ is the force exerted by the TV set on the table, $\vec{\mathbf{n}}'$. The reaction to $\vec{\mathbf{F}}_g$ is the force exerted by the TV set on Earth, $\vec{\mathbf{F}}_g{}'$.

The forces $\vec{\mathbf{n}}$ and $\vec{\mathbf{n}}'$ both have the same magnitude as $\vec{\mathbf{F}}_g$. Note that the forces acting on the TV are $\vec{\mathbf{F}}_g$ and $\vec{\mathbf{n}}$, as shown in Figure 4.8b. The two reaction forces, $\vec{\mathbf{F}}_g'$ and $\vec{\mathbf{n}}'$, are exerted by the TV on objects other than the TV. Remember, the two forces in an action-reaction pair always act on two different objects.

Because the TV is not accelerating in any direction ($\vec{\mathbf{a}} = 0$), it follows from Newton's second law that $m\vec{\mathbf{a}} = 0 = \vec{\mathbf{F}}_g + \vec{\mathbf{n}}$. However, $F_g = -mg$, so $n = mg$, a useful result.

Quick Quiz 4.5

APPLICATION

Colliding Vehicles

A small sports car collides head-on with a massive truck. The greater impact force (in magnitude) acts on (a) the car, (b) the truck, (c) neither, the force is the same on both. Which vehicle undergoes the greater magnitude acceleration? (d) the car, (e) the truck, (f) the accelerations are the same.

4.5 APPLICATIONS OF NEWTON'S LAWS

This section applies Newton's laws to objects moving under the influence of constant external forces. We assume that objects behave as particles, so we need not consider the possibility of rotational motion. We also neglect any friction effects and the masses of any ropes or strings involved. With these approximations, the magnitude of the force exerted along a rope, called the **tension**, is the same at all points in the rope. This is illustrated by the rope in Figure 4.9, showing the forces $\vec{\mathbf{T}}$ and $\vec{\mathbf{T}}'$ acting on it. If the rope has mass m, then Newton's second law applied to the rope gives $T - T' = ma$. If the mass m is taken to be negligible, however, as in the upcoming examples, then $T = T'$.

Figure 4.9 Newton's second law applied to a rope gives $T - T' = ma$. However, if $m = 0$, then $T = T'$. Thus, the tension in a massless rope is the same at all points in the rope.

When we apply Newton's law to an object, we are interested only in those forces which act *on the object*. For example, in Figure 4.8b, the only external forces acting on the TV are $\vec{\mathbf{n}}$ and $\vec{\mathbf{F}}_g$. The reactions to these forces, $\vec{\mathbf{n}}'$ and $\vec{\mathbf{F}}_g'$, act on the table and on Earth, respectively, and don't appear in Newton's second law applied to the TV.

Consider a crate being pulled to the right on a frictionless, horizontal surface, as in Figure 4.10a. Suppose you wish to find the acceleration of the crate and the force the surface exerts on it. The horizontal force exerted on the crate acts through the rope. The force that the rope exerts on the crate is denoted by $\vec{\mathbf{T}}$ (because it's a tension force). The magnitude of $\vec{\mathbf{T}}$ is equal to the tension in the rope. What we mean by the words "tension in the rope" is just the force read by a spring scale when the rope in question has been cut and the scale inserted between the cut ends. A dashed circle is drawn around the crate in Figure 4.10a to emphasize the importance of isolating the crate from its surroundings.

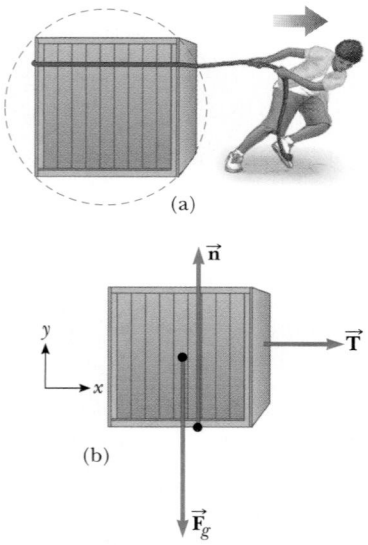

Figure 4.10 (a) A crate being pulled to the right on a frictionless surface. (b) The free-body diagram that represents the forces exerted on the crate.

Because we are interested only in the motion of the crate, we must be able to identify all forces acting on it. These forces are illustrated in Figure 4.10b. In addition to displaying the force $\vec{\mathbf{T}}$, the force diagram for the crate includes the force of gravity $\vec{\mathbf{F}}_g$ exerted by Earth and the normal force $\vec{\mathbf{n}}$ exerted by the floor. Such a force diagram is called a **free-body diagram**, because the environment is replaced by a series of forces on an otherwise free body. The construction of a correct free-body diagram is an essential step in applying Newton's laws. An incorrect diagram will most likely lead to incorrect answers!

The *reactions* to the forces we have listed—namely, the force exerted by the rope on the hand doing the pulling, the force exerted by the crate on Earth, and the force exerted by the crate on the floor—aren't included in the free-body diagram because they act on other objects and not on the crate. Consequently, they don't directly influence the crate's motion. Only forces acting directly on the crate are included.

Now let's apply Newton's second law to the crate. First we choose an appropriate coordinate system. In this case it's convenient to use the one shown in Figure 4.10b, with the x-axis horizontal and the y-axis vertical. We can apply Newton's second law in the x-direction, y-direction, or both, depending on what we're asked to find in a problem. Newton's second law applied to the crate in the x- and y-directions yields the following two equations:

$$ma_x = T \qquad ma_y = n - mg = 0$$

From these equations, we find that the acceleration in the x-direction is constant, given by $a_x = T/m$, and that the normal force is given by $n = mg$. Because the acceleration is constant, the equations of kinematics can be applied to obtain further information about the velocity and displacement of the object.

TIP 4.5 Free-Body Diagrams

The *most important step* in solving a problem by means of Newton's second law is to draw the correct free-body diagram. Include only those forces that act directly on the object of interest.

Problem-Solving Strategy Newton's Second Law

Problems involving Newton's second law can be very complex. The following protocol breaks the solution process down into smaller, intermediate goals:

1. **Read** the problem carefully at least once.
2. **Draw** a picture of the system, identify the object of primary interest, and indicate forces with arrows.
3. **Label** each force in the picture in a way that will bring to mind what physical quantity the label stands for (e.g., T for tension).
4. **Draw** a free-body diagram of the object of interest, based on the labeled picture. If additional objects are involved, draw separate free-body diagrams for them. Choose convenient coordinates for each object.
5. **Apply Newton's second law**. The x- and y-components of Newton's second law should be taken from the vector equation and written individually. This often results in two equations and two unknowns.
6. **Solve** for the desired unknown quantity, and substitute the numbers.

TIP 4.6 A Particle in Equilibrium

A zero net force on a particle does *not* mean that the particle isn't moving. It means that the particle isn't *accelerating*. If the particle has a nonzero initial velocity and is acted upon by a zero net force, it continues to move with the same velocity.

In the special case of equilibrium, the foregoing process is simplified because the acceleration is zero.

Objects in Equilibrium

Objects that are either at rest or moving with constant velocity are said to be in equilibrium. Because $\vec{\mathbf{a}} = 0$, Newton's second law applied to an object in equilibrium gives

$$\sum \vec{\mathbf{F}} = 0 \qquad [4.9]$$

This statement signifies that the *vector* sum of all the forces (the net force) acting on an object in equilibrium is zero. Equation 4.9 is equivalent to the set of component equations given by

$$\sum F_x = 0 \qquad \text{and} \qquad \sum F_y = 0 \qquad [4.10]$$

We won't consider three-dimensional problems in this book, but the extension of Equation 4.10 to a three-dimensional problem can be made by adding a third equation: $\Sigma F_z = 0$.

(i)

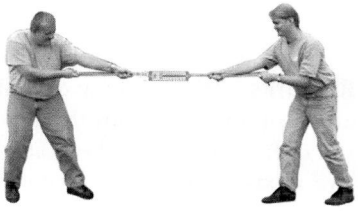

(ii)

Figure 4.11 (Quick Quiz 4.6) (i) A person pulls with a force of magnitude F on a spring scale attached to a wall. (ii) Two people pull with forces of magnitude F in opposite directions on a spring scale attached between two ropes.

Quick Quiz 4.6

Consider the two situations shown in Figure 4.11, in which there is no acceleration. In both cases, the men pull with a force of magnitude F. Is the reading on the scale in part (i) of the figure (a) greater than, (b) less than, or (c) equal to the reading in part (ii)?

EXAMPLE 4.5 A Traffic Light at Rest

Goal Use the second law in an equilibrium problem requiring two free-body diagrams.

Problem A traffic light weighing 1.00×10^2 N hangs from a vertical cable tied to two other cables that are fastened to a support, as in Figure 4.12a. The upper cables make angles of 37.0° and 53.0° with the horizontal. Find the tension in each of the three cables.

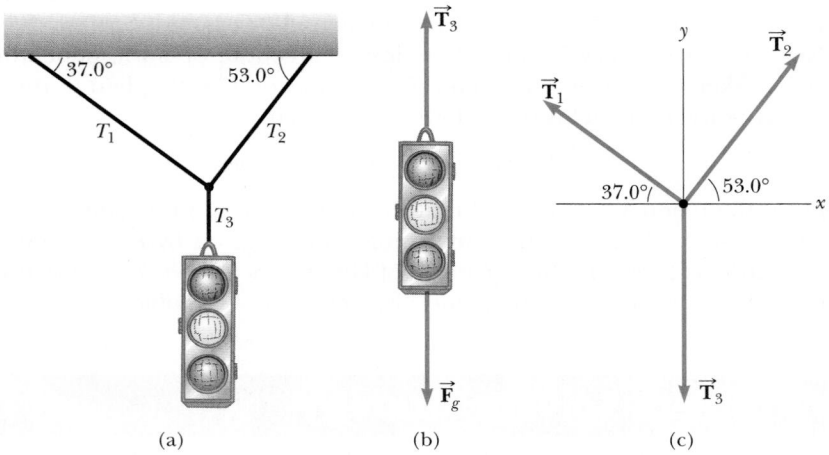

Figure 4.12 (Example 4.5) (a) A traffic light suspended by cables. (b) A free-body diagram for the traffic light. (c) A free-body diagram for the knot joining the cables.

Strategy There are three unknowns, so we need to generate three equations relating them, which can then be solved. One equation can be obtained by applying Newton's second law to the traffic light, which has forces in the y-direction only. Two more equations can be obtained by applying the second law to the knot joining the cables—one equation from the x-component and one equation from the y-component.

Solution

Find T_3 from Figure 4.12b, using the condition of equilibrium:

$$\sum F_y = 0 \quad \rightarrow \quad T_3 - F_g = 0$$
$$T_3 = F_g = 1.00 \times 10^2 \text{ N}$$

Using Figure 4.12c, resolve all three tension forces into components and construct a table for convenience:

Force	x-component	y-component
$\vec{\mathbf{T}}_1$	$-T_1 \cos 37.0°$	$T_1 \sin 37.0°$
$\vec{\mathbf{T}}_2$	$T_2 \cos 53.0°$	$T_2 \sin 53.0°$
$\vec{\mathbf{T}}_3$	0	-1.00×10^2 N

Apply the conditions for equilibrium to the knot, using the components in the table:

$$\sum F_x = -T_1 \cos 37.0° + T_2 \cos 53.0° = 0 \qquad \textbf{(1)}$$
$$\sum F_y = T_1 \sin 37.0° + T_2 \sin 53.0° - 1.00 \times 10^2 \text{ N} = 0 \qquad \textbf{(2)}$$

There are two equations and two remaining unknowns. Solve Equation (1) for T_2:

$$T_2 = T_1 \left(\frac{\cos 37.0°}{\cos 53.0°} \right) = T_1 \left(\frac{0.799}{0.602} \right) = 1.33 T_1$$

Substitute the result for T_2 into Equation (2):

$$T_1 \sin 37.0° + (1.33 T_1)(\sin 53.0°) - 1.00 \times 10^2 \text{ N} = 0$$
$$T_1 = 60.1 \text{ N}$$
$$T_2 = 1.33 T_1 = 1.33(60.0 \text{ N}) = 79.9 \text{ N}$$

Remarks It's very easy to make sign errors in this kind of problem. One way to avoid them is to always measure the angle of a vector from the positive x-direction. The trigonometric functions of the angle will then automatically give the correct signs for the components. For example, $\vec{\mathbf{T}}_1$ makes an angle of $180° - 37° = 143°$ with respect to the positive x-axis, and its x-component, $T_1 \cos 143°$, is negative, as it should be.

Exercise 4.5

Suppose the traffic light is hung so that the tensions T_1 and T_2 are both equal to 80.0 N. Find the new angles they make with respect to the x axis. (By symmetry, these angles will be the same.)

Answer Both angles are 38.7°.

EXAMPLE 4.6 Sled on a Frictionless Hill

Goal Use the second law and the normal force in an equilibrium problem.

Problem A child holds a sled at rest on a frictionless, snow-covered hill, as shown in Figure 4.13a. If the sled weighs 77.0 N, find the force exerted by the rope on the sled and the magnitude of the force $\vec{\mathbf{n}}$ exerted by the hill on the sled.

Strategy When an object is on a slope, it's convenient to use tilted coordinates, as in Figure 4.13b, so that the normal force $\vec{\mathbf{n}}$ is in the y-direction and the tension force $\vec{\mathbf{T}}$ is in the x-direction. In the absence of friction, the hill exerts no force on the sled in the x-direction. Because the sled is at rest, the conditions for equilibrium, $\Sigma F_x = 0$ and $\Sigma F_y = 0$, apply, giving two equations for the two unknowns—the tension and the normal force.

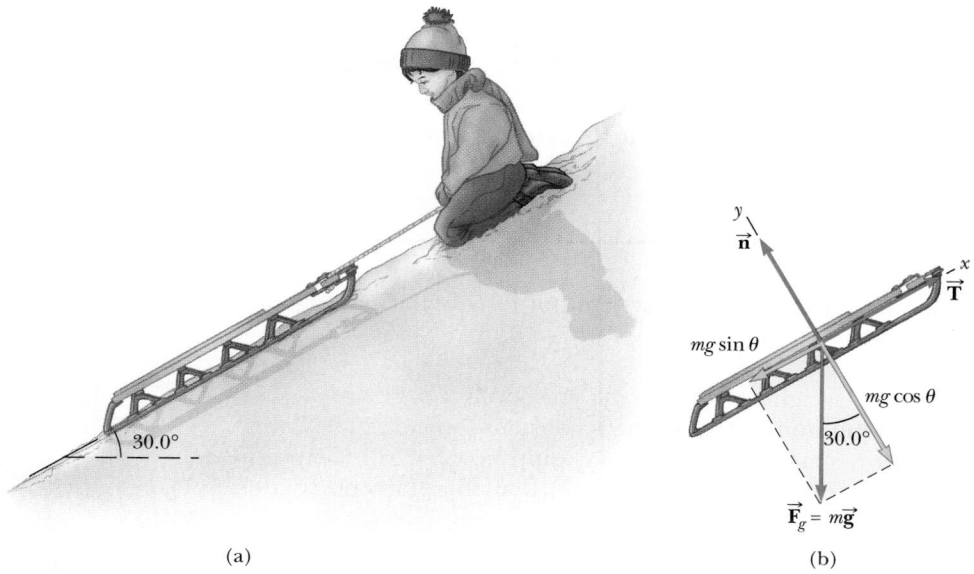

Figure 4.13 (Example 4.6) (a) A child holding a sled on a frictionless hill. (b) A free-body diagram for the sled.

Solution

Apply Newton's second law to the sled, with $\vec{\mathbf{a}} = 0$:

$$\sum \vec{\mathbf{F}} = \vec{\mathbf{T}} + \vec{\mathbf{n}} + \vec{\mathbf{F}}_g = 0$$

Extract the x-component from this equation to find T. The x-component of the normal force is zero, and the sled's weight is given by $mg = 77.0$ N.

$$\sum F_x = T + 0 - mg\sin\theta = T - (77.0\text{ N})\sin 30.0° = 0$$

$$T = 38.5\text{ N}$$

Write the y-component of Newton's second law. The y-component of the tension is zero, so this equation will give the normal force.

$$\sum F_y = 0 + n - mg\cos\theta = n - (77.0\text{ N})(\cos 30.0°) = 0$$

$$n = 66.7\text{ N}$$

Remarks Unlike its value on a horizontal surface, n is less than the weight of the sled when the sled is on the slope. This is because only part of the force of gravity (the x-component) is acting to pull the sled down the slope. The y-component of the force of gravity balances the normal force.

Exercise 4.6

Suppose another child of weight w climbs onto the sled. If the tension force is measured to be 60.0 N, find the weight of the child and the magnitude of the normal force acting on the sled.

Answers $w = 43.0$ N, $n = 104$ N

Quick Quiz 4.7

For the child being pulled forward on the toboggan in Figure 4.14, is the magnitude of the normal force exerted by the ground on the toboggan (a) equal to the total weight of the child plus the toboggan, (b) greater than the total weight, (c) less than the total weight, or (d) possibly greater than or less than the total weight, depending on the size of the weight relative to the tension in the rope?

Figure 4.14 (Quick Quiz 4.7)

Accelerating Objects and Newton's Second Law

When a net force acts on an object, the object accelerates, and we use Newton's second law to analyze the motion.

EXAMPLE 4.7 Moving a Crate

Goal Use the second law of motion for a system not in equilibrium, together with a kinematics equation.

Problem The combined weight of the crate and dolly in Figure 4.15 is 3.00×10^2 N. If the man pulls on the rope with a constant force of 20.0 N, what is the acceleration of the system (crate plus dolly), and how far will it move in 2.00 s? Assume that the system starts from rest and that there are no friction forces opposing the motion.

Strategy We can find the acceleration of the system from Newton's second law. Because the force exerted on the system is constant, its acceleration is constant. Therefore, we can apply a kinematics equation to find the distance traveled in 2.00 s.

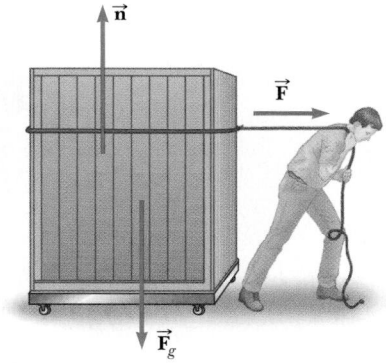

Figure 4.15 (Example 4.7)

Solution

Find the mass of the system from the definition of weight, $w = mg$:

$$m = \frac{w}{g} = \frac{3.00 \times 10^2 \text{ N}}{9.80 \text{ m/s}^2} = 30.6 \text{ kg}$$

Find the acceleration of the system from the second law:

$$a_x = \frac{F_x}{m} = \frac{20.0 \text{ N}}{30.6 \text{ kg}} = 0.654 \text{ m/s}^2$$

Use kinematics to find the distance moved in 2.00 s, with $v_0 = 0$:

$$\Delta x = \tfrac{1}{2} a_x t^2 = \tfrac{1}{2}(0.654 \text{ m/s}^2)(2.00 \text{ s})^2 = 1.31 \text{ m}$$

Remarks Note that the constant applied force of 20.0 N is assumed to act on the system at all times during its motion. If the force were removed at some instant, the system would continue to move with constant velocity and hence zero acceleration. The rollers have an effect that was neglected, here.

Exercise 4.7

A man pulls a 50.0-kg box horizontally from rest while exerting a constant horizontal force, displacing the box 3.00 m in 2.00 s. Find the force the man exerts on the box. (Ignore friction.)

Answer 75.0 N

EXAMPLE 4.8 The Runaway Car

Goal Apply the second law and kinematic equations to a problem involving a moving object on a slope.

Problem **(a)** A car of mass m is on an icy driveway inclined at an angle $\theta = 20.0°$, as in Figure 4.16a. Determine the acceleration of the car, assuming that the incline is frictionless. **(b)** If the length of the driveway is 25.0 m and the car starts from rest at the top, how long does it take to travel to the bottom? **(c)** What is the car's speed at the bottom?

Strategy Choose tilted coordinates as in Figure 4.16b, so that the normal force $\vec{\mathbf{n}}$ is in the positive y-direction, perpendicular to the driveway, and the positive x-axis is down the slope. The force of gravity $\vec{\mathbf{F}}_g$ then has an x-component, $mg \sin \theta$, and a y-component, $-mg \cos \theta$. The components of Newton's second law form a system of two equations and two unknowns for the acceleration down the slope, a_x, and the normal force. Parts (b) and (c) can be solved with the kinematics equations.

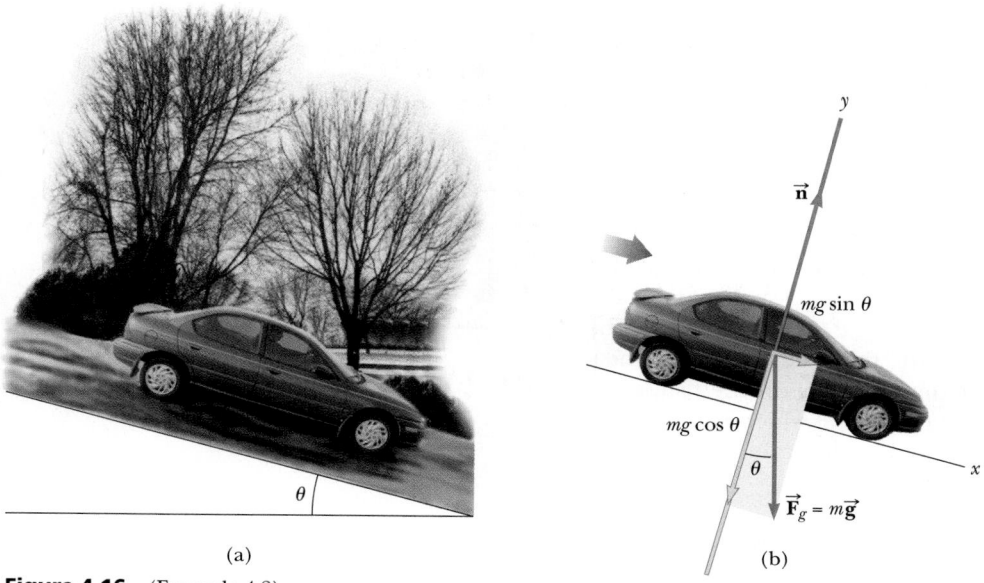

Figure 4.16 (Example 4.8)

Solution

(a) Find the acceleration of the car.

Apply Newton's second law.

$$m\vec{\mathbf{a}} = \sum \vec{\mathbf{F}} = \vec{\mathbf{F}}_g + \vec{\mathbf{n}}$$

Extract the x- and y-components from the second law:

$$ma_x = \sum F_x = mg \sin \theta \qquad \textbf{(1)}$$

$$0 = \sum F_y = -mg \cos \theta + n \qquad \textbf{(2)}$$

Divide Equation (1) by m and substitute the given values:

$$a_x = g \sin \theta = (9.80 \text{ m/s}^2) \sin 20.0° = \boxed{3.35 \text{ m/s}^2}$$

(b) Find the time taken for the car to reach the bottom.

Use Equation 2.9 for displacement, with $v_{0x} = 0$:

$$\Delta x = \tfrac{1}{2} a_x t^2 \quad \rightarrow \quad \tfrac{1}{2} \cdot (3.35 \text{ m/s}^2) t^2 = 25.0 \text{ m}$$

$$t = \boxed{3.86 \text{ s}}$$

(c) Find the speed of the car at the bottom of the driveway.

Use Equation 2.6 for velocity, again with $v_{0x} = 0$:

$$v = at = (3.35 \text{ m/s}^2)(3.86 \text{ s}) = \boxed{12.9 \text{ m/s}}$$

Remarks Notice that the final answer for the acceleration depends only on g and the angle θ, not the mass. Equation (2), which gives the normal force, isn't useful here, but is essential when friction plays a role.

Exercise 4.8

(a) Suppose a hockey puck slides down a frictionless ramp with an acceleration of 5.00 m/s^2. What angle does the ramp make with respect to the horizontal? **(b)** If the ramp has a length of 6.00 m, how long does it take the puck to reach the bottom? **(c)** Now suppose the mass of the puck is doubled. What's the puck's new acceleration down the ramp?

Answer **(a)** 30.7° **(b)** 1.55 s **(c)** unchanged, 5.00 m/s^2

EXAMPLE 4.9 Weighing a Fish in an Elevator

Goal Explore the effect of acceleration on the apparent weight of an object.

Problem A man weighs a fish with a spring scale attached to the ceiling of an elevator, as shown in Figure 4.17a. While the elevator is at rest, he measures a weight of 40.0 N. **(a)** What weight does the scale read if the elevator accelerates upward at 2.00 m/s^2? **(b)** What does the scale read if the elevator accelerates downward at 2.00 m/s^2? **(c)** If the elevator cable breaks, what does the scale read?

Strategy Write down Newton's second law for the fish, including the force $\vec{\mathbf{T}}$ exerted by the spring scale and the force of gravity, $m\vec{\mathbf{g}}$. The scale doesn't measure the true weight, it measures the force T that it exerts on the fish, so in each case solve for this force, which is the apparent weight as measured by the scale.

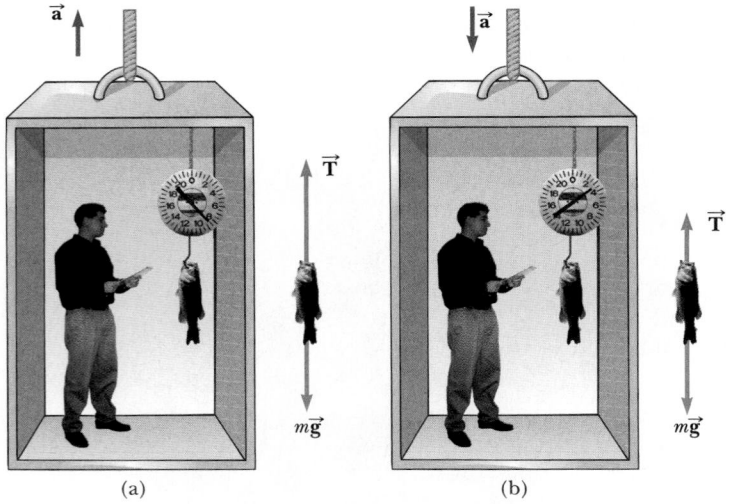

Figure 4.17 (Example 4.9)

Solution

(a) Find the scale reading as the elevator accelerates upwards.

Apply Newton's second law to the fish, taking upwards as the positive direction:

$$ma = \Sigma F = T - mg$$

Solve for T:

$$T = ma + mg = m(a + g)$$

Find the mass of the fish from its weight of 40.0 N:

$$m = \frac{w}{g} = \frac{40.0\ \text{N}}{9.80\ \text{m/s}^2} = 4.08\ \text{kg}$$

Compute the value of T, substituting $a = +2.00\ \text{m/s}^2$:

$$T = m(a + g) = (4.08\ \text{kg})(2.00\ \text{m/s}^2 + 9.80\ \text{m/s}^2) = \boxed{48.1\ \text{N}}$$

(b) Find the scale reading as the elevator accelerates downwards.

The analysis is the same, the only change being the acceleration, which is now negative: $a = -2.00\ \text{m/s}^2$.

$$T = m(a + g) = (4.08\ \text{kg})(-2.00\ \text{m/s}^2 + 9.80\ \text{m/s}^2) = \boxed{31.8\ \text{N}}$$

(c) Find the scale reading after the elevator cable breaks.

Now $a = -9.8\ \text{m/s}^2$, the acceleration due to gravity:

$$T = m(a + g) = (4.08\ \text{kg})(-9.80\ \text{m/s}^2 + 9.80\ \text{m/s}^2) = \boxed{0\ \text{N}}$$

Remarks Notice how important it is to have correct signs in this problem! Accelerations can increase or decrease the apparent weight of an object. Astronauts experience very large changes in apparent weight, from several times normal weight during ascent to weightlessness in free fall.

Exercise 4.9

Find the initial acceleration of a rocket if the astronauts on board experience eight times their normal weight during an initial vertical ascent. (*Hint*: In this exercise, the scale force is replaced by the normal force.)

Answer $68.6\ \mathrm{m/s^2}$

INTERACTIVE EXAMPLE 4.10 Atwood's Machine

Goal Use the second law to solve a simple two-body problem.

Problem Two objects of mass m_1 and m_2, with $m_2 > m_1$, are connected by a light, inextensible cord and hung over a frictionless pulley, as in Active Figure 4.18a. Both cord and pulley have negligible mass. Find the magnitude of the acceleration of the system and the tension in the cord.

Strategy The heavier mass, m_2, accelerates downwards, in the negative y-direction. Since the cord can't be stretched, the accelerations of the two masses are equal in magnitude, but *opposite* in direction, so that a_1 is positive and a_2 is negative, and $a_2 = -a_1$. Each mass is acted on by a force of tension $\vec{\mathbf{T}}$ in the upwards direction and a force of gravity in the downwards direction. Active Figure 4.18b shows free-body diagrams for the two masses. Newton's second law for each mass, together with the equation relating the accelerations, constitutes a set of three equations for the three unknowns—a_1, a_2, and T.

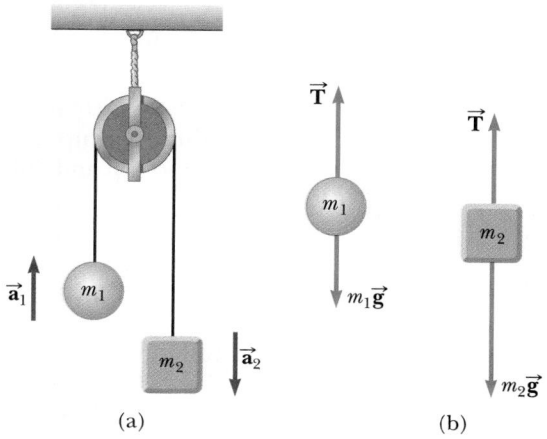

ACTIVE FIGURE 4.18
(Example 4.10) Atwood's machine. (a) Two hanging objects connected by a light string that passes over a frictionless pulley. (b) Free-body diagrams for the objects.

PhysicsNow™
Log into to PhysicsNow at **www.cp7e.com**, and go to Active Figure 4.18 to adjust the masses of objects on Atwood's machine and observe the resulting motion.

Solution

Apply the second law to each of the two masses individually:

$$m_1 a_1 = T - m_1 g \quad \textbf{(1)} \qquad\qquad m_2 a_2 = T - m_2 g \quad \textbf{(2)}$$

Substitute $a_2 = -a_1$ into the second equation, and multiply both sides by -1:

$$m_2 a_1 = -T + m_2 g$$

Add the stacked equations, and solve for a_1:

$$(m_1 + m_2)\, a_1 = m_2 g - m_1 g$$

$$a_1 = \left(\frac{m_2 - m_1}{m_1 + m_2}\right) g$$

Substitute this result into Equation (1) to find T:

$$T = \left(\frac{2 m_1 m_2}{m_1 + m_2}\right) g$$

Remarks The acceleration of the second block is the same as that of the first, but negative. When m_2 gets very large compared with m_1, the acceleration of the system approaches g, as expected, because m_2 is falling nearly freely under the influence of gravity. Indeed, m_2 is only slightly restrained by the much lighter m_1.

Exercise 4.10

Suppose that in the same Atwood setup another string is attached to the bottom of m_1 and a constant force f is applied, retarding the upward motion of m_1. If $m_1 = 5.00$ kg and $m_2 = 10.00$ kg, what value of f will reduce the acceleration of the system by 50%?

Answer 24.5 N

PhysicsNow™ Investigate the response of Atwood's machine by logging into PhysicsNow at **www.cp7e.com** and going to Interactive Example 4.10.

4.6 FORCES OF FRICTION

An object moving on a surface or through a viscous medium such as air or water encounters resistance as it interacts with its surroundings. This resistance is called **friction**. Forces of friction are essential in our everyday lives. Friction makes it possible to grip and hold things, drive a car, walk, and run. Even standing in one spot would be impossible without friction, as the slightest shift would instantly cause you to slip and fall.

Imagine that you've filled a plastic trash can with yard clippings and want to drag the can across the surface of your concrete patio. If you apply an external horizontal force $\vec{\mathbf{F}}$ to the can, acting to the right as shown in Active Figure 4.19a, the can remains stationary if $\vec{\mathbf{F}}$ is small. The force that counteracts $\vec{\mathbf{F}}$ and keeps the can from moving acts to the left, opposite the direction of $\vec{\mathbf{F}}$, and is called the **force of static friction,** $\vec{\mathbf{f}}_s$. As long as the can isn't moving, $\vec{\mathbf{f}}_s = -\vec{\mathbf{F}}$. If $\vec{\mathbf{F}}$ is increased, $\vec{\mathbf{f}}_s$ also increases. Likewise, if $\vec{\mathbf{F}}$ decreases, $\vec{\mathbf{f}}_s$ decreases. Experiments show that the friction force arises from the nature of the two surfaces: Because of their roughness, contact is made at only a few points, as shown in the magnified view of the surfaces in Active Figure 4.19a.

If we increase the magnitude of $\vec{\mathbf{F}}$, as in Active Figure 4.19b, the trash can eventually slips. When the can is on the verge of slipping, f_s is a maximum, as shown in Figure 4.19c. When F exceeds $f_{s,\max}$, the can accelerates to the right. When the can is in motion, the friction force is less than $f_{s,\max}$ (Fig. 4.19c). We call the friction force for an object in motion the **force of kinetic friction,** $\vec{\mathbf{f}}_k$. The net force $F - f_k$ in the x-direction produces an acceleration to the right, according to Newton's second law. If $F = f_k$, the acceleration is zero, and the can moves to the right with constant speed. If the applied force is removed, the friction force acting to the left provides an acceleration of the can in the $-x$-direction and eventually brings it to rest, again consistent with Newton's second law.

Experimentally, to a good approximation, both $f_{s,\max}$ and f_k for an object on a surface are proportional to the normal force exerted by the surface on the object. The experimental observations can be summarized as follows:

- The magnitude of the force of static friction between any two surfaces in contact can have the values

$$f_s \leq \mu_s n \qquad [4.11]$$

  where the dimensionless constant μ_s is called the **coefficient of static friction** and n is the magnitude of the normal force exerted by one surface on the other. Equation 4.11 also holds for $f_s = f_{s,\max} \equiv \mu_s n$ when an object is on the verge of slipping. This situation is called *impending motion.* The strict inequality holds when the component of the applied force parallel to the surfaces is less than $\mu_s n$.
- The magnitude of the force of kinetic friction acting between two surfaces is

$$f_k = \mu_k n \qquad [4.12]$$

  where μ_k is the **coefficient of kinetic friction.**

TIP 4.7 Use the Equals Sign in Limited Situations

In Equation 4.11, the equals sign is used *only* when the surfaces are just about to break free and begin sliding. Don't fall into the common trap of using $f_s = \mu_s n$ in *any* static situation.

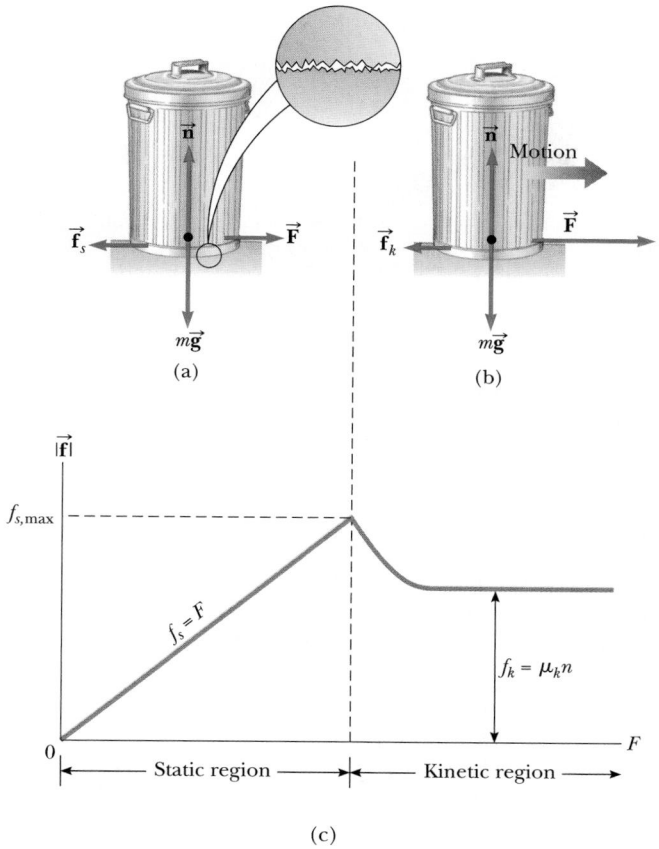

ACTIVE FIGURE 4.19 (a) The force of friction $\vec{\mathbf{f}}_s$ exerted by a concrete surface on a trash can is directed opposite the force $\vec{\mathbf{F}}$ that you exert on the can. As long as the can is not moving, the magnitude of the force of static friction equals that of the applied force $\vec{\mathbf{F}}$. (b) When the magnitude of $\vec{\mathbf{F}}$ exceeds the magnitude of $\vec{\mathbf{f}}_k$, the force of kinetic friction, the trash can accelerates to the right. (c) A graph of the magnitude of the friction force versus that of the applied force. Note that $f_{s,\max} > f_k$.

PhysicsNow™
Log into PhysicsNow at **www.cp7e.com**, and go to Active Figure 4.19 to vary the applied force on the can and practice sliding it on surfaces of varying roughness. Note the effect on the can's motion and the corresponding behavior of the graph in (c).

- The values of μ_k and μ_s depend on the nature of the surfaces, but μ_k is generally less than μ_s. Table 4.2 lists some reported values.
- The direction of the friction force exerted by a surface on an object is opposite the actual motion (kinetic friction) or the impending motion (static friction) of the object relative to the surface.
- The coefficients of friction are nearly independent of the area of contact between the surfaces.

Although the coefficient of kinetic friction varies with the speed of the object, we will neglect any such variations. The approximate nature of Equations 4.11 and 4.12

TABLE 4.2

Coefficients of Friction[a]

	μ_s	μ_k
Steel on steel	0.74	0.57
Aluminum on steel	0.61	0.47
Copper on steel	0.53	0.36
Rubber on concrete	1.0	0.8
Wood on wood	0.25–0.5	0.2
Glass on glass	0.94	0.4
Waxed wood on wet snow	0.14	0.1
Waxed wood on dry snow	—	0.04
Metal on metal (lubricated)	0.15	0.06
Ice on ice	0.1	0.03
Teflon on Teflon	0.04	0.04
Synovial joints in humans	0.01	0.003

[a]All values are approximate.

Figure 4.20 (Quick Quiz 4.10)

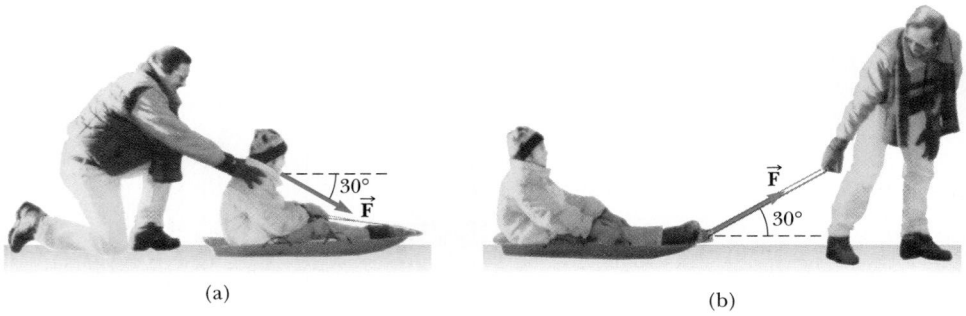

is easily demonstrated by trying to get an object to slide down an incline at constant acceleration. Especially at low speeds, the motion is likely to be characterized by alternate stick and slip episodes.

Quick Quiz 4.8

If you press a book flat against a vertical wall with your hand, in what direction is the friction force exerted by the wall on the book? (a) downward (b) upward (c) out from the wall (d) into the wall.

Quick Quiz 4.9

A crate is sitting in the center of a flatbed truck. As the truck accelerates to the east, the crate moves with it, not sliding on the bed of the truck. In what direction is the friction force exerted by the bed of the truck on the crate? (a) To the west. (b) To the east. (c) There is no friction force, because the crate isn't sliding.

Quick Quiz 4.10

Suppose you're playing with your niece in the snow. She's sitting on a sled and asks you to move her across a flat, horizontal field. You have a choice of (a) pushing her from behind by applying a force downward on her shoulders at 30° below the horizontal (Fig. 4.20a) or (b) attaching a rope to the front of the sled and pulling with a force at 30° above the horizontal (Fig 4.20b). Which option would be easier and why?

EXAMPLE 4.11 A Block on a Ramp

Goal Apply the concept of static friction to an object resting on an incline.

Problem Suppose a block with a mass of 2.50 kg is resting on a ramp. If the coefficient of static friction between the block and ramp is 0.350, what maximum angle can the ramp make with the horizontal before the block starts to slip down?

Strategy This is an application of Newton's second law involving an object in equilibrium. Choose tilted coordinates, as in Figure 4.21. Use the fact that the block is just about to slip when the force of static friction takes its maximum value, $f_s = \mu_s n$.

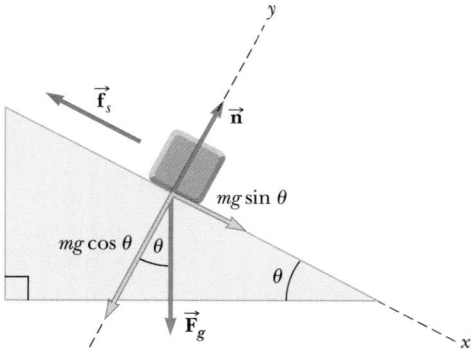

Figure 4.21 (Example 4.11)

Solution

Write Newton's laws for a static system in component form. The gravity force has two components, just as in Examples 4.6 and 4.8.

$$\sum F_x = mg \sin \theta - \mu_s n = 0 \quad (1)$$

$$\sum F_y = n - mg \cos \theta = 0 \quad (2)$$

Rearrange Equation (2) to get an expression for the normal force n:

$$n = mg\cos\theta$$

Substitute the expression for n into Equation (1) and solve for $\tan\theta$:

$$\sum F_x = mg\sin\theta - \mu_s mg\cos\theta = 0 \quad \rightarrow \quad \tan\theta = \mu_s$$

Apply the inverse tangent function to get the answer:

$$\tan\theta = 0.350 \quad \rightarrow \quad \theta = \tan^{-1}(0.350) = 19.3°$$

Remark It's interesting that the final result depends only on the coefficient of static friction. Notice also how similar Equations (1) and (2) are to the equations developed in Examples 4.6 and 4.8. Recognizing such patterns is key to solving problems successfully.

Exercise 4.11

The ramp in Example 4.11 is roughed up and the experiment repeated. **(a)** What is the new coefficient of static friction if the maximum angle turns out to be 30.0°? **(b)** Find the maximum static friction force that acts on the block.

Answer **(a)** 0.577 **(b)** 12.2 N

EXAMPLE 4.12 The Sliding Hockey Puck

Goal Apply the concept of kinetic friction.

Problem The hockey puck in Figure 4.22, struck by a hockey stick, is given an initial speed of 20.0 m/s on a frozen pond. The puck remains on the ice and slides 1.20×10^2 m, slowing down steadily until it comes to rest. Determine the coefficient of kinetic friction between the puck and the ice.

Strategy The puck slows "steadily," which means that the acceleration is constant. Consequently, we can use the kinematic equation $v^2 = v_0^2 + 2a\Delta x$ to find a, the acceleration in the x-direction. The x- and y-components of Newton's second law then give two equations and two unknowns for the coefficient of kinetic friction, μ_k, and the normal force n.

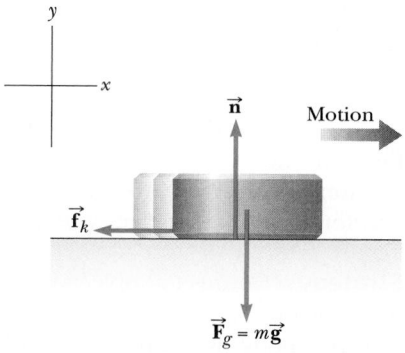

Figure 4.22 (Example 4.12) *After* the puck is given an initial velocity to the right, the external forces acting on it are the force of gravity $\vec{\mathbf{F}}_g$, the normal force $\vec{\mathbf{n}}$, and the force of kinetic friction, $\vec{\mathbf{f}}_k$.

Solution

Solve the time-independent kinematic equation for the acceleration a:

$$v^2 = v_0^2 + 2a\Delta x$$

$$a = \frac{v^2 - v_0^2}{2\Delta x}$$

Substitute $v = 0$, $v_0 = 20.0$ m/s, and $\Delta x = 1.20 \times 10^2$ m. Note the negative sign in the answer: $\vec{\mathbf{a}}$ is opposite $\vec{\mathbf{v}}$.

$$a = \frac{0 - (20.0\text{ m/s})^2}{2(1.20 \times 10^2\text{ m})} = -1.67\text{ m/s}^2$$

Find the normal force from the y-component of the second law:

$$\sum F_y = n - F_g = n - mg = 0$$

$$n = mg$$

Obtain an expression for the force of kinetic friction, and substitute it into the x-component of the second law:

$$f_k = \mu_k n = \mu_k mg$$

$$ma = \sum F_x = -f_k = -\mu_k mg$$

Solve for μ_k and substitute values:

$$\mu_k = -\frac{a}{g} = \frac{1.67\text{ m/s}^2}{9.80\text{ m/s}^2} = 0.170$$

Remarks Notice how the problem breaks down into three parts: kinematics, Newton's second law in the y-direction, and then Newton's law in the x-direction.

Exercise 4.12

An experimental rocket plane lands on skids on a dry lake bed. If it's traveling at 80.0 m/s when it touches down, how far does it slide before coming to rest? Assume the coefficient of kinetic friction between the skids and the lake bed is 0.600.

Answer 544 m

Two-body problems can often be treated as single objects and solved with a system approach. When the objects are rigidly connected—say, by a string of negligible mass that doesn't stretch—this approach can greatly simplify the analysis. When the two bodies are considered together, one or more of the forces end up becoming forces that are internal to the system, rather than external forces affecting each of the individual bodies. Both approaches will be used in Example 4.13.

EXAMPLE 4.13 Connected Objects

Goal Use both the general method and the system approach to solve a connected two-body problem involving gravity and friction.

Problem **(a)** A block with mass $m_1 = 4.00$ kg and a ball with mass $m_2 = 7.00$ kg are connected by a light string that passes over a frictionless pulley, as shown in Figure 4.23a. The coefficient of kinetic friction between the block and the surface is 0.300. Find the acceleration of the two objects and the tension in the string. **(b)** Check the answer for the acceleration by using the system approach.

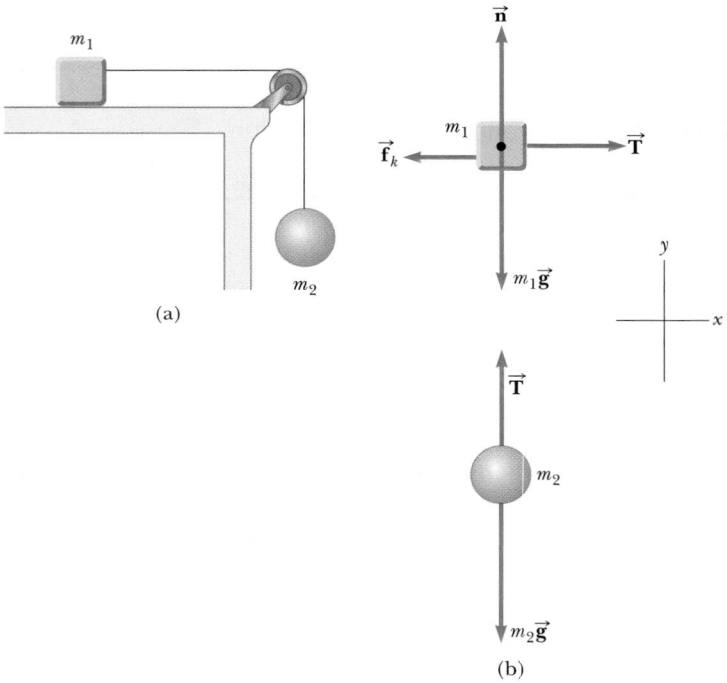

Figure 4.23 (Example 4.13) (a) Two objects connected by a light string that passes over a frictionless pulley. (b) Free-body diagrams for the objects.

Strategy Connected objects are handled by applying Newton's second law separately to each object. The free-body diagrams for the block and the ball are shown in Figure 4.23b, with the $+x$-direction to the right and the $+y$-direction upwards. The magnitude of the acceleration for both objects has the same value, $|a_1| = |a_2| = a$. The block with mass m_1 moves in the positive x-direction, and the ball with mass m_2 moves in the negative y-direction, so $a_1 = -a_2$. Using Newton's second law, we can develop two equations involving the unknowns T and a that can be solved simultaneously. In part **(b)**, treat the two masses as a single object, with the gravity force on the ball increasing the combined object's speed and the friction force on the block retarding it. The tension forces then become internal and don't appear in the second law.

Solution

(a) Find the acceleration of the objects and the tension in the string.

Write the components of Newton's second law for the cube of mass m_1:

$$\sum F_x = T - f_k = m_1 a_1 \qquad \sum F_y = n - m_1 g = 0$$

The equation for the y-component gives $n = m_1 g$. Substitute this value for n and $f_k = \mu_k n$ into the equation for the x-component:

$$T - \mu_k m_1 g = m_1 a_1 \qquad \textbf{(1)}$$

Apply Newton's second law to the ball, recalling that $a_2 = -a_1$:

$$\sum F_y = -m_2 g + T = m_2 a_2 = -m_2 a_1 \quad \textbf{(2)}$$

Subtract Equation (2) from Equation (1), eliminating T and leaving an equation that can be solved for a_1 (substitution can also be used):

$$m_2 g - \mu_k m_1 g = (m_1 + m_2) a_1$$

$$a_1 = \frac{m_2 g - \mu_k m_1 g}{m_1 + m_2}$$

Substitute the given values to obtain the acceleration.

$$a_1 = \frac{(7.00\text{ kg})(9.80\text{ m/s}^2) - (0.300)(4.00\text{ kg})(9.80\text{ m/s}^2)}{(4.00\text{ kg} + 7.00\text{ kg})}$$

$$= \boxed{5.17\text{ m/s}^2}$$

Substitute the value for a_1 into Equation (1) to find the tension T:

$$T = \boxed{32.4\text{ N}}$$

(b) Find the acceleration using the system approach, where the system consists of the two blocks.

Apply Newton's second law to the system and solve for a:

$$(m_1 + m_2) a = m_2 g - \mu_k n = m_2 g - \mu_k m_1 g$$

$$a = \boxed{\frac{m_2 g - \mu_k m_1 g}{m_1 + m_2}}$$

Remarks Although the system approach appears quick and easy, it can be applied only in special cases and can't give any information about the internal forces, such as the tension. To find the tension, you must consider the free-body diagram of one of the blocks separately.

Exercise 4.13

What if an additional mass is attached to the ball in Example 4.13? How large must this mass be to increase the downward acceleration by 50%? Why isn't it possible to add enough mass to double the acceleration?

Answer 14.0 kg. Doubling the acceleration to 10.3 m/s^2 isn't possible simply by suspending more mass, because all objects, regardless of their mass, fall freely at 9.8 m/s^2 near the Earth's surface.

EXAMPLE 4.14 Two Blocks and a Cord

Goal Apply Newton's second law and static friction in a two-body system.

Problem A block of mass 5.00 kg rides on top of a second block of mass 10.0 kg. A person attaches a string to the bottom block and pulls the system horizontally across a frictionless surface, as in Figure 4.24. Friction between the two blocks keeps the 5.00-kg block from slipping off. If the coefficient of static friction is 0.350, what maximum force can be exerted by the string on the 10.0-kg block without causing the 5.00-kg block to slip?

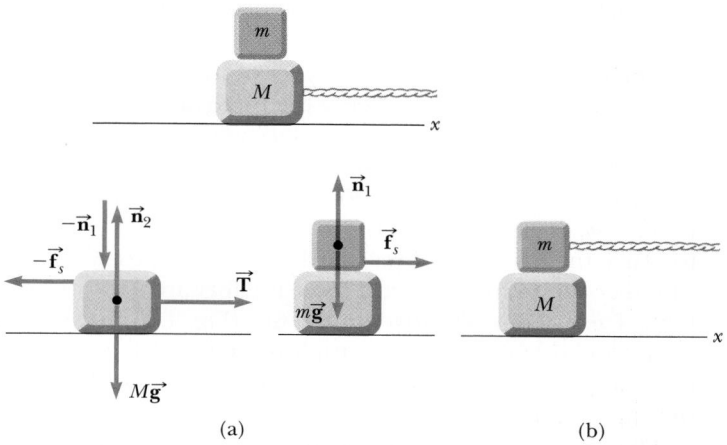

Figure 4.24 (a) (Example 4.14) (b) (Exercise 4.14)

Strategy Draw a free-body diagram for each block. The static friction force causes the top block to move horizontally, and the maximum such force corresponds to $f_s = \mu_s n$. This same static friction retards the motion of the bottom block. As long as the top block isn't slipping, the acceleration of both blocks is the same. Write Newton's second law for each block, and eliminate the acceleration a by substitution, solving for the tension T.

Solution

Write the two components of Newton's second law for the top block:

$$x\text{-component:} \quad ma = \mu_s n_1$$
$$y\text{-component:} \quad 0 = n_1 - mg$$

Solve the y-component for n, substitute the result into the x-component, and then solve for a:

$$n_1 = mg \quad \rightarrow \quad ma = \mu_s mg \quad \rightarrow \quad a = \mu_s g$$

Write the x-component of Newton's second law for the bottom block:

$$Ma = -\mu_s mg + T \qquad \textbf{(1)}$$

Substitute the expression for $a = \mu_s g$ into Equation (1) and solve for the tension T:

$$M\mu_s g = T - \mu_s mg \quad \rightarrow \quad T = (m + M)\mu_s g$$

Now evaluate to get the answer:

$$T = (5.00 \text{ kg} + 10.0 \text{ kg})(0.350)(9.80 \text{ m/s}^2) = \boxed{51.5 \text{ N}}$$

Remarks Notice that the y-component for the 10.0-kg block wasn't needed, because there was no friction between that block and the underlying surface. It's also interesting to note that the top block was accelerated by the force of static friction.

Exercise 4.14

Suppose instead that the string is attached to the top block in Example 14.4. Find the maximum force that can be exerted by the string on the block without causing the top block to slip.

Answer 25.7 N

Applying Physics 4.1 Cars and Friction

Forces of friction are important in the analysis of the motion of cars and other wheeled vehicles. How do such forces both help and hinder the motion of a car?

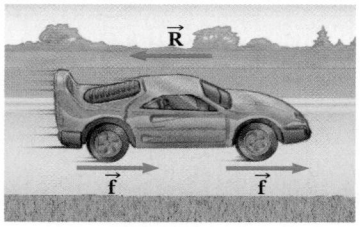

Figure 4.25 (Applying Physics 4.1) The horizontal forces acting on the car are the *forward* forces $\vec{\mathbf{f}}$ exerted by the road on each tire and the force of air resistance $\vec{\mathbf{R}}$, which acts *opposite* the car's velocity. (The car's tires exert a rearward force on the road, not shown in the diagram.)

Explanation There are several types of friction forces to consider, the main ones being the force of friction between the tires and the road surface and the retarding force produced by air resistance.

Assuming that the car is a four-wheel-drive vehicle of mass m, as each wheel turns to propel the car forward, the tire exerts a rearward force on the road. The reaction to this rearward force is a forward force $\vec{\mathbf{f}}$ exerted by the road on the tire (Fig. 4.25). If we assume that the same forward force $\vec{\mathbf{f}}$ is exerted on each tire, the net forward force on the car is $4\vec{\mathbf{f}}$, and the car's acceleration is therefore $\vec{\mathbf{a}} = 4\vec{\mathbf{f}}/m$.

The friction between the moving car's wheels and the road is normally static friction, unless the car is skidding.

When the car is in motion, we must also consider the force of air resistance, $\vec{\mathbf{R}}$, which acts in the direction opposite the velocity of the car. The net force exerted on the car is therefore $4\vec{\mathbf{f}} - \vec{\mathbf{R}}$, so the car's acceleration is $\vec{\mathbf{a}} = (4\vec{\mathbf{f}} - \vec{\mathbf{R}})/m$. At normal driving speeds, the magnitude of $\vec{\mathbf{R}}$ is proportional to the first power of the speed, $R = bv$, where b is a constant, so the force of air resistance increases with increasing speed. When R is equal to $4f$, the acceleration is zero and the car moves at a constant speed. To minimize this resistive force, racing cars often have very low profiles and streamlined contours.

Applying Physics 4.2 Air Drag

Air resistance isn't always undesirable. What are some applications that depend on it?

Explanation Consider a skydiver plunging through the air, as in Figure 4.26. Despite falling from a height of several thousand meters, she never exceeds a speed of around 120 miles per hour. This is because, aside from the downward force of gravity $m\vec{\mathbf{g}}$, there is also an upward force of air resistance, $\vec{\mathbf{R}}$. Before she reaches a final constant speed, the magnitude of $\vec{\mathbf{R}}$ is less than her weight. As her downward speed increases, the force of air resistance increases. The vector sum of the force of gravity and the force of air resistance gives a total force that decreases with time, so her acceleration decreases. Once the two forces balance each other, the net force is zero, so the acceleration is zero, and she reaches a **terminal speed**.

Terminal speed is generally still high enough to be fatal on impact, although there have been amazing stories of survival. In one case, a man fell flat on his back in a freshly plowed field and survived. (He did, however, break virtually every bone in his body). In another case, a stewardess survived a fall from thirty thousand feet into a snowbank. In neither case would the person have had any chance of surviving without the effects of air drag.

Parachutes and paragliders create a much larger drag force due to their large area and can reduce the terminal speed to a few meters per second. Some sports enthusiasts have even developed special suits with wings, allowing a long glide to the ground. In each case, a larger cross-sectional area intercepts more air, creating greater air drag, so the terminal speed is lower.

Guy Sauvage/Photo Researchers, Inc.

Figure 4.26 (Applying Physics 4.2)

Air drag is also important in space travel. Without it, returning to Earth would require a considerable amount of fuel. Air drag helps slow capsules and spaceships, and aerocapture techniques have been proposed for trips to other planets. These techniques significantly reduce fuel requirements by using air drag to slow the spacecraft down.

SUMMARY

Physics Now™ Take a practice test by logging into PhysicsNow at **www.cp7e.com** and clicking on the Pre-Test link for this chapter.

4.1 Forces

There are four known fundamental forces of nature: (1) the strong nuclear force between subatomic particles; (2) the electromagnetic forces between electric charges; (3) the weak nuclear forces, which arises in certain radioactive decay processes; and (4) the gravitational force between objects. These are collectively called field forces. Classical physics deals only with the gravitational and electromagnetic forces.

Forces such as friction or that characterizing a bat hitting a ball are called contact forces. On a more fundamental level, contact forces have an electromagnetic nature.

4.2 Newton's First Law

Newton's first law states that an object moves at constant velocity unless acted on by a force.

The tendency for an object to maintain its original state of motion is called **inertia**. **Mass** is the physical quantity that measures the resistance of an object to changes in its velocity.

4.3 Newton's Second Law

Newton's second law states that the acceleration of an object is directly proportional to the net force acting on it and inversely proportional to its mass. The net force acting on an object equals the product of its mass and acceleration:

$$\sum \vec{\mathbf{F}} = m\vec{\mathbf{a}} \qquad [4.1]$$

Newton's universal law of gravitation is

$$F_g = G\frac{m_1 m_2}{r^2} \qquad [4.5]$$

The **weight** w of an object is the magnitude of the force of gravity exerted on that object and is given by

$$w = mg \qquad [4.6]$$

where $g = F_g/m$ is the acceleration of gravity near Earth's surface.

Solving problems with Newton's second law involves finding all the forces acting on a system and writing Equation

4.1 for the x-component and y-component separately. These two equations are then solved algebraically for the unknown quantities.

4.4 Newton's Third Law

Newton's third law states that if two objects interact, the force $\vec{\mathbf{F}}_{12}$ exerted by object 1 on object 2 is equal in magnitude and opposite in direction to the force $\vec{\mathbf{F}}_{21}$ exerted by object 2 on object 1:

$$\vec{\mathbf{F}}_{12} = -\vec{\mathbf{F}}_{21}$$

An isolated force can never occur in nature.

4.5 Applications of Newton's Laws

An **object in equilibrium** has no net external force acting on it, and the second law, in component form, implies that $\Sigma F_x = 0$ and $\Sigma F_y = 0$ for such an object. These two equations are useful for solving problems in statics, in which the object is at rest or moving at constant velocity.

An object under acceleration requires the same two equations, but with the acceleration terms included: $\Sigma F_x = ma_x$ and $\Sigma F_y = ma_y$. When the acceleration is constant, the equations of kinematics can supplement Newton's second law.

4.6 Forces of Friction

The magnitude of the maximum force of static friction, $f_{s,\max}$, between an object and a surface is proportional to the magnitude of the normal force acting on the object. This maximum force occurs when the object is on the verge of slipping. In general,

$$f_s \leq \mu_s n \qquad [4.11]$$

where μ_s is the **coefficient of static friction**. When an object slides over a surface, the direction of the force of kinetic friction, $\vec{\mathbf{f}}_k$, on the object is opposite the direction of the motion of the object relative to the surface, and proportional to the magnitude of the normal force. The magnitude of $\vec{\mathbf{f}}_k$ is

$$f_k = \mu_k n \qquad [4.12]$$

where μ_k is the **coefficient of kinetic friction**. In general, $\mu_k < \mu_s$.

Solving problems that involve friction is a matter of using these two friction forces in Newton's second law. The static friction force must be handled carefully, because it refers to a maximum force, which is not always called upon in a given problem.

CONCEPTUAL QUESTIONS

1. A ball is held in a person's hand. (a) Identify all the external forces acting on the ball and the reaction to each. (b) If the ball is dropped, what force is exerted on it while it is falling? Identify the reaction force in this case. (Neglect air resistance.)
2. If a car is traveling westward with a constant speed of 20 m/s, what is the resultant force acting on it?
3. If a car moves with a constant acceleration, can you conclude that there are no forces acting on it?
4. A rubber ball is dropped onto the floor. What force causes the ball to bounce?
5. If you push on a heavy box that is at rest, you must exert some force to start its motion. However, once the box is sliding, you can apply a smaller force to maintain its motion. Why?
6. If gold were sold by weight, would you rather buy it in Denver or in Death Valley? If it were sold by mass, in which of the two locations would you prefer to buy it? Why?
7. A passenger sitting in the rear of a bus claims that she was injured as the driver slammed on the brakes, causing a suitcase to come flying toward her from the front of the bus. If you were the judge in this case, what disposition would you make? Why?
8. A space explorer is moving through space far from any planet or star. He notices a large rock, taken as a specimen from an alien planet, floating around the cabin of the ship. Should he push it gently or kick it toward the storage compartment? Why?
9. What force causes an automobile to move? A propeller-driven airplane? A rowboat?
10. Analyze the motion of a rock dropped in water in terms of its speed and acceleration as it falls. Assume that a resistive force is acting on the rock that increases as the velocity of the rock increases.
11. In the motion picture *It Happened One Night* (Columbia Pictures, 1934), Clark Gable is standing inside a stationary bus in front of Claudette Colbert, who is seated. The bus suddenly starts moving forward and Clark falls into Claudette's lap. Why did this happen?
12. A weight lifter stands on a bathroom scale. She pumps a barbell up and down. What happens to the reading on the scale? Suppose she is strong enough to actually *throw* the barbell upward. How does the reading on the scale vary now?
13. In a tug-of-war between two athletes, each pulls on the rope with a force of 200 N. What is the tension in the rope? If the rope doesn't move, what horizontal force does each athlete exert against the ground?
14. As a rocket is fired from a launching pad, its speed and acceleration increase with time as its engines continue to operate. Explain why this occurs even though the thrust of the engines remains constant.
15. Identify the action-reaction pairs in the following situations: (a) a man takes a step; (b) a snowball hits a girl in the back; (c) a baseball player catches a ball; (d) a gust of wind strikes a window.
16. The driver of a speeding empty truck slams on the brakes and skids to a stop through a distance d. (a) If the truck carried a load that doubled its mass, what would be the truck's "skidding distance"? (b) If the initial speed of the truck were halved, what would be the truck's "skidding distance"?
17. Suppose you are driving a car at a high speed. Why should you avoid slamming on your brakes when you want to stop in the shortest possible distance? (Newer cars have antilock brakes that avoid this problem.)
18. A truck loaded with sand accelerates along a highway. If the driving force on the truck remains constant, what

happens to the truck's acceleration if its trailer leaks sand at a constant rate through a hole in its bottom?

19. A large crate is placed on the bed of a truck, but is not tied down. (a) As the truck accelerates forward, the crate remains at rest relative to it. What force causes the crate to accelerate forward? (b) If the driver slams on the brakes, what could happen to the crate?

20. Draw a free-body diagram for each of the following objects: (a) a projectile in motion in the presence of air resistance, (b) a rocket leaving the launch pad with its engines operating, (c) an athlete running along a horizontal track.

PROBLEMS

1, 2, 3 = straightforward, intermediate, challenging □ = full solution available in *Student Solutions Manual/Study Guide*
PhysicsNow™ = coached solution with hints available at **www.cp7e.com** = biomedical application

Section 4.1 Forces
Section 4.2 Newton's First Law
Section 4.3 Newton's Second Law
Section 4.4 Newton's Third Law

1. A 6.0-kg object undergoes an acceleration of 2.0 m/s^2. (a) What is the magnitude of the resultant force acting on it? (b) If this same force is applied to a 4.0-kg object, what acceleration is produced?

2. A football punter accelerates a football from rest to a speed of 10 m/s during the time in which his toe is in contact with the ball (about 0.20 s). If the football has a mass of 0.50 kg, what average force does the punter exert on the ball?

3. The heaviest invertebrate is the giant squid, which is estimated to have a weight of about 2 tons spread out over its length of 70 feet. What is its weight in newtons?

4. The heaviest flying bird is the trumpeter swan, which weighs in at about 38 pounds at its heaviest. What is its weight in newtons?

5. A bag of sugar weighs 5.00 lb on Earth. What would it weigh in newtons on the Moon, where the free-fall acceleration is one-sixth that on Earth? Repeat for Jupiter, where g is 2.64 times that on Earth. Find the mass of the bag of sugar in kilograms at each of the three locations.

6. A freight train has a mass of 1.5×10^7 kg. If the locomotive can exert a constant pull of 7.5×10^5 N, how long does it take to increase the speed of the train from rest to 80 km/h?

7. The air exerts a forward force of 10 N on the propeller of a 0.20-kg model airplane. If the plane accelerates forward at 2.0 m/s^2, what is the magnitude of the resistive force exerted by the air on the airplane?

8. A 5.0-g bullet leaves the muzzle of a rifle with a speed of 320 m/s. What force (assumed constant) is exerted on the bullet while it is traveling down the 0.82-m-long barrel of the rifle?

9. A Chinook salmon has a maximum underwater speed of 3.0 m/s, and can jump out of the water vertically with a speed of 6.0 m/s. A record salmon has a length of 1.5 m and a mass of 61 kg. When swimming upward at constant speed, and neglecting buoyancy, the fish experiences three forces: an upward force F exerted by the tail fin, the downward drag force of the water, and the downward force of gravity. As the fish leaves the surface of the water, however, it experiences a net upward force causing it to accelerate from 3.0 m/s to 6.0 m/s. Assuming that the drag force disappears as soon as the head of the fish breaks the surface and that F is exerted until 2/3 of the fish's length has left the water, determine the magnitude of F.

10. Consider a solid metal sphere (S) a few centimeters in diameter and a feather (F). For each quantity in the list that follows, indicate whether the quantity is the same, greater, or lesser in the case of S or in that of F? Explain in each case why you gave the answer you did. Here is the list: (a) the gravitational force; (b) the time it will take to fall a given distance in air; (c) the time it will take to fall a given distance in vacuum; (d) the total force on the object when falling in vacuum.

11. PhysicsNow™ A boat moves through the water with two forces acting on it. One is a 2 000-N forward push by the water on the propellor, and the other is a 1 800-N resistive force due to the water around the bow. (a) What is the acceleration of the 1000-kg boat? (b) If it starts from rest, how far will the boat move in 10.0 s? (c) What will its velocity be at the end of that time?

12. Two forces are applied to a car in an effort to move it, as shown in Figure P4.12. (a) What is the resultant of these two forces? (b) If the car has a mass of 3 000 kg, what acceleration does it have? Ignore friction.

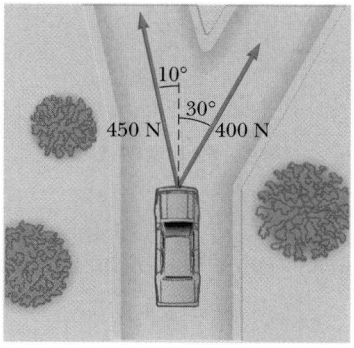

Figure P4.12

13. After falling from rest from a height of 30 m, a 0.50-kg ball rebounds upward, reaching a height of 20 m. If the contact between ball and ground lasted 2.0 ms, what average force was exerted on the ball?

14. The force exerted by the wind on the sails of a sailboat is 390 N north. The water exerts a force of 180 N east. If the boat (including its crew) has a mass of 270 kg, what are the magnitude and direction of its acceleration?

Section 4.5 Applications of Newton's Laws

15. Find the tension in each cable supporting the 600-N cat burglar in Figure P4.15.

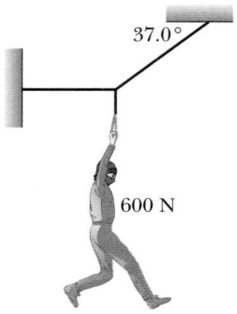

Figure P4.15

16. Find the tension in the two wires that support the 100-N light fixture in Figure P4.16.

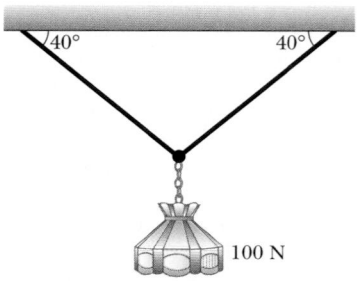

Figure P4.16

17. A 150-N bird feeder is supported by three cables as shown in Figure P4.17. Find the tension in each cable.

Figure P4.17

18. The leg and cast in Figure P4.18 weigh 220 N (w_1). Determine the weight w_2 and the angle α needed so that no force is exerted on the hip joint by the leg plus the cast.

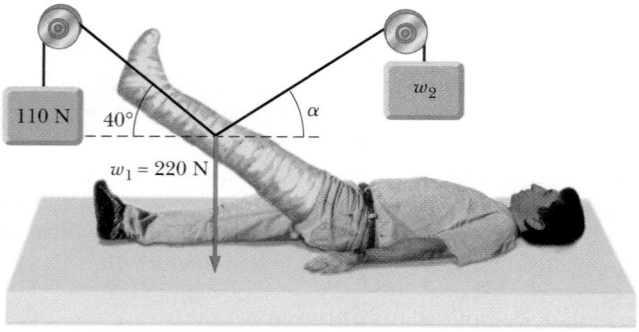

Figure P4.18

19. Two blocks are fastened to the ceiling of an elevator as in Figure P4.19. The elevator accelerates upward at 2.00 m/s^2. Find the tension in each rope.

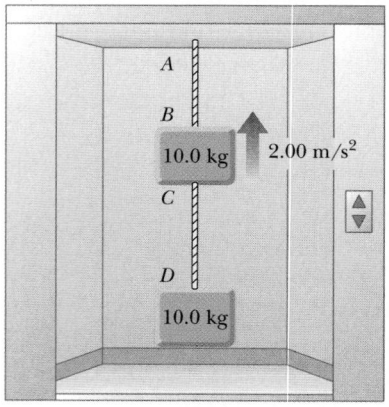

Figure P4.19

20. Two people are pulling a boat through the water as in Figure P4.20. Each exerts a force of 600 N directed at a 30.0° angle relative to the forward motion of the boat. If the boat moves with constant velocity, find the resistive force $\vec{\mathbf{F}}$ exerted by the water on the boat.

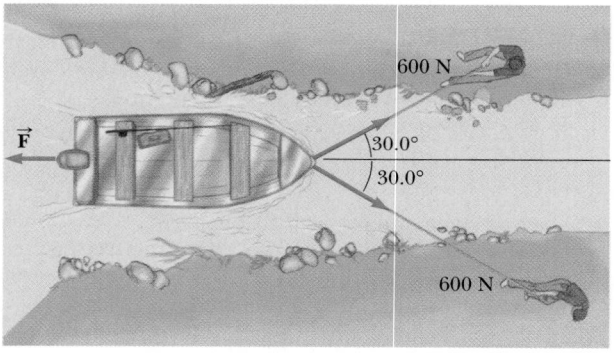

Figure P4.20

21. The distance between two telephone poles is 50.0 m. When a 1.00-kg bird lands on the telephone wire midway between the poles, the wire sags 0.200 m. Draw a free-body diagram of the bird. How much tension does the bird produce in the wire? Ignore the weight of the wire.

22. You are a judge in a children's kite-flying contest, and two children will win prizes for the kites that pull most strongly and least strongly, respectively, on their strings. To measure string tensions, you borrow a weight hanger, some slotted weights, and a protractor from your physics teacher, and use the following protocol, illustrated in Figure P4.22: Wait for a child to get her kite well controlled, hook the hanger onto the kite string about 30 cm from her hand, pile on weight until that section of string is horizontal, record the mass required, and record the angle between the horizontal and the string running up to the kite. (a) Explain how this method works. As you construct your explanation, imagine that the children's parents ask you about the method, that they might make false assumptions about your ability without concrete evidence, and that your explanation is an opportunity to give them confidence in your evaluation technique. (b) Find the tension in the string if the mass is 132 g and the angle is 46.3°.

Figure P4.22

23. Physics Now™ A 5.0-kg bucket of water is raised from a well by a rope. If the upward acceleration of the bucket is 3.0 m/s^2, find the force exerted by the rope on the bucket.

24. A shopper in a supermarket pushes a loaded cart with a horizontal force of 10 N. The cart has a mass of 30 kg. (a) How far will it move in 3.0 s, starting from rest? (Ignore friction.) (b) How far will it move in 3.0 s if the shopper places his 30-N child in the cart before he begins to push it?

25. A 2 000-kg car is slowed down uniformly from 20.0 m/s to 5.00 m/s in 4.00 s. (a) What average force acted on the car during that time, and (b) how far did the car travel during that time?

26. Two packing crates of masses 10.0 kg and 5.00 kg are connected by a light string that passes over a frictionless pulley as in Figure P4.26. The 5.00-kg crate lies on a smooth incline of angle 40.0°. Find the acceleration of the 5.00-kg crate and the tension in the string.

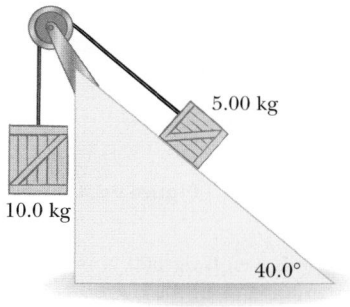

Figure P4.26

27. Assume that the three blocks portrayed in Figure P4.27 move on a frictionless surface and that a 42-N force acts as shown on the 3.0-kg block. Determine (a) the acceleration given this system, (b) the tension in the cord connecting the 3.0-kg and the 1.0-kg blocks, and (c) the force exerted by the 1.0-kg block on the 2.0-kg block.

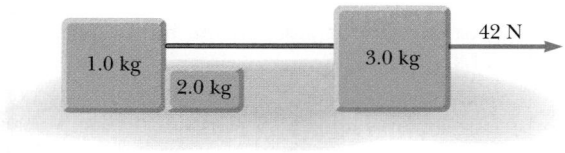

Figure P4.27

28. An object of mass 2.0 kg starts from rest and slides down an inclined plane 80 cm long in 0.50 s. What *net force* is acting on the object along the incline?

29. A 40.0-kg wagon is towed up a hill inclined at 18.5° with respect to the horizontal. The tow rope is parallel to the incline and has a tension of 140 N. Assume that the wagon starts from rest at the bottom of the hill, and neglect friction. How fast is the wagon going after moving 80.0 m up the hill?

30. An object with mass m_1 = 5.00 kg rests on a frictionless horizontal table and is connected to a cable that passes over a pulley and is then fastened to a hanging object with mass m_2 = 10.0 kg, as shown in Figure P4.30. Find the acceleration of each object and the tension in the cable.

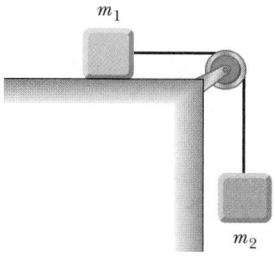

Figure P4.30 (Problems 30, 36, and 45)

31. A train has a mass of 5.22×10^6 kg and is moving at 90.0 km/h. The engineer applies the brakes, resulting in a net backward force of 1.87×10^6 N on the train. The brakes are held on for 30.0 s. (a) What is the final speed of the train? (b) How far does the train travel during this period?

32. (a) An elevator of mass m moving upward has two forces acting on it: the upward force of tension in the cable and the downward force due to gravity. When the elevator is accelerating upward, which is greater, T or w? (b) When the elevator is moving at a constant velocity upward, which is greater, T or w? (c) When the elevator is moving upward, but the acceleration is downward, which is greater, T or w? (d) Let the elevator have a mass of 1 500 kg and an upward acceleration of 2.5 m/s^2. Find T. Is your answer consistent with the answer to part (a)? (e) The elevator of part (d) now moves with a constant upward velocity of 10 m/s. Find T. Is your answer consistent with your answer to part (b)? (f) Having initially moved upward with a constant velocity, the elevator begins to accelerate downward at 1.50 m/s^2. Find T. Is your answer consistent with your answer to part (c)?

33. A 1 000-kg car is pulling a 300-kg trailer. Together, the car and trailer have an acceleration of 2.15 m/s^2 in the forward direction. Neglecting frictional forces on the trailer, determine (a) the net force on the car, (b) the net force on the trailer, (c) the force exerted by the trailer on the car, and (d) the resultant force exerted by the car on the road.

34. Two objects with masses of 3.00 kg and 5.00 kg are connected by a light string that passes over a frictionless pulley, as in Figure P4.34. Determine (a) the tension in the string, (b) the acceleration of each object, and (c) the distance each object will move in the first second of motion if both objects start from rest.

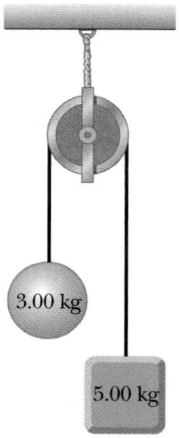

Figure P4.34

Section 4.6 Forces of Friction

35. A dockworker loading crates on a ship finds that a 20-kg crate, initially at rest on a horizontal surface, requires a 75-N horizontal force to set it in motion. However, after the crate is in motion, a horizontal force of 60 N is required to keep it moving with a constant speed. Find the coefficients of static and kinetic friction between crate and floor.

36. In Figure P4.30, $m_1 = 10$ kg and $m_2 = 4.0$ kg. The coefficient of static friction between m_1 and the horizontal surface is 0.50, and the coefficient of kinetic friction is 0.30. (a) If the system is released from rest, what will its acceleration be? (b) If the system is set in motion with m_2 moving downward, what will be the acceleration of the system?

37. A 1 000-N crate is being pushed across a level floor at a constant speed by a force $\vec{\mathbf{F}}$ of 300 N at an angle of 20.0° below the horizontal, as shown in Figure P4.37a. (a) What is the coefficient of kinetic friction between the crate and the floor? (b) If the 300-N force is instead pulling the block at an angle of 20.0° above the horizontal, as shown in Figure P4.37b, what will be the acceleration of the crate? Assume that the coefficient of friction is the same as that found in (a).

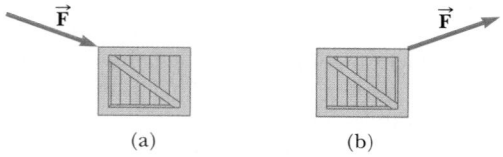

Figure P4.37

38. A hockey puck is hit on a frozen lake and starts moving with a speed of 12.0 m/s. Five seconds later, its speed is 6.00 m/s. (a) What is its average acceleration? (b) What is the average value of the coefficient of kinetic friction between puck and ice? (c) How far does the puck travel during the 5.00-s interval?

39. Consider a large truck carrying a heavy load, such as steel beams. A significant hazard for the driver is that the load may slide forward, crushing the cab, if the truck stops suddenly in an accident or even in braking. Assume, for example, that a 10 000-kg load sits on the flat bed of a 20 000-kg truck moving at 12.0 m/s. Assume the load is not tied down to the truck and has a coefficient of static friction of 0.500 with the truck bed. (a) Calculate the minimum stopping distance for which the load will not slide forward relative to the truck. (b) Is any piece of data unnecessary for the solution?

40. A woman at an airport is towing her 20.0-kg suitcase at constant speed by pulling on a strap at an angle θ above the horizontal (Fig. P4.40). She pulls on the strap with a 35.0-N force, and the friction force on the suitcase is 20.0 N. Draw a free-body diagram of the suitcase. (a) What angle does the strap make with the horizontal? (b) What normal force does the ground exert on the suitcase?

Figure P4.40

41. The coefficient of static friction between the 3.00-kg crate and the 35.0° incline of Figure P4.41 is 0.300. What minimum force $\vec{\mathbf{F}}$ must be applied to the crate perpendicular to the incline to prevent the crate from sliding down the incline?

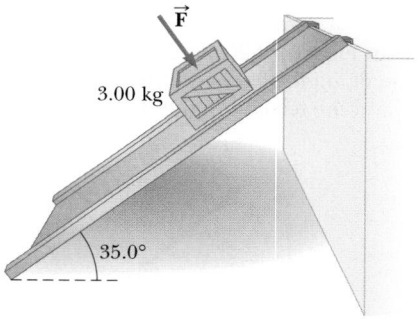

Figure P4.41

42. A box of books weighing 300 N is shoved across the floor of an apartment by a force of 400 N exerted downward at an angle of 35.2° below the horizontal. If the coefficient of kinetic friction between box and floor is 0.570, how long does it take to move the box 4.00 m, starting from rest?

43. An object falling under the pull of gravity is acted upon by a frictional force of air resistance. The magnitude of this force is approximately proportional to the speed of the object, which can be written as $f = bv$. Assume that $b = 15$ kg/s and $m = 50$ kg. (a) What is the terminal speed the object reaches while falling? (b) Does your answer to part (a) depend on the initial speed of the object? Explain.

44. A student decides to move a box of books into her dormitory room by pulling on a rope attached to the box. She pulls with a force of 80.0 N at an angle of 25.0° above the horizontal. The box has a mass of 25.0 kg, and the coefficient of kinetic friction between box and floor is 0.300. (a) Find the acceleration of the box. (b) The student now starts moving the box up a 10.0° incline, keeping her 80.0 N force directed at 25.0° above the line of the incline. If the coefficient of friction is unchanged, what is the new acceleration of the box?

45. **Physics Now™** Objects with masses $m_1 = 10.0$ kg and $m_2 = 5.00$ kg are connected by a light string that passes over a frictionless pulley as in Figure P4.30. If, when the system starts from rest, m_2 falls 1.00 m in 1.20 s, determine the coefficient of kinetic friction between m_1 and the table.

46. A car is traveling at 50.0 km/h on a flat highway. (a) If the coefficient of friction between road and tires on a rainy day is 0.100, what is the minimum distance in which the car will stop? (b) What is the stopping distance when the surface is dry and the coefficient of friction is 0.600?

47. A 3.00-kg block starts from rest at the top of a 30.0° incline and slides 2.00 m down the incline in 1.50 s. Find (a) the acceleration of the block, (b) the coefficient of kinetic friction between the block and the incline, (c) the frictional force acting on the block, and (d) the speed of the block after it has slid 2.00 m.

48. Objects of masses $m_1 = 4.00$ kg and $m_2 = 9.00$ kg are connected by a light string that passes over a frictionless pulley as in Figure P4.48. The object m_1 is held at rest on the floor, and m_2 rests on a fixed incline of $\theta = 40.0°$. The objects are released from rest, and m_2 slides 1.00 m down the incline in 4.00 s. Determine (a) the acceleration of each object, (b) the tension in the string, and (c) the coefficient of kinetic friction between m_2 and the incline.

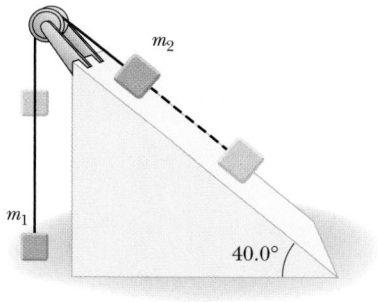

Figure P4.48

49. Find the acceleration reached by each of the two objects shown in Figure P4.49 if the coefficient of kinetic friction between the 7.00-kg object and the plane is 0.250.

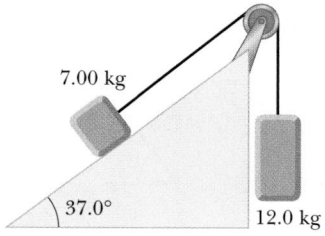

Figure P4.49

50. A 2.00-kg block is held in equilibrium on an incline of angle $\theta = 60.0°$ by a horizontal force $\vec{\mathbf{F}}$ applied in the direction shown in Figure P4.50. If the coefficient of static friction between block and incline is $\mu_s = 0.300$, determine (a) the minimum value of $\vec{\mathbf{F}}$ and (b) the normal force exerted by the incline on the block.

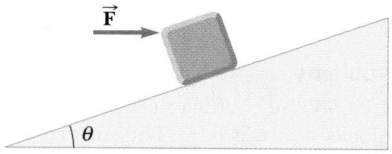

Figure P4.50

51. The person in Figure P4.51 weighs 170 lb. Each crutch makes an angle of 22.0° with the vertical (as seen from the front). Half of the person's weight is supported by the crutches, the other half by the vertical forces exerted by the ground on his feet. Assuming that he is at rest and that the force exerted by the ground on the crutches acts along the crutches, determine (a) the smallest possible coefficient of friction between crutches and ground and (b) the magnitude of the compression force supported by each crutch.

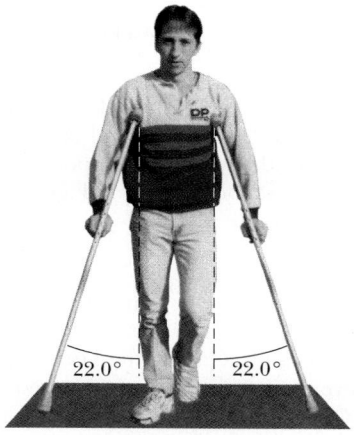

Figure P4.51

52. A block of mass $m = 2.00$ kg rests on the left edge of a block of length $L = 3.00$ m and mass $M = 8.00$ kg. The coefficient of kinetic friction between the two blocks is $\mu_k = 0.300$, and the surface on which the 8.00-kg block rests is frictionless. A constant horizontal force of magnitude $F = 10.0$ N is applied to the 2.00-kg block, setting it

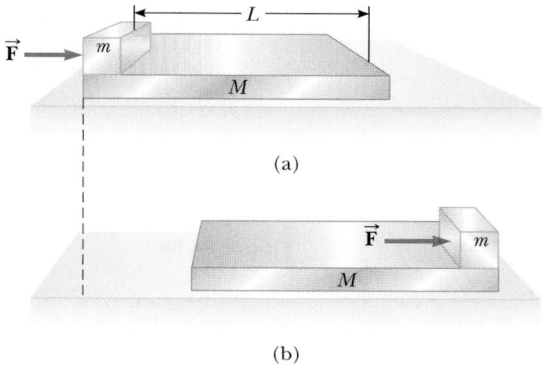

Figure P4.52

in motion as shown in Figure P4.52a. (a) How long will it take before this block makes it to the right side of the 8.00-kg block, as shown in Figure P4.52b? (*Note*: Both blocks are set in motion when the force $\vec{\mathbf{F}}$ is applied.) (b) How far does the 8.00-kg block move in the process?

Additional Problems

53. In Figure P4.53, the coefficient of kinetic friction between the two blocks shown is 0.30. The surface of the table and the pulleys are frictionless. (a) Draw a free-body diagram for each block. (b) Determine the acceleration of each block. (c) Find the tension in the strings.

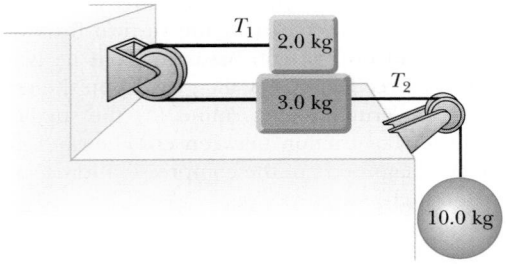

Figure P4.53

54. The force exerted by the wind on a sailboat is approximately perpendicular to the sail and proportional to the component of the wind velocity perpendicular to the sail. For the 800 kg sailboat shown in Figure P4.54, the proportionality constant is

$$F_{\text{sail}} = \left(550 \frac{\text{N}}{\text{m/s}}\right) v_{\text{wind}\perp}$$

Water exerts a force along the keel (bottom) of the boat that prevents it from moving sideways, as shown in the figure. Once the boat starts moving forward, water also exerts a drag force backwards on the boat, opposing the forward motion. If a 17-knot wind (1 knot = 0.514 m/s) is blowing to the east, what is the initial acceleration of the sailboat?

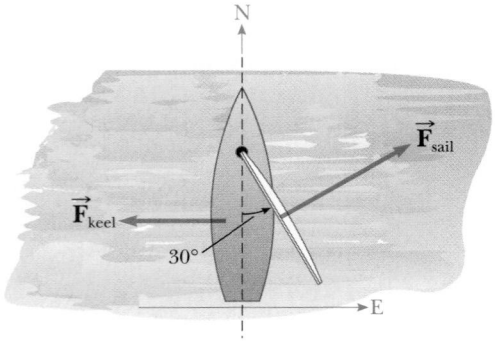

Figure P4.54

55. (a) What is the resultant force exerted by the two cables supporting the traffic light in Figure P4.55? (b) What is the weight of the light?

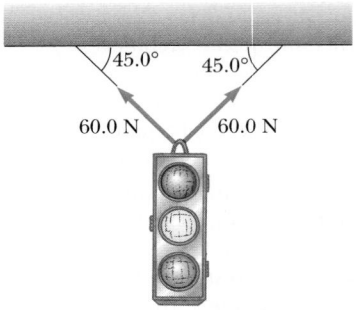

Figure P4.55

56. As a protest against the umpire's calls, a baseball pitcher throws a ball straight up into the air at a speed of 20.0 m/s. In the process, he moves his hand through a distance of 1.50 m. If the ball has a mass of 0.150 kg, find the force he exerts on the ball to give it this upward speed.

57. A boy coasts down a hill on a sled, reaching a level surface at the bottom with a speed of 7.0 m/s. If the coefficient of friction between the sled's runners and the snow is 0.050 and the boy and sled together weigh 600 N, how far does the sled travel on the level surface before coming to rest?

58. (a) What is the minimum force of friction required to hold the system of Figure P4.58 in equilibrium? (b) What coefficient of static friction between the 100-N block and the table ensures equilibrium? (c) If the coefficient of kinetic friction between the 100-N block and the table is 0.250, what hanging weight should replace the 50.0-N weight to allow the system to move at a constant speed once it is set in motion?

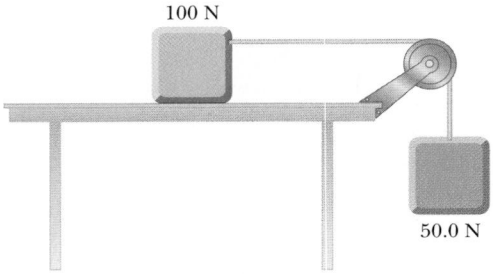

Figure P4.58

59. A box rests on the back of a truck. The coefficient of static friction between the box and the bed of the truck is 0.300. (a) When the truck accelerates forward, what force accelerates the box? (b) Find the maximum acceleration the truck can have before the box slides.

60. A 4.00-kg block is pushed along the ceiling with a constant applied force of 85.0 N that acts at an angle of 55.0°

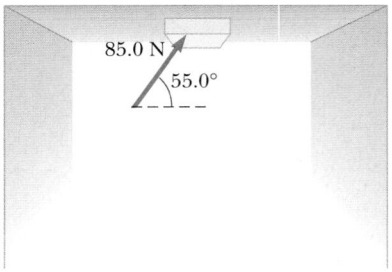

Figure P4.60

with the horizontal, as in Figure P4.60. The block accelerates to the right at 6.00 m/s^2. Determine the coefficient of kinetic friction between block and ceiling.

61. A frictionless plane is 10.0 m long and inclined at 35.0°. A sled starts at the bottom with an initial speed of 5.00 m/s up the incline. When the sled reaches the point at which it momentarily stops, a second sled is released from the top of the incline with an initial speed v_i. Both sleds reach the bottom of the incline at the same moment. (a) Determine the distance that the first sled traveled up the incline. (b) Determine the initial speed of the second sled.

62. Three objects are connected by light strings as shown in Figure P4.62. The string connecting the 4.00-kg object and the 5.00-kg object passes over a light frictionless pulley. Determine (a) the acceleration of each object and (b) the tension in the two strings.

Figure P4.62

63. A 2.00-kg aluminum block and a 6.00-kg copper block are connected by a light string over a frictionless pulley. The two blocks are allowed to move on a fixed steel block wedge (of angle $\theta = 30.0°$) as shown in Figure P4.63. Making use of Table 4.2, determine (a) the acceleration of the two blocks and (b) the tension in the string.

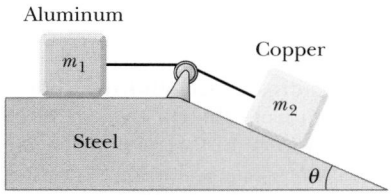

Figure P4.63

64. A 5.0-kg penguin sits on a 10-kg sled, as shown in Figure P4.64. A horizontal force of 45 N is applied to the sled, but the penguin attempts to impede the motion by holding onto a cord attached to a wall. The coefficient of kinetic friction between the sled and the snow, as well as that between the sled and the penguin, is 0.20. (a) Draw a free-body diagram for the penguin and one for the sled, and identify the reaction force for each force you include. Determine (b) the tension in the cord and (c) the acceleration of the sled.

Figure P4.64

65. Two boxes of fruit on a frictionless horizontal surface are connected by a light string as in Figure P4.65, where m_1 = 10 kg and m_2 = 20 kg. A force of 50 N is applied to the 20-kg box. (a) Determine the acceleration of each box and the tension in the string. (b) Repeat the problem for the case where the coefficient of kinetic friction between each box and the surface is 0.10.

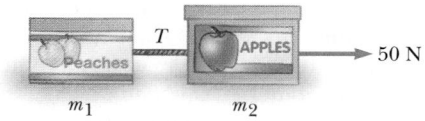

Figure P4.65

66. A high diver of mass 70.0 kg jumps off a board 10.0 m above the water. If her downward motion is stopped 2.00 s after she enters the water, what average upward force did the water exert on her?

67. Two people pull as hard as they can on ropes attached to a 200-kg boat. If they pull in the same direction, the boat has an acceleration of 1.52 m/s^2 to the right. If they pull in opposite directions, the boat has an acceleration of 0.518 m/s^2 to the left. What is the force exerted by each person on the boat? (Disregard any other forces on the boat.)

68. A 3.0-kg object hangs at one end of a rope that is attached to a support on a railroad car. When the car accelerates to the right, the rope makes an angle of 4.0° with the vertical, as shown in Figure P4.68. Find the acceleration of the car.

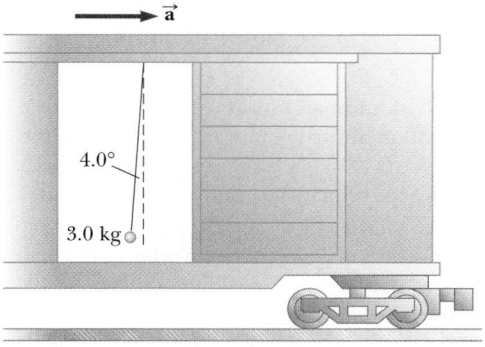

Figure P4.68

69. Three blocks of masses 10.0 kg, 5.00 kg, and 3.00 kg are connected by light strings that pass over frictionless pulleys as shown in Figure P4.69. The acceleration of the 5.00-kg block is 2.00 m/s^2 to the left, and the surfaces are

rough. Find (a) the tension in each string and (b) the coefficient of kinetic friction between blocks and surfaces. (Assume the same μ_k for both blocks that are in contact with surfaces.)

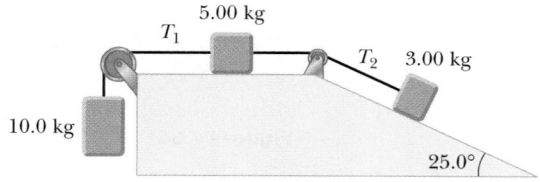

Figure P4.69

70. An inquisitive physics student, wishing to combine pleasure with scientific inquiry, rides on a rollercoaster sitting on a bathroom scale. (Do not try this yourself on a roller coaster that forbids loose heavy packages.) The bottom of the seat in the rollercoaster car is in a plane parallel to the track. The seat has a perpendicular back and a seat belt that fits around the student's chest in a plane parallel to the bottom of the seat. The student lifts his feet from the floor, so that the scale reads his weight, 200 lb, when the car is horizontal. At one point during the ride, the car zooms with negligible friction down a straight slope inclined at 30.0° below the horizontal. What does the scale read at that point?

71. A van accelerates down a hill (Fig. P4.71), going from rest to 30.0 m/s in 6.00 s. During the acceleration, a toy (m = 0.100 kg) hangs by a string from the van's ceiling. The acceleration is such that the string remains perpendicular to the ceiling. Determine (a) the angle θ and (b) the tension in the string.

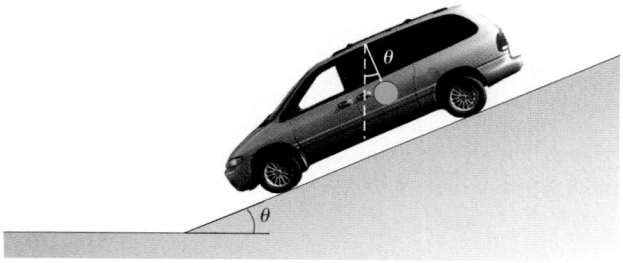

Figure P4.71

72. An 80-kg stuntman jumps from a window of a building situated 30 m above a catching net. Assuming that air resistance exerts a 100-N force on the stuntman as he falls, determine his velocity just before he hits the net.

73. The parachute on a race car of weight 8 820 N opens at the end of a quarter-mile run when the car is traveling at 35 m/s. What total retarding force must be supplied by the parachute to stop the car in a distance of 1 000 m?

74. **Physics Now™** On an airplane's takeoff, the combined action of the air around the engines and wings of an airplane exerts an 8 000-N force on the plane, directed upward at an angle of 65.0° above the horizontal. The plane rises with constant velocity in the vertical direction while continuing to accelerate in the horizontal direction. (a) What is the weight of the plane? (b) What is its horizontal acceleration?

75. A 72-kg man stands on a spring scale in an elevator. Starting from rest, the elevator ascends, attaining its maximum speed of 1.2 m/s in 0.80 s. The elevator travels with this constant speed for 5.0 s, undergoes a uniform *negative* acceleration for 1.5 s, and then comes to rest. What does the spring scale register (a) before the elevator starts to move? (b) during the first 0.80 s of the elevator's ascent? (c) while the elevator is traveling at constant speed? (d) during the elevator's negative acceleration?

76. A sled weighing 60.0 N is pulled horizontally across snow so that the coefficient of kinetic friction between sled and snow is 0.100. A penguin weighing 70.0 N rides on the sled, as in Figure P4.76. If the coefficient of static friction between penguin and sled is 0.700, find the maximum horizontal force that can be exerted on the sled before the penguin begins to slide off.

Figure P4.76

77. The board sandwiched between two other boards in Figure P4.77 weighs 95.5 N. If the coefficient of friction between the boards is 0.663, what must be the magnitude of the compression forces (assumed to be horizontal) acting on both sides of the center board to keep it from slipping?

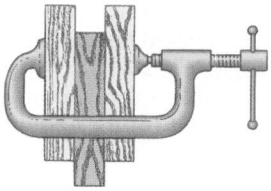

Figure P4.77

78. A magician pulls a tablecloth from under a 200-g mug located 30.0 cm from the edge of the cloth. The cloth exerts a friction force of 0.100 N on the mug and is pulled with a constant acceleration of 3.00 m/s^2. How far does the mug move relative to the horizontal tabletop before the cloth is completely out from under it? Note that the cloth must move more than 30 cm relative to the tabletop during the process.

79. An inventive child wants to reach an apple in a tree without climbing the tree. Sitting in a chair connected to a rope that passes over a frictionless pulley (Fig. P4.79), the child pulls on the loose end of the rope with such a force that the spring scale reads 250 N. The child's true weight is 320 N, and the chair weighs 160 N. (a) Show that the acceleration of the system is *upward* and find its magnitude. (b) Find the force the child exerts on the chair.

Figure P4.79

80. A fire helicopter carries a 620-kg bucket of water at the end of a 20.0-m-long cable. Flying back from a fire at a constant speed of 40.0 m/s, the cable makes an angle of 40.0° with respect to the vertical. Determine the force exerted by air resistance on the bucket.

81. A bag of cement hangs from three wires as shown in Figure P4.81. Two of the wires make angles θ_1 and θ_2, respectively, with the horizontal. (a) Show that, if the system is in equilibrium, then

$$T_1 = \frac{w \cos \theta_2}{\sin(\theta_1 + \theta_2)}$$

(b) Given that $w = 325$ N, $\theta_1 = 10.0°$, and $\theta_2 = 25.0°$, find the tensions T_1, T_2, and T_3 in the wires.

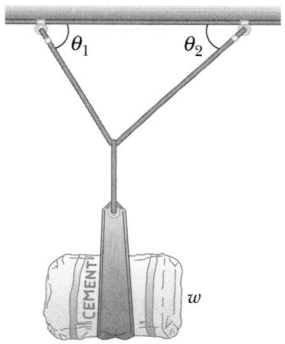

Figure P4.81

Activities

A.1. There is a simple method for measuring the coefficients of static and kinetic friction between an object and some surface. For this investigation, you will need a few coins, your textbook or some other flat surface that can be inclined, a protractor, and some double-stick tape. Place a coin at one edge of the book as it lies on a table, and lift that edge of the book until the coin just slips down the incline, as shown in Figure A4.1. When the coin starts to slip, measure the angle of incline with your protractor. Repeat the measurement five times, and find the average value of this critical angle θ_c. The coefficient of static friction between the coin and book's surface is $\mu_s = \tan \theta_c$. (You should prove this as an exercise.) Calculate the average value of μ_s. To measure the coefficient of kinetic friction, find the angle θ_c' at which the coin moves down the incline with constant speed. This angle should be less than θ_c. Measure this new angle five times, and get its average value. Calculate the average value of μ_k, using the fact that $\mu_k = \tan \theta_c'$, where $\theta_c' < \theta_c$. Repeat these measurements, using two or three stacked coins with double-stick tape between them. You should get the same results as with one coin. Why?

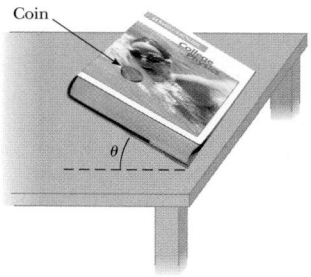

Figure A4.1

A.2. Borrow a spring scale from your instructor and use it to study some of the properties of the force of friction. (1) Attach the scale to a block of wood resting on the surface of a table, and note the force required to start the block moving. You should each take at least five trials and average your results. This measured force is the maximum value of the force of static friction between the block and surface. (2) Now use the spring scale to measure the force required to keep the block moving at constant velocity. Again, perform several trials to find the average value for this force. The force you find is the force of kinetic friction. (3) Turn the block so that a side with a different surface area is in contact with the table. Repeat the preceding experiments to see if the area of contact between the surfaces produces different values for the forces of friction.

A.3. Get a bathroom scale and stand on it while riding on an elevator. Watch carefully what happens to your apparent weight (the reading on the scale) as the elevator moves upward or downward as a function of time. What do the readings on the scale tell you about the acceleration of the elevator during the ride?

An asteroid plunges through Earth's atmosphere while pterodactyls watch. This artist's conception is of a catastrophic event thought to have led to the extinction of dinosaurs. During an impact, an asteroid only a kilometer across releases its awesome energy of motion as heat and light, delivering the explosive equivalent of one hundred million atomic bombs.

NASA

CHAPTER

5

OUTLINE

5.1 Work

5.2 Kinetic Energy and the Work–Energy Theorem

5.3 Gravitational Potential Energy

5.4 Spring Potential Energy

5.5 Systems and Energy Conservation

5.6 Power

5.7 Work Done by a Varying Force

Energy

Energy is one of the most important concepts in the world of science. In everyday use, energy is associated with the fuel needed for transportation and heating, with electricity for lights and appliances, and with the foods we consume. These associations, however, don't tell us what energy *is*, only what it *does*, and that producing it requires fuel. Our goal in this chapter, therefore, is to develop a better understanding of energy and how to quantify it.

Energy is present in the Universe in a variety of forms, including mechanical, chemical, electromagnetic, and nuclear energy. Even the inert mass of everyday matter contains a very large amount of energy. Although energy can be transformed from one kind to another, all observations and experiments to date suggest that the total amount of energy in the Universe never changes. This is also true for an isolated system—a collection of objects that can exchange energy with each other, but not with the rest of the Universe. If one form of energy in an isolated system decreases, then another form of energy in the system must increase. For example, if the system consists of a motor connected to a battery, the battery converts chemical energy to electrical energy, and the motor converts electrical energy to mechanical energy. Understanding how energy changes from one form to another is essential in all the sciences.

In this chapter the focus is mainly on *mechanical energy*, which is the sum of *kinetic energy*—the energy associated with motion—and *potential energy*—the energy associated with position. Using an energy approach to solve certain problems is often much easier than using forces and Newton's three laws. These two very different approaches are linked through the concept of *work*.

5.1 WORK

Work has a different meaning in physics than it does in everyday usage. In the physics definition, a programmer does very little work typing away at a computer. A mason, by contrast, may do a lot of work laying concrete blocks. In physics, work is done only if an object is moved through some displacement while a force is applied to it. If either the force or displacement is doubled, the work is doubled. Double them both, and the work is quadrupled. Doing work involves applying a force to an object while moving it a given distance.

Figure 5.1 shows a block undergoing a displacement $\Delta\vec{\mathbf{x}}$ along a straight line while acted on by a constant force $\vec{\mathbf{F}}$ in the same direction. We have the following definition:

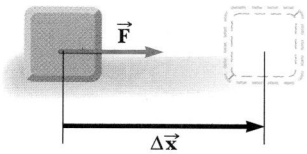

Figure 5.1 A constant force $\vec{\mathbf{F}}$ in the same direction as the displacement, $\Delta\vec{\mathbf{x}}$, does work $F\Delta x$.

◀ Work by a constant force along the displacement

> The work W done on an object by a constant force $\vec{\mathbf{F}}$ is given by
>
> $$W = F\Delta x \qquad [5.1]$$
>
> where F is the magnitude of the force, Δx is the magnitude of the displacement, and $\vec{\mathbf{F}}$ and $\Delta\vec{\mathbf{x}}$ point in the same direction.
>
> **SI unit: joule (J) = newton · meter = kg · m²/s²**

It's easy to see the difference between the physics definition and the everyday definition of work. The programmer exerts very little force on the keys of a keyboard, creating only small displacements, so relatively little physics work is done. The mason must exert much larger forces on the concrete blocks and move them significant distances, and so performs a much greater amount of work. Even very tiring tasks, however, may not constitute work according to the physics definition. A truck driver, for example, may drive for several hours, but if he doesn't exert a force, then $F = 0$ in Equation 5.1 and he doesn't do any work. Similarly, a student pressing against a wall for hours in an isometric exercise also does no work, because the displacement in Equation 5.1, Δx, is zero.[1] Atlas, of Greek mythology, bore the world on his shoulders, but that, too, wouldn't qualify as work in the physics definition.

TIP 5.1 Work is a Scalar Quantity

Work is a simple number—a scalar, not a vector—so there is no direction associated with it. Energy and energy transfer are also scalars.

Work is a scalar quantity—a number rather than a vector—and consequently is easier to handle. No direction is associated with it. Further, work doesn't depend explicitly on time, which can be an advantage in problems involving only velocities and positions. Since the units of work are those of force and distance, the SI unit is the **newton-meter** (N · m). Another name for the newton-meter is the **joule (J)** (rhymes with "pool"). The U.S. customary unit of work is the **foot-pound**, because distances are measured in feet and forces in pounds in that system.

Complications in the definition of work occur when the force exerted on an object is not in the same direction as the displacement (Figure 5.2.) The force, however, can always be split into two components—one parallel and the other perpendicular to the direction of displacement. Only the component parallel to the direction of displacement does work on the object. This fact can be expressed in the following more general definition:

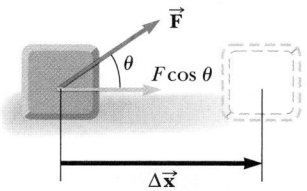

Figure 5.2 A constant force $\vec{\mathbf{F}}$ exerted at an angle θ with respect to the displacement, $\Delta\vec{\mathbf{x}}$, does work $(F\cos\theta)\Delta x$.

◀ Work by a constant force at an angle to the displacement

> The work W done on an object by a constant force $\vec{\mathbf{F}}$ is given by
>
> $$W \equiv (F\cos\theta)\Delta x \qquad [5.2]$$
>
> where F is the magnitude of the force, Δx is the magnitude of the object's displacement, and θ is the angle between the directions of $\vec{\mathbf{F}}$ and $\Delta\vec{\mathbf{x}}$.
>
> **SI unit: joule (J)**

In Figure 5.3, a man carries a bucket of water horizontally at constant velocity. The upward force exerted by the man's hand on the bucket is perpendicular to

[1]Actually, you do expend energy while doing isometric exercises, because your muscles are continuously contracting and relaxing in the process. This internal muscular movement qualifies as work according to the physics definition.

Figure 5.3 No work is done on a bucket when it is moved horizontally because the applied force $\vec{F}$ is perpendicular to the displacement.

TIP 5.2 Work is Done *by Something, on Something Else*

Work doesn't happen by itself. Work is done *by* something in the environment, *on* the object of interest.

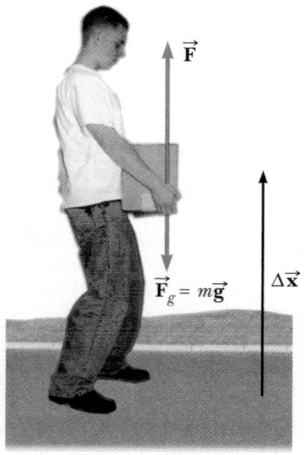

Figure 5.4 The student does positive work when he lifts the box, because the applied force $\vec{F}$ is in the same direction as the displacement. When he lowers the box to the floor, he does negative work.

the direction of motion, so it does no work on the bucket. This can also be seen from Equation 5.2, because the angle between the force exerted by the hand and the direction of motion is 90°, giving cos 90° = 0 and $W = 0$. Similarly, the force of gravity does no work on the bucket.

Work always requires a system of more than just one object. A nail, for example, can't do work on itself, but a hammer can do work on the nail by driving it into a board. In general, an object may be moving under the influence of several external forces. In that case, the total work done on the object as it undergoes some displacement is just the sum of the amount of work done by each force.

Work can be either positive or negative. In the definition of work in Equation 5.2, F and Δx are magnitudes, which are never negative. Work is therefore positive or negative depending on whether cos θ is positive or negative. This, in turn, depends on the direction of $\vec{F}$ relative the direction of $\Delta\vec{x}$. When these vectors are pointing in the same direction the angle between them is 0°, so cos 0° = +1 and the work is positive. For example, when a student lifts a box as in Figure 5.4, the work he does on the box is positive because the force he exerts on the box is upward, in the same direction as the displacement. In lowering the box slowly back down, however, the student still exerts an upward force on the box, but the motion of the box is downwards. Since the vectors $\vec{F}$ and $\Delta\vec{x}$ are now in opposite directions, the angle between them is 180°, and cos 180° = −1 and the work done by the student is negative. In general, when the part of $\vec{F}$ parallel to $\Delta\vec{x}$ points in the same direction as $\Delta\vec{x}$, the work is positive; and is otherwise negative.

Because Equations 5.1 and 5.2 assume a force constant in both direction and size, they are only special cases of a more general definition of work—that done by a varying force—treated briefly in Section 5.7.

Quick Quiz 5.1

In Active Figure 5.5 (a)–(d), a block moves to the right in the positive x-direction through the displacement Δx while under the influence of a force with the same magnitude F. Which of the following is the correct order of the amount of work done by the force F, from most positive to most negative? (A) d, c, a, b (B) c, a, b, d (C) c, a, d, b

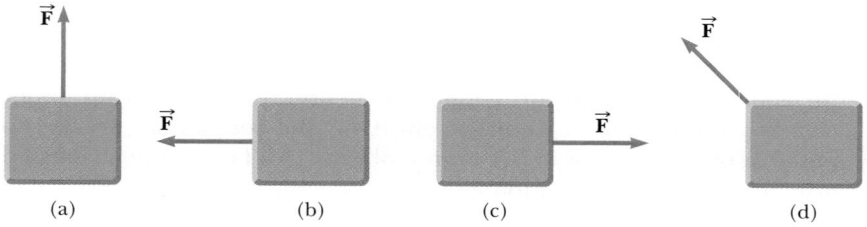

ACTIVE FIGURE 5.5
(Quick Quiz 5.1) A force $\vec{F}$ is exerted on an object that undergoes a displacement to the right. Both the magnitude of the force and the displacement are the same in all four cases.

PhysicsNow™

Log into PhysicsNow at **www.cp7e.com** and go to Active Figure 5.5, where you can move a block to see how the work done against gravity depends on position.

EXAMPLE 5.1 Sledding through the Yukon

Goal Apply the basic definitions of work done by a constant force.

Problem An Eskimo returning from a successful fishing trip pulls a sled loaded with salmon. The total mass of the sled and salmon is 50.0 kg, and the Eskimo exerts a force of 1.20×10^2 N on the sled by pulling on the rope.

(a) How much work does he do on the sled if the rope is horizontal to the ground ($\theta = 0°$ in Figure 5.6) and he pulls the sled 5.00 m? **(b)** How much work does he do on the sled if $\theta = 30.0°$ and he pulls the sled the same distance? (Treat the sled as a point particle, so details such as the point of attachment of the rope make no difference.)

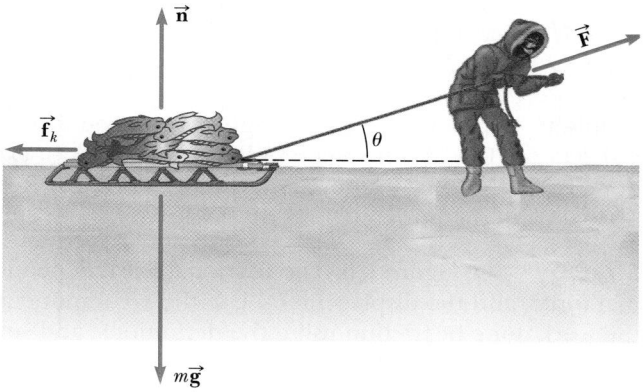

Figure 5.6 (Examples 5.1 and 5.2) An Eskimo pulling a sled with a rope at an angle θ to the horizontal.

Strategy Substitute the given values of F and Δx into the basic equations for work, Equations 5.1 and 5.2.

Solution

(a) Find the work done when the force is horizontal.

Use Equation 5.1, substituting the given values:

$$W = F\Delta x = (1.20 \times 10^2\ \text{N})(5.00\ \text{m}) = \boxed{6.00 \times 10^2\ \text{J}}$$

(b) Find the work done when the force is exerted at a 30° angle.

Use Equation 5.2, again substituting the given values:

$$W = (F\cos\theta)\Delta x = (1.20 \times 10^2\ \text{N})(\cos 30°)(5.00\ \text{m})$$
$$= \boxed{5.20 \times 10^2\ \text{J}}$$

Remarks The normal force $\vec{n}$, the gravitational force $m\vec{g}$, and the upward component of the applied force do *no* work on the sled, because they're perpendicular to the displacement. The mass of the sled didn't come into play here, but is important when the effects of friction must be calculated, and in the next section, where we introduce the work–energy theorem.

Exercise 5.1

Suppose the Eskimo is pushing the same 50.0-kg sled across level terrain with a force of 50.0 N. (a) If he does 4.00×10^2 J of work on the sled while exerting the force horizontally, through what distance must he have pushed it? (b) If he exerts the same force at an angle of 45.0° with respect to the horizontal and moves the sled through the same distance, how much work does he do on the sled?

Answers (a) 8.00 m (b) 283 J

Work and Dissipative Forces

Frictional work is extremely important in everyday life, because doing almost any other kind of work is impossible without it. The Eskimo in the last example, for instance, depends on surface friction to pull his sled. Otherwise, the rope would slip in his hands and exert no force on the sled, while his feet slid out from underneath him and he fell flat on his face. Cars wouldn't work without friction, nor could conveyor belts, nor even our muscle tissue.

The work done by pushing or pulling an object is the application of a single force. Friction, on the other hand, is a complex process caused by numerous microscopic interactions over the entire area of the surfaces in contact. Consider a metal block sliding over a metal surface. Microscopic "teeth" in the block encounter equally microscopic irregularities in the underlying surface. Pressing against each other, the teeth deform, get hot, and weld to the opposite surface. Work must be done breaking these temporary bonds, and this comes at the expense of the energy of motion of the block, to be discussed in the next section. The energy lost by the block goes into heating both the block and its environment, with some energy converted to sound.

The friction force of two objects in contact and in relative motion to each other always dissipates energy in these relatively complex ways. For our purposes, the phrase "work done by friction" will denote the effect of these processes on mechanical energy alone.

© Chris Collins/Corbis

(a)

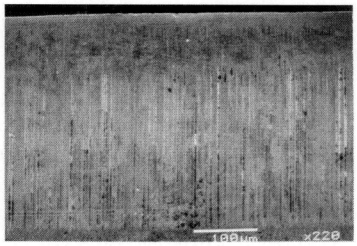

© McCrone Photo/Custom Medical Stock Photo

(b)

The edge of a razor blade looks smooth to the eye, but under a microscope proves to have numerous irregularities.

EXAMPLE 5.2 More Sledding

Goal Calculate the work done by friction when an object is acted on by an applied force.

Problem Suppose that in Example 5.1 the coefficient of kinetic friction between the loaded 50.0-kg sled and snow is 0.200. **(a)** The Eskimo again pulls the sled 5.00 m, exerting a force of 1.20×10^2 N at an angle of 0°. Find the work done on the sled by friction, and the net work. **(b)** Repeat the calculation if the applied force is exerted at an angle of 30.0° with the horizontal.

Strategy See Figure 5.6. The frictional work depends on the magnitude of the kinetic friction coefficient, the normal force, and the displacement. Use the y-component of Newton's second law to find the normal force $\vec{\mathbf{n}}$, calculate the work done by friction using the definitions, and sum with the result of Example 5.1(a) to obtain the net work on the sled. Part (b) is solved similarly, but the normal force is smaller because it has the help of the applied force $\vec{\mathbf{F}}_{\text{app}}$ in supporting the load.

Solution

(a) Find the work done by friction on the sled and the net work, if the applied force is horizontal.

First, find the normal force from the y-component of Newton's second law, which involves only the normal force and the force of gravity:

$$\Sigma F_y = n - mg = 0 \quad \rightarrow \quad n = mg$$

Use the normal force to compute the work done by friction:

$$\begin{aligned} W_{\text{fric}} &= -f_k \Delta x = -\mu_k n \Delta x = -\mu_k mg \Delta x \\ &= -(0.200)(50.0 \text{ kg})(9.80 \text{ m/s}^2)(5.00 \text{ m}) \\ &= \boxed{-4.90 \times 10^2 \text{ J}} \end{aligned}$$

Sum the frictional work with the work done by the applied force from Example 5.1 to get the net work (the normal and gravity forces are perpendicular to the displacement, so they don't contribute):

$$\begin{aligned} W_{\text{net}} &= W_{\text{app}} + W_{\text{fric}} + W_n + W_g \\ &= 6.00 \times 10^2 \text{ J} + (-4.90 \times 10^2 \text{ J}) + 0 + 0 \\ &= \boxed{1.10 \times 10^2 \text{ J}} \end{aligned}$$

(b) Recalculate the frictional work and net work if the applied force is exerted at a 30.0° angle.

Find the normal force from the y-component of Newton's second law:

$$\begin{aligned} \Sigma F_y &= n - mg + F_{\text{app}} \sin\theta = 0 \\ n &= mg - F_{\text{app}} \sin\theta \end{aligned}$$

Use the normal force to calculate the work done by friction:

$$\begin{aligned} W_{\text{fric}} &= -f_k \Delta x = -\mu_k n \Delta x = -\mu_k (mg - F_{\text{app}} \sin\theta)\, \Delta x \\ &= -(0.200)(50.0 \text{ kg} \cdot 9.80 \text{ m/s}^2 \\ &\quad - 1.20 \times 10^2 \text{ N} \sin 30.0°)(5.00 \text{ m}) \\ W_{\text{fric}} &= \boxed{-4.30 \times 10^2 \text{ J}} \end{aligned}$$

Sum this answer with the result of Example 5.1(b) to get the net work (again, the normal and gravity forces don't contribute):

$$\begin{aligned} W_{\text{net}} &= W_{\text{app}} + W_{\text{fric}} + W_n + W_g \\ &= 5.20 \times 10^2 \text{ J} - 4.30 \times 10^2 \text{ J} + 0 + 0 = \boxed{90.0 \text{ J}} \end{aligned}$$

Remark The most important thing to notice here is that exerting the applied force at different angles can dramatically affect the work done on the sled. Pulling at the optimal angle (about 15° in this case) will result in the most net work for a given amount of effort.

Exercise 5.2

(a) The Eskimo pushes the same 50.0-kg sled over level ground with a force of 1.75×10^2 N exerted horizontally, moving it a distance of 6.00 m over new terrain. If the net work done on the sled is 1.50×10^2 J, find the coefficient of kinetic friction. (b) Repeat the exercise if the applied force is upwards at a 45.0° angle with the horizontal.

Answer (a) 0.306 (b) 0.270

5.2 KINETIC ENERGY AND THE WORK–ENERGY THEOREM

Solving problems using Newton's second law can be difficult if the forces involved are complicated. An alternative is to relate the speed of an object to the net work done on it by external forces. If the net work can be calculated for a given displacement, the change in the object's speed is easy to evaluate.

Figure 5.7 shows an object of mass m moving to the right under the action of a constant net force $\vec{\mathbf{F}}_{\text{net}}$, also directed to the right. Because the force is constant, we know from Newton's second law that the object moves with constant acceleration $\vec{\mathbf{a}}$. If the object is displaced by Δx, the work done by $\vec{\mathbf{F}}_{\text{net}}$ on the object is

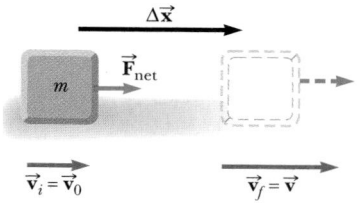

Figure 5.7 An object undergoes a displacement and a change in velocity under the action of a constant net force $\vec{\mathbf{F}}_{\text{net}}$.

$$W_{\text{net}} = F_{\text{net}}\Delta x = (ma)\Delta x \qquad \textbf{[5.3]}$$

In Chapter 2, we found that the following relationship holds when an object undergoes constant acceleration:

$$v^2 = v_0^2 + 2a\Delta x \qquad \text{or} \qquad a\Delta x = \frac{v^2 - v_0^2}{2}$$

We can substitute this expression into Equation 5.3 to get

$$W_{\text{net}} = m\left(\frac{v^2 - v_0^2}{2}\right)$$

or

$$W_{\text{net}} = \tfrac{1}{2}mv^2 - \tfrac{1}{2}mv_0^2 \qquad \textbf{[5.4]}$$

So the net work done on an object equals a change in a quantity of the form $\frac{1}{2}mv^2$. Because this term carries units of energy and involves the object's speed, it can be interpreted as energy associated with the object's motion, leading to the following definition:

◀ Kinetic energy

The **kinetic energy** ***KE*** of an object of mass m moving with a speed v is defined by

$$KE \equiv \tfrac{1}{2}mv^2 \qquad \textbf{[5.5]}$$

SI unit: joule (J) = kg · m²/s²

Like work, kinetic energy is a scalar quantity. Using this definition in Equation 5.4, we arrive at an important result known as the **work–energy theorem**:

◀ Work–energy theorem

The net work done on an object is equal to the change in the object's kinetic energy:

$$W_{\text{net}} = KE_f - KE_i = \Delta KE \qquad \textbf{[5.6]}$$

where the change in the kinetic energy is due entirely to the object's change in speed.

The proviso on the speed is necessary because work that deforms or causes the object to warm up invalidates Equation 5.6, although under most circumstances it remains approximately correct. From that equation, a positive net work W_{net} means that the final kinetic energy KE_f is greater than the initial kinetic energy KE_i. This, in turn, means that the object's final speed is greater than its initial speed. So positive net work increases an object's speed, and negative net work decreases its speed.

We can also turn the equation around and think of kinetic energy as the work a moving object can do in coming to rest. For example, suppose a hammer is on the verge of striking a nail, as in Figure 5.8. The moving hammer has kinetic energy and can therefore do work on the nail. The work done on the nail is $F\Delta x$, where F is the average net force exerted on the nail and Δx is the distance the nail is driven

Figure 5.8 The moving hammer has kinetic energy and can do work on the nail, driving it into the wall.

into the wall. This work, plus small amounts of energy carried away as heat and sound, is equal to the change in kinetic energy of the hammer, ΔKE.

For convenience, Equation 5.4 was derived under the assumption that the net force acting on the object was constant. A more general derivation, using calculus, would show that Equation 5.4 is valid under all circumstances, including the application of a variable force.

Applying Physics 5.1 Leaving Skid Marks

Suppose a car traveling at a speed v skids a distance d after its brakes lock. Estimate how far it would skid if it were traveling at speed $2v$ when its brakes locked.

Explanation Assume for simplicity that the force of kinetic friction between the car and the road surface is constant and the same at both speeds. From the work–energy theorem, the net force exerted on the car times the displacement of the car, $F_{\text{net}}\Delta x$, is equal in magnitude to its initial kinetic energy, $\frac{1}{2}mv^2$. When the speed is doubled, the kinetic energy of the car is quadrupled. So for a given applied friction force, the distance traveled must increase fourfold when the initial speed is doubled, and the estimated distance the car skids is $4d$.

EXAMPLE 5.3 Collision Analysis

Goal Apply the work–energy theorem with a known force.

Problem The driver of a 1.00×10^3 kg car traveling on the interstate at 35.0 m/s (nearly 80.0 mph) slams on his brakes to avoid hitting a second vehicle in front of him, which had come to rest because of congestion ahead. After the brakes are applied, a constant friction force of 8.00×10^3 N acts on the car. Ignore air resistance. **(a)** At what minimum distance should the brakes be applied to avoid a collision with the other vehicle? **(b)** If the distance between the vehicles is initially only 30.0 m, at what speed would the collision occur?

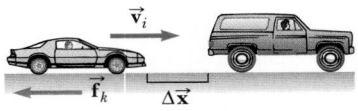

Figure 5.9 (Example 5.3) A braking vehicle just prior to an accident.

Strategy Compute the net work, which involves just the kinetic friction, because the normal and gravity forces are perpendicular to the motion. Then set the net work equal to the change in kinetic energy. To get the minimum distance in part (a), we take the final speed v_f to be zero just as the braking vehicle reaches the rear of the vehicle at rest. Solve for the unknown, Δx. For part (b) proceed similarly, except that the unknown is the final velocity v_f.

Solution

(a) Find the minimum necessary stopping distance.

Apply the work–energy theorem:

$$W_{\text{net}} = \tfrac{1}{2}mv_f^2 - \tfrac{1}{2}mv_i^2$$

Substitute an expression for the frictional work and set $v_f = 0$:

$$-f_k\Delta x = 0 - \tfrac{1}{2}mv_i^2$$

Substitute $v_i = 35.0$ m/s, $f_k = 8.00 \times 10^3$ N, and $m = 1.00 \times 10^3$ kg. Solve for Δx:

$$-(8.00 \times 10^3\ \text{N})\,\Delta x = -\tfrac{1}{2}(1.00 \times 10^3\ \text{kg})(35.0\ \text{m/s})^2$$

$$\Delta x = \boxed{76.6\ \text{m}}$$

(b) At the given distance of 30.0 m, the car is too close to the other vehicle. Find the speed at impact.

Write down the work–energy theorem:

$$W_{\text{net}} = W_{\text{fric}} = -f_k\Delta x = \tfrac{1}{2}mv_f^2 - \tfrac{1}{2}mv_i^2$$

Multiply by $2/m$ and rearrange terms, solving for the final velocity v_f:

$$v_f^2 = v_i^2 - \frac{2}{m}f_k\Delta x$$

$$v_f^2 = (35.0\ \text{m/s})^2 - \left(\frac{2}{1.00 \times 10^3\ \text{kg}}\right) \times (8.00 \times 10^3\ \text{N})(30.0\ \text{m}) = 745\ \text{m}^2/\text{s}^2$$

$$v_f = \boxed{27.3\ \text{m/s}}$$

Remarks This calculation illustrates how important it is to remain alert on the highway, allowing for an adequate stopping distance at all times. It takes about a second to react to the brake lights of the car in front of you. On a high-speed highway, your car may travel over 30 meters before you can engage the brakes. Bumper-to-bumper traffic at high speed, as often exists on the highways near big cities, is extremely unsafe.

Exercise 5.3

A police investigator measures straight skid marks 27 m long in an accident investigation. Assuming a friction force and car mass the same as in the previous problem, what was the minimum speed of the car when the brakes locked?

Answer 20.8 m/s

Conservative and Nonconservative Forces

It turns out there are two general kinds of forces. The first is called a **conservative force**. Gravity is probably the best example of a conservative force. To understand the origin of the name, think of a diver climbing to the top of a 10-meter platform. The diver has to do work against gravity in making the climb. Once at the top, however, he can recover the work—as kinetic energy—by taking a dive. His speed just before hitting the water will give him a kinetic energy equal to the work he did against gravity in climbing to the top of the platform—minus the effect of some nonconservative forces, such as air drag and internal muscular friction.

A **nonconservative force** is generally dissipative, which means it tends to randomly disperse the energy of bodies on which it acts. This dispersal of energy often takes the form of heat or sound. Kinetic friction and air drag are good examples. Propulsive forces, like the force exerted by a jet engine on a plane or by a propeller on a submarine, are also nonconservative.

Work done against a nonconservative force can't be easily recovered. Dragging objects over a rough surface requires work. When the Eskimo in Example 5.2 dragged the sled across terrain having a non-zero coefficient of friction, the net work was smaller than in the frictionless case. The missing energy went into warming the sled and its environment. As will be seen in the study of thermodynamics, such losses can't be avoided, nor all the energy recovered, so these forces are called nonconservative.

Another way to characterize conservative and nonconservative forces is to measure the work done by a force on an object traveling between two points along different paths. The work done by gravity on someone going down a frictionless slide, as in Figure 5.10, is the same as that done on someone diving into the water from

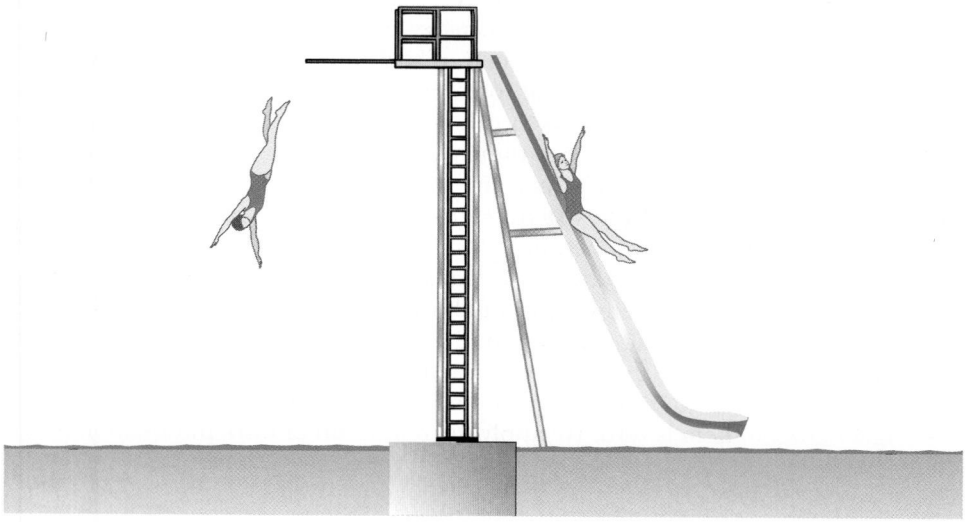

Figure 5.10 Because the gravity field is conservative, the diver regains as kinetic energy the work she did against gravity in climbing the ladder. Taking the frictionless slide gives the same result.

the same height. This equality doesn't hold for nonconservative forces. For example, sliding a book directly from point Ⓐ to point Ⓓ in Figure 5.11 requires a certain amount of work against friction, but sliding the book along the three other legs of the square, from Ⓐ to Ⓑ, Ⓑ to Ⓒ, and finally Ⓒ to Ⓓ, requires three times as much work. This observation motivates the following definition of a conservative force:

Conservative force ▶

A force is conservative if the work it does moving an object between two points is the same no matter what path is taken.

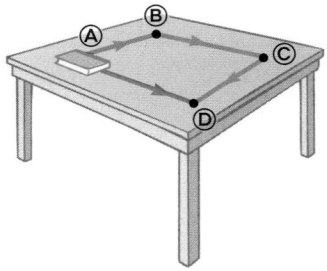

Figure 5.11 Because friction is a nonconservative force, a book pushed along the three segments Ⓐ–Ⓑ, Ⓑ–Ⓒ, and Ⓒ–Ⓓ requires three times the work as pushing the book directly from Ⓐ to Ⓓ.

Nonconservative forces, as we've seen, don't have this property. The work–energy theorem, Equation 5.6, can be rewritten in terms of the work done by conservative forces W_c and the work done by nonconservative forces W_{nc}, because the net work is just the sum of these two:

$$W_{nc} + W_c = \Delta KE \qquad [5.7]$$

It turns out that conservative forces have another useful property: The work they do can be recast as something called **potential energy**, a quantity that depends only on the beginning and end points of a curve, not the path taken.

5.3 GRAVITATIONAL POTENTIAL ENERGY

An object with kinetic energy (energy of motion) can do work on another object, just like a moving hammer can drive a nail into a wall. A brick on a high shelf can also do work: It can fall off the shelf, accelerate downwards, and hit a nail squarely, driving it into the floorboards. The brick is said to have **potential energy** associated with it, because from its location on the shelf it can potentially do work.

Potential energy is a property of a **system**, rather than of a single object, because it's due to a physical position in space relative a center of force, like the falling diver and the Earth of Figure 5.10. In this chapter, we define a system as a collection of objects interacting via forces or other processes that are internal to the system. It turns out that potential energy is another way of looking at the work done by conservative forces.

Gravitational Work and Potential Energy

Using the work–energy theorem in problems involving gravitation requires computing the work done by gravity. For most trajectories—say, for a ball traversing a parabolic arc—finding the gravitational work done on the ball requires sophisticated techniques from calculus. Fortunately, for conservative fields there's a simple alternative: potential energy.

Gravity is a conservative force, and for every conservative force a special expression called a potential energy function can be found. Evaluating that function at any two points in an object's path of motion and finding the difference will give the negative of the work done by that force between those two points. It's also advantageous that potential energy, like work and kinetic energy, is a scalar quantity.

Our first step is to find the work done by gravity on an object when it moves from one position to another. The negative of that work is the change in the gravitational potential energy of the system, and from that expression, we'll be able to identify the potential energy function.

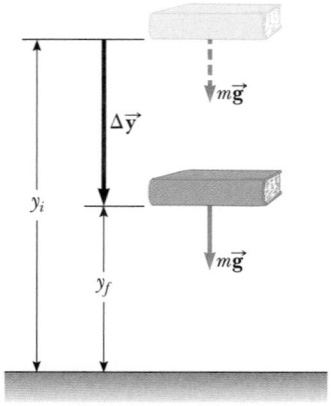

Figure 5.12 The work done by the gravitational force as the book falls from y_i to y_f equals $mgy_i - mgy_f$.

In Figure 5.12, a book of mass m falls from a height y_i to a height y_f, where the positive y-coordinate represents position above the ground. We neglect the force of air friction, so the only force acting on the book is gravitation. How much work is done? The magnitude of the force is mg and that of the displacement is $\Delta y = y_i - y_f$ (a positive number), while both $\vec{\mathbf{F}}$ and $\Delta\vec{\mathbf{y}}$ are pointing downwards, so the angle between them is zero. We apply the definition of work in Equation 5.2:

$$W_g = F\Delta y \cos\theta = mg(y_i - y_f)\cos 0° = -mg(y_f - y_i) \qquad [5.8]$$

Factoring out the minus sign was deliberate, to clarify the coming connection to potential energy. Equation 5.8 for gravitational work holds for any object, regardless of its trajectory in space, because the gravitational force is conservative. Now, W_g will appear as the work done by gravity in the work–energy theorem. For the rest of this section, assume for simplicity that we are dealing only with systems involving gravity and nonconservative forces. Then Equation 5.7 can be written as

$$W_{\text{net}} = W_{nc} + W_g = \Delta KE$$

where W_{nc} is the work done by the nonconservative forces. Substituting the expression for W_g from Equation 5.8, we obtain

$$W_{nc} - mg(y_f - y_i) = \Delta KE \qquad \textbf{[5.9a]}$$

Next, we add $mg(y_f - y_i)$ to both sides:

$$W_{nc} = \Delta KE + mg(y_f - y_i) \qquad \textbf{[5.9b]}$$

Now, by definition, we'll make the connection between gravitational work and gravitational potential energy.

◀Gravitational potential energy

The gravitational potential energy of a system consisting of the Earth and an object of mass m near the Earth's surface is given by

$$PE \equiv mgy \qquad \textbf{[5.10]}$$

where g is the acceleration of gravity and y is the vertical position of the mass relative the surface of Earth (or some other reference point).

SI unit: joule (J)

In this definition, $y = 0$ is usually taken to correspond to Earth's surface, but this is not strictly necessary, as discussed in the next subsection. It turns out that only *differences* in potential energy really matter.

So the gravitational potential energy associated with an object located near the surface of the Earth is the object's weight mg times its vertical position y above Earth. From this *definition*, we have the relationship between gravitational work and gravitational potential energy:

$$W_g = -(PE_f - PE_i) = -(mgy_f - mgy_i) \qquad \textbf{[5.11]}$$

The work done by gravity is one and the same as the negative of the change in gravitational potential energy.

Finally, using the relationship in Equation 5.11 in Equation 5.9b, we obtain an extension of the work–energy theorem:

$$W_{nc} = (KE_f - KE_i) + (PE_f - PE_i) \qquad \textbf{[5.12]}$$

This equation says that the work done by nonconservative forces, W_{nc}, is equal to the change in the kinetic energy plus the change in the gravitational potential energy.

Equation 5.12 will turn out to be true in general, even when other conservative forces besides gravity are present. The work done by these additional conservative forces will again be recast as changes in potential energy and will appear on the right-hand side along with the expression for gravitational potential energy.

TIP 5.3 Potential Energy Takes Two

Potential energy always takes a system of at least two interacting objects—for example, the Earth and a baseball interacting via the gravitational force.

Reference Levels for Gravitational Potential Energy

In solving problems involving gravitational potential energy, it's important to choose a location at which to set that energy equal to zero. Given the form of Equation 5.10, this is the same as choosing the place where $y = 0$. The choice is completely arbitrary because the important quantity is the *difference* in potential energy, and this difference will be the same regardless of the choice of zero level. However, once this position is chosen, it must remain fixed for a given problem.

Figure 5.13 Any reference level—the desktop, the floor of the room, or the ground outside the building—can be used to represent zero gravitational potential energy in the book–Earth system.

While it's always possible to choose the surface of the Earth as the reference position for zero potential energy, the statement of a problem will usually suggest a convenient position to use. As an example, consider a book at several possible locations, as in Figure 5.13. When the book is at Ⓐ, a natural zero level for potential energy is the surface of the desk. When the book is at Ⓑ, the floor might be a more convenient reference level. Finally, a location such as Ⓒ, where the book is held out a window, would suggest choosing the surface of the Earth as the zero level of potential energy. The choice, however, makes no difference: Any of the three reference levels could be used as the zero level, regardless of whether the book is at Ⓐ, Ⓑ, or Ⓒ. Example 5.4 illustrates this important point.

EXAMPLE 5.4 Wax Your Skis

Goal Calculate the change in gravitational potential energy for different choices of reference level.

Problem A 60.0-kg skier is at the top of a slope, as shown in Figure 5.14. At the initial point Ⓐ, she is 10.0 m vertically above point Ⓑ. **(a)** Setting the zero level for gravitational potential energy at Ⓑ, find the gravitational potential energy of this system when the skier is at Ⓐ and then at Ⓑ. Finally, find the change in potential energy of the skier–Earth system as the skier goes from point Ⓐ to point Ⓑ. **(b)** Repeat this problem with the zero level at point Ⓐ. **(c)** Repeat again, with the zero level 2.00 m higher than point Ⓑ.

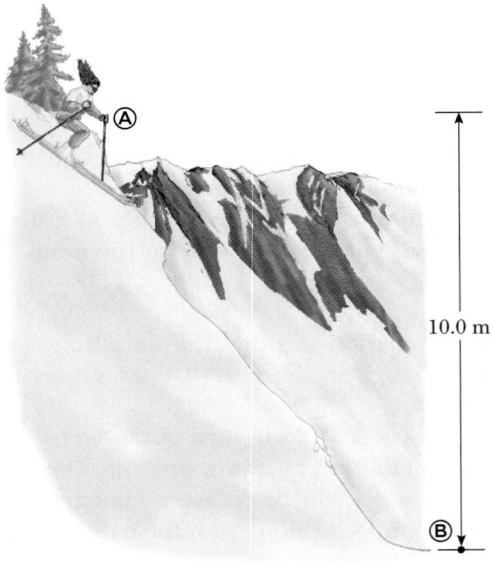

Figure 5.14 (Example 5.4)

Strategy Follow the definition, and be careful with signs. Ⓐ is the initial point, with gravitational potential energy PE_i, and Ⓑ is the final point, with gravitational potential energy PE_f. The location chosen for $y = 0$ is also the zero point for the potential energy, since $PE = mgy$.

Solution

(a) Let $y = 0$ at Ⓑ. Calculate the potential energy at Ⓐ, at Ⓑ, and the change in potential energy.

Find PE_i, the potential energy at Ⓐ, from Equation 5.10:

$$PE_i = mgy_i = (60.0\text{ kg})(9.80\text{ m/s}^2)(10.0\text{ m}) = 5.88 \times 10^3\text{ J}$$

$PE_f = 0$ at Ⓑ by choice. Find the difference in potential energy between Ⓐ and Ⓑ:

$$PE_f - PE_i = 0 - 5.88 \times 10^3\text{ J} = -5.88 \times 10^3\text{ J}$$

(b) Repeat the problem if $y = 0$ at Ⓐ, the new reference point, so that $PE = 0$ at Ⓐ.

Find PE_f, noting that point Ⓑ is now at $y = -10.0$ m:

$$PE_f = mgy_f = (60.0\text{ kg})(9.80\text{ m/s}^2)(-10.0\text{ m}) = -5.88 \times 10^3\text{ J}$$

$$PE_f - PE_i = -5.88 \times 10^3\text{ J} - 0 = -5.88 \times 10^3\text{ J}$$

(c) Repeat the problem, if $y = 0$ two meters above Ⓑ.

Find PE_i, the potential energy at Ⓐ:

$$PE_i = mgy_i = (60.0\text{ kg})(9.80\text{ m/s}^2)(8.00\text{ m}) = 4.70 \times 10^3\text{ J}$$

Find PE_f, the potential energy at Ⓑ:

$$PE_f = mgy_f = (60.0\text{ kg})(9.8\text{ m/s}^2)(-2.00\text{ m}) = -1.18 \times 10^3\text{ J}$$

Compute the change in potential energy:

$$PE_f - PE_i = -1.18 \times 10^3\text{ J} - 4.70 \times 10^3\text{ J} = -5.88 \times 10^3\text{ J}$$

Remarks These calculations show that the change in the gravitational potential energy when the skier goes from the top of the slope to the bottom is -5.88×10^3 J, *regardless of the zero level selected.*

Exercise 5.4

If the zero level for gravitational potential energy is selected to be midway down the slope, 5.00 m above point Ⓑ, find the initial potential energy, the final potential energy, and the change in potential energy as the skier goes from point Ⓐ to Ⓑ in Figure 5.14.

Answer 2.94 kJ, −2.94 kJ, −5.88 kJ

Gravity and the Conservation of Mechanical Energy

Conservation principles play a very important role in physics. **When a physical quantity is conserved the numeric value of the quantity remains the same throughout the physical process.** Although the form of the quantity may change in some way, **its final value is the same as its initial value.**

The kinetic energy KE of an object falling only under the influence of gravity is constantly changing, as is the gravitational potential energy PE. Obviously, then, these quantities aren't conserved. Because all nonconservative forces are assumed absent, however, we can set $W_{nc} = 0$ in Equation 5.12. Rearranging the equation, we arrive at the following very interesting result:

$$KE_i + PE_i = KE_f + PE_f \qquad [5.13]$$

According to this equation, **the sum of the kinetic energy and the gravitational potential energy remains constant at all times and hence is a conserved quantity.** We denote the total mechanical energy by $E = KE + PE$, and say that **the total mechanical energy is conserved.**

To show how this concept works, think of tossing a rock off a cliff, and ignore the drag forces. As the rock falls, its speed increases, so its kinetic energy increases. As the rock approaches the ground, the potential energy of the rock–Earth system decreases. Whatever potential energy is lost as the rock moves downward appears as kinetic energy, and Equation 5.13 says that in the absence of nonconservative forces like air drag, the trading of energy is exactly even. This is true for all conservative forces, not just gravity.

TIP 5.4 Conservation Principles

There are many conservation laws like the conservation of mechanical energy in isolated systems, as in Equation 5.13. For example, momentum, angular momentum, and electric charge are all conserved quantities, as will be seen later. Conserved quantities may change form during physical interactions, but their sum total for a system never changes.

In any isolated system of objects interacting only through conservative forces, the total mechanical energy $E = KE + PE$, of the system, remains the same at all times.

◀ Conservation of mechanical energy

If the force of gravity is the *only* force doing work within a system, then the principle of conservation of mechanical energy takes the form

$$\tfrac{1}{2}mv_i^2 + mgy_i = \tfrac{1}{2}mv_f^2 + mgy_f \qquad \textbf{[5.14]}$$

This form of the equation is particularly useful for solving problems involving only gravity. Further terms have to be added when other conservative forces are present, as we'll soon see.

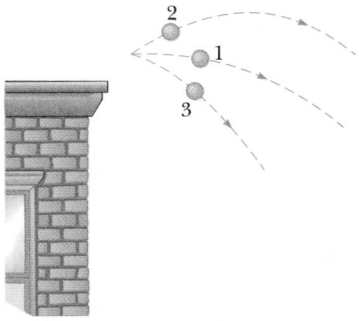

ACTIVE FIGURE 5.15
(Quick Quiz 5.2) Three identical balls are thrown with the same initial speed from the top of a building.

PhysicsNow™
Log into PhysicsNow at **www.cp7e.com**, and go to Active Figure 5.15 to throw balls at different angles from the top of the building and compare trajectories and speeds as the balls hit the ground.

Quick Quiz 5.2

Three identical balls are thrown from the top of a building, all with the same initial speed. The first ball is thrown horizontally, the second at some angle above the horizontal, and the third at some angle below the horizontal, as in Active Figure 5.15. Neglecting air resistance, rank the speeds of the balls as they reach the ground, from fastest to slowest. (a) 1, 2, 3 (b) 2, 1, 3 (c) 3, 1, 2 (d) all three balls strike the ground at the same speed.

Quick Quiz 5.3

Bob, of mass m, drops from a tree limb at the same time that Esther, also of mass m, begins her descent down a frictionless slide. If they both start at the same height above the ground, which of the following is true about their kinetic energies as they reach the ground?
(a) Bob's kinetic energy is greater than Esther's.
(b) Esther's kinetic energy is greater than Bob's.
(c) They have the same kinetic energy.
(d) The answer depends on the shape of the slide.

Problem-Solving Strategy
Applying Conservation of Mechanical Energy

Take the following steps when applying conservation of mechanical energy to problems involving gravity:

1. Define the system, including all interacting bodies. Choose a location for $y = 0$, the zero point for gravitational potential energy.
2. Select the body of interest and identify two points—one point where you have given information, and the other point where you want to find out something about the body of interest.
3. After verifying the absence of nonconservative forces, write down the conservation of energy equation, Equation 5.14, for the system. Identify the unknown quantity of interest.
4. Solve for the unknown quantity, which is usually either a speed or a position, and substitute known values.

As previously stated, it's usually best to do the algebra with symbols rather than substituting known numbers first, because it's easier to check the symbols for possible errors. The exception is when a quantity is clearly zero, in which case immediate substitution greatly simplifies the ensuing algebra.

INTERACTIVE EXAMPLE 5.5 Platform Diver

Goal Use conservation of energy to calculate the speed of a body falling straight down in the presence of gravity.

Problem A diver of mass m drops from a board 10.0 m above the water's surface, as in Figure 5.16. Neglect air resistance. **(a)** Use conservation of mechanical energy to find his speed 5.00 m above the water's surface. **(b)** Find his speed as he hits the water.

Strategy Refer to the problem-solving strategy. Step 1: The system consists of the diver and the Earth. As the diver falls, only the force of gravity acts on him (neglecting air drag), so the mechanical energy of the system is conserved and we can use conservation of energy for both (a) and (b). Choose $y = 0$ for the water's surface. Step 2: In (a), $y = 10.0$ m and $y = 5.00$ m are the points of interest, while in (b), $y = 10.0$ m and $y = 0$ m are of interest.

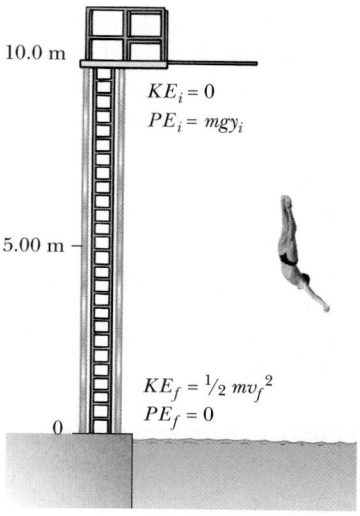

Figure 5.16 (Example 5.5) The zero of gravitational potential energy is taken to be at the water's surface.

Solution

(a) Find the diver's speed halfway down, at $y = 5.00$ m.

Step 3: we write the energy conservation equation and supply the proper terms:

$$KE_i + PE_i = KE_f + PE_f$$

$$\tfrac{1}{2}mv_i^2 + mgy_i = \tfrac{1}{2}mv_f^2 + mgy_f$$

Step 4: Substitute $v_i = 0$, cancel the mass m and solve for v_f:

$$0 + gy_i = \tfrac{1}{2}v_f^2 + gy_f$$

$$v_f = \sqrt{2g(y_i - y_f)} = \sqrt{2(9.80\ \text{m/s}^2)(10.0\,\text{m} - 5.00\ \text{m})}$$

$$v_f = 9.90\ \text{m/s}$$

(b) Find the diver's speed at the water's surface, $y = 0$.

Use the same procedure as in part (a), taking $y_f = 0$:

$$0 + mgy_i = \tfrac{1}{2}mv_f^2 + 0$$

$$v_f = \sqrt{2gy_i} = \sqrt{2(9.80\ \text{m/s}^2)(10.0\ \text{m})} = 14.0\ \text{m/s}$$

Remark Notice that the speed halfway down is not half the final speed.

Exercise 5.5

Suppose the diver vaults off the springboard, leaving it with an initial speed of 3.50 m/s upwards. Use energy conservation to find his speed when he strikes the water.

Answer 14.4 m/s

PhysicsNow™ Investigate the conservation of mechanical energy for a dropped ball by logging into PhysicsNow at **www.cp7e.com** and going to Interactive Example 5.5.

EXAMPLE 5.6 The Jumping Bug

Goal Use conservation of mechanical energy and concepts from ballistics in two dimensions to calculate a speed.

Problem A powerful grasshopper launches itself at an angle of 45° above the horizontal and rises to a maximum height of 1.00 m during the leap. (See Figure 5.17.) With what speed v_i did it leave the ground? Neglect air resistance.

Strategy This problem can be solved with conservation of energy and the relation between the initial velocity and

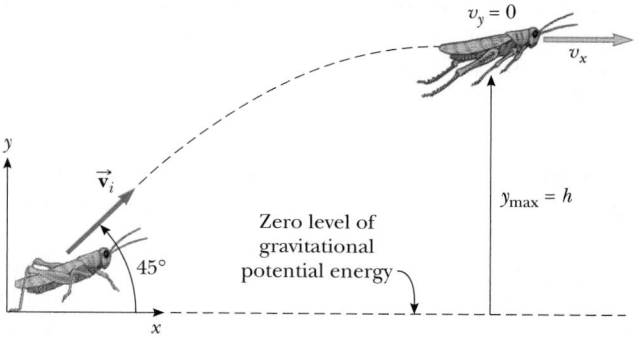

Figure 5.17 (Example 5.6)

its x-component. Aside from the origin, the other point of interest is the maximum height $y = 1.00$ m, where the grasshopper has a velocity v_x in the x-direction only. Energy conservation then gives one equation with two unknowns: the initial speed v_i and speed at maximum height, v_x. Because there are no forces in the x-direction, however, v_x is the same as the x-component of the initial velocity.

Solution

Use energy conservation:

$$\tfrac{1}{2}mv_i^2 + mgy_i = \tfrac{1}{2}mv_f^2 + mgy_f$$

Substitute $y_i = 0, v_f = v_x$, and $y_f = h$:

$$\tfrac{1}{2}mv_i^2 = \tfrac{1}{2}mv_x^2 + mgh$$

Multiply each side by $2/m$, obtaining one equation and two unknowns:

$$v_i^2 = v_x^2 + 2gh \qquad (1)$$

Eliminate v_x by substituting $v_x = v_i \sin 45°$ into equation (1), solving for v_i, and substituting known values:

$$v_i^2 = (v_i \sin 45°)^2 + 2gh = \tfrac{1}{2}v_i^2 + 2gh$$

$$v_i = 2\sqrt{gh} = 2\sqrt{(9.80\ \text{m/s}^2)(1.00\ \text{m})} = 6.26\ \text{m/s}$$

Remarks The final answer is a surprisingly high value and illustrates how strong insects are relative to their size.

Exercise 5.6

A catapult launches a rock at a 30.0° angle with respect to the horizontal. Find the maximum height attained if the speed of the rock at its highest point is 30.0 m/s.

Answer 15.3 m

TIP 5.5 Don't Use Work Done by the Force of Gravity *and* Gravitational Potential Energy!

Gravitational potential energy is just another way of including the work done by the force of gravity in the work–energy theorem. Don't use both of them in the equation at the same time, or you'll count it twice!

Gravity and Nonconservative Forces

When nonconservative forces are involved along with gravitation, the full work–energy theorem must be used, often with techniques from Chapter 4. Solving problems requires the basic procedure of the problem-solving strategy for conservation-of-energy problems in the previous section. The only difference lies in substituting Equation 5.12, the work–energy equation with potential energy, for Equation 5.14.

EXAMPLE 5.7 Der Stuka!

Goal Use the work–energy theorem with gravitational potential energy to calculate the work done by a nonconservative force.

Problem Waterslides are nearly frictionless, hence can provide bored students with high-speed thrills (Fig. 5.18). One such slide, Der Stuka, named for the terrifying German dive bombers of World War II, is 72.0 feet high (21.9 m), found at Six Flags in Dallas, Texas, and at Wet'n Wild in Orlando, Florida. **(a)** Determine the speed of a 60.0-kg woman at the bottom of such a slide, assuming no friction is present. **(b)** If the woman is clocked at 18.0 m/s at the bottom of the slide, how much mechanical energy was lost through friction?

Strategy The system consists of the woman, the Earth, and the slide. The normal force, always perpendicular to the displacement, does no work. Let $y = 0$ m represent the bottom of the slide. The two points of interest are $y = 0$ m and $y = 21.9$ m. Without friction, $W_{nc} = 0$, and we can apply conservation of mechanical energy, Equation 5.14. For part (b), use Equation 5.12, substitute two velocities and heights, and solve for W_{nc}.

Wet'n Wild Orlando

Figure 5.18 (Example 5.7) If the slide is frictionless, the speed of the thrill seeker at the bottom depends only on the height of the slide, not on the path it takes.

Solution

(a) Find the woman's speed at the bottom of the slide, assuming no friction.

Write down Equation 5.14, for conservation of energy:

$$\tfrac{1}{2}mv_i^2 + mgy_i = \tfrac{1}{2}mv_f^2 + mgy_f$$

Insert the values $v_i = 0$ and $y_f = 0$:

$$0 + mgy_i = \tfrac{1}{2}mv_f^2 + 0$$

Solve for v_f and substitute values for g and y_i:

$$v_f = \sqrt{2gy_i} = \sqrt{2(9.80\ \text{m/s}^2)(21.9\ \text{m})} = 20.7\ \text{m/s}$$

(b) Find the mechanical energy lost due to friction if $v_f = 18.0\ \text{m/s} < 20.7\ \text{m/s}$.

Write Equation 5.12, substituting expressions for the kinetic and potential energies:

$$W_{nc} = (KE_f - KE_i) + (PE_f - PE_i)$$
$$= (\tfrac{1}{2}mv_f^2 - \tfrac{1}{2}mv_f^2) + (mgy_f - mgy_i)$$

Substitute $m = 60.0$ kg, $v_f = 18.0$ m/s, and $v_i = 0$, and solve for W_{nc}:

$$W_{nc} = [\tfrac{1}{2}\cdot 60.0\ \text{kg}\cdot(18.0\ \text{m/s})^2 - 0]$$
$$+ [0 - 60.0\,\text{kg}\cdot(9.80\ \text{m/s}^2)\cdot 21.9\,\text{m}]$$

$$W_{nc} = -3.16 \times 10^3\,\text{J}$$

Remarks The speed found in part (a) is the same as if the thrill seeker fell vertically through a distance of 21.9 m, consistent with our intuition in Quick Quiz 5.3. The result of part (b) is negative because the system loses mechanical energy. Friction transforms part of the mechanical energy into thermal energy and mechanical waves, absorbed partly by the system and partly by the environment.

Exercise 5.7

Suppose a slide similar to Der Stuka is 35.0 meters high, but is a straight slope, inclined at 45.0° with respect to the horizontal. (a) Find the speed of a 60.0-kg thrill seeker at the bottom of the slide, assuming no friction. (b) If the thrill seeker has a speed of 20.0 m/s at the bottom, find the mechanical energy lost due to friction and (c) the magnitude of the force of friction, assumed constant.

Answers (a) 26.2 m/s (b) -8.58×10^3 J (c) 173 N

EXAMPLE 5.8 Hit the Ski Slopes

Goal Combine conservation of mechanical energy with the work–energy theorem involving friction on a horizontal surface.

Problem A skier starts from rest at the top of a frictionless incline of height 20.0 m, as in Figure 5.19. At the bottom of the incline, the skier encounters a horizontal surface where the coefficient of kinetic friction between skis and snow is 0.210. **(a)** Find the skier's speed at the bottom. **(b)** How far does the skier travel on the horizontal surface before coming to rest?

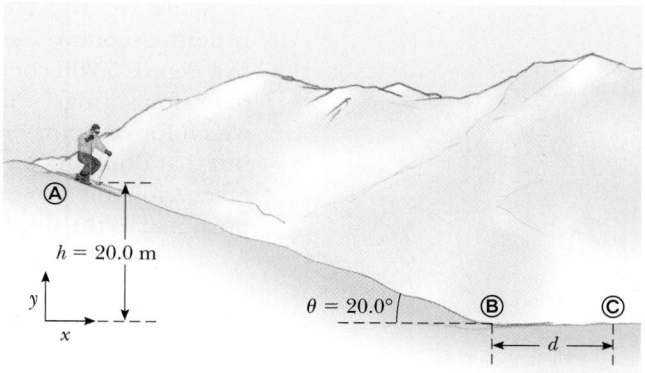

Figure 5.19 (Example 5.8) The skier slides down the slope and onto a level surface, stopping after traveling a distance d from the bottom of the hill.

Strategy Going down the frictionless incline is physically no different than going down the slide of the previous example and is handled the same way, using conservation of mechanical energy to find the speed $v_{Ⓑ}$ at the bottom. On the flat, rough surface, use the work–energy theorem, Equation 5.12, with $W_{nc} = W_{\text{fric}} = -f_k d$, where f_k is the magnitude of the force of friction and d is the distance traveled.

Solution

(a) Find the skier's speed at the bottom.

Follow the procedure used in part (a) of the previous example as the skier moves from the top, point Ⓐ, to the bottom, point Ⓑ.

$$v_{Ⓑ} = \sqrt{2gh} = \sqrt{2(9.80\ \text{m/s}^2)(20.0\ \text{m})} = 19.8\ \text{m/s}$$

(b) Find the distance traveled on the horizontal, rough surface.

Apply the work–energy theorem as the skier moves from Ⓑ to Ⓒ:

$$W_{\text{net}} = -f_k d = \Delta KE = \tfrac{1}{2}mv_{Ⓒ}^2 - \tfrac{1}{2}mv_{Ⓑ}^2$$

Substitute $v_{Ⓒ} = 0$ and $f_k = \mu_k n = \mu_k mg$:

$$-\mu_k mgd = -\tfrac{1}{2}mv_{Ⓑ}^2$$

Solve for d:

$$d = \frac{v_{Ⓑ}^2}{2\mu_k g} = \frac{(19.8\ \text{m/s})^2}{2(0.210)(9.80\ \text{m/s}^2)} = 95.2\ \text{m}$$

Remarks Substituting the symbolic expression $v_{Ⓑ} = \sqrt{2gh}$ into the equation for the distance d shows that d is linearly proportional to h: Doubling the height doubles the distance traveled.

Exercise 5.8

Find the horizontal distance the skier travels before coming to rest if the incline also has a coefficient of kinetic friction equal to 0.210.

Answer 40.3 m

5.4 SPRING POTENTIAL ENERGY

Springs are important elements in modern technology. They are found in machines of all kinds, in watches, toys, cars, and trains. Springs will be introduced here, then studied in more detail in Chapter 13.

Work done by an applied force in stretching or compressing a spring can be recovered by removing the applied force, so like gravity, the spring force is conservative. This means a potential energy function can be found and used in the work–energy theorem.

Active Figure 5.20a shows a spring in its equilibrium position, where the spring is neither compressed nor stretched. Pushing a block against the spring as in Active Figure 5.20b compresses it a distance x. While x appears to be merely a coordinate, for springs it also represents a *displacement* from the equilibrium position, which for our purposes will always be taken to be at $x = 0$. Experimentally, it turns out that doubling a given displacement requires double the force, while tripling it takes triple the force. This means the force exerted by the spring, F_s, must be proportional to the displacement x, or

$$F_s = -kx \qquad [5.15]$$

where k is a constant of proportionality, the *spring constant*, carrying units of newtons per meter. Equation 5.15 is called **Hooke's law**, after Sir Robert Hooke, who discovered the relationship. The force F_s is often called a *restoring force*, because the spring always exerts a force in a direction opposite the displacement of its end, tending to restore whatever is attached to the spring to its original position. For positive values of x, the force is negative, pointing back towards equilibrium at $x = 0$, and for negative x, the force is positive, again pointing towards $x = 0$. For a flexible spring, k is a small number (about 100 N/m), whereas for a stiff spring k is large (about 10 000 N/m). The value of the spring constant k is

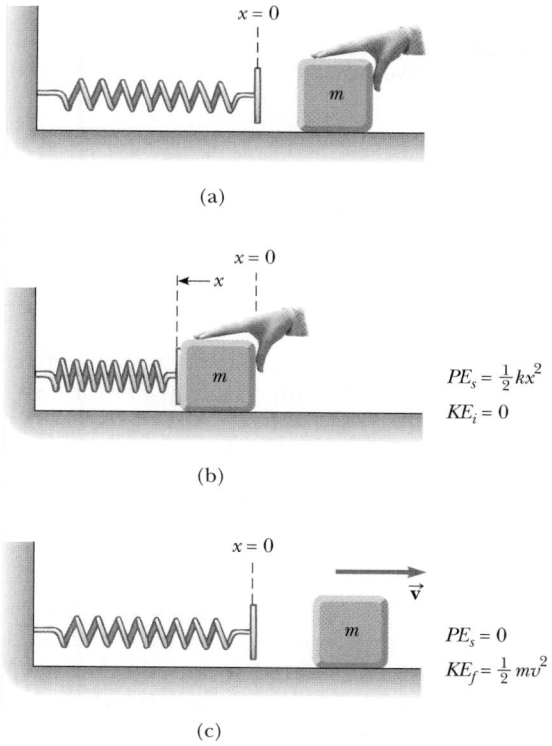

ACTIVE FIGURE 5.20
(a) A spring at equilibrium, neither compressed nor stretched. (b) A block of mass m on a frictionless surface is pushed against the spring. If x is the compression in the spring, the potential energy stored in the spring is $\frac{1}{2}kx^2$. (c) When the block is released, this energy is transferred to the block in the form of kinetic energy.

PhysicsNow™
Log into PhysicsNow at **www.cp7e.com**, and go to Active Figure 5.20 to compress the spring by varying amounts and observe the effect on the block's speed.

determined by how the spring was formed, its material composition, and the thickness of the wire. The minus sign ensures that the spring force is always directed back towards the equilibrium point.

As in the case of gravitation, a potential energy, called the **elastic potential energy**, can be associated with the spring force. Elastic potential energy is another way of looking at the work done by a spring during motion, because it is equal to the negative of the work done by the spring. It can also be considered stored energy arising from the work done to compress or stretch the spring.

Consider a horizontal spring and mass at the equilibrium position. We determine the work done by the spring when compressed by an applied force from equilibrium to a displacement x, as in Active Figure 5.20b. The spring force points in the direction opposite the motion, so we expect the work to be negative. When we studied the constant force of gravity near the Earth's surface, we found the work done on an object by multiplying the gravitational force by the vertical displacement of the object. However, this procedure can't be used with a varying force such as the spring force. Instead, we use the average force, $\bar{F}$:

$$\bar{F} = \frac{F_0 + F_1}{2} = \frac{0 - kx}{2} = -\frac{kx}{2}$$

Therefore, the work *done by the spring force* is

$$W_s = \bar{F}x = -\tfrac{1}{2}kx^2$$

In general, when the spring is stretched or compressed from x_i to x_f, the work done by the spring is

$$W_s = -\left(\tfrac{1}{2}kx_f{}^2 - \tfrac{1}{2}kx_i{}^2\right)$$

The work done by a spring can be included in the work–energy theorem. Assume Equation 5.12 now includes the work done by springs on the left-hand side. It then reads

$$W_{nc} - (\tfrac{1}{2}kx_f^2 - \tfrac{1}{2}kx_i^2) = \Delta KE + \Delta PE_g$$

where PE_g is the gravitational potential energy. We now define the elastic potential energy associated with the spring force, PE_s, by

$$PE_s \equiv \tfrac{1}{2}kx^2 \qquad [5.16]$$

Inserting this expression into the previous equation and rearranging gives the new form of the work–energy theorem, including both gravitational and elastic potential energy:

$$W_{nc} = (KE_f - KE_i) + (PE_{gf} - PF_{gi}) + (PE_{sf} - PE_{si}) \qquad [5.17]$$

where W_{nc} is the work done by nonconservative forces, KE is kinetic energy, PE_g is gravitational potential energy, and PE_s is the elastic potential energy. PE, formerly used to denote gravitational potential energy alone, will henceforth denote the total potential energy of a system, including potential energies due to all conservative forces acting on the system.

It's important to remember that the work done by gravity and springs in any given physical system is already included on the right-hand side of Equation 5.17 as potential energy and should not also be included on the left as work.

Active Figure 5.20c shows how the stored elastic potential energy can be recovered. When the block is released, the spring snaps back to its original length and the stored elastic potential energy is converted to kinetic energy of the block. The elastic potential energy stored in the spring is zero when the spring is in the equilibrium position ($x = 0$). As given by Equation 5.16, potential energy is also stored in the spring when it's stretched. Further, the elastic potential energy is a maximum when the spring has reached its maximum compression or extension. Finally, the potential energy is always positive when the spring is not in the equilibrium position, because PE_s is proportional to x^2.

In the absence of nonconservative forces, $W_{nc} = 0$, so the left-hand side of Equation 5.17 is zero, and an extended form for conservation of mechanical energy results:

$$(KE + PE_g + PE_s)_i = (KE + PE_g + PE_s)_f \qquad [5.18]$$

Problems involving springs, gravity, and other forces are handled in exactly the same way as described in problem-solving strategies 1 and 2, as will be seen in Examples 5.9–5.11.

EXAMPLE 5.9 A Horizontal Spring

Goal Use conservation of energy to calculate the speed of a block on a horizontal spring with and without friction.

Problem A block with mass of 5.00 kg is attached to a horizontal spring with spring constant $k = 4.00 \times 10^2$ N/m, as in Figure 5.21. The surface the block rests upon is frictionless. If the block is pulled out to $x_i = 0.0500$ m and released, **(a)** find the speed of the block at the equilibrium point, **(b)** find the speed when $x = 0.0250$ m, and **(c)** repeat part (a) if friction acts on the block, with coefficient $\mu_k = 0.150$.

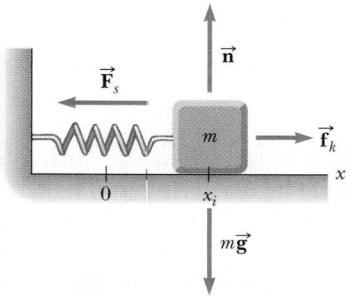

Figure 5.21 (Example 5.9) A mass attached to a spring.

Strategy In parts (a) and (b) there are no nonconservative forces, so conservation of energy, Equation 5.18, can be applied. In part (c), the definition of work and the work–energy theorem are needed to deal with the loss of mechanical energy due to friction.

Solution

(a) Find the speed of the block at equilibrium point.

Start with Equation 5.18:

$$(KE + PE_g + PE_s)_i = (KE + PE_g + PE_s)_f$$

Substitute expressions for the block's kinetic energy and the potential energy, and set the gravity terms to zero:

$$\tfrac{1}{2}mv_i^2 + \tfrac{1}{2}kx_i^2 = \tfrac{1}{2}mv_f^2 + \tfrac{1}{2}kx_f^2 \qquad (1)$$

Substitute $v_i = 0$ and $x_f = 0$, and multiply by $2/m$:

$$\frac{k}{m}x_i^2 = v_f^2$$

Solve for v_f and substitute the given values:

$$v_f = \sqrt{\frac{k}{m}}x_i = \sqrt{\frac{4.00 \times 10^2\,\text{N/m}}{5.00\,\text{kg}}}\,(0.0500\text{ m})$$

$$= \boxed{0.447\text{ m/s}}$$

(b) Find the speed of the block at the halfway point.

Set $v_i = 0$ in Equation (1), and multiply by $2/m$:

$$\frac{kx_i^2}{m} = v_f^2 + \frac{kx_f^2}{m}$$

Solve for v_f, and substitute the given values:

$$v_f = \sqrt{\frac{k}{m}(x_i^2 - x_f^2)}$$

$$= \sqrt{\frac{4.00 \times 10^2\,\text{N/m}}{5.00\,\text{kg}}((0.050\text{ m})^2 - (0.025\text{ m})^2)}$$

$$= \boxed{0.387\text{ m/s}}$$

(c) Repeat part (a), this time with friction.

Apply the work–energy theorem. The work done by the force of gravity and the normal force is zero, because these forces are perpendicular to the motion.

$$W_{\text{fric}} = \tfrac{1}{2}mv_f^2 - \tfrac{1}{2}mv_i^2 + \tfrac{1}{2}kx_f^2 - \tfrac{1}{2}kx_i^2$$

Substitute $v_i = 0$, $x_f = 0$, and $W_{\text{fric}} = -\mu_k n x_i$:

$$-\mu_k n x_i = \tfrac{1}{2}mv_f^2 - \tfrac{1}{2}kx_i^2$$

Set $n = mg$, and solve for v_f:

$$\tfrac{1}{2}mv_f^2 = \tfrac{1}{2}kx_i^2 - \mu_k mgx_i$$

$$v_f = \sqrt{\frac{k}{m}x_i^2 - 2\mu_k gx_i}$$

$$= \sqrt{\frac{4.00 \times 10^2\,\text{N/m}}{5.00\text{ kg}}(0.05.\text{ m})^2 - 2(0.150)(9.80\text{ m/s}^2)(0.050\text{ m})}$$

$$v_f = \boxed{0.230\text{ m/s}}$$

Remarks Friction or drag from immersion in a fluid damps the motion of an object attached to a spring, eventually bringing the object to rest.

Exercise 5.9

Suppose the spring system in the last example starts at $x = 0$ and the attached object is given a kick to the right, so it has an initial speed of 0.600 m/s. (a) What distance from the origin does the object travel before coming to rest, assuming the surface is frictionless? (b) How does the answer change if the coefficient of kinetic friction is $\mu_k = 0.150$? (Use the quadratic formula.)

Answer (a) 0.0671 m (b) 0.0512 m

EXAMPLE 5.10 Circus Acrobat

Goal Use conservation of mechanical energy to solve a one-dimensional problem involving gravitational potential energy and spring potential energy.

Problem A 50.0-kg circus acrobat drops from a height of 2.00 meters straight down onto a springboard with a force constant of 8.00×10^3 N/m, as in Figure 5.22. By what maximum distance does she compress the spring?

Strategy Nonconservative forces are absent, so conservation of mechanical energy can be applied. At the two points of interest, the acrobat's initial position and the point of maximum spring compression, her velocity is zero, so the kinetic energy terms will be zero. Choose $y = 0$ as the point of maximum compression, so the final gravitational potential energy is zero. This choice also means that the initial position of the acrobat is $y_i = h + d$, where h is the acrobat's initial height above the platform and d is the spring's maximum compression.

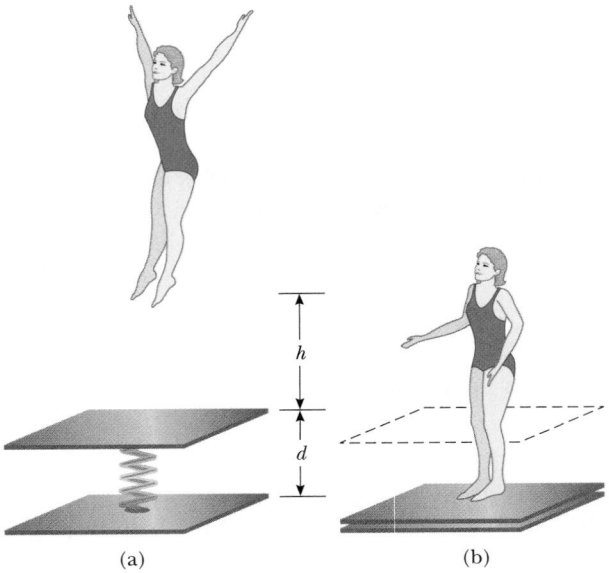

Figure 5.22 (Example 5.10) An acrobat drops onto a springboard, causing it to compress.

Solution

Use conservation of mechanical energy:

$$(KE + PE_g + PE_s)_i = (KE + PE_g + PE_s)_f \quad \textbf{(1)}$$

The only nonzero terms are the initial gravitational potential energy and the final spring potential energy.

$$0 + mg(h + d) + 0 = 0 + 0 + \tfrac{1}{2}kd^2$$

$$mg(h + d) = \tfrac{1}{2}kd^2$$

Substitute the given quantities, and rearrange the equation into standard quadratic form:

$$(50.0\text{ kg})(9.80\text{ m/s}^2)(2.00\text{ m} + d) = \tfrac{1}{2}(8.00 \times 10^3\text{ N/m})d^2$$

$$d^2 - (0.123\text{ m})d - 0.245\text{ m}^2 = 0$$

Solve with the quadratic formula (Equation A.8):

$$d = \boxed{0.560\text{ m}}$$

Remarks The other solution, $d = -0.437$ m, can be rejected because d was chosen to be a positive number at the outset. A change in the acrobat's center of mass, say, by crouching as she makes contact with the springboard, also affects the spring's compression, but that effect was neglected. Shock absorbers often involve springs, and this example illustrates how they work. The spring action of a shock absorber turns a dangerous jolt into a smooth deceleration, as excess kinetic energy is converted to spring potential energy.

Exercise 5.10

An 8.00-kg block drops straight down from a height of 1.00 m, striking a platform spring having force constant 1.00×10^3 N/m. Find the maximum compression of the spring.

Answer $d = 0.482$ m

EXAMPLE 5.11 A Block Projected up a Frictionless Incline

Goal Use conservation of mechanical energy to solve a problem involving gravitational potential energy, spring potential energy, and a ramp.

Problem A 0.500-kg block rests on a horizontal, frictionless surface as in Figure 5.23. The block is pressed back against a spring having a constant of $k = 625$ N/m, compressing the spring by 10.0 cm to point Ⓐ. Then the block is released. **(a)** Find the maximum distance d the block travels up the frictionless incline if $\theta = 30.0°$. **(b)** How fast is the block going when halfway to its maximum height?

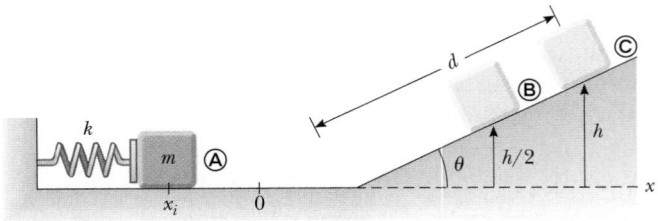

Figure 5.23 (Example 5.11)

Strategy In the absence of other forces, conservation of mechanical energy applies to parts **(a)** and **(b)**. In part **(a)**, the block starts at rest and is also instantaneously at rest at the top of the ramp, so the kinetic energies at Ⓐ and Ⓒ are both zero. Note that the question asks for a distance d along the ramp, not the height h. In part **(b)**, the system has both kinetic and gravitational potential energy at Ⓑ.

Solution

(a) Find the distance the block travels up the ramp.

Apply conservation of mechanical energy:

$$\tfrac{1}{2}mv_i^2 + mgy_i + \tfrac{1}{2}kx_i^2 = \tfrac{1}{2}mv_f^2 + mgy_f + \tfrac{1}{2}kx_f^2$$

Substitute $v_i = v_f = 0$, $y_i = 0$, $y_f = h = d\sin\theta$, and $x_f = 0$.

$$\tfrac{1}{2}kx_i^2 = mgh = mgd\sin\theta$$

Solve for the distance d and insert the known values:

$$d = \frac{\frac{1}{2}kx_i^2}{mg\sin\theta} = \frac{\frac{1}{2}(625\text{ N/m})(-0.100\text{ m})^2}{(0.500\text{ kg})(9.80\text{ m/s}^2)\sin(30.0°)}$$

$$= \boxed{1.28\text{ m}}$$

(b) Find the velocity at half the height, $h/2$. Note that $h = d\sin\theta = (1.28\text{ m})\sin 30.0° = 0.640\text{ m}$.

Use energy conservation again:

$$\tfrac{1}{2}mv_i^2 + mgy_i + \tfrac{1}{2}kx_i^2 = \tfrac{1}{2}mv_f^2 + mgy_f + \tfrac{1}{2}kx_f^2$$

Take $v_i = 0$, $y_i = 0$, $y_f = \frac{1}{2}h$, and $x_f = 0$, yielding

$$\tfrac{1}{2}kx_i^2 = \tfrac{1}{2}mv_f^2 + mg(\tfrac{1}{2}h)$$

Multiply by $2/m$, and solve for v_f:

$$\frac{k}{m}x_i^2 = v_f^2 + gh$$

$$v_f = \sqrt{\frac{k}{m}x_i^2 - gh}$$

$$= \sqrt{\left(\frac{625\text{ N/m}}{0.500\text{ kg}}\right)(-0.100\text{ m})^2 - (9.80\text{ m/s}^2)(0.640\text{ m})}$$

$$v_f = \boxed{2.50\text{ m/s}}$$

Exercise 5.11

A 1.00-kg block is shot horizontally from a spring, as in the previous example, and travels 0.500 m up along a frictionless ramp before coming to rest and sliding back down. If the ramp makes an angle of 45.0° with respect to the horizontal, and the spring was originally compressed by 0.120 m, find the spring constant.

Answer 481 N/m

Applying Physics 5.2 Accident Reconstruction

Sometimes people involved in automobile accidents make exaggerated claims of chronic pain due to subtle injuries to the neck or spinal column. The likelihood of injury can be determined by finding the change in velocity of a car during the accident. The larger the change in velocity, the more likely it is that the person suffered spinal injury resulting in chronic pain. How can reliable estimates for this change in velocity be found after the fact?

Explanation The metal and plastic of an automobile acts much like a spring, absorbing the car's kinetic energy by flexing during a collision. When the magnitude of the difference in velocity of the two cars is under five miles per hour, there is usually no visible damage, because bumpers are designed to absorb the impact and return to their original shape at such low speeds. At greater relative speeds there will be permanent damage to the vehicle. Despite the fact the structure of the car may not return to its original shape, a certain force per meter is still required to deform it, just as it takes a certain force per meter to compress a spring. The greater the original kinetic energy, the more the car is compressed during a collision, and the greater the damage. By using data obtained through crash tests, it's possible to obtain effective spring constants for all the different models of cars and determine reliable estimates of the change in velocity of a given vehicle during an accident. Medical research has established the likelihood of spinal injury for a given change in velocity, and the estimated velocity change can be used to help reduce insurance fraud.

5.5 SYSTEMS AND ENERGY CONSERVATION

Recall that the work–energy theorem can be written as

$$W_{nc} + W_c = \Delta KE$$

where W_{nc} represents the work done by nonconservative forces and W_c is the work done by conservative forces in a given physical context. As we have seen, any work done by conservative forces, such as gravity and springs, can be accounted for, by changes in potential energy. The work–energy theorem can therefore be written in the following way:

$$W_{nc} = \Delta KE + \Delta PE = (KE_f - KE_i) + (PE_f - PE_i) \quad \textbf{[5.19]}$$

where now, as previously stated, PE includes all potential energies. This equation is easily rearranged to:

$$W_{nc} = (KE_f + PE_f) - (KE_i + PE_i) \quad \textbf{[5.20]}$$

Recall, however, that the total mechanical energy is given by $E = KE + PE$. Making this substitution into Equation 5.20, we find that the work done on a system by all nonconservative forces is equal to the change in mechanical energy of that system:

$$W_{nc} = E_f - E_i = \Delta E \quad \textbf{[5.21]}$$

If the mechanical energy is changing, it has to be going somewhere. The energy either leaves the system and goes into the surrounding environment, or it stays in the system and is converted into a nonmechanical form such as thermal energy.

A simple example is a block sliding along a rough surface. Friction creates thermal energy, absorbed partly by the block and partly by the surrounding environment. When the block warms up, something called *internal energy* increases. The internal energy of a system is related to its temperature, which in turn is a consequence of the activity of its parts, such as the moving atoms of a gas or the vibration of atoms in a solid. (Internal energy will be studied in more detail in Chapter 12.)

Energy can be transferred between a nonisolated system and its environment. If positive work is done on the system, energy is transferred from the environment to the system. If negative work is done on the system, energy is transferred from the system to the environment.

So far, we have encountered three methods of storing energy in a system: kinetic energy, potential energy, and internal energy. On the other hand, we've seen only one way of transferring energy into or out of a system: through work. Other methods will be studied in later chapters, but are summarized here:

- **Work**, in the mechanical sense of this chapter, transfers energy to a system by displacing it with an applied force.
- **Heat** is the process of transferring energy through microscopic collisions between atoms or molecules. For example, a metal spoon resting in a cup of coffee becomes hot because some of the kinetic energy of the molecules in the liquid coffee is transferred to the spoon as internal energy.
- **Mechanical waves** transfer energy by creating a disturbance that propagates through air or another medium. For example, energy in the form of sound leaves your stereo system through the loudspeakers and enters your ears to stimulate the hearing process. Other examples of mechanical waves are seismic waves and ocean waves.
- **Electrical transmission** transfers energy through electric currents. This is how energy enters your stereo system or any other electrical device.
- **Electromagnetic radiation** transfers energy in the form of electromagnetic waves such as light, microwaves, and radio waves. Examples of this method of transfer include cooking a potato in a microwave oven and light energy traveling from the Sun to the Earth through space.

Conservation of Energy in General

The most important feature of the energy approach is the idea that energy is conserved; it can't be created or destroyed, only transferred from one form into another. This is the principle of **conservation of energy**.

The principle of conservation of energy is not confined to physics. In biology, energy transformations take place in myriad ways inside all living organisms. One example is the transformation of chemical energy to mechanical energy that causes flagella to move and propel an organism. Some bacteria use chemical energy to produce light. (See Figure 5.24.) Although the mechanisms that produce these light emissions are not well understood, living creatures often rely on this light for their existence. For example, certain fish have sacs beneath their eyes filled with light-emitting bacteria. The emitted light attracts creatures that become food for the fish.

APPLICATION

Flagellar Movement; Bioluminescence

Jan Hinsch/Science Photo Library/Photo Researchers, Inc.

Figure 5.24 This small plant, found in warm southern waters, exhibits bioluminescence, a process in which chemical energy is converted to light. The red areas are chlorophyll, which glows when excited by blue light.

Quick Quiz 5.4

A book of mass m is projected with a speed v across a horizontal surface. The book slides until it stops due to the friction force between the book and the surface. The surface is now tilted 30°, and the book is projected up the surface with the same initial speed v. When the book has come to rest, how does the decrease in mechanical energy of the book–Earth system compare with that when the book slid over the horizontal surface? (a) It's the same; (b) it's larger on the tilted surface; (c) it's smaller on the tilted surface; (d) more information is needed.

Applying Physics 5.3 Asteroid Impact!

An asteroid about a kilometer in radius has been blamed for the extinction of the dinosaurs 65 million years ago. How can a relatively small object, which could fit inside a college campus, inflict such injury on the vast biosphere of the Earth?

Explanation While such an asteroid is comparatively small, it travels at a very high speed relative to the Earth, typically on the order of 40 000 m/s. A roughly spherical asteroid one kilometer in radius and made mainly of rock has a mass of approximately 10 trillion kilograms—a small mountain of matter. The kinetic energy of such an asteroid would be about 10^{22} J, or 10 billion trillion joules. By contrast, the atomic bomb that devastated Hiroshima was equivalent to 15 kilotons of TNT, approximately 6×10^{13} J of energy. On striking the Earth, the asteroid's enormous kinetic energy changes into other forms, such as thermal energy, sound, and light, with a total energy release greater than 100 million Hiroshima explosions! Aside from the devastation in the immediate blast area and fires across a continent, gargantuan tidal waves would scour low-lying regions around the world and dust would block the sun for decades.

For this reason, asteroid impacts represent a threat to life on Earth. Asteroids large enough to cause widespread extinction hit Earth only every 60 million years or so. Smaller asteroids, of sufficient size to cause serious damage to civilization on a global scale, are thought to strike every five to ten thousand years. There have been several near misses by such asteroids in the last century and even in the last decade. In 1907, a small asteroid or comet fragment struck Tunguska, Siberia, annihilating a region 60 kilometers across. Had it hit northern Europe, millions of people might have perished.

Figure 5.25 is an asteroid map of the inner solar system. More asteroids are being discovered every year.

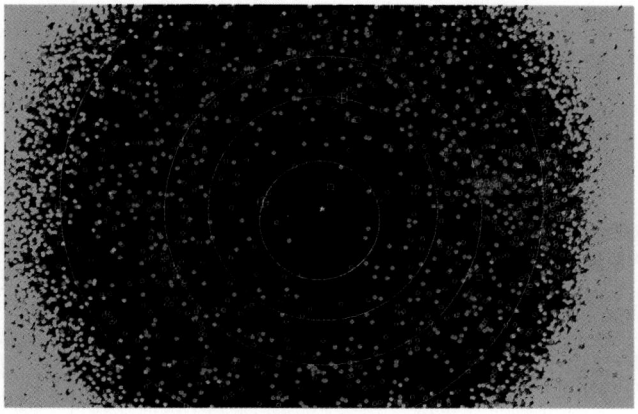

Gareth Williams, Minor Planet Center

Figure 5.25 Asteroid map of the inner solar system.

5.6 POWER

The rate at which energy is transferred is important in the design and use of practical devices, such as electrical appliances and engines of all kinds. The issue is particularly interesting for living creatures, since the maximum work per second, or power output, of an animal varies greatly with output duration. Power is defined as the rate of energy transfer with time:

Average power ▶

If an external force is applied to an object and if the work done by this force is W in the time interval Δt, then the **average power** delivered to the object during this interval is the work done divided by the time interval, or

$$\overline{\mathcal{P}} = \frac{W}{\Delta t} \qquad [5.22]$$

SI unit: watt (W = J/s)

It's sometimes useful to rewrite Equation 5.22 by substituting $W = F\Delta x$ and noticing that $\Delta x/\Delta t$ is the average speed of the object during the time Δt:

$$\overline{\mathcal{P}} = \frac{W}{\Delta t} = \frac{F\Delta x}{\Delta t} = F\overline{v} \qquad [5.23]$$

According to Equation 5.23, average power is a constant force times the average speed. The force F is the component of force in the direction of the average velocity. A more general definition can be written down with a little calculus and has the same form as Equation 5.23:

$$\mathcal{P} = Fv$$

The SI unit of power is the joule/sec, also called the **watt**, named after James Watt:

$$1\ \text{W} = 1\ \text{J/s} = 1\ \text{kg}\cdot\text{m}^2/\text{s}^3 \qquad [5.24]$$

The unit of power in the U. S. customary system is the horsepower (hp), where

$$1\ \text{hp} \equiv 550\ \frac{\text{ft}\cdot\text{lb}}{\text{s}} = 746\ \text{W} \qquad [5.25]$$

TIP 5.6 Watts the Difference?

Don't confuse the nonitalic symbol for watts, W, with the italic symbol W for work. A watt is a unit, the same as joules per second. Work is a concept, carrying units of joules.

The horsepower was first defined by Watt, who needed a large power unit to rate the power output of his new invention, the steam engine.

The watt is commonly used in electrical applications, but it can be used in other scientific areas as well. For example, European sports car engines are rated in kilowatts.

In electric power generation, it's customary to use the kilowatt-hour as a measure of energy. One kilowatt-hour (kWh) is the energy transferred in 1 h at the constant rate of 1 kW = 1 000 J/s. Therefore,

$$1\ \text{kWh} = (10^3\ \text{W})(3\ 600\ \text{s}) = (10^3\ \text{J/s})(3\ 600\ \text{s}) = 3.60 \times 10^6\ \text{J}$$

It's important to realize that a kilowatt-hour is a unit of energy, *not* power. When you pay your electric bill, you're buying energy, and that's why your bill lists a charge for electricity of about 10 cents/kWh. The amount of electricity used by an appliance can be calculated by multiplying its power rating (usually expressed in watts and valid only for normal household electrical circuits) by the length of time the appliance is operated. For example, an electric bulb rated at 100 W (= 0.100 kW) "consumes" 3.6×10^5 J of energy in 1 h.

EXAMPLE 5.12 Power Delivered by an Elevator Motor

Goal Apply the force-times-velocity definition of power.

Problem A 1.00×10^3-kg elevator carries a maximum load of 8.00×10^2 kg. A constant frictional force of 4.00×10^3 N retards its motion upward, as in Figure 5.26. What minimum power, in kilowatts and in horsepower, must the motor deliver to lift the fully loaded elevator at a constant speed of 3.00 m/s?

Strategy To solve this problem, we need to determine the force the elevator's motor must deliver through the force of tension in the cable, $\vec{\mathbf{T}}$. Substituting this force together with the given speed v into $\mathcal{P} = Fv$ gives the desired power. The tension in the cable, T, can be found with Newton's second law.

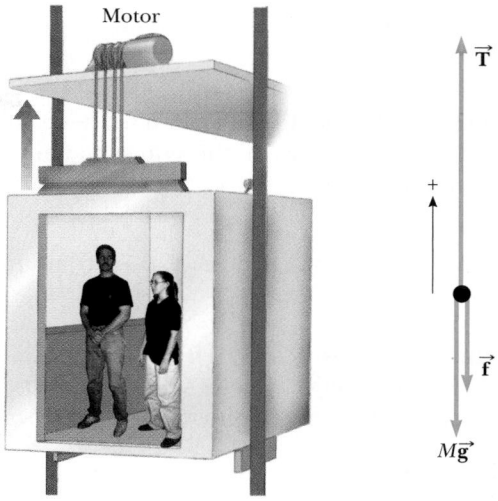

Figure 5.26 (Example 5.12) The motor exerts an upward force $\vec{\mathbf{T}}$ on the elevator. A frictional force $\vec{\mathbf{f}}$ and the force of gravity $M\vec{\mathbf{g}}$ act downwards.)

Solution

Apply Newton's second law to the elevator:

$$\Sigma \vec{\mathbf{F}} = m\vec{\mathbf{a}}$$

The velocity is constant, so the acceleration is zero. The forces acting on the elevator are the force of tension in the cable, $\vec{\mathbf{T}}$, the friction $\vec{\mathbf{f}}$, and gravity $M\vec{\mathbf{g}}$, where M is the mass of the elevator.

$$\vec{\mathbf{T}} + \vec{\mathbf{f}} + M\vec{\mathbf{g}} = 0$$

Write the equation in terms of its components:

$$T - f - Mg = 0$$

Solve this equation for the tension T and evaluate it:

$$T = f + Mg$$
$$= 4.00 \times 10^3 \text{ N} + (1.80 \times 10^3 \text{ kg})(9.80 \text{ m/s}^2)$$
$$T = 2.16 \times 10^4 \text{ N}$$

Substitute this value of T for F in the power equation:

$$\mathcal{P} = Fv = (2.16 \times 10^4 \text{ N})(3.00 \text{ m/s}) = 6.48 \times 10^4 \text{ W}$$
$$\mathcal{P} = 64.8 \text{ kW} = \boxed{86.9 \text{ hp}}$$

Remarks The friction force acts to retard the motion, requiring more power. For a descending elevator, the friction force can actually reduce the power requirement.

Exercise 5.12

Suppose the same elevator with the same load descends at 3.00 m/s. What minimum power is required? (Here, the motor removes energy from the elevator by not allowing it to fall freely.)

Answer 4.09×10^4 W = 54.9 hp

EXAMPLE 5.13 Shamu Sprint

Goal Calculate the power needed to increase an object's kinetic energy.

Problem Killer whales are known to reach 32 ft in length and have a mass of over 8 000 kg. They are also very quick, able to accelerate up to 30 mi/h in a matter of seconds. Disregarding the considerable drag force of water, calculate the average power a killer whale named Shamu with mass 8.00×10^3 kg would need to generate to reach a speed of 12.0 m/s in 6.00 s.

Strategy Find the change in kinetic energy of Shamu and use the work–energy theorem to obtain the minimum work Shamu has to do to effect this change. (Internal and external friction forces increase the necessary amount of energy.) Divide by the elapsed time to get the power.

Solution

Calculate the change in Shamu's kinetic energy. By the work–energy theorem, this equals the minimum work Shamu must do:

$$\begin{aligned}\Delta KE &= \tfrac{1}{2}mv_f^2 - \tfrac{1}{2}mv_i^2\\ &= \tfrac{1}{2}\cdot 8.00\times 10^3\,\text{kg}\cdot(12.0\,\text{m/s})^2 - 0\\ &= 5.76\times 10^5\,\text{J}\end{aligned}$$

Divide by the elapsed time (Eq. 5.22), noting that $W = \Delta KE$.

$$\mathcal{P} = \frac{W}{\Delta t} = \frac{5.76\times 10^5\,\text{J}}{6.00\,\text{s}} = 9.60\times 10^4\,\text{W}$$

Remarks This is enough power to run a moderate-sized office building! The actual requirements are larger because of friction in the water and muscular tissues. Something similar can be done with gravitational potential energy, as the exercise illustrates.

Exercise 5.13

What minimum average power must a 35-kg human boy generate climbing up the stairs to the top of the Washington monument? The trip up the nearly 170-m-tall building takes him 10 minutes. Include only work done against gravity, ignoring biological efficiency.

Answer 97 W

EXAMPLE 5.14 Speedboat Power

Goal Combine power, the work–energy theorem and nonconservative forces with one-dimensional kinematics.

Problem How much power would a 1.00×10^3-kg speedboat need to go from rest to 20.0 m/s in 5.00 s, assuming the water exerts a constant drag force of magnitude $f_d = 5.00\times 10^2$ N and the acceleration is constant.

Strategy The power is provided by the engine, which creates a nonconservative force. Use the work–energy theorem together with the work done by the engine, W_{engine}, and the work done by the drag force, W_{drag}, on the left-hand side. Use one-dimensional kinematics to find the acceleration and then the displacement Δx. Solve the work–energy theorem for W_{engine}, and divide by the elapsed time to get the power.

Solution

Write the work–energy theorem:

$$W_{\text{net}} = \Delta KE = \tfrac{1}{2}mv_f^2 - \tfrac{1}{2}mv_i^2$$

Fill in the two work terms and take $v_i = 0$:

$$W_{\text{engine}} + W_{\text{drag}} = W_{\text{engine}} - f_d\,\Delta x = \tfrac{1}{2}mv_f^2 \qquad \textbf{(1)}$$

To get the displacement Δx, first find the acceleration using the velocity equation of kinematics:

$$v_f = at + v_i \;\rightarrow\; v_f = at$$
$$(20.0\,\text{m/s}) = a(5.00\,\text{s}) \;\rightarrow\; a = 4.00\,\text{m/s}^2$$

Substitute a into the time-independent kinematics equation, and solve for Δx:

$$v_f^2 - v_i^2 = 2a\,\Delta x$$
$$(20.0\,\text{m/s})^2 - 0^2 = 2(4.00\,\text{m/s}^2)\Delta x$$
$$\Delta x = 50.0\,\text{m}$$

Now that we know Δx, we can find the mechanical energy lost due to friction.

$$W_{\text{fric}} = -f_d\,\Delta x = -(5.00\times 10^2\,\text{N})(50.0\,\text{m}) = -2.50\times 10^4\,\text{J}$$

Solve equation (1) for W_{engine}:

$$\begin{aligned}W_{\text{engine}} &= \tfrac{1}{2}mv_f^2 - f_d\,\Delta x\\ &= \tfrac{1}{2}(1.00\times 10^3\,\text{kg})(20.0\,\text{m/s})^2 - (-2.50\times 10^4\,\text{J})\\ W_{\text{engine}} &= 2.25\times 10^5\,\text{J}\end{aligned}$$

Compute the power:

$$\mathcal{P} = \frac{W_{\text{engine}}}{\Delta t} = \frac{2.25\times 10^5\,\text{J}}{5.00\,\text{s}} = 4.50\times 10^4\,\text{W} = 60.3\,\text{hp}$$

Remarks In fact, drag forces generally get larger with increasing speed.

Exercise 5.14

What power must be supplied to push a 5.00-kg block from rest to 10.0 m/s in 5.00 s when the coefficient of kinetic friction between the block and surface is 0.250? Assume the acceleration is uniform.

Answer 111 W

Energy and Power in a Vertical Jump

The stationary jump consists of two parts: extension and free flight.[2] In the extension phase the person jumps up from a crouch, straightening the legs and throwing up the arms; the free-flight phase occurs when the jumper leaves the ground. Because the body is an extended object and different parts move with different speeds, we describe the motion of the jumper in terms of the position and velocity of the **center of mass (CM)**, which is the point in the body at which all the mass may be considered to be concentrated. Figure 5.27 shows the position and velocity of the CM at different stages of the jump.

Using the principle of the conservation of mechanical energy, we can find H, the maximum increase in height of the CM, in terms of the velocity v_{CM} of the CM at liftoff. Taking PE_i, the gravitational potential energy of the jumper–Earth system just as the jumper lifts off from the ground to be zero, and noting that the kinetic energy KE_f of the jumper at the peak is zero, we have

$$PE_i + KE_i = PE_f + KE_f$$

$$\tfrac{1}{2} m v_{CM}^2 = mgH \qquad \text{or} \qquad H = \frac{v_{CM}^2}{2g}$$

We can estimate v_{CM} by assuming that the acceleration of the CM is constant during the extension phase. If the depth of the crouch is h and the time for extension is Δt, we find that $v_{CM} = 2\bar{v} = 2h/\Delta t$. Measurements on a group of male college students show typical values of $h = 0.40$ m and $\Delta t = 0.25$ s, the latter value being set by the fixed speed with which muscle can contract. Substituting, we obtain

$$v_{CM} = 2(0.40\text{ m})/(0.25\text{ s}) = 3.2\text{ m/s}$$

and

$$H = \frac{v_{CM}^2}{2g} = \frac{(3.2\text{ m/s})^2}{2(9.80\text{ m/s}^2)} = 0.52\text{ m}$$

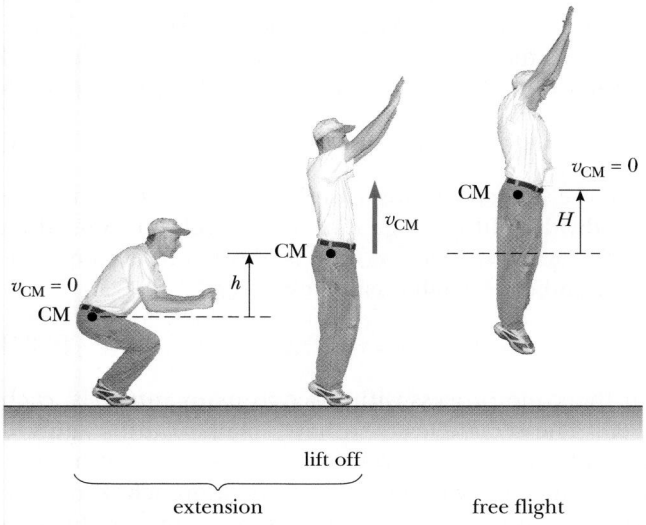

Figure 5.27 Extension and free flight in the vertical jump.

[2]For more information on this topic, see E. J. Offenbacher, *American Journal of Physics*, **38**, 829 (1969).

TABLE 5.1

Maximum Power Output from Humans over Various Periods

Power	Time
2 hp, or 1 500 W	6 s
1 hp, or 750 W	60 s
0.35 hp, or 260 W	35 min
0.2 hp, or 150 W	5 h
0.1 hp, or 75 W (safe daily level)	8 h

APPLICATION

Diet Versus Exercise in Weight-loss Programs

Measurements on this same group of students found that H was between 0.45 m and 0.61 m in all cases, confirming the basic validity of our simple calculation.

In order to relate the abstract concepts of energy, power, and efficiency to humans, it's interesting to calculate these values for the vertical jump. The kinetic energy given to the body in a jump is $KE = \frac{1}{2}mv_{CM}^2$, and for a person of mass 68 kg, the kinetic energy is

$$KE = \tfrac{1}{2}(68\text{ kg})(3.2\text{ m/s})^2 = 3.5 \times 10^2\text{ J}$$

Although this may seem like a large expenditure of energy, we can make a simple calculation to show that jumping and exercise in general are not good ways to lose weight, in spite of their many health benefits. Since the muscles are at most 25% efficient at producing kinetic energy from chemical energy (muscles always produce a lot of internal energy and kinetic energy as well as work—that's why you sweat when you work out), they use up four times the 350 J (about 1 400 J) of chemical energy in one jump. This chemical energy ultimately comes from the food we eat, with energy content given in units of food calories and one food calorie equal to 4 200 J. So the total energy supplied by the body as internal energy and kinetic energy in a vertical jump is only about one-third of a food calorie! You are a lot better off not eating that piece of cheesecake than trying to work it off by jumping.

Finally, it's interesting to calculate the mechanical power that can be generated by the body in strenuous activity for brief periods. Here we find that

$$\mathcal{P} = \frac{KE}{\Delta t} = \frac{3.5 \times 10^2\text{ J}}{0.25\text{ s}} = 1.4 \times 10^3\text{ W}$$

or $(1\,400\text{ W})(1\text{ hp}/746\text{ W}) = 1.9\text{ hp}$. So humans can produce about 2 hp of mechanical power for periods on the order of seconds. Table 5.1 shows the maximum power outputs from humans for various periods while bicycling and rowing, activities in which it is possible to measure power output accurately.

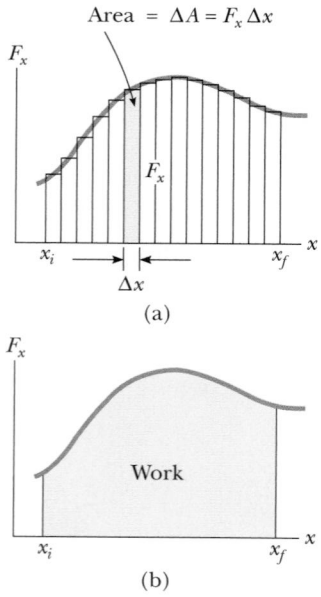

Figure 5.28 (a) The work done by the force component F_x for the small displacement Δx is $F_x \Delta x$, which equals the area of the shaded rectangle. The total work done for the displacement from x_i to x_f is approximately equal to the sum of the areas of all the rectangles. (b) The work done by the component F_x of the varying force as the particle moves from x_i to x_f is exactly equal to the area under the curve shown.

5.7 WORK DONE BY A VARYING FORCE

Suppose an object is displaced along the x-axis under the action of a force F_x that acts in the x-direction and varies with position, as shown in Figure 5.28. The object is displaced in the direction of increasing x from $x = x_i$ to $x = x_f$. In such a situation, we can't use Equation 5.1 to calculate the work done by the force because this relationship applies only when $\vec{\mathbf{F}}$ is constant in magnitude and direction. However, if we imagine that the object undergoes the *small* displacement Δx shown in Figure 5.28a, then the x-component F_x of the force is nearly constant over this interval and we can approximate the work done by the force for this small displacement as

$$W_1 \cong F_x \Delta x \qquad \textbf{[5.26]}$$

This quantity is just the area of the shaded rectangle in Figure 5.28a. If we imagine that the curve of F_x versus x is divided into a large number of such intervals, then the total work done for the displacement from x_i to x_f is approximately equal to the sum of the areas of a large number of small rectangles:

$$W \cong F_1\Delta x_1 + F_2\Delta x_2 + F_3\Delta x_3 + \cdots \qquad \textbf{[5.27]}$$

Now imagine going through the same process with twice as many intervals, each half the size of the original Δx. The rectangles then have smaller widths and will better approximate the area under the curve. Continuing the process of increasing the number of intervals while allowing their size to approach zero, the number of terms in the sum increases without limit, but the value of the sum approaches a definite value equal to the area under the curve bounded by F_x and the x-axis in Figure 5.28b. In other words, **the work done by a variable force acting on an object that undergoes a displacement is equal to the area under the graph of F_x versus x.**

A common physical system in which force varies with position consists of a block on a horizontal, frictionless surface connected to a spring, as discussed in Section 5.4. When the spring is stretched or compressed a small distance x from its equilibrium position $x = 0$, it exerts a force on the block given by $F_s = -kx$, where k is the force constant of the spring.

Now let's determine the work done by an *external agent* on the block as the spring is stretched *very slowly* from $x_i = 0$ to $x_f = x_{max}$, as in Active Figure 5.29a. This work can be easily calculated by noting that at any value of the displacement, Newton's third law tells us that the applied force $\vec{F}_{app}$ is equal in magnitude to the spring force $\vec{F}_s$ and acts in the opposite direction, so that $F_{app} = -(-kx) = kx$. A plot of F_{app} versus x is a straight line, as shown in Active Figure 5.29b. Therefore, the work done by this applied force in stretching the spring from $x = 0$ to $x = x_{max}$ is the area under the straight line in that figure, which in this case is the area of the shaded triangle:

$$W_{F_{app}} = \tfrac{1}{2}kx_{max}^2$$

During this same time the spring has done exactly the same amount of work, but that work is negative, because the spring force points in the direction opposite the motion. The potential energy of the system is exactly equal to the work done by the applied force and is the same sign, which is why potential energy is thought of as stored work.

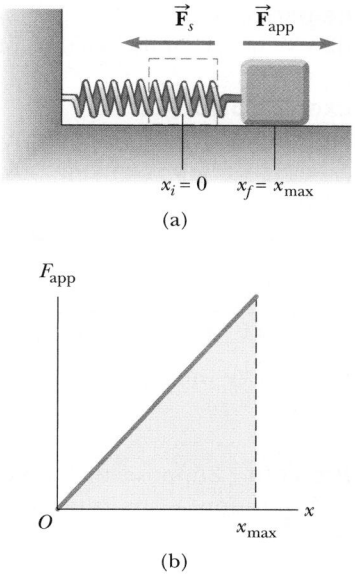

ACTIVE FIGURE 5.29
(a) A block being pulled from $x_i = 0$ to $x_f = x_{max}$ on a frictionless surface by a force $\vec{F}_{app}$. If the process is carried out very slowly, the applied force is equal in magnitude and opposite in direction to the spring force at all times. (b) A graph of F_{app} versus x.

PhysicsNow™
Log into PhysicsNow at **www.cp7e.com**, and go to Active Figure 5.29 to observe the block's motion for various maximum displacements and spring constants.

EXAMPLE 5.15 Work Required to Stretch a Spring

Goal Apply the graphical method of finding work.

Problem One end of a horizontal spring ($k = 80.0$ N/m) is held fixed while an external force is applied to the free end, stretching it slowly from $x_Ⓐ = 0$ to $x_Ⓑ = 4.00$ cm. **(a)** Find the work done by the applied force on the spring. **(b)** Find the additional work done in stretching the spring from $x_Ⓑ = 4.00$ cm to $x_Ⓒ = 7.00$ cm.

Strategy For part (a), simply find the area of the smaller triangle, using $A = \frac{1}{2}bh$, one-half the base times the height. For part (b), the easiest way to find the additional work done from $x_Ⓑ = 4.00$ cm to $x_Ⓒ = 7.00$ cm is to find the area of the new, larger triangle and subtract the area of the smaller triangle.

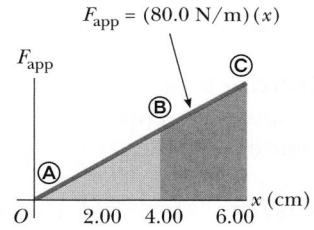

Figure 5.30 (Example 5.15) A graph of the external force required to stretch a spring that obeys Hooke's law versus the elongation of the spring.

Solution

(a) Find the work from $x_Ⓐ = 0$ cm to $x_Ⓑ = 4.00$ cm.

Compute the area of the smaller triangle:

$$W = \tfrac{1}{2}kx_B^2 = \tfrac{1}{2}(80.0\ \text{N/m})(0.040\ \text{m})^2 = \boxed{0.064\,0\ \text{J}}$$

(b) Find the work from $x_Ⓑ = 4.00$ cm to $x_Ⓒ = 7.00$ cm.

Compute the area of the large triangle, and subtract the area of the smaller triangle:

$$W = \tfrac{1}{2}kx_C^2 - \tfrac{1}{2}kx_B^2$$

$$W = \tfrac{1}{2}(80.0\ \text{N/m})(0.070\,0\ \text{m})^2 - 0.064\,0\ \text{J}$$

$$= 0.196\ \text{J} - 0.064\,0\ \text{J}$$

$$= \boxed{0.132\ \text{J}}$$

Remarks Only simple geometries—rectangles and triangles—can be solved exactly with this method. More complex shapes require calculus or the square-counting technique in the next worked example.

Exercise 5.15
How much work is required to stretch this same spring from $x_i = 5.00$ cm to $x_f = 9.00$ cm?

Answer 0.224 J

EXAMPLE 5.16 Estimating Work by Counting Boxes

Goal Use the graphical method and counting boxes to estimate the work done by a force.

Problem Suppose the force applied to stretch a thick piece of elastic changes with position as indicated in Figure 5.31a. Estimate the work done by the applied force.

Strategy To find the work, simply count the number of boxes underneath the curve, and multiply that number by the area of each box. The curve will pass through the middle of some boxes, in which case only an estimated fractional part should be counted.

Solution
There are 62 complete or nearly complete boxes under the curve, 6 boxes that are about half under the curve, and a triangular area from $x = 0$ m to $x = 0.10$ m that is equivalent to 1 box, for a total of about 66 boxes. Since the area of each box is 0.10 J, the total work done is approximately $66 \times 0.10\,\text{J} = 6.6\,\text{J}$.

Remarks Mathematically, there are a number of other methods for creating such estimates, all involving adding up regions approximating the area. To get a better estimate, make smaller boxes.

Exercise 5.16
Suppose the applied force necessary to pull the drawstring on a bow is given by Figure 5.31b. Find the approximate work done by counting boxes.

Answer About 50 J. (Individual answers may vary.)

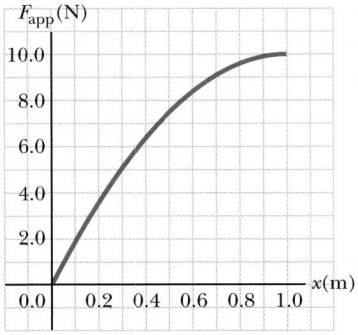

(a)

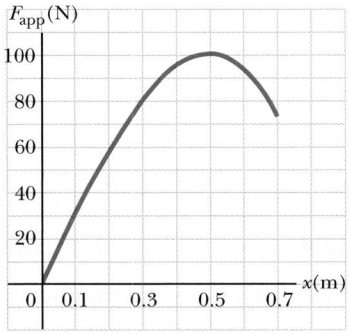

(b)

Figure 5.31 (a) (Example 5.16) (b) (Exercise 5.16)

SUMMARY

PhysicsNow™ Take a practice test by logging into PhysicsNow at **www.cp7e.com** and clicking on the Pre-Test link for this chapter.

5.1 Work

The work done on an object by a constant force is

$$W = (F\cos\theta)\Delta x \qquad [5.2]$$

where F is the magnitude of the force, Δx is the object's displacement, and θ is the angle between the direction of the force $\vec{\mathbf{F}}$ and the displacement $\Delta\vec{\mathbf{x}}$. Solving simple problems requires substituting values into this equation. More complex problems, such as those involving friction, often require using Newton's second law, $m\vec{\mathbf{a}} = \Sigma\vec{\mathbf{F}}$, to determine forces.

5.2 Kinetic Energy and the WorK–Energy Theorem

The kinetic energy of a body with mass m and speed v is given by

$$KE \equiv \tfrac{1}{2}mv^2 \qquad [5.5]$$

The work–energy theorem states that the net work done on an object of mass m is equal to the change in its kinetic energy, or

$$W_{\text{net}} = KE_f - KE_i = \Delta KE \qquad [5.6]$$

Work and energy of any kind carry units of joules. Solving problems involves finding the work done by each force acting on the object and summing them up, which is W_{net}, followed by substituting known quantities into Equation 5.6, solving for the unknown quantity.

Conservative forces are special: Work done against them can be recovered—it's conserved. An example is gravity: The work done in lifting an object through a height is effectively stored in the gravity field and can be recovered in the kinetic energy of the object simply by letting it fall. Nonconservative forces, such as surface friction and drag, dissipate energy in a form that can't be readily recovered. To account for such forces, the work–energy theorem can be rewritten as

$$W_{nc} + W_c = \Delta KE \qquad [5.7]$$

where W_{nc} is the work done by nonconservative forces and W_c is the work done by conservative forces.

5.3 Gravitational Potential Energy

The gravitational force is a conservative field. Gravitational potential energy is another way of accounting for gravitational work W_g:

$$W_g = -(PE_f - PE_i) = -(mgy_f - mgy_i) \qquad [5.11]$$

To find the change in gravitational potential energy as an object of mass m moves between two points in a gravitational field, substitute the values of the object's y-coordinates.

The work–energy theorem can be generalized to include gravitational potential energy:

$$W_{nc} = (KE_f - KE_i) + (PE_f - PE_i) \qquad [5.12]$$

Gravitational work and gravitational potential energy should not both appear in the work–energy theorem at the same time, only one or the other, because they're equivalent. Setting the work due to nonconservative forces to zero and substituting the expressions for KE and PE, a form of the conservation of mechanical energy with gravitation can be obtained:

$$\tfrac{1}{2}mv_i^2 + mgy_i = \tfrac{1}{2}mv_f^2 + mgy_f \qquad [5.14]$$

To solve problems with this equation, identify two points in the system—one where information is known and the other where information is desired. Substitute and solve for the unknown quantity.

The work done by other forces, as when frictional forces are present, isn't always zero. In that case, identify two points as before, calculate the work due to all other forces, and solve for the unknown in Equation 5.12.

5.4 Spring Potential Energy

The spring force is conservative, and its potential energy is given by

$$PE_s \equiv \tfrac{1}{2}kx^2 \qquad [5.16]$$

Spring potential energy can be put into the work–energy theorem, which then reads

$$W_{nc} = (KE_f - KE_i) + (PE_{gf} - PE_{gi}) + (PE_{sf} - PE_{si}) \qquad [5.17]$$

When nonconservative forces are absent, $W_{nc} = 0$ and mechanical energy is conserved.

5.5 Systems and Energy Conservation

The principle of the conservation of energy states that energy can't be created or destroyed. It can be transformed, but the total energy content of any isolated system is always constant. The same is true for the universe at large. The work done by all nonconservative forces acting on a system equals the change in the total mechanical energy of the system:

$$W_{nc} = (KE_f + PE_f) - (KE_i + PE_i) = E_f - E_i \qquad [5.20–21]$$

where PE represents all potential energies present.

5.6 Power

Average power is the amount of energy transferred divided by the time taken for the transfer:

$$\overline{\mathcal{P}} = \frac{W}{\Delta t} \qquad [5.22]$$

This expression can also be written

$$\overline{\mathcal{P}} = F\overline{v} \qquad [5.23]$$

where $\overline{v}$ is the object's average speed. The unit of power is the watt (W = J/s). To solve simple problems, substitute given quantities into one of these equations. More difficult problems usually require finding the work done on the object using the work–energy theorem or the definition of work.

CONCEPTUAL QUESTIONS

1. Consider a tug-of-war as in Figure Q5.1, in which two teams pulling on a rope are evenly matched, so that no motion takes place. Is work done on the rope? On the pullers? On the ground? Is work done on anything?
2. Discuss whether any work is being done by each of the following agents and, if so, whether the work is positive or negative: (a) a chicken scratching the ground, (b) a person studying, (c) a crane lifting a bucket of concrete, (d) the force of gravity on the bucket in part (c), (e) the leg muscles of a person in the act of sitting down.
3. If the height of a playground slide is kept constant, will the length of the slide or whether it has bumps make any difference in the final speed of children playing on it? Assume that the slide is slick enough to be considered

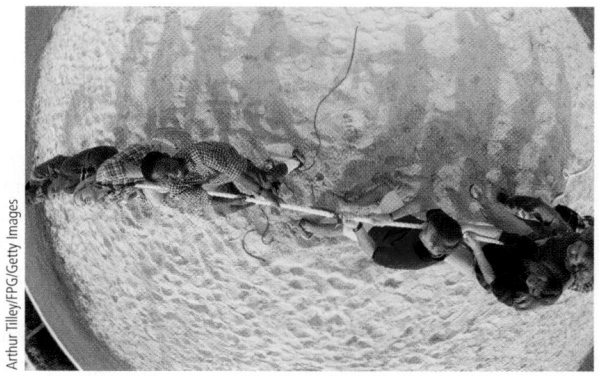

Arthur Tilley/FPG/Getty Images

Figure Q5.1

frictionless. Repeat this question, assuming that the slide is not frictionless.

4. (a) Can the kinetic energy of a system be negative? (b) Can the gravitational potential energy of a system be negative? Explain.

5. Roads going up mountains are formed into switchbacks, with the road weaving back and forth along the face of the slope such that there is only a gentle rise on any portion of the roadway. Does this configuration require any less work to be done by an automobile climbing the mountain, compared with one traveling on a roadway that is straight up the slope? Why are switchbacks used?

6. (a) If the speed of a particle is doubled, what happens to its kinetic energy? (b) If the net work done on a particle is zero, what can be said about its speed?

7. As a simple pendulum swings back and forth, the forces acting on the suspended object are the force of gravity, the tension in the supporting cord, and air resistance. (a) Which of these forces, if any, does no work on the pendulum? (b) Which of these forces does negative work at all times during the pendulum's motion? (c) Describe the work done by the force of gravity while the pendulum is swinging.

8. A bowling ball is suspended from the ceiling of a lecture hall by a strong cord. The ball is drawn away from its equilibrium position and released from rest at the tip of the demonstrator's nose, as shown in Figure Q5.8. If the demonstrator remains stationary, explain why the ball does not strike her on its return swing. Would this demonstrator be safe if the ball were given a push from its starting position at her nose?

Figure Q5.8

9. An older model car accelerates from 0 to speed v in 10 seconds. A newer, more powerful sports car accelerates from 0 to $2v$ in the same time. What is the ratio of the powers expended by the two cars? Assume the energy coming from the engine appears only as kinetic energy of the cars.

10. During a stress test of the cardiovascular system, a patient walks and runs on a treadmill. (a) Is the energy expended by the patient equivalent to the energy of walking and running on the ground? Explain. (b) What effect, if any, does tilting the treadmill upward have? Discuss.

11. When a punter kicks a football, is he doing any work on the ball while the toe of his foot is in contact with it? Is he doing any work on the ball after it loses contact with his toe? Are any forces doing work on the ball while it is in flight?

12. As a sled moves across a flat, snow-covered field at constant velocity, is any work done? How does air resistance enter into the picture?

13. A weight is connected to a spring that is suspended vertically from the ceiling. If the weight is displaced downward from its equilibrium position and released, it will oscillate up and down. If air resistance is neglected, will the total mechanical energy of the system (weight plus Earth plus spring) be conserved? How many forms of potential energy are there for this situation?

14. The driver of a car slams on her brakes to avoid colliding with a deer crossing the highway. What happens to the car's kinetic energy as it comes to rest?

15. Suppose you are reshelving books in a library. You lift a book from the floor to the top shelf. The kinetic energy of the book on the floor was zero, and the kinetic energy of the book on the top shelf is zero, so there is no change in kinetic energy. Yet you did some work in lifting the book. Is the work–energy theorem violated?

16. The feet of a standing person of mass m exert a force equal to mg on the floor, and the floor exerts an equal and opposite force upwards on the feet, which we call the normal force. During the extension phase of a vertical jump (see page 145), the feet exert a force on the floor that is greater than mg, so the normal force is greater than mg. As you learned in Chapter 4, we can use this result and Newton's second law to calculate the acceleration of the jumper: $a = F_{net}/m = (n - mg)/m$.

    Using energy ideas, we know that work is performed on the jumper to give him or her kinetic energy. But the normal force can't perform any work here, because the feet don't undergo any displacement. How is energy transferred to the jumper?

17. An Earth satellite is in a circular orbit at an altitude of 500 km. Explain why the work done by the gravitational force acting on the satellite is zero. Using the work–energy theorem, what can you say about the speed of the satellite?

18. In most circumstances, the normal force acting on an object and the force of static friction do no work on the object. However, the reason that the work is zero is different for the two cases. In each case, explain why the work done by the force is zero.

19. In most situations we have encountered in this chapter, frictional forces tend to reduce the kinetic energy of an object. However, frictional forces can sometimes increase an object's kinetic energy. Describe a few situations in which friction causes an increase in kinetic energy.

20. Discuss the energy transformations that occur as a pole vaulter runs at high speeds and attempts to clear a bar that is about 5 m from the ground. In your analysis, you must consider changes in the kinetic energy of the runner, the elastic potential energy of the pole as it bends, and the gravitational potential energy of the vaulter. Ignore rotational motion.

PROBLEMS

1, 2, 3 = straightforward, intermediate, challenging □ = full solution available in *Student Solutions Manual/Study Guide*
PhysicsNow™ = coached problem with hints available at **www.cp7e.com** = biomedical application

Section 5.1 Work

1. A weight lifter lifts a 350-N set of weights from ground level to a position over his head, a vertical distance of 2.00 m. How much work does the weight lifter do, assuming he moves the weights at constant speed?

2. If a man lifts a 20.0-kg bucket from a well and does 6.00 kJ of work, how deep is the well? Assume that the speed of the bucket remains constant as it is lifted.

3. A tugboat exerts a constant force of 5.00×10^3 N on a ship moving at constant speed through a harbor. How much work does the tugboat do on the ship if each moves a distance of 3.00 km?

4. A shopper in a supermarket pushes a cart with a force of 35 N directed at an angle of 25° downward from the horizontal. Find the work done by the shopper as she moves down a 50-m length of aisle.

5. Starting from rest, a 5.00-kg block slides 2.50 m down a rough 30.0° incline. The coefficient of kinetic friction between the block and the incline is $\mu_k = 0.436$. Determine (a) the work done by the force of gravity, (b) the work done by the friction force between block and incline, and (c) the work done by the normal force.

6. A horizontal force of 150 N is used to push a 40.0-kg packing crate a distance of 6.00 m on a rough horizontal surface. If the crate moves at constant speed, find (a) the work done by the 150-N force and (b) the coefficient of kinetic friction between the crate and surface.

□ **7.** A sledge loaded with bricks has a total mass of 18.0 kg and is pulled at constant speed by a rope inclined at 20.0° above the horizontal. The sledge moves a distance of 20.0 m on a horizontal surface. The coefficient of kinetic friction between the sledge and surface is 0.500. (a) What is the tension in the rope? (b) How much work is done by the rope on the sledge? (c) What is the mechanical energy lost due to friction?

8. A block of mass 2.50 kg is pushed 2.20 m along a frictionless horizontal table by a constant 16.0-N force directed 25.0° below the horizontal. Determine the work done by (a) the applied force, (b) the normal force exerted by the table, (c) the force of gravity, and (d) the net force on the block.

Section 5.2 Kinetic Energy and the WorK–Energy Theorem

9. A mechanic pushes a 2.50×10^3-kg car from rest to a speed of v, doing 5 000 J of work in the process. During this time, the car moves 25.0 m. Neglecting friction between car and road, find (a) v and (b) the horizontal force exerted on the car.

10. A 7.00-kg bowling ball moves at 3.00 m/s. How fast must a 2.45-g Ping-Pong ball move so that the two balls have the same kinetic energy?

11. A person doing a chin-up weighs 700 N, exclusive of the arms. During the first 25.0 cm of the lift, each arm exerts an upward force of 355 N on the torso. If the upward movement starts from rest, what is the person's velocity at that point?

12. A crate of mass 10.0 kg is pulled up a rough incline with an initial speed of 1.50 m/s. The pulling force is 100 N parallel to the incline, which makes an angle of 20.0° with the horizontal. The coefficient of kinetic friction is 0.400, and the crate is pulled 5.00 m. (a) How much work is done by gravity? (b) How much mechanical energy is lost due to friction? (c) How much work is done by the 100-N force? (d) What is the change in kinetic energy of the crate? (e) What is the speed of the crate after being pulled 5.00 m?

□ **13.** A 70-kg base runner begins his slide into second base when he is moving at a speed of 4.0 m/s. The coefficient of friction between his clothes and Earth is 0.70. He slides so that his speed is zero just as he reaches the base. (a) How much mechanical energy is lost due to friction acting on the runner? (b) How far does he slide?

14. An outfielder throws a 0.150-kg baseball at a speed of 40.0 m/s and an initial angle of 30.0°. What is the kinetic energy of the ball at the highest point of its motion?

15. A 2.0-g bullet leaves the barrel of a gun at a speed of 300 m/s. (a) Find its kinetic energy. (b) Find the average force exerted by the expanding gases on the bullet as it moves the length of the 50-cm-long barrel.

16. A 0.60-kg particle has a speed of 2.0 m/s at point A and a kinetic energy of 7.5 J at point B. What is (a) its kinetic energy at A? (b) its speed at point B? (c) the total work done on the particle as it moves from A to B?

□ **17.** PhysicsNow™ A 2 000-kg car moves down a level highway under the actions of two forces: a 1 000-N forward force exerted on the drive wheels by the road and a 950-N resistive force. Use the work–energy theorem to find the speed of the car after it has moved a distance of 20 m, assuming that it starts from rest.

18. On a frozen pond, a 10-kg sled is given a kick that imparts to it an initial speed of $v_0 = 2.0$ m/s. The coefficient of kinetic friction between sled and ice is $\mu_k = 0.10$. Use the work–energy theorem to find the distance the sled moves before coming to rest.

Section 5.3 Gravitational Potential Energy

Section 5.4 Spring Potential Energy

19. Find the height from which you would have to drop a ball so that it would have a speed of 9.0 m/s just before it hits the ground.

20. A flea is able to jump about 0.5 m. It has been said that if a flea were as big as a human, it would be able to jump over a 100-story building! When an animal jumps, it converts work done in contracting muscles into gravitational potential energy (with some steps in between). The maximum force exerted by a muscle is proportional to its cross-sectional area, and the work done by the muscle is this force times the length of contraction. If we magnified a flea by a factor of 1 000, the cross section of its muscle would increase by $1\,000^2$ and the length of contraction would increase by 1 000. How high would this "superflea" be able to jump? (Don't forget that the mass of the "superflea" increases as well.)

21. An athlete on a trampoline leaps straight up into the air with an initial speed of 9.0 m/s. Find (a) the maximum height reached by the athlete relative to the trampoline and (b) the speed of the athlete when she is halfway up to her maximum height.

22. Truck suspensions often have "helper springs" that engage at high loads. One such arrangement is a leaf spring with a helper coil spring mounted on the axle, as shown in Figure P5.22. When the main leaf spring is compressed by distance y_0, the helper spring engages and then helps to support any additional load. Suppose the leaf spring constant is 5.25×10^5 N/m, the helper spring constant is 3.60×10^5 N/m, and $y_0 = 0.500$ m. (a) What is the compression of the leaf spring for a load of 5.00×10^5 N? (b) How much work is done in compressing the springs?

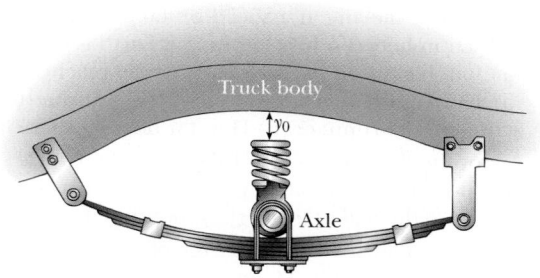

Figure P5.22

23. A daredevil on a motorcycle leaves the end of a ramp with a speed of 35.0 m/s as in Figure P5.23. If his speed is 33.0 m/s when he reaches the peak of the path, what is the maximum height that he reaches? Ignore friction and air resistance.

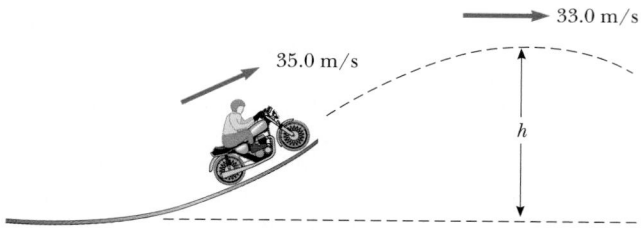

Figure P5.23

24. A softball pitcher rotates a 0.250-kg ball around a vertical circular path of radius 0.600 m before releasing it. The pitcher exerts a 30.0-N force directed parallel to the motion of the ball around the complete circular path. The speed of the ball at the top of the circle is 15.0 m/s. If the ball is released at the bottom of the circle, what is its speed upon release?

25. The chin-up is one exercise that can be used to strengthen the biceps muscle. This muscle can exert a force of approximately 800 N as it contracts a distance of 7.5 cm in a 75-kg male[3]. How much work can the biceps muscles (one in each arm) perform in a single contraction? Compare this amount of work with the energy required to lift a 75-kg person 40 cm in performing a chin-up. Do you think the biceps muscle is the only muscle involved in performing a chin-up?

Section 5.5 Systems and Energy Conservation

26. A 50-kg pole vaulter running at 10 m/s vaults over the bar. Her speed when she is above the bar is 1.0 m/s. Neglect air resistance, as well as any energy absorbed by the pole, and determine her altitude as she crosses the bar.

27. A child and a sled with a combined mass of 50.0 kg slide down a frictionless slope. If the sled starts from rest and has a speed of 3.00 m/s at the bottom, what is the height of the hill?

28. A 0.400-kg bead slides on a curved wire, starting from rest at point Ⓐ in Figure P5.28. If the wire is frictionless, find the speed of the bead (a) at Ⓑ and (b) at Ⓒ.

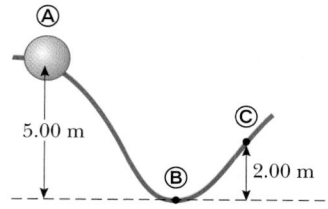

Figure P5.28 (Problems 28 and 36)

29. A 5.00-kg steel ball is dropped onto a copper plate from a height of 10.0 m. If the ball leaves a dent 3.20 mm deep in the plate, what is the average force exerted by the plate on the ball during the impact?

30. A bead of mass $m = 5.00$ kg is released from point Ⓐ and slides on the frictionless track shown in Figure P5.30. Determine (a) the bead's speed at points Ⓑ and Ⓒ and (b) the net work done by the force of gravity in moving the bead from Ⓐ to Ⓒ.

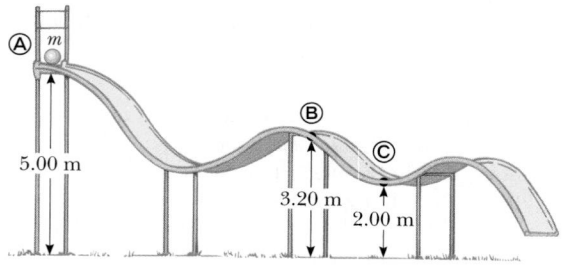

Figure P5.30

31. Tarzan swings on a 30.0-m-long vine initially inclined at an angle of 37.0° with the vertical. What is his speed at the bottom of the swing (a) if he starts from rest? (b) if he pushes off with a speed of 4.00 m/s?

32. Three objects with masses $m_1 = 5.0$ kg, $m_2 = 10$ kg, and $m_3 = 15$ kg, respectively, are attached by strings over frictionless pulleys, as indicated in Figure P5.32. The horizontal surface is frictionless and the system is released from rest. Using energy concepts, find the speed of m_3 after it moves down a distance of 4.0 m.

Figure P5.32 (Problems 32 and 89)

[3]G. P. Pappas et.al., "Nonuniform shortening in the biceps brachii during elbow flexion," *Journal of Applied Physiology* **92**, 2381, 2002.

33. **Physics Now™** The launching mechanism of a toy gun consists of a spring of unknown spring constant, as shown in Figure P5.33a. If the spring is compressed a distance of 0.120 m and the gun fired vertically as shown, the gun can launch a 20.0-g projectile from rest to a maximum height of 20.0 m above the starting point of the projectile. Neglecting all resistive forces, determine (a) the spring constant and (b) the speed of the projectile as it moves through the equilibrium position of the spring (where $x = 0$), as shown in Figure P5.33b.

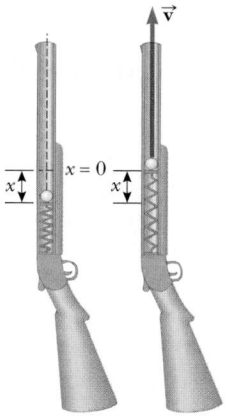

(a) (b)

Figure P5.33

34. A projectile is launched with a speed of 40 m/s at an angle of 60° above the horizontal. Use conservation of energy to find the maximum height reached by the projectile during its flight.

35. A 0.250-kg block is placed on a light vertical spring ($k = 5.00 \times 10^3$ N/m) and pushed downwards, compressing the spring 0.100 m. After the block is released, it leaves the spring and continues to travel upwards. What height above the point of release will the block reach if air resistance is negligible?

36. The wire in Problem 28 (Fig. P5.28) is frictionless between points Ⓐ and Ⓑ and rough between Ⓑ and Ⓒ. The 0.400-kg bead starts from rest at Ⓐ. (a) Find its speed at Ⓑ. (b) If the bead comes to rest at Ⓒ, find the loss in mechanical energy as it goes from Ⓑ to Ⓒ.

37. (a) A child slides down a water slide at an amusement park from an initial height h. The slide can be considered frictionless because of the water flowing down it. Can the equation for conservation of mechanical energy be used on the child? (b) Is the mass of the child a factor in determining his speed at the bottom of the slide? (c) The child drops straight down rather than following the curved ramp of the slide. In which case will he be traveling faster at ground level? (d) If friction is present, how would the conservation-of-energy equation be modified? (e) Find the maximum speed of the child when the slide is frictionless if the initial height of the slide is 12.0 m.

38. (a) A block with a mass m is pulled along a horizontal surface for a distance x by a constant force $\vec{\mathbf{F}}$ at an angle θ with respect to the horizontal. The coefficient of kinetic friction between block and table is μ_k. Is the force exerted by friction equal to $\mu_k mg$? If not, what is the force exerted by friction? (b) How much work is done by the friction force and by $\vec{\mathbf{F}}$? (Don't forget the signs.) (c) Identify all the forces that do no work on the block. (d) Let $m = 2.00$ kg, $x = 4.00$ m, $\theta = 37.0°$, $F = 15.0$ N, and $\mu_k = 0.400$, and find the answers to parts (a) and (b).

39. A 70-kg diver steps off a 10-m tower and drops from rest straight down into the water. If he comes to rest 5.0 m beneath the surface, determine the average resistive force exerted on him by the water.

40. An airplane of mass 1.5×10^4 kg is moving at 60 m/s. The pilot then revs up the engine so that the forward thrust by the air around the propeller becomes 7.5×10^4 N. If the force exerted by air resistance on the body of the airplane has a magnitude of 4.0×10^4 N, find the speed of the airplane after it has traveled 500 m. Assume that the airplane is in level flight throughout this motion.

41. A 2.1×10^3-kg car starts from rest at the top of a 5.0-m-long driveway that is inclined at 20° with the horizontal. If an average friction force of 4.0×10^3 N impedes the motion, find the speed of the car at the bottom of the driveway.

42. A 25.0-kg child on a 2.00-m-long swing is released from rest when the ropes of the swing make an angle of 30.0° with the vertical. (a) Neglecting friction, find the child's speed at the lowest position. (b) If the actual speed of the child at the lowest position is 2.00 m/s, what is the mechanical energy lost due to friction?

43. Starting from rest, a 10.0-kg block slides 3.00 m down to the bottom of a frictionless ramp inclined 30.0° from the floor. The block then slides an additional 5.00 m along the floor before coming to a stop. Determine (a) the speed of the block at the bottom of the ramp, (b) the coefficient of kinetic friction between block and floor, and (c) the mechanical energy lost due to friction.

44. A child slides without friction from a height h along a curved water slide (Fig. P5.44). She is launched from a height $h/5$ into the pool. Determine her maximum airborne height y in terms of h and the launch angle θ.

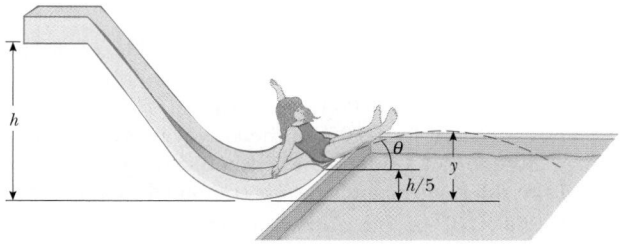

Figure P5.44

45. **Physics Now™** A skier starts from rest at the top of a hill that is inclined 10.5° with respect to the horizontal. The hillside is 200 m long, and the coefficient of friction between snow and skis is 0.075 0. At the bottom of the hill, the snow is level and the coefficient of friction is unchanged. How far does the skier glide along the horizontal portion of the snow before coming to rest?

46. In a circus performance, a monkey is strapped to a sled and both are given an initial speed of 4.0 m/s up a 20° inclined track. The combined mass of monkey and sled is 20 kg, and the coefficient of kinetic friction between sled and incline is 0.20. How far up the incline do the monkey and sled move?

47. An 80.0-kg skydiver jumps out of a balloon at an altitude of 1 000 m and opens the parachute at an altitude of

200.0 m. (a) Assuming that the total retarding force on the diver is constant at 50.0 N with the parachute closed and constant at 3 600 N with the parachute open, what is the speed of the diver when he lands on the ground? (b) Do you think the skydiver will get hurt? Explain. (c) At what height should the parachute be opened so that the final speed of the skydiver when he hits the ground is 5.00 m/s? (d) How realistic is the assumption that the total retarding force is constant? Explain.

Section 5.6 Power

48. A skier of mass 70 kg is pulled up a slope by a motor-driven cable. (a) How much work is required to pull him 60 m up a 30° slope (assumed frictionless) at a constant speed of 2.0 m/s? (b) What power must a motor have to perform this task?

49. Columnist Dave Barry poked fun at the name "The Grand Cities," adopted by Grand Forks, North Dakota, and East Grand Forks, Minnesota. Residents of the prairie towns then named a sewage pumping station for him. At the Dave Barry Lift Station No. 16, untreated sewage is raised vertically by 5.49 m in the amount of 1 890 000 liters each day. With a density of 1 050 kg/m^3, the waste enters and leaves the pump at atmospheric pressure through pipes of equal diameter. (a) Find the output power of the lift station. (b) Assume that a continuously operating electric motor with average power 5.90 kW runs the pump. Find its efficiency. In January 2002, Barry attended the outdoor dedication of the lift station and a festive potluck supper to which the residents of the different Grand Forks sewer districts brought casseroles, Jell-O® salads, and "bars" (desserts).

50. While running, a person dissipates about 0.60 J of mechanical energy per step per kilogram of body mass. If a 60-kg person develops a power of 70 W during a race, how fast is the person running? (Assume a running step is 1.5 m long.)

51. The electric motor of a model train accelerates the train from rest to 0.620 m/s in 21.0 ms. The total mass of the train is 875 g. Find the average power delivered to the train during its acceleration.

52. An electric scooter has a battery capable of supplying 120 Wh of energy. [Note that an energy of 1 Wh = (1 J/s)(3600 s) = 3600 J] If frictional forces and other losses account for 60.0% of the energy usage, what change in altitude can a rider achieve when driving in hilly terrain if the rider and scooter have a combined weight of 890 N?

53. A 1.50×10^3-kg car starts from rest and accelerates uniformly to 18.0 m/s in 12.0 s. Assume that air resistance remains constant at 400 N during this time. Find (a) the average power developed by the engine and (b) the instantaneous power output of the engine at $t = 12.0$ s, just before the car stops accelerating.

54. A 650-kg elevator starts from rest and moves upwards for 3.00 s with constant acceleration until it reaches its cruising speed, 1.75 m/s. (a) What is the average power of the elevator motor during this period? (b) How does this amount of power compare with its power during an upward trip with constant speed?

Section 5.7 Work Done by a Varying Force

55. **Physics Now™** The force acting on a particle varies as in Figure P5.55. Find the work done by the force as the particle moves (a) from $x = 0$ to $x = 8.00$ m, (b) from $x = 8.00$ m to $x = 10.0$ m, and (c) from $x = 0$ to $x = 10.0$ m.

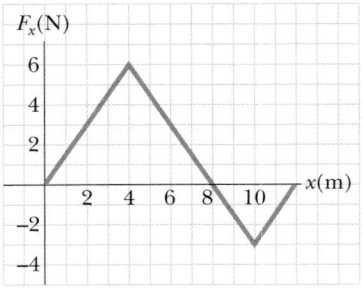

Figure P5.55

56. An object is subject to a force F_x that varies with position as in Figure P5.56. Find the work done by the force on the object as it moves (a) from $x = 0$ to $x = 5.00$ m, (b) from $x = 5.00$ m to $x = 10.0$ m, and (c) from $x = 10.0$ m to $x = 15.0$ m. (d) What is the total work done by the force over the distance $x = 0$ to $x = 15.0$ m?

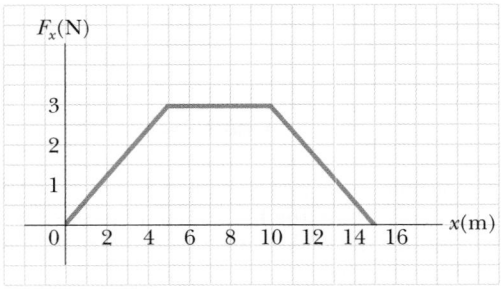

Figure P5.56

57. The force acting on an object is given by $F_x = (8x - 16)$ N, where x is in meters. (a) Make a plot of this force versus x from $x = 0$ to $x = 3.00$ m. (b) From your graph, find the net work done by the force as the object moves from $x = 0$ to $x = 3.00$ m.

ADDITIONAL PROBLEMS

58. A 2.0-m-long pendulum is released from rest when the support string is at an angle of 25° with the vertical. What is the speed of the bob at the bottom of the swing?

59. An archer pulls her bowstring back 0.400 m by exerting a force that increases uniformly from zero to 230 N. (a) What is the equivalent spring constant of the bow? (b) How much work does the archer do in pulling the bow?

60. A block of mass 12.0 kg slides from rest down a frictionless 35.0° incline and is stopped by a strong spring with $k = 3.00 \times 10^4$ N/m. The block slides 3.00 m from the point of release to the point where it comes to rest against the spring. When the block comes to rest, how far has the spring been compressed?

61. (a) A 75-kg man steps out a window and falls (from rest) 1.0 m to a sidewalk. What is his speed just before his feet strike the pavement? (b) If the man falls with his knees and ankles locked, the only cushion for his fall is an approximately 0.50-cm give in the pads of his feet. Calculate the average force exerted on him by the ground in this situation. This average force is sufficient to cause damage to cartilage in the joints or to break bones.

62. A toy gun uses a spring to project a 5.3-g soft rubber sphere horizontally. The spring constant is 8.0 N/m, the barrel of the gun is 15 cm long, and a constant frictional force of 0.032 N exists between barrel and projectile. With what speed does the projectile leave the barrel if the spring was compressed 5.0 cm for this launch?

63. Two objects are connected by a light string passing over a light, frictionless pulley as in Figure P5.63. The 5.00-kg object is released from rest at a point 4.00 m above the floor. (a) Determine the speed of each object when the two pass each other. (b) Determine the speed of each object at the moment the 5.00-kg object hits the floor. (c) How much higher does the 3.00-kg object travel after the 5.00-kg object hits the floor?

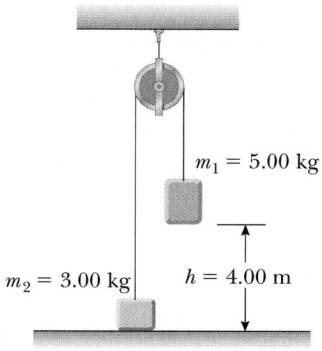

Figure P5.63

64. Two blocks, A and B (with mass 50 kg and 100 kg, respectively), are connected by a string, as shown in Figure P5.64. The pulley is frictionless and of negligible mass. The coefficient of kinetic friction between block A and the incline is $\mu_k = 0.25$. Determine the change in the kinetic energy of block A as it moves from Ⓒ to Ⓓ, a distance of 20 m up the incline if the system starts from rest.

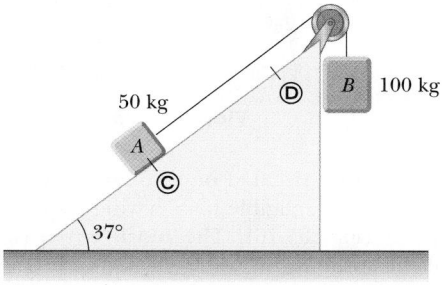

Figure P5.64

65. A 200-g particle is released from rest at point A on the inside of a smooth hemispherical bowl of radius $R = 30.0$ cm (Fig. P5.65). Calculate (a) its gravitational potential energy at A relative to B, (b) its kinetic energy at B, (c) its speed at B, (d) its potential energy at C relative to B, and (e) its kinetic energy at C.

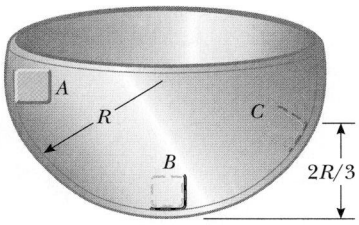

Figure P5.65

66. Energy is conventionally measured in Calories as well as in joules. One Calorie in nutrition is 1 kilocalorie, which we define in Chapter 11 as 1 kcal = 4 186 J. Metabolizing 1 gram of fat can release 9.00 kcal. A student decides to try to lose weight by exercising. She plans to run up and down the stairs in a football stadium as fast as she can and as many times as necessary. Is this in itself a practical way to lose weight? To evaluate the program, suppose she runs up a flight of 80 steps, each 0.150 m high, in 65.0 s. For simplicity, ignore the energy she uses in coming down (which is small). Assume that a typical efficiency for human muscles is 20.0%. This means that when your body converts 100 J from metabolizing fat, 20 J goes into doing mechanical work (here, climbing stairs). The remainder goes into internal energy. Assume the student's mass is 50.0 kg. (a) How many times must she run the flight of stairs to lose 1 pound of fat? (b) What is her average power output, in watts and in horsepower, as she is running up the stairs?

67. In terms of saving energy, bicycling and walking are far more efficient means of transportation than is travel by automobile. For example, when riding at 10.0 mi/h, a cyclist uses food energy at a rate of about 400 kcal/h above what he would use if he were merely sitting still. (In exercise physiology, power is often measured in kcal/h rather than in watts. Here, 1 kcal = 1 nutritionist's Calorie = 4 186 J.) Walking at 3.00 mi/h requires about 220 kcal/h. It is interesting to compare these values with the energy consumption required for travel by car. Gasoline yields about 1.30×10^8 J/gal. Find the fuel economy in equivalent miles per gallon for a person (a) walking and (b) bicycling.

68. An 80.0-N box is pulled 20.0 m up a 30° incline by an applied force of 100 N that points upwards, parallel to the incline. If the coefficient of kinetic friction between box and incline is 0.220, calculate the change in the kinetic energy of the box.

69. A ski jumper starts from rest 50.0 m above the ground on a frictionless track and flies off the track at an angle of 45.0° above the horizontal and at a height of 10.0 m above the level ground. Neglect air resistance. (a) What is her speed when she leaves the track? (b) What is the maximum altitude she attains after leaving the track? (c) Where does she land relative to the end of the track?

70. A 5.0-kg block is pushed 3.0 m up a vertical wall with constant speed by a constant force of magnitude F applied at an angle of $\theta = 30°$ with the horizontal, as shown in Figure P5.70. If the coefficient of kinetic friction between

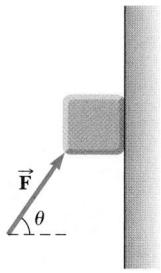

Figure P5.70

block and wall is 0.30, determine the work done by (a) $\vec{\mathbf{F}}$, (b) the force of gravity, and (c) the normal force between block and wall. (d) By how much does the gravitational potential energy increase during the block's motion?

71. The ball launcher in a pinball machine has a spring with a force constant of 1.20 N/cm (Fig. P5.71). The surface on which the ball moves is inclined 10.0° with respect to the horizontal. If the spring is initially compressed 5.00 cm, find the launching speed of a 0.100-kg ball when the plunger is released. Friction and the mass of the plunger are negligible.

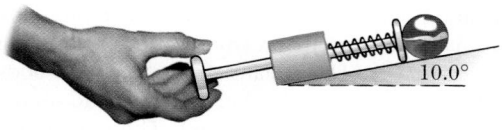

Figure P5.71

72. The masses of the javelin, discus, and shot are 0.80 kg, 2.0 kg, and 7.2 kg, respectively, and record throws in the corresponding track events are about 98 m, 74 m, and 23 m, respectively. Neglecting air resistance, (a) calculate the minimum initial kinetic energies that would produce these throws, and (b) estimate the average force exerted on each object during the throw, assuming the force acts over a distance of 2.0 m. (c) Do your results suggest that air resistance is an important factor?

73. Jane, whose mass is 50.0 kg, needs to swing across a river filled with crocodiles in order to rescue Tarzan, whose mass is 80.0 kg. However, she must swing into a *constant* horizontal wind force $\vec{\mathbf{F}}$ on a vine that is initially at an angle of θ with the vertical. (See Fig. P5.73.) In the figure, $D = 50.0$ m, $F = 110$ N, $L = 40.0$ m, and $\theta = 50.0°$. (a) With what minimum speed must Jane begin her swing in order to just make it to the other side? (*Hint:* First determine the potential energy that can be associated with the wind force. Because the wind force is constant, use an analogy with the constant gravitational force.) (b) Once the rescue is complete, Tarzan and Jane must swing back across the river. With what minimum speed must they begin their swing?

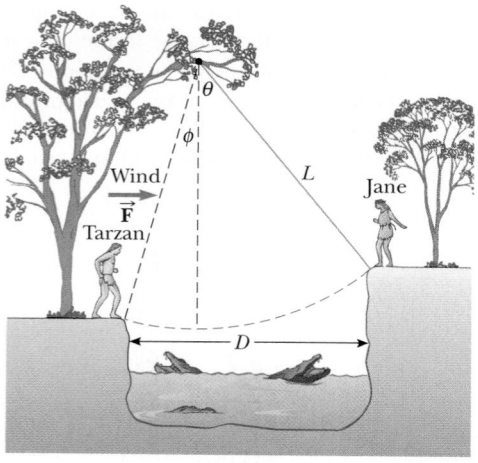

Figure P5.73

74. A hummingbird is able to hover because, as the wings move downwards, they exert a downward force on the air. Newton's third law tells us that the air exerts an equal and opposite force (upwards) on the wings. The average of this force must be equal to the weight of the bird when it hovers. If the wings move through a distance of 3.5 cm with each stroke, and the wings beat 80 times per second, determine the work performed by the wings on the air in 1 minute if the mass of the hummingbird is 3.0 grams.

75. A child's pogo stick (Fig. P5.75) stores energy in a spring ($k = 2.50 \times 10^4$ N/m). At position Ⓐ ($x_1 = -0.100$ m), the spring compression is a maximum and the child is momentarily at rest. At position Ⓑ ($x = 0$), the spring is relaxed and the child is moving upwards. At position Ⓒ, the child is again momentarily at rest at the top of the jump. Assuming that the combined mass of child and pogo stick is 25.0 kg, (a) calculate the total energy of the system if both potential energies are zero at $x = 0$, (b) determine x_2, (c) calculate the speed of the child at $x = 0$, (d) determine the value of x for which the kinetic energy of the system is a maximum, and (e) obtain the child's maximum upward speed.

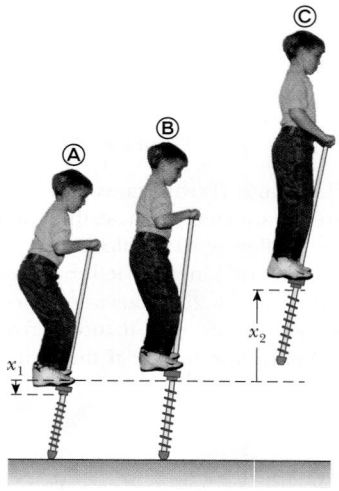

Figure P5.75

76. A 2.00-kg block situated on a rough incline is connected to a spring of negligible mass having a spring constant of 100 N/m (Fig. P5.76). The block is released from rest when the spring is unstretched, and the pulley is frictionless. The block moves 20.0 cm down the incline before coming to rest. Find the coefficient of kinetic friction between block and incline.

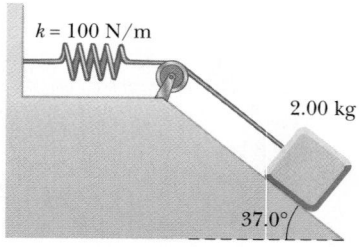

Figure P5.76

77. In the dangerous "sport" of bungee jumping, a daring student jumps from a hot-air balloon with a specially designed elastic cord attached to his waist, as shown in Figure P5.77. The unstretched length of the cord is 25.0 m, the student weighs 700 N, and the balloon is 36.0 m above the surface of a river below. Calculate the required force constant of the cord if the student is to stop safely 4.00 m above the river.

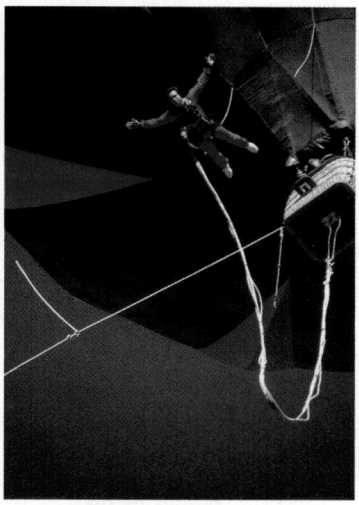

© Jamie Budge/Corbis

Figure P5.77 Bungee jumping. (Problems 77 and 82)

78. An object of mass m is suspended from the top of a cart by a string of length L as in Figure P5.78a. The cart and object are initially moving to the right at a constant speed v_0. The cart comes to rest after colliding and sticking to a bumper, as in Figure P5.78b, and the suspended object swings through an angle θ. (a) Show that the initial speed is $v_0 = \sqrt{2gL(1 - \cos\theta)}$. (b) If $L = 1.20$ m and $\theta = 35.0°$, find the initial speed of the cart. (*Hint:* The force exerted by the string on the object does no work on the object.)

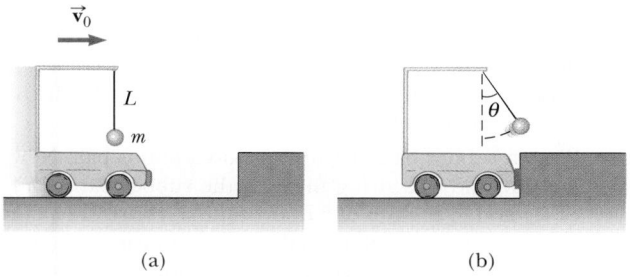

Figure P5.78

79. A truck travels uphill with constant velocity on a highway with a 7.0° slope. A 50-kg package sits on the floor of the back of the truck and does not slide, due to a static frictional force. During an interval in which the truck travels 340 m, what is the net work done on the package? What is the work done on the package by the force of gravity, the normal force, and the friction force?

80. As part of a curriculum unit on earthquakes, suppose that 375 000 British schoolchildren stand on their chairs and simultaneously jump down to the floor. Seismographers around the country see whether they can detect the resulting ground tremor. (This experiment was actually based on a suggestion by the children themselves.) (a) Find the energy released in the experiment. Model the children as having average mass 36.0 kg and as stepping from chair seats 38.0 cm above the floor. (b) Most of the energy is converted very rapidly into internal energy within the bodies of the children and the floors of the school buildings. Assume that 1% of the energy is carried away by a seismic wave. The magnitude of an earthquake on the Richter scale is given by

$$M = \frac{\log E - 4.8}{1.5}$$

where E is the seismic wave energy in joules. According to this model, what is the magnitude of the demonstration quake?

81. A loaded ore car has a mass of 950 kg and rolls on rails with negligible friction. It starts from rest and is pulled up a mine shaft by a cable connected to a winch. The shaft is inclined at 30.0° above the horizontal. The car accelerates uniformly to a speed of 2.20 m/s in 12.0 s and then continues at constant speed. (a) What power must the winch motor provide when the car is moving at constant speed? (b) What maximum power must the motor provide? (c) What total energy transfers out of the motor by work by the time the car moves off the end of the track, which is of length 1 250 m?

82. A daredevil wishes to bungee-jump from a hot-air balloon 65.0 m above a carnival midway (Fig. P5.77). He will use a piece of uniform elastic cord tied to a harness around his body to stop his fall at a point 10.0 m above the ground. Model his body as a particle and the cord as having negligible mass and a tension force described by Hooke's force law. In a preliminary test, hanging at rest from a 5.00-m length of the cord, the jumper finds that his body weight stretches it by 1.50 m. He will drop from rest at the point where the top end of a longer section of the cord is attached to the stationary balloon. (a) What length of cord should he use? (b) What maximum acceleration will he experience?

83. The system shown in Figure P5.83 consists of a light, inextensible cord, light frictionless pulleys, and blocks of equal mass. Initially, the blocks are at rest the same height above the ground. The blocks are then released. Find the speed of block A at the moment when the vertical separation of the blocks is h.

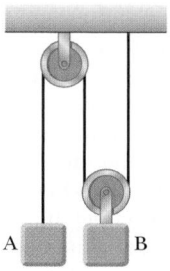

Figure P5.83

84. A cafeteria tray dispenser supports a stack of trays on a shelf that hangs from four identical spiral springs under tension, one near each corner of the shelf. Each tray has a mass of 580 g and is rectangular, 45.3 cm by 35.6 cm, and 0.450 cm thick. (a) Show that the top tray in the stack can

always be at the same height above the floor, however many trays are in the dispenser. (b) Find the spring constant each spring should have in order for the dispenser to function in this convenient way. Is any piece of data unnecessary for this determination?

85. In bicycling for aerobic exercise, a woman wants her heart rate to be between 136 and 166 beats per minute. Assume that her heart rate is directly proportional to her mechanical power output. Ignore all forces on the woman-plus-bicycle system, except for static friction forward on the drive wheel of the bicycle and an air resistance force proportional to the square of the bicycler's speed. When her speed is 22.0 km/h, her heart rate is 90.0 beats per minute. In what range should her speed be so that her heart rate will be in the range she wants?

86. In a needle biopsy, a narrow strip of tissue is extracted from a patient with a hollow needle. Rather than being pushed by hand, to ensure a clean cut the needle can be fired into the patient's body by a spring. Assume the needle has mass 5.60 g, the light spring has force constant 375 N/m, and the spring is originally compressed 8.10 cm to project the needle horizontally without friction. The tip of the needle then moves through 2.40 cm of skin and soft tissue, which exerts a resistive force of 7.60 N on it. Next, the needle cuts 3.50 cm into an organ, which exerts a backward force of 9.20 N on it. Find (a) the maximum speed of the needle and (b) the speed at which a flange on the back end of the needle runs into a stop, set to limit the penetration to 5.90 cm.

87. The power of sunlight reaching each square meter of the Earth's surface on a clear day in the tropics is close to 1 000 W. On a winter day in Manitoba, the power concentration of sunlight can be 100 W/m^2. Many human activities are described by a power-per-footprint-area on the order of 10^2 W/m^2 or less. (a) Consider, for example, a family of four paying \$80 to the electric company every 30 days for 600 kWh of energy carried by electric transmission to their house, with floor area 13.0 m by 9.50 m. Compute the power-per-area measure of this energy use. (b) Consider a car 2.10 m wide and 4.90 m long traveling at 55.0 mi/h using gasoline having a "heat of combustion" of 44.0 MJ/kg with fuel economy 25.0 mi/gallon. One gallon of gasoline has a mass of 2.54 kg. Find the power-per-area measure of the car's energy use. It can be similar to that of a steel mill where rocks are melted in blast furnaces. (c) Explain why the direct use of solar energy is not practical for a conventional automobile.

88. In 1887 in Bridgeport, Connecticut, C. J. Belknap built the water slide shown in Figure P5.88. A rider on a small sled, of total mass 80.0 kg, pushed off to start at the top of the slide (point Ⓐ) with a speed of 2.50 m/s. The chute was 9.76 m high at the top, 54.3 m long, and 0.51 m wide. Along its length, 725 wheels made friction negligible. Upon leaving the chute horizontally at its bottom end (point Ⓒ), the rider skimmed across the water of Long Island Sound for as much as 50 m, "skipping along like a flat pebble," before at last coming to rest and swimming ashore, pulling his sled after him. (a) Find the speed of the sled and rider at point Ⓒ. (b) Model the force of water friction as a constant retarding force acting on a particle. Find the work done by water friction in stopping the sled and rider. (c) Find the magnitude of the force the water exerts on the sled. (d) Find the magnitude of the force the chute exerts on the sled at point Ⓑ.

Engraving from *Scientific American*, July 1888.

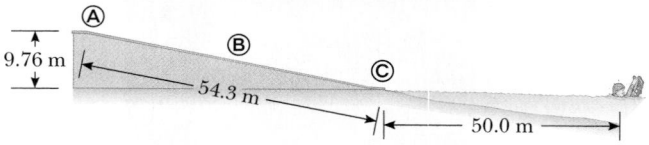

Figure P5.88

89. Three objects with masses $m_1 = 5.0$ kg, $m_2 = 10$ kg, and $m_3 = 15$ kg, respectively, are attached by strings over frictionless pulleys as indicated in Figure P5.32. The horizontal surface exerts a force of friction of 30 N on m_2. If the system is released from rest, use energy concepts to find the speed of m_3 after it moves down 4.0 m.

ACTIVITIES

A.1. Suspend a rubber band from a support and borrow some weights from your instructor to measure the rubber band's spring constant for small extensions. Calculate how much elastic potential energy is stored in the rubber band for a given extension. Use conservation of energy to predict how high a paper wad will go into the air when released from a given extension of the band. Try it to test your prediction.

A.2. Wrap a rubber band tightly around a tennis ball. Now fasten one end of a string through the rubber band and the other end to the top of a doorframe to construct a pendulum. Pull the pendulum to the side at a variety of angles to observe that the energy of the pendulum–Earth system is always conserved as the pendulum swings (almost) to the same height of its arc as the height from which it was released. (The word "almost" in the last sentence applies because some energy is lost to friction at the point of support and to air resistance. You can observe this slight loss of energy by pulling the ball to the side and letting it go from a point about half an inch from your chin. Let it go—but don't push it!—and test your belief in conservation of energy by seeing if you can avoid flinching when the ball swings back toward your chin.)

A.3. While you have your pendulum from the last activity set up, predict what will happen in the following situation and then test your guess: When a pendulum is released from a given height, it swings to the same height at the other end of its arc as you noted above. However, suppose you place a meterstick across the door opening such that the pendulum string strikes the stick about halfway up the string when it moves through the opening. How high will the ball swing in this case? Will it return to the same height as that at which it started, swing to a lower height, or swing to a greater height? Explain your answer.

A.4. Measure your pulse rate while at rest. Now slowly walk up a flight of stairs and measure your pulse rate at the top of the stairs. Repeat this activity, starting with about the same rest pulse rate at the bottom of the stairs. This time, run up the stairs. Based on your pulse rate readings, what can you conclude about the amount of work and power expended in each case? Repeat this experiment with a series of 10 push-ups.

A.5. Many fitness centers have stepper machines that enable a person to climb continuously without actually moving, because the steps move downwards as the person climbs. The work that the climber performs on the step is determined by the force exerted on the step times the distance the step moves. Since the net force on the climber is zero, the force exerted on the step must equal the climber's weight. A reasonably strenuous workout on this machine is 90 steps per minute, with each step being 8 inches (15.2 cm) high. What is the rate (in watts) at which a 130-lb (60 kg) climber does work on the stair steps? The energy actually expended by the climber is approximately five times the work done. (You may notice that a lot of heat is generated!) The machines usually report this rate in Calories/hour (1 Calorie = 1 kcal = 4186 J). Determine the rate, in Cal/h, at which energy is expended by the 130-lb climber.

Activity: If you have access to a stepper, find the ratio used by the manufacturer to determine the energy expended from the work performed. In some cases, this ratio may vary as the step speed changes. If so, generate a graph of the ratio as a function of step speed.

A small buck from the massive bull transfers a large amount of momentum to the cowboy, resulting in an involuntary dismount.

© Reuters/Corbis

CHAPTER

6

Momentum and Collisions

OUTLINE

6.1 Momentum and Impulse
6.2 Conservation of Momentum
6.3 Collisions
6.4 Glancing Collisions
6.5 Rocket Propulsion

What happens when two automobiles collide? How does the impact affect the motion of each vehicle, and what basic physical principles determine the likelihood of serious injury? How do rockets work, and what mechanisms can be used to overcome the limitations imposed by exhaust speed? Why do we have to brace ourselves when firing small projectiles at high velocity? Finally, how can we use physics to improve our golf game?

To begin answering such questions, we introduce *momentum*. Intuitively, anyone or anything that has a lot of momentum is going to be hard to stop. In politics, the term is metaphorical. Physically, the more momentum an object has, the more force has to be applied to stop it in a given time. This concept leads to one of the most powerful principles in physics: *conservation of momentum*. Using this law, complex collision problems can be solved without knowing much about the forces involved during contact. We'll also be able to derive information about the average force delivered in an impact. With conservation of momentum, we'll have a better understanding of what choices to make when designing an automobile or a moon rocket, or when addressing a golf ball on a tee.

6.1 MOMENTUM AND IMPULSE

In physics, momentum has a precise definition. A slowly moving brontosaurus has a lot of momentum, but so does a little hot lead shot from the muzzle of a gun. We therefore expect that momentum will depend on an object's mass and velocity.

Linear momentum ▶

The linear momentum $\vec{\mathbf{p}}$ of an object of mass m moving with velocity $\vec{\mathbf{v}}$ is the product of its mass and velocity :

$$\vec{\mathbf{p}} \equiv \vec{\mathbf{v}} \qquad [6.1]$$

SI unit: kilogram-meter per second (kg · m/s)

Doubling either the mass or the velocity of an object doubles its momentum; doubling both quantities quadruples its momentum. Momentum is a vector quantity with the same direction as the object's velocity. Its components are given in two dimensions by

$$p_x = mv_x \qquad p_y = mv_y$$

where p_x is the momentum of the object in the x-direction and p_y its momentum in the y-direction.

Quick Quiz 6.1

Two objects with masses m_1 and m_2 have equal kinetic energy. How do the magnitudes of their momenta compare? (a) $p_1 < p_2$ (b) $p_1 = p_2$ (c) $p_1 > p_2$ (d) not enough information is given

Changing the momentum of an object requires the application of a force. This is, in fact, how Newton originally stated his second law of motion. Starting from the more common version of the second law, we have

$$\vec{\mathbf{F}}_{\text{net}} = m\vec{\mathbf{a}} = m\frac{\Delta\vec{\mathbf{v}}}{\Delta t} = \frac{\Delta(m\vec{\mathbf{v}})}{\Delta t} \qquad [6.2]$$

where the mass m and the forces are assumed constant. The quantity in parentheses is just the momentum, so we have the following result:

The change in an object's momentum $\Delta\vec{\mathbf{p}}$ divided by the elapsed time Δt equals the constant net force $\vec{\mathbf{F}}_{\text{net}}$ acting on the object:

$$\frac{\Delta\vec{\mathbf{p}}}{\Delta t} = \frac{\text{change in momentum}}{\text{time interval}} = \vec{\mathbf{F}}_{\text{net}} \qquad [6.3]$$

◀ Newton's second law

This equation is also valid when the forces are not constant, provided the limit is taken as Δt becomes infinitesimally small. Equation 6.3 says that if the net force on an object is zero, the object's momentum doesn't change. In other words, the linear momentum of an object is *conserved* when $\vec{\mathbf{F}}_{\text{net}} = 0$. Equation 6.3 also tells us that changing an object's momentum requires the continuous application of a force over a period of time Δt, leading to the definition of *impulse:*

If a constant force $\vec{\mathbf{F}}$ acts on an object, the **impulse** $\vec{\mathbf{I}}$ delivered to the object over a time interval Δt is given by

$$\vec{\mathbf{I}} \equiv \vec{\mathbf{F}}\Delta t \qquad [6.4]$$

SI unit: kilogram meter per second (kg · m/s)

Impulse is a vector quantity with the same direction as the constant force acting on the object. When a single constant force $\vec{\mathbf{F}}$ acts on an object, Equation 6.3 can be written as

$$\vec{\mathbf{F}}\Delta t = \Delta\vec{\mathbf{p}} = m\vec{\mathbf{v}}_f - m\vec{\mathbf{v}}_i \qquad [6.5]$$

◀ Impulse–momentum theorem

This is a special case of the **impulse–momentum theorem**. Equation 6.5 shows that **the impulse of the force acting on an object equals the change in momentum of that object**. This equality is true even if the force is not constant, as long as the time interval Δt is taken to be arbitrarily small. (The proof of the general case requires concepts from calculus.)

In real-life situations, the force on an object is only rarely constant. For example, when a bat hits a baseball, the force increases sharply, reaches some maximum value, and then decreases just as rapidly. Figure 6.1(a) shows a typical graph of

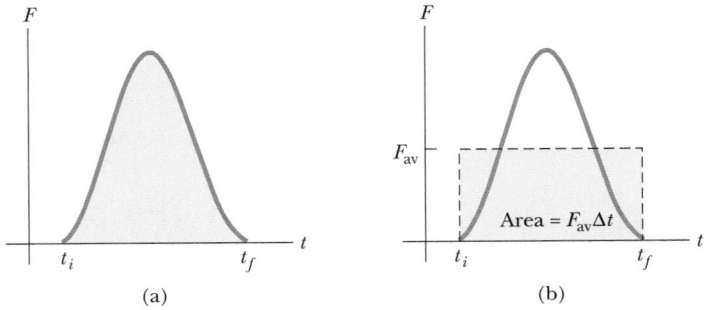

Figure 6.1 (a) A force acting on an object may vary in time. The impulse is the area under the force vs. time curve. (b) The average force (horizontal dashed line) gives the same impulse to the object in the time interval Δt as the real time-varying force described in (a).

force versus time for such incidents. The force starts out small as the bat comes in contact with the ball, rises to a maximum value when they are firmly in contact, and then drops off as the ball leaves the bat. In order to analyze this rather complex interaction, it's useful to define an **average force** $\vec{\mathbf{F}}_{av}$, shown as the dashed line in Figure 6.1b. This average force is the constant force delivering the same impulse to the object in the time interval Δt as the actual time-varying force. We can then write the impulse–momentum theorem as

$$\vec{\mathbf{F}}_{av}\Delta t = \Delta\vec{\mathbf{p}} \qquad [6.6]$$

The magnitude of the impulse delivered by a force during the time interval Δt is equal to the area under the force vs. time graph as in Figure 6.1a or, equivalently, to $F_{av}\Delta t$ as shown in Figure 6.1b. The brief collision between a bullet and an apple is illustrated in Figure 6.2.

Applying Physics 6.1 Boxing and Brain Injury

In boxing matches of the 19th century, bare fists were used. In modern boxing, fighters wear padded gloves. How do gloves protect the brain of the boxer from injury? Also, why do boxers often "roll with the punch"?

Explanation The brain is immersed in a cushioning fluid inside the skull. If the head is struck suddenly by a bare fist, the skull accelerates rapidly. The brain matches this acceleration only because of the large impulsive force exerted by the skull on the brain. This large and sudden force (large F_{av} and small Δt) can cause severe brain injury. Padded gloves extend the time Δt over which the force is applied to the head. For a given impulse $F_{av}\Delta t$, a glove results in a longer time interval than a bare fist, decreasing the average force. Because the average force is decreased, the acceleration of the skull is decreased, reducing (but not eliminating) the chance of brain injury. The same argument can be made for "rolling with the punch": If the head is held steady while being struck, the time interval over which the force is applied is relatively short and the average force is large. If the head is allowed to move in the same direction as the punch, the time interval is lengthened and the average force reduced.

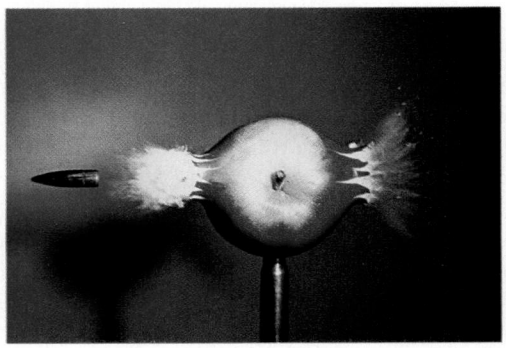

© Harold and Esther Edgerton Foundation, 2002, courtesy of Palm Press, Inc.

Figure 6.2 An apple being pierced by a 30-caliber bullet traveling at a supersonic speed of 900 m/s. This collision was photographed with a microflash stroboscope using an exposure time of 0.33 μs. Shortly after the photograph was taken, the apple disintegrated completely. Note that the points of entry and exit of the bullet are visually explosive.

EXAMPLE 6.1 Teeing Off

Goal Use the impulse–momentum theorem to estimate the average force exerted during an impact.

Problem A golf ball with mass 5.0×10^{-2} kg is struck with a club as in Figure 6.3. The force on the ball varies from zero when contact is made up to some maximum value (when the ball is maximally deformed) and then back to zero when the ball leaves the club, as in the graph of force vs. time in Figure 6.1. Assume that the ball leaves the club face with a velocity of $+44$ m/s. **(a)** Find the magnitude of the impulse due to the collision. **(b)** Estimate the duration of the collision and the average force acting on the ball.

© Harold and Esther Edgerton Foundation, 2002, Courtesy of Palm Press, Inc.

Figure 6.3 (Example 6.1) A golf ball being struck by a club.

Strategy In part (a), use the fact that the impulse is equal to the change in momentum. The mass and the initial and final speeds are known, so this change can be computed. In part (b), the average force is just the change in momentum computed in part (a) divided by an estimate of the duration of the collision. Guess at the distance the ball travels on the face of the club (about 2 cm, roughly the same as the radius of the ball). Divide this distance by the average velocity (half the final velocity) to get an estimate of the time of contact.

Solution

(a) Find the impulse delivered to the ball.

The problem is essentially one dimensional. Note that $v_i = 0$, and calculate the change in momentum, which equals the impulse:

$$I = \Delta p = p_f - p_i = (5.0 \times 10^{-2}\ \text{kg})(44\ \text{m/s}) - 0$$
$$= +2.2\ \text{kg}\cdot\text{m/s}$$

(b) Estimate the duration of the collision and the average force acting on the ball.

Estimate the time interval of the collision, Δt, using the approximate displacement (radius of the ball) and its average speed (half the maximum speed):

$$\Delta t = \frac{\Delta x}{v_{av}} = \frac{2.0 \times 10^{-2}\ \text{m}}{22\ \text{m/s}} = 9.1 \times 10^{-4}\ \text{s}$$

Estimate the average force from Equation 6.6:

$$F_{av} = \frac{\Delta p}{\Delta t} = \frac{2.2\ \text{kg}\cdot\text{m/s}}{9.1 \times 10^{-4}\ \text{s}} = +2.4 \times 10^3\ \text{N}$$

Remarks This estimate shows just how large such contact forces can be. A good golfer achieves maximum momentum transfer by shifting weight from the back foot to the front foot, transmitting the body's momentum through the shaft and head of the club. This timing, involving a short movement of the hips, is more effective than a shot powered exclusively by the arms and shoulders. Following through with the swing ensures that the motion isn't slowed at the critical instant of impact.

Exercise 6.1

A 0.150-kg baseball, thrown with a speed of 40.0 m/s, is hit straight back at the pitcher with a speed of 50.0 m/s. (a) What is the impulse delivered by the bat to the baseball? (b) Find the magnitude of the average force exerted by the bat on the ball if the two are in contact for 2.00×10^{-3} s.

Answer (a) 13.5 kg · m/s (b) 6.75 kN

EXAMPLE 6.2 How Good Are the Bumpers?

Goal Find an impulse and estimate a force in a collision of a moving object with a stationary object.

Problem In a crash test, a car of mass 1.50×10^3 kg collides with a wall and rebounds as in Figure 6.4a. The initial and final velocities of the car are $v_i = -15.0$ m/s and $v_f = 2.60$ m/s, respectively. If the collision lasts for 0.150 s, find

(a) (b)

Figure 6.4 (Example 6.2) (a) This car's momentum changes as a result of its collision with the wall. (b) In a crash test (an inelastic collision), much of the car's initial kinetic energy is transformed into the energy it took to damage the vehicle.

(a) the impulse delivered to the car due to the collision and **(b)** the size and direction of the average force exerted on the car.

Strategy This problem is similar to the previous example, except that the initial and final momenta are both nonzero. Find the momenta and substitute into the impulse–momentum theorem, Equation 6.6, solving for F_{av}.

Solution

(a) Find the impulse delivered to the car.

Calculate the initial and final momenta of the car:

$$p_i = mv_i = (1.50 \times 10^3 \text{ kg})(-15.0 \text{ m/s}) = -2.25 \times 10^4 \text{ kg} \cdot \text{m/s}$$

$$p_f = mv_f = (1.50 \times 10^3 \text{ kg})(+2.60 \text{ m/s}) = +0.390 \times 10^4 \text{ kg} \cdot \text{m/s}$$

The impulse is just the difference between the final and initial momenta:

$$I = p_f - p_i = +0.390 \times 10^4 \text{ kg} \cdot \text{m/s} - (-2.25 \times 10^4 \text{ kg} \cdot \text{m/s})$$

$$I = 2.64 \times 10^4 \text{ kg} \cdot \text{m/s}$$

(b) Find the average force exerted on the car.

Apply Equation 6.6, the impulse–momentum theorem:

$$F_{av} = \frac{\Delta p}{\Delta t} = \frac{2.64 \times 10^4 \text{ kg} \cdot \text{m/s}}{0.150 \text{ s}} = +1.76 \times 10^5 \text{ N}$$

Remarks When the car doesn't rebound off the wall, the average force exerted on the car is smaller than the value just calculated. With a final momentum of zero, the car undergoes a smaller change in momentum.

Exercise 6.2

Suppose the car doesn't rebound off the wall, but the time interval of the collision remains at 0.150 s. In this case, the final velocity of the car is zero. Find the average force exerted on the car.

Answer $+1.50 \times 10^5$ N

Injury in Automobile Collisions

APPLICATION
Injury to Passengers in Car Collisions

The main injuries that occur to a person hitting the interior of a car in a crash are brain damage, bone fracture, and trauma to the skin, blood vessels, and internal organs. Here, we compare the rather imprecisely known thresholds for human injury with typical forces and accelerations experienced in a car crash.

A force of about 90 kN (20 000 lb) compressing the tibia can cause fracture. Although the breaking force varies with the bone considered, we may take this value as the threshold force for fracture. It's well known that rapid acceleration of the head, even without skull fracture, can be fatal. Estimates show that head accelerations of $150g$ experienced for about 4 ms or $50g$ for 60 ms are fatal 50% of the time. Such injuries from rapid acceleration often result in nerve damage to the spinal cord where the nerves enter the base of the brain. The threshold for damage to skin, blood vessels, and internal organs may be estimated from whole-body impact data, where the force is uniformly distributed over the entire front surface area of $0.7\ \text{m}^2$ to $0.9\ \text{m}^2$. These data show that if the collision lasts for less than about 70 ms, a person will survive if the whole-body impact pressure (force per unit area) is less than $1.9 \times 10^5\ \text{N/m}^2$ ($28\ \text{lb/in.}^2$). Death results in 50% of cases in which the whole-body impact pressure reaches $3.4 \times 10^5\ \text{N/m}^2$ ($50\ \text{lb/in.}^2$).

Armed with the data above, we can estimate the forces and accelerations in a typical car crash and see how seat belts, air bags, and padded interiors can reduce the chance of death or serious injury in a collision. Consider a typical collision involving a 75-kg passenger not wearing a seat belt, traveling at 27 m/s (60 mi/h) who comes to rest in about 0.010 s after striking an unpadded dashboard. Using $F_{\text{av}}\Delta t = mv_f - mv_i$, we find that

$$F_{\text{av}} = \frac{mv_f - mv_i}{\Delta t} = \frac{0 - (75\ \text{kg})(27\ \text{m/s})}{0.010\ \text{s}} = -2.0 \times 10^5\ \text{N}$$

and

$$a = \left|\frac{\Delta v}{\Delta t}\right| = \frac{27\ \text{m/s}}{0.010\ \text{s}} = 2\ 700\ \text{m/s}^2 = \frac{2\ 700\ \text{m/s}^2}{9.8\ \text{m/s}^2}\,g = 280g$$

If we assume the passenger crashes into the dashboard and windshield so that the head and chest, with a combined surface area of $0.5\ \text{m}^2$, experience the force, we find a whole-body pressure of

$$\frac{F_{\text{av}}}{A} = \frac{2.0 \times 10^5\ \text{N}}{0.5\ \text{m}^2} \cong 4 \times 10^5\ \text{N/m}^2$$

We see that the force, the acceleration, and the whole-body pressure all *exceed* the threshold for fatality or broken bones and that an unprotected collision at 60 mi/h is almost certainly fatal.

What can be done to reduce or eliminate the chance of dying in a car crash? The most important factor is the collision time, or the time it takes the person to come to rest. If this time can be increased by 10 to 100 times the value of 0.01 s for a hard collision, the chances of survival in a car crash are much higher, because the increase in Δt makes the contact force 10 to 100 times smaller. Seat belts restrain people so that they come to rest in about the same amount of time it takes to stop the car, typically about 0.15 s. This increases the effective collision time by an order of magnitude. Figure 6.5 shows the measured force on a car versus time for a car crash.

Air bags also increase the collision time, absorb energy from the body as they rapidly deflate, and spread the contact force over an area of the body of about $0.5\ \text{m}^2$, preventing penetration wounds and fractures. Air bags must deploy very rapidly (in less than 10 ms) in order to stop a human traveling at 27 m/s before he or she comes to rest against the steering column about 0.3 m away. In order to achieve this rapid deployment, accelerometers send a signal to discharge a bank of capacitors (devices that store electric charge), which then ignites an explosive, thereby filling the air bag with gas very quickly. The electrical charge for ignition is

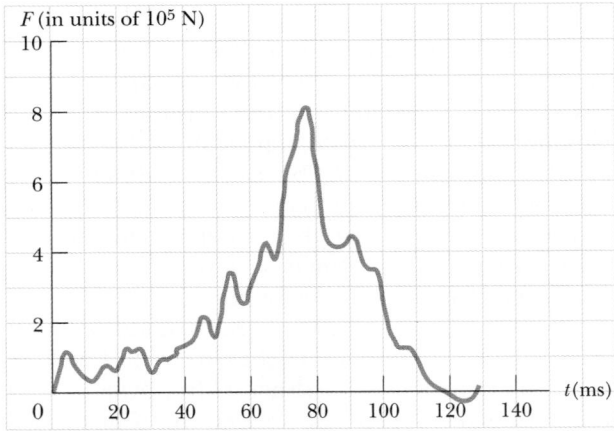

Figure 6.5 Force on a car versus time for a typical collision.

stored in capacitors to ensure that the air bag continues to operate in the event of damage to the battery or the car's electrical system in a severe collision.

The important reduction in potentially fatal forces, accelerations, and pressures to tolerable levels by the simultaneous use of seat belts and air bags is summarized as follows: If a 75-kg person traveling at 27 m/s is stopped by a seat belt in 0.15 s, the person experiences an average force of 9.8 kN, an average acceleration of 18*g*, and a whole-body pressure of 2.8×10^4 N/m^2 for a contact area of 0.5 m^2. These values are about one order of magnitude less than the values estimated earlier for an unprotected person and well below the thresholds for life-threatening injuries.

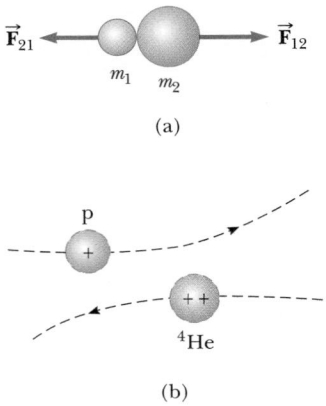

Figure 6.6 (a) A collision between two objects resulting from direct contact. (b) A collision between two charged objects (in this case, a proton and a helium nucleus).

6.2 CONSERVATION OF MOMENTUM

When a collision occurs in an isolated system, the total momentum of the system doesn't change with the passage of time. Instead, it remains constant both in magnitude and in direction. The momenta of the individual objects in the system may change, but the vector sum of *all* the momenta will not change. The total momentum, therefore, is said to be *conserved*. In this section, we will see how the laws of motion lead us to this important conservation law.

A collision may be the result of physical contact between two objects, as illustrated in Figure 6.6a. This is a common macroscopic event, as when a pair of billiard balls or a baseball and a bat strike each other. By contrast, because contact on a submicroscopic scale is hard to define accurately, the notion of *collision* must be generalized to that scale. Forces between two objects arise from the electrostatic interaction of the electrons in the surface atoms of the objects. As will be discussed in Chapter 15, electric charges are either positive or negative. Charges with the same sign repel each other, while charges with opposite sign attract each other. To understand the distinction between macroscopic and microscopic collisions, consider the collision between two positive charges, as shown in Figure 6.6b. Because the two particles in the figure are both positively charged, they repel each other. During such a microscopic collision, particles need not touch in the normal sense in order to interact and transfer momentum.

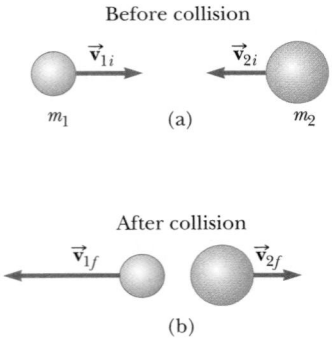

ACTIVE FIGURE 6.7
Before and after a head-on collision between two objects. The momentum of each object changes as a result of the collision, but the total momentum of the system remains constant.

Physics Now™
Log into PhysicsNow at **www.cp7e.com**, and go to Active Figure 6.7 to adjust the masses and velocities of the colliding objects and see the effect on the final velocities.

Active Figure 6.7 shows an isolated system of two particles before and after they collide. By "isolated," we mean that no external forces, such as the gravitational force or friction, act on the system. Before the collision, the velocities of the two particles are $\vec{v}_{1i}$ and $\vec{v}_{2i}$; after the collision, the velocities are $\vec{v}_{1f}$ and $\vec{v}_{2f}$. The impulse–momentum theorem applied to m_1 becomes

$$\vec{F}_{21}\Delta t = m_1\vec{v}_{1f} - m_1\vec{v}_{1i}$$

Likewise, for m_2, we have

$$\vec{F}_{12}\Delta t = m_2\vec{v}_{2f} - m_2\vec{v}_{2i}$$

where $\vec{\mathbf{F}}_{21}$ is the average force exerted by m_2 on m_1 during the collision and $\vec{\mathbf{F}}_{12}$ is the average force exerted by m_1 on m_2 during the collision, as in Figure 6.6a.

We use average values for $\vec{\mathbf{F}}_{21}$ and $\vec{\mathbf{F}}_{12}$ even though the actual forces may vary in time in a complicated way, as is the case in Figure 6.8. Newton's third law states that at all times these two forces are equal in magnitude and opposite in direction: $\vec{\mathbf{F}}_{21} = -\vec{\mathbf{F}}_{12}$. In addition, the two forces act over the same time interval. As a result, we have

$$\vec{\mathbf{F}}_{21}\Delta t = -\vec{\mathbf{F}}_{12}\Delta t$$

or

$$m_1\vec{\mathbf{v}}_{1f} - m_1\vec{\mathbf{v}}_{1i} = -(m_2\vec{\mathbf{v}}_{2f} - m_2\vec{\mathbf{v}}_{2i})$$

after substituting the expressions obtained for $\vec{\mathbf{F}}_{21}$ and $\vec{\mathbf{F}}_{12}$. This equation can be rearranged to give the following important result:

$$m_1\vec{\mathbf{v}}_{1i} + m_2\vec{\mathbf{v}}_{2i} = m_1\vec{\mathbf{v}}_{1f} + m_2\vec{\mathbf{v}}_{2f} \qquad [6.7]$$

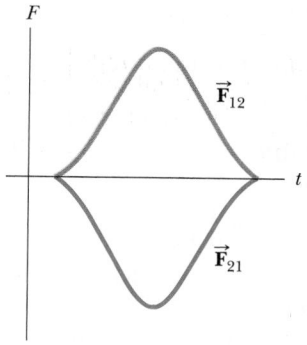

Figure 6.8 Force as a function of time for the two colliding particles in Figures 6.6(a) and 6.7. Note that $\vec{\mathbf{F}}_{21} = -\vec{\mathbf{F}}_{12}$.

This result is a special case of the law of **conservation of momentum** and is true of isolated systems containing any number of interacting objects.

When no net external force acts on a system, the total momentum of the system remains constant in time.

◀ Conservation of momentum

Defining the isolated system is an important feature of applying this conservation law. A cheerleader jumping upwards from rest might appear to violate conservation of momentum, because initially her momentum is zero and suddenly she's leaving the ground with velocity $\vec{\mathbf{v}}$. The flaw in this reasoning lies in the fact that the cheerleader isn't an isolated system. In jumping, she exerts a downward force on the Earth, changing its momentum. This change in the Earth's momentum isn't noticeable, however, because of the Earth's gargantuan mass compared to the cheerleader's. When we define the system to be *the cheerleader and the Earth*, momentum is conserved.

TIP 6.1 Momentum Conservation Applies to a *System*!

The momentum of an isolated system is conserved, but not necessarily the momentum of one particle within that system, because other particles in the system may be interacting with it. Apply conservation of momentum to an isolated system *only*.

Action and reaction, together with the accompanying exchange of momentum between two objects, is responsible for the phenomenon known as *recoil*. Everyone knows that throwing a baseball while standing straight up, without bracing your feet against the Earth, is a good way to fall over backwards. This reaction, an example of recoil, also happens when you fire a gun or shoot an arrow. Conservation of momentum provides a straightforward way to calculate such effects, as the next example shows.

Courtesy of NASA

(a)

Mike Severns/Stone/Getty Images

(b)

Conservation of momentum is the principle behind these two propulsion systems. (a) The force from a nitrogen-propelled, hand-controlled device allows an astronaut to move about freely in space without restrictive tethers. (b) A squid propels itself by expelling water at a high velocity.

INTERACTIVE EXAMPLE 6.3 The Archer

Goal Calculate recoil velocity using conservation of momentum

Problem An archer stands at rest on frictionless ice and fires a 0.500-kg arrow horizontally at 50.0 m/s. (See Fig. 6.9.) The combined mass of the archer and bow is 60.0 kg. With what velocity does the archer move across the ice after firing the arrow?

Figure 6.9 (Interactive Example 6.3) An archer fires an arrow horizontally to the right. Because he is standing on frictionless ice, he will begin to slide to the left across the ice.

Strategy Set up the conservation of momentum equation in the horizontal direction and solve for the final velocity of the archer. The system of the archer (including the bow) and the arrow is *not* isolated, because the gravitational and normal forces act on it. These forces, however, are perpendicular to the motion of the system and hence do no work on it.

Solution

Write the conservation of momentum equation. Let v_{1f} be the archer's velocity and v_{2f} the arrow's velocity.

$$p_i = p_f$$

$$0 = m_1 v_{1f} + m_2 v_{2f}$$

Substitute $m_1 = 60.0$ kg, $m_2 = 0.500$ kg, and $v_{2f} = 50.0$ m/s, and solve for v_{1f}:

$$v_{1f} = -\frac{m_2}{m_1} v_{2f} = -\left(\frac{0.500 \text{ kg}}{60.0 \text{ kg}}\right)(50.0 \text{ m/s})$$

$$= \boxed{-0.417 \text{ m/s}}$$

Remarks The negative sign on v_{1f} indicates that the archer is moving opposite the direction of motion of the arrow, in accordance with Newton's third law. Because the archer is much more massive than the arrow, his acceleration and consequent velocity are much smaller than the acceleration and velocity of the arrow.

Newton's second law, $\Sigma F = ma$, can't be used in this problem because we have no information about the force on the arrow or its acceleration. An energy approach can't be used either, because we don't know how much work is done in pulling the string back or how much potential energy is stored in the bow. Conservation of momentum, however, readily solves the problem.

Exercise 6.3

A 70.0-kg man and a 55.0-kg woman on ice skates stand facing each other. If the woman pushes the man backwards so that his final speed is 1.50 m/s, at what speed does she recoil?

Answer 1.91 m/s

PhysicsNow™ You can change the mass of the archer and the mass and speed of the arrow by logging into PhysicsNow at **www.cp7e.com** and going to Interactive Example 6.3.

Quick Quiz 6.2

A boy standing at one end of a floating raft that is stationary relative to the shore walks to the opposite end of the raft, away from the shore. As a consequence, the raft (a) remains stationary, (b) moves away from the shore, or (c) moves toward the shore. (*Hint*: Use conservation of momentum.)

6.3 COLLISIONS

We have seen that for any type of collision, the total momentum of the system just before the collision equals the total momentum just after the collision as long as the system may be considered isolated. The total kinetic energy, on the other hand, is generally not conserved in a collision because some of the kinetic energy is converted to internal energy, sound energy, and the work needed to permanently deform the objects involved, such as cars in a car crash. **We define an inelastic collision as a collision in which momentum is conserved, but kinetic energy is not.** The collision of a rubber ball with a hard surface is inelastic, because some of the kinetic energy is lost when the ball is deformed during contact with the surface. **When two objects collide and stick together, the collision is called *perfectly inelastic*.** For example, if two pieces of putty collide, they stick together and move with some common velocity after the collision. If a meteorite collides head on with the Earth, it becomes buried in the Earth and the collision is considered perfectly inelastic. Only in very special circumstances is all the initial kinetic energy lost in a perfectly inelastic collision.

An elastic collision is defined as one in which both momentum and kinetic energy are conserved. Billiard ball collisions and the collisions of air molecules with the walls of a container at ordinary temperatures are highly elastic. Macroscopic collisions such as those between billiard balls are only approximately elastic, because some loss of kinetic energy takes place—for example, in the clicking sound when two balls strike each other. Perfectly elastic collisions do occur, however, between atomic and subatomic particles. Elastic and perfectly inelastic collisions are *limiting* cases; most actual collisions fall into a range in between them.

As a practical application, an inelastic collision is used to detect glaucoma, a disease in which the pressure inside the eye builds up and leads to blindness by damaging the cells of the retina. In this application, medical professionals use a device called a *tonometer* to measure the pressure inside the eye. This device releases a puff of air against the outer surface of the eye and measures the speed of the air after reflection from the eye. At normal pressure, the eye is slightly spongy, and the pulse is reflected at low speed. As the pressure inside the eye increases, the outer surface becomes more rigid, and the speed of the reflected pulse increases. In this way, the speed of the reflected puff of air can measure the internal pressure of the eye.

We can summarize the types of collisions as follows:

- In an elastic collision, both momentum and kinetic energy are conserved.
- In an inelastic collision, momentum is conserved but kinetic energy is not.
- In a *perfectly* inelastic collision, momentum is conserved, kinetic energy is not, and the two objects stick together after the collision, so their final velocities are the same.

In the remainder of this section, we will treat perfectly inelastic collisions and elastic collisions in one dimension.

TIP 6.2 Momentum and Kinetic Energy in Collisions

The momentum of an isolated system is conserved in all collisions. However, the kinetic energy of an isolated system is conserved only when the collision is elastic.

TIP 6.3 Inelastic vs. Perfectly Inelastic Collisions

If the colliding particles stick together, the collision is perfectly inelastic. If they bounce off each other (and kinetic energy is not conserved), the collision is inelastic.

APPLICATION

Glaucoma Testing

◀ Elastic collision

◀ Inelastic collision

Quick Quiz 6.3

A car and a large truck traveling at the same speed collide head-on and stick together. Which vehicle experiences the larger change in the magnitude of its momentum? (a) the car (b) the truck (c) the change in the magnitude of momentum is the same for both (d) impossible to determine

Perfectly Inelastic Collisions

Consider two objects having masses m_1 and m_2 moving with known initial velocity components v_{1i} and v_{2i} along a straight line, as in Active Figure 6.10. If the two objects collide head-on, stick together, and move with a common velocity component v_f after the collision, then the collision is perfectly inelastic. Because the total momentum of the two-object isolated system before the collision equals the total momentum of the combined-object system after the collision, we can solve for the final velocity using conservation of momentum alone:

$$m_1 v_{1i} + m_2 v_{2i} = (m_1 + m_2) v_f \qquad [6.8]$$

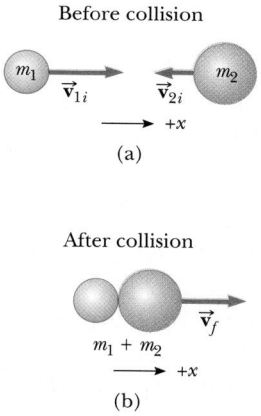

ACTIVE FIGURE 6.10
(a) Before and (b) after a perfectly inelastic head-on collision between two objects.

Physics Now™
Log into PhysicsNow at **www.cp7e.com**, and go to Active Figure 6.10 to adjust the masses and velocities of the colliding objects and see the effect on the final velocity.

Final velocity of two objects in a one-dimensional perfectly inelastic collision ▶

$$v_f = \frac{m_1 v_{1i} + m_2 v_{2i}}{m_1 + m_2} \qquad [6.9]$$

It's important to notice that v_{1i}, v_{2i}, and v_f represent the x-components of the velocity vectors, so care is needed in entering their known values, particularly with regard to signs. For example, in Active Figure 6.10, v_{1i} would have a positive value (m_1 moving to the right), whereas v_{2i} would have a negative value (m_2 moving to the left). Once these values are entered, Equation 6.9 can be used to find the correct final velocity, as shown in Examples 6.4 and 6.5.

EXAMPLE 6.4 An SUV Versus a Compact

Goal Apply conservation of momentum to a one-dimensional inelastic collision.

Problem An SUV with mass 1.80×10^3 kg is traveling eastbound at $+15.0$ m/s, while a compact car with mass 9.00×10^2 kg is traveling westbound at -15.0 m/s. (See Fig. 6.11.) The cars collide head-on, becoming entangled. **(a)** Find the speed of the entangled cars after the collision. **(b)** Find the change in the velocity of each car. **(c)** Find the change in the kinetic energy of the system consisting of both cars.

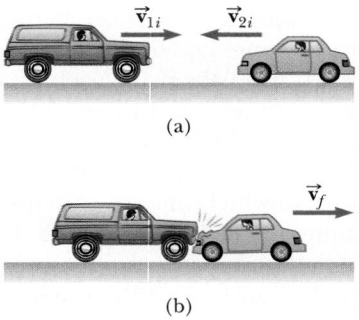

Figure 6.11 (Example 6.4)

Strategy The total momentum of the cars before the collision, p_i, equals the total momentum of the cars after the collision, p_f, if we ignore friction and assume the two cars form an isolated system. (This is called the "impulse approximation.") Solve the momentum conservation equation for the final velocity of the entangled cars. Once the velocities are in hand, the other parts can be solved by substitution.

Solution

(a) Find the final speed after collision.

Let m_1 and v_{1i} represent the mass and initial velocity of the SUV, while m_2 and v_{2i} pertain to the compact. Apply conservation of momentum:

$$p_i = p_f$$
$$m_1 v_{1i} + m_2 v_{2i} = (m_1 + m_2) v_f$$

Substitute the values and solve for the final velocity, v_f:

$$(1.80 \times 10^3 \text{ kg})(15.0 \text{ m/s}) + (9.00 \times 10^2 \text{ kg})(-15.0 \text{ m/s})$$
$$= (1.80 \times 10^3 \text{ kg} + 9.00 \times 10^2 \text{ kg}) v_f$$
$$v_f = \boxed{+5.00 \text{ m/s}}$$

(b) Find the change in velocity for each car.

Change in velocity of the SUV:

$$\Delta v_1 = v_f - v_{1i} = 5.00 \text{ m/s} - 15.0 \text{ m/s} = \boxed{-10.0 \text{ m/s}}$$

Change in velocity of the compact car:

$$\Delta v_2 = v_f - v_{2i} = 5.00 \text{ m/s} - (-15.0 \text{ m/s}) = \boxed{20.0 \text{ m/s}}$$

(c) Find the change in kinetic energy of the system.

Calculate the initial kinetic energy of the system:

$$KE_i = \tfrac{1}{2} m_1 v_{1i}^2 + \tfrac{1}{2} m_2 v_{2i}^2 = \tfrac{1}{2}(1.80 \times 10^3 \text{ kg})(15.0 \text{ m/s})^2$$
$$+ \tfrac{1}{2}(9.00 \times 10^2 \text{ kg})(-15.0 \text{ m/s})^2$$
$$= 3.04 \times 10^5 \text{ J}$$

Calculate the final kinetic energy of the system and the change in kinetic energy, ΔKE.

$$KE_f = \tfrac{1}{2}(m_1 + m_2) v_f^2$$
$$= \tfrac{1}{2}(1.80 \times 10^3 \text{ kg} + 9.00 \times 10^2 \text{ kg})(5.00 \text{ m/s})^2$$
$$= 3.38 \times 10^4 \text{ J}$$
$$\Delta KE = KE_f - KE_i = \boxed{-2.70 \times 10^5 \text{ J}}$$

Remarks During the collision, the system lost almost 90% of its kinetic energy. The change in velocity of the SUV was only 10.0 m/s, compared to twice that for the compact car. This example underscores perhaps the most important safety feature of any car: its mass. Injury is caused by a change in velocity, and the more massive vehicle undergoes a smaller velocity change in a typical accident.

Exercise 6.4

Suppose the same two vehicles are both traveling eastward, the compact car leading the SUV. The driver of the compact car slams on the brakes suddenly, slowing the vehicle to 6.00 m/s. If the SUV traveling at 18.0 m/s crashes into the compact car, find (a) the speed of the system right after the collision, assuming the two vehicles become entangled, (b) the change in velocity for both vehicles, and (c) the change in kinetic energy of the system, from the instant before impact (when the compact car is traveling at 6.00 m/s) to the instant right after the collision.

Answers (a) 14.0 m/s (b) SUV: $\Delta v_1 = -4.0$ m/s Compact car: $\Delta v_2 = 8.0$ m/s (c) -4.32×10^4 J

EXAMPLE 6.5 The Ballistic Pendulum

Goal Combine the concepts of conservation of energy and conservation of momentum in inelastic collisions.

Problem The ballistic pendulum (Fig. 6.12a) is a device used to measure the speed of a fast-moving projectile such as a bullet. The bullet is fired into a large block of wood suspended from some light wires. The bullet is stopped by the block, and the entire system swings up to a height h. It is possible to obtain the initial speed of the bullet by measuring h and the two masses. As an example of the technique, assume that the mass of the bullet, m_1, is 5.00 g, the mass of the pendulum, m_2, is 1.000 kg, and h is 5.00 cm. Find the initial speed of the bullet, v_{1i}.

Strategy First, use conservation of momentum and the properties of perfectly inelastic collisions to find the initial speed of the bullet, v_{1i}, in terms of the final velocity of the block–bullet system, v_f. Second, use conservation of energy and the height reached by the pendulum to find v_f. Finally, substitute this value of v_f into the previous result to obtain the initial speed of the bullet.

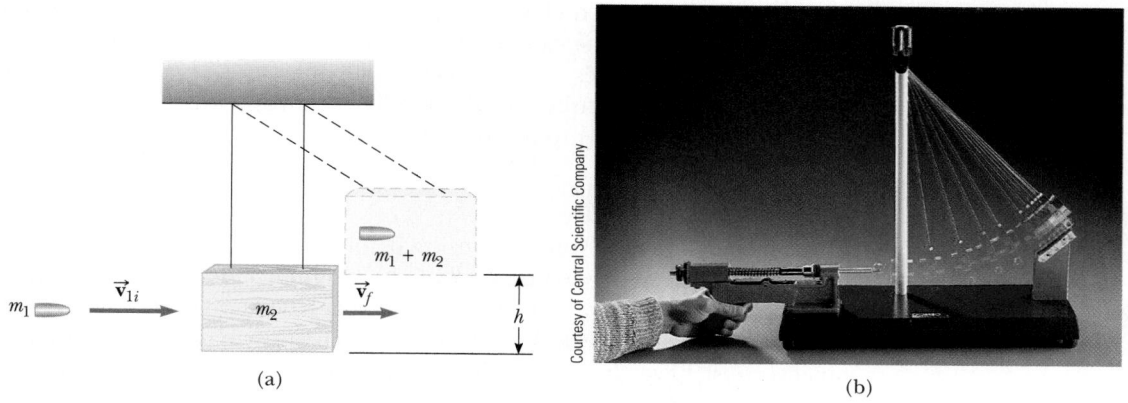

Figure 6.12 (Example 6.5) (a) Diagram of a ballistic pendulum. Note that $\vec{v}_f$ is the velocity of the system just *after* the perfectly inelastic collision. (b) Multiflash photograph of a laboratory ballistic pendulum.

Solution

Use conservation of momentum, and substitute the known masses. Note that $v_{2i} = 0$ and v_f is the velocity of the system (block + bullet) just after the collision.

$$p_i = p_f$$

$$m_1 v_{1i} + m_2 v_{2i} = (m_1 + m_2) v_f$$

$$(5.00 \times 10^{-3}\ \text{kg})\, v_{1i} + 0 = (1.005\ \text{kg})\, v_f \qquad (1)$$

Apply conservation of energy to the block–bullet system after the collision:

$$(KE + PE)_{\text{after collision}} = (KE + PE)_{\text{top}}$$

Both the potential energy at the bottom and the kinetic energy at the top are zero. Solve for the final velocity of the block–bullet system, v_f:

$$\tfrac{1}{2}(m_1 + m_2)v_f^2 + 0 = 0 + (m_1 + m_2)gh$$
$$v_f^2 = 2gh$$
$$v_f = \sqrt{2gh} = \sqrt{2(9.80\ \text{m/s}^2)(5.00 \times 10^{-2}\ \text{m})}$$
$$v_f = 0.990\ \text{m/s}$$

Finally, substitute v_f into Equation 1 to find v_{1i}, the initial speed of the bullet:

$$v_{1i} = \frac{(1.005\ \text{kg})(0.990\ \text{m/s})}{5.00 \times 10^{-3}\ \text{kg}} = \boxed{199\ \text{m/s}}$$

Remarks Because the impact is inelastic, it would be incorrect to equate the initial kinetic energy of the incoming bullet to the final gravitational potential energy associated with the bullet–block combination. The energy isn't conserved!

Exercise 6.5

A bullet with mass 5.00 g is fired horizontally into a 2.000-kg block attached to a horizontal spring. The spring has a constant 6.00×10^2 N/m and reaches a maximum compression of 6.00 cm. (a) Find the initial speed of the bullet–block system. (b) Find the speed of the bullet.

Answer (a) 1.04 m/s (b) 417 m/s

Quick Quiz 6.4

An object of mass m moves to the right with a speed v. It collides head-on with an object of mass $3m$ moving with speed $v/3$ in the opposite direction. If the two objects stick together, what is the speed of the combined object, of mass $4m$, after the collision?

(a) 0 (b) $v/2$ (c) v (d) $2v$

Quick Quiz 6.5

A skater is using very low friction rollerblades. A friend throws a Frisbee® at her, on the straight line along which she is coasting. Describe each of the following events as an elastic, an inelastic, or a perfectly inelastic collision between the skater and the Frisbee: (a) She catches the Frisbee and holds it. (b) She tries to catch the Frisbee, but it bounces off her hands and falls to the ground in front of her. (c) She catches the Frisbee and immediately throws it back with the same speed (relative to the ground) to her friend.

Quick Quiz 6.6

In a perfectly inelastic one-dimensional collision between two objects, what condition alone is necessary so that *all* of the original kinetic energy of the system is gone after the collision? (a) The objects must have momenta with the same magnitude but opposite directions. (b) The objects must have the same mass. (c) The objects must have the same velocity. (d) The objects must have the same speed, with velocity vectors in opposite directions.

Elastic Collisions

Now consider two objects that undergo an **elastic head-on collision** (Active Fig. 6.13). In this situation, **both the momentum and the kinetic energy of the system of two objects are conserved.** We can write these conditions as

$$m_1 v_{1i} + m_2 v_{2i} = m_1 v_{1f} + m_2 v_{2f} \qquad \textbf{[6.10]}$$

and

$$\tfrac{1}{2}m_1 v_{1i}^2 + \tfrac{1}{2}m_2 v_{2i}^2 = \tfrac{1}{2}m_1 v_{1f}^2 + \tfrac{1}{2}m_2 v_{2f}^2 \qquad \textbf{[6.11]}$$

where v is positive if an object moves to the right and negative if it moves to the left.

In a typical problem involving elastic collisions, there are two unknown quantities, and Equations 6.10 and 6.11 can be solved simultaneously to find them. These two equations are linear and quadratic, respectively. An alternate approach simplifies the quadratic equation to another linear equation, facilitating solution. Canceling the factor $\frac{1}{2}$ in Equation 6.11, we rewrite the equation as

$$m_1(v_{1i}^2 - v_{1f}^2) = m_2(v_{2f}^2 - v_{2i}^2)$$

Here we have moved the terms containing m_1 to one side of the equation and those containing m_2 to the other. Next, we factor both sides of the equation:

$$m_1(v_{1i} - v_{1f})(v_{1i} + v_{1f}) = m_2(v_{2f} - v_{2i})(v_{2f} + v_{2i}) \qquad \textbf{[6.12]}$$

Now we separate the terms containing m_1 and m_2 in the equation for the conservation of momentum (Equation 6.10) to get

$$m_1(v_{1i} - v_{1f}) = m_2(v_{2f} - v_{2i}) \qquad \textbf{[6.13]}$$

To obtain our final result, we divide Equation 6.12 by Equation 6.13, producing

$$v_{1i} + v_{1f} = v_{2f} + v_{2i}$$

Gathering initial and final values on opposite sides of the equation gives

$$v_{1i} - v_{2i} = -(v_{1f} - v_{2f}) \qquad \textbf{[6.14]}$$

This equation, in combination with Equation 6.10, will be used to solve problems dealing with perfectly elastic head-on collisions. According to Equation 6.14, the relative velocity of the two objects before the collision, $v_{1i} - v_{2i}$, equals the negative of the relative velocity of the two objects after the collision, $-(v_{1f} - v_{2f})$. To better understand the equation, imagine that you are riding along on one of the objects. As you measure the velocity of the other object from your vantage point, you will be measuring the relative velocity of the two objects. In your view of the collision, the other object comes toward you and bounces off, leaving the collision with the same speed, but in the opposite direction. This is just what Equation 6.14 states.

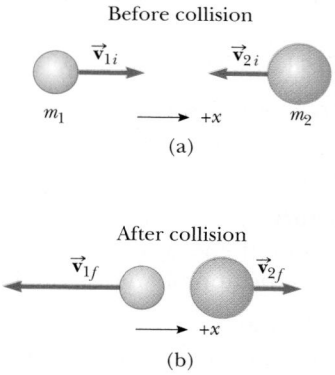

ACTIVE FIGURE 6.13
(a) Before and (b) after an elastic head-on collision between two hard spheres.

PhysicsNow™
Log into PhysicsNow at **www.cp7e.com**, and go to Active Figure 6.13 to adjust the masses and velocities of the colliding objects and see the effect on the final velocities.

Problem-Solving Strategy
One-Dimensional Collisions

The following procedure is recommended for solving one-dimensional problems involving collisions between two objects:

1. **Coordinates**. Choose a coordinate axis that lies along the direction of motion.
2. **Diagram**. Sketch the problem, representing the two objects as blocks and labeling velocity vectors and masses.
3. **Conservation of Momentum**. Write a general expression for the *total* momentum of the system of two objects *before* and *after* the collision, and equate the two, as in Equation 6.10. On the next line, fill in the known values.
4. **Conservation of Energy**. If the collision is elastic, write a general expression for the total energy before and after the collision, and equate the two quantities, as in Equation 6.11 or (preferably) Equation 6.14. Fill in the known values. (***Skip* this step if the collision is *not* perfectly elastic**.)
5. **Solve** the equations simultaneously. Equations 6.10 and 6.14 form a system of two linear equations and two unknowns. If you have forgotten Equation 6.14, use Equation 6.11 instead.

Steps 1 and 2 of the problem-solving strategy are generally carried out in the process of sketching and labeling a diagram of the problem. This is clearly the case in our next example, which makes use of Figure 6.13. Other steps are pointed out as they are applied.

EXAMPLE 6.6 Let's Play Pool

Goal Solve an elastic collision in one dimension.

Problem Two billiard balls of identical mass move toward each other as in Active Figure 6.13. Assume that the collision between them is perfectly elastic. If the initial velocities of the balls are + 30.0 cm/s and − 20.0 cm/s, what is the velocity of each ball after the collision? Assume friction and rotation are unimportant.

Strategy Solution of this problem is a matter of solving two equations, the conservation of momentum and conservation of energy equations, for two unknowns, the final velocities of the two balls. Instead of using Equation 6.11 for conservation of energy, use Equation 6.14, which is linear, hence easier to handle.

Solution

Write the conservation of momentum equation. Because $m_1 = m_2$, we can cancel the masses, then substitute $v_{1i} = +30.0$ m/s and $v_{2i} = -20.0$ cm/s (Step 3).

$$m_1 v_{1i} + m_2 v_{2i} = m_1 v_{1f} + m_2 v_{2f}$$

$$30.0 \text{ cm/s} + (-20.0 \text{ cm/s}) = v_{1f} + v_{2f}$$

$$10.0 \text{ cm/s} = v_{1f} + v_{2f} \qquad (1)$$

Next, apply conservation of energy in the form of Equation 6.14 (Step 4):

$$v_{1i} - v_{2i} = -(v_{1f} - v_{2f})$$

$$30.0 \text{ cm/s} - (-20.0 \text{ cm/s}) = v_{2f} - v_{1f}$$

$$50.0 \text{ cm/s} = v_{2f} - v_{1f} \qquad (2)$$

Now solve (1) and (2) simultaneously (Step 5): $v_{1f} = -20.0$ cm/s $\quad v_{2f} = +30.0$ cm/s

Remarks Notice the balls exchanged velocities—almost as if they'd passed through each other. This is always the case when two objects of equal mass undergo an elastic head-on collision.

Exercise 6.6

Find the final velocity of the two balls if the ball with initial velocity $v_{2i} = -20.0$ cm/s has a mass equal to half that of the ball with initial velocity $v_{1i} = +30.0$ cm/s.

Answer $v_{1f} = -3.33$ cm/s; $v_{2f} = +46.7$ cm/s

INTERACTIVE EXAMPLE 6.7 Two Blocks and a Spring

Goal Solve an elastic collision involving spring potential energy.

Problem A block of mass $m_1 = 1.60$ kg, initially moving to the right with a velocity of + 4.00 m/s on a frictionless horizontal track, collides with a massless spring attached to a second block of mass $m_2 = 2.10$ kg moving to the left with a velocity of − 2.50 m/s, as in Figure 6.14a. The spring has a spring constant of 6.00×10^2 N/m. **(a)** Determine the velocity of block 2 at the instant when block 1 is moving to the right with a velocity of + 3.00 m/s, as in Figure 6.14b. **(b)** Find the compression of the spring.

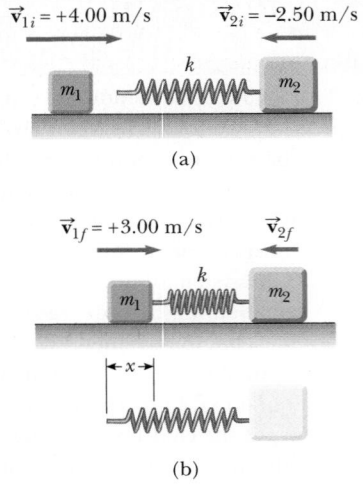

Figure 6.14 (Example 6.7)

Strategy We identify the system as the two blocks and the spring. Write down the conservation of momentum equations, and solve for the final velocity of block 2, v_{2f}. Then use conservation of energy to find the compression of the spring.

Solution

(a) Find the velocity v_{2f} when block 1 has velocity + 3.00 m/s.

$$m_1 v_{1i} + m_2 v_{2i} = m_1 v_{1f} + m_2 v_{2f}$$

$$v_{2f} = \frac{m_1 v_{1i} + m_2 v_{2i} - m_1 v_{1f}}{m_2}$$

$$= \frac{(1.60 \text{ kg})(4.00 \text{ m/s}) + (2.10 \text{ kg})(-2.50 \text{ m/s}) - (1.60 \text{ kg})(3.00 \text{ m/s})}{2.10 \text{ kg}}$$

Write the conservation of momentum equation for the system and solve for v_{2f}: $v_{2f} = -1.74$ m/s

(b) Find the compression of the spring.

Use energy conservation for the system, noticing that potential energy is stored in the spring when it is compressed a distance x:

$$E_i = E_f$$

$$\tfrac{1}{2}m_1v_{1i}^2 + \tfrac{1}{2}m_2v_{2i}^2 + 0 = \tfrac{1}{2}m_1v_{1f}^2 + \tfrac{1}{2}m_2v_{2f}^2 + \tfrac{1}{2}kx^2$$

Substitute the given values and the result of part (a) into the preceding expression, solving for x.

$$x = 0.173 \text{ m}$$

Remarks The initial velocity component of block 2 is -2.50 m/s because the block is moving to the left. The negative value for v_{2f} means that block 2 is still moving to the left at the instant under consideration.

Exercise 6.7

Find (a) the velocity of block 1 and (b) the compression of the spring at the instant that block 2 is at rest.

Answer (a) 0.719 m/s to the right (b) 0.251 m

PhysicsNow™ You can change the masses and speeds of the blocks and freeze the motion at the maximum compression of the spring by logging into PhysicsNow at **www.cp7e.com** and going to Interactive Example 6.7.

6.4 GLANCING COLLISIONS

In Section 6.2 we showed that the total linear momentum of a system is conserved when the system is isolated (that is, when no external forces act on the system). For a general collision of two objects in three-dimensional space, the conservation of momentum principle implies that the total momentum of the system in each direction is conserved. However, an important subset of collisions takes place in a plane. The game of billiards is a familiar example involving multiple collisions of objects moving on a two-dimensional surface. We restrict our attention to a single two-dimensional collision between two objects that takes place in a plane, and ignore any possible rotation. For such collisions, we obtain two component equations for the conservation of momentum:

$$m_1v_{1ix} + m_2v_{2ix} = m_1v_{1fx} + m_2v_{2fx}$$

$$m_1v_{1iy} + m_2v_{2iy} = m_1v_{1fy} + m_2v_{2fy}$$

We must use three subscripts in this general equation, to represent, respectively, (1) the object in question, and (2) the initial and final values of the components of velocity.

Now, consider a two-dimensional problem in which an object of mass m_1 collides with an object of mass m_2 that is initially at rest, as in Active Figure 6.15. After the collision, object 1 moves at an angle θ with respect to the horizontal, and object 2 moves at an angle ϕ with respect to the horizontal. This is called a *glancing* collision. Applying the law of conservation of momentum in component form, and noting that the initial y-component of momentum is zero, we have

$$x\text{-component:} \quad m_1v_{1i} + 0 = m_1v_{1f}\cos\theta + m_2v_{2f}\cos\phi \qquad \textbf{[6.15]}$$

$$y\text{-component:} \quad 0 + 0 = m_1v_{1f}\sin\theta - m_2v_{2f}\sin\phi \qquad \textbf{[6.16]}$$

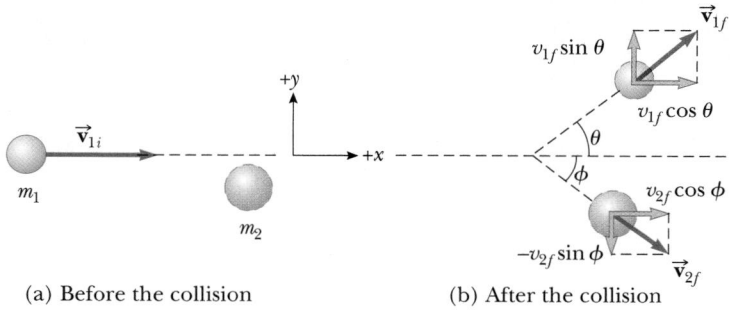

(a) Before the collision (b) After the collision

ACTIVE FIGURE 6.15
(a) Before and (b) after a glancing collision between two balls.

PhysicsNow™
Log into PhysicsNow at **www.cp7e.com**, and go to Active Figure 6.15 to adjust the speed and position of the blue particle, adjust the masses of both particles, and see the effects.

If the collision is elastic, we can write a third equation, for conservation of energy, in the form

$$\tfrac{1}{2}m_1v_{1i}^2 = \tfrac{1}{2}m_1v_{1f}^2 + \tfrac{1}{2}m_2v_{2f}^2 \qquad \textbf{[6.17]}$$

If we know the initial velocity v_{1i} and the masses, we are left with four unknowns (v_{1f}, v_{2f}, θ, and ϕ). Because we have only three equations, one of the four remaining quantities must be given in order to determine the motion after the collision from conservation principles alone.

If the collision is inelastic, the kinetic energy of the system is *not* conserved, and Equation 6.17 does *not* apply.

Problem-Solving Strategy
Two-Dimensional Collisions

To solve two-dimensional collisions, follow this procedure:

1. **Coordinate Axes.** Use both x- and y-coordinates. It's convenient to have either the x-axis or the y-axis coincide with the direction of one of the initial velocities.
2. **Diagram**. Sketch the problem, labeling velocity vectors and masses.
3. **Conservation of Momentum**. Write a separate conservation of momentum equation for each of the x- and y-directions. In each case, the total initial momentum in a given direction equals the total final momentum in that direction.
4. **Conservation of Energy**. If the collision is elastic, write a general expression for the total energy before and after the collision, and equate the two expressions, as in Equation 6.11. Fill in the known values. **(*Skip* this step if the collision is *not* perfectly elastic.)** The energy equation can't be simplified as in the one-dimensional case, so a quadratic expression such as Equation 6.11 or 6.17 must be used when the collision is elastic.
5. **Solve** the equations simultaneously. There are two equations for inelastic collisions and three for elastic collisions.

EXAMPLE 6.8 Collision at an Intersection

Goal Analyze a two-dimensional inelastic collision.

Problem A car with mass 1.50×10^3 kg traveling east at a speed of 25.0 m/s collides at an intersection with a 2.50×10^3-kg van traveling north at a speed of 20.0 m/s, as shown in Figure 6.16. Find the magnitude and direction of the velocity of the wreckage after the collision, assuming that the vehicles undergo a perfectly inelastic collision (that is, they stick together) and assuming that friction between the vehicles and the road can be neglected.

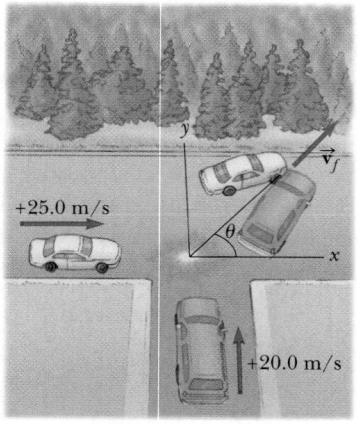

Figure 6.16 (Example 6.8) A top view of a perfectly inelastic collision between a car and a van.

Strategy Use conservation of momentum in two dimensions. (Kinetic energy is *not* conserved.) Choose coordinates as in Figure 6.16. Before the collision, the only object having momentum in the x-direction is the car, while the van carries all the momentum in the y-direction. After the totally inelastic collision, both vehicles move together at some common speed v_f and angle θ. Solve for these two unknowns, using the two components of the conservation of momentum equation.

Solution

Find the x-components of the initial and final total momenta:

$$\Sigma p_{xi} = m_{\text{car}}v_{\text{car}} = (1.50 \times 10^3 \text{ kg})(25.0 \text{ m/s}) = 3.75 \times 10^4 \text{ kg}\cdot\text{m/s}$$

$$\Sigma p_{xf} = (m_{\text{car}} + m_{\text{van}})v_f \cos\theta = (4.00 \times 10^3 \text{ kg})v_f \cos\theta$$

Set the initial x-momentum equal to the final x-momentum:

$$3.75 \times 10^4 \text{ kg}\cdot\text{m/s} = (4.00 \times 10^3 \text{ kg})\, v_f \cos\theta \qquad \textbf{(1)}$$

Find the y-components of the initial and final total momenta:

$$\Sigma p_{iy} = m_{\text{van}} v_{\text{van}} = (2.50 \times 10^3 \text{ kg})(20.0 \text{ m/s}) = 5.00 \times 10^4 \text{ kg} \cdot \text{m/s}$$

$$\Sigma p_{fy} = (m_{\text{car}} + m_{\text{van}}) v_f \sin\theta = (4.00 \times 10^3 \text{ kg}) v_f \sin\theta$$

Set the initial y-momentum equal to the final y-momentum:

$$5.00 \times 10^4 \text{ kg} \cdot \text{m/s} = (4.00 \times 10^3 \text{ kg}) v_f \sin\theta \qquad \textbf{(2)}$$

Divide Equation (2) by Equation (1) and solve for θ:

$$\tan\theta = \frac{5.00 \times 10^4 \text{ kg} \cdot \text{m/s}}{3.75 \times 10^4 \text{ kg} \cdot \text{m}} = 1.33$$

$$\theta = 53.1°$$

Substitute this angle back into Equation (2) to find v_f:

$$v_f = \frac{5.00 \times 10^4 \text{ kg} \cdot \text{m/s}}{(4.00 \times 10^3 \text{ kg}) \sin 53.1°} = 15.6 \text{ m/s}$$

Remark It's also possible to first find the x- and y-components v_{fx} and v_{fy} of the resultant velocity. The magnitude and direction of the resultant velocity can then be found with the Pythagorean theorem, $v_f = \sqrt{v_{fx}^2 + v_{fy}^2}$, and the inverse tangent function $\theta = \tan^{-1}(v_{fy}/v_{fx})$. Setting up this alternate approach is a simple matter of substituting $v_{fx} = v_f \cos\theta$ and $v_{fy} = v_f \sin\theta$ in Equations (1) and (2).

Exercise 6.8

A 3.00-kg object initially moving in the positive x-direction with a velocity of $+5.00$ m/s collides with and sticks to a 2.00-kg object initially moving in the negative y-direction with a velocity of -3.00 m/s. Find the final components of velocity of the composite object.

Answer $v_{fx} = 3.00$ m/s; $v_{fy} = -1.20$ m/s

6.5 ROCKET PROPULSION

When ordinary vehicles such as cars and locomotives move, the driving force of the motion is friction. In the case of the car, this driving force is exerted by the road on the car, a reaction to the force exerted by the wheels against the road. Similarly, a locomotive "pushes" against the tracks; hence, the driving force is the reaction force exerted by the tracks on the locomotive. However, a rocket moving in space has no road or tracks to push against. How can it move forward?

In fact, reaction forces also propel a rocket. (You should review Newton's third law, discussed in Chapter 4.) To illustrate this point, we model our rocket with a spherical chamber containing a combustible gas, as in Figure 6.17a. When an explosion occurs in the chamber, the hot gas expands and presses against all sides of the chamber, as indicated by the arrows. Because the sum of the forces exerted on the rocket is zero, it doesn't move. Now suppose a hole is drilled in the bottom of the chamber, as in Figure 6.17b. When the explosion occurs, the gas presses against the chamber in all directions, but can't press against anything at the hole, where it simply escapes into space. Adding the forces on the spherical chamber now results in a net force upwards. Just as in the case of cars and locomotives, this is a reaction force. A car's wheels press against the ground, and the reaction force of the ground on the car pushes it forward. The wall of the rocket's combustion chamber exerts a force on the gas expanding against it. The reaction force of the gas on the wall then pushes the rocket upward.

In a now infamous article in *The New York Times*, rocket pioneer Robert Goddard was ridiculed for thinking that rockets would work in space, where, according to the *Times*, there was nothing to push against. The *Times* retracted, rather belatedly, during the first Apollo moon landing mission in 1969. The hot gases are not pushing against anything external, but against the rocket itself—and ironically, rockets actually work *better* in a vacuum. In an atmosphere, the gases have to do work against the outside air pressure to escape the combustion chamber, slowing the exhaust velocity and reducing the reaction force.

(a)

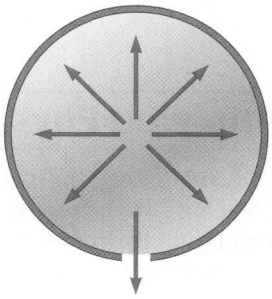

(b)

Figure 6.17 (a) A rocket reaction chamber without a nozzle has reaction forces pushing equally in all directions, so no motion results. (b) An opening at the bottom of the chamber removes the downward reaction force, resulting in a net upward reaction force.

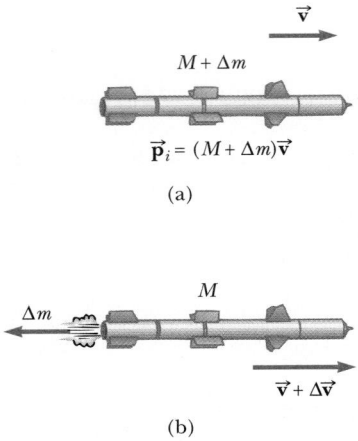

Figure 6.18 Rocket propulsion. (a) The initial mass of the rocket and fuel is $M + \Delta m$ at a time t, and the rocket's speed is v. (b) At a time $t + \Delta t$, the rocket's mass has been reduced to M, and an amount of fuel Δm has been ejected. The rocket's speed increases by an amount Δv.

At the microscopic level, this process is complicated, but it can be simplified by applying conservation of momentum to the rocket and its ejected fuel. In principle, the solution is similar to that in Example 6.3, with the archer representing the rocket and the arrows the exhaust gases.

Suppose that at some time t, the momentum of the rocket plus the fuel is $(M + \Delta m)v$, where Δm is an amount of fuel about to be burned (Fig. 6.18a). This fuel is traveling at a speed v relative to, say, the Earth, just like the rest of the rocket. During a short time interval Δt, the rocket ejects fuel of mass Δm, and the rocket's speed increases to $v + \Delta v$ (Fig. 6.18b). If the fuel is ejected with exhaust speed v_e *relative to the rocket*, the speed of the fuel relative to the Earth is $v - v_e$. Equating the total initial momentum of the system with the total final momentum, we have

$$(M + \Delta m)v = M(v + \Delta v) + \Delta m(v - v_e)$$

Simplifying this expression gives

$$M\Delta v = v_e \Delta m$$

The increase Δm in the mass of the exhaust corresponds to an equal decrease in the mass of the rocket, so that $\Delta m = -\Delta M$. Using this fact, we have

$$M\Delta v = -v_e \Delta M \qquad \textbf{[6.18]}$$

This result, together with the methods of calculus, can be used to obtain the following equation:

$$v_f - v_i = v_e \ln\left(\frac{M_i}{M_f}\right) \qquad \textbf{[6.19]}$$

where M_i is the initial mass of the rocket plus fuel and M_f is the final mass of the rocket plus its remaining fuel. This is the basic expression for rocket propulsion; it tells us that the increase in velocity is proportional to the exhaust speed v_e and to the natural logarithm of M_i/M_f. Because the maximum ratio of M_i to M_f for a single-stage rocket is about 10:1, the increase in speed can reach $v_e \ln 10 = 2.3v_e$ or about twice the exhaust speed! For best results, therefore, the exhaust speed should be as high as possible. Currently, typical rocket exhaust speeds are several kilometers per second.

The **thrust** on the rocket is defined as the force exerted on the rocket by the ejected exhaust gases. We can obtain an expression for the instantaneous thrust by dividing Equation 6.18 by Δt:

Rocket thrust ▶

$$\text{Instantaneous thrust} = Ma = M\frac{\Delta v}{\Delta t} = \left|v_e \frac{\Delta M}{\Delta t}\right| \qquad \textbf{[6.20]}$$

The absolute value signs are used for clarity: In Equation 6.18, $-\Delta M$ is a positive quantity (as is v_e, a speed). Here we see that the thrust increases as the exhaust velocity increases and as the rate of change of mass $\Delta M/\Delta t$ (the burn rate) increases.

Applying Physics 6.2 Multistage Rockets

The current maximum exhaust speed of $v_e = 4\,500$ m/s can be realized with rocket engines fueled with liquid hydrogen and liquid oxygen. But this means that the maximum speed attainable for a given rocket with a mass ratio of 10 is $v_e \ln 10 \approx 10\,000$ m/s. To reach the moon, however, requires a change in velocity of over 11 000 m/s. Further, this change must occur while working against gravity and atmospheric friction. How can that be managed without developing better engines?

Explanation The answer is the multistage rocket. By dropping stages, the spacecraft becomes lighter, so that fuel burned later in the mission doesn't have to accelerate mass that no longer serves any purpose. Strap-on boosters, as used by the Space Shuttle and a number of other rockets, such as the Titan 4 or Russian Proton, is a similar concept. The boosters are jettisoned after their fuel is exhausted, so the rocket is no longer burdened by their weight.

EXAMPLE 6.9 Single Stage to Orbit (SSTO)

Goal Apply the velocity and thrust equations of a rocket.

Problem A rocket has a total mass of 1.00×10^5 kg and a burnout mass of 1.00×10^4 kg, including engines, shell, and payload. The rocket blasts off from Earth and exhausts all its fuel in 4.00 min, burning the fuel at a steady rate with an exhaust velocity of $v_e = 4.50 \times 10^3$ m/s. **(a)** If air friction and gravity are neglected, what is the speed of the rocket at burnout? **(b)** What thrust does the engine develop at liftoff? **(c)** What is the initial acceleration of the rocket if gravity is not neglected? **(d)** Estimate the speed at burnout if gravity isn't neglected.

Strategy Although it sounds sophisticated, this problem is mainly a matter of substituting values into the appropriate equations. Part (a) requires substituting values into Equation 6.19 for the velocity. For part (b), divide the change in the rocket's mass by the total time, getting $\Delta M/\Delta t$, then substitute into Equation 6.20 to find the thrust. (c) Using Newton's second law, the force of gravity, and the result of (b), we can find the initial acceleration. For part (d), the acceleration of gravity is approximately constant over the few kilometers involved, so the velocity found in (b) will be reduced by roughly $\Delta v_g = -gt$. Add this loss to the result of part (a).

Solution

(a) Calculate the velocity at burnout.

Substitute $v_i = 0$, $v_e = 4.50 \times 10^3$ m/s, $M_i = 1.00 \times 10^5$ kg, and $M_f = 1.00 \times 10^4$ kg into Equation 6.19:

$$v_f = v_i + v_e \ln\left(\frac{M_i}{M_f}\right)$$

$$= 0 + (4.5 \times 10^3 \text{ m/s}) \ln\left(\frac{1.00 \times 10^5 \text{ kg}}{1.00 \times 10^4 \text{ kg}}\right)$$

$$v_f = 1.04 \times 10^4 \text{ m/s}$$

(b) Find the thrust at liftoff.

Compute the change in the rocket's mass:

$$\Delta M = M_f - M_i = 1.00 \times 10^4 \text{ kg} - 1.00 \times 10^5 \text{ kg}$$
$$= -9.00 \times 10^4 \text{ kg}$$

Calculate the rate at which rocket mass changes by dividing the change in mass by the time (4.00 min, converted to seconds):

$$\frac{\Delta M}{\Delta t} = \frac{-9.00 \times 10^4 \text{ kg}}{2.40 \times 10^2 \text{ s}} = -3.75 \times 10^2 \text{ kg/s}$$

Substitute this rate into Equation 6.20, obtaining the thrust:

$$\text{Thrust} = \left| v_e \frac{\Delta M}{\Delta t} \right| = (4.50 \times 10^3 \text{ m/s})(3.75 \times 10^2 \text{ kg/s})$$
$$= 1.69 \times 10^6 \text{ N}$$

(c) Find the initial acceleration.

Write Newton's second law, where T stands for thrust, and solve for the acceleration a:

$$Ma = \Sigma F = T - Mg$$

$$a = \frac{T}{M} - g = \frac{1.69 \times 10^6 \text{ N}}{1.00 \times 10^5 \text{ kg}} - 9.80 \text{ m/s}^2$$
$$= 7.10 \text{ m/s}^2$$

(d) Estimate the speed at burnout when gravity is not neglected.

Find the approximate loss of speed due to gravity:

$$\Delta v_g = -g\Delta t = -(9.80 \text{ m/s}^2)(2.40 \times 10^2 \text{ s})$$
$$= -2.35 \times 10^3 \text{ m/s}$$

Add this loss to the result of part (b):

$$v_f = 1.04 \times 10^4 \text{ m/s} - 2.35 \times 10^3 \text{ m/s}$$
$$= 8.05 \times 10^3 \text{ m/s}$$

Remarks Even taking gravity into account, the speed is sufficient to attain orbit. Some additional boost may be required to overcome air drag.

Exercise 6.9

A spaceship with a mass of 5.00×10^4 kg is traveling at 6.00×10^3 m/s relative a space station. What mass will the ship have after it fires its engines in order to reach a speed of 8.00×10^3 m/s? Assume an exhaust velocity of 4.50×10^3 m/s.

Answer 3.21×10^4 kg

SUMMARY

PhysicsNow™ Take a practice test by logging into PhysicsNow at **www.cp7e.com** and clicking on the Pre-Test link for this chapter.

6.1 Momentum and Impulse

The **linear momentum** $\vec{\mathbf{p}}$ of an object of mass m moving with velocity $\vec{\mathbf{v}}$ is defined as

$$\vec{\mathbf{p}} \equiv m\vec{\mathbf{v}} \qquad [6.1]$$

Momentum carries units of kg·m/s. The **impulse** $\vec{\mathbf{I}}$ of a constant force $\vec{\mathbf{F}}$ delivered to an object is equal to the product of the force and the time interval during which the force acts:

$$\vec{\mathbf{I}} \equiv \vec{\mathbf{F}}\Delta t \qquad [6.4]$$

These two concepts are unifed in the **impulse–momentum theorem**, which states that the impulse of a constant force delivered to an object is equal to the change in momentum of the object:

$$\vec{\mathbf{F}}\Delta t = \Delta\vec{\mathbf{p}} \equiv m\vec{\mathbf{v}}_f - m\vec{\mathbf{v}}_i \qquad [6.5]$$

Solving problems with this theorem often involves estimating speeds or contact times (or both), leading to an average force.

6.2 Conservation of Momentum

When no net external force acts on an isolated system, the total momentum of the system is constant. This principle is called **conservation of momentum**. In particular, if the isolated system consists of two objects undergoing a collision, the total momentum of the system is the same before and after the collision. Conservation of momentum can be written mathematically for this case as

$$m_1\vec{\mathbf{v}}_{1i} + m_2\vec{\mathbf{v}}_{2i} = m_1\vec{\mathbf{v}}_{1f} + m_2\vec{\mathbf{v}}_{2f} \qquad [6.7]$$

Collision and recoil problems typically require finding unknown velocities in one or two dimensions. Each vector component gives an equation, and the resulting equations are solved simultaneously.

6.3 Collisions

In an **inelastic collision**, the momentum of the system is conserved, but kinetic energy is not. In a **perfectly inelastic collision**, the colliding objects stick together. In an **elastic collision**, both the momentum and the kinetic energy of the system are conserved.

A one-dimensional **elastic collision** between two objects can be solved by using the conservation of momentum and conservation of energy equations:

$$m_1v_{1i} + m_2v_{2i} = m_1v_{1f} + m_2v_{2f} \qquad [6.10]$$

$$\tfrac{1}{2}m_1v_{1i}^2 + \tfrac{1}{2}m_2v_{2i}^2 = \tfrac{1}{2}m_1v_{1f}^2 + \tfrac{1}{2}m_2v_{2f}^2 \qquad [6.11]$$

The following equation, derived from Equations 6.10 and 6.11, is usually more convenient to use than the original conservation of energy equation:

$$v_{1i} - v_{2i} = -(v_{1f} - v_{2f}) \qquad [6.14]$$

These equations can be solved simultaneously for the unknown velocities. Energy is not conserved in **inelastic collisions**, so such problems must be solved with Equation 6.10 alone.

6.4 Glancing Collisions

In glancing collisions, conservation of momentum can be applied along two perpendicular directions: an x-axis and a y-axis. Problems can be solved by using the x- and y-components of Equation 6.7. Elastic two-dimensional collisions will usually require Equation 6.11 as well. (Equation 6.14 doesn't apply to two dimensions.) Generally, one of the two objects is taken to be traveling along the x-axis, undergoing a deflection at some angle θ after the collision. The final velocities and angles can be found with elementary trigonometry.

CONCEPTUAL QUESTIONS

1. A batter bunts a pitched baseball, blocking the ball without swinging. (a) Can the baseball deliver more kinetic energy to the bat and batter than the ball carries initially? (b) Can the baseball deliver more momentum to the bat and batter than the ball carries initially? Explain each of your answers.
2. America will never forget the terrorist attack on September 11, 2001. One commentator remarked that the force of the explosion at the Twin Towers of the World Trade Center was strong enough to blow glass and parts of the steel structure to small fragments. Yet the television coverage showed thousands of sheets of paper floating down, many still intact. Explain how that could be.
3. In perfectly inelastic collisions between two objects, there are events in which all of the original kinetic energy is transformed to forms other than kinetic. Give an example of such an event.
4. If two objects collide and one is initially at rest, is it possible for both to be at rest after the collision? Is it possible for only one to be at rest after the collision? Explain.

5. A ball of clay of mass m is thrown with a speed v against a brick wall. The clay sticks to the wall and stops. Is the principle of conservation of momentum violated in this example?
6. A skater is standing still on a frictionless ice rink. Her friend throws a Frisbee straight at her. In which of the following cases is the largest momentum transferred to the skater? (a) The skater catches the Frisbee and holds onto it. (b) The skater catches the Frisbee momentarily, but then drops it vertically downward. (c) The skater catches the Frisbee, holds it momentarily, and throws it back to her friend.
7. You are standing perfectly still and then you take a step forward. Before the step your momentum was zero, but afterwards you have some momentum. Is the conservation of momentum violated in this case?
8. If two particles have equal kinetic energies, are their momenta necessarily equal? Explain.
9. A more ordinary example of conservation of momentum than a rocket ship occurs in a kitchen dishwashing machine. In this device, water at high pressure is forced out of small holes on the spray arms. Use conservation of momentum to explain why the arms rotate, directing water to all the dishes.
10. If two automobiles collide, they usually do not stick together. Does this mean the collision is elastic? Explain why a head-on collision is likely to be more dangerous than other types of collisions.
11. An open box slides across a frictionless, icy surface of a frozen lake. What happens to the speed of the box as water from a rain shower collect in it, assuming that the rain falls vertically downward into the box? Explain.
12. Consider a perfectly inelastic collision between a car and a large truck. Which vehicle loses more kinetic energy as a result of the collision?
13. Your physical education teacher throws you a tennis ball at a certain velocity, and you catch it. You are now given the following choice: The teacher can throw you a medicine ball (which is much more massive than the tennis ball) with the same velocity, the same momentum, or the same kinetic energy as the tennis ball. Which option would you choose in order to make the easiest catch, and why?
14. While watching a movie about a superhero, you notice that the superhero hovers in the air and throws a piano at some bad guys while remaining stationary in the air. What's wrong with this scenario?
15. In golf, novice players are often advised to be sure to "follow through" with their swing. Why does this make the ball travel a longer distance? If a shot is taken near the green, very little follow-through is required. Why?
16. An air bag inflates when a collision occurs, protecting a passenger (the dummy in Figure Q6.16) from serious injury. Why does the air bag soften the blow? Discuss the physics involved in this dramatic photograph.

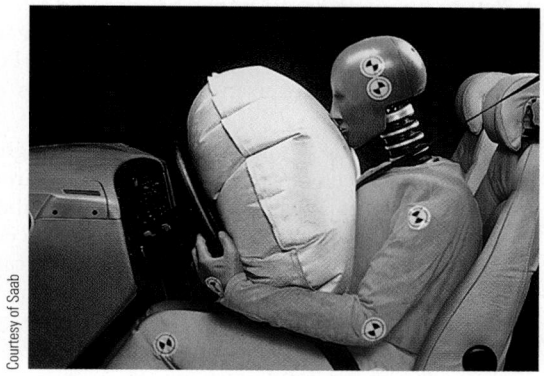

Courtesy of Saab

Figure Q6.16

17. A sharpshooter fires a rifle while standing with the butt of the gun against his shoulder. If the forward momentum of a bullet is the same as the backward momentum of the gun, why isn't it as dangerous to be hit by the gun as by the bullet?
18. A large bedsheet is held vertically by two students. A third student, who happens to be the star pitcher on the baseball team, throws a raw egg at the sheet. Explain why the egg doesn't break when it hits the sheet, regardless of its initial speed. (If you try this, make sure the pitcher hits the sheet near its center, and don't allow the egg to fall on the floor after being caught.)

PROBLEMS

1, 2, 3 = straightforward, intermediate, challenging ☐ = full solution available in *Student Solutions Manual/Study Guide*
PhysicsNow™ = coached problem with hints available at **www.cp7e.com** = biomedical application

Section 6.1 Momentum and Impulse

1. A ball of mass 0.150 kg is dropped from rest from a height of 1.25 m. It rebounds from the floor to reach a height of 0.960 m. What impulse was given to the ball by the floor?
2. A tennis player receives a shot with the ball (0.060 0 kg) traveling horizontally at 50.0 m/s and returns the shot with the ball traveling horizontally at 40.0 m/s in the opposite direction. (a) What is the impulse delivered to the ball by the racquet? (b) What work does the racquet do on the ball?
3. Calculate the magnitude of the linear momentum for the following cases: (a) a proton with mass 1.67×10^{-27} kg, moving with a speed of 5.00×10^6 m/s; (b) a 15.0-g bullet moving with a speed of 300 m/s; (c) a 75.0-kg sprinter running with a speed of 10.0 m/s; (d) the Earth (mass = 5.98×10^{24} kg) moving with an orbital speed equal to 2.98×10^4 m/s.
4. A 0.10-kg ball is thrown straight up into the air with an initial speed of 15 m/s. Find the momentum of the ball (a) at its maximum height and (b) halfway to its maximum height.
5. A pitcher claims he can throw a 0.145-kg baseball with as much momentum as a 3.00-g bullet moving with a speed of 1.50×10^3 m/s. (a) What must the baseball's speed be

if the pitcher's claim is valid? (b) Which has greater kinetic energy, the ball or the bullet?

6. A stroboscopic photo of a club hitting a golf ball, such as the photo shown in Figure 6.3, was made by Harold Edgerton in 1933. The ball was initially at rest, and the club was shown to be in contact with the club for about 0.002 0 s. Also, the ball was found to end up with a speed of 2.0×10^2 ft/s. Assuming that the golf ball had a mass of 55 g, find the average force exerted by the club on the ball.

7. Physics Now™ A professional diver performs a dive from a platform 10 m above the water surface. Estimate the order of magnitude of the average impact force she experiences in her collision with the water. State the quantities you take as data and their values.

8. A 75.0-kg stuntman jumps from a balcony and falls 25.0 m before colliding with a pile of mattresses. If the mattresses are compressed 1.00 m before he is brought to rest, what is the average force exerted by the mattresses on the stuntman?

9. A car is stopped for a traffic signal. When the light turns green, the car accelerates, increasing its speed from 0 to 5.20 m/s in 0.832 s. What are the magnitudes of the linear impulse and the average total force experienced by a 70.0-kg passenger in the car during the time the car accelerates?

10. A 0.500-kg football is thrown toward the east with a speed of 15.0 m/s. A stationary receiver catches the ball and brings it to rest in 0.020 0 s. (a) What is the impulse delivered to the ball as it's caught? (b) What is the average force exerted on the receiver?

11. The force shown in the force vs. time diagram in Figure P6.11 acts on a 1.5-kg object. Find (a) the impulse of the force, (b) the final velocity of the object if it is initially at rest, and (c) the final velocity of the object if it is initially moving along the x-axis with a velocity of -2.0 m/s.

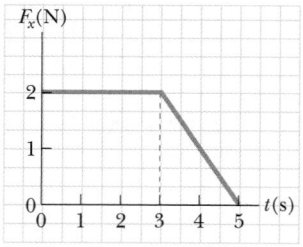

Figure P6.11

12. A force of magnitude F_x acting in the x-direction on a 2.00-kg particle varies in time as shown in Figure P6.12. Find (a) the impulse of the force, (b) the final velocity of the particle if it is initially at rest, and (c) the final velocity of the particle if it is initially moving along the x-axis with a velocity of -2.00 m/s.

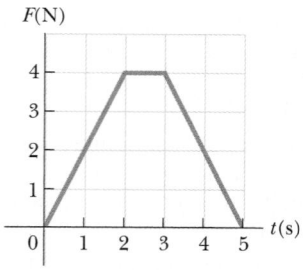

Figure P6.12

13. The forces shown in the force vs. time diagram in Figure P6.13 act on a 1.5-kg particle. Find (a) the impulse for the interval from $t = 0$ to $t = 3.0$ s and (b) the impulse for the interval from $t = 0$ to $t = 5.0$ s. (c) If the forces act on a 1.5-kg particle that is initially at rest, find the particle's speed at $t = 3.0$ s and at $t = 5.0$ s.

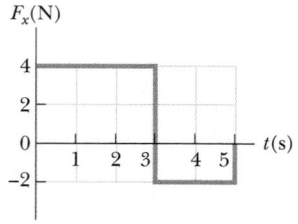

Figure P6.13

14. A 3.00-kg steel ball strikes a massive wall at 10.0 m/s at an angle of 60.0° with the plane of the wall. It bounces off the wall with the same speed and angle (Fig. P6.14). If the ball is in contact with the wall for 0.200 s, what is the average force exerted by the wall on the ball?

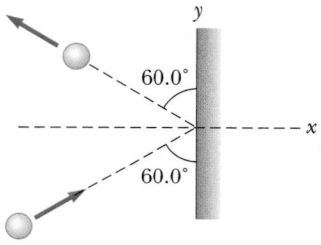

Figure P6.14

15. The front 1.20 m of a 1 400-kg car is designed as a "crumple zone" that collapses to absorb the shock of a collision. If a car traveling 25.0 m/s stops uniformly in 1.20 m, (a) how long does the collision last, (b) what is the magnitude of the average force on the car, and (c) what is the acceleration of the car? Express the acceleration as a multiple of the acceleration of gravity.

16. A pitcher throws a 0.15-kg baseball so that it crosses home plate horizontally with a speed of 20 m/s. The ball is hit straight back at the pitcher with a final speed of 22 m/s. (a) What is the impulse delivered to the ball? (b) Find the average force exerted by the bat on the ball if the two are in contact for 2.0×10^{-3} s.

17. A car of mass 1.6×10^3 kg is traveling east at a speed of 25 m/s along a horizontal roadway. When its brakes are applied, the car stops in 6.0 s. What is the average horizontal force exerted on the car while it is braking?

Section 6.2 Conservation of Momentum

18. A 730-N man stands in the middle of a frozen pond of radius 5.0 m. He is unable to get to the other side because of a lack of friction between his shoes and the ice. To overcome this difficulty, he throws his 1.2-kg physics textbook horizontally toward the north shore at a speed of 5.0 m/s. How long does it take him to reach the south shore?

19. High-speed stroboscopic photographs show that the head of a 200-g golf club is traveling at 55 m/s just before it strikes a 46-g golf ball at rest on a tee. After the collision, the club head travels (in the same direction) at 40 m/s. Find the speed of the golf ball just after impact.

20. A rifle with a weight of 30 N fires a 5.0-g bullet with a speed of 300 m/s. (a) Find the recoil speed of the rifle. (b) If a 700-N man holds the rifle firmly against his shoulder, find the recoil speed of the man and rifle.

21. **Physics Now™** A 45.0-kg girl is standing on a 150-kg plank. The plank, originally at rest, is free to slide on a frozen lake, which is a flat, frictionless surface. The girl begins to walk along the plank at a constant velocity of 1.50 m/s to the right relative to the plank. (a) What is her velocity relative to the surface of the ice? (b) What is the velocity of the plank relative to the surface of the ice?

22. A 65.0-kg person throws a 0.045 0-kg snowball forward with a ground speed of 30.0 m/s. A second person, with a mass of 60.0 kg, catches the snowball. Both people are on skates. The first person is initially moving forward with a speed of 2.50 m/s, and the second person is initially at rest. What are the velocities of the two people after the snowball is exchanged? Disregard friction between the skates and the ice.

23. In Section 6.2, we implied that the kinetic energy of the Earth can be ignored when considering the energy of a system consisting of the Earth and a dropped ball of mass m_b. Verify this statement by first setting up a ratio of the kinetic energy of the Earth to that of the ball as they collide. Then use conservation of momentum to show that

$$\frac{v_E}{v_b} = -\frac{m_b}{m_E} \quad \text{and} \quad \frac{KE_E}{KE_b} = \frac{m_b}{m_E}$$

Find the order of magnitude of the ratio of the kinetic energies, based on data that you specify.

24. Two ice skaters are holding hands at the center of a frozen pond when an argument ensues. Skater A shoves skater B along a horizontal direction. Identify (a) the horizontal forces acting on A and (b) those acting on B. (c) Which force is greater, the force on A or the force on B? (d) Can conservation of momentum be used for the system of A and B? Defend your answer. (e) If A has a mass of 0.900 times that of B, and B begins to move away with a speed of 2.00 m/s, find the speed of A.

Section 6.3 Collisions
Section 6.4 Glancing Collisions

25. An archer shoots an arrow toward a 300-g target that is sliding in her direction at a speed of 2.50 m/s on a smooth, slippery surface. The 22.5-g arrow is shot with a speed of 35.0 m/s and passes through the target, which is stopped by the impact. What is the speed of the arrow after passing through the target?

26. A 75.0-kg ice skater moving at 10.0 m/s crashes into a stationary skater of equal mass. After the collision, the two skaters move as a unit at 5.00 m/s. Suppose the average force a skater can experience without breaking a bone is 4 500 N. If the impact time is 0.100 s, does a bone break?

27. A railroad car of mass 2.00×10^4 kg moving at 3.00 m/s collides and couples with two coupled railroad cars, each of the same mass as the single car and moving in the same direction at 1.20 m/s. (a) What is the speed of the three coupled cars after the collision? (b) How much kinetic energy is lost in the collision?

28. A 7.0-g bullet is fired into a 1.5-kg ballistic pendulum. The bullet emerges from the block with a speed of 200 m/s, and the block rises to a maximum height of 12 cm. Find the initial speed of the bullet.

29. A 0.030-kg bullet is fired vertically at 200 m/s into a 0.15-kg baseball that is initially at rest. How high does the combined bullet and baseball rise after the collision, assuming the bullet embeds itself in the ball?

30. An 8.00-g bullet is fired into a 250-g block that is initially at rest at the edge of a table of height 1.00 m (Fig. P6.30). The bullet remains in the block, and after the impact the block lands 2.00 m from the bottom of the table. Determine the initial speed of the bullet.

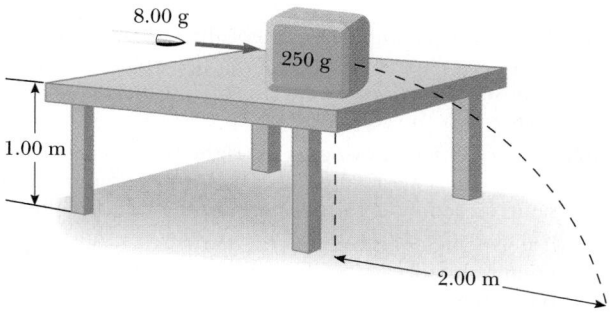

Figure P6.30

31. Gayle runs at a speed of 4.00 m/s and dives on a sled, initially at rest on the top of a frictionless, snow-covered hill. After she has descended a vertical distance of 5.00 m, her brother, who is initially at rest, hops on her back, and they continue down the hill together. What is their speed at the bottom of the hill if the total vertical drop is 15.0 m? Gayle's mass is 50.0 kg, the sled has a mass of 5.00 kg, and her brother has a mass of 30.0 kg.

32. A 1 200-kg car traveling initially with a speed of 25.0 m/s in an easterly direction crashes into the rear end of a 9 000-kg truck moving in the same direction at 20.0 m/s (Fig. P6.32). The velocity of the car right after the collision is 18.0 m/s to the east. (a) What is the velocity of the truck right after the collision? (b) How much mechanical energy is lost in the collision? Account for this loss in energy.

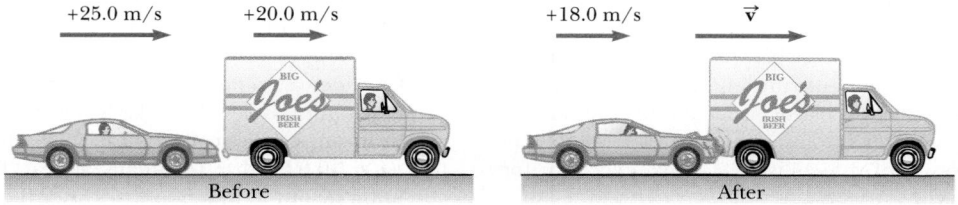

Figure P6.32

33. A 12.0-g bullet is fired horizontally into a 100-g wooden block that is initially at rest on a frictionless horizontal surface and connected to a spring having spring constant 150 N/m. The bullet becomes embedded in the block. If the bullet–block system compresses the spring by a maximum of 80.0 cm, what was the speed of the bullet at impact with the block?

34. (a) Three carts of masses 4.0 kg, 10 kg, and 3.0 kg move on a frictionless horizontal track with speeds of 5.0 m/s, 3.0 m/s, and 4.0 m/s, as shown in Figure P6.34. The carts stick together after colliding. Find the final velocity of the three carts. (b) Does your answer require that all carts collide and stick together at the same time?

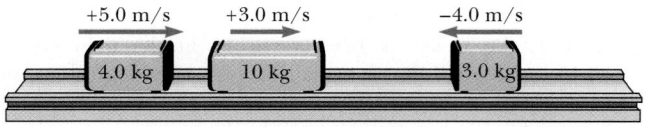

Figure P6.34

35. A 5.00-g object moving to the right at 20.0 cm/s makes an elastic head-on collision with a 10.0-g object that is initially at rest. Find (a) the velocity of each object after the collision and (b) the fraction of the initial kinetic energy transferred to the 10.0-g object.

36. A 10.0-g object moving to the right at 20.0 cm/s makes an elastic head-on collision with a 15.0-g object moving in the opposite direction at 30.0 cm/s. Find the velocity of each object after the collision.

37. A 25.0-g object moving to the right at 20.0 cm/s overtakes and collides elastically with a 10.0-g object moving in the same direction at 15.0 cm/s. Find the velocity of each object after the collision.

38. Four railroad cars, each of mass 2.50×10^4 kg, are coupled together and coasting along horizontal tracks at speed v_i toward the south. A very strong but foolish movie actor riding on the second car uncouples the front car and gives it a big push, increasing its speed to 4.00 m/s south. The remaining three cars continue moving south, now at 2.00 m/s. (a) Find the initial speed of the cars. (b) How much work did the actor do?

39. When fired from a gun into a 1.00-kg block of wood held in a vise, a 7.00-g bullet penetrates the block to a depth of 8.00 cm. The block is then placed on a frictionless, horizontal surface, and a second 7.00-g bullet is fired from the gun into the block. To what depth does the bullet penetrate the block in this case?

40. A billiard ball rolling across a table at 1.50 m/s makes a head-on elastic collision with an identical ball. Find the speed of each ball after the collision (a) when the second ball is initially at rest, (b) when the second ball is moving toward the first at a speed of 1.00 m/s, and (c) when the second ball is moving away from the first at a speed of 1.00 m/s.

41. A 90-kg fullback moving east with a speed of 5.0 m/s is tackled by a 95-kg opponent running north at 3.0 m/s. If the collision is perfectly inelastic, calculate (a) the velocity of the players just after the tackle and (b) the kinetic energy lost as a result of the collision. Can you account for the missing energy?

42. An 8.00-kg object moving east at 15.0 m/s on a frictionless horizontal surface collides with a 10.0-kg object that is initially at rest. After the collision, the 8.00-kg object moves south at 4.00 m/s. (a) What is the velocity of the 10.0-kg object after the collision? (b) What percentage of the initial kinetic energy is lost in the collision?

43. A 2 000-kg car moving east at 10.0 m/s collides with a 3 000-kg car moving north. The cars stick together and move as a unit after the collision, at an angle of 40.0° north of east and a speed of 5.22 m/s. Find the speed of the 3 000-kg car before the collision.

44. Two automobiles of equal mass approach an intersection. One vehicle is traveling with velocity 13.0 m/s toward the east, and the other is traveling north with speed v_{2i}. Neither driver sees the other. The vehicles collide in the intersection and stick together, leaving parallel skid marks at an angle of 55.0° north of east. The speed limit for both roads is 35 mi/h, and the driver of the northward-moving vehicle claims he was within the limit when the collision occurred. Is he telling the truth?

45. **PhysicsNow™** A billiard ball moving at 5.00 m/s strikes a stationary ball of the same mass. After the collision, the first ball moves at 4.33 m/s at an angle of 30° with respect to the original line of motion. (a) Find the velocity (magnitude and direction) of the second ball after collision. (b) Was the collision inelastic or elastic?

ADDITIONAL PROBLEMS

46. In research in cardiology and exercise physiology, it is often important to know the mass of blood pumped by a person's heart in one stroke. This information can be obtained by means of a *ballistocardiograph*. The instrument works as follows: The subject lies on a horizontal pallet floating on a film of air. Friction on the pallet is negligible. Initially, the momentum of the system is zero. When the heart beats, it expels a mass m of blood into the aorta with speed v, and the body and platform move in the opposite direction with speed V. The speed of the blood can be determined independently (for example, by observing an ultrasound Doppler shift). Assume that the blood's speed is 50.0 cm/s in one typical trial. The mass of the subject plus the pallet is 54.0 kg. The pallet moves 6.00×10^{-5} m in 0.160 s after one heartbeat. Calculate the mass of blood that leaves the heart. Assume that the mass of blood is negligible compared with the total mass of the person. This simplified example illustrates the principle of ballistocardiography, but in practice a more sophisticated model of heart function is used.

47. A 0.50-kg object is at rest at the origin of a coordinate system. A 3.0-N force in the $+x$-direction acts on the object for 1.50 s. (a) What is the velocity at the end of this interval? (b) At the end of the interval, a constant force of 4.0 N is applied in the $-x$-direction for 3.0 s. What is the velocity at the end of the 3.0 s?

48. Consider a frictionless track as shown in Figure P6.48. A block of mass $m_1 = 5.00$ kg is released from Ⓐ. It makes a head-on elastic collision at Ⓑ with a block of mass $m_2 = 10.0$ kg that is initially at rest. Calculate the maximum height to which m_1 rises after the collision.

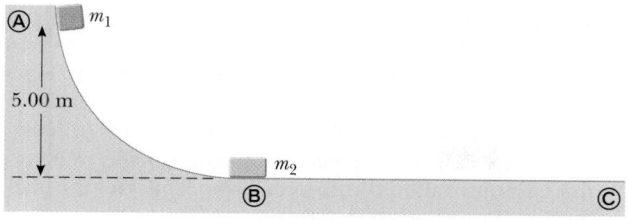

Figure P6.48

49. Most of us know intuitively that in a head-on collision between a large dump truck and a subcompact car, you are better off being in the truck than in the car. Why is this? Many people imagine that the collision force exerted on the car is much greater than that exerted on the truck. To substantiate this view, they point out that the car is crushed, whereas the truck is only dented. This idea of unequal forces, of course, is false; Newton's third law tells us that both objects are acted upon by forces of the same magnitude. The truck suffers less damage because it is made of stronger metal. But what about the two drivers? Do they experience the same forces? To answer this question, suppose that each vehicle is initially moving at 8.00 m/s and that they undergo a perfectly inelastic head-on collision. Each driver has mass 80.0 kg. Including the masses of the drivers, the total masses of the vehicles are 800 kg for the car and 4 000 kg for the truck. If the collision time is 0.120 s, what force does the seat belt exert on each driver?

50. A bullet of mass m and speed v passes completely through a pendulum bob of mass M as shown in Figure P6.50. The bullet emerges with a speed of $v/2$. The pendulum bob is suspended by a stiff rod of length ℓ and negligible mass. What is the minimum value of v such that the bob will barely swing through a complete vertical circle?

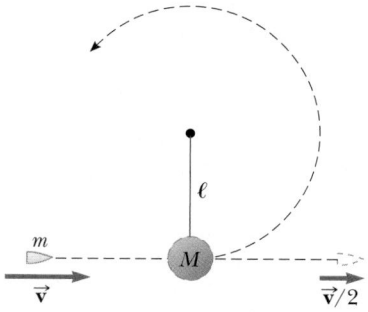

Figure P6.50

51. A 2.0-g particle moving at 8.0 m/s makes a perfectly elastic head-on collision with a resting 1.0-g object. (a) Find the speed of each particle after the collision. (b) Find the speed of each particle after the collision if the stationary particle has a mass of 10 g. (c) Find the final kinetic energy of the incident 2.0-g particle in the situations described in (a) and (b). In which case does the incident particle lose more kinetic energy?

52. A 0.400-kg green bead slides on a curved frictionless wire, starting from rest at point Ⓐ in Figure P6.52. At point Ⓑ, the bead collides elastically with a 0.600-kg blue ball at rest. Find the maximum height the blue ball rises as it moves up the wire.

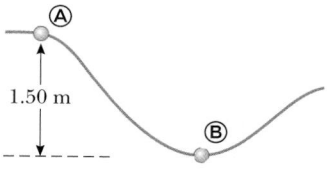

Figure P6.52

53. An 80-kg man standing erect steps off a 3.0-m-high diving platform and begins to fall from rest. The man again comes to rest 2.0 s after reaching the water. What average force did the water exert on him?

54. A 12.0-g bullet is fired horizontally into a 100-g wooden block initially at rest on a horizontal surface. After impact, the block slides 7.5 m before coming to rest. If the coefficient of kinetic friction between block and surface is 0.650, what was the speed of the bullet immediately before impact?

55. A 60.0-kg person running at an initial speed of 4.00 m/s jumps onto a 120-kg cart that is initially at rest (Figure P6.55). The person slides on the cart's top surface and finally comes to rest relative to the cart. The coefficient of kinetic friction between the person and the cart is 0.400. Friction between the cart and ground can be neglected. (a) Find the final speed of the person and cart relative to the ground. (b) Find the frictional force acting on the person while he is sliding across the top surface of the cart. (c) How long does the frictional force act on the person? (d) Find the change in momentum of the person and the change in momentum of the cart. (e) Determine the displacement of the person relative to the ground while he is sliding on the cart. (f) Determine the displacement of the cart relative to the ground while the person is sliding. (g) Find the change in kinetic energy of the person. (h) Find the change in kinetic energy of the cart. (i) Explain why the answers to (g) and (h) differ. (What kind of collision is this, and what accounts for the loss of mechanical energy?)

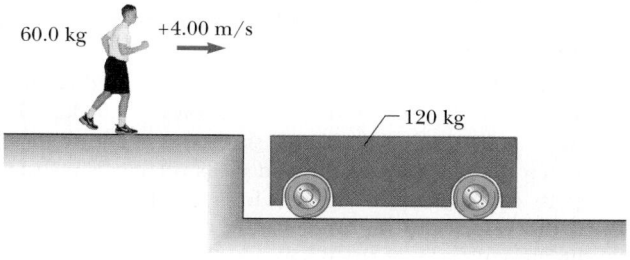

Figure P6.55

56. Two blocks of masses $m_1 = 2.00$ kg and $m_2 = 4.00$ kg are each released from rest at a height of 5.00 m on a frictionless track, as shown in Figure P6.56, and undergo an elastic head-on collision. (a) Determine the velocity of each block just before the collision. (b) Determine the velocity of each block immediately after the collision. (c) Determine the maximum heights to which m_1 and m_2 rise after the collision.

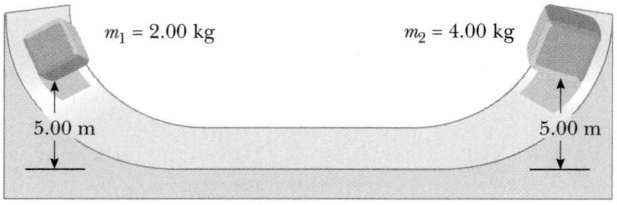

Figure P6.56

57. A 0.500-kg block is released from rest at the top of a frictionless track 2.50 m above the top of a table. It then collides elastically with a 1.00-kg object that is initially at rest on the table, as shown in Figure P6.57. (a) Determine the velocities of the two objects just after the collision. (b) How high up the track does the 0.500-kg object travel back after the collision? (c) How far away from the bottom of the table does the 1.00-kg object land, given that the table is 2.00 m high? (d) How far away from the bottom of the table does the 0.500-kg object eventually land?

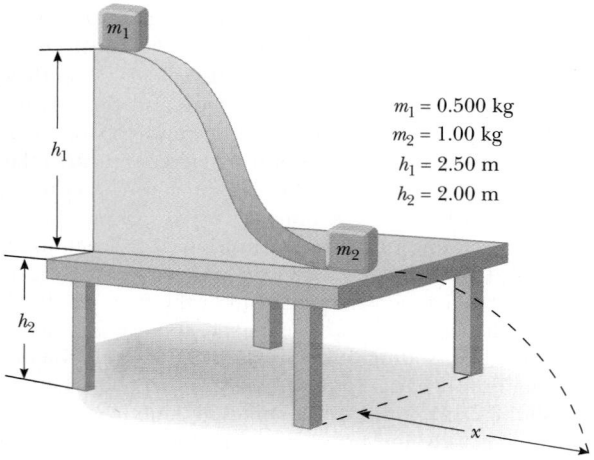

Figure P6.57

58. Tarzan, whose mass is 80.0 kg, swings from a 3.00-m vine that is horizontal when he starts. At the bottom of his arc, he picks up 60.0-kg Jane in a perfectly inelastic collision. What is the height of the highest tree limb they can reach on their upward swing?

59. A small block of mass $m_1 = 0.500$ kg is released from rest at the top of a curved wedge of mass $m_2 = 3.00$ kg, which sits on a frictionless horizontal surface as in Figure P6.59a. When the block leaves the wedge, its velocity is measured to be 4.00 m/s to the right, as in Figure P6.59b. (a) What is the velocity of the wedge after the block reaches the horizontal surface? (b) What is the height h of the wedge?

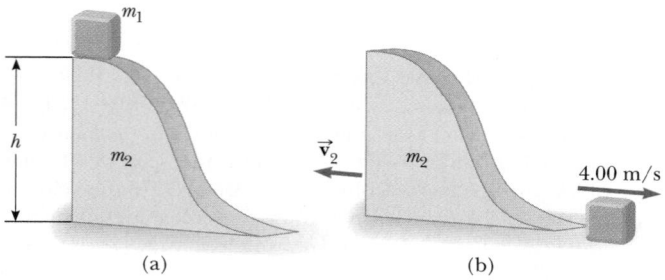

Figure P6.59

60. Two carts of equal mass $m = 0.250$ kg are placed on a frictionless track that has a light spring of force constant $k = 50.0$ N/m attached to one end of it, as in Figure P6.60. The red cart is given an initial velocity of $\vec{v}_0 = 3.00$ m/s to the right, and the blue cart is initially at rest. If the carts collide elastically, find (a) the velocity of the carts just after the first collision and (b) the maximum compression of the spring.

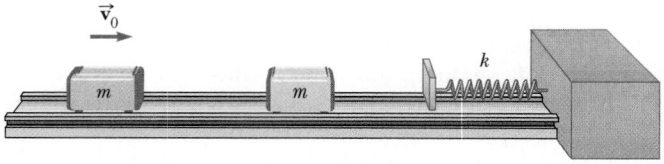

Figure P6.60

61. A cannon is rigidly attached to a carriage, which can move along horizontal rails, but is connected to a post by a large spring, initially unstretched and with force constant $k = 2.00 \times 10^4$ N/m, as in Figure P6.61. The cannon fires a 200-kg projectile at a velocity of 125 m/s directed 45.0° above the horizontal. (a) If the mass of the cannon and its carriage is 5 000 kg, find the recoil speed of the cannon. (b) Determine the maximum extension of the spring. (c) Find the maximum force the spring exerts on the carriage. (d) Consider the system consisting of the cannon, the carriage, and the shell. Is the momentum of this system conserved during the firing? Why or why not?

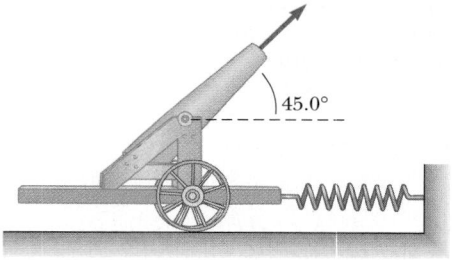

Figure P6.61

62. Two objects of masses m and $3m$ are moving toward each other along the x-axis with the same initial speed v_0. The object with mass m is traveling to the left, and the object with mass $3m$ is traveling to the right. They undergo an elastic glancing collision such that m is moving downward after the collision at right angles from its initial direction. (a) Find the final speeds of the two objects. (b) What is the angle θ at which the object with mass $3m$ is scattered?

63. **Physics Now™** A neutron in a reactor makes an elastic head-on collision with a carbon atom that is initially at rest. (The mass of the carbon nucleus is about 12 times that of the neutron.) (a) What fraction of the neutron's kinetic energy is transferred to the carbon nucleus? (b) If the neutron's initial kinetic energy is 1.6×10^{-13} J, find its final kinetic energy and the kinetic energy of the carbon nucleus after the collision.

64. A cue ball traveling at 4.00 m/s makes a glancing, elastic collision with a target ball of equal mass that is initially at rest. The cue ball is deflected so that it makes an angle of 30.0° with its original direction of travel. Find (a) the angle between the velocity vectors of the two balls after the collision and (b) the speed of each ball after the collision.

65. A block of mass m lying on a rough horizontal surface is given an initial velocity of $\vec{v}_0$. After traveling a distance d, it makes a head-on elastic collision with a block of mass $2m$. How far does the second block move before coming to rest? (Assume that the coefficient of friction, μ_k, is the same for both blocks.)

66. The "force platform" is a tool that is used to analyze the performance of athletes by measuring the vertical force as a function of time that the athlete exerts on the ground in performing various activities. A simplified force vs. time graph for an athlete performing a standing high jump is shown in Figure P6.66. The athlete started the jump at $t = 0.0$ s. How high did this athlete jump?

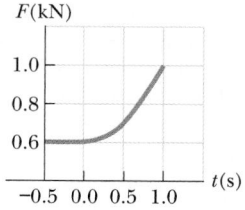

Figure P6.66

67. (a) A car traveling due east strikes a car traveling due north at an intersection, and the two move together as a unit. A property owner on the southeast corner of the intersection claims that his fence was torn down in the collision. Should he be awarded damages by the insurance company? Defend your answer. (b) Let the eastward-moving car have a mass of 1 300 kg and a speed of 30.0 km/h and the northward-moving car a mass of 1 100 kg and a speed of 20.0 km/h. Find the velocity after the collision. Are the results consistent with your answer to part (a)?

68. Two blocks collide on a frictionless surface. After the collision, the blocks stick together. Block A has a mass M and is initially moving to the right at speed v. Block B has a mass $2M$ and is initially at rest. System C is composed of both blocks. (a) Draw a free-body diagram for each block at an instant *during* the collision. (b) Rank the magnitudes of the horizontal forces in your diagram. Explain your reasoning. (c) Calculate the change in momentum of block A, block B, and system C. (d) Is kinetic energy conserved in this collision? Explain your answer. (This problem is courtesy of Edward F. Redish. For more such problems, visit http://www.physics.umd.edu/perg.)

69. A tennis ball of mass 57.0 g is held just above a basketball of mass 590 g. With their centers vertically aligned, both balls are released from rest at the same time, to fall through a distance of 1.20 m, as shown in Figure P6.69. (a) Find the magnitude of the downward velocity with which the basketball reaches the ground. (b) Assume that an elastic collision with the ground instantaneously reverses the velocity of the basketball while the tennis ball is still moving down. Next, the two balls meet in an elastic collision. (b) To what height does the tennis ball rebound?

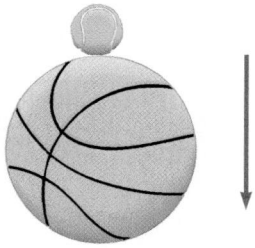

Figure P6.69

70. A 60-kg soccer player jumps vertically upwards and heads the 0.45-kg ball as it is descending vertically with a speed of 25 m/s. If the player was moving upward with a speed of 4.0 m/s just before impact, what will be the speed of the ball immediately after the collision if the ball rebounds vertically upwards and the collision is elastic? If the ball is in contact with the player's head for 20 ms, what is the average acceleration of the ball? (Note that the force of gravity may be ignored during the brief collision time.)

71. Small ice cubes, each of mass 5.00 g, slide down a frictionless ski-jump track in a steady stream, as shown in Figure P6.71. Starting from rest, each cube moves down through a net vertical distance of 1.50 m and leaves the bottom end of the track at an angle of 40.0° above the horizontal. At the highest point of its subsequent trajectory, the cube strikes a vertical wall and rebounds with half the speed it had upon impact. If 10.0 cubes strike the wall per second, what average force is exerted on the wall?

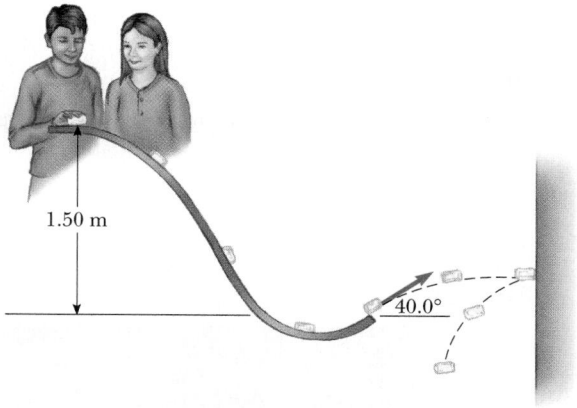

Figure P6.71

72. A 0.30-kg puck, initially at rest on a frictionless horizontal surface, is struck by a 0.20-kg puck that is initially moving along the x-axis with a velocity of 2.0 m/s. After the collision, the 0.20-kg puck has a speed of 1.0 m/s at an angle of $\theta = 53°$ to the positive x-axis. (a) Determine the velocity of the 0.30-kg puck after the collision. (b) Find the fraction of kinetic energy lost in the collision.

73. A cannon initially resting on a frictionless surface of mass $m_1 = 800$ kg (when unloaded) is loaded with a "shot" of mass $m_2 = 10.0$ kg. The cannon is aimed at mass $m_3 = 7\,990$ kg, which is connected to a massless spring of force constant $k = 4\,500$ N/m, as in Figure P6.73a. The cannon is then fired, and the shot inelastically collides with mass m_3 and sticks in it, as shown in Figure P6.73b. The combined system compresses the spring a maximum distance of $d = 0.500$ m, as in Figure P6.73c. (a) Determine the speed of m_2 just before it collides with m_3. (You may assume that m_2 travels in a straight line.) (b) Determine the recoil speed of the cannon. (c) The cannon recoils towards the right, and when it passes point A there is friction (with $\mu_k = 0.600$) between the cannon and the ground. How far to the right of A does the cannon slide before coming to rest?

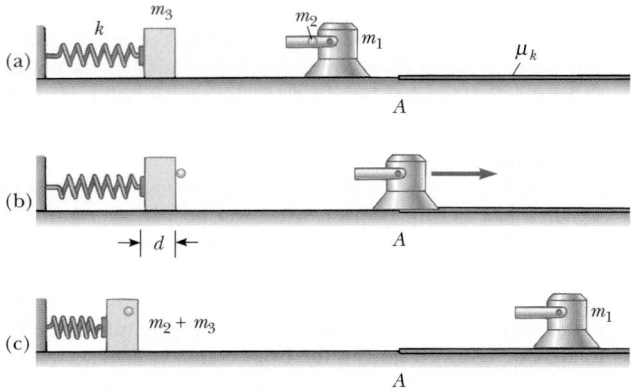

Figure P6.73

74. A flying squid (family Ommastrephidae) is able to "jump" off the surface of the sea by taking water into its body cavity and then ejecting the water vertically downward. A 0.85-kg squid is able to eject 0.30 kg of water with a speed of 20 m/s. (a) What will be the speed of the squid immediately after ejecting the water. (b) How high in the air will the squid rise?

ACTIVITIES

A.1. Tie a string to two tennis balls to make identical pendulums. Support the two such that they are of the same length and just touching. Now model an inelastic collision by attaching a piece of putty or double-stick tape to one of the balls so that when the balls collide they will stick together and move as a unit. Move one pendulum upward to some starting amplitude, and observe the amplitude to which the combination rises after the collision. Discuss and explain the result. (Don't forget that *after* the collision, the energy of the system (the two tennis balls and the Earth) is conserved, and you use this fact to determine the height to which the combination rises.)

A.2. A fun device that illustrates conservation of momentum and conservation of kinetic energy is the so-called executive stress reliever shown in Figure A6.2. It consists of five identical hard balls supported by strings of equal lengths. When one ball is pulled out and released, an almost-elastic collision causes one ball to move out on the opposite side with the same speed as the incoming ball. If two balls are pulled out and released, two balls swing out on the opposite side, and so forth—but how do the balls "know"? For example, is it possible that, on occasion, when one ball is released, two will swing out on the opposite side traveling with half the speed of the incoming ball? If you have access to this toy, try the experiment to see if you can get the latter collision to occur. When you have convinced yourself that it is not going to happen, find the kinetic energy of the system with one incoming ball of mass m and speed v and the kinetic energy of two outgoing balls with mass m and speed $v/2$ to show that kinetic energy is not conserved in such a collision.

Courtesy of CENCO

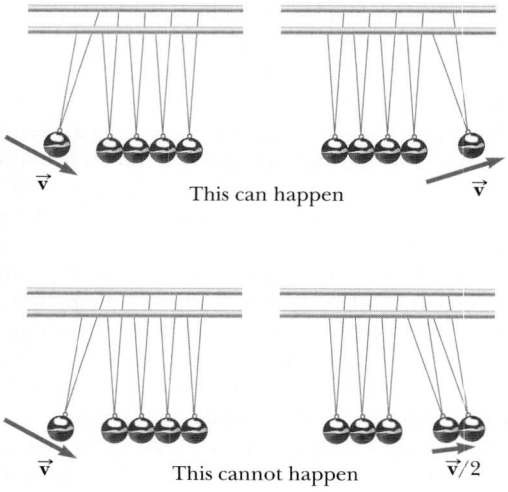

Figure A6.2 An executive stress-reliever

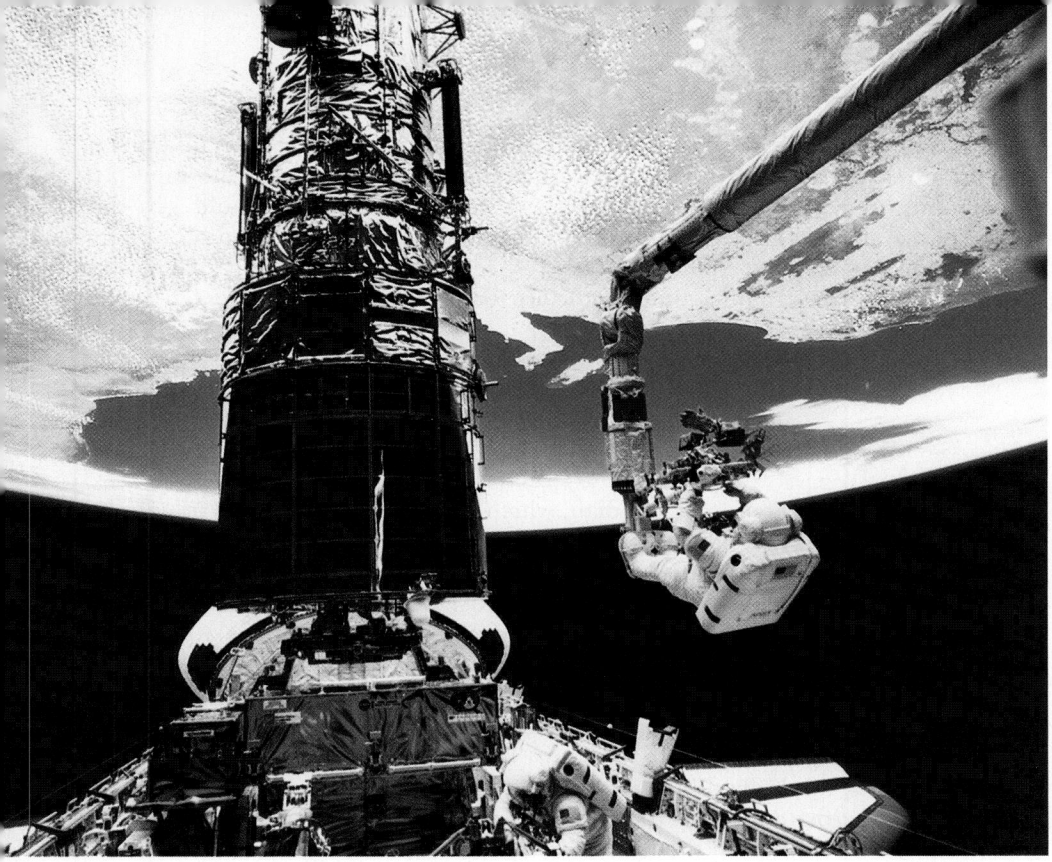

Courtesy NASA

Astronauts F. Story Musgrave and Jeffrey A. Hoffman, along with the Hubble Space Telescope and the Space Shuttle *Endeavor,* are all "falling" around Earth.

CHAPTER 7

Rotational Motion and the Law of Gravity

7.1 Angular Speed and Angular Acceleration

7.2 Rotational Motion under Constant Angular Acceleration

7.3 Relations between Angular and Linear Quantities

7.4 Centripetal Acceleration

7.5 Newtonian Gravitation

7.6 Kepler's Laws

Rotational motion is an important part of everyday life. The rotation of the Earth creates the cycle of day and night, the rotation of wheels enables easy vehicular motion, and modern technology depends on circular motion in a variety of contexts, from the tiny gears in a Swiss watch to the operation of lathes and other machinery. The concepts of *angular speed, angular acceleration*, and *centripetal acceleration* are central to understanding the motions of a diverse range of phenomena, from a car moving around a circular race track to clusters of galaxies orbiting a common center.

Rotational motion, when combined with Newton's law of universal gravitation and his laws of motion, can also explain certain facts about space travel and satellite motion, such as where to place a satellite so it will remain fixed in position over the same spot on the Earth. The generalization of gravitational potential energy and energy conservation offers an easy route to such results as planetary escape speed. Finally, we present Kepler's three laws of planetary motion, which formed the foundation of Newton's approach to gravity.

7.1 ANGULAR SPEED AND ANGULAR ACCELERATION

In the study of linear motion, the important concepts are *displacement* Δx, *velocity* v, and *acceleration* a. Each of these concepts has its analog in rotational motion: *angular displacement* $\Delta\theta$, *angular velocity* ω, and *angular acceleration* α.

The *radian*, a unit of angular measure, is essential to the understanding of these concepts. Recall that the distance s around a circle is given by $s = 2\pi r$, where r is

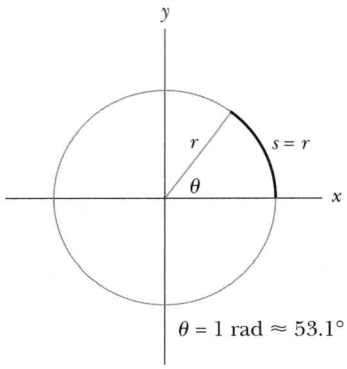

Figure 7.1 For a circle of radius r, one radian is the angle subtended by an arc length equal to r.

the radius of the circle. Dividing both sides by r results in $s/r = 2\pi$. This quantity is dimensionless, because both s and r have dimensions of length, but the value 2π corresponds to a displacement around a circle. A half circle would give an answer of π, a quarter circle an answer of $\pi/2$. The numbers 2π, π, and $\pi/2$ correspond to angles of 360°, 180°, and 90°, respectively, so a new unit of angular measure, the **radian**, can be defined as **the arc length s along a circle divided by the radius r**:

$$\theta = \frac{s}{r} \qquad [7.1]$$

Figure 7.1 illustrates the size of 1 radian, which is approximately 53°. For conversions, we use the fact that 360° = 2π radians (or 180° = π radians). For example, 45° (2π rad/360°) = ($\pi/4$) rad.

Generally, angular quantities in physics must be expressed in radians. Be sure to set your calculator to radian mode; neglecting to do this is a common error.

TIP 7.1 Remember the Radian

Equation 7.1 defines an angle expressed in *radians*. Angles expressed in terms of degrees must first be converted to radians. Also, be sure to check whether your calculator is in degree or radian mode when solving problems involving rotation.

Armed with the concept of the radian, we can now discuss angular concepts in physics. Consider Figure 7.2a, a top view of a rotating compact disc. Such a disk is an example of a "rigid body," with each part of the body fixed in position relative all other parts of the body. When a rigid body rotates through a given angle, all parts of the body rotate through the same angle at the same time. For the compact disk, the axis of rotation is at the center of the disc, O. A point P on the disc is at a distance r from the origin and moves about O in a circle of radius r. We set up a *fixed* reference line, as shown in Figure 7.2a, and assume that at time $t = 0$ the point P is on that reference line. After a time interval Δt has elapsed, P has advanced to a new position (Fig. 7.2b). In this interval, the line OP has moved through the angle θ with respect to the reference line. The angle θ, measured in radians, is called the **angular position** and is analogous to the linear position variable x. Likewise, P has moved an arc length s measured along the circumference of the circle.

In Figure 7.3, as a point on the rotating disc moves from Ⓐ to Ⓑ in a time Δt, it starts at an angle θ_i and ends at an angle θ_f. The difference $\theta_f - \theta_i$ is called the **angular displacement**.

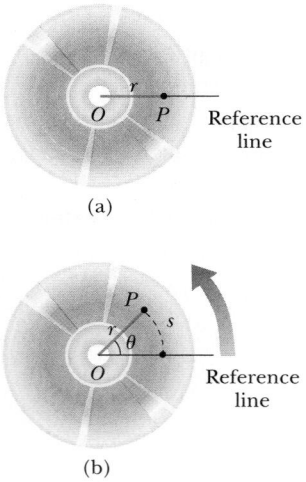

Figure 7.2 (a) The point P on a rotating compact disc at $t = 0$. (b) As the disc rotates, P moves through an arc length s.

An object's angular displacement, $\Delta\theta$, is the difference in its final and initial angles:

$$\Delta\theta = \theta_f - \theta_i \qquad [7.2]$$

SI unit: radian (rad)

For example, if a point on a disk is at $\theta_i = 4$ rad and rotates to angular position $\theta_f = 7$ rad, the angular displacement is $\Delta\theta = \theta_f - \theta_i = 7 \text{ rad} - 4 \text{ rad} = 3$ rad. Note that we use angular variables to describe the rotating disc because **each point on the disc undergoes the same angular displacement in any given time interval.**

Having defined angular displacements, it's natural to define an angular speed:

Average angular speed ▶

The average angular speed ω_{av} of a rotating rigid object during the time interval Δt is defined as the angular displacement $\Delta\theta$ divided by Δt:

$$\omega_{av} \equiv \frac{\theta_f - \theta_i}{t_f - t_i} = \frac{\Delta\theta}{\Delta t} \qquad [7.3]$$

SI unit: radian per second (rad/s)

For very short time intervals, the average angular speed approaches the instantaneous angular speed, just as in the linear case.

The **instantaneous angular speed** ω of a rotating rigid object is defined as the limit of the average speed $\Delta\theta/\Delta t$ as the time interval Δt approaches zero:

$$\omega \equiv \lim_{\Delta t \to 0} \frac{\Delta\theta}{\Delta t} \qquad [7.4]$$

SI unit: radian per second (rad/s)

We take ω to be positive when θ is increasing (counterclockwise motion) and negative when θ is decreasing (clockwise motion). When the angular speed is constant, the instantaneous angular speed is equal to the average angular speed.

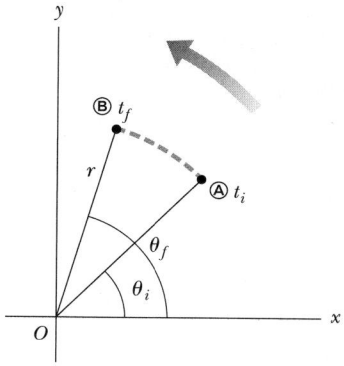

Figure 7.3 As a point on the compact disc moves from Ⓐ to Ⓑ, the disc rotates through the angle $\Delta\theta = \theta_f - \theta_i$.

EXAMPLE 7.1 Whirlybirds

Goal Convert an angular speed in revolutions per minute to radians per second.

Problem The rotor on a helicopter turns at an angular speed of 3.20×10^2 revolutions per minute. (In this book, we sometimes use the abbreviation rpm, but in most cases we use rev/min.) Express this angular speed in radians per second.

Strategy During one revolution, the rotor turns through an angle of 2π radians. Use this relationship as a conversion factor.

Solution

Apply the conversion factors 1 rev = 2π rad and 60 s = 1 min:

$$\omega = 3.20 \times 10^2 \frac{\text{rev}}{\text{min}}$$

$$= 3.20 \times 10^2 \frac{\cancel{\text{rev}}}{\cancel{\text{min}}}\left(\frac{2\pi \text{ rad}}{\cancel{\text{rev}}}\right)\left(\frac{1.00 \cancel{\text{min}}}{60.0 \text{ s}}\right)$$

$$= 33.5 \text{ rad/s}$$

Exercise 7.1

A waterwheel turns at 1 500 revolutions per hour. Express this figure in radians per second.

Answer 2.6 rad/s

Quick Quiz 7.1

A rigid body is rotating counterclockwise about a fixed axis. Each of the following pairs of quantities represents an initial angular position and a final angular position of the rigid body. Which of the sets can occur *only* if the rigid body rotates through more than 180°? (a) 3 rad, 6 rad; (b) − 1 rad, 1 rad; (c) 1 rad, 5 rad.

Quick Quiz 7.2

Suppose that the change in angular position for each of the pairs of values in Quick Quiz 7.1 occurred in 1 s. Which choice represents the lowest average angular speed?

Figure 7.4 shows a bicycle turned upside down so that a repair technician can work on the rear wheel. The bicycle pedals are turned so that at time t_i the wheel has angular speed ω_i (Fig. 7.4a) and at a later time t_f it has angular speed ω_f

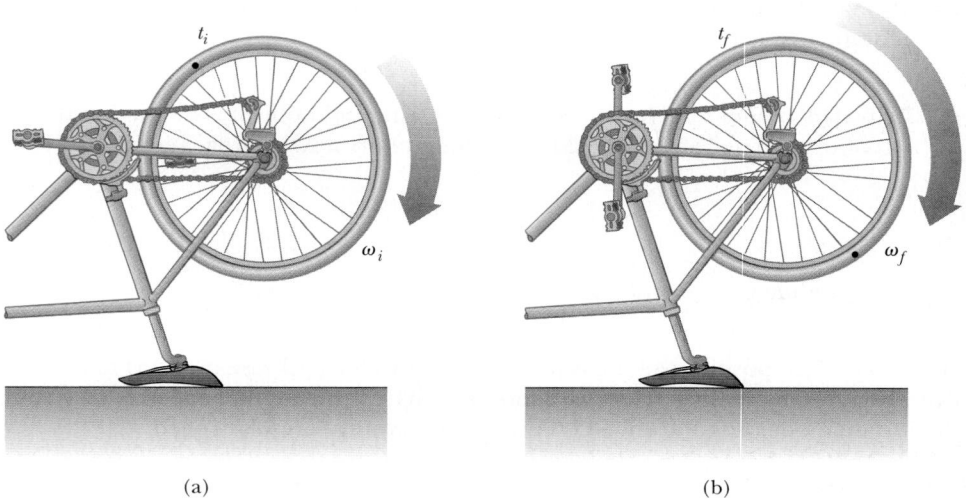

Figure 7.4 An accelerating bicycle wheel rotates with (a) angular speed ω_i at time t_i and (b) angular speed ω_f at time t_f.

(Fig. 7.4b). Just as a changing speed leads to the concept of an acceleration, a changing angular speed leads to the concept of an angular acceleration.

Average angular acceleration ▶

An object's average angular acceleration α_{av} during the time interval Δt is defined as the change in its angular speed $\Delta\omega$ divided by Δt:

$$\alpha_{av} \equiv \frac{\omega_f - \omega_i}{t_f - t_i} = \frac{\Delta\omega}{\Delta t} \qquad [7.5]$$

SI unit: radian per second squared (rad/s^2)

As with angular velocity, positive angular accelerations are in the counterclockwise direction, negative angular accelerations in the clockwise direction. If the angular speed goes from 15 rad/s to 9.0 rad/s in 3.0 s, then the average angular acceleration during that time interval is

$$\alpha_{av} = \frac{\Delta\omega}{\Delta t} = \frac{9.0 \text{ rad/s} - 15 \text{ rad/s}}{3.0 \text{ s}} = -2.0 \text{ rad/s}$$

The negative sign indicates that the angular acceleration is clockwise (though the angular speed, still positive but slowing down, is in the counterclockwise direction). There is also an instantaneous version of angular acceleration:

Instantaneous angular acceleration ▶

The instantaneous angular acceleration α is defined as the limit of the average angular acceleration $\Delta\omega/\Delta t$ as the time interval Δt approaches zero:

$$\alpha \equiv \lim_{\Delta t \to 0} \frac{\Delta\omega}{\Delta t} \qquad [7.6]$$

SI unit: radian per second squared (rad/s^2)

When a rigid object rotates about a fixed axis, as does the bicycle wheel, every portion of the object has the same angular speed and the same angular acceleration. This fact is what makes these variables so useful for describing rotational motion. In contrast, the tangential (linear) speed and acceleration of the object take different values that depend on the distance from a given point to the axis of rotation.

7.2 ROTATIONAL MOTION UNDER CONSTANT ANGULAR ACCELERATION

A number of parallels exist between the equations for rotational motion and those for linear motion. For example, compare the defining equation for the average angular speed,

$$\omega_{av} \equiv \frac{\theta_f - \theta_i}{t_f - t_i} = \frac{\Delta\theta}{\Delta t}$$

with that of the average linear speed,

$$v_{av} \equiv \frac{x_f - x_i}{t_f - t_i} = \frac{\Delta x}{\Delta t}$$

In these equations, ω takes the place of v and θ takes the place of x, so the equations differ only in the names of the variables. In the same way, every linear quantity we have encountered so far has a corresponding "twin" in rotational motion.

The procedure used in Section 2.5 to develop the kinematic equations for linear motion under constant acceleration can be used to derive a similar set of equations for rotational motion under constant angular acceleration. The resulting equations of rotational kinematics, along with the corresponding equations for linear motion, are as follows:

Linear Motion with a Constant (Variables: x and v)	Rotational Motion about a Fixed Axis with α Constant (Variables: θ and ω)	
$v = v_i + at$	$\omega = \omega_i + \alpha t$	**[7.7]**
$\Delta x = v_i t + \frac{1}{2}at^2$	$\Delta\theta = \omega_i t + \frac{1}{2}\alpha t^2$	**[7.8]**
$v^2 = v_i^2 + 2a\Delta x$	$\omega^2 = \omega_i^2 + 2\alpha\Delta\theta$	**[7.9]**

Notice that every term in a given linear equation has a corresponding term in the analogous rotational equation.

Quick Quiz 7.3

Consider again the pairs of angular positions for the rigid object in Quick Quiz 7.1. If the object starts from rest at the initial angular position, moves counterclockwise with constant angular acceleration, and arrives at the final angular position with the same angular speed in all three cases, for which choice is the angular acceleration the highest?

EXAMPLE 7.2 A Rotating Wheel

Goal Apply the rotational kinematic equations.

Problem A wheel rotates with a constant angular acceleration of 3.50 rad/s^2. If the angular speed of the wheel is 2.00 rad/s at $t_i = 0$, **(a)** through what angle does the wheel rotate between $t = 0$ and $t = 2.00$ s? Give your answer in radians and in revolutions. **(b)** What is the angular speed of the wheel at $t = 2.00$ s?

Strategy The angular acceleration is constant, so this problem just requires substituting given values into Equations 7.7 and 7.8.

Solution

(a) Find the angular displacement after 2.00 s, in both radians and revolutions.

Use Equation 7.8, setting $\omega_i = 2.00$ rad/s, $\alpha = 3.5$ rad/s^2, and $t = 2.00$ s:

$$\begin{aligned}\Delta\theta &= \omega_i t + \tfrac{1}{2}\alpha t^2 \\ &= (2.00 \text{ rad/s})(2.00 \text{ s}) + \tfrac{1}{2}(3.50 \text{ rad/s}^2)(2.00 \text{ s})^2 \\ &= 11.0 \text{ rad}\end{aligned}$$

Convert radians to revolutions.

$$\Delta\theta = (11.0 \text{ rad})(1.00 \text{ rev}/2\pi \text{ rad}) = 1.75 \text{ rev}$$

(b) What is the angular speed of the wheel at $t = 2.00$ s?

Substitute the same values into Equation 7.7:

$$\omega = \omega_i + \alpha t = 2.00 \text{ rad/s} + (3.50 \text{ rad/s}^2)(2.00 \text{ s})$$
$$= 9.00 \text{ rad/s}$$

Remarks The result of part (b) could also be obtained from Equation 7.9 and the results of part (a).

Exercise 7.2
(a) Find the angle through which the wheel rotates between $t = 2.00$ s and $t = 3.00$ s. (b) Find the angular speed when $t = 3.00$ s.

Answer (a) 10.8 rad (b) 12.5 rad/s

EXAMPLE 7.3 Slowing Propellers

Goal Apply the time-independent rotational kinematic equation.

Problem An airplane propeller slows from an initial angular speed of 12.5 rev/s to a final angular speed of 5.00 rev/s. During this process, the propeller rotates through 21.0 revolutions. Find the angular acceleration of the propeller in radians per second squared, assuming it's constant.

Strategy The given quantities are the angular speeds and the displacement, which suggests applying Equation 7.9, the time-independent rotational kinematic equation, to find α.

Solution
First, convert the angular displacement to radians and the angular speeds to rad/s:

$$\Delta\theta = (21.0 \text{ rev})(2\pi \text{ rad/rev}) = 42.0\pi \text{ rad}$$
$$\omega_i = (12.5 \text{ rev/s})(2\pi \text{ rad/rev}) = 25.0\pi \text{ rad/s}$$
$$\omega = (5.00 \text{ rev/s})(2\pi \text{ rad/rev}) = 10.0\pi \text{ rad/s}$$

Substitute these values into Equation 7.9 to find the angular acceleration α:

$$\omega^2 = \omega_i^2 + 2\alpha\Delta\theta$$
$$(10.0\pi \text{ rad/s})^2 = (25.0\pi \text{ rad/s})^2 + 2\alpha(42\pi \text{ rad})$$

Solve for α:

$$\alpha = -6.25\pi \text{ rad/s}^2$$

Remark Waiting until the end to convert revolutions to radians is also possible and requires only one conversion instead of three.

Exercise 7.3
Suppose, instead, the engine speeds up so that the propeller goes through 28.0 revolutions while the angular speed increases uniformly from 5.00 rev/s to 15.0 rev/s. Find the angular acceleration.

Answer 7.14π rad/s^2

7.3 RELATIONS BETWEEN ANGULAR AND LINEAR QUANTITIES

Angular variables are closely related to linear variables. Consider the arbitrarily shaped object in Active Figure 7.5 rotating about the z-axis through the point O. Assume that the object rotates through the angle $\Delta\theta$, and hence point P moves through the arc length Δs, in the interval Δt. We know from the defining equation for radian measure that

$$\Delta\theta = \frac{\Delta s}{r}$$

Dividing both sides of this equation by Δt, the time interval during which the rotation occurs, yields

$$\frac{\Delta\theta}{\Delta t} = \frac{1}{r}\frac{\Delta s}{\Delta t}$$

When Δt is very small, the angle $\Delta\theta$ through which the object rotates is also small and the ratio $\Delta\theta/\Delta t$ is close to the instantaneous angular speed ω. On the other side of the equation, similarly, the ratio $\Delta s/\Delta t$ approaches the instantaneous linear speed v for small values of Δt. Hence, when Δt gets arbitrarily small, the preceding equation is equivalent to

$$\omega = \frac{v}{r}$$

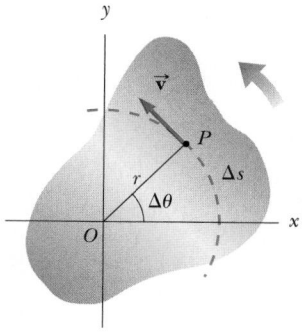

ACTIVE FIGURE 7.5
Rotation of an object about an axis through O (the z-axis) that is perpendicular to the plane of the figure. Note that a point P on the object rotates in a circle of radius r centered at O.

Physics Now™
Log into PhysicsNow at **www.cp7e.com,** and go to Active Figure 7.5 to move point P and observe the tangential velocity as the object rotates.

In Active Figure 7.5, the point P traverses a distance Δs along a circular arc during the time interval Δt at a linear speed of v. The direction of P's velocity vector $\vec{\mathbf{v}}$ is *tangent to the circular path.* The magnitude of $\vec{\mathbf{v}}$ is the linear speed $v = v_t$, called the **tangential speed** of a particle moving in a circular path, written

$$v_t = r\omega \qquad [7.10]$$

◀ Tangential speed

The tangential speed of a point on a rotating object equals the distance of that point from the axis of rotation multiplied by the angular speed. Equation 7.10 shows that the linear speed of a point on a rotating object increases as that point is moved outward from the center of rotation toward the rim, as expected; however, **every point on the rotating object has the same *angular* speed.**

Equation 7.10, derived using the defining equation for radian measure, is valid only when ω is measured in radians per unit time. Other measures of angular speed, such as degrees per second and revolutions per second, shouldn't be used.

To find a second equation relating linear and angular quantities, refer again to Figure 7.5, and suppose the rotating object changes its angular speed by $\Delta\omega$ in the time interval Δt. At the end of this interval, the speed of a point on the object, such as P, has changed by the amount Δv_t. From Equation 7.10 we have

$$\Delta v_t = r\,\Delta\omega$$

Dividing by Δt gives

$$\frac{\Delta v_t}{\Delta t} = r\frac{\Delta\omega}{\Delta t}$$

As the time interval Δt is taken to be arbitrarily small, $\Delta\omega/\Delta t$ approaches the instantaneous angular acceleration. On the left-hand side of the equation, the ratio $\Delta v_t/\Delta t$ tends to the instantaneous linear acceleration, called the tangential acceleration of that point, given by

$$a_t = r\alpha \qquad [7.11]$$

◀ Tangential acceleration

The tangential acceleration of a point on a rotating object equals the distance of that point from the axis of rotation multiplied by the angular acceleration. Again, radian measure must be used for the angular acceleration term in this equation.

One last equation that relates linear quantities to angular quantities will be derived in the next section.

Quick Quiz 7.4

Andrea and Chuck are riding on a merry-go-round. Andrea rides on a horse at the outer rim of the circular platform, twice as far from the center of the circular platform as Chuck, who rides on an inner horse. When the merry-go-round is rotating at a constant angular speed, Andrea's angular speed is (a) twice Chuck's (b) the same as Chuck's (c) half of Chuck's (d) impossible to determine.

Quick Quiz 7.5

When the merry-go-round of Quick Quiz 7.4 is rotating at a constant angular speed, Andrea's tangential speed is (a) twice Chuck's (b) the same as Chuck's (c) half of Chuck's (d) impossible to determine.

Applying Physics 7.1 ESA Launch Site

Why is the launch area for the European Space Agency in South America and not in Europe?

Explanation Satellites are boosted into orbit on top of rockets, which provide the large tangential speed necessary to achieve orbit. Due to its rotation, the surface of Earth is already traveling toward the east at a tangential speed of nearly 1 700 m/s at the equator. This tangential speed is steadily reduced further north, because the distance to the axis of rotation is decreasing. It finally goes to zero at the North Pole. Launching eastward from the equator gives the satellite a starting initial tangential speed of 1 700 m/s, whereas a European launch provides roughly half that speed (depending on the exact latitude).

EXAMPLE 7.4 Compact Discs

Goal Apply the rotational kinematics equations in tandem with tangential acceleration and speed.

Problem A compact disc rotates from rest up to an angular speed of 31.4 rad/s in a time of 0.892 s. **(a)** What is the angular acceleration of the disc, assuming the angular acceleration is uniform? **(b)** Through what angle does the disc turn while coming up to speed? **(c)** If the radius of the disc is 4.45 cm, find the final tangential speed of a microbe riding on the rim of the disc. **(d)** What is the magnitude of the tangential acceleration of the microbe at the given time?

Strategy We can solve parts (a) and (b) by applying the kinematic equations for angular speed and angular displacement (Equations 7.7 and 7.8). Multiplying the radius by the angular acceleration yields the tangential acceleration at the rim, while multiplying the radius by the angular speed gives the tangential speed at that point.

Solution

(a) Find the angular acceleration.

Apply the angular velocity equation $\omega = \omega_i + \alpha t$, taking $\omega_i = 0$ at $t = 0$:

$$\alpha = \frac{\omega}{t} = \frac{31.4 \text{ rad/s}}{0.892 \text{ s}} = 35.2 \text{ rad/s}^2$$

(b) Through what angle does the disc turn?

Use Equation 7.8 for angular displacement, with $t = 0.892$ s and $\omega_i = 0$:

$$\Delta\theta = \omega_i t + \tfrac{1}{2}\alpha t^2 = \tfrac{1}{2}(35.2 \text{ rad/s}^2)(0.892 \text{ s})^2 = 14.0 \text{ rad}$$

(c) Find the final tangential speed of a microbe at $r = 4.45$ cm.

Substitute into Equation 7.10:

$$v_t = r\omega = (0.0445 \text{ m})(31.4 \text{ rad/s}) = 1.40 \text{ m/s}$$

(d) Find the tangential acceleration of the microbe at $r = 4.45$ cm.

Substitute into Equation 7.11:

$$a_t = r\alpha = (0.0445 \text{ m})(35.2 \text{ rad/s}^2) = 1.57 \text{ m/s}^2$$

Remarks Because 2π rad = 1 rev, the angular displacement in part (**b**) corresponds to 2.23 rev. In general, dividing the number of radians by six gives a good approximation to the number of revolutions, because $2\pi \sim 6$.

Exercise 7.4

(a) What are the angular speed and angular displacement of the disc 0.300 s after it begins to rotate? (b) Find the tangential speed at the rim at this time.

Answers (a) 10.6 rad/s; 1.58 rad (b) 0.472 m/s

APPLICATION

Phonograph Records and Compact Discs

Before compact discs became the medium of choice for recorded music, phonographs were popular. There are similarities and differences between the rotational motion of phonograph records and that of compact discs. A phonograph record rotates at a constant angular speed. Popular angular speeds were $33\frac{1}{3}$ rev/min for long-playing albums (hence the nickname "LP"), 45 rev/min for "singles," and 78 rev/min used in very early recordings. At the outer edge of the record, the pickup needle (stylus) moves over the vinyl material at a faster tangential speed than when the needle is close to the center of the record. As a result, the sound information is compressed into a smaller length of track near the center of the record than near the outer edge.

CDs, on the other hand, are designed so that the disc moves under the laser pickup at a constant tangential speed. Because the pickup moves radially as it follows the tracks of information, the angular speed of the compact disc must vary according to the radial position of the laser. Since the tangential speed is fixed, the information density (per length of track) anywhere on the disc is the same. Example 7.5 demonstrates numerical calculations for both compact discs and phonograph records.

EXAMPLE 7.5 Track Length of a Compact Disc

Goal Relate angular to linear variables.

Problem In a compact disc player, as the read head moves out from the center of the disc, the angular speed of the disc changes so that the linear speed at the position of the head remains at a constant value of about 1.3 m/s. **(a)** Find the angular speed of the compact disc when the read head is at $r = 2.0$ cm and again at $r = 5.6$ cm. **(b)** An old-fashioned record player rotates at a constant angular speed, so the linear speed of the record groove moving under the detector (the stylus) changes. Find the linear speed of a 45.0-rpm record at points 2.0 and 5.6 cm from the center. **(c)** In both the CD and phonograph record, information is recorded in a continuous spiral track. Calculate the total length of the track for a CD designed to play for 1.0 h.

Strategy This problem is just a matter of substituting numbers into the appropriate equations. Part (a) requires relating angular and linear speed with Equation 7.10, $v_t = r\omega$, solving for ω and substituting given values. In part (b), convert from rev/min to rad/s and substitute straight into Equation 7.10 to obtain the linear speeds. In part (c), linear speed multiplied by time gives the total distance.

Solution

(a) Find the angular speed when the read head is at $r = 2.0$ cm and $r = 5.6$ cm.

Solve $v_t = r\omega$ for ω and substitute the numbers. At $r = 2.0$ cm:

$$\omega = \frac{v_t}{r} = \frac{1.3 \text{ m/s}}{2.0 \times 10^{-2} \text{ m}} = 65 \text{ rad/s}$$

Likewise, find the angular speed at $r = 5.6$ cm:

$$\omega = \frac{v_t}{r} = \frac{1.3 \text{ m/s}}{5.6 \times 10^{-2} \text{ m}} = 23 \text{ rad/s}$$

(b) Find the linear speed in m/s of a 45.0-rpm record at points 2.0 cm and 5.6 cm from the center.

Convert rev/min to rad/s:

$$45.0\,\frac{\text{rev}}{\text{min}} = 45.0\,\frac{\cancel{\text{rev}}}{\cancel{\text{min}}}\left(\frac{2\pi\text{ rad}}{\cancel{\text{rev}}}\right)\left(\frac{1.00\,\cancel{\text{min}}}{60.0\text{ s}}\right) = 4.71\,\frac{\text{rad}}{\text{s}}$$

Calculate the linear speed at $r = 2.0$ cm:

$$v_t = r\omega = (2.0 \times 10^{-2}\text{ m})(4.71\text{ rad/s}) = \boxed{0.094\text{ m/s}}$$

Calculate the linear speed at $r = 5.6$ cm:

$$v_t = r\omega = (5.6 \times 10^{-2}\text{ m})(4.71\text{ rad/s}) = \boxed{0.26\text{ m/s}}$$

(c) Calculate the total length of the track for a CD designed to play for 1.0 h.

Multiply the linear speed of the read head by the time in seconds.

$$\Delta x = v_t t = (1.3\text{ m/s})(3\,600\text{ s}) = \boxed{4\,700\text{ m}}$$

Remark Notice that for the record player in part (b), even though the angular speed is constant at all points along a radial line, the tangential speed steadily increases with increasing r. The calculation for a CD in part (c) is easy only because the linear (tangential) speed is constant. It would be considerably more difficult for a record player, where the tangential speed depends on the distance from the center.

Exercise 7.5

Compute the linear speed of a record playing at $33\frac{1}{3}$ revolutions per minute (a) at $r = 2.00$ cm and (b) at $r = 5.60$ cm.

Answers (a) 0.069 8 m/s (b) 0.195 m/s

7.4 CENTRIPETAL ACCELERATION

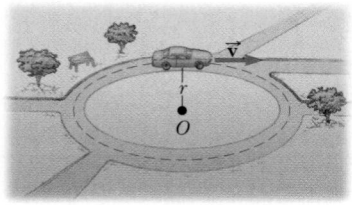

(a)

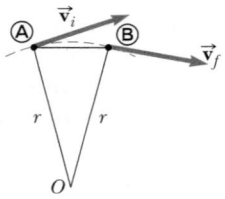

(b)

Figure 7.6 (a) Circular motion of a car moving with constant speed. (b) As the car moves along the circular path from Ⓐ to Ⓑ, the direction of its velocity vector changes, so the car undergoes a centripetal acceleration.

Figure 7.6a shows a car moving in a circular path with *constant linear speed* v. **Even though the car moves at a constant speed, it still has an acceleration**. To understand this, consider the defining equation for average acceleration:

$$\vec{\mathbf{a}}_{\text{av}} = \frac{\vec{\mathbf{v}}_f - \vec{\mathbf{v}}_i}{t_f - t_i} \qquad [7.12]$$

The numerator represents the difference between the velocity vectors $\vec{\mathbf{v}}_f$ and $\vec{\mathbf{v}}_i$. These vectors may have the same *magnitude*, corresponding to the same speed, but if they have different *directions*, their difference can't equal zero. The direction of the car's velocity as it moves in the circular path is continually changing, as shown in Figure 7.6b. For circular motion at constant speed, the acceleration vector always points toward the center of the circle. Such an acceleration is called a **centripetal** (center-seeking) **acceleration**. Its magnitude is given by

$$a_c = \frac{v^2}{r} \qquad [7.13]$$

To derive Equation 7.13, consider Figure 7.7a. An object is first at point Ⓐ with velocity $\vec{\mathbf{v}}_i$ at time t_i and then at point Ⓑ with velocity $\vec{\mathbf{v}}_f$ at a later time t_f. We assume that $\vec{\mathbf{v}}_i$ and $\vec{\mathbf{v}}_f$ differ only in direction; their magnitudes are the same ($v_i = v_f = v$). To calculate the acceleration, we begin with Equation 7.12,

$$\vec{\mathbf{a}}_{\text{av}} = \frac{\vec{\mathbf{v}}_f - \vec{\mathbf{v}}_i}{t_f - t_i} = \frac{\Delta\vec{\mathbf{v}}}{\Delta t} \qquad [7.14]$$

where $\Delta\vec{\mathbf{v}} = \vec{\mathbf{v}}_f - \vec{\mathbf{v}}_i$ is the change in velocity. When Δt is very small, Δs and $\Delta\theta$ are also very small. In Figure 7.7b, $\vec{\mathbf{v}}_f$ is almost parallel to $\vec{\mathbf{v}}_i$, and the vector $\Delta\vec{\mathbf{v}}$ is approximately perpendicular to them, pointing toward the center of the circle. In the limiting case when Δt becomes vanishingly small, $\Delta\vec{\mathbf{v}}$ points exactly toward the center of the circle, and the average acceleration $\vec{\mathbf{a}}_{\text{av}}$ becomes the instantaneous acceleration $\vec{\mathbf{a}}$. From Equation 7.14, $\vec{\mathbf{a}}$ and $\Delta\vec{\mathbf{v}}$ point in the same direction (in this limit), so the instantaneous acceleration points to the center of the circle.

The triangle in Figure 7.7a, which has sides Δs and r, is similar to the one formed by the vectors in Figure 7.7b, so the ratios of their sides are equal:

$$\frac{\Delta v}{v} = \frac{\Delta s}{r}$$

or

$$\Delta v = \frac{v}{r}\,\Delta s \qquad \textbf{[7.15]}$$

Substituting the result of Equation 7.15 into $a_{\text{av}} = \Delta v/\Delta t$ gives

$$a_{\text{av}} = \frac{v}{r}\frac{\Delta s}{\Delta t} \qquad \textbf{[7.16]}$$

But Δs is the distance traveled along the arc of the circle in time Δt, and in the limiting case when Δt becomes very small, $\Delta s/\Delta t$ approaches the instantaneous value of the tangential speed, v. At the same time, the average acceleration a_{av} approaches a_c, the instantaneous centripetal acceleration, so Equation 7.16 reduces to Equation 7.13:

$$a_c = \frac{v^2}{r}$$

Because the tangential speed is related to the angular speed through the relation $v_t = r\omega$ (Eq. 7.10), an alternate form of Equation 7.13 is

$$a_c = \frac{r^2\omega^2}{r} = r\omega^2 \qquad \textbf{[7.17]}$$

Dimensionally, $[r] = \text{L}$ and $[\omega] = 1/\text{T}$, so the units of centripetal acceleration are L/T^2, as they should be. This is a geometric result relating the centripetal acceleration to the angular speed, but physically an acceleration can occur only if some force is present. For example, if a car travels in a circle on flat ground, the force of static friction between the tires and the ground provides the necessary centripetal force.

Note that a_c in Equations 7.13 and 7.17 represents only the *magnitude* of the centripetal acceleration. The acceleration itself is always directed towards the center of rotation.

The foregoing derivations concern circular motion at constant speed. When an object moves in a circle but is speeding up or slowing down, a tangential component of acceleration, $a_t = r\alpha$, is also present. Because the tangential and centripetal components of acceleration are perpendicular to each other, we can find the magnitude of the **total acceleration** with the Pythagorean theorem:

$$a = \sqrt{a_t^{\,2} + a_c^{\,2}} \qquad \textbf{[7.18]}$$

◀ Total acceleration

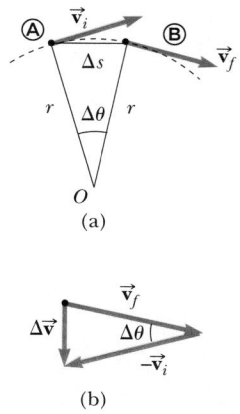

Figure 7.7 (a) As the particle moves from Ⓐ to Ⓑ, the direction of its velocity vector changes from $\vec{v}_i$ to $\vec{v}_f$. (b) The construction for determining the direction of the change in velocity $\Delta\vec{v}$, which is toward the center of the circle.

Quick Quiz 7.6

A race track is constructed such that two arcs of radius 80 m at Ⓐ and 40 m at Ⓑ are joined by two stretches of straight track as in Figure 7.8. In a particular trial run, a driver travels at a constant speed of 50 m/s for one complete lap.

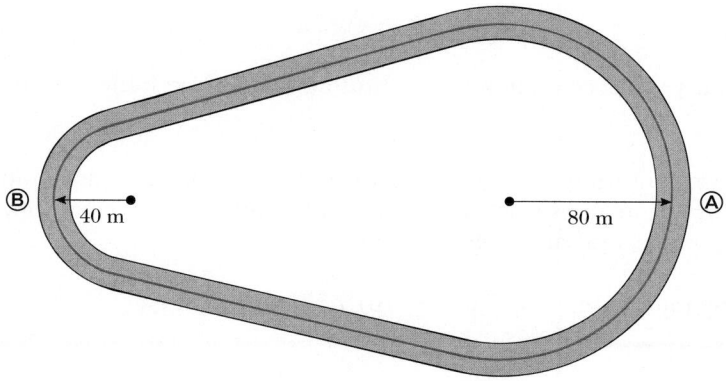

Figure 7.8 (Quick Quiz 7.6)

1. The ratio of the tangential acceleration at Ⓐ to that at Ⓑ is
   (a) $\frac{1}{2}$ (b) $\frac{1}{4}$ (c) 2 (d) 4 (e) The tangential acceleration is zero at both points.
2. The ratio of the centripetal acceleration at Ⓐ to that at Ⓑ is
   (a) $\frac{1}{2}$ (b) $\frac{1}{4}$ (c) 2 (d) 4 (e) The centripetal acceleration is zero at both points.
3. The angular speed is greatest at
   (a) Ⓐ (b) Ⓑ (c) It is equal at both Ⓐ and Ⓑ.

Quick Quiz 7.7

An object moves in a circular path with constant speed v. Which of the following statements is true concerning the object? (a) Its velocity is constant, but its acceleration is changing. (b) Its acceleration is constant, but its velocity is changing. (c) Both its velocity and acceleration are changing. (d) Its velocity and acceleration remain constant.

EXAMPLE 7.6 At the Race Track

Goal Apply the concepts of centripetal acceleration and tangential speed.

Problem A race car accelerates uniformly from a speed of 40.0 m/s to a speed of 60.0 m/s in 5.00 s while traveling counterclockwise around a circular track of radius 4.00×10^2 m. When the car reaches a speed of 50.0 m/s, find **(a)** the magnitude of the car's centripetal acceleration **(b)** the angular speed **(c)** the tangential acceleration, and **(d)** the magnitude of the total acceleration.

Strategy Substitute values into the definitions of centripetal acceleration (Equation 7.13), tangential speed (Equation 7.10), and total acceleration (Equation 7.18). Dividing the change in linear speed by the time yields the tangential acceleration.

Solution

(a) Find the magnitude of the centripetal acceleration when $v = 50.0$ m/s.

Substitute into Equation 7.13:

$$a_c = \frac{v^2}{r} = \frac{(50.0 \text{ m/s})^2}{(4.00 \times 10^2 \text{ m})} = 6.25 \text{ m/s}^2$$

(b) Find the angular speed.

Solve Equation 7.10 for ω and substitute:

$$\omega = \frac{v}{r} = \frac{50.0 \text{ m/s}}{4.00 \times 10^2 \text{ m}} = 0.125 \text{ rad/s}$$

(c) Find the tangential acceleration.

Divide the change in linear speed by the time:

$$a_t = \frac{v_f - v_i}{\Delta t} = \frac{60.0 \text{ m/s} - 40.0 \text{ m/s}}{5.00 \text{ s}} = 4.00 \text{ m/s}^2$$

(d) Find the magnitude of the total acceleration.

Substitute into Equation 7.18:

$$a = \sqrt{a_t^2 + a_c^2} = \sqrt{(4.00 \text{ m/s}^2)^2 + (6.25 \text{ m/s}^2)^2}$$

$$a = 7.42 \text{ m/s}^2$$

Remarks We can also find the centripetal acceleration by substituting the derived value of ω into Equation 7.17.

Exercise 7.6

Suppose the race car now slows down uniformly from 60.0 m/s to 30.0 m/s in 4.50 s to avoid an accident, while still traversing a circular path 4.00×10^2 m in radius. Find the car's (a) centripetal acceleration, (b) angular speed, (c) tangential acceleration, and (d) total acceleration when the speed is 40.0 m/s.

Answers (a) 4.00 m/s^2 (b) 0.100 rad/s (c) -6.67 m/s^2 (d) 7.77 m/s^2

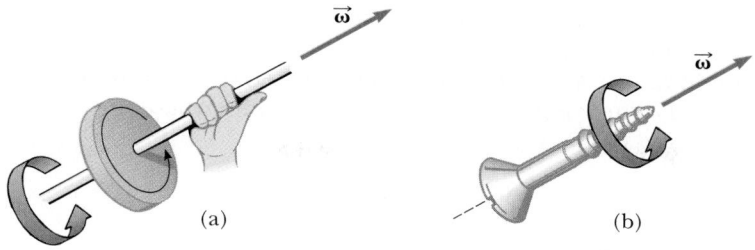

Figure 7.9 (a) The right-hand rule for determining the direction of the angular velocity vector $\vec{\omega}$. (b) The direction of $\vec{\omega}$ is in the direction of advance of a right-handed screw.

Angular Quantities Are Vectors

When we discussed linear motion in Chapter 2, we emphasized the fact that displacement, velocity, and acceleration are all vector quantities. In describing rotational motion, angular displacement, angular velocity, and angular acceleration are also vector quantities.

The direction of the angular velocity vector $\vec{\omega}$ can be found with the **right-hand rule**, as illustrated in Figure 7.9a. Grasp the axis of rotation with your right hand so that your fingers wrap in the direction of rotation. Your extended thumb then points in the direction of $\vec{\omega}$. Figure 7.9b shows that $\vec{\omega}$ is also in the direction of advance of a rotating right-handed screw.

We can apply this rule to a rotating disk viewed along the axis of rotation, as in Figure 7.10. When the disk rotates clockwise (Fig. 7.10a), the right-hand rule shows that the direction of $\vec{\omega}$ is into the page. When the disk rotates counterclockwise (Fig. 7.10b), the direction of $\vec{\omega}$ is out of the page.

Finally, the directions of the angular acceleration $\vec{\alpha}$ and the angular velocity $\vec{\omega}$ are the same if the angular speed ω (the magnitude of $\vec{\omega}$) is increasing with time, and opposite each other if the angular speed is decreasing with time.

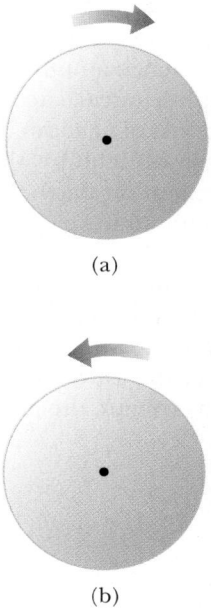

Figure 7.10 A top view of a disk rotating about an axis through its center perpendicular to the page. (a) When the disk rotates clockwise, $\vec{\omega}$ points into the page. (b) When the disk rotates counterclockwise, $\vec{\omega}$ points out of the page.

Forces Causing Centripetal Acceleration

An object can have a centripetal acceleration *only* if some external force acts on it. For a ball whirling in a circle at the end of a string, that force is the tension in the string. In the case of a car moving on a flat circular track, the force is friction between the car and track. A satellite in circular orbit around Earth has a centripetal acceleration due to the gravitational force between the satellite and Earth.

Some books use the term "centripetal force," which can give the mistaken impression that it is a new force of nature. This is not the case: The adjective "centripetal" in "centripetal force" simply means that the force in question acts toward a center. The gravitational force and the force of tension in the string of a yo-yo whirling in a circle are examples of centripetal forces, as is the force of gravity on a satellite circling the Earth.

Consider a ball of mass m that is tied to a string of length r and is being whirled at constant speed in a horizontal circular path, as illustrated in Figure 7.11. Its weight is supported by a frictionless table. Why does the ball move in a circle? Because of its inertia, the tendency of the ball is to move in a straight line; however, the string prevents motion along a straight line by exerting a radial force on the ball—a tension force—that makes it follow the circular path. The tension is directed along the string toward *the center of the circle, as shown* in the figure.

In general, applying Newton's second law along the radial direction yields the equation relating the net centripetal force F_c—the sum of the radial components of all forces acting on a given object—with the centripetal acceleration:

$$F_c = ma_c = m\frac{v^2}{r} \qquad [7.19]$$

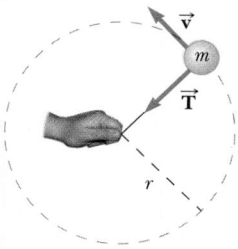

Figure 7.11 A ball attached to a string of length r, rotating in a circular path at constant speed.

TIP 7.2 Centripetal Force is a *Type* of Force, not a Force in Itself!

"Centripetal force" is a classification that includes forces acting toward a central point, like string tension on a tetherball or gravity on a satellite. A centripetal force must be *supplied* by some actual, physical force.

A net force causing a centripetal acceleration acts toward the center of the circular path and effects a change in the direction of the velocity vector. If that force should vanish, the object would immediately leave its circular path and move along a straight line tangent to the circle at the point where the force vanished.

Applying Physics 7.2 Artificial Gravity

Astronauts spending lengthy periods of time in space experience a number of negative effects due to weightlessness, such as weakening of muscle tissue and loss of calcium in bones. These effects may make it very difficult for them to return to their usual environment on Earth. How could artificial gravity be generated in space to overcome such complications?

Solution A rotating cylindrical space station creates an environment of artificial gravity. The normal force of the rigid walls provides the centripetal force, which keeps the astronauts moving in a circle (Fig. 7.12). To an astronaut, the normal force can't be easily distinguished from a gravitational force as long as the radius of the station is large compared with the astronaut's height. (Otherwise there are unpleasant inner ear effects.) This same principle is used in certain amusement park rides in which passengers are pressed against the inside of a rotating cylinder as it tilts in various directions. The visionary physicist Gerard O'Neill proposed creating a giant space colony a kilometer in radius that rotates slowly, creating Earth-normal artificial gravity for the inhabitants in its interior. These inside-out artificial worlds could enable safe transport on a several-thousand-year journey to another star system.

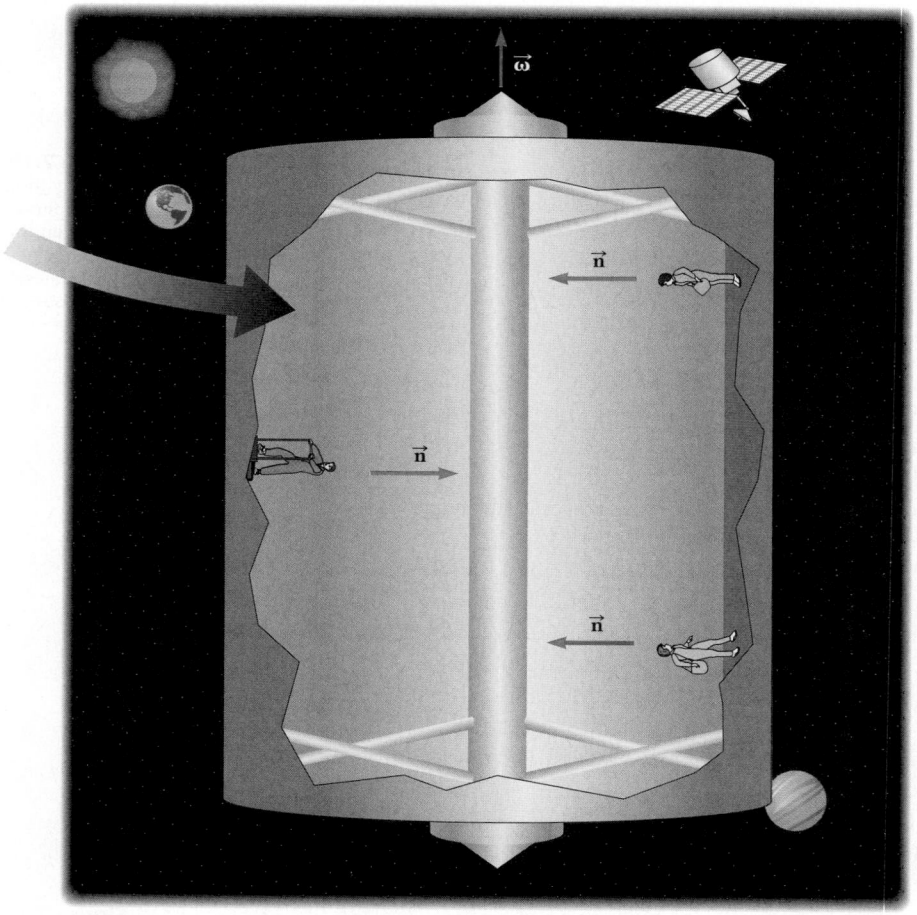

Figure 7.12 Artificial gravity inside a spinning cylinder is provided by the normal force.

Problem Solving Strategy Forces that Cause Centripetal Acceleration

Use the following steps in dealing with centripetal accelerations and the forces that produce them:

1. **Draw a free-body diagram** of the object under consideration, labeling all forces that act on it.
2. **Choose a coordinate system** that has one axis perpendicular to the circular path followed by the object (the radial direction) and one axis tangent to the circular path (the tangential, or angular, direction). The normal direction, perpendicular to the plane of motion, is also often needed.
3. **Find the net force F_c toward the center** of the circular path, $F_c = \Sigma F_r$, where ΣF_c is the sum of the radial components of the forces. This net radial force causes the centripetal acceleration.
4. **Use Newton's second law for the radial, tangential, and normal directions**, as required, writing $\Sigma F_r = ma_c$, $\Sigma F_t = ma_t$, and $\Sigma F_n = ma_n$. Remember that the magnitude of the centripetal acceleration for uniform circular motion can always be written $a_c = v_t{}^2/r$.
5. **Solve** for the unknown quantities.

INTERACTIVE EXAMPLE 7.7 Buckle Up for Safety

Goal Calculate the frictional force that causes an object to have a centripetal acceleration.

Problem A car travels at a constant speed of 30.0 mi/h (13.4 m/s) on a level circular turn of radius 50.0 m, as shown in the bird's-eye view in Figure 7.13a. What minimum coefficient of static friction, μ_s, between the tires and roadway will allow the car to make the circular turn without sliding?

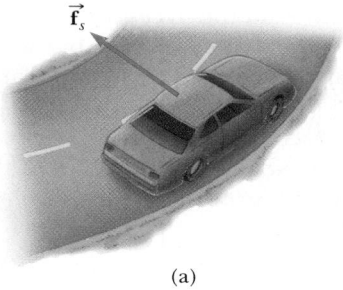

(a)

Strategy In the car's free-body diagram (Fig. 7.13b) the normal direction is vertical and the tangential direction is into the page (step 2). Use Newton's second law. The net force acting on the car in the radial direction is the force of static friction toward the center of the circular path, which causes the car to have a centripetal acceleration. Calculating the maximum static friction force requires the normal force, obtained from the normal component of the second law.

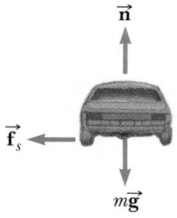

(b)

Figure 7.13 (Example 7.7) (a) Top view of a car on a curved path. (b) A free-body diagram of the car, showing an end view.

Solution

(Step 3, 4) Write the components of Newton's second law. The radial component involves only the maximum static friction force, $f_{s,\max}$:

$$m\frac{v^2}{r} = f_{s,\max} = \mu_s n$$

In the vertical component of the second law, the gravity force and the normal force are in equilibrium:

$$n - mg = 0 \quad \rightarrow \quad n = mg$$

(Step 5) Substitute the expression for n into the first equation and solve for μ_s:

$$m\frac{v^2}{r} = \mu_s mg$$

$$\mu_s = \frac{v^2}{rg} = \frac{(13.4\ \text{m/s})^2}{(50.0\ \text{m})(9.80\ \text{m/s}^2)} = \boxed{0.366}$$

Remarks The value of μ_s for rubber on dry concrete is very close to 1, so the car can negotiate the curve with ease. If the road were wet or icy, however, the value for μ_s could be 0.2 or lower. Under such conditions, the radial force provided by static friction wouldn't be great enough to keep the car on the circular path, and it would slide off on a tangent, leaving the roadway.

Exercise 7.7

At what maximum speed can a car negotiate a turn on a wet road with coefficient of static friction 0.230 without sliding out of control? The radius of the turn is 25.0 m.

Answer 7.51 m/s

PhysicsNow™ Investigate the motion of a car around a level curve by logging into PhysicsNow at **www.cp7e.com** and going to Interactive Example 7.7.

EXAMPLE 7.8 Daytona International Speedway

Goal Solve a centripetal force problem involving two dimensions.

Problem The Daytona International Speedway in Daytona Beach, Florida, is famous for its races, especially the Daytona 500, held every spring. Both of its courses feature four-story, 31.0° banked curves, with maximum radius of 316 m. If a car negotiates the curve too slowly, it tends to slip down the incline of the turn, whereas if it's going too fast, it may begin to slide up the incline. **(a)** Find the necessary centripetal acceleration on this banked curve so the car won't slip down or slide up the incline. (Neglect friction.) **(b)** Calculate the speed of the race car.

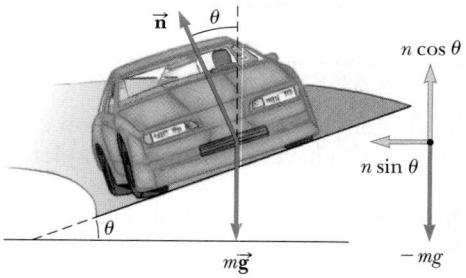

Figure 7.14 (Example 7.8) Front view of a car rounding a banked roadway. Vector components are shown to the right.

Strategy Two forces act on the race car: the force of gravity and the normal force $\vec{n}$. (See Fig. 7.14.) Use Newton's second law in the upward and radial directions to find the centripetal acceleration a_c. Solving $a_c = v^2/r$ for v then gives the race car's speed.

Solution

(a) Find the centripetal acceleration.

Write Newton's second law for the car:

$$m\vec{a} = \sum \vec{F} = \vec{n} + m\vec{g}$$

Use the y-component of Newton's second law to solve for the normal force n:

$$n \cos \theta - mg = 0$$

$$n = \frac{mg}{\cos \theta}$$

Obtain an expression for the horizontal component of $\vec{n}$, which is the centripetal force F_c in this example:

$$F_c = n \sin \theta = \frac{mg \sin \theta}{\cos \theta} = mg \tan \theta$$

Substitute this expression for F_c into the radial component of Newton's second law and divide by m to get the centripetal acceleration:

$$ma_c = F_c$$

$$a_c = \frac{F_c}{m} = \frac{mg \tan \theta}{m} = g \tan \theta$$

$$a_c = (9.80 \text{ m/s}^2)(\tan 31.0°) = 5.89 \text{ m/s}^2$$

(b) Find the speed of the race car.

Apply Equation 7.13:

$$\frac{v^2}{r} = a_c$$

$$v = \sqrt{ra_c} = \sqrt{(316 \text{ m})(5.89 \text{ m/s}^2)} = 43.1 \text{ m/s}$$

Remarks Both banking and friction assist in keeping the race car on the track.

APPLICATION
Banked Roadways

Exercise 7.8
A race track is to have a banked curve with radius of 245 m. What should be the angle of the bank if the normal force alone is to allow safe travel around the curve at 58.0 m/s?

Answer 54.5°

EXAMPLE 7.9 Riding the Tracks

Goal Combine centripetal force with conservation of energy.

Problem Figure 7.15a shows a roller-coaster car moving around a circular loop of radius R. **(a)** What speed must the car have so that it will just make it over the top without any assistance from the track? **(b)** What speed will the car subsequently have at the bottom of the loop? **(c)** What will be the normal force on a passenger at the bottom of the loop if the loop has a radius of 10.0 m?

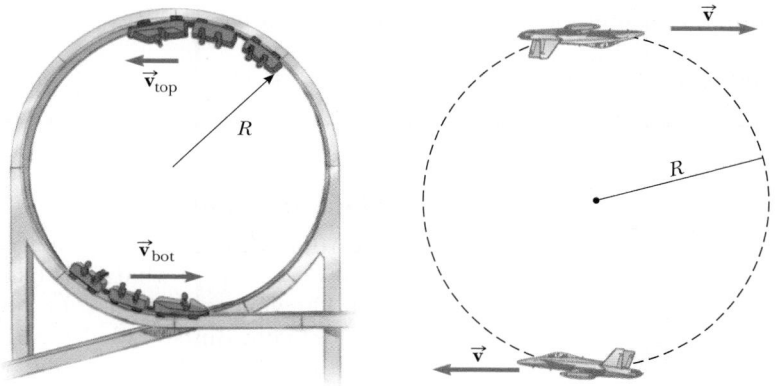

(a) (b)

Figure 7.15 (a) (Example 7.9) A roller coaster traveling around a nearly circular track. (b) (Exercise 7.9) A jet executing a vertical loop.

Strategy This problem requires Newton's second law and centripetal acceleration to find an expression for the car's speed at the top of the loop, followed by conservation of energy to find its speed at the bottom. If the car just makes it over the top, the force $\vec{\mathbf{n}}$ must become zero there, so the only force exerted on the car at that point is the force of gravity, $m\vec{\mathbf{g}}$. At the bottom of the loop, the normal force acts up toward the center and the gravity force acts down, away from the center. The difference of these two is the centripetal force. The normal force can then be calculated from Newton's second law.

Solution
(a) Find the speed at the top of the loop.

Write Newton's second law for the car:

$$m\vec{\mathbf{a}}_c = \vec{\mathbf{n}} + m\vec{\mathbf{g}} \qquad \textbf{(1)}$$

At the top of the loop, set $n = 0$. The force of gravity acts toward the center and provides the centripetal acceleration $a_c = v^2/R$.

$$m\frac{v_{top}^2}{R} = mg$$

Solve the foregoing equation for v_{top}:

$$v_{top} = \sqrt{gR}$$

(b) Find the speed at the bottom of the loop.

Apply conservation of mechanical energy to find the total mechanical energy at the top of the loop:

$$E_{top} = \tfrac{1}{2}mv_{top}^2 + mgh = \tfrac{1}{2}mgR + mg(2R) = 2.5mgR$$

Find the total mechanical energy at the bottom of the loop:

$$E_{bot} = \tfrac{1}{2}mv_{bot}^2$$

Energy is conserved, so these two energies may be equated and solved for v_{bot}:

$$\tfrac{1}{2}mv_{bot}^2 = 2.5mgR$$

$$v_{bot} = \sqrt{5gR}$$

(c) Find the normal force on a passenger at the bottom. (This is the passenger's perceived weight.)

Use Equation (1). The net centripetal force is $n - mg$:

$$m\frac{v_{\text{bot}}^2}{R} = n - mg$$

Solve for n:

$$n = mg + m\frac{v_{\text{bot}}^2}{R} = mg + m\frac{5gR}{R} = 6\,mg$$

Remarks The final answer for n shows that the rider experiences a force six times normal weight at the bottom of the loop! Astronauts experience a similar force during space launches.

Exercise 7.9

A jet traveling at a speed of 1.20×10^2 m/s executes a vertical loop with a radius of 5.00×10^2 m. (See Fig. 7.15b.) Find the magnitude of the force of the seat on a 70.0-kg pilot at (a) the top and (b) the bottom of the loop.

Answers (a) 1.33×10^3 N (b) 2.70×10^3 N

Fictitious Forces

Anyone who has ridden a merry-go-round as a child (or as a fun-loving grown-up) has experienced what feels like a "center-fleeing" force. Holding onto the railing and moving toward the center feels like a walk up a steep hill.

Actually, this so-called centrifugal force is *fictitious*. In reality, the rider is exerting a centripetal force on his body with his hand and arm muscles. In addition, a smaller centripetal force is exerted by the static friction between his feet and the platform. If the rider's grip slipped, he wouldn't be flung radially away; rather, he would go off on a straight line, tangent to the point in space where he let go of the railing. The rider lands at a point that is further away from the center, but not by "fleeing the center" along a radial line. Instead, he travels perpendicular to a radial line, traversing an angular displacement while increasing his radial displacement. (See Fig. 7.16.)

TIP 7.3 Centrifugal Force

A so-called centrifugal force is most often just the *absence* of an adequate *centripetal force*, arising from measuring phenomena from a noninertial (accelerating) frame of reference such as a merry-go-round.

7.5 NEWTONIAN GRAVITATION

Prior to 1686, a great deal of data had been collected on the motions of the Moon and planets, but no one had a clear understanding of the forces affecting them. In that year, Isaac Newton provided the key that unlocked the secrets of the heavens. He knew from the first law that a net force had to be acting on the Moon. If it were not, the Moon would move in a straight-line path rather than in its almost circular orbit around Earth. Newton reasoned that this force arose as a result of an attractive force between Moon and Earth, called the force of gravity, and that it was the same kind of force that attracted objects—such as apples—close to the surface of the Earth.

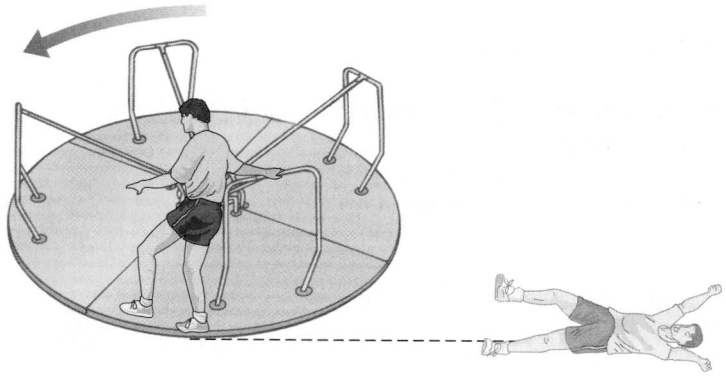

Figure 7.16 A fun-loving student loses his grip and falls along a line tangent to the rim of the merry-go-round.

In 1687, Newton published his work on the law of universal gravitation:

◀ Law of universal gravitation

If two particles with masses m_1 and m_2 are separated by a distance r, then a gravitational force acts along a line joining them, with magnitude given by

$$F = G\frac{m_1 m_2}{r^2} \qquad [7.20]$$

where $G = 6.673 \times 10^{-11}\ \text{kg}^{-1}\ \text{m}^3\ \text{s}^{-2}$ is a constant of proportionality called the **constant of universal gravitation**. The gravitational force is always attractive.

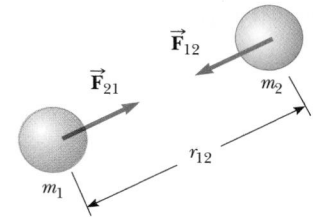

ACTIVE FIGURE 7.17
The gravitational force between two particles is attractive and acts along the line joining the particles. Note that according to Newton's third law, $\vec{F}_{12} = -\vec{F}_{21}$.

PhysicsNow™
Log into PhysicsNow at **www.cp7e.com**, and go to Active Figure 7.17 to change the masses of the particles and the separation between them to see the effect on the gravitational force.

This force law is an example of an **inverse-square law**, in that it varies as one over the square of the separation of the particles. From Newton's third law, we know that the force exerted by m_1 on m_2, designated $\vec{F}_{12}$ in Active Figure 7.17, is equal in magnitude but opposite in direction to the force $\vec{F}_{21}$ exerted by m_2 on m_1, forming an action–reaction pair.

Another important fact is that **the gravitational force exerted by a uniform sphere on a particle outside the sphere is the same as the force exerted if the entire mass of the sphere were concentrated at its center**. This is called Gauss's law, after the German mathematician and astronomer Karl Friedrich Gauss, and is also true of electric fields, which we will encounter in Chapter 15. Gauss's law is a mathematical result, true because the force falls off as an inverse square of the separation between the particles.

Near the surface of the Earth, the expression $F = mg$ is valid. As shown in Table 7.1, however, the free-fall acceleration g varies considerably with altitude above the Earth.

Quick Quiz 7.8

A ball falls to the ground. Which of the following statements are false? (a) The force that the ball exerts on Earth is equal in magnitude to the force that Earth exerts on the ball. (b) The ball undergoes the same acceleration as Earth. (c) Earth pulls much harder on the ball than the ball pulls on Earth, so the ball falls while Earth remains stationary.

Quick Quiz 7.9

A planet has two moons with identical mass. Moon 1 is in a circular orbit of radius r. Moon 2 is in a circular orbit of radius $2r$. The magnitude of the gravitational force exerted by the planet on Moon 2 is (a) four times as large (b) twice as large (c) the same (d) half as large (e) one-fourth as large as the gravitational force exerted by the planet on Moon 1.

TABLE 7.1

Free-Fall Acceleration g at Various Altitudes

Altitude (km)[a]	g (m/s^2)
1 000	7.33
2 000	5.68
3 000	4.53
4 000	3.70
5 000	3.08
6 000	2.60
7 000	2.23
8 000	1.93
9 000	1.69
10 000	1.49
50 000	0.13

[a]All figures are distances above Earth's surface.

Measurement of the Gravitational Constant

The gravitational constant G in Equation 7.20 was first measured in an important experiment by Henry Cavendish in 1798. His apparatus consisted of two small spheres, each of mass m, fixed to the ends of a light horizontal rod suspended by a thin metal wire, as in Figure 7.18a (see page 208). Two large spheres, each of mass M, were placed near the smaller spheres. The attractive force between the smaller and larger spheres caused the rod to rotate in a horizontal plane and the wire to twist. The angle through which the suspended rod rotated was measured with a light beam reflected from a mirror attached to the vertical suspension. (Such a moving spot of light is an effective technique for amplifying the motion.) The experiment was carefully repeated with different masses at various separations. In addition to providing a value for G, the results showed that the force is attractive, proportional to the product mM, and inversely proportional to the square of the distance r. Modern forms of such experiments are carried out regularly today, in an effort to determine G with greater precision.

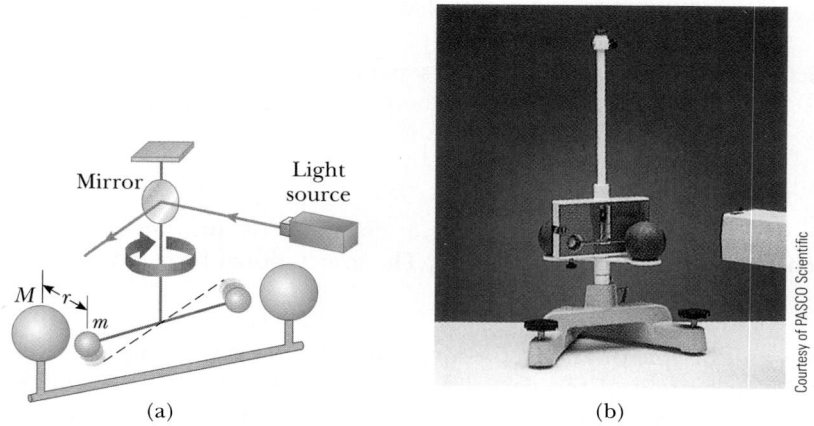

Figure 7.18 (a) A schematic diagram of the Cavendish apparatus for measuring G. The smaller spheres of mass m are attracted to the large spheres of mass M, and the rod rotates through a small angle. A light beam reflected from a mirror on the rotating apparatus measures the angle of rotation. (b) A student Cavendish apparatus.

EXAMPLE 7.10 Billiards, Anyone?

Goal Use vectors to find the net gravitational force on an object.

Problem **(a)** Three 0.300-kg billiard balls are placed on a table at the corners of a right triangle, as shown from overhead in Figure 7.19. Find the net gravitational force on the cue ball (designated as m_1) resulting from the forces exerted by the other two balls. **(b)** Find the components of the gravitational force of m_2 on m_3.

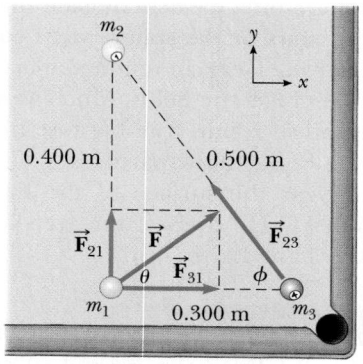

Figure 7.19 (Example 7.10)

Strategy **(a)** To find the net gravitational force on the cue ball of mass m_1, we first calculate the force $\vec{\mathbf{F}}_{21}$ exerted by m_2 on m_1. This force is the y-component of the net force acting on m_1. Then we find the force $\vec{\mathbf{F}}_{31}$ exerted by m_3 on m_1, which is the x-component of the net force acting on m_1. With these two components, we can find the magnitude and direction of the net force on the cue ball. (b) In this case, we must use trigonometry to find the components of the force $\vec{\mathbf{F}}_{23}$.

Solution

(a) Find the net gravitational force on the cue ball. Find the magnitude of the force $\vec{\mathbf{F}}_{21}$ exerted by m_2 on m_1 using the law of gravitation, Equation 7.20:

$$F_{21} = \frac{m_2 m_1}{r_{21}^2}$$

$$= (6.67 \times 10^{-11}\ \mathrm{N \cdot m^2/kg^2}) \frac{(0.300\ \mathrm{kg})(0.300\ \mathrm{kg})}{(0.400\ \mathrm{m})^2}$$

$$F_{21} = 3.75 \times 10^{-11}\ \mathrm{N}$$

Find the magnitude of the force $\vec{\mathbf{F}}_{31}$ exerted by m_3 on m_1, again using Newton's law of gravity:

$$F_{31} = G\frac{m_3 m_1}{r_{31}^2}$$

$$= (6.67 \times 10^{-11}\ \mathrm{N \cdot m^2/kg^2}) \frac{(0.300\ \mathrm{kg})(0.300\ \mathrm{kg})}{(0.300\ \mathrm{m})^2}$$

$$F_{31} = 6.67 \times 10^{-11}\ \mathrm{N}$$

The net force has components $F_x = F_{31}$ and $F_y = F_{21}$. Compute the magnitude of this net force:

$$F = \sqrt{F_x^2 + F_y^2} = \sqrt{(6.67)^2 + (3.75)^2} \times 10^{-11}\ \mathrm{N}$$

$$= 7.65 \times 10^{-11}\ \mathrm{N}$$

Use the inverse tangent to obtain the direction of $\vec{\mathbf{F}}$:

$$\theta = \tan^{-1}\left(\frac{F_y}{F_x}\right) = \tan^{-1}\left(\frac{3.75 \times 10^{-11}\ \mathrm{N}}{6.67 \times 10^{-11}\ \mathrm{N}}\right) = 29.3°$$

(b) Find the components of the force of m_2 on m_3. First, compute the magnitude of $\vec{\mathbf{F}}_{23}$:

$$F_{23} = G\frac{m_2 m_1}{r_{23}^2}$$

$$= (6.67 \times 10^{-11}\ \mathrm{kg^{-1}m^3s^{-2}}) \frac{(0.300\ \mathrm{kg})(0.300\ \mathrm{kg})}{(0.500\ \mathrm{m})^2}$$

$$= 2.40 \times 10^{-11}\ \mathrm{N}$$

To obtain the x- and y-components of F_{23}, we need $\cos\varphi$ and $\sin\varphi$. Use the sides of the large triangle in Figure 7.19:

$$\cos\varphi = \frac{\text{adj}}{\text{hyp}} = \frac{0.300\text{ m}}{0.500\text{ m}} = 0.600$$

$$\sin\varphi = \frac{\text{opp}}{\text{hyp}} = \frac{0.400\text{ m}}{0.500\text{ m}} = 0.800$$

Compute the components of $\vec{\mathbf{F}}_{23}$. A minus sign must be supplied for the x-component, because it's in the negative x-direction.

$$F_{23x} = -F_{23}\cos\varphi = -(2.40\times10^{-11}\text{ N})(0.600)$$
$$= -1.44\times10^{-11}\text{ N}$$
$$F_{23y} = F_{23}\sin\varphi = (2.40\times10^{-11}\text{ N})(0.800)$$
$$= 1.92\times10^{-11}\text{ N}$$

Remarks Notice how small the gravity forces are between everyday objects. Nonetheless, such forces can be measured directly with torsion balances.

Exercise 7.10

Find magnitude and direction of the force exerted by m_1 and m_3 on m_2.

Answers 5.85×10^{-11} N, $-75.8°$

EXAMPLE 7.11 Ceres

Goal Relate Newton's universal law of gravity to mg, and show how g changes with position.

Problem An astronaut standing on the surface of Ceres, the largest asteroid, drops a rock from a height of 10.0 m. It takes 8.06 s to hit the ground. **(a)** Calculate the acceleration of gravity on Ceres. **(b)** Find the mass of Ceres, given that the radius of Ceres is $R_C = 5.10\times10^2$ km. **(c)** Calculate the gravitational acceleration 50.0 km from the surface of Ceres.

Strategy Part (a) is a review of one-dimensional kinematics. In part (b), the weight of an object, $w = mg$, is the same as the magnitude of the force given by the universal law of gravity. Solve for the unknown mass of Ceres, after which the answer for (c) can be found by substitution into the universal law of gravity, Equation 7.20.

Solution

(a) Calculate the acceleration of gravity, g_C, on Ceres.

Apply the kinematics displacement equation to the falling rock:

$$\Delta x = \tfrac{1}{2}at^2 + v_0 t$$

Substitute $\Delta x = -10.0$ m, $v_0 = 0$, $a = -g_C$, and $t = 8.06$ s, and solve for the gravitational acceleration on Ceres, g_C:

$$-10.0\text{ m} = -\tfrac{1}{2}g_C(8.06\text{ s})^2 \quad\rightarrow\quad g_C = 0.308\text{ m/s}^2$$

(b) Find the mass of Ceres.

Equate the weight of the rock on Ceres to the gravitational force acting on the rock:

$$mg_C = G\frac{M_C m}{R_C^2}$$

Solve for the mass of Ceres, M_C:

$$M_C = \frac{g_C R_C^2}{G} = 1.20\times10^{21}\text{ kg}$$

(c) Calculate the acceleration of gravity at a height of 50.0 km above the surface of Ceres.

Equate the weight at 50.0 km to the gravitational force:

$$mg'_C = G\frac{mM_C}{r^2}$$

Cancel m, and substitute the mass of Ceres and $r = 5.60 \times 10^5$ m:

$$g'_C = G\frac{M_C}{r^2}$$

$$= (6.67 \times 10^{-11}\ \text{kg}^{-1}\text{m}^3\text{s}^{-2})\frac{1.20 \times 10^{21}\ \text{kg}}{(5.60 \times 10^5\ \text{m})^2}$$

$$= 0.255\ \text{m/s}^2$$

Remarks This is the standard method of finding the mass of a planetary body: study the motion of a falling (or orbiting) object.

Exercise 7.11

An object takes 2.40 s to fall 5.00 m on a certain planet. (a) Find the acceleration due to gravity on the planet. (b) Find the planet's mass if its radius is 5 250 km.

Solution (a) $1.74\ \text{m/s}^2$ (b) 7.19×10^{23} kg

Gravitational Potential Energy Revisited

In Chapter 5 we introduced the concept of gravitational potential energy and found that the potential energy associated with an object could be calculated from the equation $PE = mgh$, where h is the height of the object above or below some reference level. This equation, however, is valid *only* when the object is near Earth's surface. For objects high above Earth's surface, such as a satellite, an alternative must be used, because g varies with distance from the surface, as shown in Table 7.1.

General form of gravitational potential energy ▶

The gravitational potential energy associated with an object of mass m at a distance r from the center of Earth is

$$PE = -G\frac{M_E m}{r} \qquad [7.21]$$

where M_E and R_E are the mass and radius of Earth, respectively, with $r > R_E$.

SI units: Joules (J)

As before, gravitational potential energy is a property of a system, in this case the object of mass m and Earth. Equation 7.21 is valid for the special case where the zero level for potential energy is at an infinite distance from the center of Earth. Recall that the gravitational potential energy associated with an object is nothing more than the negative of the work done by the force of gravity in moving the object. If an object falls under the force of gravity from a great distance (effectively infinity), the change in gravitational potential energy is negative, which corresponds to a positive amount of gravitational work done on the system. This positive work is equal to the (also positive) change in kinetic energy, as the next example shows.

EXAMPLE 7.12 A Near-Earth Asteroid

Goal Use gravitational potential energy to calculate the work done by gravity on a falling object.

Problem An asteroid with mass $m = 1.00 \times 10^9$ kg comes from deep space, effectively from infinity, and falls toward Earth. **(a)** Find the change in potential energy when it reaches a point 4.00×10^8 m from Earth (just beyond the Moon), assuming it falls from rest at infinity. In addition, find the work done by the force of gravity. **(b)** Calculate the speed of the asteroid at that point. **(c)** How much work would have to be done on the asteroid by some other agent so the asteroid would be traveling at only half the speed found in (b) at the same point?

Strategy Part (a) requires simple substitution into the definition of gravitational potential energy. To find the work done by the force of gravity, recall that the work done on an object by a conservative force is just the negative of the change in potential energy. Part (b) can be solved with conservation of energy, and part (c) is an application of the work–energy theorem.

Solution

(a) Find the change in potential energy and the work done by the force of gravity.

Apply Equation 7.21:

$$\Delta PE = PE_f - PE_i = -\frac{GM_E m}{r_f} - \left(-\frac{GM_E m}{r_i}\right)$$
$$= GM_E m\left(-\frac{1}{r_f} + \frac{1}{r_i}\right)$$

Substitute known quantities. The asteroid's initial position is effectively infinity, so $1/r_i$ is zero.

$$\Delta PE = (6.67 \times 10^{-11}\ \text{kg}^{-1}\text{m}^3/\text{s}^2)(5.98 \times 10^{24}\ \text{kg})$$
$$\times (1.00 \times 10^9\ \text{kg})\left(-\frac{1}{4.00 \times 10^8\ \text{m}} + 0\right)$$
$$\Delta PE = -9.97 \times 10^{14}\ \text{J}$$

Compute the work done by the force of gravity:

$$W_{\text{grav}} = -\Delta PE = 9.97 \times 10^{14}\ \text{J}$$

(b) Find the speed of the asteroid when it reaches $r_f = 4.00 \times 10^8$ m.

Use conservation of energy:

$$\Delta KE + \Delta PE = 0$$
$$(\tfrac{1}{2}mv^2 - 0) - 9.97 \times 10^{14}\ \text{J} = 0$$
$$v = 1.41 \times 10^3\ \text{m/s}$$

(c) Find the work needed to reduce the speed to 7.05×10^2 m/s (half the value just found) at this point.

Apply the work–energy theorem:

$$W = \Delta KE + \Delta PE$$

The change in potential energy remains the same as in part (a), but substitute only half the speed in the kinetic-energy term:

$$W = (\tfrac{1}{2}mv^2 - 0) - 9.97 \times 10^{14}\ \text{J}$$
$$W = \tfrac{1}{2}(1.00 \times 10^9\ \text{kg})(7.05 \times 10^2\ \text{m/s})^2 - 9.97 \times 10^{14}\ \text{J}$$
$$= -7.48 \times 10^{14}\ \text{J}$$

Remark The amount of work calculated in part (c) is negative because an external agent must exert a force against the direction of motion of the asteroid. It would take a thruster with a megawatt of output about 24 years to slow down the asteroid to half its original speed. An asteroid endangering Earth need not be slowed that much: A small change in its speed, if applied early enough, will cause it to miss Earth. Timeliness of the applied thrust, however, is important. By the time you can look over your shoulder and see the Earth, it's already far too late, despite how these scenarios play out in Hollywood. Last minute rescues won't work!

Exercise 7.12

Suppose the asteroid starts from rest at a great distance (effectively infinity), falling toward Earth. How much work would have to be done on the asteroid to slow it to 425 m/s by the time it reached a distance of 2.00×10^8 m from Earth?

Answer -1.90×10^{15} J

Applying Physics 7.3 Why is the Sun Hot?

Explanation The Sun formed when particles in a cloud of gas in coalesced, due to gravitational attraction, into a massive astronomical object. Before this occurred, the particles in the cloud were widely scattered, representing a large amount of gravitational potential energy. As the particles fell closer together, their kinetic energy increased, but the gravitational potential energy of the system decreased, as required by the conservation of energy. With further slow collapse, the cloud became more dense, and the average kinetic energy of the particles increased. This kinetic energy is the internal energy of the cloud, which is proportional to the temperature. If enough particles come together, the temperature can rise to a point at which nuclear fusion occurs, and the ball of gas becomes a star. Otherwise, the temperature may rise, but not enough to ignite fusion reactions, and the object becomes a brown dwarf (a failed star) or a planet.

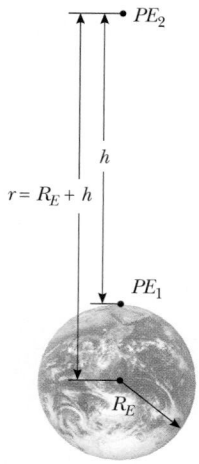

Figure 7.20 Relating the general form of gravitational potential energy to mgh.

On inspecting Equation 7.21, some may wonder what happened to mgh, the gravitational potential energy expression introduced in Chapter 5. That expression is still valid when h is small compared with the Earth's radius. To see this, we write the change in potential energy as an object is raised from the ground to height h, using the general form for gravitational potential energy (see Fig. 7.20):

$$PE_2 - PE_1 = -G\frac{M_E m}{(R_E + h)} - \left(-G\frac{M_E m}{R_E}\right)$$

$$= -GM_E m\left[\frac{1}{(R_E + h)} - \frac{1}{R_E}\right]$$

After finding a common denominator and applying some algebra, we obtain

$$PE_2 - PE_1 = \frac{GM_E mh}{R_E(R_E + h)}$$

When the height h is very small compared with R_E, h can be dropped from the second factor in the denominator, yielding

$$\frac{1}{R_E(R_E + h)} \cong \frac{1}{R_E^2}$$

Substituting this into the previous expression, we have

$$PE_2 - PE_1 \cong \frac{GM_E}{R_E^2}\, mh$$

Now recall from Chapter 4 that the free-fall acceleration at the surface of Earth is given by $g = GM_E/R_E^2$, giving

$$PE_2 - PE_1 \cong mgh$$

Escape Speed

If an object is projected upward from Earth's surface with a large enough speed, it can soar off into space and never return. This speed is called Earth's **escape speed**. (It is also commonly called the *escape velocity*, but in fact is more properly a speed.)

Earth's escape speed can be found by applying conservation of energy. Suppose an object of mass m is projected vertically upward from Earth's surface with an initial speed v_i. The initial mechanical energy (kinetic plus potential energy) of the object–Earth system is given by

$$KE_i + PE_i = \tfrac{1}{2}mv_i^2 - \frac{GM_E m}{R_E}$$

We neglect air resistance and assume that the initial speed is just large enough to allow the object to reach infinity with a speed of zero. This value of v_i is the escape speed v_{esc}. When the object is at an infinite distance from Earth, its kinetic energy is zero, because $v_f = 0$, and the gravitational potential energy is also zero, because $1/r$ goes to zero as r goes to infinity. Hence the total mechanical energy is zero, and the law of conservation of energy gives

$$\tfrac{1}{2}mv_{\text{esc}}^2 - \frac{GM_E m}{R_E} = 0$$

so that

$$v_{\text{esc}} = \sqrt{\frac{2GM_E}{R_E}} \qquad [7.22]$$

The escape speed for Earth is about 11.2 km/s, which corresponds to about 25 000 mi/h. (See Example 7.13.) Note that the expression for v_{esc} doesn't depend on the mass of the object projected from Earth, so a spacecraft has the same escape speed as a molecule. Escape speeds for the planets, the Moon, and the Sun are listed in Table 7.2. Escape speed and temperature determine to a large extent whether a world has an atmosphere and, if so, what the constituents of the atmosphere are. Planets with low escape speeds, such as Mercury, generally don't have atmospheres because the average speed of gas molecules is close to the escape speed. Venus has a very thick atmosphere, but it's almost entirely carbon dioxide, a heavy gas. The atmosphere of Earth has very little hydrogen or helium, but has retained the much heavier nitrogen and oxygen molecules.

TABLE 7.2

Escape Speeds for the Planets and the Moon

Planet	v_e (km/s)
Mercury	4.3
Venus	10.3
Earth	11.2
Moon	2.3
Mars	5.0
Jupiter	60.0
Saturn	36.0
Uranus	22.0
Neptune	24.0
Pluto	1.1

EXAMPLE 7.13 From the Earth to the Moon

Goal Apply conservation of energy with the general form of Newton's universal law of gravity.

Problem In Jules Verne's classic novel, *From the Earth to the Moon*, a giant cannon dug into the Earth in Florida fired a spacecraft all the way to the Moon. **(a)** If the spacecraft leaves the cannon at escape speed, at what speed is it moving when 1.50×10^5 km from the center of Earth? Neglect any friction effects. **(b)** Approximately what constant acceleration is needed to propel the spacecraft to this speed through a cannon bore a kilometer long?

Strategy For part (a), use conservation of energy and solve for the final speed v_f. Part (b) is an application of the time-independent kinematic equation: solve for the acceleration a.

Solution

(a) Find the speed at $r = 1.50 \times 10^5$ km.

Apply conservation of energy:

$$\tfrac{1}{2}mv_i^2 - \frac{GM_E m}{R_E} = \tfrac{1}{2}mv_f^2 - \frac{GM_E m}{r_f}$$

Multiply by $2/m$ and rearrange, solving for v_f^2. Then substitute known values and take the square root.

$$v_f^2 = v_i^2 + \frac{2GM_E}{r_f} - \frac{2GM_E}{R_E} = v_i^2 + 2GM_E\left(\frac{1}{r_f} - \frac{1}{R_E}\right)$$

$$v_f^2 = (1.12 \times 10^4\ \text{m/s})^2 + 2(6.67 \times 10^{-11}\ \text{kg}^{-1}\ \text{m}^3\text{s}^{-2}) \times (5.98 \times 10^{24}\ \text{kg})\left(\frac{1}{1.50 \times 10^8\ \text{m}} - \frac{1}{6.37 \times 10^6\ \text{m}}\right)$$

$$v_f = 2.35 \times 10^3\ \text{m/s}$$

(b) Find the acceleration through the cannon bore, assuming that it's constant.

Use the time-independent kinematics equation:

$$v^2 - v_0^2 = 2a\Delta x$$

$$(1.12 \times 10^4\ \text{m/s})^2 - 0 = 2a(1.00 \times 10^3\ \text{m})$$

$$a = 6.27 \times 10^4\ \text{m/s}^2$$

Remark This result corresponds to an acceleration of over 6 000 times the free-fall acceleration on Earth. Such a huge acceleration is far beyond what the human body can tolerate.

Exercise 7.13

Using the data in Table 7.3 (see page 216), find (a) the escape speed from the surface of Mars and (b) the speed of a space vehicle when it is 1.25×10^7 m from the center of Mars if it leaves the surface at the escape speed.

Answer (a) 5.04×10^3 m/s (b) 2.62×10^3 m/s

7.6 KEPLER'S LAWS

The movements of the planets, stars, and other celestial bodies have been observed for thousands of years. In early history, scientists regarded Earth as the center of the Universe. This **geocentric model** was developed extensively by the Greek astronomer Claudius Ptolemy in the second century A.D. and was accepted for the next 1 400 years. In 1543, the Polish astronomer Nicolaus Copernicus (1473–1543) showed that Earth and the other planets revolve in circular orbits around the Sun (the **heliocentric model**).

The Danish astronomer Tycho Brahe (pronounced Brah or BRAH–huh; 1546–1601) made accurate astronomical measurements over a period of 20 years, providing the data for the currently accepted model of the Solar System. Brahe's precise observations of the planets and 777 stars were carried out with nothing more elaborate than a large sextant and compass; the telescope had not yet been invented.

The German astronomer Johannes Kepler, who was Brahe's assistant, acquired Brahe's astronomical data and spent about 16 years trying to deduce a mathematical model for the motions of the planets. After many laborious calculations, he found that Brahe's precise data on the motion of Mars about the Sun provided the answer. Kepler's analysis first showed that the concept of circular orbits about the Sun had to be abandoned. He eventually discovered that the orbit of Mars could be accurately described by an ellipse with the Sun at one focus. He then generalized this analysis to include the motions of all planets. The complete analysis is summarized in three statements known as **Kepler's laws**:

1. All planets move in elliptical orbits with the Sun at one of the focal points.
2. A line drawn from the Sun to any planet sweeps out equal areas in equal time intervals.
3. The square of the orbital period of any planet is proportional to the cube of the average distance from the planet to the Sun.

Newton later demonstrated that these laws are consequences of the gravitational force that exists between any two objects. Newton's law of universal gravitation, together with his laws of motion, provides the basis for a full mathematical description of the motions of planets and satellites.

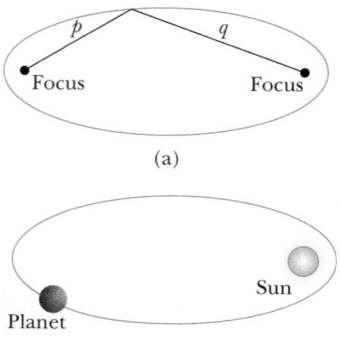

ACTIVE FIGURE 7.21
(a) The sum $p + q$ is the same for every point on the ellipse. (b) In the Solar System, the Sun is at one focus of the elliptical orbit of each planet and the other focus is empty.

PhysicsNow™
Log into PhysicsNow at **www.cp7e.com**, and go to Active Figure 7.21a to move the focal points, or enter values for *a*, *b*, *c*, and *e* and see the resulting elliptical shape.

Kepler's First Law

The first law arises as a natural consequence of the inverse-square nature of Newton's law of gravitation. Any object bound to another by a force that varies as $1/r^2$ will move in an elliptical orbit. As shown in Active Figure 7.21a, an ellipse is a curve drawn so that the sum of the distances from any point on the curve to two internal points called *focal points* or *foci* (singular, *focus*) is always the same. The semimajor axis a is half the length of the line that goes across the ellipse and contains both foci. For the Sun–planet configuration (Active Fig. 7.21b), the Sun is at one focus and the other focus is empty. Because the orbit is an ellipse, the distance from the Sun to the planet continuously changes.

Kepler's Second Law

Kepler's second law states that a line drawn from the Sun to any planet sweeps out equal areas in equal time intervals. Consider a planet in an elliptical orbit about the Sun, as in Figure 7.22. In a given period Δt, the planet moves from point Ⓐ to point Ⓑ. The planet moves more slowly on that side of the orbit because it's farther away from the sun. On the opposite side of its orbit, the planet moves from point Ⓒ to point Ⓓ in the same amount of time, Δt, moving faster because it's closer to the sun. Kepler's second law says that any two wedges formed as in Figure 7.22 will always have the same area. As we will see in Chapter 8, Kepler's second law is related to a physical principle known as conservation of angular momentum.

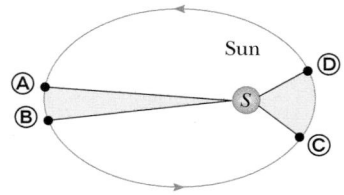

Figure 7.22 The two areas swept out by the planet in its elliptical orbit about the Sun are equal if the time interval between points Ⓐ and Ⓑ is equal to the time interval between points Ⓒ and Ⓓ.

Kepler's Third Law

The derivation of Kepler's third law is simple enough to carry out for the special case of a circular orbit. Consider a planet of mass M_p moving around the Sun, which has a mass of M_S, in a circular orbit. Because the orbit is circular, the planet moves at a constant speed v. Newton's second law, his law of gravitation, and centripetal acceleration then give the following equation:

$$M_p a_c = \frac{M_p v^2}{r} = \frac{G M_S M_p}{r^2}$$

The speed v of the planet in its orbit is equal to the circumference of the orbit divided by the time required for one revolution, T, called the **period** of the planet, so $v = 2\pi r/T$. Substituting, the preceding expression becomes

$$\frac{G M_S}{r^2} = \frac{(2\pi r/T)^2}{r}$$

$$T^2 = \left(\frac{4\pi^2}{G M_S}\right) r^3 = K_S r^3 \qquad [7.23]$$

◀ Kepler's third law

where K_S is a constant given by

$$K_S = \frac{4\pi^2}{G M_S} = 2.97 \times 10^{-19}\ \mathrm{s^2/m^3}$$

Equation 7.23 is Kepler's third law for a circular orbit. The orbits of most of the planets are very nearly circular. Comets and asteroids, however, usually have elliptical orbits. For these orbits, the radius r must be replaced with a, the semimajor axis—half the longest distance across the elliptical orbit. (This is also the average distance of the comet or asteroid from the sun.) A more detailed calculation shows that K_S actually depends on the sum of both the mass of a given planet and the Sun's mass. The masses of the planets, however, are negligible compared with the Sun's mass; hence can be neglected, meaning Equation 7.23 is valid for any planet in the Sun's family. If we consider the orbit of a satellite such as the Moon around Earth, then the constant has a different value, with the mass of the Sun replaced by the mass of Earth. In that case, K_E equals $4\pi^2/GM_E$.

The mass of the Sun can be determined from Kepler's third law, because the constant K_S in Equation 7.23 includes the mass of the Sun and the other variables and constants can be easily measured. The value of this constant can be found by substituting the values of a planet's period and orbital radius and solving for K_S. The mass of the Sun is then

$$M_S = \frac{4\pi^2}{G K_S}$$

This same process can be used to calculate the mass of Earth (by considering the period and orbital radius of the Moon) and the mass of other planets in the Solar System that have satellites.

TABLE 7.3

Useful Planetary Data

Body	Mass (kg)	Mean Radius (m)	Period (s)	Mean Distance from Sun (m)	$\frac{T^2}{r^3}\ 10^{-19}\left(\frac{s^2}{m^3}\right)$
Mercury	3.18×10^{23}	2.43×10^6	7.60×10^6	5.79×10^{10}	2.97
Venus	4.88×10^{24}	6.06×10^6	1.94×10^7	1.08×10^{11}	2.99
Earth	5.98×10^{24}	6.38×10^6	3.156×10^7	1.496×10^{11}	2.97
Mars	6.42×10^{23}	3.37×10^6	5.94×10^7	2.28×10^{11}	2.98
Jupiter	1.90×10^{27}	6.99×10^7	3.74×10^8	7.78×10^{11}	2.97
Saturn	5.68×10^{26}	5.85×10^7	9.35×10^8	1.43×10^{12}	2.99
Uranus	8.68×10^{25}	2.33×10^7	2.64×10^9	2.87×10^{12}	2.95
Neptune	1.03×10^{26}	2.21×10^7	5.22×10^9	4.50×10^{12}	2.99
Pluto	1.27×10^{23}	1.14×10^6	7.82×10^9	5.91×10^{12}	2.96
Moon	7.36×10^{22}	1.74×10^6	—	—	—
Sun	1.991×10^{30}	6.96×10^8	—	—	—

The last column in Table 7.3 confirms the fact that T^2/r^3 is very nearly constant. When time is measured in Earth years and the semimajor axis in astronomical units (1 AU = the distance from the Earth to the Sun), Kepler's law takes the following simple form:

$$T^2 = a^3$$

This equation can be easily checked: The Earth has a semimajor axis of one astronomical unit (by definition), and it takes one year to circle the sun. This equation, of course, is valid only for the sun and its planets, asteroids, and comets.

Quick Quiz 7.10

Suppose an asteroid has a semimajor axis of 4 AU How long does it take the asteroid to go around the sun? (a) 2 years (b) 4 years (c) 6 years (d) 8 years

EXAMPLE 7.14 Geosynchronous Orbit and Telecommunications Satellites

Goal Apply Kepler's third law to an Earth satellite

Problem From a telecommunications point of view, it's advantageous for satellites to remain at the same location relative to a location on the Earth. This can occur only if the satellite's orbital period is the same as the Earth's period of rotation, 24.0 h. **(a)** At what distance from the center of the Earth can this geosynchronous orbit be found? **(b)** What's the orbital speed of the satellite?

Strategy This problem can be solved with the same method that was used to derive a special case of Kepler's third law, with Earth's mass replacing the Sun's mass. There's no need to repeat the analysis, just replace the Sun's mass with Earth's mass in Kepler's third law, substitute the period T (converted to seconds), and solve for r. For part (b), find the circumference of the circular orbit and divide by the elapsed time.

Solution

(a) Find the distance r to geosynchronous orbit.

Apply Kepler's third law:

$$T^2 = \left(\frac{4\pi^2}{GM_E}\right)r^3$$

Substitute the period in seconds, $T = 86\,400$ s, the gravity constant $G = 6.67 \times 10^{-11}\ \text{kg}^{-1}\ \text{m}^3/\text{s}^2$, and the mass of the Earth, $M_E = 5.98 \times 10^{24}$ kg. Solve for r:

$$r = 4.23 \times 10^7\ \text{m}$$

(b) Find the orbital speed:

$$v = \frac{d}{T} = \frac{2\pi r}{T} = \frac{2\pi(4.23 \times 10^7\ \text{m})}{8.64 \times 10^4\ \text{s}} = 3.08 \times 10^3\ \text{m/s}$$

Remarks Both these results are independent of the mass of the satellite. Notice that Earth's mass could be found by substituting the Moon's distance and period into this form of Kepler's third law.

Exercise 7.14

Mars rotates on its axis once every 1.02 days (almost the same as Earth does). (a) Find the distance from Mars at which a satellite would remain in one spot over the Martian surface. (b) Find the speed of the satellite.

Answer (a) 2.03×10^7 m (b) 1.45×10^3 m/s

SUMMARY

PhysicsNow™ Take a practice test by logging into PhysicsNow at **www.cp7e.com** and clicking on the Pre-Test link for this chapter.

7.1 Angular Speed and Angular Acceleration

The **average angular speed** ω_{av} of a rigid object is defined as the ratio of the angular displacement $\Delta\theta$ to the time interval Δt, or

$$\omega_{\text{av}} = \frac{\theta_f - \theta_i}{t_f - t_i} = \frac{\Delta\theta}{\Delta t} \qquad [7.3]$$

where ω_{av} is in radians per second (rad/s).

The **average angular acceleration** α_{av} of a rotating object is defined as the ratio of the change in angular speed $\Delta\omega$ to the time interval Δt, or

$$\alpha_{\text{av}} \equiv \frac{\omega_f - \omega_i}{t_f - t_i} = \frac{\Delta\omega}{\Delta t} \qquad [7.5]$$

where α_{av} is in radians per second per second (rad/s^2).

7.2 Rotational Motion under Constant Angular Acceleration

If an object undergoes rotational motion about a fixed axis under a constant angular acceleration α, its motion can be described with the following set of equations:

$$\omega = \omega_i + \alpha t \qquad [7.7]$$

$$\Delta\theta = \omega_i t + \tfrac{1}{2}\alpha t^2 \qquad [7.8]$$

$$\omega^2 = \omega_i^2 + 2\alpha\Delta\theta \qquad [7.9]$$

Problems are solved as in one-dimensional kinematics.

7.3 Relations between Angular and Linear Quantities

When an object rotates about a fixed axis, the angular speed and angular acceleration are related to the tangential speed and tangential acceleration through the relationships

$$v_t = r\omega \qquad [7.10]$$

and

$$a_t = r\alpha \qquad [7.11]$$

7.4 Centripetal Acceleration

Any object moving in a circular path has an acceleration directed toward the center of the circular path, called a **centripetal acceleration**. Its magnitude is given by

$$a_c = \frac{v^2}{r} = r\omega^2 \qquad [7.13, 7.17]$$

Any object moving in a circular path must have a net force exerted on it that is directed toward the center of the path. Some examples of forces that cause centripetal acceleration are the force of gravity (as in the motion of a satellite) and the force of tension in a string.

7.5 Newtonian Gravitation

Newton's law of universal gravitation states that every particle in the Universe attracts every other particle with a force that is directly proportional to the product of their masses and inversely proportional to the square of the distance r between them:

$$F = G\frac{m_1 m_2}{r^2} \qquad [7.20]$$

where $G = 6.673 \times 10^{-11}\ \text{N}\cdot\text{m}^2/\text{kg}^2$ is the **constant of universal gravitation**. A general expression for gravitational potential energy is

$$PE = -G\frac{M_E m}{r} \qquad [7.21]$$

This expression reduces to $PE = mgh$ close to the surface of Earth and holds for other worlds through replacement of the mass M_E. Problems such as finding the escape velocity from Earth can be solved by using Equation 7.21 in the conservation of energy equation.

7.6 Kepler's Laws

Kepler derived the following three laws of planetary motion:

1. All planets move in elliptical orbits with the Sun at one of the focal points.
2. A line drawn from the Sun to any planet sweeps out equal areas in equal time intervals.
3. The square of the orbital period of a planet is proportional to the cube of the average distance from the planet to the Sun:

$$T^2 = \left(\frac{4\pi^2}{GM_S}\right)r^3 \qquad [7.23]$$

The third law can be applied to any large body and its system of satellites by replacing the Sun's mass with the body's mass. In particular, it can be used to determine the mass of the central body once the average distance to a satellite and its period are known.

CONCEPTUAL QUESTIONS

1. In a race like the Indianapolis 500, a driver circles the track counterclockwise and feels his head pulled toward one shoulder. To relieve his neck muscles from having to hold his head erect, the driver fastens a strap to one wall of the car and the other to his helmet. The length of the strap is adjusted to keep his head vertical. (a) Which shoulder does his head tend to lean toward? (b) What force or forces produce the centripetal acceleration when there is no strap? (c) What force or forces do so when there is a strap?
2. Two schoolmates, Romeo and Juliet, catch each other's eye across a crowded dance floor at a school dance. Find the order of magnitude of the gravitational attraction that Juliet exerts on Romeo and that Romeo exerts on Juliet. State the quantities you take as data and the values you measure or estimate for them.
3. An object executes circular motion with a constant speed whenever a net force of constant magnitude acts perpendicular to its velocity. What happens to the speed if the force is not perpendicular to the velocity?
4. Explain why Earth is not spherical in shape, but bulges at the equator.
5. If a car's wheels are replaced with wheels of greater diameter, will the reading of the speedometer change? Explain.
6. At night, you are farther away from the Sun than during the day. What's more, the force exerted by the Sun on you is downward into Earth at night, and upward into the sky during the day. If you had a sensitive enough bathroom scale, would you appear to weigh more at night than during the day?
7. Correct the following statement: "The race car rounds the turn at a constant velocity of 90 miles per hour."
8. Why does an astronaut in a spacecraft orbiting Earth experience a feeling of weightlessness?
9. Explain why it's easier to determine the mass of a planet when it has a moon.
10. Because of Earth's rotation about its axis, you weigh slightly less at the equator than at the poles. Why?
11. It has been suggested that rotating cylinders about 10 miles long and 5 miles in diameter be placed in space for colonies. The purpose of their rotation is to simulate gravity for the inhabitants. Explain the concept behind this proposal.
12. Describe the path of a moving object in the event that the object's acceleration is constant in magnitude at all times and (a) perpendicular to its velocity; (b) parallel to its velocity.
13. A pail of water can be whirled in a vertical circular path such that no water is spilled. Why does the water remain in the pail, even when the pail is upside down above your head?
14. The orbital planes of all the planets must pass through the center of the Sun. Why?
15. Is it possible for a car to move in a circular path in such a way that it has a tangential acceleration but no centripetal acceleration?
16. Use Kepler's second law to convince yourself that Earth must move faster in its orbit during the northern-hemisphere winter, when it is closest to the Sun, than during the summer, when it is farthest from the Sun.
17. If the mass of Earth were doubled at the same time its radius were doubled, the free-fall acceleration would (a) increase, (b) decrease, or (c) stay the same?
18. A satellite in orbit is not truly traveling through a vacuum—it's moving through very thin air. Does the resulting air friction cause the satellite to slow down?

PROBLEMS

1, 2, 3 = straightforward, intermediate, challenging □ = full solution available in *Student Solutions Manual/Study Guide*
PhysicsNow™ = coached problem with hints available at **www.cp7e.com** = biomedical application

Section 7.1 Angular Speed and Angular Acceleration

1. The tires on a new compact car have a diameter of 2.0 ft and are warranted for 60 000 miles. (a) Determine the angle (in radians) through which one of these tires will rotate during the warranty period. (b) How many revolutions of the tire are equivalent to your answer in (a)?
2. A wheel has a radius of 4.1 m. How far (path length) does a point on the circumference travel if the wheel is rotated through angles of 30°, 30 rad, and 30 rev, respectively?
3. Find the angular speed of Earth about the Sun in radians per second and degrees per day.

4. A potter's wheel moves from rest to an angular speed of 0.20 rev/s in 30 s. Find its angular acceleration in radians per second per second.

Section 7.2 Rotational Motion under Constant Angular Acceleration

Section 7.3 Relations between Angular and Linear Quantities

5. A dentist's drill starts from rest. After 3.20 s of constant angular acceleration, it turns at a rate of 2.51×10^4 rev/min. (a) Find the drill's angular acceleration. (b) Determine the angle (in radians) through which the drill rotates during this period.

6. A centrifuge in a medical laboratory rotates at an angular speed of 3 600 rev/min. When switched off, it rotates through 50.0 revolutions before coming to rest. Find the constant angular acceleration of the centrifuge.

7. A machine part rotates at an angular speed of 0.60 rad/s; its speed is then increased to 2.2 rad/s at an angular acceleration of 0.70 rad/s^2. Find the angle through which the part rotates before reaching this final speed.

8. A tire placed on a balancing machine in a service station starts from rest and turns through 4.7 revolutions in 1.2 s before reaching its final angular speed. Calculate its angular acceleration.

9. **Physics Now™** The diameters of the main rotor and tail rotor of a single-engine helicopter are 7.60 m and 1.02 m, respectively. The respective rotational speeds are 450 rev/min and 4 138 rev/min. Calculate the speeds of the tips of both rotors. Compare these speeds with the speed of sound, 343 m/s.

10. The tub of a washer goes into its spin-dry cycle, starting from rest and reaching an angular speed of 5.0 rev/s in 8.0 s. At this point, the person doing the laundry opens the lid, and a safety switch turns off the washer. The tub slows to rest in 12.0 s. Through how many revolutions does the tub turn during the entire 20-s interval? Assume constant angular acceleration while it is starting and stopping.

11. A standard cassette tape is placed in a standard cassette player. Each side lasts for 30 minutes. The two tape wheels of the cassette fit onto two spindles in the player. Suppose that a motor drives one spindle at constant angular velocity of approximately 1 rad/s and the other spindle is free to rotate at any angular speed. Find the order of magnitude of the tape's thickness. Specify any other quantities you estimate and the values you take for them.

12. A coin with a diameter of 2.40 cm is dropped on edge onto a horizontal surface. The coin starts out with an initial angular speed of 18.0 rad/s and rolls in a straight line without slipping. If the rotation slows with an angular acceleration of magnitude 1.90 rad/s^2, how far does the coin roll before coming to rest?

13. A rotating wheel requires 3.00 s to rotate 37.0 revolutions. Its angular velocity at the end of the 3.00-s interval is 98.0 rad/s. What is the constant angular acceleration of the wheel?

Section 7.4 Centripetal Acceleration

14. It has been suggested that rotating cylinders about 10 mi long and 5.0 mi in diameter be placed in space and used as colonies. What angular speed must such a cylinder have so that the centripetal acceleration at its surface equals the free-fall acceleration on Earth?

15. Find the centripetal accelerations of (a) a point on the equator of Earth and (b) the North Pole, due to the rotation of Earth about its axis.

16. A tire 2.00 ft in diameter is placed on a balancing machine, where it is spun so that its tread is moving at a constant speed of 60.0 mi/h. A small stone is stuck in the tread of the tire. What is the acceleration of the stone as the tire is being balanced?

17. (a) What is the tangential acceleration of a bug on the rim of a 10-in.-diameter disk if the disk moves from rest to an angular speed of 78 rev/min in 3.0 s? (b) When the disk is at its final speed, what is the tangential velocity of the bug? (c) One second after the bug starts from rest, what are its tangential acceleration, centripetal acceleration, and total acceleration?

18. A race car starts from rest on a circular track of radius 400 m. The car's speed increases at the constant rate of 0.500 m/s^2. At the point where the magnitudes of the centripetal and tangential accelerations are equal, determine (a) the speed of the race car, (b) the distance traveled, and (c) the elapsed time.

19. A 55.0-kg ice-skater is moving at 4.00 m/s when she grabs the loose end of a rope, the opposite end of which is tied to a pole. She then moves in a circle of radius 0.800 m around the pole. (a) Determine the force exerted by the horizontal rope on her arms. (b) Compare this force with her weight.

20. A sample of blood is placed in a centrifuge of radius 15.0 cm. The mass of a red blood cell is 3.0×10^{-16} kg, and the magnitude of the force acting on it as it settles out of the plasma is 4.0×10^{-11} N. At how many revolutions per second should the centrifuge be operated?

21. A certain light truck can go around a flat curve having a radius of 150 m with a maximum speed of 32.0 m/s. With what maximum speed can it go around a curve having a radius of 75.0 m?

22. The *cornering performance* of an automobile is evaluated on a skid pad, where the maximum speed that a car can maintain around a circular path on a dry, flat surface is measured. Then the centripetal acceleration, also called the lateral acceleration, is calculated as a multiple of the free-fall acceleration g. The main factors affecting the performance of the car are its tire characteristics and suspension system. A Dodge Viper GTS can negotiate a skid pad of radius 61.0 m at 86.5 km/h. Calculate its maximum lateral acceleration.

23. **Physics Now™** A 50.0-kg child stands at the rim of a merry-go-round of radius 2.00 m, rotating with an angular speed of 3.00 rad/s. (a) What is the child's centripetal acceleration? (b) What is the minimum force between her feet and the floor of the carousel that is required to keep her in the circular path? (c) What minimum coefficient of static friction is required? Is the answer you found reasonable? In other words, is she likely to stay on the merry-go-round?

24. An engineer wishes to design a curved exit ramp for a toll road in such a way that a car will not have to rely on friction to round the curve without skidding. He does so by banking the road in such a way that the force causing the

centripetal acceleration will be supplied by the component of the normal force toward the center of the circular path. (a) Show that, for a given speed v and a radius r, the curve must be banked at the angle θ such that $\tan \theta = v^2/rg$. (b) Find the angle at which the curve should be banked if a typical car rounds it at a 50.0-m radius and a speed of 13.4 m/s.

25. An air puck of mass 0.25 kg is tied to a string and allowed to revolve in a circle of radius 1.0 m on a frictionless horizontal table. The other end of the string passes through a hole in the center of the table, and a mass of 1.0 kg is tied to it (Fig. P7.25). The suspended mass remains in equilibrium while the puck on the tabletop revolves. (a) What is the tension in the string? (b) What is the horizontal force acting on the puck? (c) What is the speed of the puck?

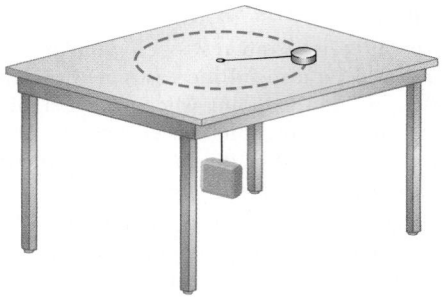

Figure P7.25

26. Tarzan ($m = 85$ kg) tries to cross a river by swinging from a 10-m-long vine. His speed at the bottom of the swing (as he just clears the water) is 8.0 m/s. Tarzan doesn't know that the vine has a breaking strength of 1 000 N. Does he make it safely across the river? Justify your answer.

27. A 40.0-kg child takes a ride on a Ferris wheel that rotates four times each minute and has a diameter of 18.0 m. (a) What is the centripetal acceleration of the child? (b) What force (magnitude and direction) does the seat exert on the child at the lowest point of the ride? (c) What force does the seat exert on the child at the highest point of the ride? (d) What force does the seat exert on the child when the child is halfway between the top and bottom?

28. A roller-coaster vehicle has a mass of 500 kg when fully loaded with passengers (Fig. P7.28). (a) If the vehicle has a speed of 20.0 m/s at point Ⓐ, what is the force of the track on the vehicle at this point? (b) What is the maximum speed the vehicle can have at point Ⓑ in order for gravity to hold it on the track?

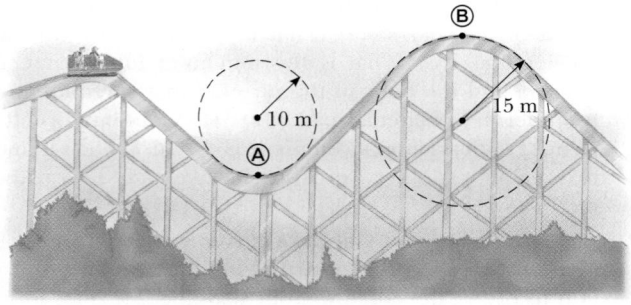

Figure P7.28

Section 7.5 Newtonian Gravitation

29. The average distance separating Earth and the Moon is 384 000 km. Use the data in Table 7.3 to find the net gravitational force exerted by Earth and the Moon on a 3.00×10^4-kg spaceship located halfway between them.

30. During a solar eclipse, the Moon, Earth, and Sun all lie on the same line, with the Moon between Earth and the Sun. (a) What force is exerted by the Sun on the Moon? (b) What force is exerted by Earth on the Moon? (c) What force is exerted by the Sun on Earth? (See Table 7.3 and Problem 29.)

31. A coordinate system (in meters) is constructed on the surface of a pool table, and three objects are placed on the table as follows: a 2.0-kg object at the origin of the coordinate system, a 3.0-kg object at (0, 2.0), and a 4.0-kg object at (4.0, 0). Find the resultant gravitational force exerted by the other two objects on the object at the origin.

32. Use the data of Table 7.3 to find the point between Earth and the Sun at which an object can be placed so that the net gravitational force exerted by Earth and the Sun on that object is zero.

33. Objects with masses of 200 kg and 500 kg are separated by 0.400 m. (a) Find the net gravitational force exerted by these objects on a 50.0-kg object placed midway between them. (b) At what position (other than infinitely remote ones) can the 50.0-kg object be placed so as to experience a net force of zero?

34. Two objects attract each other with a gravitational force of magnitude 1.00×10^{-8} N when separated by 20.0 cm. If the total mass of the objects is 5.00 kg, what is the mass of each?

Section 7.6 Kepler's Laws

35. Physics Now™ A satellite moves in a circular orbit around Earth at a speed of 5 000 m/s. Determine (a) the satellite's altitude above the surface of Earth and (b) the period of the satellite's orbit.

36. A 600-kg satellite is in a circular orbit about Earth at a height above Earth equal to Earth's mean radius. Find (a) the satellite's orbital speed, (b) the period of its revolution, and (c) the gravitational force acting on it.

37. Io, a satellite of Jupiter, has an orbital period of 1.77 days and an orbital radius of 4.22×10^5 km. From these data, determine the mass of Jupiter.

38. A satellite has a mass of 100 kg and is located at 2.00×10^6 m above the surface of Earth. (a) What is the potential energy associated with the satellite at this location? (b) What is the magnitude of the gravitational force on the satellite?

39. A satellite of mass 200 kg is launched from a site on Earth's equator into an orbit 200 km above the surface of Earth. (a) Assuming a circular orbit, what is the orbital period of this satellite? (b) What is the satellite's speed in it's orbit? (c) What is the minimum energy necessary to place the satellite in orbit, assuming no air friction?

ADDITIONAL PROBLEMS

40. Neutron stars are extremely dense objects that are formed from the remnants of supernova explosions. Many rotate very rapidly. Suppose that the mass of a certain spherical

neutron star is twice the mass of the sun and its radius is 10.0 km. Determine the greatest possible angular speed the neutron star can have so that the matter at its surface on the equator is just held in orbit by the gravitational force.

41. One method of pitching a softball is called the "windmill" delivery method, in which the pitcher's arm rotates through approximately 360° in a vertical plane before the 198-gram ball is released at the lowest point of the circular motion. An experienced pitcher can throw a ball with a speed of 98.0 mi/h. Assume that the angular acceleration is uniform throughout the pitching motion, and take the distance between the softball and the shoulder joint to be 74.2 cm. (a) Determine the angular speed of the arm in rev/s at the instant of release. (b) Find the value of the angular acceleration in rev/s^2 and the radial and tangential acceleration of the ball just before it is released. (c) Determine the force exerted on the ball by the pitcher's hand (both radial and tangential components) just before it is released.

42. The Mars probe *Pathfinder* is designed to drop the vehicle's instrument package from a height of 20 meters above the Martian surface, after the speed of the probe has been brought to zero by a combination parachute–rocket system at that height. To cushion the landing, giant air bags surround the package. The mass of Mars is 0.107 4 times that of Earth, and the radius of Mars is 0.528 2 that of Earth. Find (a) the acceleration due to gravity at the surface of Mars and (b) how long it takes for the instrument package to fall the last 20 meters.

43. An athlete swings a 5.00-kg ball horizontally on the end of a rope. The ball moves in a circle of radius 0.800 m at an angular speed of 0.500 rev/s. What are (a) the tangential speed of the ball and (b) its centripetal acceleration? (c) If the maximum tension the rope can withstand before breaking is 100 N, what is the maximum tangential speed the ball can have?

44. A digital audio compact disc carries data along a continuous spiral track from the inner circumference of the disc to the outside edge. Each bit occupies 0.6 μm of the track. A CD player turns the disc to carry the track counterclockwise above a lens at a constant speed of 1.30 m/s. Find the required angular speed (a) at the beginning of the recording, where the spiral has a radius of 2.30 cm, and (b) at the end of the recording, where the spiral has a radius of 5.80 cm. (c) A full-length recording lasts for 74 min, 33 s. Find the average angular acceleration of the disc. (d) Assuming that the acceleration is constant, find the total angular displacement of the disc as it plays. (e) Find the total length of the track.

45. The Solar Maximum Mission Satellite was placed in a circular orbit about 150 mi above Earth. Determine (a) the orbital speed of the satellite and (b) the time required for one complete revolution.

46. A car rounds a banked curve where the radius of curvature of the road is R, the banking angle is θ, and the coefficient of static friction is μ. (a) Determine the range of speeds the car can have without slipping up or down the road. (b) What is the range of speeds possible if $R = 100$ m, $\theta = 10°$, and $\mu = 0.10$ (slippery conditions)?

47. **Physics Now™** A car moves at speed v across a bridge made in the shape of a circular arc of radius r. (a) Find an expression for the normal force acting on the car when it is at the top of the arc. (b) At what minimum speed will the normal force become zero (causing the occupants of the car to seem weightless) if $r = 30.0$ m?

48. A 0.400-kg pendulum bob passes through the lowest part of its path at a speed of 3.00 m/s. (a) What is the tension in the pendulum cable at this point if the pendulum is 80.0 cm long? (b) When the pendulum reaches its highest point, what angle does the cable make with the vertical? (c) What is the tension in the pendulum cable when the pendulum reaches its highest point?

49. Because of Earth's rotation about its axis, a point on the equator has a centripetal acceleration of 0.034 0 m/s^2, while a point at the poles has no centripetal acceleration. (a) Show that, at the equator, the gravitational force on an object (the object's true weight) must exceed the object's apparent weight. (b) What are the apparent weights of a 75.0-kg person at the equator and at the poles? (Assume Earth is a uniform sphere, and take $g = 9.800$ m/s^2.)

50. A stuntman whose mass is 70 kg swings from the end of a 4.0-m-long rope along the arc of a vertical circle. Assuming that he starts from rest when the rope is horizontal, find the tensions in the rope that are required to make him follow his circular path (a) at the beginning of his motion, (b) at a height of 1.5 m above the bottom of the circular arc, and (c) at the bottom of the arc.

51. In a popular amusement park ride, a rotating cylinder of radius 3.00 m is set in rotation at an angular speed of 5.00 rad/s, as in Figure P7.51. The floor then drops away, leaving the riders suspended against the wall in a vertical position. What minimum coefficient of friction between a rider's clothing and the wall is needed to keep the rider from slipping? (*Hint:* Recall that the magnitude of the maximum force of static friction is equal to μn, where n is the normal force—in this case, the force causing the centripetal acceleration.)

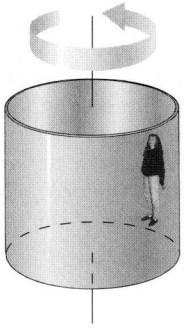

Figure P7.51

52. A 0.50-kg ball that is tied to the end of a 1.5-m light cord is revolved in a horizontal plane, with the cord making a 30° angle with the vertical. (See Fig. P7.52.) (a) Determine the ball's speed. (b) If, instead, the ball is revolved so that its speed is 4.0 m/s, what angle does the cord make with the vertical? (c) If the cord can withstand a maximum tension of 9.8 N, what is the highest speed at which the ball can move?

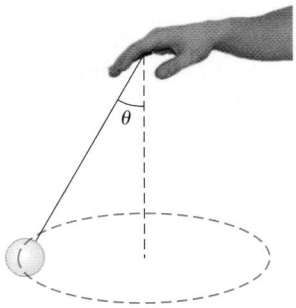

Figure P7.52

53. A skier starts at rest at the top of a large hemispherical hill (Fig. P7.53). Neglecting friction, show that the skier will leave the hill and become airborne at a distance $h = R/3$ below the top of the hill. (*Hint:* At this point, the normal force goes to zero.)

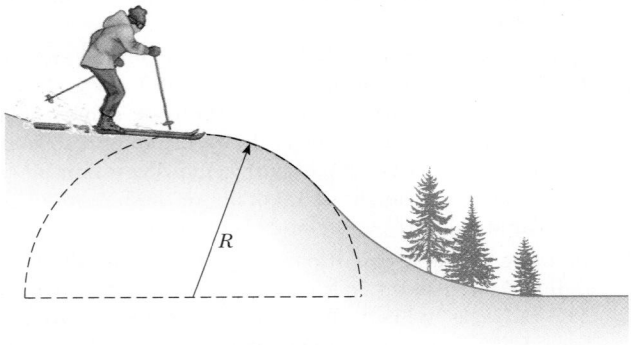

Figure P7.53

54. After consuming all of its nuclear fuel, a massive star can collapse to form a black hole, which is an immensely dense object whose escape speed is greater than the speed of light. Newton's law of universal gravitation still describes the force that a black hole exerts on objects outside it. A spacecraft in the shape of a long cylinder has a length of 100 m, and its mass with occupants is 1 000 kg. It has strayed too close to a black hole having a mass 100 times that of the Sun (Figure P7.54). If the nose of the spacecraft points toward the center of the black hole, and if the distance between the nose of the spacecraft and the black hole's center is 10 km, (a) determine the total force on the spacecraft. (b) What is the difference in the force per kilogram of mass felt by the occupants in the nose of the ship versus those in the rear of the ship farthest from the black hole?

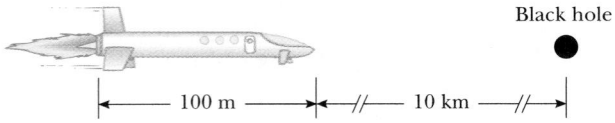

Figure P7.54

55. In Robert Heinlein's *The Moon Is a Harsh Mistress*, the colonial inhabitants of the Moon threaten to launch rocks down onto Earth if they are not given independence (or at least representation). Assuming that a gun could launch a rock of mass m at twice the lunar escape speed, calculate the speed of the rock as it enters Earth's atmosphere.

56. Show that the escape speed from the surface of a planet of uniform density is directly proportional to the radius of the planet.

57. A massless spring of constant $k = 78.4$ N/m is fixed on the left side of a level track. A block of mass $m = 0.50$ kg is pressed against the spring and compresses it a distance d, as in Figure P7.57. The block (initially at rest) is then released and travels toward a circular loop-the-loop of radius $R = 1.5$ m. The entire track and the loop-the-loop are frictionless, except for the section of track between points A and B. Given that the coefficient of kinetic friction between the block and the track along AB is $\mu_k = 0.30$, and that the length of AB is 2.5 m, determine the minimum compression d of the spring that enables the block to just make it through the loop-the-loop at point C. (*Hint:* The force exerted by the track on the block will be zero if the block barely makes it through the loop-the-loop.)

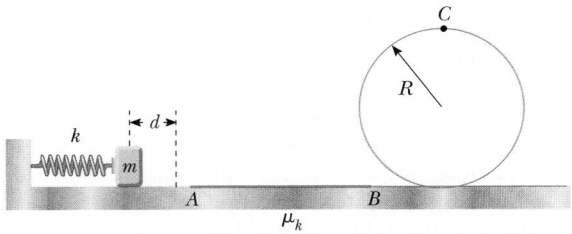

Figure P7.57

58. A small block of mass $m = 0.50$ kg is fired with an initial speed of $v_0 = 4.0$ m/s along a horizontal section of frictionless track, as shown in the top portion of Figure P7.58. The block then moves along the frictionless, semicircular, *vertical* tracks of radius $R = 1.5$ m. (a) Determine the force exerted by the track on the block at points Ⓐ and Ⓑ. (b) The bottom of the track consists of a section ($L = 0.40$ m) with friction. Determine the coefficient of kinetic friction between the block and that portion of the bottom track if the block just makes it to point Ⓒ on the first trip. (*Hint:* If the block just makes it to point Ⓒ, the force of contact exerted by the track on the block at that point is zero.)

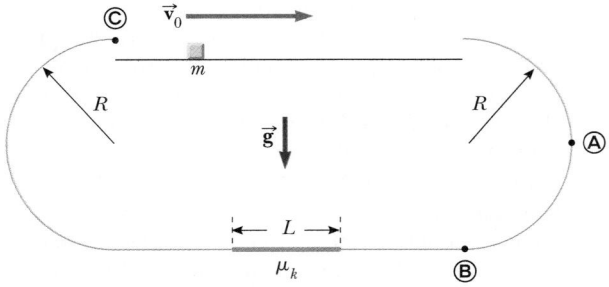

Figure P7.58

59. A frictionless roller coaster is given an initial speed v_0 at height h, as in Figure P7.59. The radius of curvature of the track at point Ⓐ is R. (a) Find the maximum value of v_0 so that the roller coaster stays on the track at Ⓐ solely because of gravity. (b) Using the value of v_0 calculated in (a), determine the value of h' that is necessary if the roller coaster just makes it to point Ⓑ.

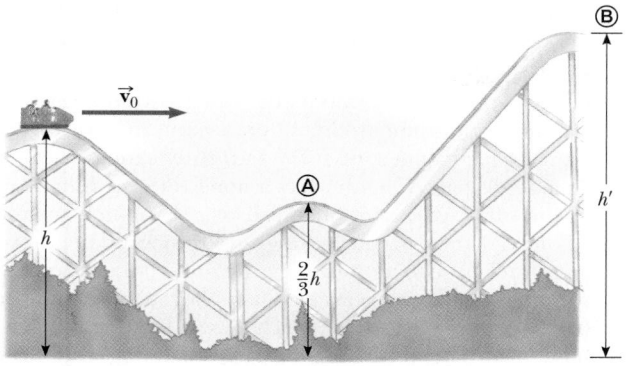

Figure P7.59

60. A roller coaster travels in a circular path. (a) Identify the forces on a passenger at the top of the circular loop that cause centripetal acceleration. Show the direction of all forces in a sketch. (b) Identify the forces on the passenger at the bottom of the loop that produce centripetal acceleration. Show these in a sketch. (c) Based on your answers to (a) and (b), at what point, top or bottom, should the seat exert the greatest force on the passenger? (d) Assume the speed of the roller coaster is 4.00 m/s at the top of the loop of radius 8.00 m. Find the force exerted by the seat on a 70.0-kg passenger at the top of the loop. Then, assume the speed remains the same at the bottom of the loop, and find the force exerted by the seat on the passenger at this point. Are your answers consistent with your choice of answers for (a) and (b)?

61. Assume that you are agile enough to run across a horizontal surface at 8.50 m/s, independently of the value of the gravitational field. What would be (a) the radius and (b) the mass of an airless spherical asteroid of uniform density 1.10×10^3 kg/m^3 on which you could launch yourself into orbit by running? (c) What would be your period?

62. Figure P7.62 shows the elliptical orbit of a spacecraft around Earth. Take the origin of your coordinate system to be at the center of Earth.

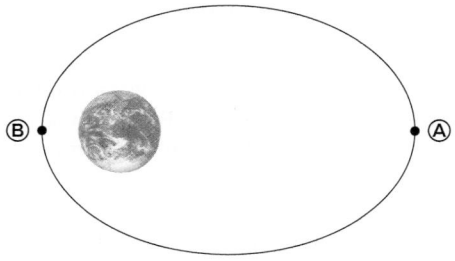

Figure P7.62

(a) On a copy of the figure (enlarged if necessary), draw vectors representing
(i) the position of the spacecraft when it is at Ⓐ and Ⓑ;
(ii) the velocity of the spacecraft when it is at Ⓐ and Ⓑ;
(iii) the acceleration of the spacecraft when it is at Ⓐ and Ⓑ.

Make sure that each type of vector can be distinguished. Provide a legend that shows how each type is represented.

(b) Have you drawn the velocity vector at Ⓐ longer than, shorter than, or the same length as the one at Ⓑ? Explain. Have you drawn the acceleration vector at Ⓐ longer than, shorter than, or the same length as the one at Ⓑ? Explain. (Problem 62 is courtesy of E. F. Redish. For more problems of this type, visit www.physics.umd.edu/perg/)

63. In a home laundry drier, a cylindrical tub containing wet clothes rotates steadily about a horizontal axis, as in Figure P7.63. The clothes are made to tumble so that they will dry uniformly. The rate of rotation of the smooth-walled tub is chosen so that a small piece of cloth loses contact with the tub when the cloth is at an angle of 68.0° above the horizontal. If the radius of the tub is 0.330 m, what rate of revolution is needed?

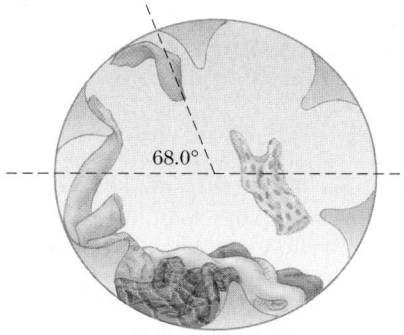

Figure P7.63

64. Casting of molten metal is important in many industrial processes. *Centrifugal casting* is used for manufacturing pipes, bearings, and many other structures. A cylindrical enclosure is rotated rapidly and steadily about a horizontal axis, as in Figure P7.64. Molten metal is poured into the rotating cylinder and then cooled, forming the finished product. Turning the cylinder at a high rotation rate forces the solidifying metal strongly to the outside. Any bubbles are displaced toward the axis so that unwanted voids will not be present in the casting.

Suppose that a copper sleeve of inner radius 2.10 cm and outer radius 2.20 cm is to be cast. To eliminate bubbles and give high structural integrity, the centripetal acceleration of each bit of metal should be $100g$. What rate of rotation is required? State the answer in revolutions per minute.

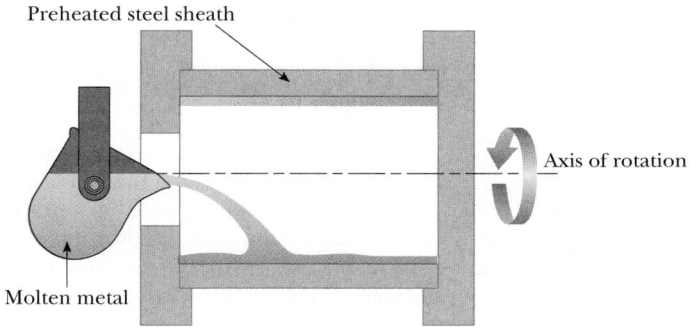

Figure P7.64

65. Suppose that a 1 800-kg car passes over a bump in a roadway that follows the arc of a circle of radius 20.4 m, as in Figure P7.65. (a) What force does the road exert on the

car as the car passes the highest point of the bump if the car travels at 8.94 m/s? (b) What is the maximum speed the car can have without losing contact with the road as it passes this highest point?

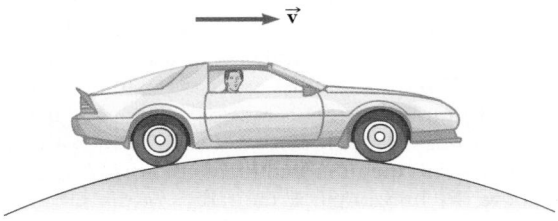

Figure P7.65

66. One popular design of a household juice machine is a conical, perforated stainless-steel basket 3.30 cm high, with a closed bottom of diameter 8.00 cm and an open top of diameter 13.70 cm, that spins at 20 000 revolutions per minute about a vertical axis (Fig. P7.66). Solid pieces of fruit or vegetables are chopped into granules by cutters on the bottom of the spinning cone. The dry pulp is ejected from the top of the cone. After passing through the perforations on the surface of the cone, the juice is collected in an enclosure immediately surrounding the cone. (a) What centripetal acceleration does a bit of fruit experience as it spins with the basket at a point midway between the top and bottom of the basket? Express the answer as a multiple of *g*. (b) Observe that the weight of the fruit is a negligible force. What is the normal force on 2.00 g of fruit at the midway point? (c) If the coefficient of kinetic friction between the fruit and the cone is 0.600, with what acceleration relative to the cone will the bit of fruit start to slide up the wall of the cone at that point, after being temporarily stuck?

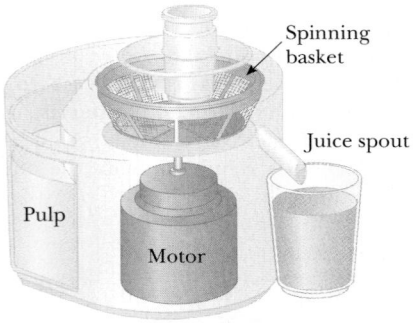

Figure P7.66

67. (a) A luggage carousel at an airport has the form of a section of a large cone, steadily rotating about its vertical axis. Its metallic surface slopes downward toward the outside, making an angle of 20.0° with the horizontal. A 30.0-kg piece of luggage is placed on the carousel, 7.46 m from the axis of rotation. The travel bag goes around once in 38.0 s. Calculate the force of static friction between the bag and the carousel. (b) The drive motor is shifted to turn the carousel at a higher constant rate of rotation, and the piece of luggage is bumped to a position 7.94 m from the axis of rotation. The bag is on the verge of slipping as it goes around once every 34.0 s. Calculate the coefficient of static friction between the bag and the carousel.

68. A merry-go-round is stationary. A dog is running on the ground just outside its circumference, moving with a constant angular speed of 0.750 rad/s. The dog does not change his pace when he sees a bone resting on the edge of the merry-go-round one-third of a revolution in front of him. At the instant the dog sees the bone, the merry-go-round begins to move in the direction the dog is running, with a constant angular acceleration of 0.015 0 rad/s^2. (a) After what time will the dog reach the bone? (b) The confused dog keeps running and passes the bone. How long after the merry-go-round starts to turn do the dog and the bone draw even with each other for the second time?

69. Figure P7.69 shows the drive train of a bicycle that has wheels 67.3 cm in diameter and pedal cranks 17.5 cm long. The cyclist pedals at a steady angular rate of 76.0 rev/min. The chain engages with a front sprocket 15.2 cm in diameter and a rear sprocket 7.00 cm in diameter. (a) Calculate the speed of a link of the chain relative to the bicycle frame. (b) Calculate the angular speed of the bicycle wheels. (c) Calculate the speed of the bicycle relative to the road. (d) What pieces of data, if any, are not necessary for the calculations?

Figure P7.69

70. The maximum lift force on a bat is proportional to the square of its flying speed v. For the hoary bat (*Lasiurus cinereus*), the magnitude of the lift force is given by

$$F_L \leq (0.018\ \mathrm{N \cdot s^2/m^2})v^2$$

The bat can fly in a horizontal circle by "banking" its wings at an angle θ, as shown in Figure P7.70. In this situation, the magnitude of the vertical component of the lift force must equal the bat's weight. The horizontal component of the force provides the centripetal acceleration. (a) What is the minimum speed that the bat can have if its mass is 0.031 kg? (b) If the maximum speed of the bat is 10 m/s, what is the maximum banking angle that allows the bat to stay in a horizontal plane? (c) What is the radius of the circle of its flight when the bat flies at its maximum speed? (d) Can the bat turn with a smaller radius by flying more slowly?

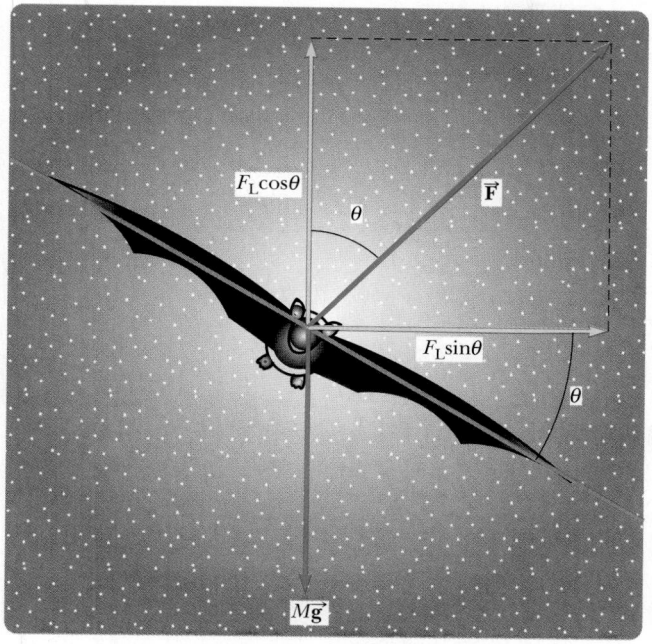

Figure P7.70

71. (a) Find the acceleration due to gravity at the surface of a neutron star of mass 1.5 solar masses and having a radius of 10.0 km. (b) Find the weight of a 0.120-kg baseball on this star. (c) Assume the equation $PE = mgh$ applies, and calculate the energy that a 70.0-kg person would expend climbing a 1.00-cm-tall mountain on the neutron star.

ACTIVITIES

A.1. Experiment with a bicycle wheel like the one in Figure 7.4. Wrap tape around the middle of one of the spokes so that you can easily see it, and draw a chalk mark on a point on the rim at the end of the spoke. Make the tire rotate by gently and steadily turning the crank by hand. Compare the linear speed of the tape with that of the chalk mark. Why do these two points move at different speeds? What happens to the speed of the tape as you slide it closer to the center of the wheel?

With the tape at a given distance from the center, measure the time it takes for the tape to make five rotations. From this time, find the angular speed and the linear speed of the wheel. Repeat at least five times and average your results. Move the tape to a different point and repeat the observations.

A.2. Tie a tennis ball to a string having a length of about 1.0 m. At a safe distance from other students, whirl the ball in a horizontal circle. Note the increasing tension in the string as you whirl it faster. Now whirl the ball in a vertical circle, and observe the difference in tension at the top and at the bottom of the path. Why is there such a difference? While whirling the ball, release it and observe that it flies off in a direction tangent to its circular path.

A.3. On a clear night, spend a few hours outside observing the stars. Choose a pattern of stars toward the north and others toward the east, the south, and the west. As time passes, notice how these patterns move. Obtain a star chart of the constellations from your instructor and also a list of the planets that are visible on your date of observation. Locate as many of these constellations and stars as possible.

Rotational motion is key in the harnessing of energy for power and propulsion, as illustrated by a steamboat. The rotating paddlewheel, driven by a steam engine, pushes water backwards, and the reaction force of the water thrusts the boat forward.

© Kevin Fleming/Corbis

CHAPTER

8

OUTLINE

8.1 Torque

8.2 Torque and the Two Conditions for Equilibrium

8.3 The Center of Gravity

8.4 Examples of Objects in Equilibrium

8.5 Relationship between Torque and Angular Acceleration

8.6 Rotational Kinetic Energy

8.7 Angular Momentum

Rotational Equilibrium and Rotational Dynamics

In the study of linear motion, objects were treated as point particles without structure. It didn't matter *where* a force was applied, only *whether* it was applied or not.

The reality is that the point of application of a force *does* matter. In football, for example, if the ball carrier is tackled near his midriff, he might carry the tackler several yards before falling. If tackled well below the waistline, however, his center of mass rotates toward the ground, and he can be brought down immediately. Tennis provides another good example. If a tennis ball is struck with a strong horizontal force acting through its center of mass, it may travel a long distance before hitting the ground, far out of bounds. Instead, the same force applied in an upward, glancing stroke will impart topspin to the ball, which can cause it to land in the opponent's court.

The concepts of rotational equilibrium and rotational dynamics are also important in other disciplines. For example, students of architecture benefit from understanding the forces that act on buildings and biology students should understand the forces at work in muscles and on bones and joints. These forces create torques, which tell us how the forces affect an object's equilibrium and rate of rotation.

We will find that an object remains in a state of uniform rotational motion unless acted on by a net torque. This principle is the equivalent of Newton's first law. Further, the angular acceleration of an object is proportional to the net torque acting on it, which is the analog of Newton's second law. A net torque acting on an object causes a change in its rotational energy.

Finally, torques applied to an object through a given time interval can change the object's angular momentum. In the absence of external torques, angular momentum is conserved, a property that explains some of the mysterious and formidable properties of pulsars—remnants of supernova explosions that rotate at equatorial speeds approaching that of light.

8.1 TORQUE

Forces cause accelerations; *torques* cause angular accelerations. There is a definite relationship, however, between the two concepts.

Figure 8.1 depicts an overhead view of a door hinged at point O. From this viewpoint, the door is free to rotate around an axis perpendicular to the page and passing through O. If a force $\vec{\mathbf{F}}$ is applied to the door, there are three factors that determine the effectiveness of the force in opening the door: the *magnitude* of the force, the *position* of application of the force, and the *angle* at which it is applied.

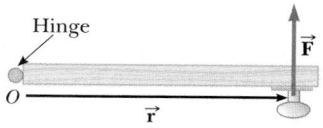

Figure 8.1 A bird's-eye view of a door hinged at O, with a force applied perpendicular to the door.

For simplicity, we restrict our discussion to position and force vectors lying in a plane. When the applied force $\vec{\mathbf{F}}$ is perpendicular to the outer edge of the door, as in Figure 8.1, the door rotates counterclockwise with constant angular acceleration. The same perpendicular force applied at a point nearer the hinge results in a smaller angular acceleration. In general, a larger radial distance r between the applied force and the axis of rotation results in a larger angular acceleration. Similarly, a larger applied force will also result in a larger angular acceleration. These considerations motivate the basic definition of **torque** for the special case of forces perpendicular to the position vector:

◀ Basic definition of torque

Let $\vec{\mathbf{F}}$ be a force acting on an object, and let $\vec{\mathbf{r}}$ be a position vector from a chosen point O to the point of application of the force, with $\vec{\mathbf{F}}$ perpendicular to $\vec{\mathbf{r}}$. The magnitude of the torque $\vec{\boldsymbol{\tau}}$ exerted by the force $\vec{\mathbf{F}}$ is given by

$$\tau = rF \qquad [8.1]$$

where r is the length of the position vector and F is the magnitude of the force.

SI unit: Newton-meter (N·m)

The vectors $\vec{\mathbf{r}}$ and $\vec{\mathbf{F}}$ lie in a plane. As discussed in detail shortly in conjunction with Figure 8.4, the torque $\vec{\boldsymbol{\tau}}$ is then perpendicular to this plane. The point O is usually chosen to coincide with the axis the object is rotating around, such as the hinge of a door or hub of a merry-go-round. (Other choices are possible as well.) In addition, we consider only forces acting in the plane perpendicular to the axis of rotation. This criterion excludes, for example, a force with upward component on a merry-go-round railing, which cannot affect the merry-go-round's rotation.

Under these conditions, an object can rotate around the chosen axis in one of two directions. By convention, counterclockwise is taken to be the positive direction, clockwise the negative direction. When an applied force causes an object to rotate counterclockwise, the torque on the object is positive. When the force causes the object to rotate clockwise, the torque on the object is negative. When two or more torques act on an object at rest, the torques are added. If the net torque isn't zero, the object starts rotating at an ever-increasing rate. If the net torque is zero, the object's rate of rotation doesn't change. These considerations lead to the rotational analog of the first law: **the rate of rotation of an object doesn't change, unless the object is acted on by a net torque.**

EXAMPLE 8.1 Battle of the Revolving Door

Goal Apply the basic definition of torque.

Problem Two disgruntled businessmen are trying to use a revolving door, as in Figure 8.2. The door has a diameter of 2.60 m. The businessman on the left exerts a force of 625 N perpendicular to the door and 1.20 m from the hub's center, while the one on the right exerts a force of 8.50×10^2 N perpendicular to the door and 0.800 m from the hub's center. Find the net torque on the revolving door.

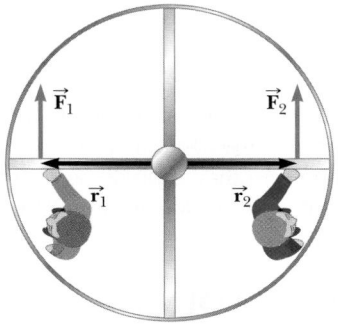

Figure 8.2 (Example 8.1)

Strategy Calculate the individual torques on the door using the definition of torque, Equation 8.1, and then sum to find the net torque on the door. The first businessman exerts a negative torque, the second a positive torque. Their positions of application also differ.

Solution

Calculate the torque exerted by the first businessman. A negative sign must be supplied, because $\vec{F}_1$, if unopposed, would cause a clockwise rotation.

$$\tau_1 = -r_1F_1 = -(1.20\text{ m})(625\text{ N}) = -7.50 \times 10^2\text{ N}\cdot\text{m}$$

Calculate the torque exerted by the second businessman. The torque is positive because $\vec{F}_2$, if unopposed, would cause a counterclockwise rotation.

$$\tau_2 = r_2F_2 = (0.800\text{ m})(8.50 \times 10^2\text{ N}) = 6.80 \times 10^2\text{ N}\cdot\text{m}$$

Sum the torques to find the net torque on the door:

$$\tau_{net} = \tau_1 + \tau_2 = -7.0 \times 10^1\text{ N}\cdot\text{m}$$

Remark The negative result here means the net torque will produce a clockwise rotation.

Exercise 8.1

A businessman enters the same revolving door on the right, pushing with 576 N of force directed perpendicular to the door and 0.700 m from the hub, while a boy exerts a force of 365 N perpendicular to the door, 1.25 m to the left of the hub. Find (a) the torques exerted by each person and (b) the net torque on the door.

Answers (a) $\tau_{boy} = -456\text{ N}\cdot\text{m}$, $\tau_{man} = 403\text{ N}\cdot\text{m}$ (b) $\tau_{net} = -53\text{ N}\cdot\text{m}$

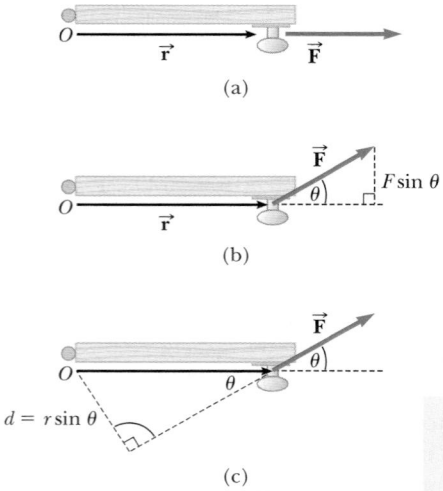

Figure 8.3 (a) A force $\vec{F}$ acting at an angle $\theta = 180°$ exerts zero torque about the pivot O. (b) The part of the force perpendicular to the door, $F\sin\theta$, exerts torque $rF\sin\theta$ about O. (c) An alternate interpretation of torque in terms of a *lever arm* $d = r\sin\theta$.

The applied force isn't always perpendicular to the position vector $\vec{r}$. Suppose the force $\vec{F}$ exerted on a door is directed away from the axis, as in Figure 8.3a, say, by someone's grasping the doorknob and pushing to the right. Exerting the force in this direction couldn't possibly open the door. However, if the applied force acts at an angle to the door as in Figure 8.3b, the component of the force *perpendicular* to the door will cause it to rotate. This figure shows that the component of the force perpendicular to the door is $F\sin\theta$, where θ is the angle between the position vector $\vec{r}$ and the force $\vec{F}$. When the force is directed away from the axis, $\theta = 0°$, $\sin(0°) = 0$, and $F\sin(0°) = 0$. When the force is directed toward the axis, $\theta = 180°$ and $F\sin(180°) = 0$. The maximum absolute value of $F\sin\theta$ is attained only when $\vec{F}$ is perpendicular to $\vec{r}$—that is, when $\theta = 90°$ or $\theta = 270°$. These considerations motivate a more general definition of torque:

Let $\vec{F}$ be a force acting on an object, and let $\vec{r}$ be a position vector from a chosen point O to the point of application of the force. The magnitude of the torque $\vec{\tau}$ exerted by the force $\vec{F}$ is

$$\tau = rF\sin\theta \qquad [8.2]$$

where r is the length of the position vector, F the magnitude of the force, and θ the angle between $\vec{r}$ and $\vec{F}$.

SI unit: Newton-meter (N · m)

Again the vectors $\vec{\mathbf{r}}$ and $\vec{\mathbf{F}}$ lie in a plane and for our purposes the chosen point O will usually correspond to an axis of rotation perpendicular to the plane.

A second way of understanding the sin θ factor is to associate it with the magnitude r of the position vector $\vec{\mathbf{r}}$. The quantity $d = r \sin \theta$ is called the **lever arm**, which is the perpendicular distance from the axis of rotation to a line drawn along the direction of the force. This alternate interpretation is illustrated in Figure 8.3c.

It's important to remember that **the value of τ depends on the chosen axis of rotation**. Torques can be computed around any axis, regardless of whether there is some actual, physical rotation axis present. Once the point is chosen, however, it must be used consistently throughout a given problem.

Torque is a vector perpendicular to the plane determined by the position and force vectors, as illustrated in Figure 8.4. The direction can be determined by the *right-hand rule*:

1. Point the fingers of your right hand in the direction of $\vec{\mathbf{r}}$.
2. Curl your fingers toward the direction of vector $\vec{\mathbf{F}}$.
3. Your thumb then points approximately in the direction of the torque, in this case out of the page.

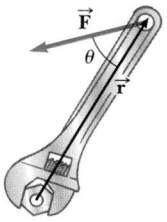

Figure 8.4 The right-hand rule: Point the fingers of your right hand along $\vec{\mathbf{r}}$ and curl them in the direction of $\vec{\mathbf{F}}$. Your thumb then points in the direction of the torque (out of the page, in this case).

Problems used in this book will be confined to objects rotating around an axis perpendicular to the plane containing $\vec{\mathbf{r}}$ and $\vec{\mathbf{F}}$, so if these vectors are in the plane of the page, the torque will always point either into or out of the page, parallel to the axis of rotation. If your right thumb is pointed in the direction of a torque, your fingers curl naturally in the direction of rotation that the torque would produce on an object at rest.

EXAMPLE 8.2 The Swinging Door

Goal Apply the more general definition of torque.

Problem **(a)** A man applies a force of $F = 3.00 \times 10^2$ N at an angle of 60.0° to the door of Figure 8.5a, 2.00 m from the hinges. Find the torque on the door, choosing the position of the hinges as the axis of rotation. **(b)** Suppose a wedge is placed 1.50 m from the hinges on the other side of the door. What minimum force must the wedge exert so that the force applied in part (a) won't open the door?

Strategy Part (a) can be solved by substitution into the general torque equation. In part (b), the hinges, the wedge, and the applied force all exert torques on the door. The door doesn't open, so the sum of these torques must be zero, a condition that can be used to find the wedge force.

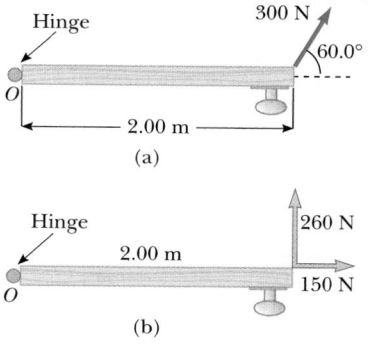

Figure 8.5 (Example 8.2a) (a) Top view of a door being pushed by a 300-N force. (b) The components of the 300-N force.

Solution

(a) Compute the torque due to the applied force exerted at 60.0°.

Substitute into the general torque equation:

$$\tau_F = rF \sin \theta = (2.00 \text{ m})(3.00 \times 10^2 \text{ N}) \sin 60.0°$$
$$= (2.00 \text{ m})(2.60 \times 10^2 \text{ N}) = \boxed{5.20 \times 10^2 \text{ N}\cdot\text{m}}$$

(b) Calculate the force exerted by the wedge on the other side of the door.

Set the sum of the torques equal to zero:

$$\tau_{\text{hinge}} + \tau_{\text{wedge}} + \tau_F = 0$$

The hinge force provides no torque because it acts at the axis ($r = 0$). The wedge force acts at an angle of $-90.0°$, opposite F_y.

$$0 + F_{\text{wedge}}(1.50 \text{ m}) \sin(-90.0°) + 5.20 \times 10^2 \text{ N}\cdot\text{m} = 0$$

$$F_{\text{wedge}} = \boxed{347 \text{ N}}$$

Remark Notice that the angle from the position vector to the wedge force is $-90°$. This is because, starting at the position vector, it's necessary to go 90° clockwise (the negative angular direction) to get to the force vector. Measuring the angle in this way automatically supplies the correct sign for the torque term and is consistent with the right-hand rule. Alternately, the magnitude of the torque can be found and the correct sign chosen based on physical intuition.

Exercise 8.2

A man ties one end of a strong rope 8.00 m long to the bumper of his truck, 0.500 m from the ground, and the other end to a vertical tree trunk at a height of 3.00 m. He uses the truck to create a tension of 8.00×10^2 N in the rope. Compute the magnitude of the torque on the tree due to the tension in the rope, with the base of the tree acting as the reference point.

Answer $2.28 \times 10^3\ \mathrm{N \cdot m}$

David Serway

This large balanced rock at the Garden of the Gods in Colorado, Springs, Colorado, is in mechanical equilibrium.

8.2 TORQUE AND THE TWO CONDITIONS FOR EQUILIBRIUM

An object in mechanical equilibrium must satisfy the following two conditions:

1. **The net external force must be zero:** $\Sigma \vec{\mathbf{F}} = 0$
2. **The net external torque must be zero:** $\Sigma \vec{\boldsymbol{\tau}} = 0$

The first condition is a statement of translational equilibrium: The sum of all forces acting on the object must be zero, so the object has no translational acceleration, $\vec{\mathbf{a}} = 0$. The second condition is a statement of rotational equilibrium: The sum of all torques on the object must be zero, so the object has no angular acceleration, $\vec{\boldsymbol{\alpha}} = 0$. For an object to be in equilibrium, it must both translate and rotate at a constant rate.

Because we can choose any location for calculating torques, it's usually best to select an axis that will make at least one torque equal to zero, just to simplify the net torque equation.

EXAMPLE 8.3 Balancing Act

Goal Apply the conditions of equilibrium and illustrate the use of different axes for calculating the net torque on an object.

Problem A woman of mass $m = 55.0$ kg sits on the left end of a seesaw—a plank of length $L = 4.00$ m, pivoted in the middle as in Figure 8.6. **(a)** First compute the torques on the seesaw about an axis that passes through the pivot point. Where should a man of mass $M = 75.0$ kg sit if the system (seesaw plus man and woman) is to be balanced? **(b)** Find the normal force exerted by the pivot if the plank has a mass of $m_{\mathrm{pl}} = 12.0$ kg. **(c)** Repeat part (b), but this time compute the torques about an axis through the left end of the plank.

Strategy In part (a), apply the second condition of equilibrium, $\Sigma\tau = 0$, computing torques around the pivot point. The mass of the plank forming the seesaw is distributed evenly on either side of the pivot point, so the torque exerted by gravity on the plank, τ_{gravity}, can be computed as if all the plank's mass is concentrated at the pivot point. Then

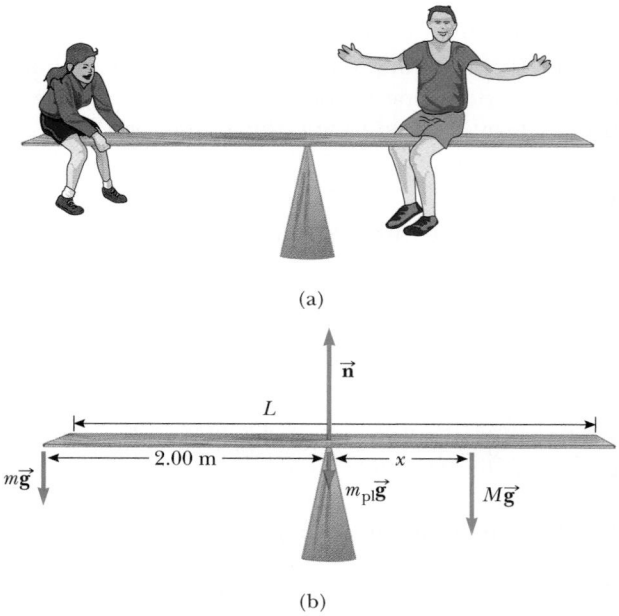

Figure 8.6 (a) (Example 8.3) Two people on a see-saw. (b) Free body diagram for the plank.

$\tau_{gravity}$ is zero, as is the torque exerted by the pivot, because their moment arms are zero. In part (b), the first condition of equilibrium, $\Sigma\vec{\mathbf{F}} = 0$, must be applied. Part (c) is a repeat of part (a) showing that choice of a different axis yields the same answer.

Solution

(a) Where should the man sit to balance the seesaw?

Apply the second condition of equilibrium to the plank by setting the sum of the torques equal to zero:

$$\tau_{pivot} + \tau_{gravity} + \tau_{man} + \tau_{woman} = 0$$

The first two torques are zero. Let x represent the man's distance from the pivot. The woman is at a distance $L/2$ from the pivot.

$$0 + 0 - Mgx + mg(L/2) = 0$$

Solve this equation for x and evaluate it:

$$x = \frac{m(L/2)}{M} = \frac{(55.0\text{ kg})(2.00\text{ m})}{75.0\text{ kg}} = \boxed{1.47\text{ m}}$$

(b) Find the normal force n exerted by the pivot on the seesaw.

Apply for first condition of equilibrium to the plank, solving the resulting equation for the unknown normal force, n:

$$-Mg - mg - m_{pl}g + n = 0$$

$$n = (M + m + m_{pl})g$$

$$= (75.0\text{ kg} + 55.0\text{ kg} + 12.0\text{ kg})(9.80\text{ m/s}^2)$$

$$n = \boxed{1.39 \times 10^3\text{ N}}$$

(c) Repeat part (a), choosing a new axis through the left end of the plank.

Compute the torques using this axis, and set their sum equal to zero. Now the pivot and gravity forces on the plank result in nonzero torques.

$$\tau_{man} + \tau_{woman} + \tau_{plank} + \tau_{pivot} = 0$$

$$-Mg(L/2 + x) + mg(0) - m_{pl}g(L/2) + n(L/2) = 0$$

Substitute all known quantities:

$$-(75.0\text{ kg})(9.80\text{ m/s}^2)(2.00\text{ m} + x) + 0$$

$$-(12.0\text{ kg})(9.80\text{ m/s}^2)(2.00\text{ m}) + n(2.00\text{ m}) = 0$$

$$-(1.47 \times 10^3\text{ N}\cdot\text{m}) - (735\text{ N})x - (235\text{ N}\cdot\text{m})$$

$$+ (2.00\text{ m})n = 0$$

Solve for x, substituting the normal force found in part (b):

$$x = \boxed{1.46\text{ m}}$$

Remarks The answers for x in parts (a) and (c) agree except for a small round-off discrepancy. This illustrates how choosing a different axis leads to the same solution.

Exercise 8.3

Suppose a 30.0-kg child sits 1.50 m to the left of center on the same seesaw. A second child sits at the end on the opposite side, and the system is balanced. (a) Find the mass of the second child. (b) Find the normal force acting at the pivot point.

Answers (a) 22.5 kg (b) 632 N

TIP 8.1 Specify Your Axis

Choose the axis of rotation, and use that axis exclusively throughout a given problem. The axis need not correspond to a physical axle or pivot point. Any convenient point will do.

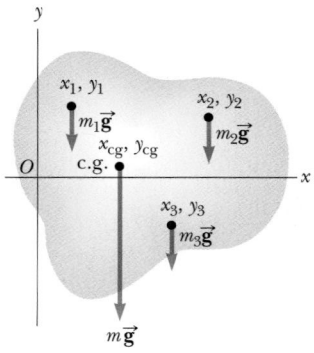

Figure 8.7 The net gravitational torque on an object is zero if computed around the center of gravity. The object will balance if supported at that point (or at any point along a vertical line above or below that point).

8.3 THE CENTER OF GRAVITY

In the example of the seesaw in the previous section, we guessed that the torque due to the force of gravity on the plank was the same as if all the plank's weight were concentrated at its center. This is a general procedure: To compute the torque on a rigid body due to the force of gravity, the body's entire weight can be thought of as concentrated at a single point. The problem then reduces to finding the location of that point. If the body is homogeneous (its mass is distributed evenly) and symmetric, it's usually possible to guess the location of that point, as in Example 8.3. Otherwise, it's necessary to calculate the point's location, as explained in this section.

Consider an object of arbitrary shape lying in the xy-plane, as in Figure 8.7. The object is divided into a large number of very small particles of weight m_1g, m_2g, m_3g, . . . having coordinates (x_1, y_1), (x_2, y_2), (x_3, y_3), ... If the object is free to rotate around the origin, each particle contributes a torque about the origin that is equal to its weight multiplied by its lever arm. For example, the torque due to the weight m_1g is m_1gx_1, and so forth.

We wish to locate the point of application of the single force of magnitude $w = F_g = Mg$ (the total weight of the object), where the effect on the rotation of the object is the same as that of the individual particles. This point is called the object's **center of gravity**. Equating the torque exerted by w at the center of gravity to the sum of the torques acting on the individual particles gives

$$(m_1g + m_2g + m_3g + \cdots)x_{cg} = m_1gx_1 + m_2gx_2 + m_3gx_3 + \cdots$$

We assume that g is the same everywhere in the object (which is true for all objects we will encounter). Then the g factors in the preceding equation cancel, resulting in

$$x_{cg} = \frac{m_1x_1 + m_2x_2 + m_3x_3 + \cdots}{m_1 + m_2 + m_3 + \cdots} = \frac{\Sigma m_i x_i}{\Sigma m_i} \qquad \textbf{[8.3a]}$$

where x_{cg} is the x-coordinate of the center of gravity. Similarly, the y-coordinate and z-coordinate of the center of gravity of the system can be found from

$$y_{cg} = \frac{\Sigma m_i y_i}{\Sigma m_i} \qquad \textbf{[8.3b]}$$

and

$$z_{cg} = \frac{\Sigma m_i z_i}{\Sigma m_i} \qquad \textbf{[8.3c]}$$

These three equations are identical to the equations for a similar concept called **center of mass**. The center of mass and center of gravity of an object are exactly the same when g doesn't vary significantly over the object.

It's often possible to guess the location of the center of gravity. **The center of gravity of a homogeneous, symmetric body must lie on the axis of symmetry**. For example, the center of gravity of a homogeneous rod lies midway between the ends of the rod, and the center of gravity of a homogeneous sphere or a homogeneous cube lies at the geometric center of the object. The center of gravity of an irregularly shaped object, such as a wrench, can be determined experimentally by suspending the wrench from two different arbitrary points (Fig. 8.8). The wrench is first hung from point A, and a vertical line AB (which can be established with a plumb bob) is drawn when the wrench is in equilibrium. The wrench is then hung from point C, and a second vertical line CD is drawn. The center of gravity coincides with the intersection of these two lines. In fact, if the wrench is hung freely from any point, the center of gravity always lies straight below the point of support, so the vertical line through that point must pass through the center of gravity.

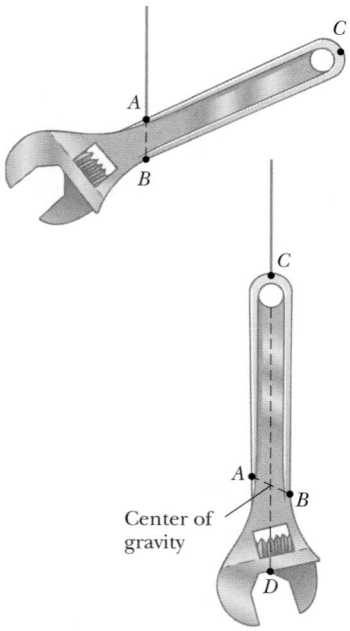

Figure 8.8 An experimental technique for determining the center of gravity of a wrench. The wrench is hung freely from two different pivots, A and C. The intersection of the two vertical lines, AB and CD, locates the center of gravity.

Several examples in Section 8.4 involve homogeneous, symmetric objects where the centers of gravity coincide with their geometric centers. A rigid object in a uniform gravitational field can be balanced by a single force equal in magnitude to the weight of the object, as long as the force is directed upward through the object's center of gravity.

EXAMPLE 8.4 Where Is the Center of Gravity?

Goal Find the center of gravity of a system of particles.

Problem Three particles are located in a coordinate system as shown in Figure 8.9. Find the center of gravity.

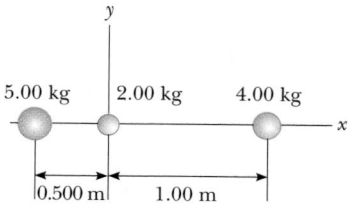

Figure 8.9 (Example 8.4) Locating the center of gravity of a system of three particles.

Strategy The y-coordinate and z-coordinate of the center of gravity are both zero because all the particles are on the x-axis. We can find the x-coordinate of the center of gravity using Equation 8.3a.

Solution

Apply Equation 8.3a to the system of three particles:

$$x_{cg} = \frac{\Sigma m_i x_i}{\Sigma m_i} = \frac{m_1 x_1 + m_2 x_2 + m_3 x_3}{m_1 + m_2 + m_3} \quad (1)$$

Compute the numerator of Equation (1):

$$\begin{aligned}\Sigma m_i x_i &= m_1 x_1 + m_2 x_2 + m_3 x_3 \\ &= (5.00\text{ kg})(-0.500\text{ m}) + (2.00\text{ kg})(0\text{ m}) \\ &\quad + (4.00\text{ kg})(1.00\text{ m}) \\ &= 1.50\text{ kg}\cdot\text{m}\end{aligned}$$

Substitute the denominator, $\Sigma m_i = 11.0$ kg, and the numerator into Equation (1).

$$x_{cg} = \frac{1.50\text{ kg}\cdot\text{m}}{11.0\text{ kg}} = \boxed{0.136\text{ m}}$$

Exercise 8.4

If a fourth particle of mass 2.00 kg is placed at $x = 0$, $y = 0.250$ m, find the x-and y-coordinates of the center of gravity for this system of four particles.

Answer $x_{cg} = 0.115$ m; $y_{cg} = 0.0385$ m

EXAMPLE 8.5 Locating Your Lab Partner's Center of Gravity

Goal Use torque to find a center of gravity.

Problem In this example, we show how to find the location of a person's center of gravity. Suppose your lab partner has a height L of 173 cm (5 ft, 8 in) and a weight w of 715 N (160 lb). You can determine the position of his center of gravity by having him stretch out on a uniform board supported at one end by a scale, as shown in Figure 8.10. If the board's weight w_b is 49 N and the scale reading F is 3.50×10^2 N, find the distance of your lab partner's center of gravity from the left end of the board.

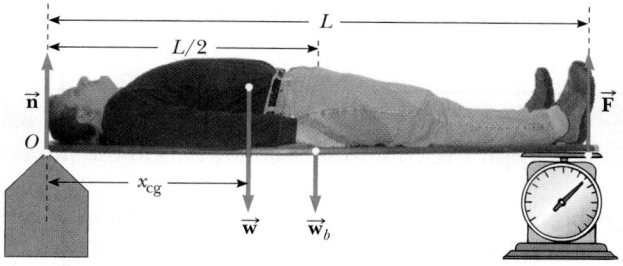

Figure 8.10 (Example 8.5) Determining your lab partner's center of gravity.

Strategy To find the position x_{cg} of the center of gravity, compute the torques using an axis through O. Set the sum of the torques equal to zero and solve for x_{cg}.

Solution

Apply the second condition of equilibrium. There is no torque due to the normal force $\vec{\mathbf{n}}$ because its moment arm is zero.

$$\Sigma \tau_O = 0$$

$$-wx_{cg} - w_b(L/2) + FL = 0$$

Solve for x_{cg} and substitute known values:

$$x_{cg} = \frac{FL - w_b(L/2)}{w}$$

$$= \frac{(350\text{ N})(173\text{ cm}) - (49\text{ N})(86.5\text{ cm})}{715\text{ N}} = \boxed{79\text{ cm}}$$

Remarks The given information is sufficient only to determine the x-coordinate of the center of gravity. The other two coordinates can be estimated, based on the body's symmetry.

Exercise 8.5

Suppose a 416-kg alligator of length 3.5 m is stretched out on a board of the same length weighing 65 N. If the board is supported on the ends as in Figure 8.10, and the scale reads 1 880 N, find the x-component of the alligator's center of gravity.

Answer 1.59 m

TIP 8.2 Rotary Motion Under Zero Torque

If a net torque of zero is exerted on an object, it will continue to rotate at a constant angular speed—which need not be zero. However, zero torque *does* imply that the angular acceleration is zero.

8.4 EXAMPLES OF OBJECTS IN EQUILIBRIUM

Recall from Chapter 4 that when an object is treated as a geometric point, equilibrium requires only that the net force on the object is zero. In this chapter, we have shown that for extended objects a second condition for equilibrium must also be satisfied: The net torque on the object must be zero. The following general procedure is recommended for solving problems that involve objects in equilibrium.

Problem-Solving Strategy Objects in Equilibrium

1. **Diagram the system.** Include coordinates and choose a convenient rotation axis for computing the net torque on the object.
2. **Draw a free-body diagram** of the object of interest, showing all external forces acting on it. For systems with more than one object, draw a *separate* diagram for each object. (Most problems will have a single object of interest.)
3. **Apply $\Sigma\tau_i = 0$, the second condition of equilibrium.** This condition yields a single equation for each object of interest. If the axis of rotation has been carefully chosen, the equation often has only one unknown and can be solved immediately.
4. **Apply $\Sigma F_x = 0$ and $\Sigma F_y = 0$, the first condition of equilibrium.** This yields two more equations per object of interest.
5. **Solve the system of equations.** For each object, the two conditions of equilibrium yield three equations, usually with three unknowns. Solve by substitution.

EXAMPLE 8.6 A Weighted Forearm

Goal Apply the equilibrium conditions to the human body.

Problem A 50.0-N (11-lb) weight is held in a person's hand with the forearm horizontal, as in Figure 8.11. The biceps muscle is attached 0.030 0 m from the joint, and the weight is 0.350 m from the joint. Find the upward force $\vec{\mathbf{F}}$ exerted by the biceps on the forearm (the ulna) and the downward force $\vec{\mathbf{R}}$ exerted by the humerus on the forearm, acting at the joint. Neglect the weight of the forearm.

Strategy The forces acting on the forearm are equivalent to those acting on a bar of length 0.350 m, as shown in Figure 8.11b. Choose the usual x- and y-coordinates as shown and the axis at O on the left end. (This completes Steps 1 and 2.) Use the conditions of equilibrium to generate equations for the unknowns, and solve.

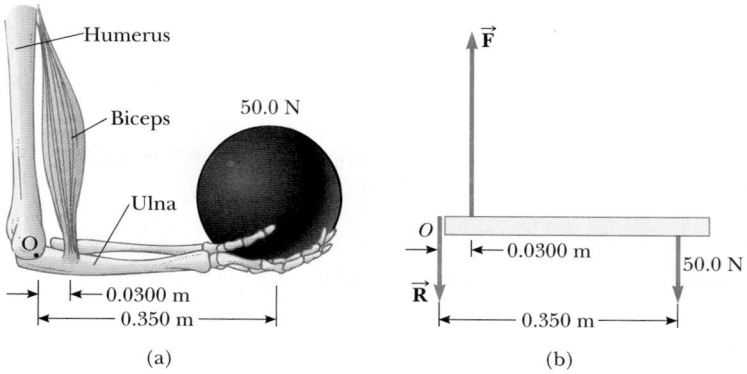

Figure 8.11 (Example 8.6) (a) A weight held with the forearm horizontal. (b) The mechanical model for the system.

Solution

Apply the second condition for equilibrium (step 3):

$$\Sigma\tau_i = \tau_R + \tau_F + \tau_{BB} = 0$$

$$R(0) + F(0.0300\text{ m}) - (50.0\text{ N})(0.350\text{ m}) = 0$$

$$F = \boxed{583\text{ N (131 lb)}}$$

Apply the first condition for equilibrium (step 4):

$$\Sigma F_y = F - R - 50.0\text{ N} = 0$$

$$R = F - 50.0\text{ N} = 583\text{ N} - 50\text{ N} = \boxed{533\text{ N (120 lb)}}$$

Exercise 8.6

Suppose you wanted to limit the force acting on your joint to a maximum value of 8.00×10^2 N. (a) Under these circumstances, what maximum weight would you attempt to lift? (b) What force would your biceps apply while lifting this weight?

Answers (a) 75.0 N (b) 875 N

EXAMPLE 8.7 Walking a Horizontal Beam

Goal Solve an equilibrium problem with nonperpendicular torques.

Problem A uniform horizontal beam 5.00 m long and weighing 3.00×10^2 N is attached to a wall by a pin connection that allows the beam to rotate. Its far end is supported by a cable that makes an angle of 53.0° with the horizontal (Fig. 8.12a). If a person weighing 6.00×10^2 N stands 1.50 m from the wall, find the magnitude of the tension $\vec{\mathbf{T}}$ in the cable and the force $\vec{\mathbf{R}}$ exerted by the wall on the beam.

Strategy

The second condition of equilibrium, $\Sigma\tau_i = 0$, with torques computed around the pin, can be solved for the tension T in the cable. The first condition of equilibrium, $\Sigma\vec{\mathbf{F}}_i = 0$, gives two equations and two unknowns for the two components of the force exerted by the wall, R_x and R_y.

Solution

From Figure 8.12, the forces causing torques are the wall force $\vec{\mathbf{R}}$, the gravity forces on the beam and the man, w_B and w_M, and the tension force $\vec{\mathbf{T}}$. Apply the condition of rotational equilibrium:

$$\Sigma\tau_i = \tau_R + \tau_B + \tau_M + \tau_T = 0$$

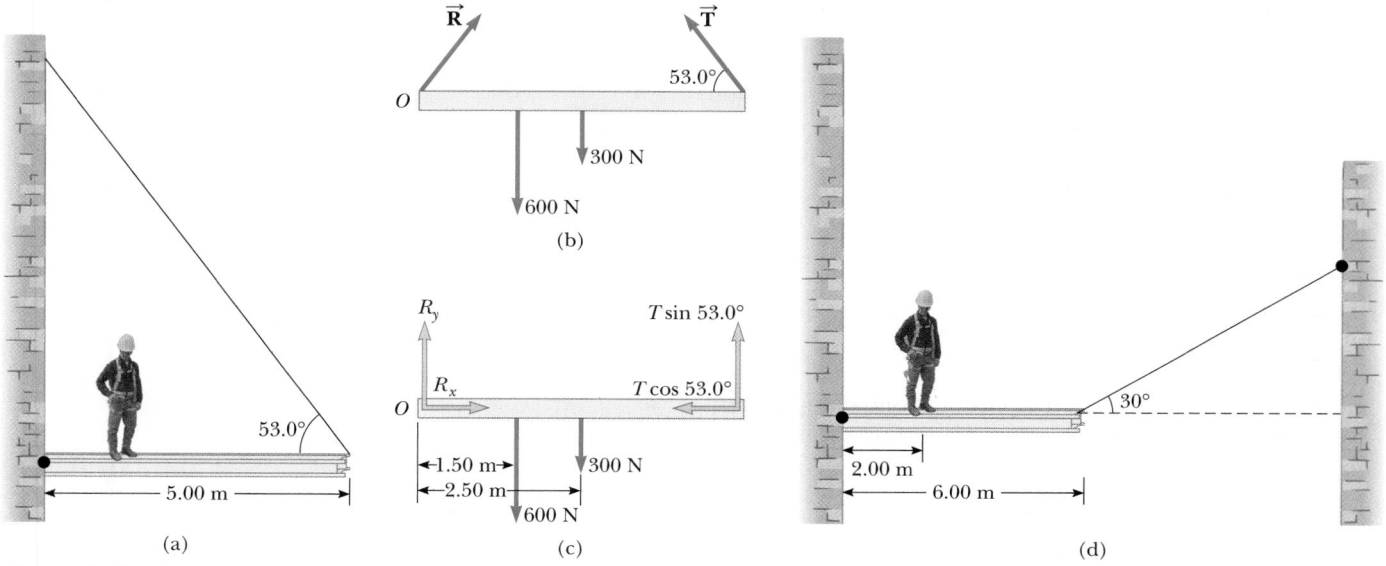

Figure 8.12 (Example 8.7) (a) A uniform beam attached to a wall and supported by a cable. (b) A free-body diagram for the beam. (c) The component form of the free-body diagram. (d) (Exercise 8.7)

Compute torques around the pin at O, so $\tau_R = 0$ (zero moment arm). The torque due to the beam's weight acts at the beam's center of gravity.

$$\Sigma\tau_i = 0 - w_B(L/2) - w_M(1.50\text{ m}) + TL\sin(53°) = 0$$

Substitute $L = 5.00$ m and the weights, solving for T:

$$-(3.00 \times 10^2\text{ N})(2.50\text{ m}) - (6.00 \times 10^2\text{ N})(1.50\text{ m}) + (T\sin 53.0°)(5.00\text{ m}) = 0$$

$$T = 413\text{ N}$$

Now apply the first condition of equilibrium to the beam:

$$\Sigma F_x = R_x - T\cos 53.0° = 0 \quad \textbf{(1)}$$

$$\Sigma F_y = R_y - w_B - w_M + T\sin 53.0° = 0 \quad \textbf{(2)}$$

Substituting the value of T found in the previous step and the weights, obtain the components of $\vec{\mathbf{R}}$:

$$R_x = 249\text{ N} \qquad R_y = 5.70 \times 10^2\text{ N}$$

Remarks Even if we selected some other axis for the torque equation, the solution would be the same. For example, if the axis were to pass through the center of gravity of the beam, the torque equation would involve both T and R_y. Together with equations (1) and (2), however, the unknowns could still be found—a good exercise.

Exercise 8.7

A person with mass 55.0 kg stands 2.00 m away from the wall on a 6.00-m beam, as shown in Figure 8.12d. The mass of the beam is 40.0 kg. Find the hinge force components and the tension in the wire.

Answers $T = 751$ N, $R_x = -6.50 \times 10^2$ N, $R_y = 556$ N

INTERACTIVE EXAMPLE 8.8 Don't Climb the Ladder

Goal Apply the two conditions of equilibrium.

Problem A uniform ladder 10.0 m long and weighing 50.0 N rests against a smooth vertical wall as in Figure 8.13a. If the ladder is just on the verge of slipping when it makes a 50.0° angle with the ground, find the coefficient of static friction between the ladder and ground.

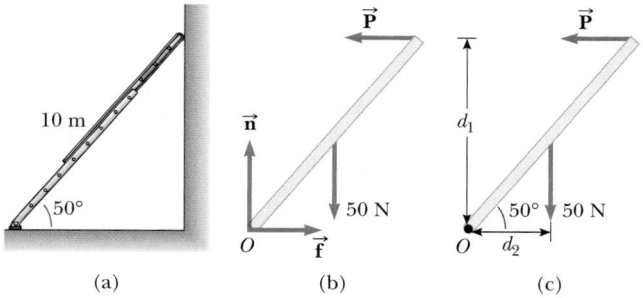

Figure 8.13 (Interactive Example 8.8) (a) A ladder leaning against a frictionless wall. (b) A free-body diagram of the ladder. (c) Lever arms for the force of gravity and $\vec{\mathbf{P}}$.

Strategy Figure 8.13b is the free-body diagram for the ladder. The first condition of equilibrium, $\Sigma\vec{\mathbf{F}}_i = 0$, gives two equations for three unknowns: the magnitudes of the static friction force f and the normal force n, both acting on the base of the ladder, and the magnitude of the force of the wall, P, acting on the top of the ladder. The second condition of equilibrium, $\Sigma\tau_i = 0$, gives a third equation (for P), so all three quantities can be found. The definition of static friction then allows computation of the coefficient of static friction.

Solution

Apply the first condition of equilibrium to the ladder:

$$\Sigma F_x = f - P = 0 \quad \rightarrow \quad f = P \quad \textbf{(1)}$$

$$\Sigma F_y = n - 50.0\text{ N} = 0 \quad \rightarrow \quad n = 50.0\text{ N} \quad \textbf{(2)}$$

Apply the second condition of equilibrium, computing torques around the base of the ladder, with τ_{grav} standing for the torque due to the ladder's 50.0-N weight:

$$\Sigma\tau_i = \tau_f + \tau_n + \tau_{\text{grav}} + \tau_P = 0$$

The torques due to friction and the normal force are zero about O because their moment arms are zero. (Moment arms can be found from Figure 8.13c.)

$$0 + 0 - (50.0\text{ N})(5.00\text{ m})\sin 40.0° + P(10.0\text{ m})\sin 50.0° = 0$$

$$P = 21.0\text{ N}$$

From Equation (1), we now have $f = P = 21.0$ N. The ladder is on the verge of slipping, so write an expression for the maximum force of static friction and solve for μ_s:

$$21.0 \text{ N} = f = f_{s,\max} = \mu_s n = \mu_s(50.0 \text{ N})$$

$$\mu_s = \frac{21.0 \text{ N}}{50.0 \text{ N}} = 0.420$$

Remarks Note that torques were computed around an axis through the bottom of the ladder so that only $\vec{\mathbf{P}}$ and the force of gravity contributed nonzero torques. This choice of axis reduces the complexity of the torque equation, often resulting in an equation with only one unknown.

Exercise 8.8

If the coefficient of static friction is 0.360, and the same ladder makes a 60.0° angle with respect to the horizontal, how far along the length of the ladder can a 70.0-kg painter climb before the ladder begins to slip?

Answer 6.33 m

PhysicsNow™ You can adjust the angle of the ladder and watch what happens when it is released by logging into PhysicsNow at **www.cp7e.com** and going to Interactive Example 8.8.

8.5 RELATIONSHIP BETWEEN TORQUE AND ANGULAR ACCELERATION

When a rigid object is subject to a net torque, it undergoes an angular acceleration that is directly proportional to the net torque. This result, which is analogous to Newton's second law, is derived as follows.

The system shown in Figure 8.14 consists of an object of mass m connected to a very light rod of length r. The rod is pivoted at the point O, and its movement is confined to rotation on a frictionless *horizontal* table. Assume that a force F_t acts perpendicular to the rod and hence is tangent to the circular path of the object. Because there is no force to oppose this tangential force, the object undergoes a tangential acceleration a_t in accordance with Newton's second law:

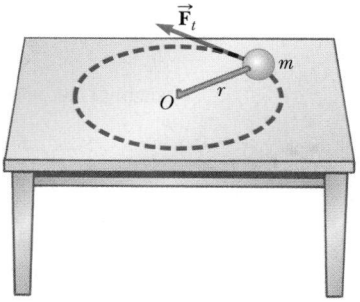

Figure 8.14 An object of mass m attached to a light rod of length r moves in a circular path on a frictionless horizontal surface while a tangential force $\vec{\mathbf{F}}_t$ acts on it.

$$F_t = ma_t$$

Multiply both sides of this equation by r:

$$F_t r = mra_t$$

Substituting the equation $a_t = r\alpha$ relating tangential and angular acceleration into the above expression gives

$$F_t r = mr^2\alpha \qquad [8.4]$$

The left side of Equation 8.4 is the torque acting on the object about its axis of rotation, so we can rewrite it as

$$\tau = mr^2\alpha \qquad [8.5]$$

Equation 8.5 shows that the torque on the object is proportional to the angular acceleration of the object, where the constant of proportionality mr^2 is called the **moment of inertia** of the object of mass m. (Because the rod is very light, its moment of inertia can be neglected.)

Quick Quiz 8.1

Using a screwdriver, you try to remove a screw from a piece of furniture, but can't get it to turn. To increase the chances of success, you should use a screwdriver that (a) is longer, (b) is shorter, (c) has a narrower handle, or (d) has a wider handle.

Torque on a Rotating Object

Consider a solid disk rotating about its axis as in Figure 8.15a. The disk consists of many particles at various distances from the axis of rotation. (See Fig. 8.15b.) The

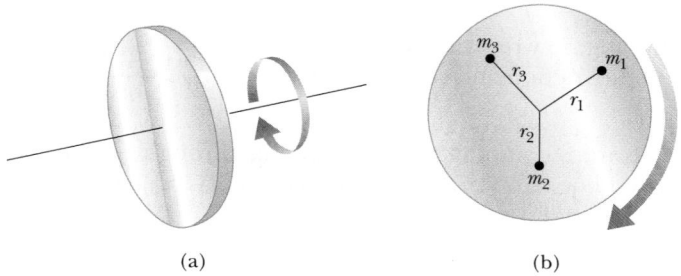

Figure 8.15 (a) A solid disk rotating about its axis. (b) The disk consists of many particles, all with the same angular acceleration.

torque on each one of these particles is given by Equation 8.5. The *net* torque on the disk is given by the sum of the individual torques on all the particles:

$$\Sigma\tau = (\Sigma mr^2)\alpha \qquad \textbf{[8.6]}$$

Because the disk is rigid, all of its particles have the *same* angular acceleration, so α is not involved in the sum. If the masses and distances of the particles are labeled with subscripts as in Figure 8.15b, then

$$\Sigma mr^2 = m_1r_1^2 + m_2r_2^2 + m_3r_3^2 + \cdots$$

This quantity is the moment of inertia, I, of the whole body:

Moment of inertia ▶

$$I \equiv \sum mr^2 \qquad \textbf{[8.7]}$$

The moment of inertia has the SI units $\text{kg}\cdot\text{m}^2$. Using this result in Equation 8.6, we see that the net torque on a rigid body rotating about a fixed axis is given by

Rotational analog of Newton's second law ▶

$$\sum\tau = I\alpha \qquad \textbf{[8.8]}$$

Equation 8.8 says that **the angular acceleration of an extended rigid object is proportional to the net torque acting on it**. This equation is the rotational analog of Newton's second law of motion, with torque replacing force, moment of inertia replacing mass, and angular acceleration replacing linear acceleration. Although the moment of inertia of an object is related to its mass, there is an important difference between them. The mass m depends only on the quantity of matter in an object while the moment of inertia, I, depends on both the quantity of matter and its distribution (through the r^2 term in $I = \Sigma mr^2$) in the rigid object.

Quick Quiz 8.2

A constant net torque is applied to an object. Which one of the following will *not* be constant? (a) angular acceleration, (b) angular velocity, (c) moment of inertia, or (d) center of gravity.

Quick Quiz 8.3

The two rigid objects shown in Figure 8.16 have the same mass, radius, and angular speed. If the same braking torque is applied to each, which takes longer to stop? (a) A (b) B (c) more information is needed

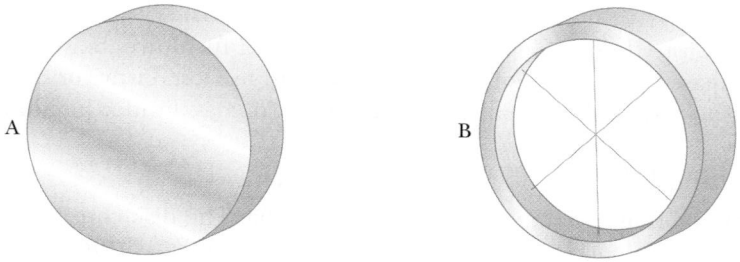

Figure 8.16 (Quick Quiz 8.3)

The gear system on a bicycle provides an easily visible example of the relationship between torque and angular acceleration. Consider first a five-speed gear system in which the drive chain can be adjusted to wrap around any of five gears attached to the back wheel (Fig. 8.17). The gears, with different radii, are concentric with the wheel hub. When the cyclist begins pedaling from rest, the chain is attached to the largest gear. Because it has the largest radius, this gear provides the largest torque to the drive wheel. A large torque is required initially, because the bicycle starts from rest. As the bicycle rolls faster, the tangential speed of the chain increases, eventually becoming too fast for the cyclist to maintain by pushing the pedals. The chain is then moved to a gear with a smaller radius, so the chain has a smaller tangential speed that the cyclist can more easily maintain. This gear doesn't provide as much torque as the first, but the cyclist needs to accelerate only to a somewhat higher speed. This process continues as the bicycle moves faster and faster and the cyclist shifts through all five gears. The fifth gear supplies the lowest torque, but now the main function of that torque is to counter the frictional torque from the rolling tires, which tends to reduce the speed of the bicycle. The small radius of the fifth gear allows the cyclist to keep up with the chain's movement by pushing the pedals.

A 15-speed bicycle has the same gear structure on the drive wheel, but has three gears on the sprocket connected to the pedals. By combining different positions of the chain on the rear gears and the sprocket gears, 15 different torques are available.

APPLICATION

Bicycle Gears

George Semple

Figure 8.17 The drive wheel and gears of a bicycle.

More on the Moment of Inertia

As we have seen, a small object (or a particle) has a moment of inertia equal to mr^2 about some axis. The moment of inertia of a *composite* object about some axis is just the sum of the moments of inertia of the object's components. For example, suppose a majorette twirls a baton as in Figure 8.18. Assume that the baton can be modeled as a very light rod of length 2ℓ with a heavy object at each end. (The rod of a real baton has a significant mass relative to its ends.) Because we are neglecting the mass of the rod, the moment of inertia of the baton about an axis through its center and perpendicular to its length is given by Equation 8.7:

$$I = \Sigma mr^2$$

Because this system consists of two objects with equal masses equidistant from the axis of rotation, $r = \ell$ for each object, and the sum is

$$I = \Sigma mr^2 = m\ell^2 + m\ell^2 = 2m\ell^2$$

If the mass of the rod were not neglected, we would have to include its moment of inertia to find the total moment of inertia of the baton.

We pointed out earlier that I is the rotational counterpart of m. However, there are some important distinctions between the two. For example, mass is an intrinsic property of an object that doesn't change, whereas **the moment of inertia of a system depends on how the mass is distributed and on the location of the axis of rotation**. Example 8.9 illustrates this point.

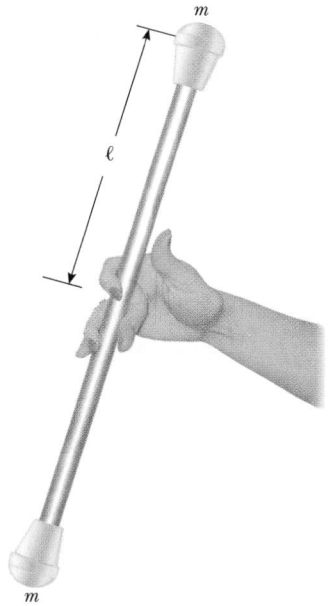

Figure 8.18 A baton of length 2ℓ and mass $2m$. (The mass of the connecting rod is neglected.) The moment of inertia about the axis through the baton's center and perpendicular to its length is $2m\ell^2$.

EXAMPLE 8.9 The Baton Twirler

Goal Calculate a moment of inertia.

Problem In an effort to be the star of the half-time show, a majorette twirls an unusual baton made up of four spheres fastened to the ends of very light rods (Fig. 8.19). Each rod is 1.0 m long. **(a)** Find the moment of inertia of the baton about an axis perpendicular to the page and passing through the point where the rods cross. **(b)** The majorette tries spinning her strange baton about the axis OO', as shown in Figure 8.20. Calculate the moment of inertia of the baton about this axis.

Strategy In Figure 8.19, all four balls contribute to the moment of inertia, whereas in Figure 8.20, with the new axis, only the two balls on the left and right contribute. Technically, the balls on the top and bottom still make a small contribution because they're not really point particles. However, their moment of inertia can be neglected, because the radius of the sphere is much smaller than the radius formed by the rods.

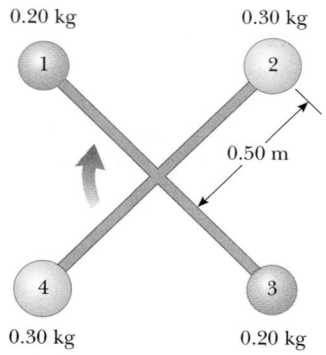

Figure 8.19 (Example 8.9a) Four objects connected to light rods rotating in the plane of the page.

Solution

(a) Calculate the moment of inertia of the baton when oriented as in Figure 8.19.

Apply Equation 8.7, neglecting the mass of the connecting rods:

$$I = \Sigma mr^2 = m_1r_1^2 + m_2r_2^2 + m_3r_3^2 + m_4r_4^2$$
$$= (0.20\text{ kg})(0.50\text{ m})^2 + (0.30\text{ kg})(0.50\text{ m})^2$$
$$+ (0.20\text{ kg})(0.50\text{ m})^2 + (0.30\text{ kg})(0.50\text{ m})^2$$
$$I = 0.25\text{ kg}\cdot\text{m}^2$$

(b) Calculate the moment of inertia of the baton when oriented as in Figure 8.20.

Apply Equation 8.7 again, neglecting the radii of the 0.20-kg spheres.

$$I = \Sigma mr^2 = m_1r_1^2 + m_2r_2^2 + m_3r_3^2 + m_4r_4^2$$
$$= (0.20\text{ kg})(0)^2 + (0.30\text{ kg})(0.50\text{ m})^2$$
$$+ (0.20\text{ kg})(0)^2 + (0.30\text{ kg})(0.50\text{ m})^2$$
$$I = 0.15\text{ kg}\cdot\text{m}^2$$

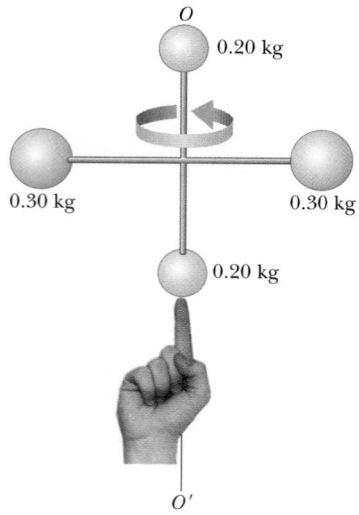

Figure 8.20 (Example 8.9b) A double baton rotating about the axis OO'.

Remarks The moment of inertia is smaller in part (b) because in this configuration the 0.20-kg spheres are essentially located on the axis of rotation.

Exercise 8.9

Yet another bizarre baton is created by taking four identical balls, each with mass 0.300 kg, and fixing them as before, except that one of the rods has a length of 1.00 m and the other has a length of 1.50 m. Calculate the moment of inertia of this baton (a) when oriented as in Figure 8.19; (b) when oriented as in Figure 8.20, with the shorter rod vertical; and (c) when oriented as in Figure 8.20, but with longer rod vertical.

Answers (a) $0.488\text{ kg}\cdot\text{m}^2$ (b) $0.338\text{ kg}\cdot\text{m}^2$ (c) $0.150\text{ kg}\cdot\text{m}^2$

Calculation of Moments of Inertia for Extended Objects

The method used for calculating moments of inertia in Example 8.9 is simple when only a few small objects rotate about an axis. When the object is an extended one, such as a sphere, a cylinder, or a cone, techniques of calculus are often required, unless some simplifying symmetry is present. One such extended object amenable to a simple solution is a hoop rotating about an axis perpendicular to its plane and passing through its center, as shown in Figure 8.21. (A bicycle tire, for example, would approximately fit into this category.)

To evaluate the moment of inertia of the hoop, we can still use the equation $I = \Sigma mr^2$ and imagine that the mass of the hoop M is divided into n small segments having masses $m_1, m_2, m_3, \ldots, m_n$, as in Figure 8.21, with $M = m_1 + m_2 + m_3 + \ldots + m_n$. This approach is just an extension of the baton problem described in the preceding examples, except that now we have a large number of small masses in rotation instead of only four.

We can express the sum for I as

$$I = \Sigma mr^2 = m_1r_1^2 + m_2r_2^2 + m_3r_3^2 + \cdots + m_nr_n^2$$

All of the segments around the hoop are at the *same distance* R from the axis of rotation, so we can drop the subscripts on the distances and factor out R^2 to obtain

$$I = (m_1 + m_2 + m_3 + \cdots + m_n)R^2 = MR^2 \qquad [8.9]$$

This expression can be used for the moment of inertia of any ring-shaped object rotating about an axis through its center and perpendicular to its plane. Note that the result is strictly valid only if the thickness of the ring is small relative to its inner radius.

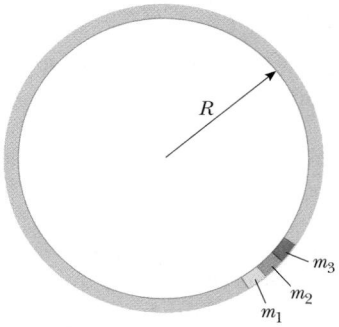

Figure 8.21 A uniform hoop can be divided into a large number of small segments that are equidistant from the center of the hoop.

The hoop we selected as an example is unique in that we were able to find an expression for its moment of inertia by using only simple algebra. Unfortunately, for most extended objects the calculation is much more difficult because the mass elements are not all located at the same distance from the axis, so the methods of integral calculus are required. The moments of inertia for some other common shapes are given without proof in Table 8.1. You can use this table as needed to determine the moment of inertia of a body having any one of the listed shapes.

TIP 8.3 No Single Moment of Inertia

Moment of inertia is analogous to mass, but there are major differences. Mass is an inherent property of an object. The moment of inertia of an object depends on the shape of the object, its mass, and the choice of rotation axis.

If mass elements in an object are redistributed parallel to the axis of rotation, the moment of inertia of the object doesn't change. Consequently, the expression $I = MR^2$ can be used equally well to find the axial moment of inertia of an embroidery hoop or of a long sewer pipe. Likewise, a door turning on its hinges is described by the same moment-of-inertia expression as that tabulated for a long thin rod rotating about an axis through its end.

TABLE 8.1

Moments of Inertia for Various Rigid Objects of Uniform Composition

Hoop or thin cylindrical shell $I = MR^2$	Solid sphere $I = \frac{2}{5}MR^2$
Solid cylinder or disk $I = \frac{1}{2}MR^2$	Thin spherical shell $I = \frac{2}{3}MR^2$
Long thin rod with rotation axis through center $I = \frac{1}{12}ML^2$	Long thin rod with rotation axis through end $I = \frac{1}{3}ML^2$

EXAMPLE 8.10 Warming Up

Goal Find a moment of inertia and apply the rotational analog of Newton's second law.

Problem A baseball player loosening up his arm before a game tosses a 0.150-kg baseball, using only the rotation of his forearm to accelerate the ball (Fig. 8.22). The forearm has a mass of 1.50 kg and a length of 0.350 m. The ball starts at rest and is released with a speed of 30.0 m/s in 0.300 s. **(a)** Find the constant angular acceleration of the arm and ball. **(b)** Calculate the moment of inertia of the system consisting of the forearm and ball. **(c)** Find the torque exerted on the ball to give it the angular acceleration found in part (a).

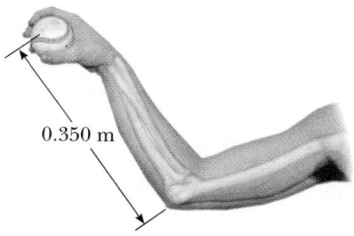

Figure 8.22 (Example 8.10) A ball being tossed by a pitcher. The forearm is used to accelerate the ball.

Strategy The angular acceleration can be found with rotational kinematic equations, while the moment of inertia of the system can be obtained by summing the separate moments of inertia of the ball and forearm. Multiplying these two results together gives the torque.

Solution

(a) Find the angular acceleration of the ball.

The angular acceleration is constant, so use the angular velocity kinematic equation with $\omega_i = 0$.

$$\omega = \omega_i + \alpha t \quad \rightarrow \quad \alpha = \frac{\omega}{t}$$

The ball accelerates along a circular arc with radius given by the length of the forearm. Solve $v = r\omega$ for ω and substitute:

$$\alpha = \frac{\omega}{t} = \frac{v}{rt} = \frac{30.0 \text{ m/s}}{(0.350 \text{ m})(0.300 \text{ s})} = 286 \text{ rad/s}^2$$

(b) Find the moment of inertia of the system (forearm plus ball).

Find the moment of inertia of the ball about an axis that passes through the elbow, perpendicular to the arm:

$$I_{\text{ball}} = mr^2 = (0.150 \text{ kg})(0.350 \text{ m})^2 = 1.84 \times 10^{-2} \text{ kg}\cdot\text{m}^2$$

Obtain the moment of inertia of the forearm, modeled as a rod, by consulting Table 8.1:

$$I_{\text{forearm}} = \tfrac{1}{3}ML^2 = \tfrac{1}{3}(1.50 \text{ kg})(0.350 \text{ m})^2 = 6.13 \times 10^{-2} \text{ kg}\cdot\text{m}^2$$

Sum the individual moments of inertia to obtain the moment of inertia of the system (ball plus forearm):

$$I_{\text{system}} = I_{\text{ball}} + I_{\text{forearm}} = 7.97 \times 10^{-2} \text{ kg}\cdot\text{m}^2$$

(c) Find the torque exerted on the ball.

Apply Equation 8.8, using the results of parts (a) and (b):

$$\tau = I_{\text{system}}\alpha = (7.97 \times 10^{-2} \text{ kg}\cdot\text{m}^2)(286 \text{ rad/s}^2) = 22.8 \text{ N}\cdot\text{m}$$

Remarks Notice that having a long forearm can greatly increase the torque and hence the acceleration of the ball. This is one reason it's advantageous for a pitcher to be tall—the pitching arm is proportionately longer. A similar advantage holds in tennis, where taller players can usually deliver faster serves.

Exercise 8.10

A catapult with a radial arm 4.00 m long accelerates a ball of mass 20.0 kg through a quarter circle. The ball leaves the apparatus at 45.0 m/s. If the mass of the arm is 25.0 kg and the acceleration is uniform, find (a) the angular acceleration, (b) the moment of inertia of the arm and ball, and (c) the net torque exerted on the ball and arm. *Hint:* Use the time-independent rotational kinematics equation to find the angular acceleration, rather than the angular velocity equation.

Answers (a) 40.3 rad/s^2 (b) 453 kg·m^2 (c) 1.83×10^4 N·m

INTERACTIVE EXAMPLE 8.11 The Falling Bucket

Goal Combine Newton's second law with its rotational analog.

Problem A solid, frictionless cylindrical reel of mass $M = 3.00$ kg and radius $R = 0.400$ m is used to draw water from a well (Fig. 8.23a). A bucket of mass $m = 2.00$ kg is attached to a cord that is wrapped around the cylinder. **(a)** Find the tension T in the cord and acceleration a of the bucket. **(b)** If the bucket starts from rest at the top of the well and falls for 3.00 s before hitting the water, how far does it fall?

Strategy This problem involves three equations and three unknowns. The three equations are Newton's second law applied to the bucket, $ma = \Sigma F_i$; the rotational version of the second law applied to the cylinder, $I\alpha = \Sigma \tau_i$; and the relationship between linear and angular acceleration, $a = r\alpha$, which connects the dynamics of the bucket and cylinder. The three unknowns are the acceleration a of the bucket, the angular acceleration a of the cylinder, and the tension T in the rope. Assemble the terms of the three equations and solve for the three unknowns by substitution. Part (b) is a review of kinematics.

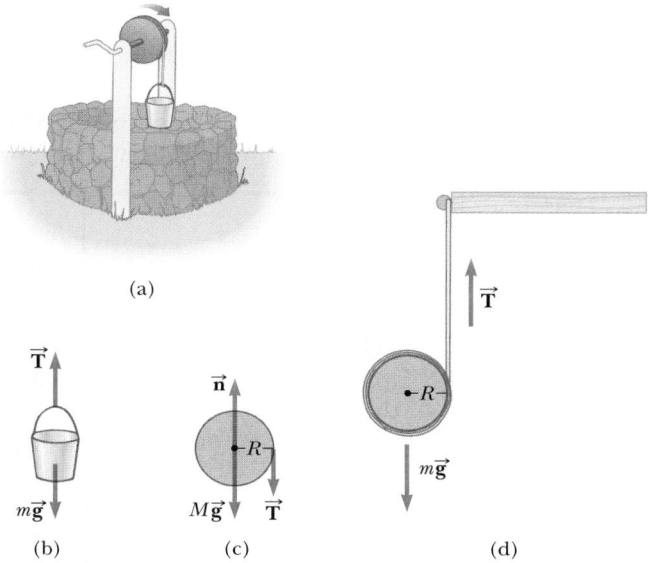

Figure 8.23 (Interactive Example 8.11) (a) A water bucket attached to a rope passing over a frictionless reel. (b) A free-body diagram for the bucket. (c) The tension produces a torque on the cylinder about its axis of rotation. (d) A falling cylinder (Exercise 8.11).

Solution

(a) Find the tension in the cord and the acceleration of the bucket.

Apply Newton's second law to the bucket in Figure 8.23b. There are two forces: the tension $\vec{\mathbf{T}}$ acting upwards and gravity $m\vec{\mathbf{g}}$ acting downwards.

$$ma = -mg + T \qquad \textbf{(1)}$$

Apply $\tau = I\alpha$ to the cylinder in Figure 8.23c:

$$\Sigma\tau = I\alpha = \tfrac{1}{2}MR^2\alpha \qquad \text{(solid cylinder)}$$

Notice the angular acceleration is clockwise, so the torque is negative. The normal and gravity forces have zero moment arm, and don't contribute any torque.

$$-TR = \tfrac{1}{2}MR^2\alpha \qquad \textbf{(2)}$$

Solve for T and substitute $\alpha = a/R$ (notice that both α and a are negative):

$$T = -\tfrac{1}{2}MR\alpha = -\tfrac{1}{2}Ma \qquad \textbf{(3)}$$

Substitute the expression for T in Equation (3) into Equation (1), and solve for the acceleration:

$$ma = -mg - \tfrac{1}{2}Ma \quad \rightarrow \quad a = -\frac{mg}{m + \frac{1}{2}M}$$

Substitute the values for m, M, and g, getting a, then substitute a into Equation (3) to get T.

$$a = -5.60 \text{ m/s}^2 \qquad T = 8.40 \text{ N}$$

(b) Find the distance the bucket falls in 3.00 s.
Apply the displacement kinematic equation for constant acceleration, with $t = 3.00$ s and $v_0 = 0$:

$$\Delta y = v_0 t + \tfrac{1}{2}at^2 = -\tfrac{1}{2}(5.60 \text{ m/s}^2)(3.00 \text{ s})^2 = -25.2 \text{ m}$$

Remarks Proper handling of signs is very important in these problems. All such signs should be chosen initially and checked mathematically and physically. In this problem, for example, both the angular acceleration α and the acceleration a are negative, so $\alpha = a/R$ applies. If the rope had been wound the other way on the cylinder, causing counterclockwise rotation, the torque would have been positive, and the relationship would have been $\alpha = -a/R$, with the double negative making the right-hand side positive, just like the left-hand side.

Exercise 8.11

A hollow cylinder of mass 0.100 kg and radius 4.00 cm has a string wrapped several times around it, as in Figure 8.23d. If the string is attached to a rigid support and the cylinder allowed to drop from rest, find (a) the acceleration of the cylinder and (b) the speed of the cylinder when a meter of string has unwound off of it.

Answers (a) $-4.90\ \text{m/s}^2$ (b) $3.13\ \text{m/s}$

PhysicsNow™ You can change the mass of the object and the mass and radius of the wheel to see the effect on how the system moves by logging into PhysicsNow at **www.cp7e.com** and going to Interactive Example 8.11.

8.6 ROTATIONAL KINETIC ENERGY

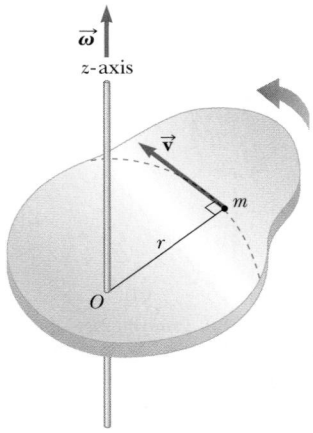

Figure 8.24 A rigid plate rotating about the z-axis with angular speed ω. The kinetic energy of a particle of mass m is $\frac{1}{2}mv^2$. The total kinetic energy of the plate is $\frac{1}{2}I\omega^2$.

In Chapter 5 we defined the kinetic energy of a particle moving through space with a speed v as the quantity $\frac{1}{2}mv^2$. Analogously, **an object rotating about some axis with an angular speed ω has rotational kinetic energy given by $\frac{1}{2}I\omega^2$.** To prove this, consider an object in the shape of a thin rigid plate rotating around some axis perpendicular to its plane, as in Figure 8.24. The plate consists of many small particles, each of mass m. All these particles rotate in circular paths around the axis. If r is the distance of one of the particles from the axis of rotation, the speed of that particle is $v = r\omega$. Because the *total* kinetic energy of the plate's rotation is the sum of all the kinetic energies associated with its particles, we have

$$KE_r = \Sigma(\tfrac{1}{2}mv^2) = \Sigma(\tfrac{1}{2}mr^2\omega^2) = \tfrac{1}{2}(\Sigma mr^2)\omega^2$$

In the last step, the ω^2 term is factored out because it's the same for every particle. Now, the quantity in parentheses on the right is the moment of inertia of the plate in the limit as the particles become vanishingly small, so

$$KE_r = \tfrac{1}{2}I\omega^2 \qquad [8.10]$$

where $I = \Sigma mr^2$ is the moment of inertia of the plate.

A system such as a bowling ball rolling down a ramp is described by three types of energy: **gravitational potential energy PE_g, translational kinetic energy KE_t,** and **rotational kinetic energy KE_r.** All these forms of energy, plus the potential energies of any other conservative forces, must be included in our equation for the conservation of mechanical energy of an isolated system:

Conservation of mechanical energy ▶

$$(KE_t + KE_r + PE)_i = (KE_t + KE_r + PE)_f \qquad [8.11]$$

where i and f refer to initial and final values, respectively, and PE includes the potential energies of all conservative forces in a given problem. This relation is true *only* if we ignore dissipative forces such as friction. In that case, it's necessary to resort to a generalization of the work–energy theorem:

Work–energy theorem including rotational energy ▶

$$W_{nc} = \Delta KE_t + \Delta KE_r + \Delta PE \qquad [8.12]$$

Problem-Solving Strategy Energy Methods and Rotation

1. **Choose two points of interest**, one where all necessary information is known, and the other where information is desired.
2. **Identify** the conservative and nonconservative forces acting on the system being analyzed.
3. **Write the general work–energy theorem**, Equation 8.12, or Equation 8.11 if all forces are conservative.
4. **Substitute general expressions** for the terms in the equation.
5. **Use $v = r\omega$** to eliminate either ω or v from the equation.
6. **Solve** for the unknown.

EXAMPLE 8.12 A Ball Rolling Down an Incline

Goal Combine gravitational, translational, and rotational energy.

Problem A ball of mass M and radius R starts from rest at a height of 2.00 m and rolls down a 30.0° slope, as in Active Figure 8.25. What is the linear speed of the ball when it leaves the incline? Assume that the ball rolls without slipping.

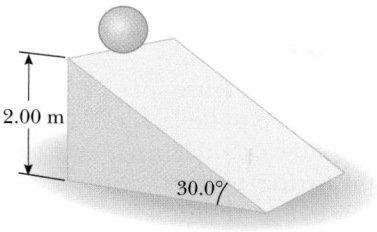

ACTIVE FIGURE 8.25
(Example 8.12) A ball starts from rest at the top of an incline and rolls to the bottom without slipping.

PhysicsNow™
Log into PhysicsNow at **www.cp7e.com**, and go to Active Figure 8.25 to roll several objects down a hill and see how the final speed depends on the shape of the object.

Strategy The two points of interest are the top and bottom of the incline, with the bottom acting as the zero point of gravitational potential energy. The force of static friction converts translational kinetic energy to rotational kinetic energy without dissipating any energy, so mechanical energy is conserved and Equation 8.11 can be applied.

Solution

Apply conservation of energy with $PE = PE_g$, the potential energy associated with gravity.

$$(KE_t + KE_r + PE_g)_i = (KE_t + KE_r + PE_g)_f$$

Substitute the appropriate general expressions, noting that $(KE_t)_i = (KE_r)_i = 0$ and $(PE_g)_f = 0$ (obtain the moment of inertia of a ball from Table 8.1):

$$0 + 0 + Mgh = \tfrac{1}{2}Mv^2 + \tfrac{1}{2}(\tfrac{2}{5}MR^2)\omega^2 + 0$$

The ball rolls without slipping, so $R\omega = v$, the "no-slip condition," can be applied:

$$Mgh = \tfrac{1}{2}Mv^2 + \tfrac{1}{5}Mv^2 = \tfrac{7}{10}Mv^2$$

Solve for v, noting that M cancels.

$$v = \sqrt{\frac{10gh}{7}} = \sqrt{\frac{10(9.80\ \text{m/s}^2)(2.00\ \text{m})}{7}} = 5.29\ \text{m/s}$$

Exercise 8.12

Repeat this example for a solid cylinder of the same mass and radius as the ball and released from the same height. In a race between the two objects on the incline, which one would win?

Answer $v = \sqrt{4gh/3} = 5.11$ m/s; the ball would win.

Quick Quiz 8.4

Two spheres, one hollow and one solid, are rotating with the same angular speed around an axis through their centers. Both spheres have the same mass and radius. Which sphere, if either, has the higher rotational kinetic energy? (a) The hollow sphere. (b) The solid sphere. (c) They have the same kinetic energy.

Quick Quiz 8.5

Which arrives at the bottom first, (a) a ball rolling without sliding down a certain incline A, (b) a solid cylinder rolling without sliding down incline A, or (c) a box of the same mass as the ball sliding down a frictionless incline B having the same dimensions as A? Assume that each object is released from rest at the top of its incline.

EXAMPLE 8.13 Blocks and Pulley

Goal Solve a system requiring rotation concepts and the work–energy theorem.

Problem Two blocks with masses $m_1 = 5.00$ kg and $m_2 = 7.00$ kg are attached by a string as in Figure 8.26a, over a pulley with mass $M = 2.00$ kg. The pulley, which turns on a frictionless axle, is a hollow cylinder with radius 0.050 0 m over which the string moves without slipping. The horizontal surface has coefficient of kinetic friction 0.350. Find the speed of the system when the block of mass m_2 has dropped 2.00 m.

Strategy This problem can be solved with the extension of the work–energy theorem, Equation 8.12. If the block of mass m_2 falls from height h to 0, then the block of mass m_1 moves the same distance, $\Delta x = h$. Apply the work–energy theorem, solve for v, and substitute. Kinetic friction the sole nonconservative force.

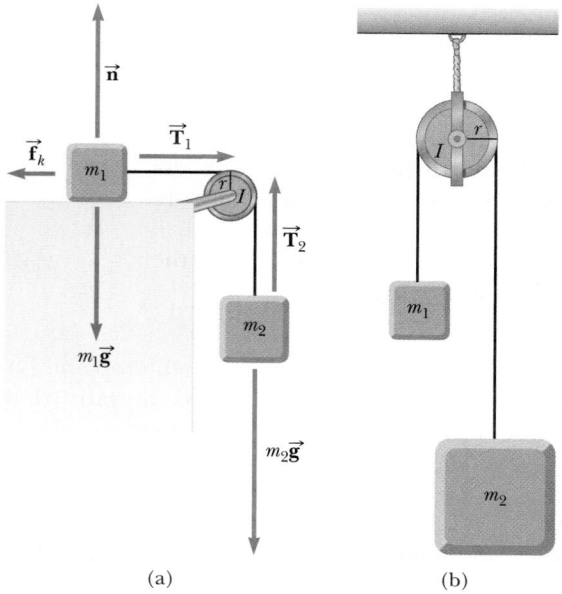

Figure 8.26 (a) (Example 8.13) (b) (Exercise 8.13) In both cases, $\vec{T}_1$ and $\vec{T}_2$ exert torques on the pulley.

Solution

Apply the work–energy theorem, with $PE = PE_g$, the potential energy associated with gravity.

$$W_{nc} = \Delta KE_t + \Delta KE_r + \Delta PE_g$$

Substitute the frictional work for W_{nc}, kinetic energy changes for the two blocks, the rotational kinetic energy change for the pulley, and the potential energy change for the second block.

$$-\mu_k n\Delta x = -\mu_k(m_1 g)\Delta x = (\tfrac{1}{2}m_1 v^2 - 0) + (\tfrac{1}{2}m_2 v^2 - 0) + (\tfrac{1}{2}I\omega^2 - 0) + (0 - m_2 gh)$$

Substitute $\Delta x = h$, and write I as $(I/r^2)r^2$:

$$-\mu_k(m_1 g)h = \tfrac{1}{2}m_1 v^2 + \tfrac{1}{2}m_2 v^2 + \tfrac{1}{2}\left(\frac{I}{r^2}\right)r^2\omega^2 - m_2 gh$$

For a hoop, $I = Mr^2$ so $(I/r^2) = M$. Substitute this quantity and $v = r\omega$:

$$-\mu_k(m_1 g)h = \tfrac{1}{2}m_1 v^2 + \tfrac{1}{2}m_2 v^2 + \tfrac{1}{2}Mv^2 - m_2 gh$$

Solve for v:

$$m_2 gh - \mu_k(m_1 g)h = \tfrac{1}{2}m_1 v^2 + \tfrac{1}{2}m_2 v^2 + \tfrac{1}{2}Mv^2 = \tfrac{1}{2}(m_1 + m_2 + M)v^2$$

$$v = \sqrt{\frac{2gh(m_2 - \mu_k m_1)}{m_1 + m_2 + M}}$$

Substitute $m_1 = 5.00$ kg, $m_2 = 7.00$ kg, $M = 2.00$ kg, $g = 9.80$ m/s^2, $h = 2.00$ m, and $\mu_k = 0.350$:

$$v = 3.83 \text{ m/s}$$

Remarks In the expression for the speed v, the mass m_1 of the first block and the mass M of the pulley all appear in the denominator, reducing the speed, as they should. In the numerator, m_2 is positive while the friction term is negative. Both assertions are reasonable, because the force of gravity on m_2 increases the speed of the system while the force of friction on m_1 slows it down. This problem can also be solved with Newton's second law together with $\tau = I\alpha$, a difficult exercise (though it can be facilitated with a system approach).

Exercise 8.13

Two blocks with masses $m_1 = 2.00$ kg and $m_2 = 9.00$ kg are attached over a pulley with mass $M = 3.00$ kg, hanging straight down as in Atwood's machine (Fig. 8.26b). The pulley is a solid cylinder with radius 0.050 0 m, and there is

some friction in the axle. The system is released from rest, and the string moves without slipping over the pulley. If the larger mass is traveling at a speed of 2.50 m/s when it has dropped 1.00 m, how much mechanical energy was lost due to friction in the pulley's axle?
[*Hint:* This exercise is slightly easier than the associated example because the friction force need not be determined.]

Answer 29.5 J

8.7 ANGULAR MOMENTUM

In Figure 8.27, an object of mass m rotates in a circular path of radius r, acted on by a net force, $\vec{\mathbf{F}}_{\text{net}}$. The resulting net torque on the object increases its angular speed from the value ω_0 to the value ω in a time interval Δt. Therefore, we can write

$$\Sigma\tau = I\alpha = I\frac{\Delta\omega}{\Delta t} = I\left(\frac{\omega - \omega_0}{\Delta t}\right) = \frac{I\omega - I\omega_0}{\Delta t}$$

If we define the product

$$L \equiv I\omega \qquad [8.13]$$

◀ Definition of angular momentum

as the **angular momentum** of the object, then we can write

$$\Sigma\tau = \frac{\text{change in angular momentum}}{\text{time interval}} = \frac{\Delta L}{\Delta t} \qquad [8.14]$$

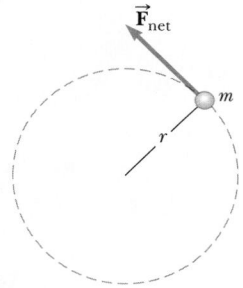

Figure 8.27 An object of mass m rotating in a circular path under the action of a constant torque.

Equation 8.14 is the rotational analog of Newton's second law in the form $F = \Delta p/\Delta t$ and states that **the net torque acting on an object is equal to the time rate of change of the object's angular momentum**. Recall that this equation also parallels the impulse–momentum theorem.

When the net external torque ($\Sigma\tau$) acting on a system is zero, Equation 8.14 gives that $\Delta L/\Delta t = 0$, which says that the time rate of change of the system's angular momentum is zero. We then have the following important result:

◀ Conservation of angular momentum

Let L_i and L_f be the angular momenta of a system at two different times, and suppose there is no net external torque, so $\Sigma\tau = 0$. Then

$$L_i = L_f \qquad [8.15]$$

and angular momentum is said to be *conserved.*

Equation 8.15 gives us a third conservation law to add to our list: **conservation of angular momentum**. We can now state that **the mechanical energy, linear momentum, and angular momentum of an isolated system all remain constant**.

If the moment of inertia of an isolated rotating system changes, the system's angular speed will change. Conservation of angular momentum then requires that

$$I_i\omega_i = I_f\omega_f \quad \text{if} \quad \Sigma\tau = 0 \qquad [8.16]$$

Note that conservation of angular momentum applies to macroscopic objects such as planets and people, as well as to atoms and molecules. There are many examples of conservation of angular momentum; one of the most dramatic is that of a figure skater spinning in the finale of her act. In Figure 8.28a, the skater has pulled her arms and legs close to her body, reducing their distance from her axis of rotation and hence also reducing her moment of inertia. By conservation of angular momentum, a reduction in her moment of inertia must increase her angular velocity. Coming out of the spin in Figure 8.28b, she needs to reduce her angular velocity, so she extends her arms and legs again, increasing her moment of inertia and thereby slowing her rotation.

APPLICATION

Figure Skating

© Benson Krista Hicks/Corbis Sygma

(a) (b)

Figure 8.28 Michelle Kwan controls her moment of inertia. (a) By pulling in her arms and legs, she reduces her moment of inertia and increases her angular velocity (rate of spin). (b) Upon landing, extending her arms and legs increases her moment of inertia and helps slow her spin.

APPLICATION

Aerial Somersaults

© Patrick Giardino/Corbis

Tightly curling her body, a diver decreases her moment of inertia, increasing her angular velocity.

Similarly, when a diver or an acrobat wishes to make several somersaults, she pulls her hands and feet close to the trunk of her body in order to rotate at a greater angular speed. In this case, the external force due to gravity acts through her center of gravity and hence exerts no torque about her axis of rotation, so the angular momentum about her center of gravity is conserved. For example, when a diver wishes to double her angular speed, she must reduce her moment of inertia to half its initial value.

An interesting astrophysical example of conservation of angular momentum occurs when a massive star, at the end of its lifetime, uses up all its fuel and collapses under the influence of gravitational forces, causing a gigantic outburst of energy called a supernova. The best-studied example of a remnant of a supernova explosion is the Crab Nebula, a chaotic, expanding mass of gas (Fig. 8.29). In a supernova, part of the star's mass is ejected into space, where it eventually condenses into new stars and planets. Most of what is left behind typically collapses into a **neutron star**—an extremely dense sphere of matter with a diameter of about 10 km, greatly reduced from the 10^6-km diameter of the original star and containing a large fraction of the star's original mass. In a neutron star, pressures become so great that atomic electrons combine with protons, becoming neutrons. As the moment of inertia of the system decreases

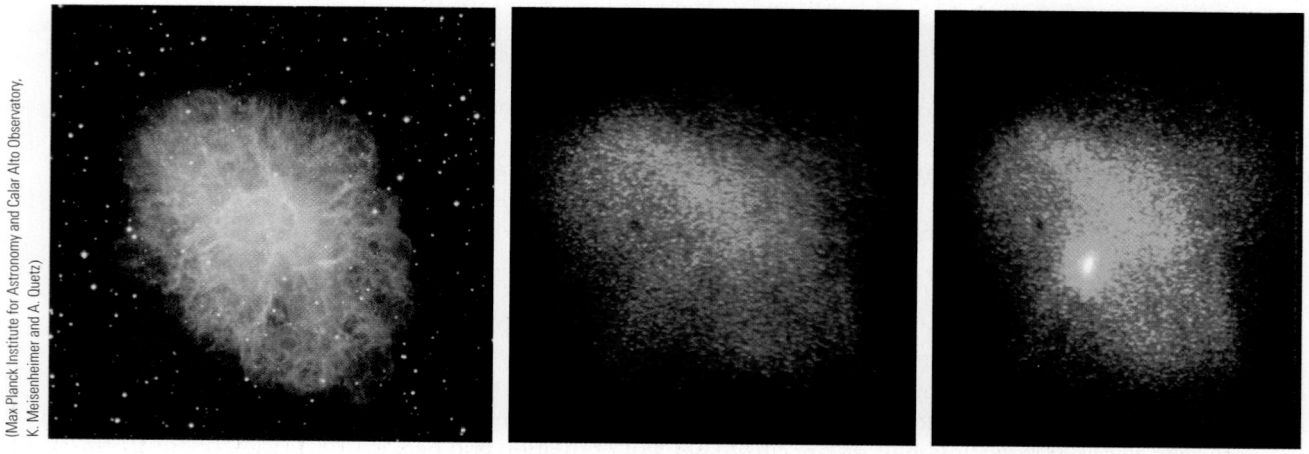

(Max Planck Institute for Astronomy and Calar Alto Observatory, K. Meisenheimer and A. Quetz)

© Smithsonian Institute/Photo Researchers, Inc.

(a) (b) (c)

Figure 8.29 (a) The Crab Nebula in the constellation Taurus. This nebula is the remnant of a supernova seen on Earth in A.D. 1054. It is located some 6 300 lightyears away and is approximately 6 lightyears in diameter, still expanding outward. A pulsar deep inside the nebula flashes 30 times every second. (b) Pulsar off. (c) Pulsar on.

during the collapse, the star's rotational speed increases. More than 700 rapidly rotating neutron stars have been identified since their first discovery in 1967, with periods of rotation ranging from a millisecond to several seconds. The neutron star is an amazing system—an object with a mass greater than the Sun, fitting comfortably within the space of a small county and rotating so fast that the tangential speed of the surface approaches a sizeable fraction of the speed of light!

APPLICATION
Rotating Neutron Stars

Quick Quiz 8.6

A horizontal disk with moment of inertia I_1 rotates with angular speed ω_1 about a vertical frictionless axle. A second horizontal disk, with moment of inertia I_2 and initially not rotating, drops onto the first. Because their surfaces are rough, the two eventually reach the same angular speed ω. The ratio ω/ω_1 is equal to (a) I_1/I_2 (b) I_2/I_1 (c) $I_1/(I_1 + I_2)$ (d) $I_2/(I_1 + I_2)$

Quick Quiz 8.7

If global warming continues, it's likely that some ice from the polar ice caps of the Earth will melt and the water will be distributed closer to the Equator. If this occurs, would the length of the day (one revolution) (a) increase, (b) decrease, or (c) remain the same?

EXAMPLE 8.14 The Spinning Stool

Goal Apply conservation of angular momentum to a simple system.

Problem A student sits on a pivoted stool while holding a pair of weights. (See Fig. 8.30.) The stool is free to rotate about a vertical axis with negligible friction. The moment of inertia of student, weights, and stool is 2.25 kg · m^2. The student is set in rotation with arms outstretched, making one complete turn every 1.26 s, arms outstretched. **(a)** What is the initial angular speed of the system? **(b)** As he rotates, he pulls the weights inward so that the new moment of inertia of the system (student, objects, and stool) becomes 1.80 kg · m^2. What is the new angular speed of the system? **(c)** Find the work done by the student on the system while pulling in the weights. (Ignore energy lost through dissipation in his muscles.)

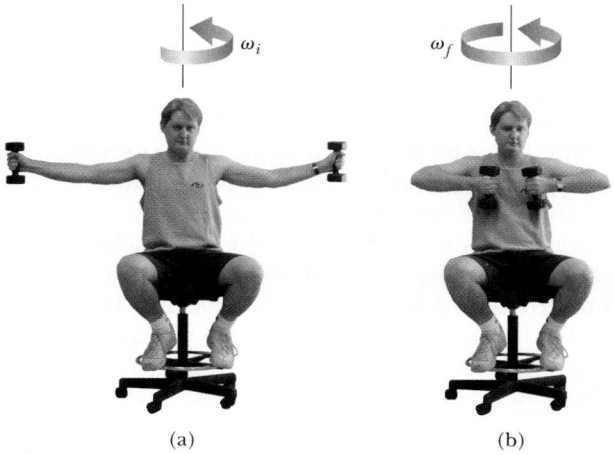

Figure 8.30 (Example 8.14) (a) The student is given an initial angular speed while holding two weights out. (b) The angular speed increases as the student draws the weights inwards.

Strategy (a) The angular frequency can be obtained from the frequency, which is the inverse of the period. (b) There are no external torques acting on the system, so the new angular speed can be found with the principle of conservation of angular momentum. (c) The work done on the system during this process is the same as the system's change in rotational kinetic energy.

Solution

(a) Find the initial angular speed of the system.

Invert the period to get the frequency, and multiply by 2π: $\omega_i = 2\pi f = 2\pi/T = 4.99 \text{ rad/s}$

(b) After he pulls the weights in, what's the system's new angular speed?

Equate the initial and final angular momenta of the system: $L_i = L_f \rightarrow I_i\omega_i = I_f\omega_f$ **(1)**

Substitute and solve for the final angular speed ω_f:

$$(2.25\ \text{kg}\cdot\text{m}^2)(4.99\ \text{rad/s}) = (1.80\ \text{kg}\cdot\text{m}^2)\omega_f \qquad \textbf{(2)}$$

$$\omega_f = \boxed{6.24\ \text{rad/s}}$$

(c) Find the work the student does on the system.

Apply the work–energy theorem:

$$\begin{aligned} W_{\text{student}} &= \Delta K_r = \tfrac{1}{2}I_f\omega_f^2 - \tfrac{1}{2}I_i\omega_i^2 \\ &= \tfrac{1}{2}(1.80\ \text{kg}\cdot\text{m}^2)(6.24\ \text{rad/s})^2 \\ &\quad - \tfrac{1}{2}(2.25\ \text{kg}\cdot\text{m}^2)(4.99\ \text{rad/s})^2 \end{aligned}$$

$$W_{\text{student}} = \boxed{7.03\ \text{J}}$$

Remarks Although the angular momentum of the system is conserved, mechanical energy is not conserved because the student does work on the system.

Exercise 8.14

A star with an initial radius of 1.0×10^8 m and period of 30.0 days collapses suddenly to a radius of 1.0×10^4 m. (a) Find the period of rotation after collapse. (b) Find the work done by gravity during the collapse if the mass of the star is 2.0×10^{30} kg. (c) What is the speed of an indestructible person standing on the equator of the collapsed star? (Neglect any relativistic or thermal effects, and assume the star is spherical before and after it collapses.)

Answers (a) 2.6×10^{-2} s (b) 2.3×10^{42} J (c) 2.4×10^6 m/s

EXAMPLE 8.15 The Merry-Go-Round

Goal Apply conservation of angular momentum while combining two moments of inertia.

Problem A merry-go-round modeled as a disk of mass $M = 1.00 \times 10^2$ kg and radius $R = 2.00$ m is rotating in a horizontal plane about a frictionless vertical axle (Fig. 8.31). **(a)** After a student with mass $m = 60.0$ kg jumps onto the merry-go-round, the system's angular speed decreases to 2.00 rad/s. If the student walks slowly from the edge toward the center, find the angular speed of the system when she reaches a point 0.500 m from the center. **(b)** Find the change in the system's rotational kinetic energy caused by her movement to the center. **(c)** Find the work done on the student as she walks to $r = 0.500$ m.

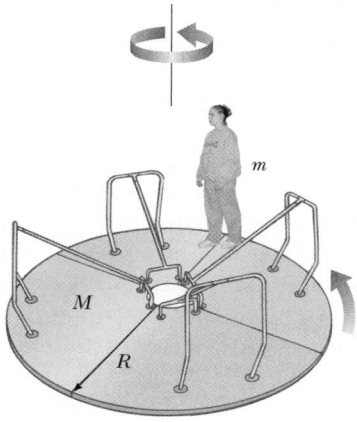

Figure 8.31 (Example 8.15) As the student walks toward the center of the rotating platform, the moment of inertia of the system (student plus platform) decreases. Because angular momentum is conserved, the angular speed of the system must increase.

Strategy This problem can be solved with conservation of angular momentum by equating the system's initial angular momentum when the student stands at the rim to the angular momentum when the student has reached $r = 0.500$ m. The key is to find the different moments of inertia.

Solution

(a) Find the angular speed when the student reaches a point 0.500 m from the center.

Calculate the moment of inertia of the disk, I_D:

$$\begin{aligned} I_D &= \tfrac{1}{2}MR^2 = \tfrac{1}{2}(1.00 \times 10^2\ \text{kg})(2.00\ \text{m})^2 \\ &= 2.00 \times 10^2\ \text{kg}\cdot\text{m}^2 \end{aligned}$$

Calculate the initial moment of inertia of the student. This is the same as the moment of inertia of a mass a distance R from the axis:

$$I_S = mR^2 = (60.0\ \text{kg})(2.00\ \text{m})^2 = 2.40 \times 10^2\ \text{kg}\cdot\text{m}^2$$

Sum the two moments of inertia and multiply by the initial angular speed to find L_i, the initial angular momentum of the system:

$$\begin{aligned} L_i &= (I_D + I_S)\omega_i \\ &= (2.00 \times 10^2\ \text{kg}\cdot\text{m}^2 + 2.40 \times 10^2\ \text{kg}\cdot\text{m}^2)(2.00\ \text{rad/s}) \\ &= 8.80 \times 10^2\ \text{kg}\cdot\text{m}^2/\text{s} \end{aligned}$$

Calculate the student's final moment of inertia, I_{Sf}, when she is 0.500 m from the center:

$$I_{Sf} = mr_f^2 = (60.0\text{ kg})(0.50\text{ m})^2 = 15.0\text{ kg}\cdot\text{m}^2$$

The moment of inertia of the platform is unchanged. Add it to the student's final moment of inertia, and multiply by the unknown final angular speed to find L_f:

$$L_f = (I_D + I_{Sf})\omega_f = (2.00 \times 10^2\text{ kg}\cdot\text{m}^2 + 15.0\text{ kg}\cdot\text{m}^2)\omega_f$$

Equate the initial and final angular momenta and solve for the final angular speed of the system:

$$L_i = L_f$$
$$(8.80 \times 10^2\text{ kg}\cdot\text{m}^2/\text{s}) = (2.15 \times 10^2\text{ kg}\cdot\text{m}^2)\omega_f$$
$$\omega_f = \boxed{4.09\text{ rad/s}}$$

(b) Find the change in the rotational kinetic energy of the system.

Calculate the initial kinetic energy of the system:

$$KE_i = \tfrac{1}{2}I_i\omega_i^2 = \tfrac{1}{2}(4.40 \times 10^2\text{ kg}\cdot\text{m}^2)(2.00\text{ rad/s})^2 = 8.80 \times 10^2\text{ J}$$

Calculate the final kinetic energy of the system:

$$KE_f = \tfrac{1}{2}I_f\omega_f^2 = \tfrac{1}{2}(215\text{ kg}\cdot\text{m}^2)(4.09\text{ rad/s})^2 = 1.80 \times 10^3\text{ J}$$

Calculate the change in kinetic energy of the system.

$$KE_f - KE_i = \boxed{920\text{ J}}$$

(c) Find the work done on the student.

The student undergoes a change in kinetic energy that equals the work done on her. Apply the work–energy theorem:

$$W = \Delta KE_{\text{student}} = \tfrac{1}{2}I_{Sf}\omega_f^2 - \tfrac{1}{2}I_S\omega_i^2 = \tfrac{1}{2}(15.0\text{ kg}\cdot\text{m}^2)(4.09\text{ rad/s})^2 - \tfrac{1}{2}(2.40 \times 10^2\text{ kg}\cdot\text{m}^2)(2.00\text{ rad/s})^2$$
$$W = \boxed{-355\text{ J}}$$

Remarks The angular momentum is unchanged by internal forces; however, the kinetic energy increases, because the student must perform positive work in order to walk toward the center of the platform.

Exercise 8.15

(a) Find the angular speed of the merry-go-round before the student jumped on, assuming the student didn't transfer any momentum or energy as she jumped on the merry-go-round. (b) By how much did the kinetic energy of the system change when the student jumped on? Notice that energy is lost in this process, as should be expected, since it is essentially a perfectly inelastic collision.

Answers (a) 4.4 rad/s (b) $KE_f - KE_i = -1.06 \times 10^3$ J.

SUMMARY

PhysicsNow™ Take a practice test by logging into PhysicsNow at **www.cp7e.com** and clicking on the Pre-Test link for this chapter.

8.1 Torque

Let $\vec{\mathbf{F}}$ be a force acting on an object, and let $\vec{\mathbf{r}}$ be a position vector from a chosen point O to the point of application of the force. Then the magnitude of the torque $\vec{\boldsymbol{\tau}}$ of the force $\vec{\mathbf{F}}$ is given by

$$\tau = rF\sin\theta \qquad [8.2]$$

where r is the length of the position vector, F the magnitude of the force, and θ the angle between $\vec{\mathbf{F}}$ and $\vec{\mathbf{r}}$.

The quantity $d = r\sin\theta$ is called the *lever arm* of the force.

8.2 Torque and the Two Conditions for Equilibrium

An object in mechanical equilibrium must satisfy the following two conditions:

1. The net external force must be zero: $\Sigma\vec{\mathbf{F}} = 0$.
2. The net external torque must be zero: $\Sigma\vec{\boldsymbol{\tau}} = 0$.

These two conditions, used in solving problems involving rotation in a plane—result in three equations and three unknowns—two from the first condition (corresponding to the x- and y-components of the force) and one from the second condition, on torques. These equations must be solved simultaneously.

8.5 Relationship Between Torque and Angular Acceleration

The **moment of inertia** of a group of particles is

$$I \equiv \Sigma mr^2 \qquad [8.7]$$

If a rigid object free to rotate about a fixed axis has a net external torque $\Sigma\tau$ acting on it, then the object undergoes an angular acceleration α, where

$$\Sigma\tau = I\alpha \qquad [8.8]$$

This equation is the rotational equivalent of the second law of motion.

Problems are solved by using Equation 8.8 together with Newton's second law and solving the resulting equations simultaneously. The relation $a = r\alpha$ is often key in relating the translational equations to the rotational equations.

8.6 Rotational Kinetic Energy

If a rigid object rotates about a fixed axis with angular speed ω, its **rotational kinetic energy** is

$$KE_r \equiv \tfrac{1}{2}I\omega^2 \qquad [8.10]$$

where I is the moment of inertia of the object around the axis of rotation.

A system involving rotation is described by three types of energy: potential energy PE, translational kinetic energy KE_t, and rotational kinetic energy KE_r. All these forms of energy must be included in the equation for conservation of mechanical energy for an isolated system:

$$(KE_t + KE_r + PE)_i = (KE_t + KE_r + PE)_f \qquad [8.11]$$

where i and f refer to initial and final values, respectively. When non-conservative forces are present, it's necessary to use a generalization of the work–energy theorem:

$$W_{nc} = \Delta KE_t + \Delta KE_r + \Delta PE \qquad [8.12]$$

8.7 Angular Momentum

The **angular momentum** of a rotating object is given by

$$L \equiv I\omega \qquad [8.13]$$

Angular momentum is related to torque in the following equation:

$$\Sigma\tau = \frac{\text{change in angular momentum}}{\text{time interval}} = \frac{\Delta L}{\Delta t} \qquad [8.14]$$

If the net external torque acting on a system is zero, then the total angular momentum of the system is constant,

$$L_i = L_f \qquad [8.15]$$

and is said to be conserved. Solving problems usually involves substituting into the expression

$$I_i\omega_i = I_f\omega_f \qquad [8.16]$$

and solving for the unknown.

CONCEPTUAL QUESTIONS

1. Why can't you put your heels firmly against a wall and then bend over without falling?
2. Why does a tall athlete have an advantage over a smaller one when the two are competing in the high jump?
3. Both torque and work are products of force and distance. How are they different? Do they have the same units?
4. Is it possible to calculate the torque acting on a rigid object without specifying an origin? Is the torque independent of the location of the origin?
5. Can an object be in equilibrium when only one force acts on it? If you believe the answer is yes, give an example to support your conclusion.
6. The polar ice caps contain about 2.3×10^{19} kg of ice. This mass contributes almost nothing to the moment of inertia of the Earth because it is located at the poles, close to the Earth's axis of rotation. *Estimate* the change in the length of the day that would be expected if the polar ice caps were to melt and the water were distributed uniformly over the surface of the Earth. (Note that the moment of inertia of a thin spherical shell of radius r and mass m is $2mr^2/3$.) (Question 6 is courtesy of Edward F. Redish. For more questions of this type, see www.physics.umd.edu/perg/.)
7. In some motorcycle races, the riders drive over small hills, and the motorcycle becomes airborne for a short time. If the motorcycle racer keeps the throttle open while leaving the hill and going into the air, the motorcycle's nose tends to rise upwards. Why does this happen?
8. In the movie *Jurassic Park*, there is a scene in which some members of the visiting group are trapped in the kitchen with dinosaurs outside. The paleontologist is pressing against the center of the door, trying to keep out the dinosaurs on the other side. The botanist throws herself against the door at the edge near the hinge. A pivotal point in the film is that she cannot reach a gun on the floor because she is trying to hold the door closed. If the paleontologist is pressing at the center of the door, and the botanist is pressing at the edge about 8 cm from the hinge, estimate how far the paleontologist would have to relocate in order to have a greater effect on keeping the door closed than both of them pushing together have in their original positions. (Question 8 is courtesy of Edward F. Redish. For more questions of this type, see www.physics.umd.edu/perg/.)
9. Suppose you are designing a car for a coasting race—a race in which the cars have no engines, but simply coast downhill. Do you want large wheels or small wheels? Do you want solid, disklike wheels or hooplike wheels? Should the wheels be heavy or light?
10. If you toss a textbook into the air, rotating it each time about one of the three axes perpendicular to it, you will find that it will not rotate smoothly about one of those axes. (Try placing a strong rubber band around the book

before the toss so that it will stay closed.) The book's rotation is stable about those axes having the largest and smallest moments of inertia, but unstable about the axis of intermediate moment. Try this on your own to find the axis that has this intermediate moment of inertia.

11. Stars originate as large bodies of slowly rotating gas. Because of gravity, these clumps of gas slowly decrease in size. What happens to the angular speed of a star as it shrinks? Explain.

12. If a high jumper positions his body correctly when going over the bar, the center of gravity of the athlete may actually pass under the bar. (See Fig. Q8.12.) Explain how this is possible.

© Eye Ubiquitous/Corbis

Figure Q8.12

13. In a tape recorder, the tape is pulled past the read–write heads at a constant speed by the drive mechanism. Consider the reel from which the tape is pulled: As the tape is pulled off, the radius of the roll of remaining tape decreases. How does the torque on the reel change with time? How does the angular speed of the reel change with time? If the tape mechanism is suddenly turned on so that the tape is quickly pulled with a large force, is the tape more likely to break when pulled from a nearly full reel or from a nearly empty reel?

14. (a) Give an example in which the net force acting on an object is zero, yet the net torque is nonzero. (b) Give an example in which the net torque acting on an object is zero, yet the net force is nonzero.

15. A mouse is initially at rest on a horizontal turntable mounted on a frictionless vertical axle. If the mouse begins to walk clockwise around the perimeter of the table, what happens to the turntable? Explain.

16. A cat usually lands on its feet regardless of the position from which it is dropped. A slow-motion film of a cat falling shows that the upper half of its body twists in one direction while the lower half twists in the opposite direction. (See Fig. Q8.16.) Why does this type of rotation occur?

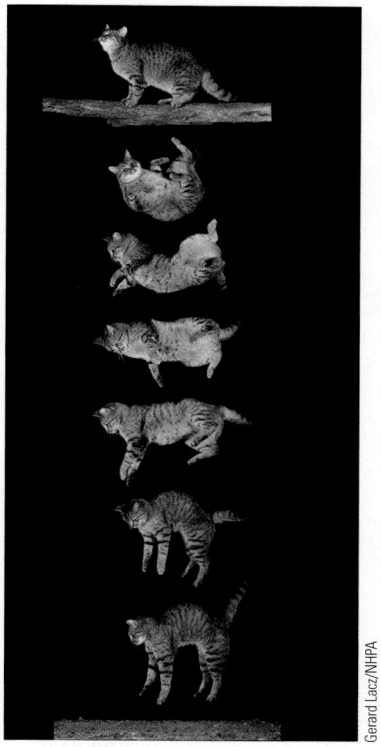

Gerard Lacz/NHPA

Figure Q8.16 A falling, twisting cat.

17. A ladder rests inclined against a wall. Would you feel safer climbing up the ladder if you were told that the floor was frictionless, but the wall was rough, or that the wall was frictionless, but the floor was rough? Justify your answer.

18. Two solid spheres—a large, massive sphere and a small sphere with low mass—are rolled down a hill. Which one reaches the bottom of the hill first? Next, we roll a large, low-density sphere and a small, high-density sphere, both with the same mass. Which one wins the race?

PROBLEMS

1, 2, 3 = straightforward, intermediate, challenging □ = full solution available in *Student Solutions Manual/Study Guide*
PhysicsNow™ = coached problem with hints available at **www.cp7e.com** = biomedical application

Section 8.1 Torque

1. If the torque required to loosen a nut that is holding a flat tire in place on a car has a magnitude of 40.0 N · m, what *minimum* force must be exerted by the mechanic at the end of a 30.0-cm lug wrench to accomplish the task?

2. A steel band exerts a horizontal force of 80.0 N on a tooth at point *B* in Figure P8.2. What is the torque on the root of the tooth about point *A*?

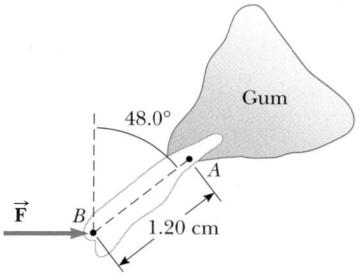

Figure P8.2

3. Calculate the net torque (magnitude and direction) on the beam in Figure P8.3 about (a) an axis through O perpendicular to the page and (b) an axis through C perpendicular to the page.

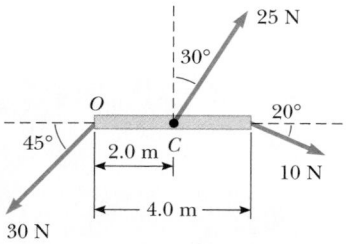

Figure P8.3

4. Write the necessary equations of equilibrium of the object shown in Figure P8.4. Take the origin of the torque equation about an axis perpendicular to the page through the point O.

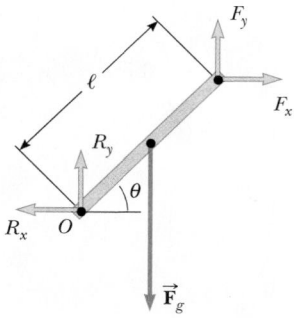

Figure P8.4

5. A simple pendulum consists of a small object of mass 3.0 kg hanging at the end of a 2.0-m-long light string that is connected to a pivot point. Calculate the magnitude of the torque (due to the force of gravity) about this pivot point when the string makes a 5.0° angle with the vertical.

6. A fishing pole is 2.00 m long and inclined to the horizontal at an angle of 20.0° (Fig. P8.6). What is the torque exerted by the fish about an axis perpendicular to the page and passing through the hand of the person holding the pole?

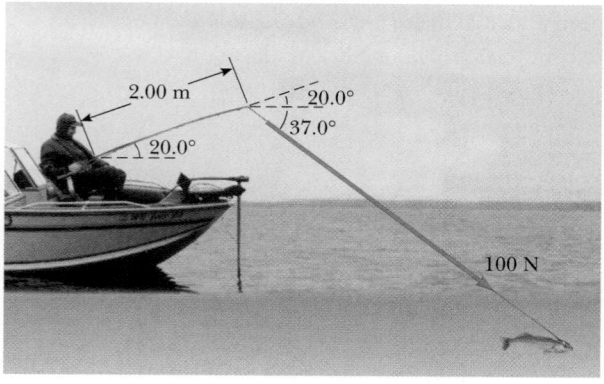

Figure P8.6

Section 8.2 Torque and the Two Conditions for Equilibrium
Section 8.3 The Center of Gravity
Section 8.4 Examples of Objects in Equilibrium

7. The arm in Figure P8.7 weighs 41.5 N. The force of gravity acting on the arm acts through point A. Determine the magnitudes of the tension force $\vec{F}_t$ in the deltoid muscle and the force $\vec{F}_s$ exerted by the shoulder on the humerus (upper-arm bone) to hold the arm in the position shown.

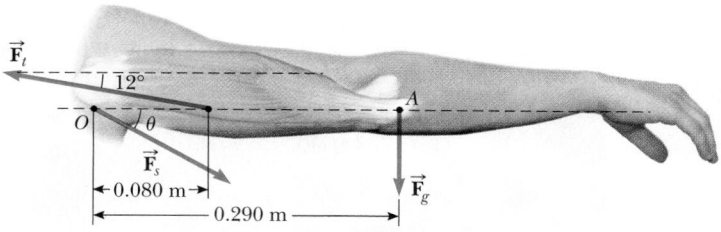

Figure P8.7

8. A water molecule consists of an oxygen atom with two hydrogen atoms bound to it as shown in Figure P8.8. The bonds are 0.100 nm in length, and the angle between the two bonds is 106°. Use the coordinate axes shown, and determine the location of the center of gravity of the molecule. Take the mass of an oxygen atom to be 16 times the mass of a hydrogen atom.

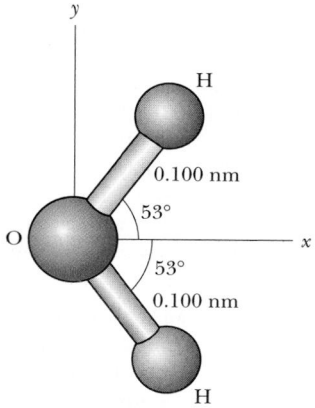

Figure P8.8

9. **Physics Now™** A cook holds a 2.00-kg carton of milk at arm's length (Fig. P8.9). What force $\vec{F}_B$ must be exerted by the biceps muscle? (Ignore the weight of the forearm.)

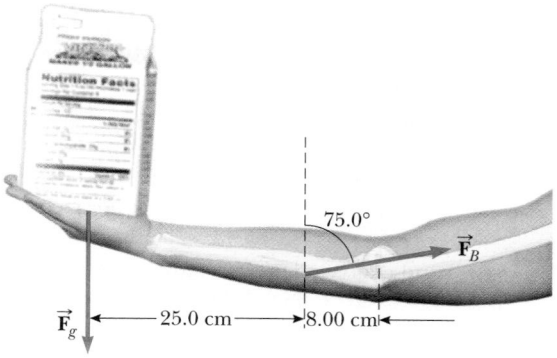

Figure P8.9

10. A meterstick is found to balance at the 49.7-cm mark when placed on a fulcrum. When a 50.0-gram mass is attached at the 10.0-cm mark, the fulcrum must be moved to the 39.2-cm mark for balance. What is the mass of the meter stick?

11. Find the x- and y-coordinates of the center of gravity of a 4.00-ft by 8.00-ft uniform sheet of plywood with the upper right quadrant removed as shown in Figure P8.11.

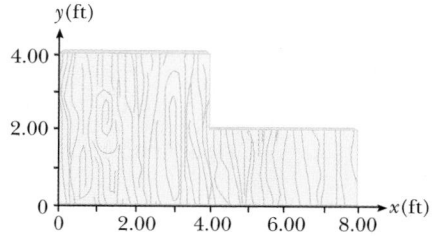

Figure P8.11

12. Consider the following mass distribution, where x- and y-coordinates are given in meters: 5.0 kg at (0.0, 0.0) m, 3.0 kg at (0.0, 4.0) m, and 4.0 kg at (3.0, 0.0) m. Where should a fourth object of 8.0 kg be placed so that the center of gravity of the four-object arrangement will be at (0.0, 0.0) m?

13. Many of the elements in horizontal-bar exercises can be modeled by representing the gymnast by four segments consisting of arms, torso (including the head), thighs, and lower legs, as shown in Figure P8.13a. Inertial parameters for a particular gymnast are as follows:

Segment	Mass (kg)	Length (m)	r_{cg} (m)	I (kg-m^2)
Arms	6.87	0.548	0.239	0.205
Torso	33.57	0.601	0.337	1.610
Thighs	14.07	0.374	0.151	0.173
Legs	7.54	—	0.227	0.164

Note that in Figure P8.13a r_{cg} is the distance to the center of gravity measured from the joint closest to the bar and the masses for the arms, thighs, and legs include both appendages. I is the moment of inertia of each segment about its center of gravity. Determine the distance from the bar to the center of gravity of the gymnast for the two positions shown in Figures P8.13b and P8.13c.

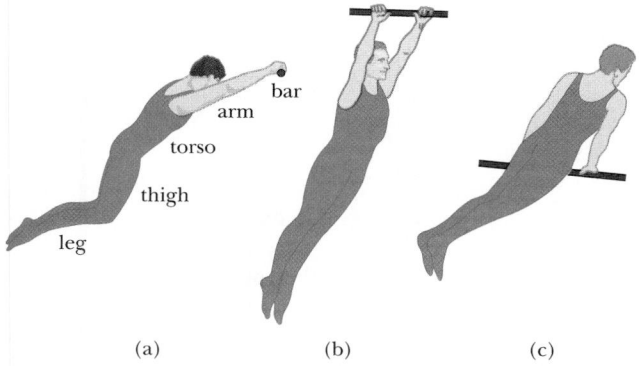

Figure P8.13

14. Using the data given in Problem 13 and the coordinate system shown in Figure P8.14b, calculate the position of the center of gravity of the gymnast shown in Figure P8.14a. Pay close attention to the definition of r_{cg} in the table.

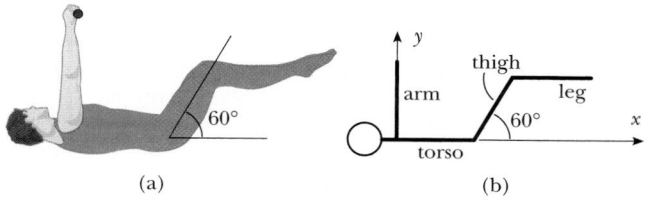

Figure P8.14

15. A person bending forward to lift a load "with his back" (Fig. P8.15a) rather than "with his knees" can be injured by large forces exerted on the muscles and vertebrae. The spine pivots mainly at the fifth lumbar vertebra, with the principal supporting force provided by the erector spinalis muscle in the back. To see the magnitude of the forces involved, and to understand why back problems are common among humans, consider the model shown in Fig. P8.15b of a person bending forward to lift a 200-N object. The spine and upper body are represented as a uniform horizontal rod of weight 350 N, pivoted at the base of the spine. The erector spinalis muscle, attached at a point two-thirds of the way up the spine, maintains the position of the back. The angle between the spine and this muscle is 12.0°. Find the tension in the back muscle and the compressional force in the spine.

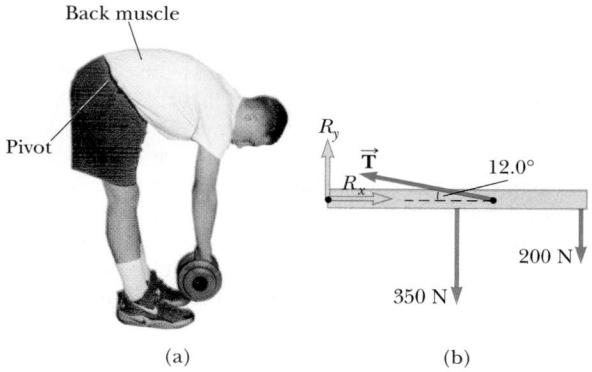

Figure P8.15

16. When a person stands on tiptoe (a strenuous position), the position of the foot is as shown in Figure P8.16a. The total gravitational force on the body, $\vec{F}_g$, is supported by the force $\vec{n}$ exerted by the floor on the toes of one foot. A mechanical model of the situation is shown in Figure P8.16b, where $\vec{T}$ is the force exerted by the Achilles tendon on the foot and $\vec{R}$ is the force exerted by the tibia on the foot. Find the values of T, R, and θ when $F_g = 700$ N.

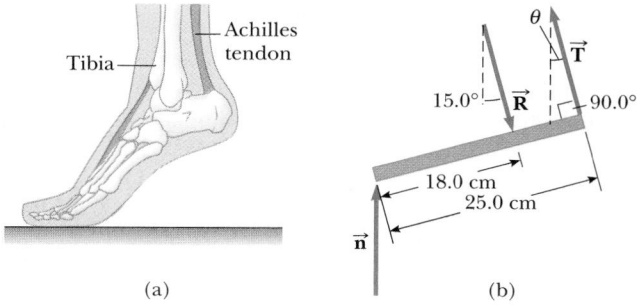

Figure P8.16

17. A 500-N uniform rectangular sign 4.00 m wide and 3.00 m high is suspended from a horizontal, 6.00-m-long, uniform, 100-N rod as indicated in Figure P8.17. The left end of the rod is supported by a hinge, and the right end is supported by a thin cable making a 30.0° angle with the vertical. (a) Find the tension T in the cable. (b) Find the horizontal and vertical components of force exerted on the left end of the rod by the hinge.

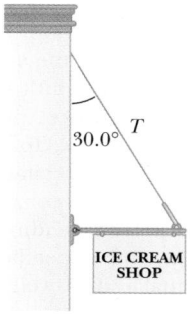

Figure P8.17

18. A window washer is standing on a scaffold supported by a vertical rope at each end. The scaffold weighs 200 N and is 3.00 m long. What is the tension in each rope when the 700-N worker stands 1.00 m from one end?

19. The chewing muscle, the masseter, is one of the strongest in the human body. It is attached to the mandible (lower jawbone) as shown in Figure P8.19a. The jawbone is pivoted about a socket just in front of the auditory canal. The forces acting on the jawbone are equivalent to those acting on the curved bar in Figure P8.19b: $\vec{\mathbf{F}}_c$ is the force exerted by the food being chewed against the jawbone, $\vec{\mathbf{T}}$ is the force of tension in the masseter, and $\vec{\mathbf{R}}$ is the force exerted by the socket on the mandible. Find $\vec{\mathbf{T}}$ and $\vec{\mathbf{R}}$ for a person who bites down on a piece of steak with a force of 50.0 N.

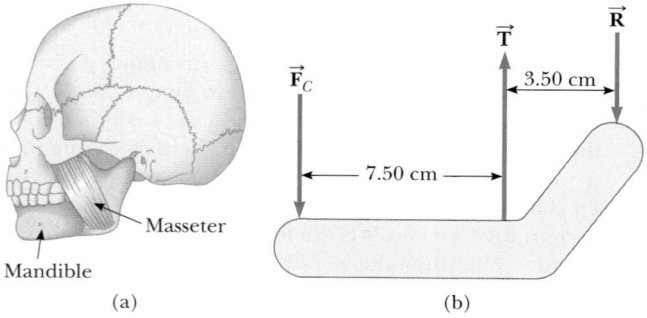

Figure P8.19

20. A hungry 700-N bear walks out on a beam in an attempt to retrieve some "goodies" hanging at the end (Fig. P8.20). The beam is uniform, weighs 200 N, and is 6.00 m long; the goodies weigh 80.0 N. (a) Draw a free-body diagram of the beam. (b) When the bear is at $x = 1.00$ m, find the tension in the wire and the components of the reaction force at the hinge. (c) If the wire can withstand a maximum tension of 900 N, what is the maximum distance the bear can walk before the wire breaks?

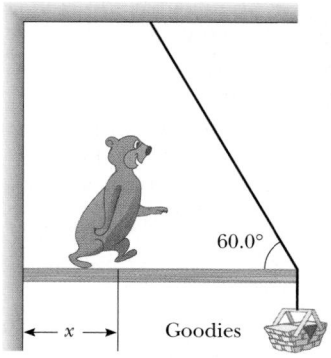

Figure P8.20

21. A uniform semicircular sign 1.00 m in diameter and of weight w is supported by two wires as shown in Figure P8.21. What is the tension in each of the wires supporting the sign?

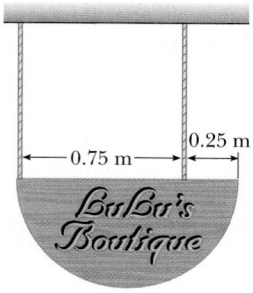

Figure P8.21

22. A 20.0-kg floodlight in a park is supported at the end of a horizontal beam of negligible mass that is hinged to a pole, as shown in Figure P8.22. A cable at an angle of 30.0° with the beam helps to support the light. Find (a) the tension in the cable and (b) the horizontal and vertical forces exerted on the beam by the pole.

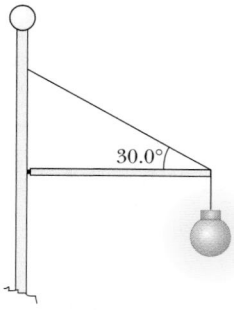

Figure P8.22

23. A uniform plank of length 2.00 m and mass 30.0 kg is supported by three ropes, as indicated by the blue vectors in Figure P8.23. Find the tension in each rope when a 700-N person is 0.500 m from the left end.

24. A 15.0-m, 500-N uniform ladder rests against a frictionless wall, making an angle of 60.0° with the horizontal. (a) Find the horizontal and vertical forces exerted on the base of the ladder by the Earth when an 800-N firefighter is 4.00 m from the bottom. (b) If the ladder is just on the

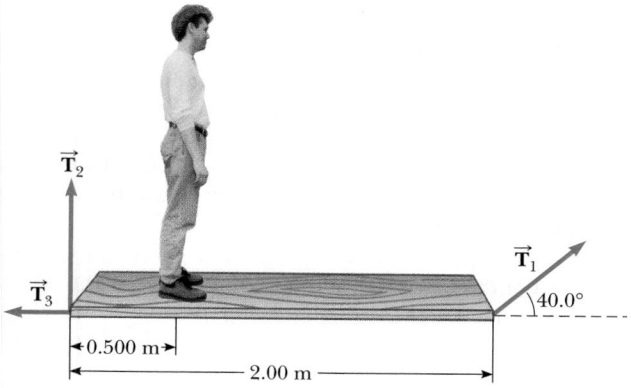

Figure P8.23

verge of slipping when the firefighter is 9.00 m up, what is the coefficient of static friction between ladder and ground?

25. An 8.00-m, 200-N uniform ladder rests against a smooth wall. The coefficient of static friction between the ladder and the ground is 0.600, and the ladder makes a 50.0° angle with the ground. How far up the ladder can an 800-N person climb before the ladder begins to slip?

26. A 1 200-N uniform boom is supported by a cable perpendicular to the boom as in Figure P8.26. The boom is hinged at the bottom, and a 2 000-N weight hangs from its top. Find the tension in the supporting cable and the components of the reaction force exerted on the boom by the hinge.

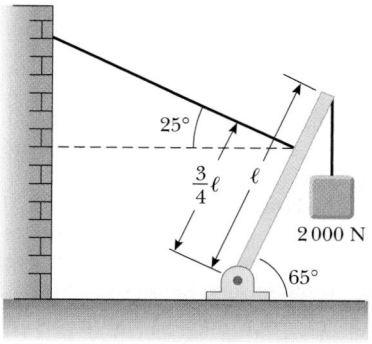

Figure P8.26

27. The large quadriceps muscle in the upper leg terminates at its lower end in a tendon attached to the upper end of the tibia (Fig. P8.27a). The forces on the lower leg when the leg is extended are modeled as in Figure P8.27b, where $\vec{\mathbf{T}}$ is the force of tension in the tendon, $\vec{\mathbf{w}}$ is the force of gravity acting on the lower leg, and $\vec{\mathbf{F}}$ is the force of gravity acting on the foot. Find $\vec{\mathbf{T}}$ when the tendon is at an angle of 25.0° with the tibia, assuming that $w = 30.0$ N, $F = 12.5$ N, and the leg is extended at an angle θ of 40.0° with the vertical. Assume that the center of gravity of the lower leg is at its center and that the tendon attaches to the lower leg at a point one-fifth of the way down the leg.

28. One end of a uniform 4.0-m-long rod of weight w is supported by a cable. The other end rests against a wall, where it is held by friction. (See Fig. P8.28.) The coefficient of static friction between the wall and the rod is

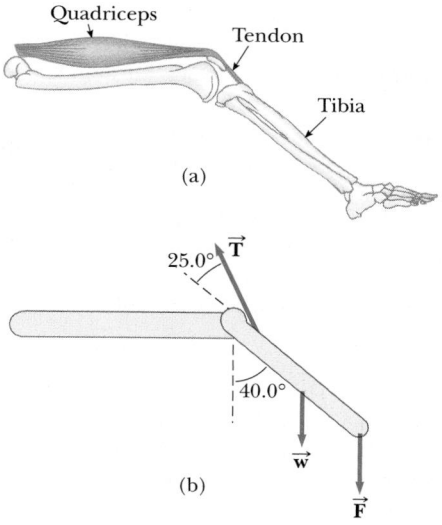

Figure P8.27

$\mu_s = 0.50$. Determine the minimum distance x from point A at which an additional weight w (the same as the weight of the rod) can be hung without causing the rod to slip at point A.

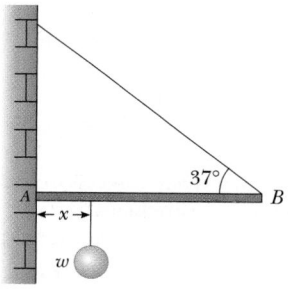

Figure P8.28

Section 8.5 Relationship Between Torque and Angular Acceleration

29. Four objects are held in position at the corners of a rectangle by light rods as shown in Figure P8.29. Find the moment of inertia of the system about (a) the x-axis, (b) the y-axis, and (c) an axis through O and perpendicular to the page.

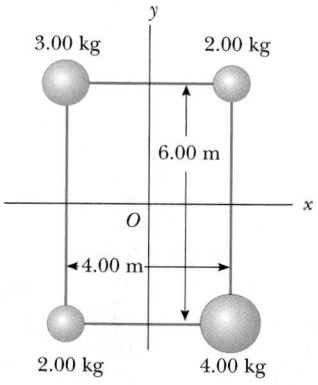

Figure P8.29 (Problems 29 and 30)

30. If the system shown in Figure P8.29 is set in rotation about each of the axes mentioned in Problem 29, find the torque that will produce an angular acceleration of 1.50 rad/s^2 in each case.

31. A model airplane with mass 0.750 kg is tethered by a wire so that it flies in a circle 30.0 m in radius. The airplane engine provides a net thrust of 0.800 N perpendicular to the tethering wire. (a) Find the torque the net thrust produces about the center of the circle. (b) Find the angular acceleration of the airplane when it is in level flight. (c) Find the linear acceleration of the airplane tangent to its flight path.

32. A potter's wheel having a radius of 0.50 m and a moment of inertia of 12 kg · m^2 is rotating freely at 50 rev/min. The potter can stop the wheel in 6.0 s by pressing a wet rag against the rim and exerting a radially inward force of 70 N. Find the effective coefficient of kinetic friction between the wheel and the wet rag.

33. A cylindrical fishing reel has a moment of inertia $I = 6.8 \times 10^{-4}$ kg · m^2 and a radius of 4.0 cm. A friction clutch in the reel exerts a restraining torque of 1.3 N · m if a fish pulls on the line. The fisherman gets a bite, and the reel begins to spin with an angular acceleration of 66 rad/s^2. (a) What is the force exerted by the fish on the line? (b) How much line unwinds in 0.50 s?

34. A bicycle wheel has a diameter of 64.0 cm and a mass of 1.80 kg. Assume that the wheel is a hoop with all the mass concentrated on the outside radius. The bicycle is placed on a stationary stand, and a resistive force of 120 N is applied tangent to the rim of the tire. (a) What force must be applied by a chain passing over a 9.00-cm-diameter sprocket in order to give the wheel an acceleration of 4.50 rad/s^2? (b) What force is required if you shift to a 5.60-cm-diameter sprocket?

35. A 150-kg merry-go-round in the shape of a uniform, solid, horizontal disk of radius 1.50 m is set in motion by wrapping a rope about the rim of the disk and pulling on the rope. What constant force must be exerted on the rope to bring the merry-go-round from rest to an angular speed of 0.500 rev/s in 2.00 s?

36. A 5.00-kg cylindrical reel with a radius of 0.600 m and a frictionless axle starts from rest and speeds up uniformly as a 3.00-kg bucket falls into a well, making a light rope unwind from the reel (Fig. P8.36). The bucket starts from rest and falls for 4.00 s. (a) What is the linear acceleration of the falling bucket? (b) How far does it drop? (c) What is the angular acceleration of the reel?

37. An airliner lands with a speed of 50.0 m/s. Each wheel of the plane has a radius of 1.25 m and a moment of inertia of 110 kg · m^2. At touchdown, the wheels begin to spin under the action of friction. Each wheel supports a weight of 1.40×10^4 N, and the wheels attain their angular speed in 0.480 s while rolling without slipping. What is the coefficient of kinetic friction between the wheels and the runway? Assume that the speed of the plane is constant.

Section 8.6 Rotational Kinetic Energy

38. A constant torque of 25.0 N · m is applied to a grindstone whose moment of inertia is 0.130 kg · m^2. Using energy principles and neglecting friction, find the angular speed after the grindstone has made 15.0 revolutions. [*Hint:* The angular equivalent of $W_{\text{net}} = F\Delta x = \frac{1}{2}mv_f^2 - \frac{1}{2}mv_i^2$ is $W_{\text{net}} = \tau\Delta\theta = \frac{1}{2}I\omega_f^2 - \frac{1}{2}I\omega_i^2$. You should convince yourself that this relationship is correct.)

39. **Physics Now**™ A 10.0-kg cylinder rolls without slipping on a rough surface. At an instant when its center of gravity has a speed of 10.0 m/s, determine (a) the translational kinetic energy of its center of gravity, (b) the rotational kinetic energy about its center of gravity, and (c) its total kinetic energy.

40. Use conservation of energy to determine the angular speed of the spool shown in Figure P8.36 after the 3.00-kg bucket has fallen 4.00 m, starting from rest. The light string attached to the bucket is wrapped around the spool and does not slip as it unwinds.

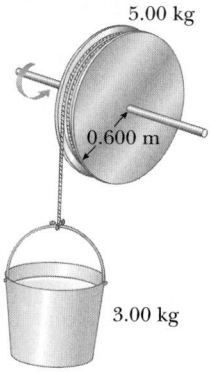

Figure P8.36 (Problems 36 and 40)

41. A horizontal 800-N merry-go-round of radius 1.50 m is started from rest by a constant horizontal force of 50.0 N applied tangentially to the merry-go-round. Find the kinetic energy of the merry-go-round after 3.00 s. (Assume it is a solid cylinder.)

42. A car is designed to get its energy from a rotating flywheel with a radius of 2.00 m and a mass of 500 kg. Before a trip, the flywheel is attached to an electric motor, which brings the flywheel's rotational speed up to 5 000 rev/min. (a) Find the kinetic energy stored in the flywheel. (b) If the flywheel is to supply energy to the car as a 10.0-hp motor would, find the length of time the car could run before the flywheel would have to be brought back up to speed.

43. The top in Figure P8.43 has a moment of inertia of 4.00×10^{-4} kg · m^2 and is initially at rest. It is free to rotate about a stationary axis AA'. A string wrapped around a peg along the axis of the top is pulled in such a manner as to maintain a constant tension of 5.57 N in the string. If the string does not slip while wound around the peg, what

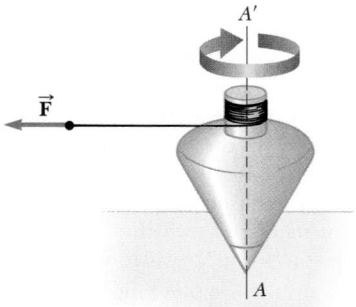

Figure P8.43

is the angular speed of the top after 80.0 cm of string has been pulled off the peg? [*Hint:* Consider the work that is done.]

44. A 240-N sphere 0.20 m in radius rolls without slipping 6.0 m down a ramp that is inclined at 37° with the horizontal. What is the angular speed of the sphere at the bottom of the slope if it starts from rest?

Section 8.7 Angular Momentum

45. A light rigid rod 1.00 m in length rotates about an axis perpendicular to its length and through its center, as shown in Figure P8.45. Two particles of masses 4.00 kg and 3.00 kg are connected to the ends of the rod. What is the angular momentum of the system if the speed of each particle is 5.00 m/s? (Neglect the rod's mass.)

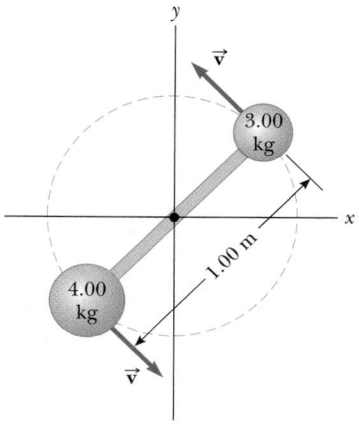

Figure P8.45

46. Halley's comet moves about the Sun in an elliptical orbit, with its closest approach to the Sun being 0.59 A.U. and its greatest distance being 35 A.U. (1 A.U. is the Earth–Sun distance). If the comet's speed at closest approach is 54 km/s, what is its speed when it is farthest from the Sun? You may neglect any change in the comet's mass and assume that its angular momentum about the Sun is conserved.

47. The system of small objects shown in Figure P8.47 is rotating at an angular speed of 2.0 rev/s. The objects are connected by light, flexible spokes that can be lengthened or shortened. What is the new angular speed if the spokes are shortened to 0.50 m? (An effect similar to that illustrated in this problem occurred in the early stages of the formation of our galaxy. As the massive cloud of dust and gas that was the source of the stars and planets contracted, an initially small angular speed increased with time.)

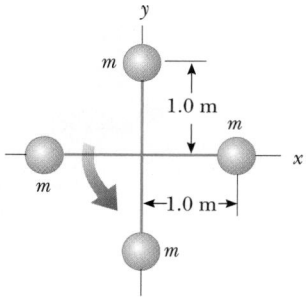

Figure P8.47

48. A playground merry-go-round of radius 2.00 m has a moment of inertia $I = 275\ \text{kg}\cdot\text{m}^2$ and is rotating about a frictionless vertical axle. As a child of mass 25.0 kg stands at a distance of 1.00 m from the axle, the system (merry-go-round and child) rotates at the rate of 14.0 rev/min. The child then proceeds to walk toward the edge of the merry-go-round. What is the angular speed of the system when the child reaches the edge?

49. A solid, horizontal cylinder of mass 10.0 kg and radius 1.00 m rotates with an angular speed of 7.00 rad/s about a fixed vertical axis through its center. A 0.250-kg piece of putty is dropped vertically onto the cylinder at a point 0.900 m from the center of rotation and sticks to the cylinder. Determine the final angular speed of the system.

50. A student sits on a rotating stool holding two 3.0-kg objects. When his arms are extended horizontally, the objects are 1.0 m from the axis of rotation and he rotates with an angular speed of 0.75 rad/s. The moment of inertia of the student plus stool is $3.0\ \text{kg}\cdot\text{m}^2$ and is assumed to be constant. The student then pulls in the objects horizontally to 0.30 m from the rotation axis. (a) Find the new angular speed of the student. (b) Find the kinetic energy of the student before and after the objects are pulled in.

51. The puck in Figure P8.51 has a mass of 0.120 kg. Its original distance from the center of rotation is 40.0 cm, and it moves with a speed of 80.0 cm/s. The string is pulled downward 15.0 cm through the hole in the frictionless table. Determine the work done on the puck. [*Hint:* Consider the change in kinetic energy of the puck.]

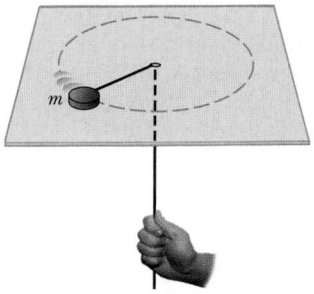

Figure P8.51

52. A merry-go-round rotates at the rate of 0.20 rev/s with an 80-kg man standing at a point 2.0 m from the axis of rotation. (a) What is the new angular speed when the man walks to a point 1.0 m from the center? Assume that the merry-go-round is a solid 25-kg cylinder of radius 2.0 m. (b) Calculate the change in kinetic energy due to the man's movement. How do you account for this change in kinetic energy?

53. A 60.0-kg woman stands at the rim of a horizontal turntable having a moment of inertia of $500\ \text{kg}\cdot\text{m}^2$ and a radius of 2.00 m. The turntable is initially at rest and is free to rotate about a frictionless, vertical axle through its center. The woman then starts walking around the rim clockwise (as viewed from above the system) at a constant speed of 1.50 m/s relative to the Earth. (a) In what direction and with what angular speed does the turntable rotate? (b) How much work does the woman do to set herself and the turntable into motion?

54. A space station shaped like a giant wheel has a radius of 100 m and a moment of inertia of 5.00×10^8 kg·m^2. A crew of 150 lives on the rim, and the station is rotating so that the crew experiences an apparent acceleration of $1g$ (Fig. P8.54). When 100 people move to the center of the station for a union meeting, the angular speed changes. What apparent acceleration is experienced by the managers remaining at the rim? Assume the average mass of a crew member is 65.0 kg.

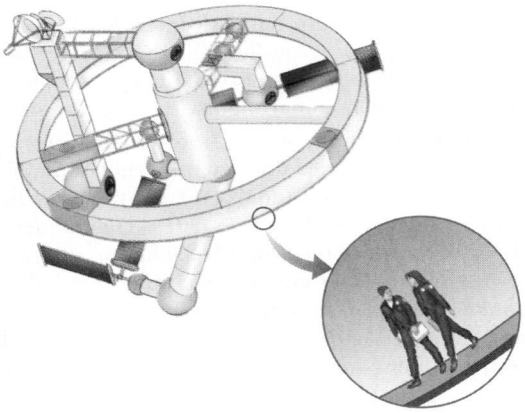

Figure P8.54

ADDITIONAL PROBLEMS

55. A cylinder with moment of inertia I_1 rotates with angular velocity ω_0 about a frictionless vertical axle. A second cylinder, with moment of inertia I_2, initially not rotating, drops onto the first cylinder (Fig. P8.55). Since the surfaces are rough, the two cylinders eventually reach the same angular speed ω. (a) Calculate ω. (b) Show that kinetic energy is lost in this situation, and calculate the ratio of the final to the initial kinetic energy.

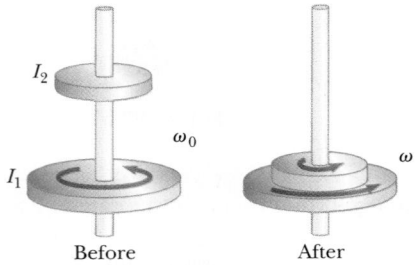

Figure P8.55

56. A new General Electric stove has a mass of 68.0 kg and the dimensions shown in Figure P8.56. The stove comes with a warning that it can tip forward if a person stands or sits on the oven door when it is open. What can you conclude about the weight of such a person? Could it be a child? List the assumptions you make in solving this problem. (The stove is supplied with a wall bracket to prevent the accident.)

57. A 40.0-kg child stands at one end of a 70.0-kg boat that is 4.00 m long (Fig. P8.57). The boat is initially 3.00 m from the pier. The child notices a turtle on a rock beyond the far end of the boat and proceeds to walk to that end to catch the turtle. (a) Neglecting friction between the boat and water, describe the motion of the system (child plus boat). (b) Where will the child be relative to the pier when he reaches the far end of the boat? (c) Will he catch the turtle? (Assume that he can reach out 1.00 m from the end of the boat.)

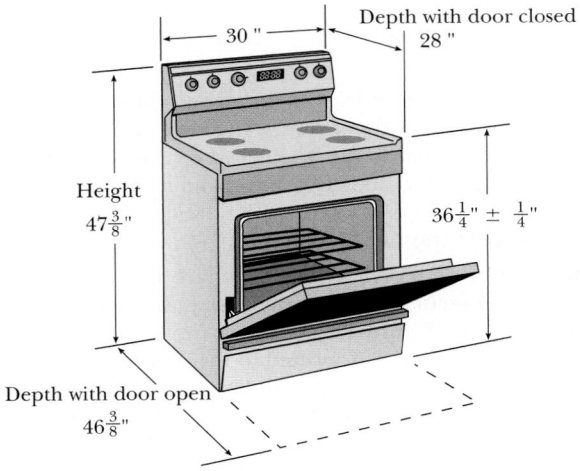

Figure P8.56

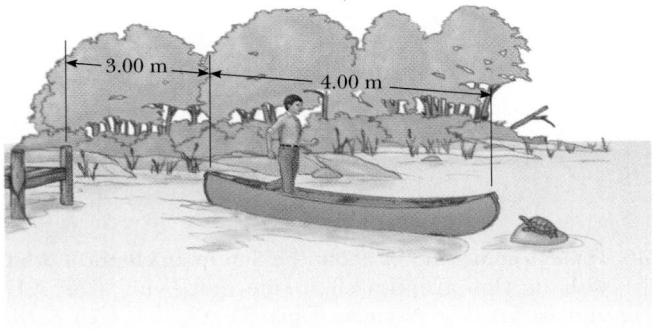

Figure P8.57

58. Figure P8.58 shows a clawhammer as it is being used to pull a nail out of a horizontal board. If a force of magnitude 150 N is exerted horizontally as shown, find (a) the

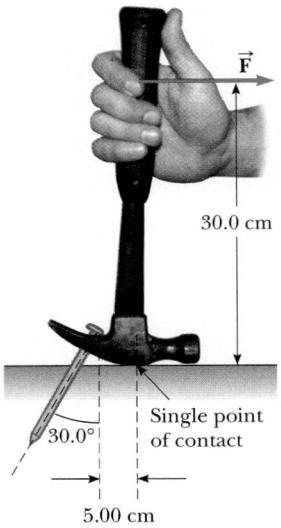

Figure P8.58

force exerted by the hammer claws on the nail and (b) the force exerted by the surface at the point of contact with the hammer head. Assume that the force the hammer exerts on the nail is parallel to the nail.

59. The pulley in Figure P8.59 has a moment of inertia of $5.0\ \text{kg}\cdot\text{m}^2$ and a radius of 0.50 m. The cord supporting the masses m_1 and m_2 does not slip, and the axle is frictionless. (a) Find the acceleration of each mass when $m_1 = 2.0$ kg and $m_2 = 5.0$ kg. (b) Find the tension in the cable supporting m_1 and the tension in the cable supporting m_2. [*Note:* The two tensions are different].

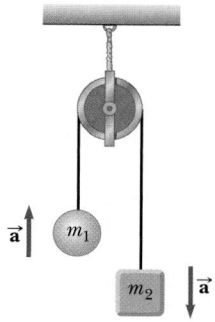

Figure P8.59

60. A 12.0-kg object is attached to a cord that is wrapped around a wheel of radius $r = 10.0$ cm (Fig. P8.60). The acceleration of the object down the frictionless incline is measured to be $2.00\ \text{m/s}^2$. Assuming the axle of the wheel to be frictionless, determine (a) the tension in the rope, (b) the moment of inertia of the wheel, and (c) the angular speed of the wheel 2.00 s after it begins rotating, starting from rest.

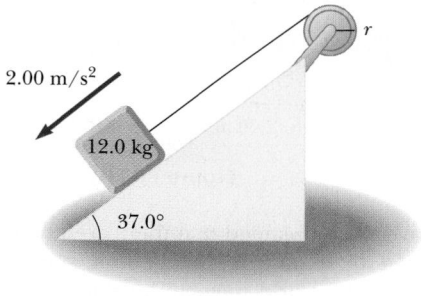

Figure P8.60

61. A uniform ladder of length L and weight w is leaning against a vertical wall. The coefficient of static friction between the ladder and the floor is the same as that between the ladder and the wall. If this coefficient of static friction is $\mu_s = 0.500$, determine the smallest angle the ladder can make with the floor without slipping.

62. A uniform 10.0-N picture frame is supported as shown in Figure P8.62. Find the tension in the cords and the magnitude of the horizontal force at P that are required to hold the frame in the position shown.

63. **Physics Now™** A solid 2.0-kg ball of radius 0.50 m starts at a height of 3.0 m above the surface of the Earth and *rolls* down a 20° slope. A solid disk and a ring start at the same time and the same height. The ring and disk each have the same mass and radius as the ball. Which of the three wins the race to the bottom if all roll without slipping?

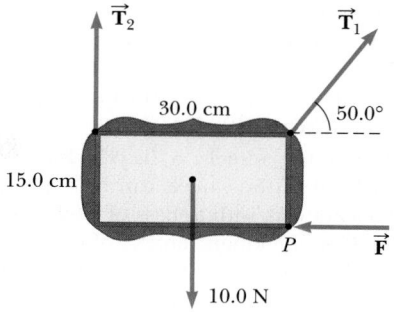

Figure P8.62

64. A common physics demonstration (Fig. P8.64) consists of a ball resting at the end of a board of length ℓ that is elevated at an angle θ with the horizontal. A light cup is attached to the board at r_c so that it will catch the ball when the support stick is suddenly removed. Show that (a) the ball will lag behind the falling board when $\theta < 35.3°$ and (b) the ball will fall into the cup when the board is supported at this limiting angle and the cup is placed at

$$r_c = \frac{2\ell}{3\cos\theta}$$

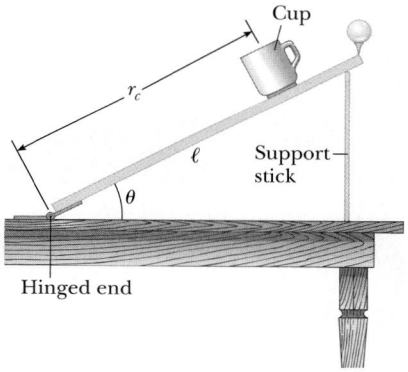

Figure P8.64

65. In Figure P8.65, the sliding block has a mass of 0.850 kg, the counterweight has a mass of 0.420 kg, and the pulley is a uniform solid cylinder with a mass of 0.350 kg and an outer radius of 0.0300 m. The coefficient of kinetic friction between the block and the horizontal surface is 0.250. The pulley turns without friction on its axle. The light cord does not stretch and does not slip on the pulley. The block has a

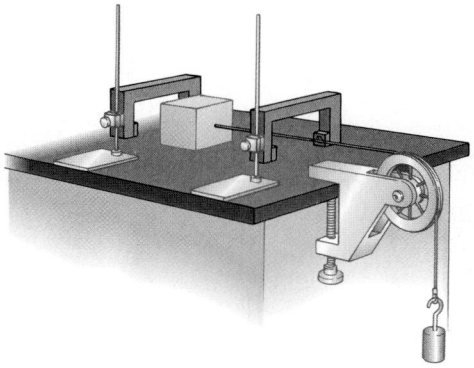

Figure P8.65

velocity of 0.820 m/s toward the pulley when it passes through a photogate. (a) Use energy methods to predict the speed of the block after it has moved to a second photogate 0.700 m away. (b) Find the angular speed of the pulley at the same moment.

66. (a) Without the wheels, a bicycle frame has a mass of 8.44 kg. Each of the wheels can be roughly modeled as a uniform solid disk with a mass of 0.820 kg and a radius of 0.343 m. Find the kinetic energy of the whole bicycle when it is moving forward at 3.35 m/s. (b) Before the invention of a wheel turning on an axle, ancient people moved heavy loads by placing rollers under them. (Modern people use rollers, too: Any hardware store will sell you a roller bearing for a lazy Susan.) A stone block of mass 844 kg moves forward at 0.335 m/s, supported by two uniform cylindrical tree trunks, each of mass 82.0 kg and radius 0.343 m. There is no slipping between the block and the rollers or between the rollers and the ground. Find the total kinetic energy of the moving objects.

67. In exercise physiology studies, it is sometimes important to determine the location of a person's center of gravity. This can be done with the arrangement shown in Figure P8.67. A light plank rests on two scales that read $F_{g1} = 380$ N and $F_{g2} = 320$ N. The scales are separated by a distance of 2.00 m. How far from the woman's feet is her center of gravity?

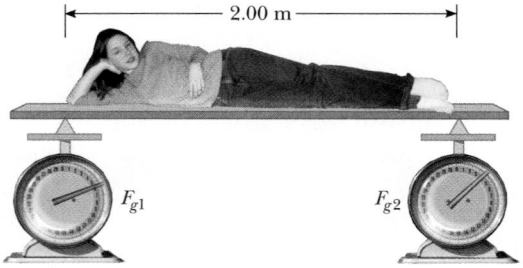

Figure P8.67

68. Two astronauts (Fig. P8.68), each having a mass of 75.0 kg, are connected by a 10.0-m rope of negligible mass. They are isolated in space, moving in circles around the point halfway between them at a speed of 5.00 m/s. Treating the astronauts as particles, calculate (a) the magnitude of the angular momentum and (b) the rotational energy of the system. By pulling on the rope, the astronauts shorten the distance between them to 5.00 m. (c) What is the new angular momentum of the system? (d) What are their new speeds? (e) What is the new rotational energy of the system? (f) How much work is done by the astronauts in shortening the rope?

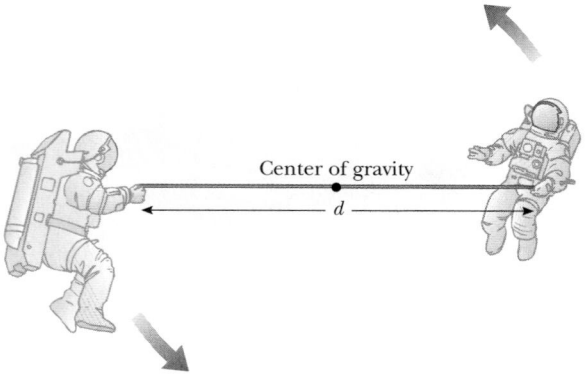

Figure P8.68 (Problems 68 and 69)

69. Two astronauts (Fig. P8.68), each having a mass M, are connected by a rope of length d having negligible mass. They are isolated in space, moving in circles around the point halfway between them at a speed v. (a) Calculate the magnitude of the angular momentum of the system by treating the astronauts as particles. (b) Calculate the rotational energy of the system. By pulling on the rope, the astronauts shorten the distance between them to $d/2$. (c) What is the new angular momentum of the system? (d) What are their new speeds? (e) What is the new rotational energy of the system? (f) How much work is done by the astronauts in shortening the rope?

70. Two window washers, Bob and Joe, are on a 3.00-m-long, 345-N scaffold supported by two cables attached to its ends. Bob weighs 750 N and stands 1.00 m from the left end, as shown in Figure P8.70. Two meters from the left end is the 500-N washing equipment. Joe is 0.500 m from the right end and weighs 1 000 N. Given that the scaffold is in rotational and translational equilibrium, what are the forces on each cable?

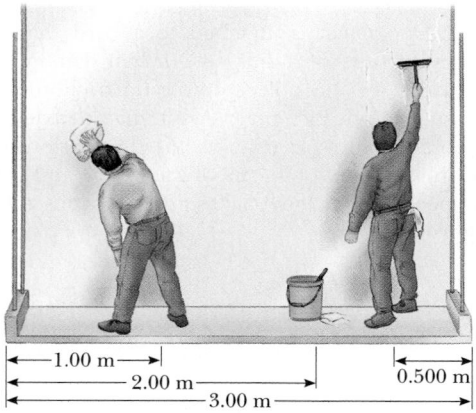

Figure P8.70

71. We have all complained that there aren't enough hours in a day. In an attempt to change that, suppose that all the people in the world lined up at the equator and started running east at 2.5 m/s relative to the surface of the Earth. By how much would the length of a day increase? (Assume that there are 5.5×10^9 people in the world with an average mass of 70 kg each and that the Earth is a solid, homogeneous sphere. In addition, you may use the result $1/(1 - x) \approx 1 + x$ for small x.)

72. In a circus performance, a large 5.0-kg hoop of radius 3.0 m rolls without slipping. If the hoop is given an angular speed of 3.0 rad/s while rolling on the horizontal ground and is then allowed to roll up a ramp inclined at 20° with the horizontal, how far along the incline does the hoop roll?

73. **PhysicsNow™** A uniform solid cylinder of mass M and radius R rotates on a frictionless horizontal axle (Fig. P8.73). Two objects with equal masses m hang from light cords wrapped around the cylinder. If the system is released from rest, find (a) the tension in each cord and (b) the acceleration of each object after the objects have descended a distance h.

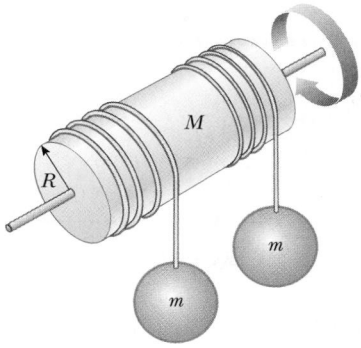

Figure P8.73

74. Figure P8.74 shows a vertical force applied tangentially to a uniform cylinder of weight w. The coefficient of static friction between the cylinder and all surfaces is 0.500. Find, in terms of w, the maximum force $\vec{\mathbf{F}}$ that can be applied without causing the cylinder to rotate. [*Hint:* When the cylinder is on the verge of slipping, both friction forces are at their maximum values. Why?]

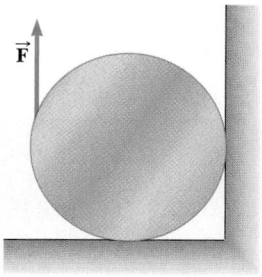

Figure P8.74

75. Due to a gravitational torque exerted by the Moon on the Earth, our planet's period of rotation slows at a rate on the order of 1 ms/century. (a) Determine the order of magnitude of Earth's angular acceleration. (b) Find the order of magnitude of the torque. (c) Find the order of magnitude of the size of the wrench an ordinary person would need to exert such a torque, as in Figure P8.75. Assume the person can brace his feet against a solid firmament.

Figure P8.75

76. A uniform pole is propped between the floor and the ceiling of a room. The height of the room is 7.80 ft, and the coefficient of static friction between the pole and the ceiling is 0.576. The coefficient of static friction between the pole and the floor is greater than that. What is the length of the longest pole that can be propped between the floor and the ceiling?

77. A *war-wolf*, or *trebuchet*, is a device used during the Middle Ages to throw rocks at castles and now sometimes used to fling pumpkins and pianos. A simple trebuchet is shown in Figure P8.77. Model it as a stiff rod of negligible mass 3.00 m long and joining particles of mass 60.0 kg and 0.120 kg at its ends. It can turn on a frictionless horizontal axle perpendicular to the rod and 14.0 cm from the particle of larger mass. The rod is released from rest in a horizontal orientation. Find the maximum speed that the object of smaller mass attains.

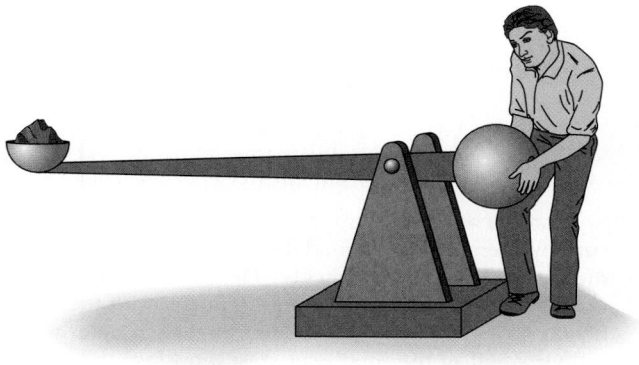

Figure P8.77

78. A painter climbs a ladder leaning against a smooth wall. At a certain height, the ladder is on the verge of slipping. (a) Explain why the force exerted by the vertical wall on the ladder is horizontal. (b) If the ladder of length L leans at an angle θ with the horizontal, what is the lever arm for this horizontal force with the axis of rotation taken at the base of the ladder? (c) If the ladder is uniform, what is the lever arm for the force of gravity acting on the ladder? (d) Let the mass of the painter be 80 kg, $L = 4.0$ m, the ladder's mass be 30 kg, $\theta = 53°$, and the coefficient of friction between ground and ladder be 0.45. Find the maximum distance the painter can climb up the ladder.

79. A 4.00-kg mass is connected by a light cord to a 3.00-kg mass on a smooth surface (Fig. P8.79). The pulley rotates about a frictionless axle and has a moment of inertia of $0.500\ \text{kg}\cdot\text{m}^2$ and a radius of 0.300 m. Assuming that the cord does not slip on the pulley, find (a) the acceleration of the two masses and (b) the tensions T_1 and T_2.

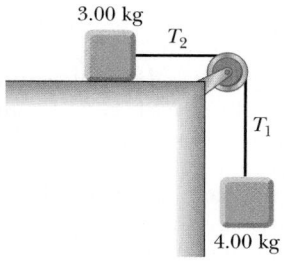

Figure P8.79

80. A string is wrapped around a uniform cylinder of mass M and radius R. The cylinder is released from rest with the string vertical and its top end tied to a fixed bar (Fig. P8.80). Show that (a) the tension in the string is one-third the weight of the cylinder, (b) the magnitude of the acceleration of the center of gravity is $2g/3$, and (c) the speed of the center of gravity is $(4gh/3)^{1/2}$ after the cylinder has descended through distance h. Verify your answer to (c) with the energy approach.

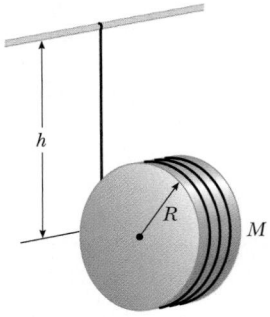

Figure P8.80

81. A person in a wheelchair wishes to roll up over a sidewalk curb by exerting a horizontal force $\vec{\mathbf{F}}$ to the top of each of the wheelchair's main wheels (Fig. P8.81a). The main wheels have radius r and come in contact with a curb of height h (Fig. P8.81b). (a) Assume that each main wheel supports half of the total load, and show that the magnitude of the minimum force necessary to raise the wheelchair from the street is given by

$$F = \frac{mg\sqrt{2rh - h^2}}{2(2r - h)}$$

where mg is the combined weight of the wheelchair and person. (b) Estimate the value of F, taking $mg = 1\,400$ N, $r = 30$ cm, and $h = 10$ cm.

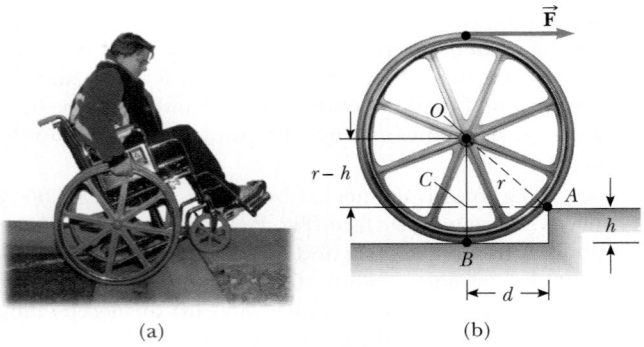

Figure P8.81

82. The truss structure in Figure P8.82 represents part of a bridge. Assume that the structural components are connected by pin joints and that the entire structure is free to slide horizontally at each end. Assume furthermore that the mass of the structure is negligible compared with the load it must support. In this situation, the force exerted by each of the bars (struts) on the pin joints is either a force of tension or one of compression and must be along the length of the bar. Calculate the force in each strut when the bridge supports a 7 200-N load at its center.

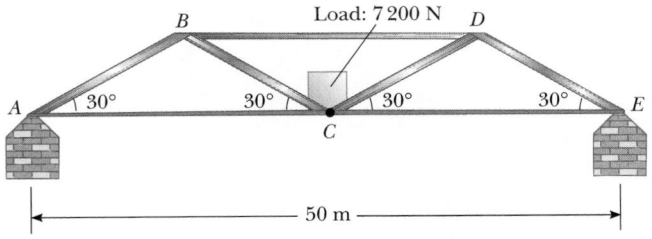

Figure P8.82

83. **The Iron Cross** When a gymnast weighing 750 N executes the iron cross as in Figure P8.83a, the primary muscles involved in supporting this position are the latissimus dorsi ("lats") and the pectoralis major ("pecs"). The rings exert an upward force on the arms and support the weight of the gymnast. The force exerted by the shoulder joint on the arm is labeled $\vec{\mathbf{F}}_s$ while the two muscles exert a total force $\vec{\mathbf{F}}_m$ on the arm. Estimate the magnitude of the force $\vec{\mathbf{F}}_m$. Note that one ring supports half the weight of the gymnast, which is 375 N as indicated in Figure P8.83b. Assume that the force $\vec{\mathbf{F}}_m$ acts at an angle of 45° below the horizontal at a distance of 4.0 cm from the shoulder joint. In your estimate, take the distance from the shoulder joint to the hand to be 70 cm and ignore the weight of the arm.

© Ed Bock/Corbis

(a)

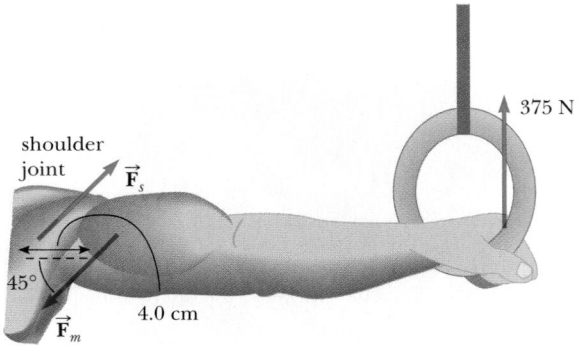

(b)

Figure P8.83

84. **Swinging on a high bar** The gymnast shown in Figure P8.84 is performing a backwards giant swing on the high bar. Starting from rest in a near-vertical orientation, he rotates around the bar in a counterclockwise direction,

keeping his body and arms straight. Friction between the bar and the gymnast's hands exerts a constant torque opposing the rotational motion. If the angular velocity of the gymnast at position 2 is measured to be 4.0 rad/s, determine his angular velocity at position 3. (Note that this maneuver is called a backwards giant swing, even though the motion of the gymnast would seem to be forwards.)

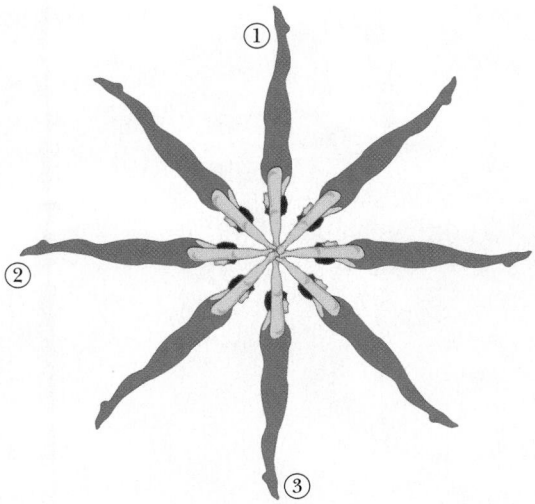

Figure P8.84

ACTIVITIES

A.1. Compare the motion of an empty soup can and a filled soup can down the same incline, such as a tilted table. If they are released from rest at the same height on the incline, which one reaches the bottom first? Repeat your observations with different kinds of soup (tomato, chicken noodle, etc.) and a can of beans. Compare their motions, and try to explain your observations. Finally, compare the motion of a filled soup can with a tennis ball, and explain your results to a friend.

A.2. Before attempting this exercise, review Example 8.14, dealing with the spinning stool. The techniques used here are similar to those used there. (a) First, make an estimate of the moment of inertia of your body. One way to do this would be to model your body as a solid cylinder and find I from $I = \frac{1}{2}MR^2$. You would have to determine your mass and estimate your average "radius" for this approach. Can you think of an alternative way to estimate I? (b) Now use the approach of Example 8.13 to measure I: Sit on a rotating stool, hold two weights (say, two books), and determine the angular speed of rotation with the books extended and after they are pulled in. The angular speed can be found by estimating the time taken for a given number of rotations. Use conservation of angular momentum to determine I. Do this five times to determine an average value for I. How well do your results for (a) and (b) compare, and if they differ greatly, what might cause the discrepancy?

A.3. This experiment demonstrates a simple way to find the center of gravity of an irregularly shaped object. Cut out an irregular shape from a piece of cardboard, and punch three to five holes around the edge of the shape. Put a pushpin through one of the holes, and tack the shape to a corkboard so that the shape can rotate freely. Now tie a weight to one end of a string, and hang the other end of the string from the pushpin. When the string stops moving, trace a line on the cardboard that follows the string. Repeat this for each of the holes in the cardboard. You will find that there is a point where all the lines intersect. This point is the center of gravity of the object.

In the Dead Sea, a lake between Jordan and Israel, the high percentage of salt dissolved in the water raises the fluid's density, dramatically increasing the buoyant force. Bathers can kick back and enjoy a good read, dispensing with the floating lounge chairs.

© Alison Wright/Corbis

CHAPTER

9

OUTLINE

9.1 States of Matter
9.2 The Deformation of Solids
9.3 Density and Pressure
9.4 Variation of Pressure with Depth
9.5 Pressure Measurements
9.6 Buoyant Forces and Archimedes's Principle
9.7 Fluids in Motion
9.8 Other Applications of Fluid Dynamics
9.9 Surface Tension, Capillary Action, and Viscous Fluid Flow
9.10 Transport Phenomena

Solids and Fluids

There are four known states of matter: solids, liquids, gases, and plasmas. In the universe at large, plasmas—systems of charged particles interacting electromagnetically—are the most common. In our environment on Earth, solids, liquids, and gases predominate.

An understanding of the fundamental properties of these different states of matter is important in all the sciences, in engineering, and in medicine. Forces put stresses on solids, and stresses can strain, deform, and break those solids, whether they are steel beams or bones. Fluids under pressure can perform work, or they can carry nutrients and essential solutes, like the blood flowing through our arteries and veins. Flowing gases cause pressure differences that can lift a massive cargo plane or the roof off a house in a hurricane. High-temperature plasmas created in fusion reactors may someday allow humankind to harness the energy source of the sun.

The study of any one of these states of matter is itself a vast discipline. Here, we'll introduce basic properties of solids and liquids, the latter including some properties of gases. In addition, we'll take a brief look at surface tension, viscosity, osmosis, and diffusion.

9.1 STATES OF MATTER

Matter is normally classified as being in one of three states: **solid**, **liquid**, or **gas**. Often this classification system is extended to include a fourth state of matter, called a **plasma**.

Everyday experience tells us that a solid has a definite volume and shape. A brick, for example, maintains its familiar shape and size day in and day out. A liquid has a definite volume but no definite shape. When you fill the tank on a lawn mower, the gasoline changes its shape from that of the original container to that of the tank on the mower, but the original volume is unchanged. A gas differs from solids and liquids in that it has neither definite volume nor definite shape. Because gas can flow, however, it shares many properties with liquids.

All matter consists of some distribution of atoms or molecules. The atoms in a solid, held together by forces that are mainly electrical, are located at specific positions with respect to one another and vibrate about those positions. At low temperatures, the vibrating motion is slight and the atoms can be considered essentially fixed. As energy is added to the material, the amplitude of the vibrations increases. A vibrating atom can be viewed as being bound in its equilibrium position by springs attached to neighboring atoms. A collection of such atoms and imaginary springs is shown in Figure 9.1. We can picture applied external forces as compressing these tiny internal springs. When the external forces are removed, the solid tends to return to its original shape and size. Consequently, a solid is said to have *elasticity*.

Charles D. Winters

Crystals of natural quartz (SiO_2), one of the most common minerals on Earth. Quartz crystals are used to make special lenses and prisms and are employed in certain electronic applications.

Solids can be classified as either crystalline or amorphous. In a **crystalline solid** the atoms have an ordered structure. For example, in the sodium chloride crystal (common table salt), sodium and chlorine atoms occupy alternate corners of a cube, as in Figure 9.2a. In an **amorphous solid**, such as glass, the atoms are arranged almost randomly, as in Figure 9.2b.

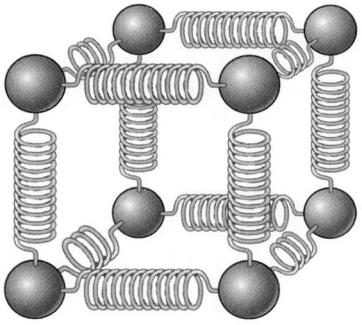

Figure 9.1 A model of a portion of a solid. The atoms (spheres) are imagined as being attached to each other by springs, which represent the elastic nature of the interatomic forces. A solid consists of trillions of segments like this, with springs connecting all of them.

For any given substance, the liquid state exists at a higher temperature than the solid state. The intermolecular forces in a liquid aren't strong enough to keep the molecules in fixed positions, and they wander through the liquid in random fashion (Fig. 9.2c). Solids and liquids both have the property that when an attempt is made to compress them, strong repulsive atomic forces act internally to resist the compression.

In the gaseous state, molecules are in constant random motion and exert only weak forces on each other. The average distance between the molecules of a gas is quite large compared with the size of the molecules. Occasionally the molecules collide with each other, but most of the time they move as nearly free, noninteracting particles. As a result, unlike solids and liquids, gases can be easily compressed. We'll say more about gases in subsequent chapters.

When a gas is heated to high temperature, many of the electrons surrounding each atom are freed from the nucleus. The resulting system is a collection of free, electrically charged particles—negatively charged electrons and positively charged ions. Such a highly ionized state of matter containing equal amounts of positive and negative charges is called a **plasma**. Unlike a neutral gas, the long-range electric and magnetic forces allow the constituents of a plasma to interact with each other. Plasmas are found inside stars and in accretion disks around black holes, for example, and are far more common than the solid, liquid, and gaseous states because there are far more stars around than any other form of celestial matter,

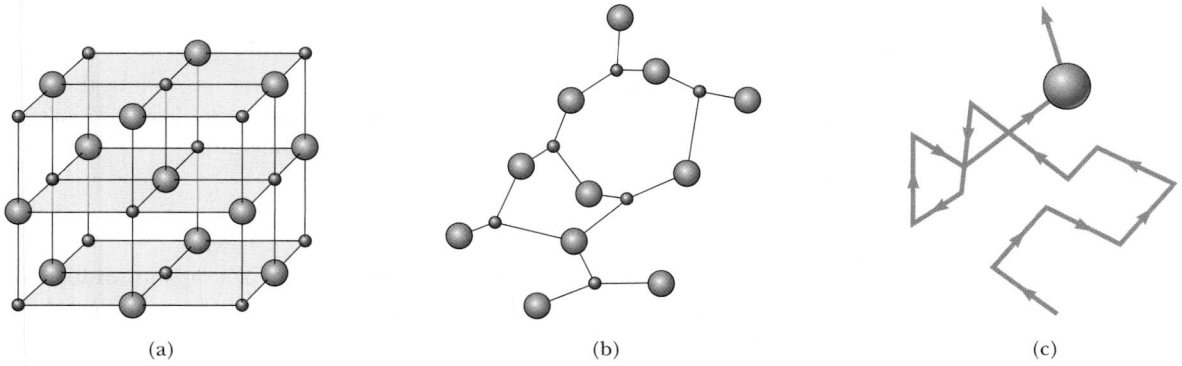

Figure 9.2 (a) The NaCl structure, with the Na^+ (gray) and Cl^- (green) ions at alternate corners of a cube. (b) In an amorphous solid, the atoms are arranged randomly. (c) Erratic motion of a molecule in a liquid.

except possibly **dark matter**. Dark matter, inferred by observations of the motion of stars around the galaxy, makes up about 90% of the matter in the universe and is of unknown composition. In this chapter, however, we ignore plasmas and dark matter and concentrate on the more familiar solid, liquid, and gaseous forms that make up the environment of our planet.

9.2 THE DEFORMATION OF SOLIDS

While a solid may be thought of as having a definite shape and volume, it's possible to change its shape and volume by applying external forces. A sufficiently large force will permanently deform or break an object, but otherwise, when the external forces are removed, the object tends to return to its original shape and size. This is called *elastic behavior.*

The elastic properties of solids are discussed in terms of stress and strain. **Stress** is the force per unit area causing a deformation; **strain** is a measure of the amount of the deformation. For sufficiently small stresses, **stress is proportional to strain**, with the constant of proportionality depending on the material being deformed and on the nature of the deformation. We call this proportionality constant the **elastic modulus**:

$$\text{stress} = \text{elastic modulus} \times \text{strain} \qquad [9.1]$$

The elastic modulus is analogous to a spring constant. It can be taken as the stiffness of a material: A material having a large elastic modulus is very stiff and difficult to deform. There are three relationships having the form of Equation 9.1, corresponding to tensile, shear, and bulk deformation, and all of them satisfy an equation similar to Hooke's law for springs:

$$F = k\Delta x \qquad [9.2]$$

where F is the applied force, k is the spring constant, and Δx is the amount by which the spring is compressed.

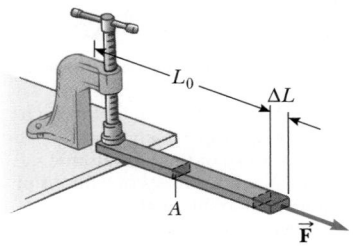

ACTIVE FIGURE 9.3
A long bar clamped at one end is stretched by the amount ΔL under the action of a force $\vec{F}$.

PhysicsNow™
Log into PhysicsNow at **www.cp7e.com**, and go to Active Figure 9.3 to adjust the values of the applied force and Young's modulus and observe the change in length of the bar.

Young's Modulus: Elasticity in Length

Consider a long bar of cross-sectional area A and length L_0, clamped at one end (Active Fig. 9.3). When an external force $\vec{F}$ is applied along the bar, perpendicular to the cross section, internal forces in the bar resist the distortion ("stretching") that $\vec{F}$ tends to produce. Nevertheless, the bar attains an equilibrium in which (1) its length is greater than L_0 and (2) the external force is balanced by internal forces. Under these circumstances, the bar is said to be *stressed.* We define the **tensile stress** as the ratio of the magnitude of the external force F to the cross-sectional area A. The word "tensile" has the same root as the word "tension" and is used because the bar is under tension. The SI unit of stress is the newton per square meter (N/m^2), called the **pascal** (Pa):

The pascal ▶

$$1 \text{ Pa} \equiv 1 \text{ N/m}^2$$

The **tensile strain** in this case is defined as the ratio of the change in length ΔL to the original length L_0 and is therefore a dimensionless quantity. Using Equation 9.1, we can write an equation relating tensile stress to tensile strain:

$$\frac{F}{A} = Y\frac{\Delta L}{L_0} \qquad [9.3]$$

In this equation, Y is the constant of proportionality, called **Young's modulus**. Notice that Equation 9.3 could be solved for F and put in the form $F = k\Delta L$, where $k = YA/L_0$, making it look just like Hooke's law, Equation 9.2.

A material having a large Young's modulus is difficult to stretch or compress. This quantity is typically used to characterize a rod or wire stressed under either tension or compression. Because strain is a dimensionless quantity, Y is in pascals.

TABLE 9.1

Typical Values for the Elastic Modulus

Substance	Young's Modulus (Pa)	Shear Modulus (Pa)	Bulk Modulus (Pa)
Aluminum	7.0×10^{10}	2.5×10^{10}	7.0×10^{10}
Bone	1.8×10^{10}	8.0×10^{10}	—
Brass	9.1×10^{10}	3.5×10^{10}	6.1×10^{10}
Copper	11×10^{10}	4.2×10^{10}	14×10^{10}
Steel	20×10^{10}	8.4×10^{10}	16×10^{10}
Tungsten	35×10^{10}	14×10^{10}	20×10^{10}
Glass	$6.5–7.8 \times 10^{10}$	$2.6–3.2 \times 10^{10}$	$5.0–5.5 \times 10^{10}$
Quartz	5.6×10^{10}	2.6×10^{10}	2.7×10^{10}
Rib Cartilage	1.2×10^{7}	—	—
Rubber	0.1×10^{7}	—	—
Tendon	2×10^{7}	—	—
Water	—	—	0.21×10^{10}
Mercury	—	—	2.8×10^{10}

Typical values are given in Table 9.1. Experiments show that (1) the change in length for a fixed external force is proportional to the original length and (2) the force necessary to produce a given strain is proportional to the cross-sectional area. The value of Young's modulus for a given material depends on whether the material is stretched or compressed. A human femur, for example, is stronger under tension than compression.

It's possible to exceed the **elastic limit** of a substance by applying a sufficiently great stress (Fig. 9.4). At the elastic limit, the stress–strain curve departs from a straight line. A material subjected to a stress beyond this limit ordinarily doesn't return to its original length when the external force is removed. As the stress is increased further, it surpasses the **ultimate strength**: the greatest stress the substance can withstand without breaking. The **breaking point** for brittle materials is just beyond the ultimate strength. For ductile metals like copper and gold, after passing the point of ultimate strength, the metal thins and stretches at a lower stress level before breaking.

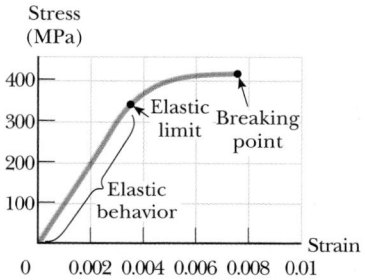

Figure 9.4 Stress-versus-strain curve for an elastic solid.

Shear Modulus: Elasticity of Shape

Another type of deformation occurs when an object is subjected to a force $\vec{F}$ *parallel* to one of its faces while the opposite face is held fixed by a second force (Active Fig. 9.5a). If the object is originally a rectangular block, such a parallel force results in a shape with the cross section of a parallelogram. This kind of stress is called a **shear stress**. A book pushed sideways, as in Active Figure 9.5b, is being subjected to a shear stress. There is no change in volume with this kind of deformation. It's important to remember that in shear stress, the applied force is *parallel* to the cross-sectional area, whereas in tensile stress the force is *perpendicular* to the cross-sectional area. We define **the shear stress as *F/A*, the ratio of the magnitude of the**

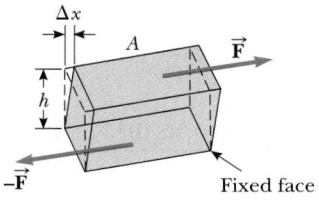

(a)

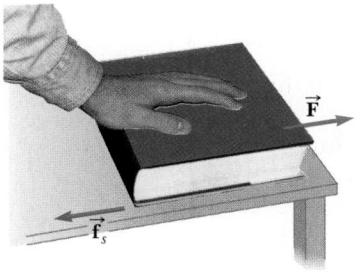

(b)

ACTIVE FIGURE 9.5
(a) A shear deformation in which a rectangular block is distorted by forces applied tangent to two of its faces. (b) A book under shear stress.

PhysicsNow™
Log into PhysicsNow at **www.cp7e.com**, and go to Active Figure 9.5 to adjust the values of the applied force and the shear modulus and observe the change in shape of the block in part (a).

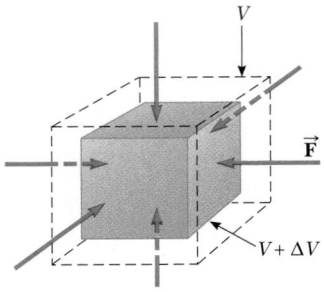

ACTIVE FIGURE 9.6
When a solid is under uniform pressure, it undergoes a change in volume, but no change in shape. This cube is compressed on all sides by forces normal to its six faces.

PhysicsNow™
Log into PhysicsNow at **www.cp7e.com**, and go to Active Figure 9.6 to adjust the values of the applied force and the bulk modulus and observe the change in volume of the cube.

parallel force to the area A of the face being sheared. The shear strain is the ratio $\Delta x/h$, where Δx is the horizontal distance the sheared face moves and h is the height of the object. The shear stress is related to the shear strain according to

$$\frac{F}{A} = S\frac{\Delta x}{h} \qquad [9.4]$$

where S is the **shear modulus** of the material, with units of pascals (force per unit area). Once again, notice the similarity to Hooke's law.

A material having a large shear modulus is difficult to bend. Shear moduli for some representative materials are listed in Table 9.1.

Bulk Modulus: Volume Elasticity

The bulk modulus characterizes the response of a substance to uniform squeezing. Suppose that the external forces acting on an object are all perpendicular to the surface on which the force acts and are distributed uniformly over the surface of the object (Active Fig. 9.6). This occurs when an object is immersed in a fluid. An object subject to this type of deformation undergoes a change in volume but no change in shape. **The volume stress ΔP is defined as the ratio of the magnitude of the change in the applied force ΔF to the surface area A.** (In dealing with fluids, we'll refer to the quantity F/A as the **pressure**, to be defined and discussed more formally in the next section.) The volume strain is equal to the change in volume ΔV divided by the original volume V. Again using Equation 9.1, we can relate a volume stress to a volume strain by the formula

Bulk modulus ▶

$$\Delta P = -B\frac{\Delta V}{V} \qquad [9.5]$$

A material having a large bulk modulus doesn't compress easily. Note that a negative sign is included in this defining equation so that B is always positive. An increase in pressure (positive ΔP) causes a decrease in volume (negative ΔV) and vice versa.

Table 9.1 lists bulk modulus values for some materials. If you look up such values in a different source, you may find that the reciprocal of the bulk modulus, called the **compressibility** of the material, is listed. Note from the table that both solids and liquids have bulk moduli. There is neither a Young's modulus nor shear modulus for liquids, however, because liquids simply flow when subjected to a tensile or shearing stress.

EXAMPLE 9.1 Built to Last

Goal Calculate a compression due to tensile stress, and maximum load.

Problem A vertical steel beam in a building supports a load of 6.0×10^4 N. **(a)** If the length of the beam is 4.0 m and its cross-sectional area is 8.0×10^{-3} m^2, find the distance the beam is compressed along its length. **(b)** What maximum load in newtons could the steel beam support before failing?

Strategy Equation 9.3 pertains to compressive stress and strain and can be solved for ΔL, followed by substitution of known values. For part (b), set the compressive stress equal to the ultimate strength of steel from Table 9.2. Solve for the magnitude of the force, which is the total weight the structure can support.

Solution

(a) Find the amount of compression in the beam.

Solve Equation 9.3 for ΔL and substitute, using the value of Young's modulus from Table 9.1:

$$\frac{F}{A} = Y\frac{\Delta L}{L_0}$$

$$\Delta L = \frac{FL_0}{YA} = \frac{(6.0 \times 10^4\ \text{N})(4.0\ \text{m})}{(2.0 \times 10^{11}\ \text{Pa})(8.0 \times 10^{-3}\ \text{m}^2)}$$

$$= 1.5 \times 10^{-4}\ \text{m}$$

(b) Find the maximum load that the beam can support.

Set the compressive stress equal to the ultimate compressive strength from Table 9.2, and solve for F:

$$\frac{F}{A} = \frac{F}{8.0 \times 10^{-3}\ \text{m}^2} = 5.0 \times 10^8\ \text{Pa}$$

$$F = 4.0 \times 10^6\ \text{N}$$

Remarks In designing load-bearing structures of any kind, it's always necessary to build in a safety factor. No one would drive a car over a bridge that had been designed to supply the minimum necessary strength to keep it from collapsing.

Exercise 9.1

A cable used to lift heavy materials like steel I-beams must be strong enough to resist breaking even under a load of 1.0×10^6 N. For safety, the cable must support twice that load. (a) What cross-sectional area should the cable have if it's to be made of steel? (b) By how much will an 8.0-m length of this cable stretch when subject to the 1.0×10^6-N load?

Answers (a) $4.0 \times 10^{-3}\ \text{m}^2$ (b) 1.0×10^{-2} m

TABLE 9.2

Ultimate Strength of Materials

Material	Tensile Strength (N/m^2)	Compressive Strength (N/m^2)
Iron	1.7×10^8	5.5×10^8
Steel	5.0×10^8	5.0×10^8
Aluminum	2.0×10^8	2.0×10^8
Bone	1.2×10^8	1.5×10^8
Marble	—	8.0×10^7
Brick	1×10^6	3.5×10^7
Concrete	2×10^6	2×10^7

EXAMPLE 9.2 Explosive Bolts

Goal Calculate the maximum shear stress supported by a set of bolts.

Problem Until launch, rockets are generally held to the launch pad by explosive bolts. Such bolts are also used in escape hatches and to secure different stages of the rocket, external tanks, and strap-on boosters, allowing rapid release when a part needs to be separated from the rest of the vehicle. Suppose a rocket has a strap-on booster supported by eight horizontal steel bolts, each 9.00 cm in diameter and oriented horizontally. (Bolts similar to these would be used in rockets like the Titan IV, shown at right.) **(a)** What maximum load can be placed on these bolts before they are sheared off? Assume the load is shared equally by the eight bolts. The ultimate shear strength of steel is 2.50×10^8 Pa. **(b)** If the booster has a mass of 3.00×10^5 kg, calculate the shear deformation of one of the bolts if the length of the bolt between rocket and booster is 8.00 cm.

NASA

The Titan IV launch vehicle, with its two solid rocket boosters, is capable of placing large satellites in geosynchronous orbit.

Strategy (a) The total force required to shear off the bolts increases with the number of bolts, but the necessary shear stress does not. Set the ultimate shear strength equal to the shear stress and solve for the force, multiplying the answer by eight to find the total shear force that can be applied. Part (b) can be solved by substituting values into Equation 9.4, which relates shear stress to shear strain.

Solution

(a) Find the maximum load the bolts can support.

Set the shear stress for one bolt equal to its ultimate shear strength:

$$\frac{F_1}{A} = \frac{F_1}{\pi r^2} = 2.50 \times 10^8\ \text{N/m}^2$$

Solve for the force needed to shear off one bolt:

$$F_1 = \pi r^2(2.50 \times 10^8 \text{ Pa})$$
$$= \pi(4.50 \times 10^{-2} \text{ m})^2(2.50 \times 10^8 \text{ Pa})$$
$$= 1.59 \times 10^6 \text{ N}$$

Multiply this result by eight to get the total shear force that the eight bolts can support:

$$F_{tot} = 1.27 \times 10^7 \text{ N}$$

(b) Calculate the shear deformation of one bolt at half the maximum load.

The bolts must support the booster against the force of gravity. The reaction to this force is the force exerted by the booster on the bolts, equal in magnitude to the weight of the booster.

$$F_{tot} = mg = (3.00 \times 10^5 \text{ kg})(9.80 \text{ m/s}^2) = 2.94 \times 10^6 \text{ N}$$

This force is shared among eight bolts. Use Equation 9.4, taking $F = F_{tot}/8$ and solving for Δx:

$$\frac{F}{A} = S\frac{\Delta x}{h}$$

$$\Delta x = \frac{(F_{tot}/8)h}{AS} = \frac{(3.68 \times 10^5 \text{ N})(0.080 \text{ m})}{\pi(4.50 \times 10^{-2} \text{ m})^2(8.40 \times 10^{10} \text{ Pa})}$$
$$= 5.51 \times 10^{-5} \text{ m}$$

Remarks The Titan IV launch vehicle, which is capable of putting large satellites into geosynchronous orbit, has a pair of strap-on boosters similar to this one. Notice that the bolts in this example are capable of supporting about four times as much weight as needed. Because all materials either have or develop microscopic defects, a safety factor has to be built in, so structures are designed to tolerate several times the maximum stresses they are expected to undergo.

Exercise 9.2

Calculate the diameter of a single steel horizontal bolt if it is expected to support a maximum load having a mass of 2.00×10^3 kg, but for safety reasons must be designed to support three times that load.

Answer 1.73 cm

EXAMPLE 9.3 Stressing a Lead Ball

Goal Apply the concepts of bulk stress and strain.

Problem A solid lead sphere of volume 0.50 m^3, dropped in the ocean, sinks to a depth of 2.0×10^3 m (about 1 mile), where the pressure increases by 2.0×10^7 Pa. Lead has a bulk modulus of 4.2×10^{10} Pa. What is the change in volume of the sphere?

Strategy Solve Equation 9.5 for ΔV and substitute the given quantities.

Solution

Start with the definition of bulk modulus:

$$B = -\frac{\Delta P}{\Delta V/V}$$

Solve for ΔV:

$$\Delta V = -\frac{V\Delta P}{B}$$

Substitute the known values:

$$\Delta V = -\frac{(0.50 \text{ m}^3)(2.0 \times 10^7 \text{ Pa})}{4.2 \times 10^{10} \text{ Pa}} = -2.4 \times 10^{-4} \text{ m}^3$$

Remarks The negative sign indicates a *decrease* in volume. The following exercise shows that even water can be compressed, though not by much, despite the depth.

Exercise 9.3

(a) By what percentage does a similar globe of water shrink at that same depth? (b) What is the ratio of the new radius to the initial radius?

Answer (a) 0.95% (b) 0.997

Arches and the Ultimate Strength of Materials

As we have seen, the ultimate strength of a material is the maximum force per unit area the material can withstand before it breaks or fractures. Such values are of great importance, particularly in the construction of buildings, bridges, and roads. Table 9.2 gives the ultimate strength of a variety of materials under both tension and compression. Note that bone and a variety of building materials (concrete, brick, and marble) are stronger under compression than under tension. The greater ability of brick and stone to resist compression is the basis of the semicircular arch, developed and used extensively by the Romans in everything from memorial arches to expansive temples and aqueduct supports.

APPLICATION

Arch Structures in Buildings

Before the development of the arch, the principal method of spanning a space was the simple post-and-beam construction (Fig. 9.7a), in which a horizontal beam is supported by two columns. This type of construction was used to build the great Greek temples. The columns of these temples were closely spaced because of the limited length of available stones and the low ultimate tensile strength of a sagging stone beam.

The semicircular arch (Fig. 9.7b) developed by the Romans was a great technological achievement in architectural design. It effectively allowed the heavy load of a wide roof span to be channeled into horizontal and vertical forces on narrow supporting columns. The stability of this arch depends on the compression between its wedge-shaped stones. The stones are forced to squeeze against each other by the uniform loading, as shown in the figure. This compression results in horizontal outward forces at the base of the arch where it starts curving away from the vertical. These forces must then be balanced by the stone walls shown on the sides of the arch. It's common to use very heavy walls (buttresses) on either side of the arch to provide horizontal stability. If the foundation of the arch should move, the compressive forces between the wedge-shaped stones may decrease to the extent that the arch collapses. The stone surfaces used in the arches constructed by the Romans were cut to make very tight joints; mortar was usually not used. The resistance to slipping between stones was provided by the compression force and the friction between the stone faces.

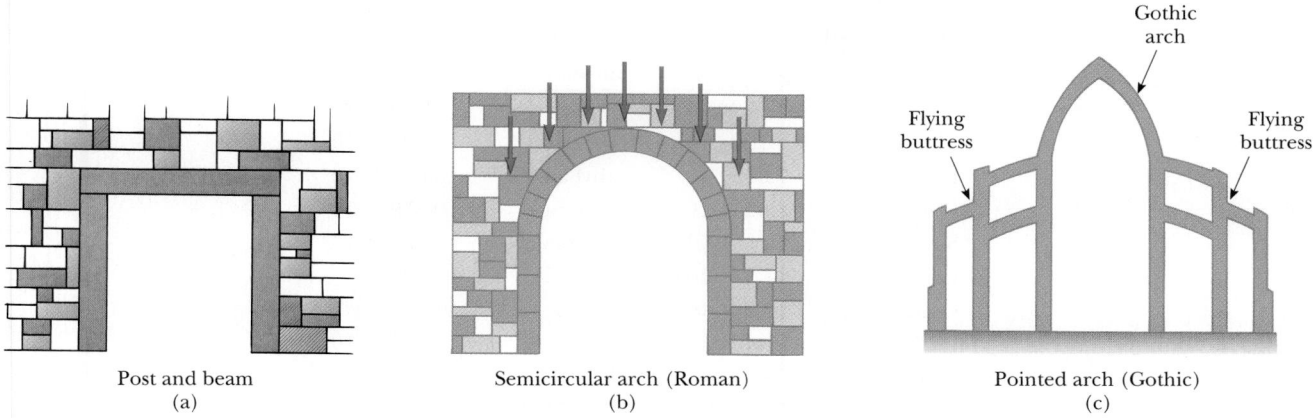

Figure 9.7 (a) A simple post-and-beam structure. (b) The semicircular arch developed by the Romans. (c) Gothic arch with flying buttresses to provide lateral support.

Another important architectural innovation was the pointed Gothic arch, shown in Figure 9.7c. This type of structure was first used in Europe beginning in the 12th century, followed by the construction of several magnificent Gothic cathedrals in France in the 13th century. One of the most striking features of these cathedrals is their extreme height. For example, the cathedral at Chartres rises to 118 ft, and the one at Reims has a height of 137 ft. Such magnificent buildings evolved over a very short time, without the benefit of any mathematical theory of structures. However, Gothic arches required flying buttresses to prevent the spreading of the arch supported by the tall, narrow columns.

9.3 DENSITY AND PRESSURE

Equal masses of aluminum and gold have an important physical difference: The aluminum takes up over seven times as much space as the gold. While the reasons for the difference lie at the atomic and nuclear levels, a simple measure of this difference is the concept of *density*.

Density ▶

The **density** ρ of an object having uniform composition is defined as its mass M divided by its volume V:

$$\rho \equiv \frac{M}{V} \qquad [9.6]$$

SI unit: kilogram per meter cubed (kg/m^3)

The most common units used for density are kilograms per cubic meter in the SI system and grams per cubic centimeter in the cgs system. Table 9.3 lists the densities of some substances. The densities of most liquids and solids vary slightly with changes in temperature and pressure; the densities of gases vary greatly with such changes. Under normal conditions, the densities of solids and liquids are about 1 000 times greater than the densities of gases. This difference implies that the average spacing between molecules in a gas under such conditions is about ten times greater than in a solid or liquid.

The **specific gravity** of a substance is the ratio of its density to the density of water at 4°C, which is 1.0×10^3 kg/m^3. (The size of the kilogram was originally defined to make the density of water 1.0×10^3 kg/m^3 at 4°C.) By definition, specific gravity is a dimensionless quantity. For example, if the specific gravity of a substance is 3.0, its density is $3.0(1.0 \times 10^3 \text{ kg/m}^3) = 3.0 \times 10^3 \text{ kg/m}^3$.

TABLE 9.3

Densities of Some Common Substances

Substance	ρ(kg/m^3)[a]	Substance	ρ(kg/m^3)[a]
Ice	0.917×10^3	Water	1.00×10^3
Aluminum	2.70×10^3	Glycerin	1.26×10^3
Iron	7.86×10^3	Ethyl alcohol	0.806×10^3
Copper	8.92×10^3	Benzene	0.879×10^3
Silver	10.5×10^3	Mercury	13.6×10^3
Lead	11.3×10^3	Air	1.29
Gold	19.3×10^3	Oxygen	1.43
Platinum	21.4×10^3	Hydrogen	8.99×10^{-2}
Uranium	18.7×10^3	Helium	1.79×10^{-1}

[a]All values are at standard atmospheric temperature and pressure (STP), defined as 0°C (273 K) and 1 atm (1.013×10^5 Pa). To convert to grams per cubic centimeter, multiply by 10^{-3}.

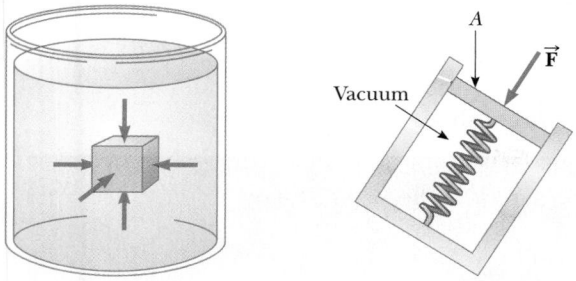

FIGURE 9.8 (a) The force exerted by a fluid on a submerged object at any point is perpendicular to the surface of the object. The force exerted by the fluid on the walls of the container is perpendicular to the walls at all points and increases with depth. (b) A simple device for measuring pressure in a fluid.

Quick Quiz 9.1

Suppose you have one cubic meter of gold, two cubic meters of silver, and six cubic meters of aluminum. Rank them by mass, from smallest to largest. (a) gold, aluminum, silver (b) gold, silver, aluminum (c) aluminum, gold, silver (d) silver, aluminum, gold

Fluids don't sustain shearing stresses, so the only stress that a fluid can exert on a submerged object is one that tends to compress it, which is bulk stress. The force exerted by the fluid on the object is always perpendicular to the surfaces of the object, as shown in Figure 9.8a.

The pressure at a specific point in a fluid can be measured with the device pictured in Figure 9.8b: an evacuated cylinder enclosing a light piston connected to a spring that has been previously calibrated with known weights. As the device is submerged in a fluid, the fluid presses down on the top of the piston and compresses the spring until the inward force exerted by the fluid is balanced by the outward force exerted by the spring. Let F be the magnitude of the force on the piston and A the area of the top surface of the piston. Notice that the force that compresses the spring is spread out over the entire area, motivating our formal definition of pressure:

TIP 9.1 Force and Pressure

Equation 9.7 makes a clear distinction between force and pressure. Another important distinction is that *force is a vector* and *pressure is a scalar.* There is no direction associated with pressure, but the direction of the force associated with the pressure is perpendicular to the surface of interest.

◀ Average pressure

If F is the magnitude of a force exerted perpendicular to a given surface of area A, then the pressure P is the force divided by the area:

$$P \equiv \frac{F}{A} \qquad [9.7]$$

SI unit: pascal (Pa)

Because pressure is defined as force per unit area, it has units of pascals (newtons per square meter). The English customary unit for pressure is the pound per inch squared. Atmospheric pressure at sea level is $14.7\ \text{lb/in}^2$, which in SI units is 1.01×10^5 Pa.

As we see from Equation 9.7, the effect of a given force depends critically on the area to which it's applied. A 700-N man can stand on a vinyl-covered floor in regular street shoes without damaging the surface, but if he wears golf shoes, the metal cleats protruding from the soles can do considerable damage to the floor. With the cleats, the same force is concentrated into a smaller area, greatly elevating the pressure in those areas, resulting in a greater likelihood of exceeding the ultimate strength of the floor material.

Snowshoes use the same principle (Fig. 9.9). The snow exerts an upward normal force on the shoes to support the person's weight. According to Newton's third law, this upward force is accompanied by a downward force exerted by the shoes on the snow. If the person is wearing snowshoes, that force is distributed over the very large area of each snowshoe, so that the pressure at any given point is relatively low and the person doesn't penetrate very deeply into the snow.

© Royalty-Free/Corbis

Figure 9.9 Snowshoes prevent the person from sinking into the soft snow because the force on the snow is spread over a larger area, reducing the pressure on the snow's surface.

Applying Physics 9.1 Bed of Nails Trick

After an exciting but exhausting lecture, a physics professor stretches out for a nap on a bed of nails, as in Figure 9.10, suffering no injury and only moderate discomfort. How is this possible?

Explanation If you try to support your entire weight on a single nail, the pressure on your body is your weight divided by the very small area of the end of the nail. The resulting pressure is large enough to penetrate the skin. If you distribute your weight over several hundred nails, however, as demonstrated by the professor, the pressure is considerably reduced because the area that supports your weight is the total area of all nails in contact with your body. (Why is lying on a bed of nails more comfortable than sitting on the same bed? Extend the logic to show that it would be more uncomfortable yet to stand on a bed of nails without shoes.)

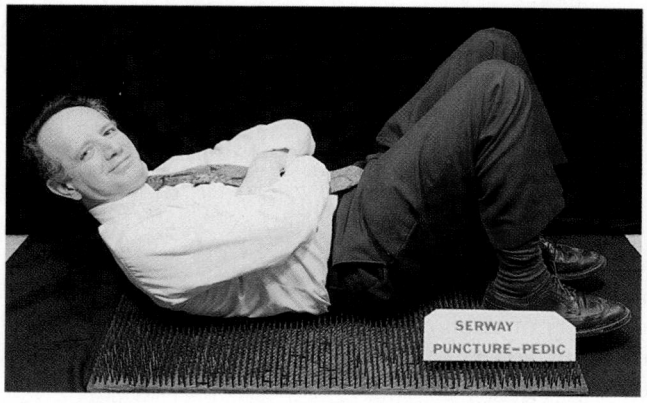

Raymond A. Serway

Figure 9.10 (Applying Physics 9.1) Does anyone have a pillow?

EXAMPLE 9.4 The Water Bed

Goal Calculate a density and a pressure from a weight.

Problem A water bed is 2.00 m on a side and 30.0 cm deep. **(a)** Find its weight. **(b)** Find the pressure that the water bed exerts on the floor. Assume that the entire lower surface of the bed makes contact with the floor.

Strategy Density is mass per unit volume: first, find the volume of the bed and multiply it by the density of water to get the bed's mass. Multiplying by the acceleration of gravity then gives the weight of the bed. The weight divided by the area of floor the bed rests upon gives the pressure exerted on the floor.

Solution

(a) Find the weight of the water bed.

First, find the volume of the bed:

$$V = lwh = (2.00\text{ m})(2.00\text{ m})(0.300\text{ m}) = 1.20\text{ m}^3$$

Solve the density equation for the mass and substitute, then multiply the result by g to get the weight:

$$\rho = \frac{M}{V}$$

$$M = \rho V = (1.00 \times 10^3\text{ kg/m}^3)(1.20\text{ m}^3) = 1.20 \times 10^3\text{ kg}$$

$$w = Mg = (1.20 \times 10^3\text{ kg})(9.80\text{ m/s}^2) = 1.18 \times 10^4\text{ N}$$

(b) Find the pressure that the bed exerts on the floor.

Use the cross-sectional area $A = 4.00\text{ m}^2$ and the value of w from part (a) to get the pressure:

$$P = \frac{F}{A} = \frac{w}{A} = \frac{1.18 \times 10^4\text{ N}}{4.00\text{ m}^2} = 2.95 \times 10^3\text{ Pa}$$

Remarks Notice that the answer to part (b) is far less than atmospheric pressure. Water is heavier than air for a given volume, but the air is stacked up considerably higher (100 km!). The total pressure exerted on the floor would include the pressure of the atmosphere.

Exercise 9.4

Calculate the pressure exerted by the water bed on the floor if the bed rests on its side.

Answer 1.97×10^4 Pa

9.4 VARIATION OF PRESSURE WITH DEPTH

When a fluid is at rest in a container, **all portions of the fluid must be in static equilibrium**—at rest with respect to the observer. Furthermore, **all points at the same depth must be at the same pressure**. If this were not the case, fluid would flow from the higher pressure region to the lower pressure region. For example, consider the small block of fluid shown in Figure 9.11a. If the pressure were greater on the left side of the block than on the right, $\vec{\mathbf{F}}_1$ would be greater than $\vec{\mathbf{F}}_2$, and the block would accelerate to the right and thus would not be in equilibrium.

Next, let's examine the fluid contained within the volume indicated by the darker region in Figure 9.11b. This region has cross-sectional area A and extends from position y_1 to position y_2 below the surface of the liquid. Three external forces act on this volume of fluid: the force of gravity, Mg; the upward force P_2A exerted by the liquid below it; and a downward force P_1A exerted by the fluid above it. Because the given volume of fluid is in equilibrium, these forces must add to zero, so we get

$$P_2A - P_1A - Mg = 0 \qquad [9.8]$$

From the definition of density, we have

$$M = \rho V = \rho A(y_1 - y_2) \qquad [9.9]$$

Substituting Equation 9.9 into Equation 9.8, canceling the area A, and rearranging terms, we get

$$P_2 = P_1 + \rho g(y_1 - y_2) \qquad [9.10]$$

Notice that $(y_1 - y_2)$ is positive, because $y_2 < y_1$. The force P_2A is greater than the force P_1A by exactly the weight of water between the two points. This is the same principle experienced by the person at the bottom of a pileup in football or rugby.

Atmospheric pressure is also caused by a piling up of fluid—in this case, the fluid is the gas of the atmosphere. The weight of all the air from sea level to the edge of space results in an atmospheric pressure of $P_0 = 1.013 \times 10^5$ Pa (equivalent to 14.7 lb/in.^{2}) at sea level. This result can be adapted to find the pressure P at any depth $h = (y_1 - y_2) = (0 - y_2)$ below the surface of the water:

$$P = P_0 + \rho gh \qquad [9.11]$$

According to Equation 9.11, **the pressure P at a depth h below the surface of a liquid open to the atmosphere is greater than atmospheric pressure by the amount ρgh.** Moreover, the pressure isn't affected by the shape of the vessel, as shown in Figure 9.12.

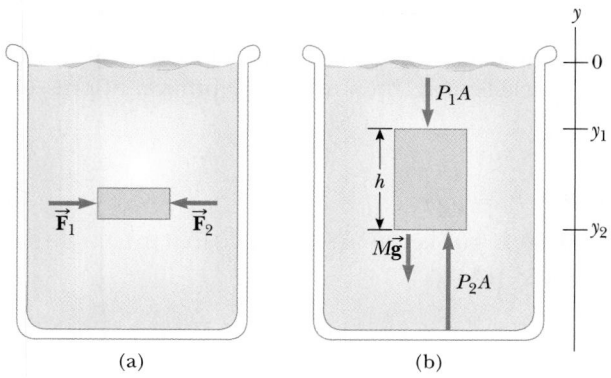

Figure 9.11 (a) If the block of fluid is to be in equilibrium, the force $\vec{\mathbf{F}}_1$ must balance the force $\vec{\mathbf{F}}_2$. (b) The net force on the volume of liquid within the darker region must be zero.

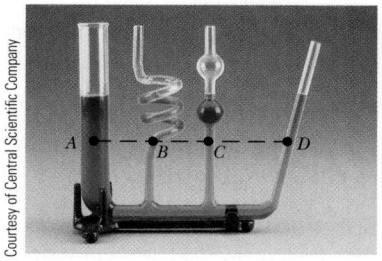

Courtesy of Central Scientific Company

Figure 9.12 This photograph illustrates the fact that the pressure in a liquid is the same at all points lying at the same elevation. For example, the pressure is the same at points A, B, C, and D. Note that the shape of the vessel does not affect the pressure.

Quick Quiz 9.2

The pressure at the bottom of a glass filled with water ($\rho = 1\,000\ \text{kg/m}^3$) is P. The water is poured out and the glass is filled with ethyl alcohol ($\rho = 806\ \text{kg/m}^3$). The pressure at the bottom of the glass is now (a) smaller than P (b) equal to P (c) larger than P (d) indeterminate.

EXAMPLE 9.5 Oil and Water

Goal Calculate pressures created by layers of different fluids.

Problem In a huge oil tanker, salt water has flooded an oil tank to a depth of 5.00 m. On top of the water is a layer of oil 8.00 m deep, as in the cross-sectional view of the tank in Figure 9.13. The oil has a density of $0.700\ \text{g/cm}^3$. Find the pressure at the bottom of the tank. (Take $1\,025\ \text{kg/m}^3$ as the density of salt water.)

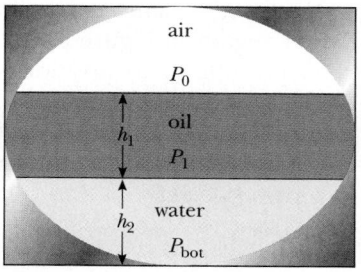

Figure 9.13 (Example 9.5)

Strategy Equation 9.11 must be used twice. First, use it to calculate the pressure P_1 at the bottom of the oil layer. Then use this pressure in place of P_0 in Equation 9.11 and calculate the pressure P_{bot} at the bottom of the water layer.

Solution

Use Equation 9.11 to calculate the pressure at the bottom of the oil layer:

$$P_1 = P_0 + \rho g h_1 \quad (1)$$
$$= 1.01 \times 10^5\ \text{Pa} + (7.00 \times 10^2\ \text{kg/m}^3)(9.80\ \text{m/s}^2)(8.00\ \text{m})$$
$$P_1 = 1.56 \times 10^5\ \text{Pa}$$

Now adapt Equation 9.11 to the new starting pressure, and use it to calculate the pressure at the bottom of the water layer:

$$P_{bot} = P_1 + \rho g h_2 \quad (2)$$
$$= 1.56 \times 10^5\ \text{Pa} + (1.025 \times 10^3\ \text{kg/m}^3)(9.80\ \text{m/s}^2)(5.00\ \text{m})$$
$$P_{bot} = 2.06 \times 10^5\ \text{Pa}$$

Remark The weight of the atmosphere results in P_0 at the surface of the oil layer. Then the weight of the oil and the weight of the water combine to create the pressure at the bottom.

Exercise 9.5

Calculate the pressure on the top lid of a chest buried under 4.00 meters of mud with density $1.75 \times 10^3\ \text{kg/m}^3$ at the bottom of a 10.0-m-deep lake.

Answer $2.68 \times 10^5\ \text{Pa}$

EXAMPLE 9.6 A Pain in the Ear

Goal Calculate a pressure difference at a given depth, and estimate a force.

Problem Estimate the net force exerted on your eardrum due to the water above when you are swimming at the bottom of a pool that is 5.0 m deep.

Strategy Use Equation 9.11 to find the pressure difference across the eardrum at the given depth. The air inside the ear is generally at atmospheric pressure. Estimate the eardrum's surface area, then use the definition of pressure to get the net force exerted on the eardrum.

Solution

Use Equation 9.11 to calculate the difference between the water pressure at the depth h and the pressure inside the ear:

$$\Delta P = P - P_0 = \rho g h$$
$$= (1.00 \times 10^3\ \text{kg/m}^3)(9.80\ \text{m/s}^2)(5.0\ \text{m})$$
$$= 4.9 \times 10^4\ \text{Pa}$$

Mutliply by area A to get the net force on the eardrum associated with this pressure difference, estimating the area of the eardrum as $1\ \text{cm}^2$.

$$F_{net} = A\Delta P \approx (1 \times 10^{-4}\ \text{m}^2)(4.9 \times 10^4\ \text{Pa}) \approx 5\ \text{N}$$

Remarks Because a force on the eardrum of this magnitude is uncomfortable, swimmers often "pop their ears" by swallowing or expanding their jaws while underwater, an action that pushes air from the lungs into the middle ear. Using this technique equalizes the pressure on the two sides of the eardrum and relieves the discomfort.

Exercise 9.6

An airplane takes off at sea level and climbs to a height of 425 m. Estimate the net outward force on a passenger's eardrum assuming the density of air is approximately constant at 1.3 kg/m^3 and that the inner ear pressure hasn't been equalized.

Answer 0.54 N

In view of the fact that the pressure in a fluid depends on depth and on the value of P_0, any increase in pressure at the surface must be transmitted to every point in the fluid. This was first recognized by the French scientist Blaise Pascal (1623–1662) and is called **Pascal's principle**:

> A change in pressure applied to an enclosed fluid is transmitted undiminished to every point of the fluid and to the walls of the container.

APPLICATION
Hydraulic Lifts

An important application of Pascal's principle is the hydraulic press (Fig. 9.14a). A downward force $\vec{\mathbf{F}}_1$ is applied to a small piston of area A_1. The pressure is transmitted through a fluid to a larger piston of area A_2. As the pistons move and the fluids in the left and right cylinders change their relative heights, there are slight differences in the pressures at the input and output pistons. Neglecting these small differences, the fluid pressure on each of the pistons may be taken to be the same; $P_1 = P_2$. From the definition of pressure, it then follows that $F_1/A_1 = F_2/A_2$. Therefore, the magnitude of the force $\vec{\mathbf{F}}_2$ is larger than the magnitude of $\vec{\mathbf{F}}_1$ by the factor A_2/A_1. That's why a large load, such as a car, can be moved on the large piston by a much smaller force on the smaller piston. Hydraulic brakes, car lifts, hydraulic jacks, forklifts, and other machines make use of this principle.

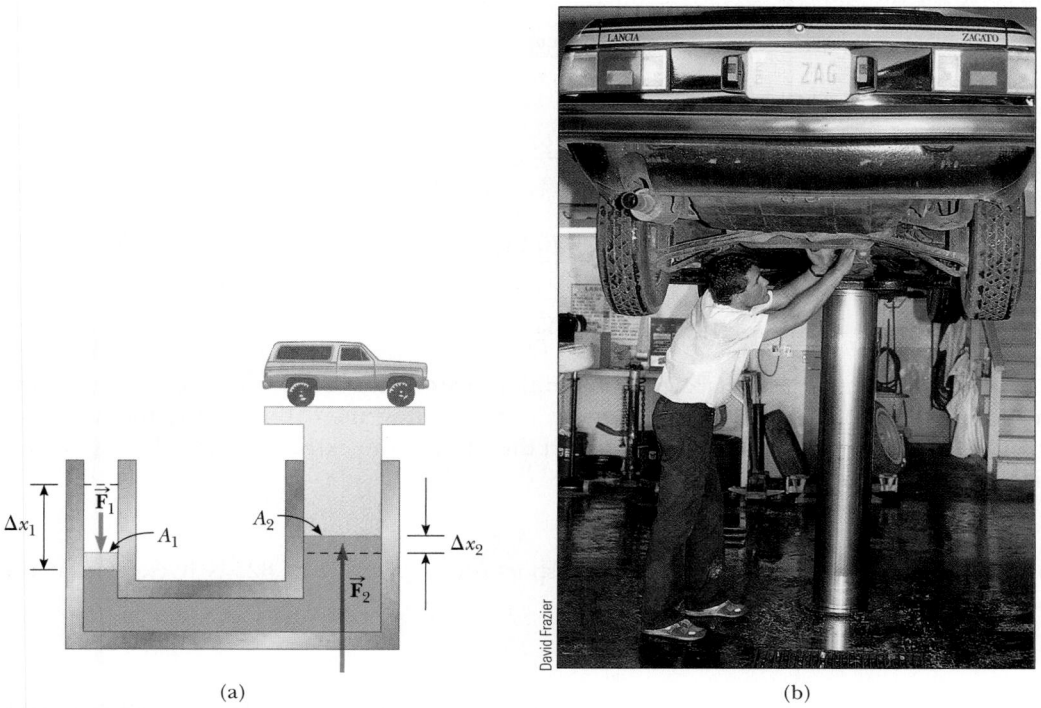

Figure 9.14 (a) Diagram of a hydraulic press (Example 9.7). Because the pressure is the same at the left and right sides, a small force $\vec{\mathbf{F}}_1$ at the left produces a much larger force $\vec{\mathbf{F}}_2$ at the right. (b) A vehicle under repair is supported by a hydraulic lift in a garage.

INTERACTIVE EXAMPLE 9.7 The Car Lift

Goal Apply Pascal's principle to a car lift, and show that the input work is the same as the output work.

Problem In a car lift used in a service station, compressed air exerts a force on a small piston of circular cross section having a radius of $r_1 = 5.00$ cm. This pressure is transmitted by an incompressible liquid to a second piston of radius $r_2 = 15.0$ cm. **(a)** What force must the compressed air exert on the small piston in order to lift a car weighing 13 300 N? Neglect the weights of the pistons. **(b)** What air pressure will produce a force of that magnitude? **(c)** Show that the work done by the input and output pistons is the same.

Strategy Substitute into Pascal's principle in part (a), while recognizing that the magnitude of the output force, F_2, must be equal to the car's weight in order to support it. Use the definition of pressure in part (b). In part (c), use $W = F\Delta x$ to find the ratio W_1/W_2, showing that it must equal 1. This requires combining Pascal's principle with the fact that the input and output pistons move through the same volume.

Solution

(a) Find the necessary force on the small piston.

Substitute known values into Pascal's principle, using $A = \pi r^2$ for the area.

$$F_1 = \left(\frac{A_1}{A_2}\right)F_2 = \frac{\pi r_1^2}{\pi r_2^2}F_2$$
$$= \frac{\pi(5.00 \times 10^{-2}\text{ m})^2}{\pi(15.0 \times 10^{-2}\text{ m})^2}(1.33 \times 10^4\text{ N})$$
$$= 1.48 \times 10^3\text{ N}$$

(b) Find the air pressure producing F_1.

Substitute into the definition of pressure:

$$P = \frac{F_1}{A_1} = \frac{1.48 \times 10^3\text{ N}}{\pi(5.00 \times 10^{-2}\text{ m})^2} = 1.88 \times 10^5\text{ Pa}$$

(c) Show that the work done by the input and output pistons is the same.

First equate the volumes, and solve for the ratio of A_2 to A_1:

$$V_1 = V_2 \rightarrow A_1\Delta x_1 = A_2\Delta x_2$$
$$\frac{A_2}{A_1} = \frac{\Delta x_1}{\Delta x_2}$$

Now use Pascal's principle to get a relationship for F_1/F_2:

$$\frac{F_1}{A_1} = \frac{F_2}{A_2} \rightarrow \frac{F_1}{F_2} = \frac{A_1}{A_2}$$

Evaluate the work ratio, substituting the preceding two results:

$$\frac{W_1}{W_2} = \frac{F_1\Delta x_1}{F_2\Delta x_2} = \left(\frac{F_1}{F_2}\right)\left(\frac{\Delta x_1}{\Delta x_2}\right) = \left(\frac{A_1}{A_2}\right)\left(\frac{A_2}{A_1}\right) = 1$$
$$W_1 = W_2$$

Remark In this problem, we didn't address the effect of possible differences in the heights of the pistons. If the column of fluid is higher in the small piston, the fluid weight assists in supporting the car, reducing the necessary applied force. If the column of fluid is higher in the large piston, both the car and the extra fluid must be supported, so additional applied force is required.

Exercise 9.7

A hydraulic lift has pistons with diameters 8.00 cm and 36.0 cm, respectively. If a force of 825 N is exerted at the input piston, what maximum mass can be lifted by the output piston?

Answer 1.70×10^3 kg

PhysicsNow™ You can adjust the weight of the truck in Figure 9.14a by logging into PhysicsNow at **www.cp7e.com** and going to Interactive Example 9.7.

Applying Physics 9.2 Building the Pyramids

A corollary to the statement that pressure in a fluid increases with depth is that water always seeks its own level. This means that if a vessel is filled with water, then regardless of the vessel's shape the surface of the water is perfectly flat and at the same height at all points. The ancient Egyptians used this fact to make the pyramids level. Devise a scheme showing how this could be done.

Explanation There are many ways it could be done, but Figure 9.15 shows the scheme used by the Egyptians. The builders cut grooves in the base of the pyramid as in (a) and partially filled the grooves with water. The height of the water was marked as in (b), and the rock was chiseled down to the mark, as in (c). Finally, the groove was filled with crushed rock and gravel, as in (d).

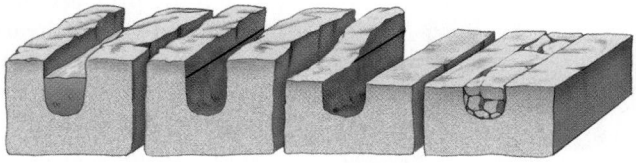

(a) (b) (c) (d)

Figure 9.15 (Applying Physics 9.2)

9.5 PRESSURE MEASUREMENTS

A simple device for measuring pressure is the open-tube manometer (Fig. 9.16a). One end of a U-shaped tube containing a liquid is open to the atmosphere, and the other end is connected to a system of unknown pressure P. The pressure at point B equals $P_0 + \rho gh$, where ρ is the density of the fluid. The pressure at B, however, equals the pressure at A, which is also the unknown pressure P. We conclude that $P = P_0 + \rho gh$.

The pressure P is called the **absolute pressure**, and $P - P_0$ is called the **gauge pressure**. If P in the system is greater than atmospheric pressure, h is positive. If P is less than atmospheric pressure (a partial vacuum), h is negative, meaning that the right-hand column in Figure 9.16a is lower than the left-hand column.

Another instrument used to measure pressure is the **barometer** (Fig. 9.16b), invented by Evangelista Torricelli (1608–1647). A long tube closed at one end is filled with mercury and then inverted into a dish of mercury. The closed end of the tube is nearly a vacuum, so its pressure can be taken to be zero. It follows that $P_0 = \rho gh$, where ρ is the density of the mercury and h is the height of the mercury column. Note that the barometer measures the pressure of the atmosphere, whereas the manometer measures pressure in an enclosed fluid.

One atmosphere of pressure is defined to be the pressure equivalent of a column of mercury that is exactly 0.76 m in height at 0°C with $g = 9.806\,65\ \text{m/s}^2$. At this temperature, mercury has a density of $13.595 \times 10^3\ \text{kg/m}^3$; therefore,

$$P_0 = \rho gh = (13.595 \times 10^3\ \text{kg/m}^3)(9.806\,65\ \text{m/s}^2)(0.760\,0\ \text{m})$$
$$= 1.013 \times 10^5\ \text{Pa} = 1\ \text{atm}$$

It is interesting to note that the force of the atmosphere on our bodies (assuming a body area of $2\,000\ \text{in}^2$) is extremely large, on the order of 30 000 lb! If it were not for the fluids permeating our tissues and body cavities, our bodies would collapse. The fluids provide equal and opposite forces. In the upper atmosphere or in space, sudden decompression can lead to serious injury and death. Air retained in the lungs can damage the tiny alveolar sacs, and intestinal gas can even rupture internal organs.

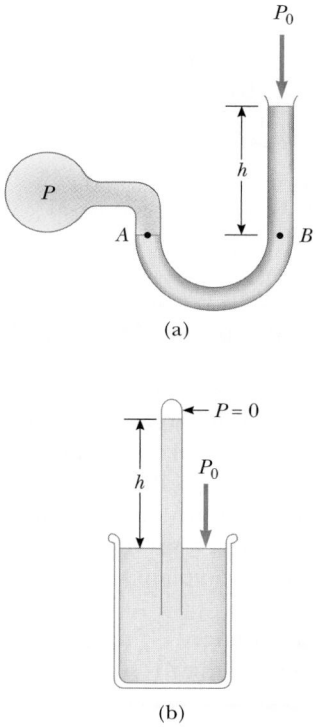

Figure 9.16 Two devices for measuring pressure: (a) an open-tube manometer and (b) a mercury barometer.

Quick Quiz 9.3

Several common barometers are built using a variety of fluids. For which fluid will the column of fluid in the barometer be the highest? (Refer to Table 9.3.) (a) mercury (b) water (c) ethyl alcohol (d) benzene.

APPLICATION

Measuring Blood Pressure

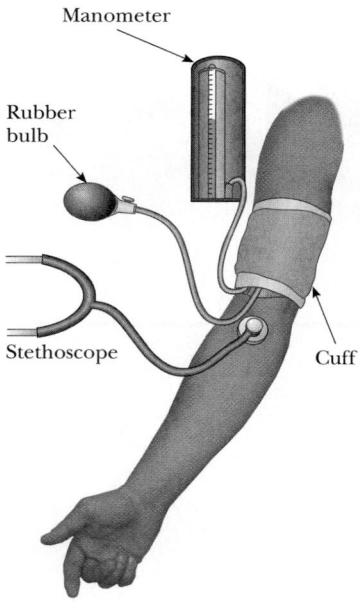

Figure 9.17 A sphygmomanometer can be used to measure blood pressure.

Blood Pressure Measurements

A specialized manometer (called a sphygmomanometer) is often used to measure blood pressure. In this application, a rubber bulb forces air into a cuff wrapped tightly around the upper arm and simultaneously into a manometer, as in Figure 9.17. The pressure in the cuff is increased until the flow of blood through the brachial artery in the arm is stopped. A valve on the bulb is then opened, and the measurer listens with a stethoscope to the artery at a point just below the cuff. When the pressure in the cuff and brachial artery is just below the maximum value produced by the heart (the systolic pressure), the artery opens momentarily on each beat of the heart. At this point, the velocity of the blood is high and turbulent, and the flow is noisy and can be heard with the stethoscope. The manometer is calibrated to read the pressure in millimeters of mercury, and the value obtained is about 120 mm for a normal heart. Values of 130 mm or above are considered high, and medication to lower the blood pressure is often prescribed for such patients. As the pressure in the cuff is lowered further, intermittent sounds are still heard until the pressure falls just below the minimum heart pressure (the diastolic pressure). At this point, continuous sounds are heard. In the normal heart, this transition occurs at about 80 mm of mercury, and values above 90 require medical intervention. Blood pressure readings are usually expressed as the ratio of the systolic pressure to the diastolic pressure, which is 120/80 for a healthy heart.

Quick Quiz 9.4

Blood pressure is normally measured with the cuff of the sphygmomanometer around the arm. Suppose that the blood pressure is measured with the cuff around the calf of the leg of a standing person. Would the reading of the blood pressure be (a) the same here as it is for the arm? (b) greater than it is for the arm? or (c) less than it is for the arm?

Applying Physics 9.3 Ballpoint Pens

In a ballpoint pen, ink moves down a tube to the tip, where it is spread on a sheet of paper by a rolling stainless steel ball. Near the top of the ink cartridge, there is a small hole open to the atmosphere. If you seal this hole, you will find that the pen no longer functions. Use your knowledge of how a barometer works to explain this behavior.

Explanation If the hole is sealed, or if it were not present, the pressure of the air above the ink would decrease as the ink gets used. Consequently, atmospheric pressure exerted against the ink at the bottom of the cartridge would prevent some of the ink from flowing out. The hole allows the pressure above the ink to remain at atmospheric pressure. Why does a ballpoint pen seem to run out of ink when you write on a vertical surface?

9.6 BUOYANT FORCES AND ARCHIMEDES'S PRINCIPLE

A fundamental principle affecting objects submerged in fluids was discovered by the Greek mathematician and natural philosopher Archimedes. **Archimedes's principle** can be stated as follows:

Archimedes's principle ▶

> Any object completely or partially submerged in a fluid is buoyed up by a force with magnitude equal to the weight of the fluid displaced by the object.

Many historians attribute the concept of buoyancy to Archimedes's "bathtub epiphany," when he noticed an apparent change in his weight upon lowering himself into a tub of water. As will be seen in Example 9.8, buoyancy yields a method of determining density.

Everyone has experienced Archimedes's principle. It's relatively easy, for example, to lift someone if you're both standing in a swimming pool, whereas lifting that same individual on dry land may be a difficult task. Water provides partial support to any object placed in it. We often say that an object placed in a fluid is buoyed up by the fluid, so we call this upward force the **buoyant force**.

The buoyant force is *not* a mysterious new force that arises in fluids. In fact, the physical cause of the buoyant force is the pressure difference between the upper and lower sides of the object, which can be shown to be equal to the weight of the displaced fluid. In Figure 9.18a, the fluid inside the indicated sphere, colored darker blue, is pressed on all sides by the surrounding fluid. Arrows indicate the forces arising from the pressure. Because pressure increases with depth, the arrows on the underside are larger than those on top. Adding them all up, the horizontal components cancel, but there is a net force upwards. This force, due to differences in pressure, is the buoyant force $\vec{\mathbf{B}}$. The sphere of water neither rises nor falls, so the vector sum of the buoyant force and the force of gravity on the sphere of fluid must be zero, and it follows that $B = Mg$, where M is the mass of the fluid.

Replacing the shaded fluid with a bowling ball of the same volume, as in Figure 9.18b, changes only the mass on which the pressure acts, so the buoyant force is the same: $B = Mg$, where M is the mass of the displaced fluid, *not* the mass of the bowling ball. The force of gravity on the heavier ball is greater than it was on the fluid, so the bowling ball sinks.

Archimedes's principle can also be obtained from Equation 9.8, relating pressure and depth, using Figure 9.11b. Horizontal forces from the pressure cancel, but in the vertical direction P_2A acts upwards on the bottom of the block of fluid and P_1A and the gravity force on the fluid, Mg, act downwards, giving

$$B = P_2A - P_1A = Mg \qquad \textbf{[9.12a]}$$

where the buoyancy force has been identified as a difference in pressure equal in magnitude to the weight of the displaced fluid. This buoyancy force remains the same regardless of the material occupying the volume in question because it's due to the *surrounding* fluid. Using the definition of density, Equation 9.12a becomes

$$B = \rho_{\text{fluid}} V_{\text{fluid}} g \qquad \textbf{[9.12b]}$$

where ρ_{fluid} is the density of the fluid and V_{fluid} is the volume of the displaced fluid. This result applies equally to all shapes, because any irregular shape can be approximated by a large number of infinitesimal cubes.

It's instructive to compare the forces on a totally submerged object with those on a floating object.

Case I: A Totally Submerged Object. When an object is *totally* submerged in a fluid of density ρ_{fluid}, the upward buoyant force acting on the object has a magnitude of $B = \rho_{\text{fluid}} V_{\text{obj}} g$, where V_{obj} is the volume of the object. If the object has density ρ_{obj}, the downward gravitational force acting on the object has a magni-

North Wind Picture Archives

ARCHIMEDES: Greek Mathematician, Physicist, and Engineer (287–212 B.C.)

Archimedes was perhaps the greatest scientist of antiquity. He is well known for discovering the nature of the buoyant force and was a gifted inventor. According to legend, Archimedes was asked by King Hieron to determine whether the king's crown was made of pure gold or merely a gold alloy. The task was to be performed without damaging the crown. Archimedes allegedly arrived at a solution while taking a bath, noting a partial loss of weight after submerging his arms and legs in the water. As the story goes, he was so excited about his great discovery that he ran naked through the streets of Syracuse, shouting, "Eureka!" which is Greek for "I have found it."

TIP 9.2 Buoyant Force is Exerted by the Fluid

The buoyant force on an object is exerted by the fluid and is the same, regardless of the density of the object. Objects more dense than the fluid sink; objects less dense rise.

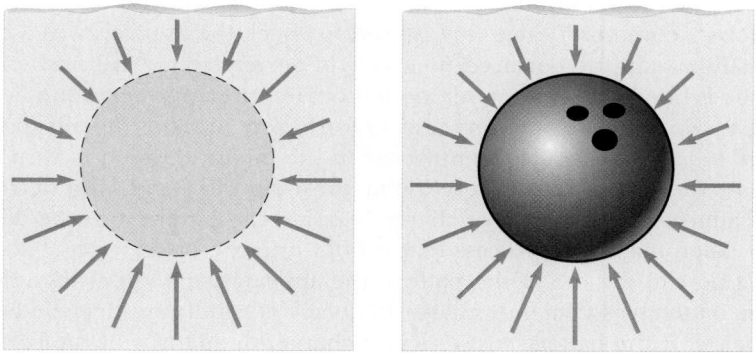

(a) (b)

Figure 9.18 (a) The arrows indicate forces on the sphere of fluid due to pressure, larger on the underside because pressure increases with depth. The net upward force is the buoyant force. (b) The buoyant force, which is caused by the *surrounding* fluid, is the same on any object of the same volume, including this bowling ball. The magnitude of the buoyant force is equal to the weight of the displaced fluid.

© Royalty-Free/Corbis

Hot-air balloons. Because hot air is less dense than cold air, there is a net upward force on the balloons.

ACTIVE FIGURE 9.19
(a) A totally submerged object that is less dense than the fluid in which it is submerged is acted upon by a net upward force. (b) A totally submerged object that is denser than the fluid sinks.

Physics Now™
Log into PhysicsNow at **www.cp7e.com**, and go to Active Figure 9.19 to move the object to new positions, as well as change the density of the object, and see the results.

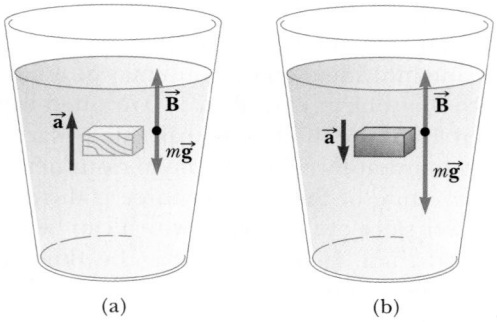

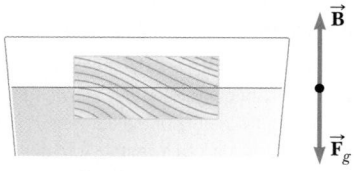

ACTIVE FIGURE 9.20
An object floating on the surface of a fluid is acted upon by two forces: the gravitational force $\vec{F}_g$ and the buoyant force $\vec{B}$. These two forces are equal in magnitude and opposite in direction.

Physics Now™
Log into PhysicsNow at **www.cp7e.com**, and go to Active Figure 9.20 to change the densities of the object and the fluid and see the results.

tude equal to $w = mg = \rho_{obj} V_{obj} g$, and the net force on it is $B - w = (\rho_{fluid} - \rho_{obj}) V_{obj} g$. Therefore, if the density of the object is *less* than the density of the fluid, as in Active Figure 9.19a, the net force exerted on the object is *positive* (upward) and the object accelerates *upward*. If the density of the object is *greater* than the density of the fluid, as in Active Figure 9.19b, the net force is *negative* and the object accelerates *downwards*.

Case II: A Floating Object. Now consider a partially submerged object in static equilibrium floating in a fluid, as in Active Figure 9.20. In this case, the upward buoyant force is balanced by the downward force of gravity acting on the object. If V_{fluid} is the volume of the fluid displaced by the object (which corresponds to the volume of the part of the object beneath the fluid level), then the magnitude of the buoyant force is given by $B = \rho_{fluid} V_{fluid} g$. Because the weight of the object is $w = mg = \rho_{obj} V_{obj} g$, and because $w = B$, it follows that $\rho_{fluid} V_{fluid} g = \rho_{obj} V_{obj} g$, or

$$\frac{\rho_{obj}}{\rho_{fluid}} = \frac{V_{fluid}}{V_{obj}} \qquad [9.13]$$

Equation 9.13 neglects the buoyant force of the air, which is slight, because the density of air is only 1.29 kg/m^3 at sea level.

Under normal circumstances, the average density of a fish is slightly greater than the density of water, so it would sink if it didn't have a mechanism for adjusting its density. By changing the size of an internal swim bladder, fish maintain neutral buoyancy as they swim to various depths.

APPLICATION

Cerebrospinal Fluid

The human brain is immersed in a fluid (the cerebrospinal fluid) of density 1 007 kg/m^3, which is slightly less than the average density of the brain, 1 040 kg/m^3. Consequently, most of the weight of the brain is supported by the buoyant force of the surrounding fluid. In some clinical procedures, a portion of this fluid must be removed for diagnostic purposes. During such procedures, the nerves and blood vessels in the brain are placed under great strain, which in turn can cause extreme discomfort and pain. Great care must be exercised with such patients until the initial volume of brain fluid has been restored by the body.

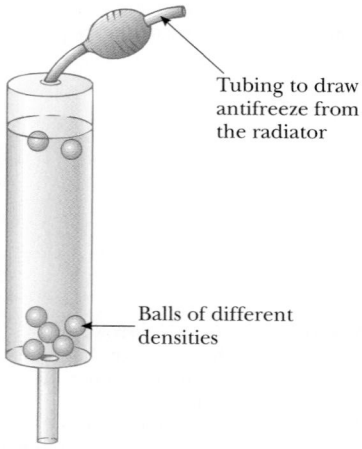

Figure 9.21 The number of balls that float in this device is a measure of the density of the antifreeze solution in a vehicle's radiator and, consequently, a measure of the temperature at which freezing will occur.

When service station attendants check the antifreeze in your car or the condition of your battery, they often use devices that apply Archimedes's principle. Figure 9.21 shows a common device that is used to check the antifreeze in a car radiator. The small balls in the enclosed tube vary in density, so that all of them float when the tube is filled with pure water, none float in pure antifreeze, one floats in a 5% mixture, two in a 10% mixture, and so forth. The number of balls that float is a measure of the percentage of antifreeze in the mixture, which in turn is used to determine the lowest temperature the mixture can withstand without freezing.

Similarly, the degree of charge in some car batteries can be determined with a so-called magic-dot process that is built into the battery (Fig. 9.22). Inside a viewing port in the top of the battery, the appearance of an orange dot indicates that the battery is sufficiently charged; a black dot indicates that the battery has lost its charge. If the battery has sufficient charge, the density of the battery fluid is high enough to cause the orange ball to float. As the battery loses its charge, the density

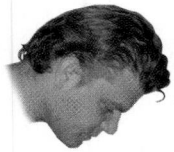

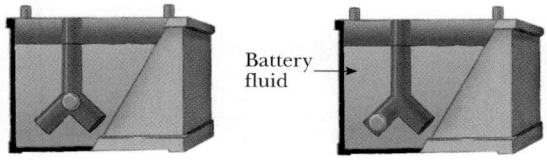

Figure 9.22 The orange ball in the plastic tube inside the battery serves as an indicator of whether the battery is (a) charged or (b) discharged. As the battery loses its charge, the density of the battery fluid decreases, and the ball sinks out of sight.

APPLICATION

Checking the Battery Charge

of the battery fluid decreases and the ball sinks beneath the surface of the fluid, leaving the dot to appear black.

Quick Quiz 9.5

Atmospheric pressure varies from day to day. The level of a floating ship on a high-pressure day is (a) higher (b) lower, or (c) no different than on a low-pressure day.

Quick Quiz 9.6

The density of lead is greater than iron, and both metals are denser than water. Is the buoyant force on a solid lead object (a) greater than, (b) equal to, or (c) less than the buoyant force acting on a solid iron object of the same dimensions?

© Royalty-Free/Corbis

Most of the volume of this iceberg is beneath the water. Can you determine what fraction of the total volume is under water?

EXAMPLE 9.8 A Red-Tag Special on Crowns

Goal Apply Archimedes's principle to a submerged object.

Problem A bargain hunter purchases a "gold" crown at a flea market. After she gets home, she hangs it from a scale and finds its weight to be 7.84 N (Fig. 9.23a). She then weighs the crown while it is immersed in water, as in Figure 9.23b, and now the scale reads 6.86 N. Is the crown made of pure gold?

Strategy The goal is to find the density of the crown and compare it to the density of gold. We already have the weight of the crown in air, so we can get the mass by dividing by the acceleration of gravity. If we can find the volume of the crown, we can obtain the desired density by dividing the mass by this volume.

When the crown is fully immersed, the displaced water is equal to the volume of the crown. This same volume is used in calculating the buoyant force. So our strategy is as follows: (1) Apply Newton's second law to the crown, both in the water and in the air to find the buoyant force. (2) Use the buoyant force to find the crown's volume. (3) Divide the crown's scale weight in air by the acceleration of gravity to get the mass, then by the volume to get the density.

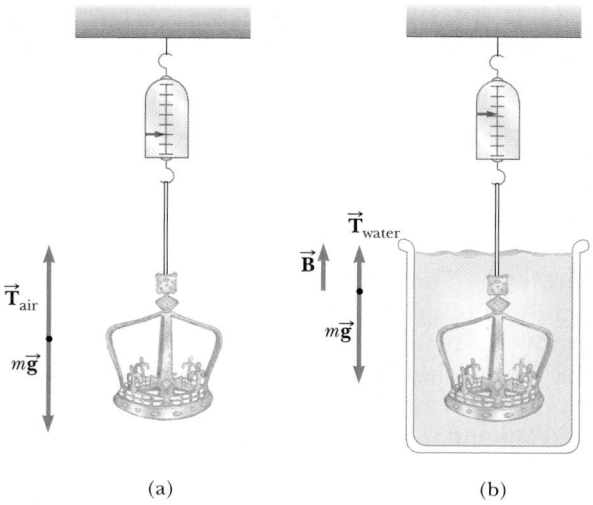

Figure 9.23 (Example 9.8) (a) When the crown is suspended in air, the scale reads $T_{air} = mg$, the crown's true weight. (b) When the crown is immersed in water, the buoyant force $\vec{B}$ reduces the scale reading by the magnitude of the buoyant force, $T_{water} = mg - B$.

Solution

Apply Newton's second law to the crown when it's weighed in air. There are two forces on the crown—gravity $m\vec{g}$ and $\vec{T}_{air}$, the force exerted by the scale on the crown, with magnitude equal to the reading on the scale.

$$T_{air} - mg = 0 \qquad (1)$$

When the crown is immersed in water, the scale force is $\vec{\mathbf{T}}_{\text{water}}$, with magnitude equal to the scale reading, and there is an upward buoyant force $\vec{\mathbf{B}}$ and the force of gravity.

$$T_{\text{water}} - mg + B = 0 \qquad \textbf{(2)}$$

Solve Equation (1) for mg, substitute into Equation (2), and solve for the buoyant force, which equals the difference in scale readings:

$$T_{\text{water}} - T_{\text{air}} + B = 0$$

$$B = T_{\text{air}} - T_{\text{water}} = 7.84\text{ N} - 6.86\text{ N} = 0.980\text{ N}$$

Find the volume of the displaced water, using the fact that the magnitude of the buoyant force equals the weight of the displaced water:

$$B = \rho_{\text{water}} g V_{\text{water}} = 0.980\text{ N}$$

$$V_{\text{water}} = \frac{0.980\text{ N}}{g\rho_{\text{water}}} = \frac{0.980\text{ N}}{(9.80\text{ m/s}^2)(1.00 \times 10^3\text{ kg/m}^3)}$$

$$= 1.00 \times 10^{-4}\text{ m}^3$$

The crown is totally submerged, so $V_{\text{crown}} = V_{\text{water}}$. From Equation (1), the mass is the crown's weight in air, T_{air}, divided by g:

$$m = \frac{T_{\text{air}}}{g} = \frac{7.84\text{ N}}{9.80\text{ m/s}^2} = 0.800\text{ kg}$$

Find the density of the crown:

$$\rho_{\text{crown}} = \frac{m}{V_{\text{crown}}} = \frac{0.800\text{ kg}}{1.00 \times 10^{-4}\text{ m}^3} = 8.00 \times 10^3\text{ kg/m}^3$$

Remarks Because the density of gold is 19.3×10^3 kg/m^3, the crown is either hollow, made of an alloy, or both. Despite the mathematical complexity, it is certainly conceivable that this was the method that occurred to Archimedes. Conceptually, it's a matter of realizing (or guessing) that equal weights of gold and a silver–gold alloy would have different scale readings when immersed in water, because their densities and hence their volumes are different, leading to differing buoyant forces.

Exercise 9.8

The weight of a metal bracelet is measured to be 0.100 N in air and 0.092 N when immersed in water. Find its density.

Answer 1.25×10^4 kg/m^3

EXAMPLE 9.9 Floating down the River

Goal Apply Archimedes's principle to a partially submerged object.

Problem A raft is constructed of wood having a density of 6.00×10^2 kg/m^3. Its surface area is 5.70 m^2, and its volume is 0.60 m^3. When the raft is placed in fresh water as in Figure 9.24, to what depth h is the bottom of the raft submerged?

Strategy There are two forces acting on the raft: the buoyant force of magnitude B, acting upwards, and the force of gravity, acting downwards. Because the raft is in equilibrium, the sum of these forces is zero. The buoyant force depends on the submerged volume $V_{\text{water}} = Ah$. Set up Newton's second law and solve for h, the depth reached by the bottom of the raft.

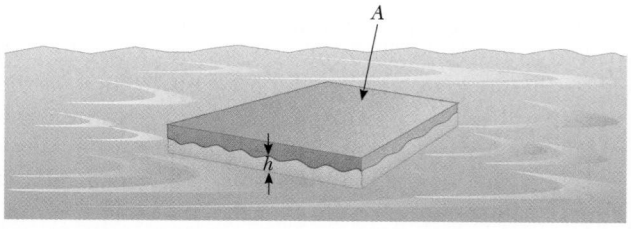

Figure 9.24 (Example 9.9) A raft partially submerged in water.

Solution

Apply Newton's second law to the raft, which is in equilibrium:

$$B - m_{\text{raft}} g = 0 \quad \rightarrow \quad B = m_{\text{raft}} g$$

The volume of the raft submerged in water is given by $V_{\text{water}} = Ah$. The magnitude of the buoyant force is equal to the weight of this displaced volume of water:

$$B = m_{\text{water}} g = (\rho_{\text{water}} V_{\text{water}}) g = (\rho_{\text{water}} Ah) g$$

Now rewrite the gravity force on the raft using the raft's density and volume:

$$m_{\text{raft}} g = (\rho_{\text{raft}} V_{\text{raft}}) g$$

Substitute these two expressions into Newton's second law, $B = m_{\text{raft}}g$, and solve for h (note that g cancels):

$$(\rho_{\text{water}}Ah)\cancel{g} = (\rho_{\text{raft}}V_{\text{raft}})\cancel{g}$$

$$h = \frac{\rho_{\text{raft}}V_{\text{raft}}}{\rho_{\text{water}}A}$$

$$= \frac{(6.00 \times 10^2\ \text{kg/m}^3)(0.600\ \text{m}^3)}{(1.00 \times 10^3\ \text{kg/m}^3)(5.70\ \text{m}^2)}$$

$$= \boxed{0.063\ 2\ \text{m}}$$

Remarks How low the raft rides in the water depends on the density of the raft. The same is true of the human body: Fat is less dense than muscle and bone, so those with a higher percentage of body fat float better.

Exercise 9.9

Calculate how much of an iceberg is beneath the surface of the ocean, given that the density of ice is 917 kg/m^3, and salt water has density 1 025 kg/m^3.

Answer 89.5%

EXAMPLE 9.10 Floating in Two Fluids

Goal Apply Archimedes's principle to an object floating in a fluid having two layers with different densities.

Problem A 1.00×10^3-kg cube of aluminum is placed in a tank. Water is then added to the tank until half the cube is immersed. **(a)** What is the normal force on the cube? (See Fig. 9.25a.) **(b)** Mercury is now slowly poured into the tank until the normal force on the cube goes to zero. (See Fig. 9.25b.) How deep is the layer of mercury?

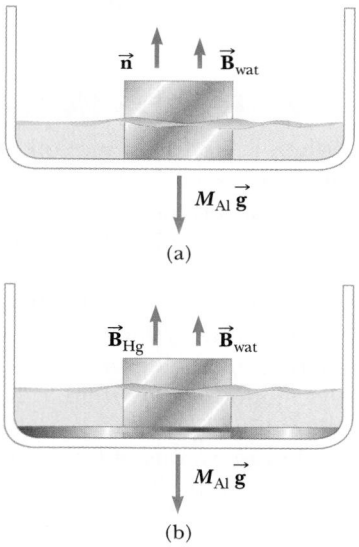

Figure 9.25 (Example 9.10)

Strategy Both parts of this problem involve applications of Newton's second law for a body in equilibrium, together with the concept of a buoyant force. In part (a), the normal, gravitational, and buoyant force of water act on the cube. In part (b), there is an additional buoyant force of mercury, while the normal force goes to zero. Using $V_{\text{Hg}} = Ah$, solve for the height of mercury, h.

Solution

(a) Find the normal force on the cube when half-immersed in water.

Calculate the volume V of the cube and the length d of one side, for future reference (both quantities will be needed for what follows):

$$V_{\text{Al}} = \frac{M_{\text{Al}}}{\rho_{\text{Al}}} = \frac{1.00 \times 10^3\ \text{kg}}{2.70 \times 10^3\ \text{kg/m}^3} = 0.370\ \text{m}^3$$

$$d = V_{\text{Al}}^{1/3} = 0.718\ \text{m}$$

Write Newton's second law for the cube, and solve for the normal force. The buoyant force is equal to the weight of the displaced water (half the volume of the cube).

$$n - M_{\text{Al}}g + B_{\text{wat}} = 0$$

$$n = M_{\text{Al}}g - B_{\text{wat}} = M_{\text{Al}}g - \rho_{\text{wat}}(V/2)g$$

$$= (1.00 \times 10^3\ \text{kg})(9.80\ \text{m/s}^2)$$

$$- (1.00 \times 10^3\ \text{kg/m}^3)(0.370\ \text{m}^3/2.00)(9.80\ \text{m/s}^2)$$

$$n = 9.80 \times 10^3\ \text{N} - 1.81 \times 10^3\ \text{N} = \boxed{7.99 \times 10^3\ \text{N}}$$

(b) Calculate the level h of added mercury.

Apply Newton's second law to the cube:

$$n - M_{Al}g + B_{wat} + B_{Hg} = 0$$

Set $n = 0$ and solve for the buoyant force of mercury:

$$B_{Hg} = (\rho_{Hg} Ah)g = M_{Al}g - B_{wat} = 7.99 \times 10^3 \text{ N}$$

Solve for h, noting that $A = d^2$:

$$h = \frac{M_{Al}g - B_{wat}}{\rho_{Hg}Ag} = \frac{7.99 \times 10^3 \text{ N}}{(13.6 \times 10^3 \text{ kg/m}^3)(0.718 \text{ m})^2(9.80 \text{ m/s}^2)}$$

$$h = 0.116 \text{ m}$$

Remarks Notice that the buoyant force of mercury calculated in part (b) is the same as the normal force in part (a). This is naturally the case, because enough mercury was added to exactly cancel out the normal force. We could have used this fact to take a shortcut, simply writing $B_{Hg} = 7.99 \times 10^3$ N immediately, solving for h, and avoiding a second use of Newton's law. Most of the time, however, we won't be so lucky! Try calculating the normal force when the level of mercury is 4.00 cm.

Exercise 9.10

A cube of aluminum 1.00 m on a side is immersed one-third in water and two-thirds in glycerin. What is the normal force on the cube?

Answer 1.50×10^4 N

9.7 FLUIDS IN MOTION

When a fluid is in motion, its flow can be characterized in one of two ways. The flow is said to be **streamline**, or **laminar**, if every particle that passes a particular point moves along exactly the same smooth path followed by previous particles passing that point. This path is called a *streamline* (Fig. 9.26). Different streamlines can't cross each other under this steady-flow condition, and the streamline at any point coincides with the direction of the velocity of the fluid at that point.

In contrast, the flow of a fluid becomes irregular, or **turbulent**, above a certain velocity or under any conditions that can cause abrupt changes in velocity. Irregular motions of the fluid, called *eddy currents*, are characteristic in turbulent flow, as shown in Figure 9.27.

In discussions of fluid flow, the term **viscosity** is used for the degree of internal friction in the fluid. This internal friction is associated with the resistance between

Andy Sacks/Stone/Getty Images

Figure 9.26 An illustration of streamline flow around an automobile in a test wind tunnel. The streamlines in the airflow are made visible by smoke particles.

two adjacent layers of the fluid moving relative to each other. A fluid such as kerosene has a lower viscosity than does crude oil or molasses.

Many features of fluid motion can be understood by considering the behavior of an **ideal fluid**, which satisfies the following conditions:

1. **The fluid is nonviscous**, which means there is no internal friction force between adjacent layers.
2. **The fluid is incompressible**, which means its density is constant.
3. **The fluid motion is steady**, meaning that the velocity, density, and pressure at each point in the fluid don't change with time.
4. **The fluid moves without turbulence**. This implies that each element of the fluid has zero angular velocity about its center, so there can't be any eddy currents present in the moving fluid. A small wheel placed in the fluid would translate but not rotate.

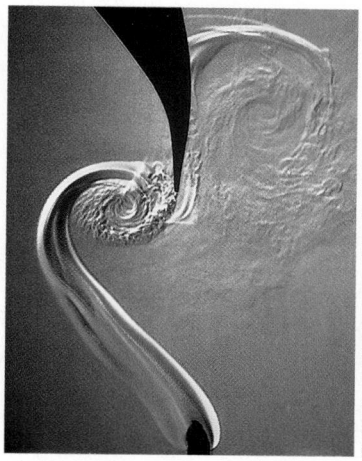

Kim Vandiver and Harold Edgerton. © Harold and Esther Edgerton Foundation, 2002, courtesy of Palm Press, Inc.

Figure 9.27 Turbulent flow: The tip of a rotating blade (the dark region at the top) forms a vortex in air that is being heated by an alcohol lamp. (The wick is at the bottom.) Note the air turbulence on both sides of the blade.

Equation of Continuity

Figure 9.28a represents a fluid flowing through a pipe of nonuniform size. The particles in the fluid move along the streamlines in steady-state flow. In a small time interval Δt, the fluid entering the bottom end of the pipe moves a distance $\Delta x_1 = v_1 \Delta t$, where v_1 is the speed of the fluid at that location. If A_1 is the cross-sectional area in this region, then the mass contained in the bottom blue region is $\Delta M_1 = \rho_1 A_1 \Delta x_1 = \rho_1 A_1 v_1 \Delta t$, where ρ_1 is the density of the fluid at A_1. Similarly, the fluid that moves out of the upper end of the pipe in the same time interval Δt has a mass of $\Delta M_2 = \rho_2 A_2 v_2 \Delta t$. However, **because mass is conserved and because the flow is steady**, the mass that flows into the bottom of the pipe through A_1 in the time Δt must equal the mass that flows out through A_2 in the same interval. Therefore, $\Delta M_1 = \Delta M_2$, or

$$\rho_1 A_1 v_1 = \rho_2 A_2 v_2 \qquad \textbf{[9.14]}$$

For the case of an incompressible fluid, $\rho_1 = \rho_2$ and Equation 9.14 reduces to

$$A_1 v_1 = A_2 v_2 \qquad \textbf{[9.15]}$$

◀ Equation of continuity

This expression is called the **equation of continuity**. From this result, we see that **the product of the cross-sectional area of the pipe and the fluid speed at that cross section is a constant**. Therefore, the speed is high where the tube is constricted and low where the tube has a larger diameter. The product Av, which has dimensions of volume per unit time, is called the **flow rate. The condition Av = constant is equivalent to the fact that the volume of fluid that enters one end of the tube in a given time interval equals the volume of fluid leaving the tube in the same interval, assuming that the fluid is incompressible and there are no leaks.** Figure 9.28b

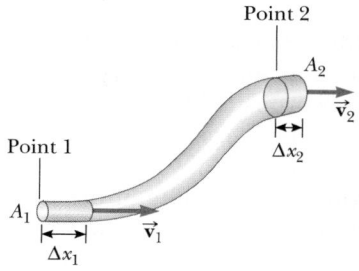

(a)

George Semple

(b)

Figure 9.28 (a) A fluid moving with streamline flow through a pipe of varying cross-sectional area. The volume of fluid flowing through A_1 in a time interval Δt must equal the volume flowing through A_2 in the same time interval. Therefore, $A_1 v_1 = A_2 v_2$. (b) Water flowing slowly out of a faucet. The width of the stream narrows as the water falls and speeds up in accord with the continuity equation.

is an example of an application of the equation of continuity: As the stream of water flows continuously from a faucet, the width of the stream narrows as it falls and speeds up.

There are many instances in everyday experience that involve the equation of continuity. Reducing the cross-sectional area of a garden hose by putting a thumb over the open end makes the water spray out with greater speed; hence the stream goes farther. Similar reasoning explains why smoke from a smoldering piece of wood first rises in a streamline pattern, getting thinner with height, eventually breaking up into a swirling, turbulent pattern. The smoke rises because it's less dense than air and the buoyant force of the air accelerates it upward. As the speed of the smoke stream increases, the cross-sectional area of the stream decreases, in accordance with the equation of continuity. The stream soon reaches a speed so great that streamline flow is not possible. We will study the relationship between speed of fluid flow and turbulence in a later discussion on the Reynolds number.

TIP 9.3 Continuity Equations

The rate of flow of fluid into a system equals the rate of flow out of the system. The incoming fluid occupies a certain volume and can enter the system only if the fluid already inside goes out, thereby making room.

EXAMPLE 9.11 Niagara Falls

Goal Apply the equation of continuity.

Problem Each second, 5 525 m^3 of water flows over the 670-m-wide cliff of the Horseshoe Falls portion of Niagara Falls. The water is approximately 2 m deep as it reaches the cliff. Estimate its speed at that instant?

Strategy This is an estimate, so only one significant figure will be retained in the answer. The volume flow rate is given, and according to the equation of continuity, is a constant equal to Av. Find the cross-sectional area, substitute, and solve for the speed.

Solution

Calculate the cross-sectional area of the water as it reaches the edge of the cliff:

$$A = (670\text{ m})(2\text{ m}) = 1\,340\text{ m}^2$$

Multiply this result by the speed and set it equal to the flow rate. Then solve for v.

$$Av = \text{volume flow rate}$$

$$(1340\text{ m}^2)v = 5\,525\text{ m}^3/\text{s} \quad \rightarrow \quad v \approx 4\text{ m/s}$$

Exercise 9.11

The Garfield Thomas water tunnel at Pennsylvania State University has a circular cross section that constricts from a diameter of 3.6 m to the test section, which is 1.2 m in diameter. If the speed of flow is 3.0 m/s in the larger-diameter pipe, determine the speed of flow in the test section.

Answer 27 m/s

EXAMPLE 9.12 Watering a Garden

Goal Combine the equation of continuity with concepts of flow rate and kinematics.

Problem A water hose 2.50 cm in diameter is used by a gardener to fill a 30.0-liter bucket. (One liter = 1 000 cm^3.) The gardener notices that it takes 1.00 min to fill the bucket. A nozzle with an opening of cross-sectional area 0.500 cm^2 is then attached to the hose. The nozzle is held so that water is projected horizontally from a point 1.00 m above the ground. Over what horizontal distance can the water be projected?

Strategy We can find the volume flow rate through the hose by dividing the volume of the bucket by the time it takes to fill it. After finding the flow rate, apply the equation of continuity to find the speed at which the water shoots horizontally out the nozzle. The rest of the problem is an application of two-dimensional kinematics. The answer obtained is the same as would be found for a ball having the same initial velocity and height.

Solution

Calculate the volume flow rate into the bucket, and convert to m^3/s:

$$\text{volume flow rate} = \frac{30.0\text{ L}}{1.00\text{ min}}\left(\frac{1.00\times10^3\text{ cm}^3}{1.00\text{ L}}\right)\left(\frac{1.00\text{ m}}{100.0\text{ cm}}\right)^3\left(\frac{1.00\text{ min}}{60.0\text{ s}}\right)$$
$$= 5.00\times10^{-4}\text{ m}^3/\text{s}$$

Solve the equation of continuity for v_{0x}, the x-component of the initial velocity of the stream exiting the hose:

$$A_1v_1 = A_2v_2 = A_2v_{0x}$$
$$v_{0x} = \frac{A_1v_1}{A_2} = \frac{5.00\times10^{-4}\text{ m}^3/\text{s}}{0.500\times10^{-4}\text{ m}^2} = 10.0\text{ m/s}$$

Calculate the time for the stream to fall 1.00 m, using kinematics. Initially, the stream is horizontal, so v_{0y} is zero

$$\Delta y = v_{0y}t - \tfrac{1}{2}gt^2$$

Set $v_{0y} = 0$ in the preceding equation and solve for t noting that $\Delta y = -1.00$ m:

$$t = \sqrt{\frac{-2\Delta y}{g}} = \sqrt{\frac{-2(-1.00\text{ m})}{9.80\text{ m/s}^2}} = 0.452\text{ s}$$

Find the horizontal distance the stream travels:

$$x = v_{0x}t = (10.0\text{ m/s})(0.452\text{ s}) = \boxed{4.52\text{ m}}$$

Remark It's interesting that the motion of fluids can be treated with the same kinematics equations as individual objects.

Exercise 9.12

The nozzle is replaced with a Y-shaped fitting that splits the flow in half. Garden hoses are connected to each end of the Y, with each hose having a 0.400 cm^2 nozzle. (a) How fast does the water come out of one of the nozzles? (b) How far would one of the nozzles squirt water if both were operated simultaneously and held horizontally 1.00 m off the ground? [*Hint:* Find the volume flow rate through each 0.400-cm^2 nozzle, then follow the same steps as before.]

Answer (a) 6.25 m/s (b) 2.82 m

Bernoulli's Equation

As a fluid moves through a pipe of varying cross section and elevation, the pressure changes along the pipe. In 1738 the Swiss physicist Daniel Bernoulli (1700–1782) derived an expression that relates the pressure of a fluid to its speed and elevation. Bernoulli's equation is not a freestanding law of physics; rather, it's **a consequence of energy conservation as applied to an ideal fluid**.

In deriving Bernoulli's equation, we again assume that the fluid is incompressible, nonviscous, and flows in a nonturbulent, steady-state manner. Consider the flow through a nonuniform pipe in the time Δt, as in Figure 9.29. The force on the lower end of the fluid is P_1A_1, where P_1 is the pressure at the lower end. The work done on the lower end of the fluid by the fluid behind it is

$$W_1 = F_1\,\Delta x_1 = P_1A_1\,\Delta x_1 = P_1V$$

where V is the volume of the lower blue region in the figure. In a similar manner, the work done on the fluid on the upper portion in the time Δt is

$$W_2 = -P_2A_2\Delta x_2 = -P_2V$$

The volume is the same because by the equation of continuity, the volume of fluid that passes through A_1 in the time Δt equals the volume that passes through A_2 in the same interval. The work W_2 is negative because the force on the fluid at the top is opposite its displacement. The net work done by these forces in the time Δt is

$$W_{\text{fluid}} = P_1V - P_2V$$

Part of this work goes into changing the fluid's kinetic energy, and part goes into changing the gravitational potential energy of the fluid–Earth system. If m is the

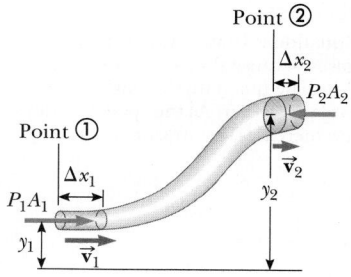

Figure 9.29 A fluid flowing through a constricted pipe with streamline flow. The fluid in the section with a length of Δx_1 moves to the section with a length of Δx_2. The volumes of fluid in the two sections are equal.

Corbis-Bettmann

DANIEL BERNOULLI, Swiss Physicist and Mathematician (1700–1782)

In his most famous work, *Hydrodynamica*, Bernoulli showed that, as the velocity of fluid flow increases, its pressure decreases. In this same publication, Bernoulli also attempted the first explanation of the behavior of gases with changing pressure and temperature; this was the beginning of the kinetic theory of gases.

mass of the fluid passing through the pipe in the time interval Δt, then the change in kinetic energy of the volume of fluid is

$$\Delta KE = \tfrac{1}{2}mv_2^2 - \tfrac{1}{2}mv_1^2$$

The change in the gravitational potential energy is

$$\Delta PE = mgy_2 - mgy_1$$

Because the net work done by the fluid on the segment of fluid shown in Figure 9.29 changes the kinetic energy and the potential energy of the nonisolated system, we have

$$W_{\text{fluid}} = \Delta KE + \Delta PE$$

The three terms in this equation are those we have just evaluated. Substituting expressions for each of the terms gives

$$P_1V - P_2V = \tfrac{1}{2}mv_2^2 - \tfrac{1}{2}mv_1^2 + mgy_2 - mgy_1$$

If we divide each term by V and recall that $\rho = m/V$, this expression becomes

$$P_1 - P_2 = \tfrac{1}{2}\rho v_2^2 - \tfrac{1}{2}\rho v_1^2 + \rho gy_2 - \rho gy_1$$

Rearrange the terms as follows:

$$P_1 + \tfrac{1}{2}\rho v_1^2 + \rho gy_1 = P_2 + \tfrac{1}{2}\rho v_2^2 + \rho gy_2 \qquad \textbf{[9.16]}$$

Bernoulli's equation ▶

This is **Bernoulli's equation**, often expressed as

$$P + \tfrac{1}{2}\rho v^2 + \rho gy = \text{constant} \qquad \textbf{[9.17]}$$

Bernoulli's equation states that the sum of the pressure P, the kinetic energy per unit volume, $\frac{1}{2}\rho v^2$, and the potential energy per unit volume, ρgy, has the same value at all points along a streamline.

TIP 9.4 Bernoulli's Principle for Gases

Equation 9.16 isn't strictly true for gases because they aren't incompressible. The qualitative behavior is the same, however: As the speed of the gas increases, its pressure decreases.

An important consequence of Bernoulli's equation can be demonstrated by considering Figure 9.30, which shows water flowing through a horizontal constricted pipe from a region of large cross-sectional area into a region of smaller cross-sectional area. This device, called a **Venturi tube**, can be used to measure the speed of fluid flow. Because the pipe is horizontal, $y_1 = y_2$, and Equation 9.16 applied to points 1 and 2 gives

$$P_1 + \tfrac{1}{2}\rho v_1^2 = P_2 + \tfrac{1}{2}\rho v_2^2 \qquad \textbf{[9.18]}$$

Because the water is not backing up in the pipe, its speed v_2 in the constricted region must be greater than its speed v_1 in the region of greater diameter. From Equation 9.18, we see that P_2 must be less than P_1 because $v_2 > v_1$. This result is often expressed by the statement that **swiftly moving fluids exert less pressure than do slowly moving fluids**. This important fact enables us to understand a wide range of everyday phenomena.

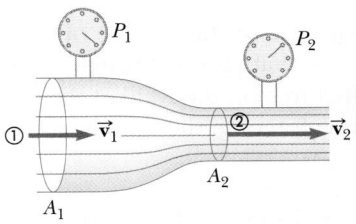

(a)

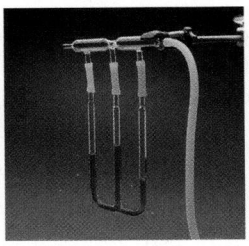

Courtesy of Central Scientific Company

(b)

Figure 9.30 (a) The pressure P_1 is greater than the pressure P_2, because $v_1 < v_2$. This device can be used to measure the speed of fluid flow. (b) A Venturi tube, located at the top of the photograph. The higher level of fluid in the middle column shows that the pressure at the top of the column, which is in the constricted region of the Venturi tube, is lower than the pressure elsewhere in the column.

Quick Quiz 9.7

You observe two helium balloons floating next to each other at the ends of strings secured to a table. The facing surfaces of the balloons are separated by 1–2 cm. You blow through the opening between the balloons. What happens to the balloons? (a) They move toward each other; (b) they move away from each other; (c) they are unaffected.

INTERACTIVE EXAMPLE 9.13 Shoot-Out at the Old Water Tank

Goal Apply Bernoulli's equation to find the speed of a fluid.

Problem A nearsighted sheriff fires at a cattle rustler with his trusty six-shooter. Fortunately for the rustler, the bullet misses him and penetrates the town water tank, causing a leak (Fig. 9.31). **(a)** If the top of the tank is open to the atmosphere, determine the speed at which the water leaves the hole when the water level is 0.500 m above the hole. **(b)** Where does the stream hit the ground if the hole is 3.00 m above the ground?

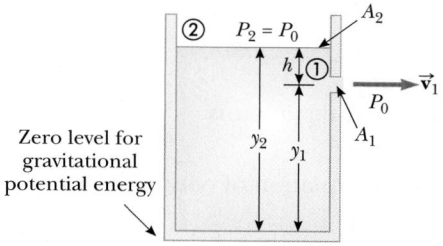

Figure 9.31 (Example 9.13) The water speed v_1 from the hole in the side of the container is given by $v_1 = \sqrt{2gh}$.

Strategy (a) Assume the tank's cross-sectional area is large compared to the hole's ($A_2 \gg A_1$), so the water level drops very slowly and $v_2 \approx 0$. Apply Bernoulli's equation to points ① and ② in Figure 9.31, noting that P_1 equals atmospheric pressure P_0 at the hole and is approximately the same at the top of the water tank. Part (b) can be solved with kinematics, just as if the water were a ball thrown horizontally.

Solution

(a) Find the speed of the water leaving the hole.

Substitute $P_1 = P_2 = P_0$ and $v_2 \approx 0$ into Bernoulli's equation, and solve for v_1:

$$P_0 + \tfrac{1}{2}\rho v_1^2 + \rho g y_1 = P_0 + \rho g y_2$$

$$v_1 = \sqrt{2g(y_2 - y_1)} = \sqrt{2gh}$$

$$v_1 = \sqrt{2(9.80\ \text{m/s}^2)(0.500\ \text{m})} = \boxed{3.13\ \text{m/s}}$$

(b) Find where the stream hits the ground.

Use the displacement equation to find the time of the fall, noting that the stream is initially horizontal, so $v_{0y} = 0$.

$$\Delta y = -\tfrac{1}{2}gt^2 + v_{0y}t$$

$$-3.00\ \text{m} = -(4.90\ \text{m/s}^2)t^2$$

$$t = 0.782\ \text{s}$$

Compute the horizontal distance the stream travels in this time:

$$x = v_{0x}t = (3.13\ \text{m/s})(0.782\ \text{s}) = \boxed{2.45\ \text{m}}$$

Remarks As the analysis of part (a) shows, the speed of the water emerging from the hole is equal to the speed acquired by an object falling freely through the vertical distance h. This is known as **Torricelli's law**.

Exercise 9.13

Suppose, in a similar situation, the water hits the ground 4.20 m from the hole in the tank. If the hole is 2.00 m above the ground, how far above the hole is the water level?

Answer 2.20 m above the hole

PhysicsNow™ You can move the hole vertically and see where the water lands by logging into PhysicsNow at **www.cp7e.com** and going to Interactive Example 9.13.

Example 9.14 Fluid Flow in a Pipe

Goal Solve a problem combining Bernoulli's equation and the equation of continuity.

Problem A large pipe with a cross-sectional area of 1.00 m^2 descends 5.00 m and narrows to 0.500 m^2, where it terminates in a valve (Fig. 9.32). If the pressure at point ② is atmospheric pressure, and the valve is opened wide and water allowed to flow freely, find the speed of the water leaving the pipe.

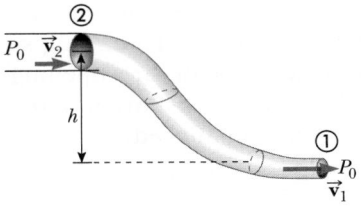

Figure 9.32 (Example 9.14)

Strategy The equation of continuity, together with Bernoulli's equation, constitute two equations in two unknowns: the speeds v_1 and v_2. Eliminate v_2 from Bernoulli's equation with the equation of continuity, and solve for v_1.

Solution

Bernoulli's equation is

$$P_1 + \tfrac{1}{2}\rho v_1^2 + \rho g y_1 = P_2 + \tfrac{1}{2}\rho v_2^2 + \rho g y_2 \qquad (1)$$

Solve the equation of continuity for v_2:

$$A_2 v_2 = A_1 v_1$$

$$v_2 = \frac{A_1}{A_2} v_1 \qquad (2)$$

In Equation 1, set $P_1 = P_2 = P_0$, and substitute the expression for v_2. Then solve for v_1.

$$P_0 + \tfrac{1}{2}\rho v_1^2 + \rho g y_1 = P_0 + \tfrac{1}{2}\rho\left(\frac{A_1}{A_2} v_1\right)^2 + \rho g y_2$$

$$v_1^2\left(1 - \left(\frac{A_1}{A_2}\right)^2\right) = 2g(y_2 - y_1) = 2gh$$

$$v_1 = \frac{\sqrt{2gh}}{\sqrt{1 - (A_1/A_2)^2}}$$

Substitute the given values:

$$v_1 = 11.4 \text{ m/s}$$

Remarks This speed is slightly higher than the speed predicted by Torricelli's law, because the narrowing pipe squeezes the fluid.

Exercise 9.14

Water flowing in a horizontal pipe is at a pressure of 1.4×10^5 Pa at a point where its cross-sectional area is 1.00 m^2. When the pipe narrows to 0.400 m^2, the pressure drops to 1.16×10^5 Pa. Find the water's speed (a) in the wider pipe and (b) in the narrower pipe.

Answer (a) 3.02 m/s (b) 7.56 m/s

9.8 OTHER APPLICATIONS OF FLUID DYNAMICS

In this section we describe some common phenomena that can be explained, at least in part, by Bernoulli's equation.

In general, an object moving through a fluid is acted upon by a net upward force as the result of any effect that causes the fluid to change its direction as it flows past the object. For example, a golf ball struck with a club is given a rapid backspin, as shown in Figure 9.33. The dimples on the ball help entrain the air along the curve of the ball's surface. The figure shows a thin layer of air wrapping partway around the ball and being deflected downward as a result. Because the

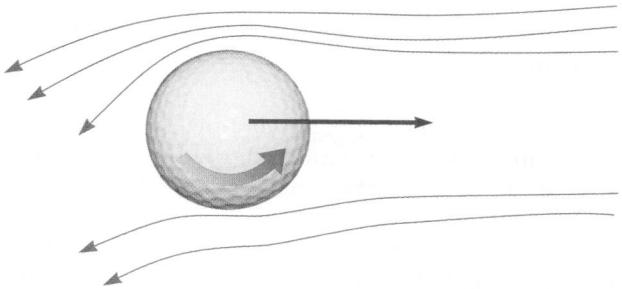

FIGURE 9.33 A spinning golf ball is acted upon by a lifting force that allows it to travel much further than it would if it were not spinning.

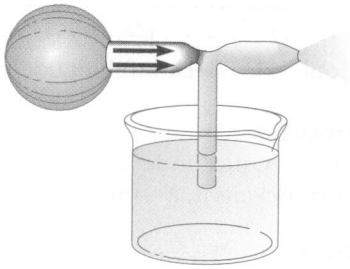

Figure 9.34 A stream of air passing over a tube dipped in a liquid causes the liquid to rise in the tube. This effect is used in perfume bottles and paint sprayers.

ball pushes the air down, by Newton's third law the air must push up on the ball and cause it to rise. Without the dimples, the air isn't as well entrained, so the golf ball doesn't travel as far. A tennis ball's fuzz performs a similar function, though the desired result is ball placement rather than greater distance.

APPLICATION

"Atomizers" in Perfume Bottles and Paint Sprayers

Many devices operate in the manner illustrated in Figure 9.34. A stream of air passing over an open tube reduces the pressure above the tube, causing the liquid to rise into the airstream. The liquid is then dispersed into a fine spray of droplets. You might recognize that this so-called atomizer is used in perfume bottles and paint sprayers. The same principle is used in the carburetor of a gasoline engine. In that case, the low-pressure region in the carburetor is produced by air drawn in by the piston through the air filter. The gasoline vaporizes, mixes with the air, and enters the cylinder of the engine for combustion.

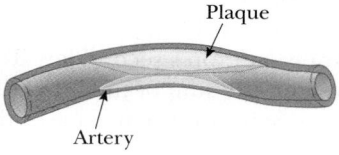

Figure 9.35 Blood must travel faster than normal through a constricted region of an artery.

In a person with advanced arteriosclerosis, the Bernoulli effect produces a symptom called **vascular flutter**. In this condition, the artery is constricted as a result of accumulated plaque on its inner walls, as shown in Figure 9.35. To maintain a constant flow rate, the blood must travel faster than normal through the constriction. If the speed of the blood is sufficiently high in the constricted region, the blood pressure is low, and the artery may collapse under external pressure, causing a momentary interruption in blood flow. During the collapse there is no Bernoulli effect, so the vessel reopens under arterial pressure. As the blood rushes through the constricted artery, the internal pressure drops and the artery closes again. Such variations in blood flow can be heard with a stethoscope. If the plaque becomes dislodged and ends up in a smaller vessel that delivers blood to the heart, it can cause a heart attack.

APPLICATION

Vascular Flutter and Aneurysms

An **aneurysm** is a weakened spot on an artery where the artery walls have ballooned outward. Blood flows more slowly though this region, as can be seen from the equation of continuity, resulting in an increase in pressure in the vicinity of the aneurysm relative to the pressure in other parts of the artery. This condition is dangerous because the excess pressure can cause the artery to rupture.

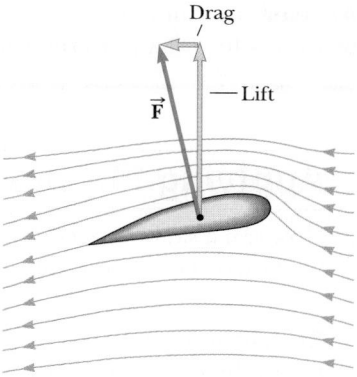

Figure 9.36 Streamline flow around an airplane wing. The pressure above is less than the pressure below, and there is a dynamic upward lift force.

The lift on an aircraft wing can also be explained in part by the Bernoulli effect. Airplane wings are designed so that the air speed above the wing is greater than the speed below. As a result, the air pressure above the wing is less than the pressure below, and there is a net upward force on the wing, called the *lift*. (There is also a horizontal component called the *drag*.) Another factor influencing the lift on a wing, shown in Figure 9.36, is the slight upward tilt of the wing. This causes air molecules striking the bottom to be deflected downward, producing a reaction force upward by Newton's third law. Finally, turbulence also has an effect. If the wing is tilted too much, the flow of air across the upper surface becomes turbulent, and the pressure difference across the wing is not as great as that predicted by the Bernoulli effect. In an extreme case, this turbulence may cause the aircraft to stall.

APPLICATION

Lift on Aircraft Wings

EXAMPLE 9.15 Lift on an Airfoil

Goal Use Bernoulli's equation to calculate the lift on an airplane wing.

Problem An airplane has wings, each with area $4.00\ \text{m}^2$, designed so that air flows over the top of the wing at 245 m/s and underneath the wing at 222 m/s. Find the mass of the airplane such that the lift on the plane will support its weight, assuming the force from the pressure difference across the wings is directed straight upwards.

Strategy This problem can be solved by substituting values into Bernoulli's equation to find the pressure difference between the air under the wing and the air over the wing, followed by applying Newton's second law to find the mass the airplane can lift.

Solution

Apply Bernoulli's equation to the air flowing under the wing (point 1) and over the wing (point 2). Gravitational potential energy terms are small compared with the other terms, and can be neglected.

$$P_1 + \tfrac{1}{2}\rho v_1^2 = P_2 + \tfrac{1}{2}\rho v_2^2$$

Solve this equation for the pressure difference:

$$\Delta P = P_1 - P_2 = \tfrac{1}{2}\rho v_2^2 - \tfrac{1}{2}\rho v_1^2 = \tfrac{1}{2}\rho(v_2^2 - v_1^2)$$

Substitute the given speeds and $\rho = 1.29\ \text{kg/m}^3$, the density of air:

$$\Delta P = \tfrac{1}{2}(1.29\ \text{kg/m}^3)(245^2\ \text{m}^2/\text{s}^2 - 222^2\ \text{m}^2/\text{s}^2)$$

$$\Delta P = 6.93 \times 10^3\ \text{Pa}$$

Apply Newton's second law. To support the plane's weight, the sum of the lift and gravity forces must equal zero. Solve for the mass m of the plane.

$$2A\Delta P - mg = 0 \quad \rightarrow \quad m = 5.66 \times 10^3\ \text{kg}$$

Remarks Note the factor of two in the last equation, needed because the airplane has two wings. The density of the atmosphere drops steadily with increasing height, reducing the lift. As a result, all aircraft have a maximum operating altitude.

Exercise 9.15

Approximately what size wings would an aircraft need on Mars if its engine generates the same differences in speed as in the example and the total mass of the craft is 400 kg? The density of air on the surface of Mars is approximately one percent Earth's density at sea level, and the acceleration of gravity on the surface of Mars is about $3.8\ \text{m/s}^2$.

Answer Rounding to one significant digit, each wing would have to have an area of about $10\ \text{m}^2$. There have been proposals for solar-powered robotic Mars aircraft, which would have to be gossamer-light with large wings.

Applying Physics 9.4 Sailing Upwind

How can a sailboat accomplish the seemingly impossible task of sailing into the wind?

Explanation As shown in Figure 9.37, the wind blowing in the direction of the arrow causes the sail to billow out and take on a shape similar to that of an airplane wing. By Bernoulli's equation, just as for an airplane wing, there is a force on the sail in the direction shown. The component of force perpendicular to the boat tends to make the boat move sideways in the water, but the keel prevents this sideways motion. The component of the force in the forward direction drives the boat almost against the wind. The word *almost* is used because a sailboat can move forward only

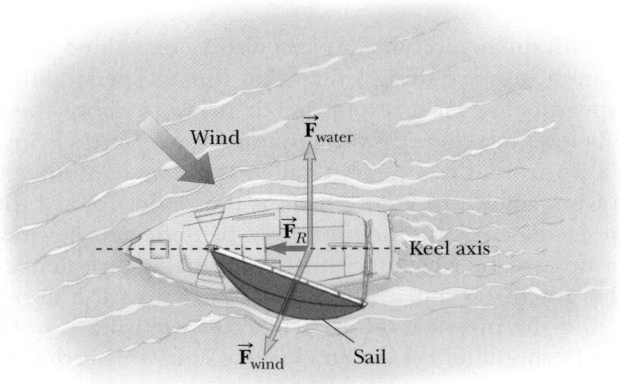

Figure 9.37 (Applying Physics 9.4)

when the wind direction is about 10 to 15° with respect to the forward direction. This means that in order to sail directly against the wind, a boat must follow a zigzag path, a procedure called *tacking*, so that the wind is always at some angle with respect to the direction of travel.

Applying Physics 9.5 Home Plumbing

Consider the portion of a home plumbing system shown in Figure 9.38. The water trap in the pipe below the sink captures a plug of water that prevents sewer gas from finding its way from the sewer pipe, up the sink drain, and into the home. Suppose the dishwasher is draining, so that water is moving to the left in the sewer pipe. What is the purpose of the vent, which is open to the air above the roof of the house? In which direction is air moving at the opening of the vent, upwards or downwards?

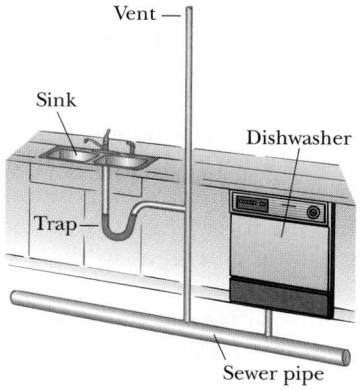

Figure 9.38 (Applying Physics 9.5)

Explanation Imagine that the vent isn't present, so that the drainpipe for the sink is simply connected through the trap to the sewer pipe. As water from the dishwasher moves to the left in the sewer pipe, the pressure in the sewer pipe is reduced below atmospheric pressure, in accordance with Bernoulli's principle. The pressure at the drain in the sink is still at atmospheric pressure. This pressure difference can push the plug of water in the water trap of the sink down the drainpipe and into the sewer pipe, removing it as a barrier to sewer gas. With the addition of the vent to the roof, the reduced pressure of the dishwasher water will result in air entering the vent pipe at the roof. This inflow of air will keep the pressure in the vent pipe and the right-hand side of the sink drainpipe close to atmospheric pressure, so that the plug of water in the water trap will remain in place.

APPLICATION
Rocket Engines

The exhaust speed of a rocket engine can also be understood qualitatively with Bernoulli's equation, though, in actual practice a large number of additional variables need to be taken into account. Rockets actually work better in vacuum than in the atmosphere, contrary to an early *New York Times* article criticizing rocket pioneer Robert Goddard, which held that they wouldn't work at all, having no air to push against. The pressure inside the combustion chamber is P, and the pressure just outside the nozzle is the ambient atmospheric pressure, P_{atm}. Differences in height between the combustion chamber and the end of the nozzle result in negligible contributions of gravitational potential energy. In addition, the gases inside the chamber flow at negligible speed compared to gases going through the nozzle. The exhaust speed can be found from Bernoulli's equation,

$$v_{ex} = \sqrt{\frac{2(P - P_{atm})}{\rho}}$$

This equation shows that the exhaust speed is reduced in the atmosphere, so rockets are actually more effective in the vacuum of space. Also of interest is the appearance of the density ρ in the denominator. A lower density working fluid or gas will give a higher exhaust speed, which partly explains why liquid hydrogen, which has a very low density, is a fuel of choice.

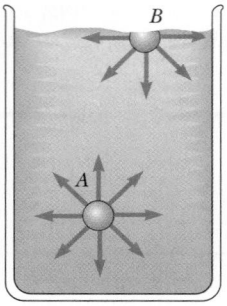

Figure 9.39 The net force on a molecule at A is zero because such a molecule is completely surrounded by other molecules. The net force on a surface molecule at B is downward because it isn't completely surrounded by other molecules.

9.9 SURFACE TENSION, CAPILLARY ACTION, AND VISCOUS FLUID FLOW

If you look closely at a dewdrop sparkling in the morning sunlight, you will find that the drop is spherical. The drop takes this shape because of a property of liquid surfaces called **surface tension**. In order to understand the origin of surface tension, consider a molecule at point A in a container of water, as in Figure 9.39. Although nearby molecules exert forces on this molecule, the net force on it is zero because it's completely surrounded by other molecules and hence is attracted equally in all directions. The molecule at B, however, is not attracted equally in all directions. Because there are no molecules above it to exert upward forces, the molecule at B is pulled toward the interior of the liquid. The contraction at the surface of the liquid ceases when the inward pull exerted on the surface molecules is balanced by the outward repulsive forces that arise from collisions with molecules in the interior of the liquid. **The net effect of this pull on all the surface molecules is to make the surface of the liquid contract and, consequently, to make the surface area of the liquid as small as possible**. Drops of water take on a spherical shape because a sphere has the smallest surface area for a given volume.

If you place a sewing needle very carefully on the surface of a bowl of water, you will find that the needle floats even though the density of steel is about eight times that of water. This phenomenon can also be explained by surface tension. A close examination of the needle shows that it actually rests in a depression in the liquid surface as shown in Figure 9.40. The water surface acts like an elastic membrane under tension. The weight of the needle produces a depression, increasing the surface area of the film. Molecular forces now act at all points along the depression, tending to restore the surface to its original horizontal position. The vertical components of these forces act to balance the force of gravity on the needle. The floating needle can be sunk by adding a little detergent to the water, which reduces the surface tension.

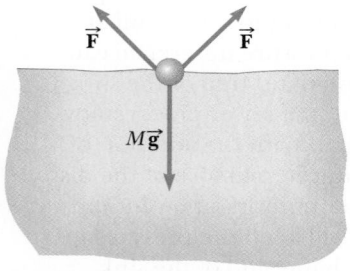

Figure 9.40 End view of a needle resting on the surface of water. The components of surface tension balance the force of gravity.

The **surface tension** γ in a film of liquid is defined as the magnitude of the surface tension force F divided by the length L along which the force acts:

$$\gamma \equiv \frac{F}{L} \qquad [9.19]$$

The SI unit of surface tension is the newton per meter, and values for a few representative materials are given in Table 9.4.

Surface tension can be thought of as the energy content of the fluid at its surface per unit surface area. To see that this is reasonable, we can manipulate the units of surface tension γ as follows:

$$\frac{\text{N}}{\text{m}} = \frac{\text{N}\cdot\text{m}}{\text{m}^2} = \frac{\text{J}}{\text{m}^2}$$

In general, in **any equilibrium configuration of an object, the energy is a minimum**. Consequently, a fluid will take on a shape such that its surface area is as small as possible. For a given volume, a spherical shape has the smallest surface area; therefore, a drop of water takes on a spherical shape.

TABLE 9.4

Surface Tensions for Various Liquids

Liquid	T (°C)	Surface Tension (N/m)
Ethyl alcohol	20	0.022
Mercury	20	0.465
Soapy water	20	0.025
Water	20	0.073
Water	100	0.059

An apparatus used to measure the surface tension of liquids is shown in Figure 9.41. A circular wire with a circumference L is lifted from a body of liquid. The surface film clings to the inside and outside edges of the wire, holding back the wire and causing the spring to stretch. If the spring is calibrated, the force required to overcome the surface tension of the liquid can be measured. In this case, the surface tension is given by

$$\gamma = \frac{F}{2L}$$

We use $2L$ for the length because the surface film exerts forces on both the inside and outside of the ring.

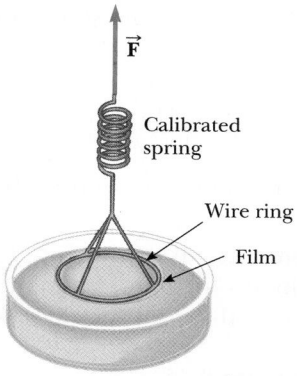

Figure 9.41 An apparatus for measuring the surface tension of liquids. The force on the wire ring is measured just before the ring breaks free of the liquid.

The surface tension of liquids decreases with increasing temperature, because the faster moving molecules of a hot liquid aren't bound together as strongly as are those in a cooler liquid. In addition, certain ingredients called surfactants decrease surface tension when added to liquids. For example, soap or detergent decreases the surface tension of water, making it easier for soapy water to penetrate the cracks and crevices of your clothes to clean them better than plain water does. A similar effect occurs in the lungs. The surface tissue of the air sacs in the lungs contains a fluid that has a surface tension of about 0.050 N/m. A liquid with a surface tension this high would make it very difficult for the lungs to expand during inhalation. However, as the area of the lungs increases with inhalation, the body secretes into the tissue a substance that gradually reduces the surface tension of the liquid. At full expansion, the surface tension of the lung fluid can drop to as low as 0.005 N/m.

APPLICATION

Air Sac Surface Tension

EXAMPLE 9.16 Walking on Water

Goal Apply the surface tension equation.

Problem Many insects can literally walk on water, using surface tension for their support. To show this is feasible, assume that the insect's "foot" is spherical. When the insect steps onto the water with all six legs, a depression is formed in the water around each foot, as shown in Figure 9.42a. The surface tension of the water produces upward forces on the water that tend to restore the water surface to its normally flat shape. If the insect has a mass of 2.0×10^{-5} kg and if the radius of each foot is 1.5×10^{-4} m, find the angle θ.

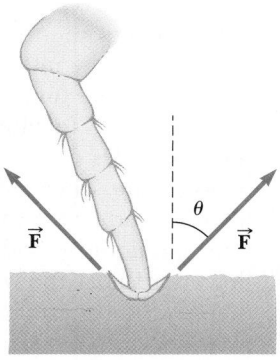

(a)

Herman Eisenbeiss/Photo Researchers, Inc.

(b)

Figure 9.42 (Example 9.16) (a) One foot of an insect resting on the surface of water. (b) This water strider resting on the surface of a lake remains on the surface, rather than sinking, because an upward surface tension force acts on each leg, balancing the force of gravity on the insect.

Strategy Find an expression for the magnitude of the net force F directed tangentially to the depressed part of the water surface, and obtain the part that is acting vertically, in opposition to the downward force of gravity. Assume that the radius of depression is the same as the radius of the insect's foot. Because the insect has six legs, one-sixth of the insect's weight must be supported by one of the legs, assuming the weight is distributed evenly. The length L is just the distance around a circle. Using Newton's second law for a body in equilibrium (zero acceleration), solve for θ.

Solution

Start with the surface tension equation:

$$F = \gamma L$$

Focus on one circular foot, substituting $L = 2\pi r$. Multiply by $\cos\theta$ to get the vertical component F_v:

$$F_v = \gamma(2\pi r)\cos\theta$$

Write Newton's second law for the insect's one foot, which supports one-sixth of the insect's weight:

$$\Sigma F = F_v - F_{\text{grav}} = 0$$

$$\gamma 2\pi r\cos\theta - \tfrac{1}{6}mg = 0$$

Solve for cos θ and substitute:

$$\cos\theta = \frac{mg}{12\pi r\gamma}$$

$$= \frac{(2.0 \times 10^{-5}\ \text{kg})(9.80\ \text{m/s}^2)}{12\pi(1.5 \times 10^{-4}\ \text{m})(0.073\ \text{N/m})} = 0.47 \qquad \textbf{(1)}$$

Take the inverse cosine of both sides to find the angle θ: $\theta = \cos^{-1}(0.47) = 62°$

Remarks If the weight of the insect were great enough to make the right side of Equation (1) greater than one, a solution for θ would be impossible because the cosine of an angle can never be greater than one. In this circumstance, the insect would sink.

Exercise 9.16

A typical sewing needle floats on water. Create an estimate for the needle's maximum possible mass. Assume the sewing needle is two inches long. [*Hint:* The cosine of an angle is never larger than 1.]

Answer 0.8 g

The Surface of Liquid

If you have ever closely examined the surface of water in a glass container, you may have noticed that the surface of the liquid near the walls of the glass curves upwards as you move from the center to the edge, as shown in Figure 9.43a. However, if mercury is placed in a glass container, the mercury surface curves downwards, as in Figure 9.43b. These surface effects can be explained by considering the forces between molecules. In particular, we must consider the forces that the molecules of the liquid exert on one another and the forces that the molecules of the glass surface exert on those of the liquid. In general terms, forces between like molecules, such as the forces between water molecules, are called **cohesive forces**, and forces between unlike molecules, such as those exerted by glass on water, are called **adhesive forces**.

Water tends to cling to the walls of the glass because the adhesive forces between the molecules of water and the glass molecules are *greater* than the cohesive forces between the water molecules. In effect, the water molecules cling to the surface of the glass rather than fall back into the bulk of the liquid. When this condition prevails, the liquid is said to "wet" the glass surface. The surface of the mercury curves downward near the walls of the container because the cohesive forces between the mercury atoms are greater than the adhesive forces between mercury and glass. A mercury atom near the surface is pulled more strongly toward other mercury atoms than toward the glass surface, so mercury doesn't wet the glass surface.

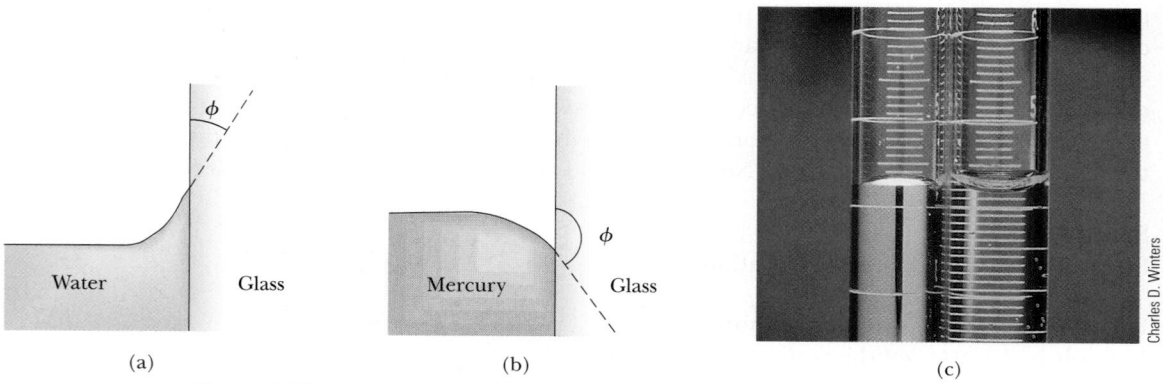

Figure 9.43 A liquid in contact with a solid surface. (a) For water, the adhesive force is greater than the cohesive force. (b) For mercury, the adhesive force is less than the cohesive force. (c) The surface of mercury (*left*) curves downwards in a glass container, whereas the surface of water (*right*) curves upwards, as you move from the center to the edge.

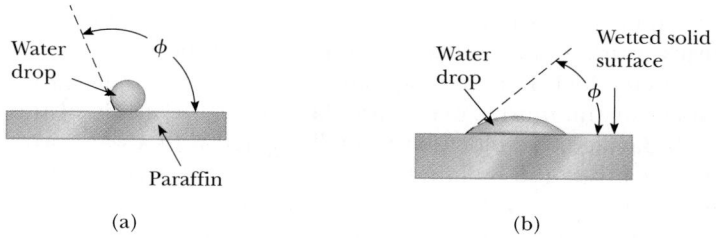

Figure 9.44 (a) The contact angle between water and paraffin is about 107°. In this case, the cohesive force is greater than the adhesive force. (b) When a chemical called a wetting agent is added to the water, it wets the paraffin surface, and $f < 90°$. In this case, the adhesive force is greater than the cohesive force.

The angle ϕ between the solid surface and a line drawn tangent to the liquid at the surface is called the **contact angle** (Fig. 9.44). The angle ϕ is less than 90° for any substance in which adhesive forces are stronger than cohesive forces and greater than 90° if cohesive forces predominate. For example, if a drop of water is placed on paraffin, the contact angle is approximately 107° (Fig. 9.44a). If certain chemicals, called wetting agents or detergents, are added to the water, the contact angle becomes less than 90°, as shown in Figure 9.44b. The addition of such substances to water ensures that the water makes intimate contact with a surface and penetrates it. For this reason, detergents are added to water to wash clothes or dishes.

APPLICATION

Detergents and Waterproofing Agents

On the other hand, it is sometimes necessary to *keep* water from making intimate contact with a surface, as in waterproof clothing, where a situation somewhat the reverse of that shown in Figure 9.44 is called for. The clothing is sprayed with a waterproofing agent, which changes ϕ from less than 90° to greater than 90°. The water beads up on the surface and doesn't easily penetrate the clothing.

Capillary Action

In capillary tubes the diameter of the opening is very small, on the order of a hundredth of a centimeter. In fact, the word *capillary* means "hairlike." If such a tube is inserted into a fluid for which adhesive forces dominate over cohesive forces, the liquid rises into the tube, as shown in Figure 9.45. The rising of the liquid in the tube can be explained in terms of the shape of the liquid's surface and surface tension effects. At the point of contact between liquid and solid, the upward force of surface tension is directed as shown in the figure. From Equation 9.19, the magnitude of this force is

$$F = \gamma L = \gamma(2\pi r)$$

(We use $L = 2\pi r$ here because the liquid is in contact with the surface of the tube at all points around its circumference.) The vertical component of this force due to surface tension is

$$F_v = \gamma(2\pi r)(\cos\phi) \qquad \textbf{[9.20]}$$

In order for the liquid in the capillary tube to be in equilibrium, this upward force must be equal to the weight of the cylinder of water of height h inside the capillary tube. The weight of this water is

$$w = Mg = \rho V g = \rho g \pi r^2 h \qquad \textbf{[9.21]}$$

Equating F_v in Equation 9.20 to w in Equation 9.21 (applying Newton's second law for equilibrium), we have

$$\gamma(2\pi r)(\cos\phi) = \rho g \pi r^2 h$$

Solving for h gives the height to which water is drawn into the tube:

$$h = \frac{2\gamma}{\rho g r}\cos\phi \qquad \textbf{[9.22]}$$

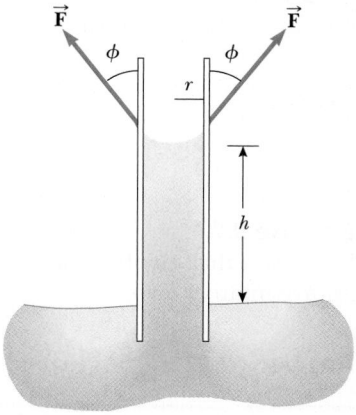

Figure 9.45 A liquid rises in a narrow tube because of capillary action, a result of surface tension and adhesive forces.

If a capillary tube is inserted into a liquid in which cohesive forces dominate over adhesive forces, the level of the liquid in the capillary tube will be below the surface of the surrounding fluid, as shown in Figure 9.46. An analysis similar to the above would show that the distance h to the depressed surface is given by Equation 9.22.

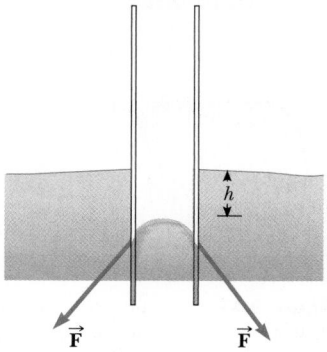

Figure 9.46 When cohesive forces between molecules of a liquid exceed adhesive forces, the level of the liquid in the capillary tube is below the surface of the surrounding fluid.

Capillary tubes are often used to draw small samples of blood from a needle prick in the skin. Capillary action must also be considered in the construction of concrete-block buildings, because water seepage through capillary pores in the blocks or the mortar may cause damage to the inside of the building. To prevent such damage, the blocks are usually coated with a waterproofing agent either outside or inside the building. Water seepage through a wall is an undesirable effect of capillary action, but there are many useful effects. Plants depend on capillary action to transport water and nutrients, and sponges and paper towels use capillary action to absorb spilled fluids.

EXAMPLE 9.17 Rising Water

Goal Apply surface tension to capillary action.

Problem Find the height to which water would rise in a capillary tube with a radius equal to 5.0×10^{-5} m. Assume that the contact angle between the water and the material of the tube is small enough to be considered zero.

Strategy This problem requires substituting values into Equation 9.22.

Solution

Substitute the known values into Equation 9.22:

$$h = \frac{2\gamma \cos 0°}{\rho g r}$$

$$= \frac{2(0.073 \text{ N/m})}{(1.00 \times 10^3 \text{ kg/m}^3)(9.80 \text{ m/s}^2)(5.0 \times 10^{-5} \text{ m})}$$

$$= 0.30 \text{ m}$$

Exercise 9.17

Suppose ethyl alcohol rises 0.250 m in a thin tube. Estimate the radius of the tube, assuming the contact angle is approximately zero.

Answer 2.23×10^{-5} m

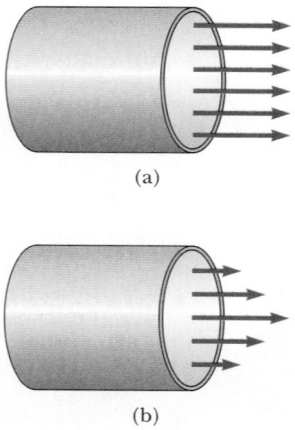

Figure 9.47 (a) The particles in an ideal (nonviscous) fluid all move through the pipe with the same velocity. (b) In a viscous fluid, the velocity of the fluid particles is zero at the surface of the pipe and increases to a maximum value at the center of the pipe.

Viscous Fluid Flow

It is considerably easier to pour water out of a container than to pour honey. This is because honey has a higher viscosity than water. In a general sense, **viscosity refers to the internal friction of a fluid**. It's very difficult for layers of a viscous fluid to slide past one another. Likewise, it's difficult for one solid surface to slide past another if there is a highly viscous fluid, such as soft tar, between them.

When an ideal (nonviscous) fluid flows through a pipe, the fluid layers slide past one another with no resistance. If the pipe has a uniform cross section, each layer has the same velocity, as shown in Figure 9.47a. In contrast, the layers of a viscous fluid have different velocities, as Figure 9.47b indicates. The fluid has the greatest velocity at the center of the pipe, whereas the layer next to the wall doesn't move because of adhesive forces between molecules and the wall surface.

To better understand the concept of viscosity, consider a layer of liquid between two solid surfaces, as in Figure 9.48. The lower surface is fixed in position, and the top surface moves to the right with a velocity $\vec{\mathbf{v}}$ under the action of an external force $\vec{\mathbf{F}}$. Because of this motion, a portion of the liquid is distorted from its original shape, *ABCD*, at one instant to the shape *AEFD* a moment later. The force required to move the upper plate and distort the liquid is proportional to both the area A in contact with the fluid and the speed v of the fluid. Furthermore, the

TABLE 9.5

Viscosities of Various Fluids

Fluid	T (°C)	Viscosity η (N·s/m^2)
Water	20	1.0×10^{-3}
Water	100	0.3×10^{-3}
Whole blood	37	2.7×10^{-3}
Glycerin	20	1500×10^{-3}
10-wt motor oil	30	250×10^{-3}

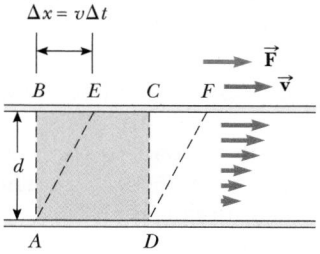

Figure 9.48 A layer of liquid between two solid surfaces in which the lower surface is fixed and the upper surface moves to the right with a velocity $\vec{v}$.

force is inversely proportional to the distance d between the two plates. We can express these proportionalities as $F \propto Av/d$. The force required to move the upper plate at a fixed speed v is therefore

$$F = \eta \frac{Av}{d} \qquad [9.23]$$

where η (the lowercase Greek letter *eta*) is the **coefficient of viscosity** of the fluid. ◀ Coefficient of viscosity

The SI units of viscosity are N·s/m^2. The units of viscosity in many reference sources are often expressed in dyne·s/cm^2, called 1 **poise**, in honor of the French scientist J. L. Poiseuille (1799–1869). The relationship between the SI unit of viscosity and the poise is

$$1 \text{ poise} = 10^{-1} \text{ N}\cdot\text{s/m}^2 \qquad [9.24]$$

Small viscosities are often expressed in centipoise (cp), where 1 cp = 10^{-2} poise. The coefficients of viscosity for some common substances are listed in Table 9.5.

Poiseuille's Law

Figure 9.49 shows a section of a tube of length L and radius R containing a fluid under a pressure P_1 at the left end and a pressure P_2 at the right. Because of this pressure difference, the fluid flows through the tube. The rate of flow (volume per unit time) depends on the pressure difference $(P_1 - P_2)$, the dimensions of the tube, and the viscosity of the fluid. The result, known as **Poiseuille's law**, is

$$\text{Rate of flow} = \frac{\Delta V}{\Delta t} = \frac{\pi R^4 (P_1 - P_2)}{8\eta L} \qquad [9.25]$$

◀ Poiseuille's law

where η is the coefficient of viscosity of the fluid. We won't attempt to derive this equation here, because the methods of integral calculus are required. However, it is reasonable that the rate of flow should increase if the pressure difference across the tube or the tube radius increases. Likewise, the flow rate should decrease if the viscosity of the fluid or the length of the tube increases. So the presence of R and the pressure difference in the numerator of Equation 9.25 and of L and η in the denominator makes sense.

From Poiseuille's law, we see that in order to maintain a constant flow rate, the pressure difference across the tube has to increase if the viscosity of the fluid increases. This fact is important in understanding the flow of blood through the circulatory system. The viscosity of blood increases as the number of red blood cells rises. Blood with a high concentration of red blood cells requires greater pumping pressure from the heart to keep it circulating than does blood of lower red blood cell concentration.

Note that the flow rate varies as the radius of the tube raised to the fourth power. Consequently, if a constriction occurs in a vein or artery, the heart will have to work considerably harder in order to produce a higher pressure drop and hence maintain the required flow rate.

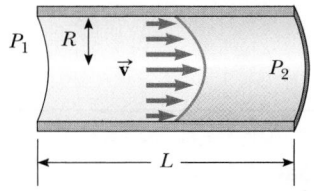

Figure 9.49 Velocity profile of a fluid flowing through a uniform pipe of circular cross section. The rate of flow is given by Poiseuille's law. Note that the fluid velocity is greatest at the middle of the pipe.

EXAMPLE 9.18 A Blood Transfusion

Goal Apply Poiseuille's law.

Problem A patient receives a blood transfusion through a needle of radius 0.20 mm and length 2.0 cm. The density of blood is 1 050 kg/m^3. The bottle supplying the blood is 0.50 m above the patient's arm. What is the rate of flow through the needle?

Strategy Find the pressure difference between the level of the blood and the patient's arm. Substitute into Poiseuille's law, using the value for the viscosity of whole blood in Table 9.5.

Solution

Calculate the pressure difference:

$$P_1 - P_2 = \rho g h = (1\,050\ \text{kg/m}^3)(9.80\ \text{m/s}^2)(0.50\ \text{m}) = 5.15 \times 10^3\ \text{Pa}$$

Substitute into Poiseuille's law:

$$\frac{\Delta V}{\Delta t} = \frac{\pi R^4 (P_1 - P_2)}{8\eta L} = \frac{\pi(2.0 \times 10^{-4}\ \text{m})^4 (5.15 \times 10^3\ \text{Pa})}{8(2.7 \times 10^{-3}\ \text{N}\cdot\text{s/m}^2)(2.0 \times 10^{-2}\ \text{m})} = 6.0 \times 10^{-8}\ \text{m}^3/\text{s}$$

Remarks Compare this to the volume flow rate in the absence of any viscosity. Using Bernoulli's equation, the calculated volume flow rate is approximately five times as great. As expected, viscosity greatly reduces flow rate.

Exercise 9.18

A pipe carrying water from a tank 20.0 m tall must cross 3.00×10^2 km of wilderness to reach a remote town. Find the radius of pipe so that the volume flow rate is at least 0.0500 m^3/s. (Use the viscosity of water at 20°C.)

Answer 0.118 m

Reynolds Number

At sufficiently high velocities, fluid flow changes from simple streamline flow to turbulent flow, characterized by a highly irregular motion of the fluid. Experimentally, the onset of turbulence in a tube is determined by a dimensionless factor called the **Reynolds number**, RN, given by

Reynolds number ▶

$$RN = \frac{\rho v d}{\eta} \qquad [9.26]$$

where ρ is the density of the fluid, v is the average speed of the fluid along the direction of flow, d is the diameter of the tube, and η is the viscosity of the fluid. If RN is below about 2 000, the flow of fluid through a tube is streamline; turbulence occurs if RN is above 3 000. In the region between 2 000 and 3 000, the flow is unstable, meaning that the fluid can move in streamline flow, but any small disturbance will cause its motion to change to turbulent flow.

EXAMPLE 9.19 Turbulent Flow of Blood

Goal Use the Reynolds number to determine a speed associated with the onset of turbulence.

Problem Determine the speed at which blood flowing through an artery of diameter 0.20 cm will become turbulent.

Strategy The solution requires only the substitution of values into Equation 9.26 giving the Reynolds number and then solving it for the speed v.

Solution

Solve Equation 9.26 for v, and substitute the viscosity and density of blood from Example 9.18, the diameter d of the artery, and a Reynolds number of 3.00×10^3:

$$v = \frac{\eta(RN)}{\rho d} = \frac{(2.7 \times 10^{-3}\ \mathrm{N \cdot s/m^2})(3.00 \times 10^3)}{(1.05 \times 10^3\ \mathrm{kg/m^3})(0.20 \times 10^{-2}\ \mathrm{m})}$$

$$v = 3.9\ \mathrm{m/s}$$

Remark The exercise shows that rapid ingestion of soda through a straw may create a turbulent state.

Exercise 9.19

Determine the speed v at which water at 20° sucked up a straw would become turbulent. The straw has a diameter of 0.0060 m.

Answer $v = 0.50\ \mathrm{m/s}$

9.10 TRANSPORT PHENOMENA

When a fluid flows through a tube, the basic mechanism that produces the flow is a difference in pressure across the ends of the tube. This pressure difference is responsible for the transport of a mass of fluid from one location to another. The fluid may also move from place to place because of a second mechanism—one that depends on a difference in *concentration* between two points in the fluid, as opposed to a pressure difference. When the concentration (the number of molecules per unit volume) is higher at one location than at another, molecules will flow from the point where the concentration is high to the point where it is lower. The two fundamental processes involved in fluid transport resulting from concentration differences are called *diffusion* and *osmosis*.

Diffusion

In a diffusion process, molecules move from a region where their concentration is high to a region where their concentration is lower. To understand why diffusion occurs, consider Figure 9.50, which depicts a container in which a high concentration of molecules has been introduced into the left side. The dashed line in the figure represents an imaginary barrier separating the two regions. Because the molecules are moving with high speeds in random directions, many of them will cross the imaginary barrier moving from left to right. Very few molecules will pass through moving from right to left, simply because there are very few of them on the right side of the container at any instant. As a result, there will always be a *net* movement from the region with many molecules to the region with fewer molecules. For this reason, the concentration on the left side of the container will decrease, and that on the right side will increase with time. Once a concentration equilibrium has been reached, there will be no *net* movement across the cross-sectional area: The rate of movement of molecules from left to right will equal the rate from right to left.

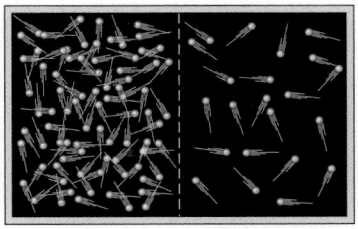

Figure 9.50 When the concentration of gas molecules on the left side of the container exceeds the concentration on the right side, there will be a net motion (diffusion) of molecules from left to right.

The basic equation for diffusion is **Fick's law**,

$$\text{Diffusion rate} = \frac{\text{mass}}{\text{time}} = \frac{\Delta M}{\Delta t} = DA\left(\frac{C_2 - C_1}{L}\right) \qquad [9.27]$$

◀ Fick's law

where D is a constant of proportionality. The left side of this equation is called the *diffusion rate* and is a measure of the mass being transported per unit time. The equation says that the rate of diffusion is proportional to the cross-sectional area A and to the change in concentration per unit distance, $(C_2 - C_1)/L$, which is called the *concentration gradient*. The concentrations C_1 and C_2 are measured in kilograms per cubic meter. The proportionality constant D is called the **diffusion coefficient** and has units of square meters per second. Table 9.6 (page 306) lists diffusion coefficients for a few substances.

TABLE 9.6

Diffusion Coefficients of Various Substances at 20°C

Substance	D (m^2/s)
Oxygen through air	6.4×10^{-5}
Oxygen through tissue	1×10^{-11}
Oxygen through water	1×10^{-9}
Sucrose through water	5×10^{-10}
Hemoglobin through water	76×10^{-11}

The Size of Cells and Osmosis

Diffusion through cell membranes is vital in carrying oxygen to the cells of the body and in removing carbon dioxide and other waste products from them. Cells require oxygen for those metabolic processes in which substances are either synthesized or broken down. In such processes, the cell uses up oxygen and produces carbon dioxide as a by-product. A fresh supply of oxygen diffuses from the blood, where its concentration is high, into the cell, where its concentration is low. Likewise, carbon dioxide diffuses from the cell into the blood, where it is in lower concentration. Water, ions, and other nutrients also pass into and out of cells by diffusion.

A cell can function properly only if it can transport nutrients and waste products rapidly across the cell membrane. The surface area of the cell should be large enough so that the exposed membrane area can exchange materials effectively while the volume should be small enough so that materials can reach or leave particular locations rapidly. This requires a large surface-area-to-volume ratio.

Model a cell as a cube, each side with length L. The total surface area is $6L^2$ and the volume is L^3. The surface area to volume is then

$$\frac{\text{surface area}}{\text{volume}} = \frac{6L^2}{L^3} = \frac{6}{L}$$

Because L is in the denominator, a smaller L means a larger ratio. This shows that the smaller the size of a body, the more efficiently it can transport nutrients and waste products across the cell membrane. Cells range in size from a millionth of a meter to several millionths, so a good estimate of a typical cell's surface-to-volume ratio is 10^6.

The diffusion of material through a membrane is partially determined by the size of the pores (holes) in the membrane wall. Small molecules, such as water, may pass through the pores easily, while larger molecules, such as sugar, may pass through only with difficulty or not at all. A membrane that allows passage of some molecules but not others is called a **selectively permeable** membrane.

Osmosis is the diffusion of water across a selectively permeable membrane from a high water concentration to a low water concentration. As in the case of diffusion, osmosis continues until the concentrations on the two sides of the membrane are equal.

APPLICATION

Effect of Osmosis on Living Cells

To understand the effect of osmosis on living cells, consider a particular cell in the body that contains a sugar concentration of 1%. (A 1% solution is 1 g of sugar dissolved in enough water to make 100 ml of solution; "ml" is the abbreviation for milliliters, so 10^{-3} L = 1 cm^3.) Assume this cell is immersed in a 5% sugar solution (5 g of sugar dissolved in enough water to make 100 ml). Compared to the 1% solution, there are five times as many sugar molecules per unit volume in the 5% sugar solution, so there must be fewer water molecules. Accordingly, water will diffuse from inside the cell, where its concentration is higher, across the cell membrane to the outside solution, where the concentration of water is lower. This loss of water from the cell would cause it to shrink and perhaps become damaged through dehydration. If the concentrations were reversed, water would diffuse *into* the cell, causing it to swell and perhaps burst. If solutions are introduced into the body intravenously, care must be taken to ensure that they don't disturb the osmotic balance of the body, else cell damage can occur. For example, if a 9% saline solution surrounds a red blood cell, the cell will shrink. By contrast, if the solution is about 1%, the cell will eventually burst.

APPLICATION

Kidney Function and Dialysis

In the body, blood is cleansed of impurities by osmosis as it flows through the kidneys. (See Fig. 9.51a.) Arterial blood first passes through a bundle of capillaries known as a *glomerulus*, where most of the waste products and some essential salts and minerals are removed. From the glomerulus, a narrow tube emerges that is in intimate contact with other capillaries throughout its length. As blood passes through the tubules, most of the essential elements are returned to it; waste products are not allowed to reenter and are eventually removed in urine.

If the kidneys fail, an artificial kidney or a dialysis machine can filter the blood. Figure 9.51b shows how this is done. Blood from an artery in the arm is mixed

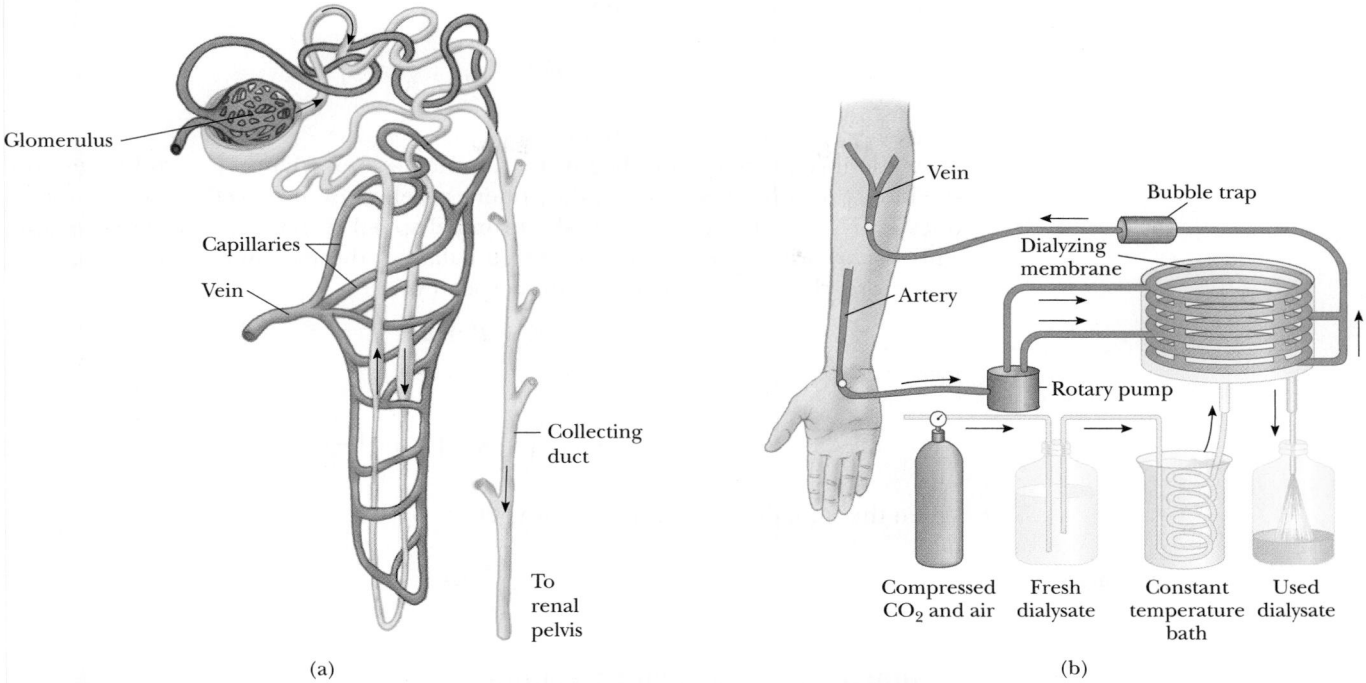

Figure 9.51 (a) Diagram of a single nephron in the human excretory system. (b) An artificial kidney.

with heparin, a blood thinner, and allowed to pass through a tube covered with a semipermeable membrane. The tubing is immersed in a bath of a dialysate fluid with the same chemical composition as purified blood. Waste products from the blood enter the dialysate by diffusion through the membrane. The filtered blood is then returned to a vein.

Motion through a Viscous Medium

When an object falls through air, its motion is impeded by the force of air resistance. In general, this force is dependent on the shape of the falling object and on its velocity. The force of air resistance acts on all falling objects, but the exact details of the motion can be calculated only for a few cases in which the object has a simple shape, such as a sphere. In this section, we will examine the motion of a tiny spherical object falling slowly through a viscous medium.

In 1845 a scientist named George Stokes found that the magnitude of the resistive force on a very small spherical object of radius r falling slowly through a fluid of viscosity η with speed v is given by

$$F_r = 6\pi\eta r v \qquad \textbf{[9.28]}$$

This equation, called **Stokes's law**, has many important applications. For example, it describes the sedimentation of particulate matter in blood samples. It was used by Robert Millikan (1886–1953) to calculate the radius of charged oil droplets falling through air. From this, Millikan was ultimately able to determine the charge on the electron, and was awarded the Nobel prize in 1923 for his pioneering work on elemental charges.

As a sphere falls through a viscous medium, three forces act on it, as shown in Figure 9.52: $\vec{\mathbf{F}}_r$, the force of friction; $\vec{\mathbf{B}}$, the buoyant force of the fluid; and $\vec{\mathbf{w}}$, the force of gravity acting on the sphere. The magnitude of $\vec{\mathbf{w}}$ is given by

$$w = \rho g V = \rho g\left(\frac{4}{3}\pi r^3\right)$$

where ρ is the density of the sphere and $\frac{4}{3}\pi r^3$ is its volume. According to Archimedes's principle, the magnitude of the buoyant force is equal to the weight

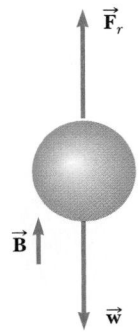

Figure 9.52 A sphere falling through a viscous medium. The forces acting on the sphere are the resistive frictional force $\vec{\mathbf{F}}_r$, the buoyant force $\vec{\mathbf{B}}$, and the force of gravity $\vec{\mathbf{w}}$ acting on the sphere.

of the fluid displaced by the sphere,

$$B = \rho_f g V = \rho_f g \left(\frac{4}{3}\pi r^3\right)$$

where ρ_f is the density of the fluid.

At the instant the sphere begins to fall, the force of friction is zero because the speed of the sphere is zero. As the sphere accelerates, its speed increases and so does $\vec{\mathbf{F}}_r$. Finally, **at a speed called the terminal speed v_t, the net force goes to zero.** This occurs when the net upward force balances the downward force of gravity. Therefore, the sphere reaches terminal speed when

$$F_r + B = w$$

or

$$6\pi\eta r v_t + \rho_f g \left(\frac{4}{3}\pi r^3\right) = \rho g \left(\frac{4}{3}\pi r^3\right)$$

When this equation is solved for v_t, we get

Terminal speed ▶

$$v_t = \frac{2r^2 g}{9\eta}(\rho - \rho_f) \qquad \textbf{[9.29]}$$

Sedimentation and Centrifugation

If an object isn't spherical, we can still use the basic approach just described to determine its terminal speed. The only difference is that we can't use Stokes's law for the resistive force. Instead, we assume that the resistive force has a magnitude given by $F_r = kv$, where k is a coefficient that must be determined experimentally. As discussed previously, the object reaches its terminal speed when the downward force of gravity is balanced by the net upward force, or

$$w = B + F_r \qquad \textbf{[9.30]}$$

where $B = \rho_f g V$ is the buoyant force. The volume V of the displaced fluid is related to the density ρ of the falling object by $V = m/\rho$. Hence, we can express the buoyant force as

$$B = \frac{\rho_f}{\rho} mg$$

We substitute this expression for B and $F_r = kv_t$ into Equation 9.30 (terminal speed condition):

$$mg = \frac{\rho_f}{\rho} mg + kv_t$$

or

$$v_t = \frac{mg}{k}\left(1 - \frac{\rho_f}{\rho}\right) \qquad \textbf{[9.31]}$$

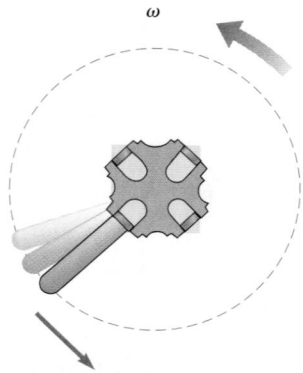

Figure 9.53 Simplified diagram of a centrifuge (top view).

The terminal speed for particles in biological samples is usually quite small. For example, the terminal speed for blood cells falling through plasma is about 5 cm/h in the gravitational field of the Earth. The terminal speeds for the molecules that make up a cell are many orders of magnitude smaller than this because of their much smaller mass. The speed at which materials fall through a fluid is called the **sedimentation rate** and is important in clinical analysis.

The sedimentation rate in a fluid can be increased by increasing the effective acceleration g that appears in Equation 9.31. A fluid containing various biological molecules is placed in a centrifuge and whirled at very high angular speeds (Fig. 9.53). Under these conditions, the particles gain a large radial acceleration

$a_c = v^2/r = \omega^2 r$ that is much greater than the free-fall acceleration, so we can replace g in Equation 9.31 by $\omega^2 r$ and obtain

$$v_t = \frac{m\omega^2 r}{k}\left(1 - \frac{\rho_f}{\rho}\right) \qquad [9.32]$$

This equation indicates that the sedimentation rate is enormously speeded up in a centrifuge ($\omega^2 r \gg g$) and that those particles with the greatest mass will have the largest terminal speed. Consequently the most massive particles will settle out on the bottom of a test tube first.

SUMMARY

PhysicsNow™ Take a practice test by logging into PhysicsNow at **www.cp7e.com** and clicking on the Pre-Test link for this chapter.

9.1 States of Matter

Matter is normally classified as being in one of three states: solid, liquid, or gaseous. The fourth state of matter is called a plasma, which consists of a neutral system of charged particles interacting electromagnetically.

9.2 The Deformation of Solids

The elastic properties of a solid can be described using the concepts of stress and strain. **Stress** is related to the force per unit area producing a deformation; **strain** is a measure of the amount of deformation. Stress is proportional to strain, and the constant of proportionality is the **elastic modulus**:

$$\text{Stress} = \text{Elastic modulus} \times \text{strain} \qquad [9.1]$$

Three common types of deformation are (1) the resistance of a solid to elongation or compression, characterized by **Young's modulus** Y; (2) the resistance to displacement of the faces of a solid sliding past each other, characterized by the shear modulus S; and (3) the resistance of a solid or liquid to a change in volume, characterized by the bulk modulus B.

All three types of deformation obey laws similar to Hooke's law for springs. Solving problems is usually a matter of identifying the given physical variables and solving for the unknown variable.

9.3 Density and Pressure

The **density** ρ of a substance of uniform composition is its mass per unit volume—kilograms per cubic meter (kg/m^3) in the SI system:

$$\rho \equiv \frac{M}{V} \qquad [9.6]$$

The **pressure** P in a fluid, measured in pascals (Pa), is the force per unit area that the fluid exerts on an object immersed in it:

$$P \equiv \frac{F}{A} \qquad [9.7]$$

9.4 Variation of Pressure with Depth

The pressure in an incompressible fluid varies with depth h according to the expression

$$P = P_0 + \rho g h \qquad [9.11]$$

where P_0 is atmospheric pressure (1.013×10^5 Pa) and ρ is the density of the fluid.

Pascal's principle states that when pressure is applied to an enclosed fluid, the pressure is transmitted undiminished to every point of the fluid and to the walls of the containing vessel.

9.6 Buoyant Forces and Archimedes's Principle

When an object is partially or fully submerged in a fluid, the fluid exerts an upward force, called the **buoyant force**, on the object. This force is, in fact, just the net difference in pressure between the top and bottom of the object. It can be shown that the magnitude of the buoyant force B is equal to the weight of the fluid displaced by the object, or

$$B = \rho_{\text{fluid}} V_{\text{fluid}} g \qquad [9.12b]$$

Equation 9.12b is known as **Archimedes's principle**.

Solving a buoyancy problem usually involves putting the buoyant force into Newton's second law and then proceeding as in Chapter 4.

9.7 Fluids in Motion

Certain aspects of a fluid in motion can be understood by assuming that the fluid is nonviscous and incompressible and that its motion is in a steady state with no turbulence:

1. The flow rate through the pipe is a constant, which is equivalent to stating that the product of the cross-sectional area A and the speed v at any point is constant. At any two points, therefore, we have

$$A_1 v_1 = A_2 v_2 \qquad [9.15]$$

   This relation is referred to as the **equation of continuity**.
2. The sum of the pressure, the kinetic energy per unit volume, and the potential energy per unit volume is the same at any two points along a streamline:

$$P_1 + \tfrac{1}{2}\rho v_1^2 + \rho g y_1 = P_2 + \tfrac{1}{2}\rho v_2^2 + \rho g y_2 \qquad [9.16]$$

Equation 9.16 is known as **Bernoulli's equation**. Solving problems with Bernoulli's equation is similar to solving problems with the work–energy theorem, whereby two points are chosen, one point where a quantity is unknown and another where all quantities are known. Equation 9.16 is then solved for the unknown quantity.

CONCEPTUAL QUESTIONS

1. Baseball home-run hitters like to play in Denver, but curveball pitchers do not. Why?

2. The density of air is $1.3\ \text{kg/m}^3$ at sea level. From your knowledge of air pressure at ground level, estimate the height of the atmosphere. As a simplifying assumption, take the atmosphere to be of uniform density up to some height, after which the density rapidly falls to zero. (In reality, the density of the atmosphere decreases as we go up.) (Question 2 is courtesy of Edward F. Redish. For more questions of this type, see http://www.physics.umd.edu/perg/.)

3. A woman wearing high-heeled shoes is invited into a home in which the kitchen has vinyl floor covering. Why should the homeowner be concerned?

4. Figure Q9.4 shows aerial views from directly above two dams. Both dams are equally long (the vertical dimension in the diagram) and equally deep (into the page in the diagram). The dam on the left holds back a very large lake, while the dam on the right holds back a narrow river. Which dam has to be built more strongly?

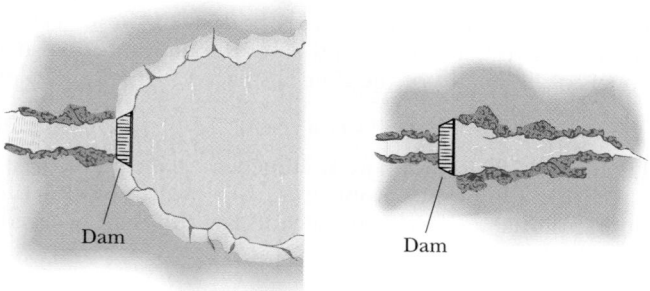

Figure Q9.4

5. A typical silo on a farm has many bands wrapped around its perimeter, as shown in Figure Q9.5. Why is the spacing between successive bands smaller at the lower portions of the silo?

Henry Leap and Jim Lehman

Figure Q9.5

6. During inhalation, the pressure in the lungs is slightly less than external pressure and the muscles controlling exhalation are relaxed. Under water, the body equalizes internal and external pressures. Discuss the condition of the muscles if a person under water is breathing through a snorkel. Would a snorkel work in deep water?

7. Colloquially, we say a nurse uses a syringe to "draw" blood. Why is it more accurate to say that a syringe is used to "accept" blood?

8. Many people believe that a vacuum created inside a vacuum cleaner causes particles of dirt to be drawn in. Actually, the dirt is pushed in. Explain.

9. Suppose a damaged ship just barely floats in the ocean after a hole in its hull has been sealed. It is pulled by a tugboat toward shore and into a river, heading toward a dry dock for repair. As the boat is pulled up the river, it sinks. Why?

10. Will an ice cube float higher in water or in an alcoholic beverage?

11. A pound of Styrofoam and a pound of lead have the same weight. If they are placed on a sensitive equal-arm balance, will the scales balance?

12. An ice cube is placed in a glass of water. What happens to the level of the water as the ice melts?

13. Place two cans of soft drinks, one regular and one diet, in a container of water. You will find that the diet drink floats while the regular one sinks. Use Archimedes' principle to devise an explanation. [*Broad Hint:* The artificial sweetener used in diet drinks is less dense than sugar.]

14. Prairie dogs live in underground burrows with at least two entrances. They ventilate their burrows by building a mound over one entrance, as shown in Figure Q9.14. This entrance is open to a stream of air when a breeze blows from any direction. A second entrance at ground level is open to almost stagnant air. How does this construction create an airflow through the burrow?

Pamela Zilly/The Image Bank/Getty Images

Figure Q9.14

15. When you are driving a small car on the freeway and a truck passes you at high speed, why do you feel pulled toward the truck?

16. A barge is carrying a load of gravel along a river. It approaches a low bridge, and the captain realizes that the top of the pile of gravel is not going to make it under the bridge. The captain orders the crew to quickly shovel gravel from the pile into the water. Is this a good decision?

17. Tornadoes and hurricanes often lift the roofs of houses. Use the Bernoulli effect to explain why. Why should you keep your windows open under these conditions?

18. Water is poured to the same level in each of the three vessels shown in Figure Q9.18. Each vessel has the same base area. Because the water fills each to the same depth, each vessel will have the same pressure at the bottom. Because the area and pressure at each base is the same, each liquid should exert the same force on the base of the vessel. Yet, if the vessels are weighed, three different values are obtained. (The one in the center clearly holds less liquid than the one at the left, so it will weigh less.) How can you resolve this apparent contradiction? (Question 18 is courtesy of Edward F. Redish. For more questions of this type, see http://www.physics.umd.edu/perg/.)

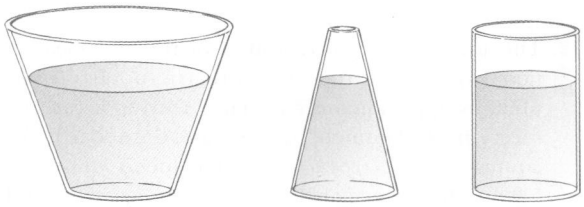

Figure Q9.18

PROBLEMS

1, 2, 3 = straightforward, intermediate, challenging □ = full solution available in *Student Solutions Manual/Study Guide*
PhysicsNow™ = coached problem with hints available at **www.cp7e.com** = biomedical application

Section 9.1 States of Matter

Section 9.2 The Deformation of Solids

1. If the elastic limit of steel is 5.0×10^8 Pa, determine the minimum diameter a steel wire can have if it is to support a 70-kg circus performer without its elastic limit being exceeded.

2. If the shear stress in steel exceeds about 4.00×10^8 N/m^2, the steel ruptures. Determine the shearing force necessary to (a) shear a steel bolt 1.00 cm in diameter and (b) punch a 1.00-cm-diameter hole in a steel plate 0.500 cm thick.

3. The heels on a pair of women's shoes have radii of 0.50 cm at the bottom. If 30% of the weight of a woman weighing 480 N is supported by each heel, find the stress on each heel.

4. When water freezes, it expands about 9.00%. What would be the pressure increase inside your automobile engine block if the water in it froze? The bulk modulus of ice is 2.00×10^9 N/m^2.

5. For safety in climbing, a mountaineer uses a nylon rope that is 50 m long and 1.0 cm in diameter. When supporting a 90-kg climber, the rope elongates 1.6 m. Find its Young's modulus.

6. A stainless-steel orthodontic wire is applied to a tooth, as in Figure P9.6. The wire has an unstretched length of 3.1 cm and a diameter of 0.22 mm. If the wire is stretched 0.10 mm, find the magnitude and direction of the force on the tooth. Disregard the width of the tooth, and assume that Young's modulus for stainless steel is 18×10^{10} Pa.

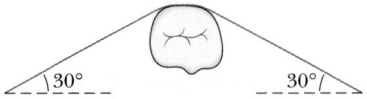

Figure P9.6

7. Bone has a Young's modulus of about 18×10^9 Pa. Under compression, it can withstand a stress of about 160×10^6 Pa before breaking. Assume that a femur (thighbone) is 0.50 m long, and calculate the amount of compression this bone can withstand before breaking.

8. The distortion of the Earth's crustal plates is an example of shear on a large scale. A particular crustal rock has a shear modulus of 1.5×10^{10} Pa. What shear stress is involved when a 10-km layer of this rock is sheared through a distance of 5.0 m?

9. A child slides across a floor in a pair of rubber-soled shoes. The friction force acting on each foot is 20 N, the footprint area of each foot is 14 cm^2, and the thickness of the soles is 5.0 mm. Find the horizontal distance traveled by the sheared face of the sole. The shear modulus of the rubber is 3.0×10^6 Pa.

10. A high-speed lifting mechanism supports an 800-kg object with a steel cable that is 25.0 m long and 4.00 cm^2 in cross-sectional area. (a) Determine the elongation of the cable. (b) By what additional amount does the cable increase in length if the object is accelerated upwards at a rate of 3.0 m/s^2? (c) What is the greatest mass that can be accelerated upwards at 3.0 m/s^2 if the stress in the cable is not to exceed the elastic limit of the cable, which is 2.2×10^8 Pa?

11. Determine the elongation of the rod in Figure P9.11 if it is under a tension of 5.8×10^3 N.

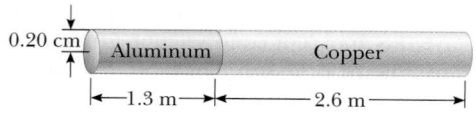

Figure P9.11

12. The total cross-sectional area of the load-bearing calcified portion of the two forearm bones (radius and ulna) is approximately 2.4 cm^2. During a car crash, the forearm is slammed against the dashboard. The arm comes to rest from an initial speed of 80 km/h in 5.0 ms. If the arm has an effective mass of 3.0 kg and bone material can withstand a maximum compressional stress of 16×10^7 Pa, is the arm likely to withstand the crash?

Section 9.3 Density and Pressure

13. A 50.0-kg ballet dancer stands on her toes during a performance with four square inches (26.0 cm^2) in contact with the floor. What is the pressure exerted by the floor over the area of contact (a) if the dancer is stationary and (b) if the dancer is leaping upwards with an acceleration of 4.00 m/s^2?

14. The four tires of an automobile are inflated to a gauge pressure of 2.0×10^5 Pa. Each tire has an area of 0.024 m^2 in contact with the ground. Determine the weight of the automobile.

15. Air is trapped above liquid ethyl alcohol in a rigid container, as shown in Figure P9.15. If the air pressure above the liquid is 1.10 atm, determine the pressure inside a bubble 4.0 m below the surface of the liquid.

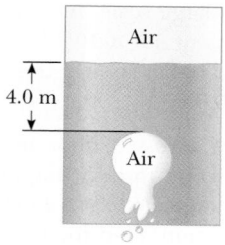

Figure P9.15

16. A 70-kg man in a 5.0-kg chair tilts back so that all the weight is balanced on two legs of the chair. Assume that each leg makes contact with the floor over a circular area with a radius of 1.0 cm, and find the pressure exerted by each leg on the floor.

17. If 1.0 m^3 of concrete weighs 5.0×10^4 N, what is the height of the tallest cylindrical concrete pillar that will not collapse under its own weight? The compression strength of concrete (the maximum pressure that can be exerted on the base of the structure) is 1.7×10^7 Pa.

Section 9.3 Density and Pressure

Section 9.4 Variation of Pressure with Depth

Section 9.5 Pressure Measurements

18. The deepest point in the ocean is in the Mariana Trench, about 11 km deep. The pressure at the ocean floor is huge, about 1.13×10^8 N/m^2. (a) Calculate the change in volume of 1.00 m^3 of water carried from the surface to the bottom of the Pacific. (b) The density of water at the surface is 1.03×10^3 kg/m^3. Find its density at the bottom. (c) Is it a good approximation to think of water as incompressible?

19. A collapsible plastic bag (Figure P9.19) contains a glucose solution. If the average gauge pressure in the vein is 1.33×10^3 Pa, what must be the minimum height h of the bag in order to infuse glucose into the vein? Assume that the specific gravity of the solution is 1.02.

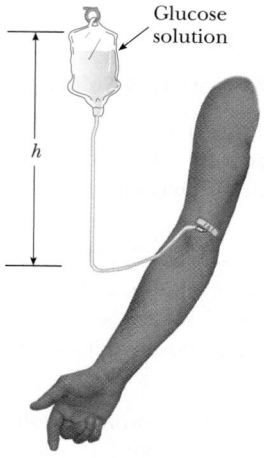

Figure P9.19

20. (a) A very powerful vacuum cleaner has a hose 2.86 cm in diameter. With no nozzle on the hose, what is the weight of the heaviest brick it can lift? (b) A very powerful octopus uses one sucker, of diameter 2.86 cm, on each of the two shells of a clam, in an attempt to pull the shells apart. Find the greatest force the octopus can exert in salt water 32.3 m deep.

21. For the cellar of a new house, a hole is dug in the ground, with vertical sides going down 2.40 m. A concrete foundation wall is built all the way across the 9.60-m width of the excavation. The foundation wall is 0.183 m away from the front of the cellar hole. During a rainstorm, drainage from the street fills up the space in front of the concrete wall, but not the cellar behind the wall. The water does not soak into the clay soil. Find the force the water

exerts on the foundation wall. For comparison, the weight of the water is $2.40\ \text{m} \times 9.60\ \text{m} \times 0.183\ \text{m} \times 1\,000\ \text{kg/m}^3 \times 9.80\ \text{m/s}^2 = 41.3\ \text{kN}$.

22. Blaise Pascal duplicated Torricelli's barometer using a red Bordeaux wine of density $984\ \text{kg/m}^3$ as the working liquid (Fig. P9.22). What was the height h of the wine column for normal atmospheric pressure? Would you expect the vacuum above the column to be as good as for mercury?

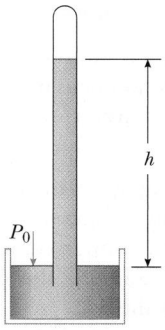

Figure P9.22

23. Physics Now™ A container is filled to a depth of 20.0 cm with water. On top of the water floats a 30.0-cm-thick layer of oil with specific gravity 0.700. What is the absolute pressure at the bottom of the container?

24. Piston ① in Figure P9.24 has a diameter of 0.25 in.; piston ② has a diameter of 1.5 in. In the absence of friction, determine the force $\vec{\mathbf{F}}$ necessary to support the 500-lb weight.

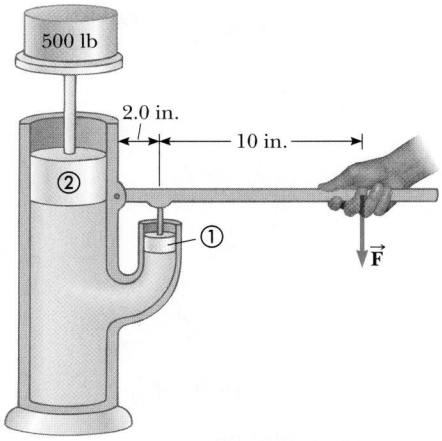

Figure P9.24

25. Figure P9.25 shows the essential parts of a hydraulic brake system. The area of the piston in the master cylinder is $6.4\ \text{cm}^2$, and that of the piston in the brake cylinder is $1.8\ \text{cm}^2$. The coefficient of friction between shoe and wheel drum is 0.50. If the wheel has a radius of 34 cm, determine the frictional torque about the axle when a force of 44 N is exerted on the brake pedal.

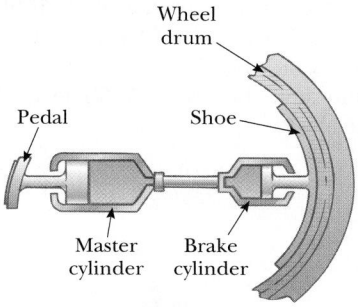

Figure P9.25

Section 9.6 Buoyant Forces and Archimedes' Principle

26. A frog in a hemispherical pod finds that he just floats without sinking in a fluid of density $1.35\ \text{g/cm}^3$. If the pod has a radius of 6.00 cm and negligible mass, what is the mass of the frog? (See Fig. P9.26.)

Figure P9.26

27. A small ferryboat is 4.00 m wide and 6.00 m long. When a loaded truck pulls onto it, the boat sinks an additional 4.00 cm into the river. What is the weight of the truck?

28. The density of ice is $920\ \text{kg/m}^3$, and that of sea water is $1\,030\ \text{kg/m}^3$. What fraction of the total volume of an iceberg is exposed?

29. As a first approximation, the Earth's continents may be thought of as granite blocks floating in a denser rock (called peridotite) in the same way that ice floats in water. (a) Show that a formula describing this phenomenon is

$$\rho_g t = \rho_p d$$

where ρ_g is the density of granite ($2.8 \times 10^3\ \text{kg/m}^3$), ρ_p is the density of peridotite ($3.3 \times 10^3\ \text{kg/m}^3$), t is the thickness of a continent, and d is the depth to which a continent floats in the peridotite. (b) If a continent sinks 5.0 km into the peridotite layer (this surface may be thought of as the ocean floor), what is the thickness of the continent?

30. A 10.0-kg block of metal is suspended from a scale and immersed in water, as in Figure P9.30. The dimensions of the block are 12.0 cm × 10.0 cm × 10.0 cm. The 12.0-cm dimension is vertical, and the top of the block is 5.00 cm below the surface of the water. (a) What are the forces exerted by the water on the top and bottom of the block? (Take $P_0 = 1.013\,0 \times 10^5\ \text{N/m}^2$.) (b) What is the reading of the spring scale? (c) Show that the buoyant force equals the difference between the forces at the top and bottom of the block.

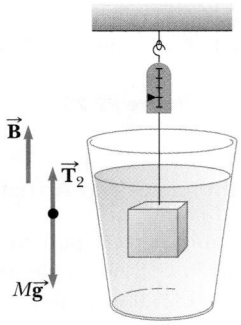

Figure P9.30

31. A bathysphere used for deep sea exploration has a radius of 1.50 m and a mass of 1.20×10^4 kg. In order to dive, the sphere takes on mass in the form of sea water. Determine the mass the bathysphere must take on so that it can descend at a constant speed of 1.20 m/s when the resistive force on it is 1 100 N upward. The density of sea water is $1.03 \times 10^3\ \text{kg/m}^3$.

32. The United States possesses the eight largest warships in the world—aircraft carriers of the *Nimitz* class—and is building one more. Suppose that, at a location where $g = 9.78\ \text{m/s}^2$, one of the ships bobs up to float 11.0 cm higher in the water when 50 fighters take off from it in 25 minutes. Bristling with bombs and missiles, each plane has an average mass of 29 000 kg. Find the horizontal area enclosed by the waterline of the $4-billion ship. By comparison, its flight deck has area of 18 000 m^2. Below decks are passageways hundreds of meters long, so narrow that two large men cannot pass each other.

33. An empty rubber balloon has a mass of 0.012 0 kg. The balloon is filled with helium at a density of $0.181\ \text{kg/m}^3$. At this density, the balloon has a radius of 0.500 m. If the filled balloon is fastened to a vertical line, what is the tension in the line?

34. A light spring of force constant $k = 160$ N/m rests vertically on the bottom of a large beaker of water (Fig. P9.34a). A 5.00-kg block of wood (density = $650\ \text{kg/m}^3$) is connected to the spring, and the block–spring system is allowed to come to static equilibrium (Fig. P9.34b). What is the elongation ΔL of the spring?

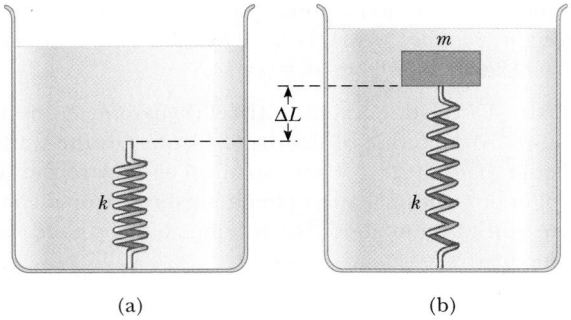

Figure P9.34

35. A sample of an unknown material appears to weigh 300 N in air and 200 N when immersed in alcohol of specific gravity 0.700. What are (a) the volume and (b) the density of the material?

36. An object weighing 300 N in air is immersed in water after being tied to a string connected to a balance. The scale now reads 265 N. Immersed in oil, the object appears to weigh 275 N. Find (a) the density of the object and (b) the density of the oil.

37. A thin spherical shell of mass 0.400 kg and diameter 0.200 m is filled with alcohol ($\rho = 806\ \text{kg/m}^3$). It is then released from rest at the bottom of a pool of water. Find the acceleration of the alcohol-filled shell as it starts to rise toward the surface of the water.

38. A rectangular air mattress is 2.0 m long, 0.50 m wide, and 0.08 m thick. If it has a mass of 2.0 kg, what additional mass can it support in water?

39. **Physics Now™** A 1.00-kg beaker containing 2.00 kg of oil (density = $916\ \text{kg/m}^3$) rests on a scale. A 2.00-kg block of iron is suspended from a spring scale and is completely submerged in the oil (Fig. P9.39). Find the equilibrium readings of both scales.

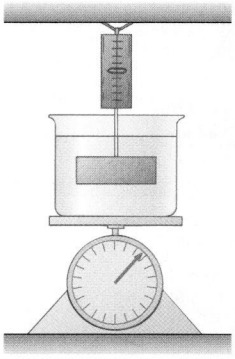

Figure P9.39

Section 9.7 Fluids in Motion

Section 9.8 Other Applications of Fluid Dynamics

40. Water is pumped into a storage tank from a well delivering 20.0 gallons of water in 30.0 seconds through a pipe of 1.00-in.^{2} cross-sectional area. What is the average velocity of the water in the pipe as the water is pumped from the well?

41. (a) Calculate the mass flow rate (in grams per second) of blood ($\rho = 1.0\ \text{g/cm}^3$) in an aorta with a cross-sectional area of $2.0\ \text{cm}^2$ if the flow speed is 40 cm/s. (b) Assume that the aorta branches to form a large number of capillaries with a combined cross-sectional area of $3.0 \times 10^3\ \text{cm}^2$. What is the flow speed in the capillaries?

42. A liquid ($\rho = 1.65\ \text{g/cm}^3$) flows through two horizontal sections of tubing joined end to end. In the first section, the cross-sectional area is $10.0\ \text{cm}^2$, the flow speed is 275 cm/s, and the pressure is 1.20×10^5 Pa. In the second section, the cross-sectional area is $2.50\ \text{cm}^2$. Calculate the smaller section's (a) flow speed and (b) pressure.

43. A hypodermic syringe contains a medicine with the density of water (Fig. P9.43). The barrel of the syringe has a cross-sectional area of $2.50 \times 10^{-5}\ \text{m}^2$. In the absence of a force on the plunger, the pressure everywhere is 1.00 atm. A force $\vec{\mathbf{F}}$ of magnitude 2.00 N is exerted on the plunger, making medicine squirt from the needle. Determine the medicine's flow speed through the needle. Assume that the pressure in the needle remains equal to 1.00 atm and that the syringe is horizontal.

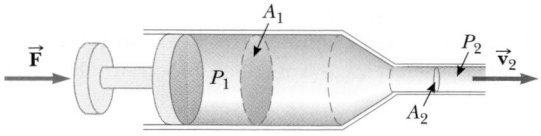

Figure P9.43

44. When a person inhales, air moves down the bronchus (windpipe) at 15 cm/s. The average flow speed of the air doubles through a constriction in the bronchus. Assuming incompressible flow, determine the pressure drop in the constriction.

45. PhysicsNow™ A jet of water squirts out horizontally from a hole near the bottom of the tank shown in Figure P9.45. If the hole has a diameter of 3.50 mm, what is the height h of the water level in the tank?

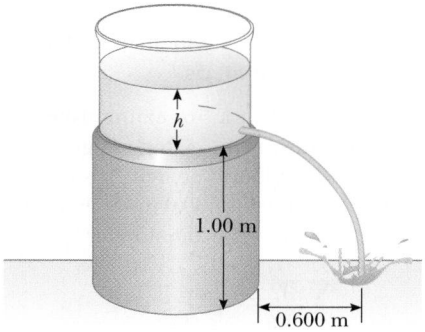

Figure P9.45

46. A large storage tank, open to the atmosphere at the top and filled with water, develops a small hole in its side at a point 16.0 m below the water level. If the rate of flow from the leak is $2.50 \times 10^{-3}\ \text{m}^3/\text{min}$, determine (a) the speed at which the water leaves the hole and (b) the diameter of the hole.

47. The inside diameters of the larger portions of the horizontal pipe depicted in Figure P9.47 are 2.50 cm. Water flows to the right at a rate of $1.80 \times 10^{-4}\ \text{m}^3/\text{s}$. Determine the inside diameter of the constriction.

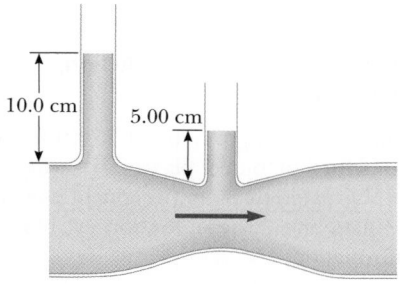

Figure P9.47

48. Water is pumped through a pipe of diameter 15.0 cm from the Colorado River up to Grand Canyon Village, on the rim of the canyon. The river is at 564 m elevation and the village is at 2 096 m. (a) At what minimum pressure must the water be pumped to arrive at the village? (b) If $4\ 500\ \text{m}^3$ are pumped per day, what is the speed of the water in the pipe? (c) What additional pressure is necessary to deliver this flow? [*Note:* You may assume that the free-fall acceleration and the density of air are constant over the given range of elevations.]

49. Old Faithful geyser in Yellowstone Park erupts at approximately 1-hour intervals, and the height of the fountain reaches 40.0 m. (a) Consider the rising stream as a series of separate drops. Analyze the free-fall motion of one of the drops to determine the speed at which the water leaves the ground. (b) Treat the rising stream as an ideal fluid in streamline flow. Use Bernoulli's equation to determine the speed of the water as it leaves ground level. (c) What is the pressure (above atmospheric pressure) in the heated underground chamber 175 m below the vent? You may assume that the chamber is large compared with the geyser vent.

50. An airplane is cruising at an altitude of 10 km. The pressure outside the craft is 0.287 atm; within the passenger compartment, the pressure is 1.00 atm and the temperature is 20°C. The density of air is $1.20\ \text{kg/m}^3$ at 20°C and 1 atm of pressure. A small leak forms in one of the window seals in the passenger compartment. Model the air as an ideal fluid to find the speed of the stream of air flowing through the leak.

51. A siphon is a device that allows a fluid to seemingly defy gravity (Fig. P9.51). The flow must be initiated by a partial vacuum in the tube, as in a drinking straw. (a) Show that the speed at which the water emerges from the siphon is given by $v = \sqrt{2gh}$. (b) For what values of y will the siphon work?

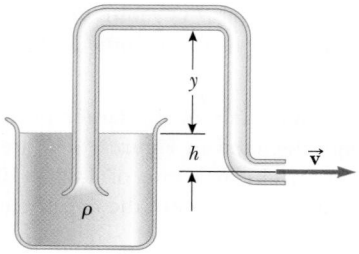

Figure P9.51

Section 9.9 Surface Tension, Capillary Action, and Viscous Fluid Flow

52. In order to lift a wire ring of radius 1.75 cm from the surface of a container of blood plasma, a vertical force of 1.61×10^{-2} N greater than the weight of the ring is required. Calculate the surface tension of blood plasma from this information.

53. A square metal sheet 3.0 cm on a side and of negligible thickness is attached to a balance and inserted into a container of fluid. The contact angle is found to be zero, as shown in Figure P9.53a, and the balance to which the metal sheet is attached reads 0.40 N. A thin veneer of oil is then spread over the sheet, and the contact angle becomes 180°, as shown in Figure P9.53b. The balance now reads 0.39 N. What is the surface tension of the fluid?

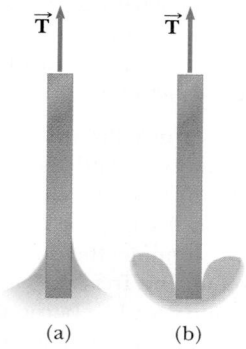

Figure P9.53

54. Whole blood has a surface tension of 0.058 N/m and a density of 1 050 kg/m^3. To what height can whole blood rise in a capillary blood vessel that has a radius of 2.0×10^{-6} m if the contact angle is zero?

55. A certain fluid has a density of 1 080 kg/m^3 and is observed to rise to a height of 2.1 cm in a 1.0-mm-diameter tube. The contact angle between the wall and the fluid is zero. Calculate the surface tension of the fluid.

56. A staining solution used in a microbiology laboratory has a surface tension of 0.088 N/m and a density 1.035 times the density of water. What must be the diameter of a capillary tube so that this solution will rise to a height of 5 cm? (Assume a contact angle of zero.)

57. The block of ice (temperature 0°C) shown in Figure P9.57 is drawn over a level surface lubricated by a layer of water 0.10 mm thick. Determine the magnitude of the force $\vec{F}$ needed to pull the block with a constant speed of 0.50 m/s. At 0°C, the viscosity of water has the value $\eta = 1.79 \times 10^{-3}$ N·s/m^2.

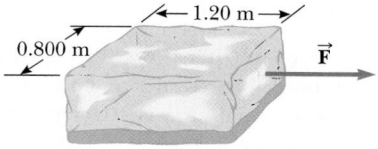

Figure P9.57

58. A thin 1.5-mm coating of glycerine has been placed between two microscope slides of width 1.0 cm and length 4.0 cm. Find the force required to pull one of the microscope slides at a constant speed of 0.30 m/s relative to the other slide.

59. A straight horizontal pipe with a diameter of 1.0 cm and a length of 50 m carries oil with a coefficient of viscosity of 0.12 N·s/m^2. At the output of the pipe, the flow rate is 8.6×10^{-5} m^3/s and the pressure is 1.0 atm. Find the gauge pressure at the pipe input.

60. The pulmonary artery, which connects the heart to the lungs, has an inner radius of 2.6 mm and is 8.4 cm long. If the pressure drop between the heart and lungs is 400 Pa, what is the average speed of blood in the pulmonary artery?

61. Spherical particles of a protein of density 1.8 g/cm^3 are shaken up in a solution of 20°C water. The solution is allowed to stand for 1.0 h. If the depth of water in the tube is 5.0 cm, find the radius of the largest particles that remain in solution at the end of the hour.

62. A hypodermic needle is 3.0 cm in length and 0.30 mm in diameter. What excess pressure is required along the needle so that the flow rate of water through it will be 1 g/s? (Use 1.0×10^{-3} Pa·s as the viscosity of water.)

63. What diameter needle should be used to inject a volume of 500 cm^3 of a solution into a patient in 30 min? Assume that the length of the needle is 2.5 cm and that the solution is elevated 1.0 m above the point of injection. Furthermore, assume the viscosity and density of the solution are those of pure water, and assume that the pressure inside the vein is atmospheric.

64. Water is forced out of a fire extinguisher by air pressure, as shown in Figure P9.64. What gauge air pressure in the tank (above atmospheric pressure) is required for the water to have a jet speed of 30.0 m/s when the water level in the tank is 0.500 m below the nozzle?

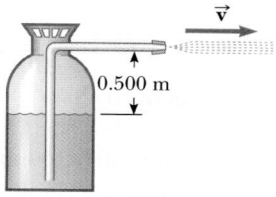

Figure P9.64

65. The aorta in humans has a diameter of about 2.0 cm, and at certain times the blood speed through it is about 55 cm/s. Is the blood flow turbulent? The density of whole blood is 1 050 kg/m^3, and its coefficient of viscosity is 2.7×10^{-3} N·s/m^2.

66. A pipe carrying 20°C water has a diameter of 2.5 cm. Estimate the maximum flow speed if the flow must be streamline.

Section 9.10 Transport Phenomena

67. Sucrose is allowed to diffuse along a 10-cm length of tubing filled with water. The tube is 6.0 cm^2 in cross-sectional area. The diffusion coefficient is equal to 5.0×10^{-10} m^2/s, and 8.0×10^{-14} kg is transported along the tube in 15 s. What is the difference in the concentration levels of sucrose at the two ends of the tube?

68. Glycerin in water diffuses along a horizontal column that has a cross-sectional area of 2.0 cm^2. The concentration gradient is 3.0×10^{-2} kg/m^4, and the diffusion rate is found to be 5.7×10^{-15} kg/s. Determine the diffusion coefficient.

69. The viscous force on an oil drop is measured to be 3.0×10^{-13} N when the drop is falling through air with a speed of 4.5×10^{-4} m/s. If the radius of the drop is 2.5×10^{-6} m, what is the viscosity of air?

70. Small spheres of diameter 1.00 mm fall through 20°C water with a terminal speed of 1.10 cm/s. Calculate the density of the spheres.

ADDITIONAL PROBLEMS

71. An iron block of volume 0.20 m^3 is suspended from a spring scale and immersed in a flask of water. Then the iron block is removed, and an aluminum block of the same volume replaces it. (a) In which case is the buoyant force the greatest, for the iron block or the aluminum block? (b) In which case does the spring scale read the largest value? (c) Use the known densities of these materials to calculate the quantities requested in parts (a) and (b). Are your calculations consistent with your previous answers to part (a) and (b)?

72. Take the density of blood to be ρ and the distance between the feet and the heart to be h_H. Ignore the flow of blood. (a) Show that the difference in blood pressure between the feet and the heart is given by $P_F - P_H = \rho g h_H$. (b) Take the density of blood to be 1.05×10^3 kg/m^3 and the distance between the heart and the feet to be 1.20 m. Find the difference in blood pressure between these two points. This problem indicates that pumping blood from the extremities is very difficult for the heart. The veins in the legs have valves in them that open when blood is pumped toward the heart and close when blood flows away from the heart. Also, pumping action produced by physical activities such as walking and breathing assists the heart.

73. PhysicsNow™ The approximate inside diameter of the aorta is 0.50 cm; that of a capillary is 10 μm. The approximate average blood flow speed is 1.0 m/s in the aorta and 1.0 cm/s in the capillaries. If all the blood in the aorta eventually flows through the capillaries, estimate the number of capillaries in the circulatory system.

74. Superman attempts to drink water through a very long vertical straw. With his great strength, he achieves maximum possible suction. The walls of the straw don't collapse. (a) Find the maximum height through which he can lift the water? (b) Still thirsty, the Man of Steel repeats his attempt on the Moon, which has no atmosphere. Find the difference between the water levels inside and outside the straw.

75. The human brain and spinal cord are immersed in the cerebrospinal fluid. The fluid is normally continuous between the cranial and spinal cavities and exerts a pressure of 100 to 200 mm of H_2O above the prevailing atmospheric pressure. In medical work, pressures are often measured in units of mm of H_2O because body fluids, including the cerebrospinal fluid, typically have nearly the same density as water. The pressure of the cerebrospinal fluid can be measured by means of a *spinal tap.* A hollow tube is inserted into the spinal column, and the height to which the fluid rises is observed, as shown in Figure P9.75. If the fluid rises to a height of 160 mm, we write its gauge pressure as 160 mm H_2O. (a) Express this pressure in pascals, in atmospheres, and in millimeters of mercury. (b) Sometimes it is necessary to determine whether an accident victim has suffered a crushed vertebra that is blocking the flow of cerebrospinal fluid in the spinal column. In other cases, a physician may suspect that a tumor or other growth is blocking the spinal column and inhibiting the flow of cerebrospinal fluid. Such conditions can be investigated by means of the *Queckensted test.* In this procedure, the veins in the patient's neck are compressed, to make the blood pressure rise in the brain. The increase in pressure in the blood vessels is transmitted to the cerebrospinal fluid. What should be the normal effect on the height of the fluid in the spinal tap? (c) Suppose that compressing the veins had no effect on the level of the fluid. What might account for this phenomenon?

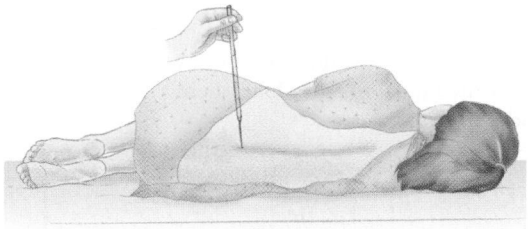

Figure P9.75

76. Determining the density of a fluid has many important applications. A car battery contains sulfuric acid, and the battery will not function properly if the acid density is too low. Similarly, the effectiveness of antifreeze in your car's engine coolant depends on the density of the mixture (usually ethylene glycol and water). When you donate blood to a blood bank, its screening includes a determination of the density of the blood, since higher density correlates with higher hemoglobin content. A *hydrometer* is an instrument used to determine the density of a liquid. A simple one is sketched in Figure P9.76. The bulb of a syringe is squeezed and released to lift a sample of the liquid of interest into a tube containing a calibrated rod of known density. (Assume the rod is cylindrical.) The rod, of length L and average density ρ_0, floats partially immersed in the liquid of density ρ. A length h of the rod protrudes above the surface of the liquid. Show that the density of the liquid is given by

$$\rho = \frac{\rho_0 L}{L - h}$$

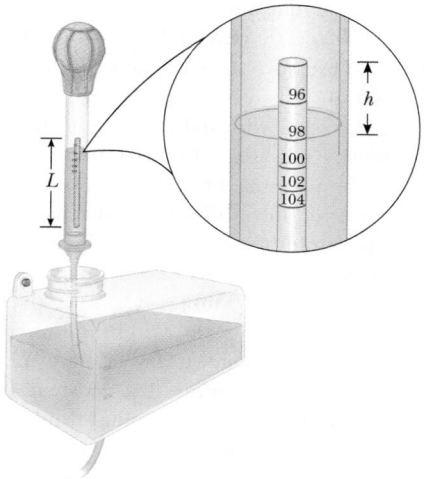

Figure P9.76

77. A 600-kg weather balloon is designed to lift a 4 000-kg package. What volume should the balloon have after being inflated with helium at standard temperature and pressure (see Table 9.3) in order that the total load can be lifted?

78. A helium-filled balloon is tied to a 2.0-m-long, 0.050-kg string. The balloon is spherical with a radius of 0.40 m. When released, it lifts a length h of the string and then remains in equilibrium, as in Figure P9.78. Determine the value of h. When deflated, the balloon has a mass of 0.25 kg. [*Hint:* Only that part of the string above the floor contributes to the load being held up by the balloon.]

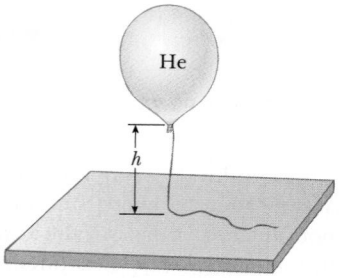

Figure P9.78

79. A block of wood weighs 50.0 N in air. A sinker is attached to the block, and the weight of the wood–sinker combination is 200 N when the sinker alone is immersed in water. Finally, the wood–sinker combination is completely immersed, and the weight is measured to be 140 N. Find the density of the block.

80. A U-tube open at both ends is partially filled with water (Fig. P9.80a). Oil ($\rho = 750\ \text{kg/m}^3$) is then poured into the right arm and forms a column $L = 5.00$ cm high (Fig. P9.80b). (a) Determine the difference h in the heights of the two liquid surfaces. (b) The right arm is then shielded from any air motion while air is blown across the top of the left arm until the surfaces of the two liquids are at the same height (Fig. 9.80c). Determine the speed of the air being blown across the left arm. Assume that the density of air is $1.29\ \text{kg/m}^3$.

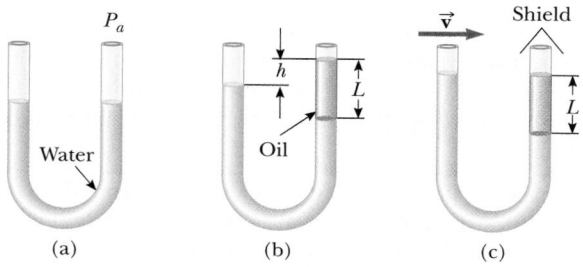

Figure P9.80

81. Figure P9.81 shows a water tank with a valve. If the valve is opened, what is the maximum height attained by the stream of water coming out of the right side of the tank? Assume that $h = 10.0$ m, $L = 2.00$ m, and $\theta = 30.0°$. Assume also that the cross-sectional area at A is very large compared with that at B.

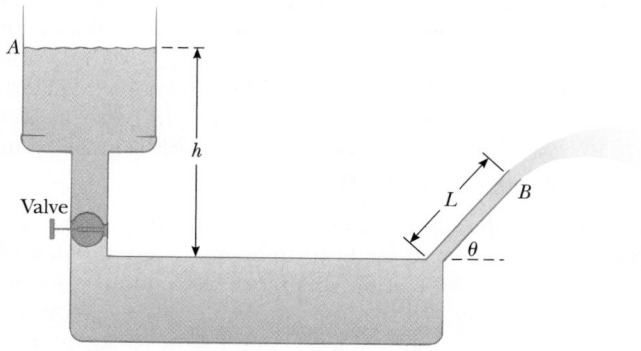

Figure P9.81

82. A solid copper ball with a diameter of 3.00 m at sea level is placed on the bottom of the ocean, at a depth of 10.0 km. If the density of sea water is $1\,030\ \text{kg/m}^3$, how much does the diameter of the ball decrease when it reaches bottom?

83. A 1.0-kg hollow ball with a radius of 0.10 m and filled with air is released from rest at the bottom of a 2.0-m-deep

pool of water. How high above the water does the ball shoot upward? Neglect all frictional effects, and neglect changes in the ball's motion when it is only partially submerged.

84. In about 1657, Otto von Guericke, inventor of the air pump, evacuated a sphere made of two brass hemispheres (Fig. P9.84). Two teams of eight horses each could pull the hemispheres apart only on some trials and even then with the greatest difficulty. (a) Show that the force required to pull the evacuated hemispheres apart is $\pi R^2(P_0 - P)$, where R is the radius of the hemispheres and P is the pressure inside the sphere, which is much less than atmospheric pressure P_0. (b) Determine the required force if $P = 0.10\ P_0$ and $R = 0.30$ m.

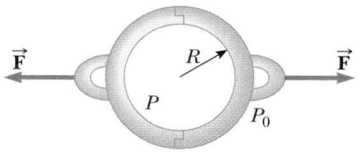

The Granger Collection

Figure P9.84 The colored engraving, dated 1672, illustrates Otto von Guericke's demonstration of the force due to air pressure, as performed before Emporer Ferdinand III in 1657.

85. A 2.0-cm-thick bar of soap is floating on a water surface so that 1.5 cm of the bar is under water. Bath oil of specific gravity 0.60 is poured into the water and floats on top of it. What is the depth of the oil layer when the top of the soap is just level with the upper surface of the oil?

86. A cube of ice whose edge is 20.0 mm is floating in a glass of ice-cold water with one of its faces parallel to the water's surface. (a) How far below the water surface is the bottom face of the block? (b) Ice-cold ethyl alcohol is gently poured onto the water surface to form a layer 5.00 mm thick above the water. When the ice cube attains hydrostatic equilibrium again, what will be the distance from the top of the water to the bottom face of the block? (c) Additional cold ethyl alcohol is poured onto the water surface until the top surface of the alcohol coincides with the top surface of the ice cube (in hydrostatic equilibrium). How thick is the required layer of ethyl alcohol?

87. A water tank open to the atmosphere at the top has two small holes punched in its side, one above the other. The holes are 5.00 cm and 12.0 cm above the floor. How high does water stand in the tank if the two streams of water hit the floor at the same place?

88. Oil having a density of 930 kg/m^3 floats on water. A rectangular block of wood 4.00 cm high and with a density of 960 kg/m^3 floats partly in the oil and partly in the water. The oil completely covers the block. How far below the interface between the two liquids is the bottom of the block?

89. A hollow object with an average density of 900 kg/m^3 floats in a pan containing 500 cm^3 of water. Ethanol is added to the water and mixed into it until the object is just on the verge of sinking. What volume of ethanol has been added? (Disregard the loss of volume caused by mixing.)

90. A walkway suspended across a hotel lobby is supported at numerous points along its edges by a vertical cable above each point and a vertical column underneath. The steel cable is 1.27 cm in diameter and is 5.75 m long before loading. The aluminum column is a hollow cylinder with an inside diameter of 16.14 cm, an outside diameter of 16.24 cm, and an unloaded length of 3.25 m. When the walkway exerts a load force of 8 500 N on one of the support points, through what distance does the point move down?

ACTIVITIES

A.1. You will need a large spring scale, a clear container partially filled with water, and a few cylinders of the same size, but made of different materials. Your instructor may be able to supply these items. Measure the volume of each cylinder. Hang a cylinder on the scale, record the reading, and start lowering the cylinder into the water. What happens to the reading on the scale as more of the cylinder is submerged? Why does the reading behave as it does? Is the effect independent of the type of material used for the cylinder? Record the reading for a cylinder when it is not submerged and when it is completely immersed. How can you use these two readings to verify Archimedes's principle?

A.2. Place an egg at the bottom of a container of fresh water. Now use a funnel to slowly add a salt solution to the water. You will observe that the egg begins to rise to the surface. Use Archimedes's principle to explain your observation.

A.3. Suppose you have the following collection of objects: a pencil, a coin, an empty plastic box for a tape cassette with its edges taped shut, a needle, an unopened can of soft drink, and an empty can of soft drink. Which of these objects do you expect will float and which will sink in water? Will it make a difference if you carefully place the object with its largest surface on the surface of the water? In which cases? Explain your reasoning. After you have written your answer, perform the experiments and compare the results with your predictions. (Activity 3 is courtesy of Edward F. Redish. For more problems of this type, see http://www.physics.umd.edu/perg/.)

© Jim Sugar/Corbis

High temperatures inside a volcano turn water into a high pressure steam. Unless the steam and other gases vent into the atmosphere, pressure can build until a catastrophic explosion results.

CHAPTER

10

OUTLINE

10.1 Temperature and the Zeroth Law of Thermodynamics

10.2 Thermometers and Temperature Scales

10.3 Thermal Expansion of Solids and Liquids

10.4 Macroscopic Description of an Ideal Gas

10.5 The Kinetic Theory of Gases

Thermal Physics

How can trapped water blow off the top of a volcano in a giant explosion? What causes a sidewalk or road to fracture and buckle spontaneously when the temperature changes? How can thermal energy be harnessed to do work, running the engines that make everything in modern living possible?

Answering these and related questions is the domain of **thermal physics**, the study of temperature, heat, and how they affect matter. Quantitative descriptions of thermal phenomena require careful definitions of the concepts of temperature, heat, and internal energy. Heat leads to changes in internal energy and thus to changes in temperature, which cause the expansion or contraction of matter. Such changes can damage roadways and buildings, create stress fractures in metal, and render flexible materials stiff and brittle, the latter resulting in compromised O-rings and the *Challenger* disaster. Changes in internal energy can also be harnessed for transportation, construction, and food preservation.

Gases are critical in the harnessing of thermal energy to do work. Within normal temperature ranges, a gas acts like a large collection of non-interacting point particles, called an ideal gas. Such gases can be studied on either a macroscopic or microscopic scale. On the macroscopic scale, the pressure, volume, temperature, and number of particles associated with a gas can be related in a single equation known as the ideal gas law. On the microscopic scale, a model called the kinetic theory of gases pictures the components of a gas as small particles. This model will enable us to understand how processes on the atomic scale affect macroscopic properties like pressure, temperature, and internal energy.

10.1 TEMPERATURE AND THE ZEROTH LAW OF THERMODYNAMICS

Temperature is commonly associated with how hot or cold an object feels when we touch it. While our senses provide us with qualitative indications of temperature, they are unreliable and often misleading. A metal ice tray feels colder to the hand, for example, than a package of frozen vegetables at the same temperature, due to the fact metals conduct thermal energy more rapidly than a cardboard package. What we need is a reliable and reproducible method of making quantitative measurements that establish the relative "hotness" or "coldness" of objects—a method related solely to temperature. Scientists have developed a variety of thermometers for making such measurements.

When placed in contact with each other, two objects at different initial temperatures will eventually reach a common intermediate temperature. If a cup of hot coffee is cooled with an ice cube, for example, the ice rises in temperature and eventually melts while the temperature of the coffee decreases.

Understanding the concept of temperature requires understanding *thermal contact* and *thermal equilibrium.* Two objects are in **thermal contact** if energy can be exchanged between them. Two objects are in **thermal equilibrium** if they are in thermal contact and there is no net exchange of energy.

The exchange of energy between two objects because of differences in their temperatures is called **heat**, a concept examined in more detail in Chapter 11.

Using these ideas, we can develop a formal definition of temperature. Consider two objects A and B that are not in thermal contact with each other, and a third object C that acts as a **thermometer**—a device calibrated to measure the temperature of an object. We wish to determine whether A and B would be in thermal equilibrium if they were placed in thermal contact. The thermometer (object C) is first placed in thermal contact with A until thermal equilibrium is reached, as in Figure 10.1a, whereupon the reading of the thermometer is recorded. The thermometer is then placed in thermal contact with B, and its reading is again recorded at equilibrium (Fig. 10.1b). If the two readings are the same, then A and B are in thermal equilibrium with each other. If A and B are placed in thermal contact with each other, as in Figure 10.1c, there is no net transfer of energy between them.

We can summarize these results in a statement known as the **zeroth law of thermodynamics (the law of equilibrium)**:

Zeroth law of thermodynamics ▶

> If objects A and B are separately in thermal equilibrium with a third object C, then A and B are in thermal equilibrium with each other.

This statement is important because it makes it possible to define **temperature**. We can think of temperature as the property that determines whether or not an object is in thermal equilibrium with other objects. **Two objects in thermal equilibrium with each other are at the same temperature.**

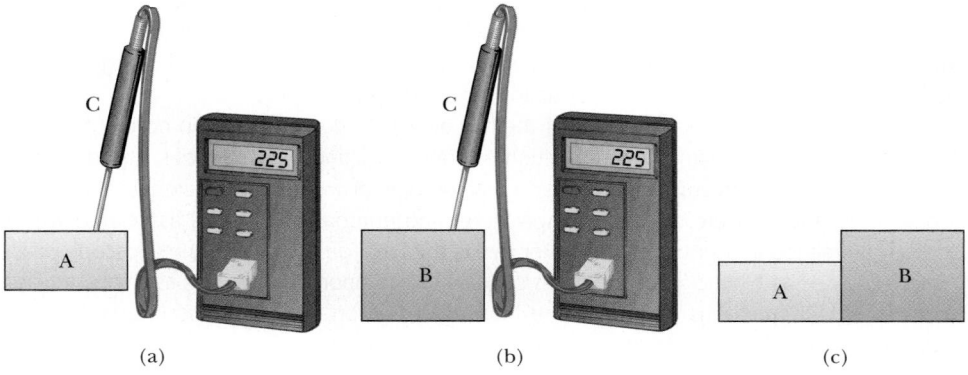

Figure 10.1 The zeroth law of thermodynamics. (a) and (b): If the temperatures of A and B are found to be the same as measured by object C (a thermometer), no energy will be exchanged between them when they are placed in thermal contact with each other, as in (c).

Quick Quiz 10.1

Two objects with different sizes, masses, and temperatures are placed in thermal contact. Choose the best answer: Energy travels (a) from the larger object to the smaller object (b) from the object with more mass to the one with less mass (c) from the object at higher temperature to the object at lower temperature.

10.2 THERMOMETERS AND TEMPERATURE SCALES

Thermometers are devices used to measure the temperature of an object or a system. When a thermometer is in thermal contact with a system, energy is exchanged until the thermometer and the system are in thermal equilibrium with each other. For accurate readings, the thermometer must be much smaller than the system, so that the energy the thermometer gains or loses doesn't significantly alter the energy content of the system. All thermometers make use of some physical property that changes with temperature and can be calibrated to make the temperature measurable. Some of the physical properties used are (1) the volume of a liquid, (2) the length of a solid, (3) the pressure of a gas held at constant volume, (4) the volume of a gas held at constant pressure, (5) the electric resistance of a conductor, and (6) the color of a very hot object.

One common thermometer in everyday use consists of a mass of liquid—usually mercury or alcohol—that expands into a glass capillary tube when its temperature rises (Fig. 10.2). In this case the physical property that changes is the volume of a liquid. To serve as an effective thermometer, the change in volume of the liquid with change in temperature must be very nearly constant over the temperature ranges of interest. When the cross-sectional area of the capillary tube is constant as well, the change in volume of the liquid varies linearly with its length along the tube. We can then define a temperature in terms of the length of the liquid column. The thermometer can be calibrated by placing it in thermal contact with environments that remain at constant temperature. One such environment is a mixture of water and ice in thermal equilibrium at atmospheric pressure. Another commonly used system is a mixture of water and steam in thermal equilibrium at atmospheric pressure.

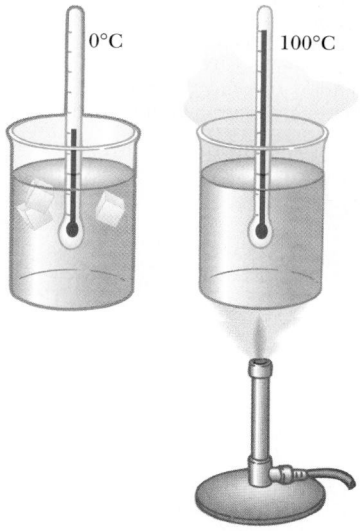

Figure 10.2 Schematic diagram of a mercury thermometer. Because of thermal expansion, the level of the mercury rises as the temperature of the mercury changes from 0°C (the ice point) to 100°C (the steam point).

Once we have marked the ends of the liquid column for our chosen environment on our thermometer, we need to define a scale of numbers associated with various temperatures. An example of such a scale is the **Celsius temperature scale**, formerly called the centigrade scale. On the Celsius scale, the temperature of the ice–water mixture is defined to be zero degrees Celsius, written 0°C and called the **ice point** or **freezing point** of water. The temperature of the water–steam mixture is defined as 100°C, called the **steam point** or **boiling point** of water. Once the ends of the liquid column in the thermometer have been marked at these two points, the distance between marks is divided into 100 equal segments, each corresponding to a change in temperature of one degree Celsius.

Thermometers calibrated in this way present problems when extremely accurate readings are needed. For example, an alcohol thermometer calibrated at the ice and steam points of water might agree with a mercury thermometer only at the calibration points. Because mercury and alcohol have different thermal expansion properties, when one indicates a temperature of 50°C, say, the other may indicate a slightly different temperature. The discrepancies between different types of thermometers are especially large when the temperatures to be measured are far from the calibration points.

The Constant-Volume Gas Thermometer and the Kelvin Scale

We can construct practical thermometers such as the mercury thermometer, but these types of thermometers don't define temperature in a fundamental way. One thermometer, however, *is* more fundamental, and offers a way to define temperature and relate it directly to internal energy: the **gas thermometer**. In a gas thermometer, the temperature readings are nearly independent of the substance used in the thermometer. One type of gas thermometer is the constant-volume unit shown in Figure 10.3.

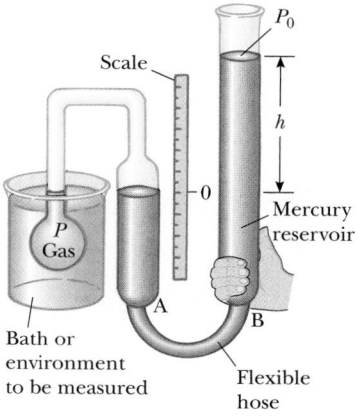

Figure 10.3 A constant-volume gas thermometer measures the pressure of the gas contained in the flask immersed in the bath. The volume of gas in the flask is kept constant by raising or lowering reservoir B to keep the mercury level constant.

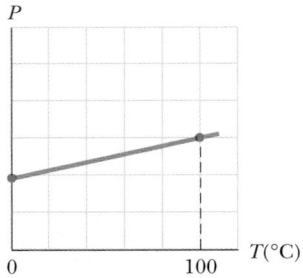

Figure 10.4 A typical graph of pressure versus temperature taken with a constant-volume gas thermometer. The dots represent known reference temperatures (the ice and the steam points of water).

The behavior observed in this device is the variation of pressure with temperature of a fixed volume of gas. When the constant-volume gas thermometer was developed, it was calibrated using the ice and steam points of water as follows (a different calibration procedure, to be discussed shortly, is now used): The gas flask is inserted into an ice–water bath, and mercury reservoir B is raised or lowered until the volume of the confined gas is at some value, indicated by the zero point on the scale. The height h, the difference between the levels in the reservoir and column A, indicates the pressure in the flask at 0°C. The flask is inserted into water at the steam point, and reservoir B is readjusted until the height in column A is again brought to zero on the scale, ensuring that the gas volume is the same as it had been in the ice bath (hence the designation "constant-volume"). A measure of the new value for h gives a value for the pressure at 100°C. These pressure and temperature values are then plotted on a graph, as in Figure 10.4. The line connecting the two points serves as a calibration curve for measuring unknown temperatures. If we want to measure the temperature of a substance, we place the gas flask in thermal contact with the substance and adjust the column of mercury until the level in column A returns to zero. The height of the mercury column tells us the pressure of the gas, and we could then find the temperature of the substance from the calibration curve.

Now suppose that temperatures are measured with various gas thermometers containing different gases. Experiments show that the thermometer readings are nearly independent of the type of gas used, as long as the gas pressure is low and the temperature is well above the point at which the gas liquifies.

We can also perform the temperature measurements with the gas in the flask at different starting pressures at 0°C. As long as the pressure is low, we will generate straight-line calibration curves for each starting pressure, as shown for three experimental trials (solid lines) in Figure 10.5.

If the curves in Figure 10.5 are extended back toward negative temperatures, we find a startling result: In every case, regardless of the type of gas or the value of the low starting pressure, **the pressure extrapolates to zero when the temperature is − 273.15°C**. This fact suggests that this particular temperature is universal in its importance, because it doesn't depend on the substance used in the thermometer. In addition, because the lowest possible pressure is $P = 0$, a perfect vacuum, the temperature − 273.15°C must represent a lower bound for physical processes. We define this temperature as **absolute zero**.

Absolute zero is used as the basis for the **Kelvin temperature scale**, which sets − 273.15°C as its zero point (0 K). The size of a "degree" on the Kelvin scale is chosen to be identical to the size of a degree on the Celsius scale. The relationship between these two temperature scales is

$$T_C = T - 273.15 \qquad [10.1]$$

where T_C is the Celsius temperature and T is the Kelvin temperature (sometimes called the **absolute temperature**).

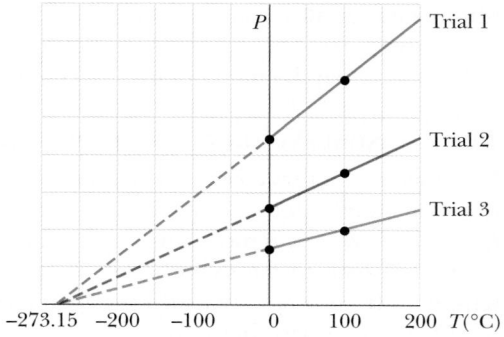

Figure 10.5 Pressure versus temperature for experimental trials in which gases have different pressures in a constant-volume gas thermometer. Note that for all three trials the pressure extrapolates to zero at the temperature − 273.15°C.

Technically, Equation 10.1 should have units on the right-hand side so that it reads $T_C = T\ °C/K - 273.15\ °C$. The units are rather cumbersome in this context, so we will usually suppress them in such calculations except in the final answer. (This will also be the case when discussing the Celsius and Fahrenheit scales).

Early gas thermometers made use of ice and steam points according to the procedure just described. These points are experimentally difficult to duplicate, however, because they are pressure-sensitive. Consequently, a procedure based on two new points was adopted in 1954 by the International Committee on Weights and Measures. The first point is absolute zero. The second point is **the triple point of water, which is the single temperature and pressure at which water, water vapor, and ice can coexist in equilibrium**. This point is a convenient and reproducible reference temperature for the Kelvin scale; it occurs at a temperature of 0.01°C and a pressure of 4.58 mm of mercury. The temperature at the triple point of water on the Kelvin scale occurs at 273.16 K. Therefore, **the SI unit of temperature, the kelvin, is defined as 1/273.16 of the temperature of the triple point of water**. Figure 10.6 shows the Kelvin temperatures for various physical processes and structures. Absolute zero has been closely approached but never achieved.

◀ The kelvin

What would happen to a substance if its temperature could reach 0 K? As Figure 10.5 indicates, the substance would exert zero pressure on the walls of its container (assuming that the gas doesn't liquefy or solidify on the way to absolute zero). In Section 10.5 we show that the pressure of a gas is proportional to the kinetic energy of the molecules of that gas. According to classical physics, therefore, the kinetic energy of the gas would go to zero, and there would be no motion at all of the individual components of the gas. According to quantum theory, however (to be discussed in Chapter 27), the gas would always retain some residual energy, called the *zero-point energy*, at that low temperature.

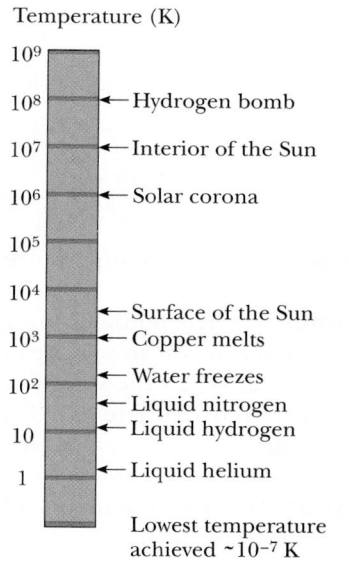

Figure 10.6 Absolute temperatures at which various selected physical processes take place. Note that the scale is logarithmic.

The Celsius, Kelvin, and Fahrenheit Temperature Scales

Equation 10.1 shows that the Celsius temperature T_C is shifted from the absolute (Kelvin) temperature T by 273.15. Because the size of a Celsius degree is the same as a kelvin, a temperature difference of 5°C is equal to a temperature difference of 5 K. The two scales differ only in the choice of zero point. The ice point (273.15 K) corresponds to 0.00°C, and the steam point (373.15 K) is equivalent to 100.00°C.

The most common temperature scale in use in the United States is the Fahrenheit scale. It sets the temperature of the ice point at 32°F and the temperature of the steam point at 212°F. The relationship between the Celsius and Fahrenheit temperature scales is

$$T_F = \tfrac{9}{5}\,T_C + 32 \qquad \textbf{[10.2a]}$$

For example, a temperature of 50.0°F corresponds to a Celsius temperature of 10.0°C and an absolute temperature of 283 K.

Equation 10.2 can be inverted to give Celsius temperatures in terms of Fahrenheit temperatures:

$$T_C = \tfrac{5}{9}(T_F - 32) \qquad \textbf{[10.2b]}$$

Equation 10.2 can also be used to find a relationship between changes in temperature on the Celsius and Fahrenheit scales. In a problem at the end of the chapter you will be asked to show that if the Celsius temperature changes by ΔT_C, the Fahrenheit temperature changes by the amount

$$\Delta T_F = \tfrac{9}{5}\,\Delta T_C \qquad \textbf{[10.3]}$$

Figure 10.7 (page 326) compares the three temperature scales we have discussed.

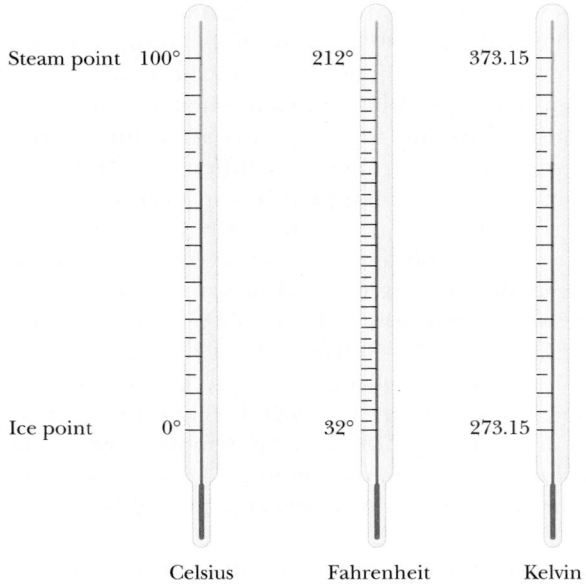

Figure 10.7 A comparison of the Celsius, Fahrenheit, and Kelvin temperature scales.

EXAMPLE 10.1 Skin Temperature

Goal Apply the temperature conversion formulas.

Problem The temperature gradient between the skin and the air is regulated by cutaneous (skin) blood flow. If the cutaneous blood vessels are constricted, the skin temperature and the temperature of the environment will be about the same. When the vessels are dilated, more blood is brought to the surface. Suppose that during dilation the skin warms from 72.0°F to 84.0°F. **(a)** Convert these temperatures to Celsius and find the difference. **(b)** Convert the temperatures to Kelvin, again finding the difference.

Strategy This is a matter of applying the conversion formulas, Equations 10.1 and 10.2. For part (b) it's easiest to use the answers for Celsius, rather than develop another set of conversion equations.

Solution

(a) Convert the temperatures from Fahrenheit to Celsius and find the difference.

Convert the lower temperature, using Equation 10.2b. $T_C = \frac{5}{9}(T_F - 32) = \frac{5}{9}(72 - 32) = 22°\text{C}$

Convert the upper temperature: $T_C = \frac{5}{9}(T_F - 32) = \frac{5}{9}(84 - 32) = 29°\text{C}$

Find the difference of the two temperatures: $\Delta T_C = 29°\text{C} - 22°\text{C} = 7°\text{C}$

(b) Convert the temperatures from Fahrenheit to Kelvin and find their difference.

Convert the lower temperature, using the answers for Celsius found in part (a): $T_C = T - 273.15 \quad \rightarrow \quad T = T_C + 273.15$

$T = 22 + 273.15 = 285\ \text{K}$

Convert the upper temperature: $T = 29 + 273.15 = 292\ \text{K}$

Find the difference of the two temperatures: $\Delta T = 292\ \text{K} - 285\ \text{K} = 7\ \text{K}$

Remark The change in temperature in Kelvin and Celsius is the same, as it should be.

Exercise 10.1

Core body temperature can rise from 98.6°F to 107°F during extreme exercise, such as a marathon run. Such elevated temperatures can also be caused by viral or bacterial infections or tumors and are dangerous if sustained. (a) Convert the given temperatures to Celsius and find the difference. (b) Convert the temperatures to Kelvin, again finding the difference.

Answer (a) 37.0°C, 41.7°C, 4.7°C (b) 310.2 K, 314.9 K, 4.7 K

EXAMPLE 10.2 Extraterrestrial Temperature Scale

Goal Understand how to relate different temperature scales.

Problem An extraterrestrial scientist invents a temperature scale such that water freezes at − 75°E and boils at 325°E, where E stands for an extraterrestrial scale. Find an equation that relates temperature in °E to temperature in °C.

Strategy Using the given data, find the ratio of the number of °E between the two temperatures to the number of °C. This ratio will be the same as a similar ratio for any other such process—say, from the freezing point to an unknown temperature—corresponding to T_E and T_C. Setting the two ratios equal and solving for T_E in terms of T_C yields the desired relationship.

Solution

Find the change in temperature in °E between the freezing and boiling points of water:

$$\Delta T_E = 325°\text{E} - (-75°\text{E}) = 400°\text{E}$$

Find the change in temperature in °C between the freezing and boiling points of water:

$$\Delta T_C = 100°\text{C} - 0°\text{C} = 100°\text{C}$$

Form the ratio of these two quantities.

$$\frac{\Delta T_E}{\Delta T_C} = \frac{400°\text{E}}{100°\text{C}} = 4\,\frac{°\text{E}}{°\text{C}}$$

This ratio is the same between any other two temperatures—say, from the freezing point to an unknown final temperature. Set the two ratios equal to each other:

$$\frac{\Delta T_E}{\Delta T_C} = \frac{T_E - (-75°\text{E})}{T_C - 0°\text{C}} = 4\,\frac{°\text{E}}{°\text{C}}$$

Solve for T_E:

$$T_E - (-75°\text{E}) = 4(°\text{E}/°\text{C})(T_C - 0°\text{C})$$

$$T_E = 4T_C - 75$$

Remark The relationship between any other two temperatures scales can be derived in the same way.

Exercise 10.2

Find the equation converting °F to °E.

Answer $T_E = \frac{20}{9}T_F - 146$

10.3 THERMAL EXPANSION OF SOLIDS AND LIQUIDS

Our discussion of the liquid thermometer made use of one of the best-known changes that occur in most substances: As temperature of the substance increases, its volume increases. This phenomenon, known as **thermal expansion**, plays an important role in numerous applications. Thermal expansion joints, for example,

George Semple

(a)

George Semple

(b)

Figure 10.8 (a) Thermal expansion joints are used to separate sections of roadways on bridges. Without these joints, the surfaces would buckle due to thermal expansion on very hot days or crack due to contraction on very cold days. (b) The long, vertical joint is filled with a soft material that allows the wall to expand and contract as the temperature of the bricks changes.

must be included in buildings, concrete highways, and bridges to compensate for changes in dimensions with variations in temperature (Fig. 10.8).

The overall thermal expansion of an object is a consequence of the change in the average separation between its constituent atoms or molecules. To understand this idea, consider how the atoms in a solid substance behave. These atoms are located at fixed equilibrium positions; if an atom is pulled away from its position, a restoring force pulls it back. We can imagine that the atoms are particles connected by springs to their neighboring atoms. (See Fig. 9.1 in the previous chapter.) If an atom is pulled away from its equilibrium position, the distortion of the springs provides a restoring force.

At ordinary temperatures, the atoms vibrate around their equilibrium positions with an amplitude (maximum distance from the center of vibration) of about 10^{-11} m, with an average spacing between the atoms of about 10^{-10} m. As the temperature of the solid increases, the atoms vibrate with greater amplitudes and the average separation between them increases. Consequently, the solid as a whole expands.

If the thermal expansion of an object is sufficiently small compared with the object's initial dimensions, then the change in any dimension is, to a good approximation, proportional to the first power of the temperature change. Suppose an object has an initial length L_0 along some direction at some temperature T_0. Then the length increases by ΔL for a change in temperature ΔT. So for small changes in temperature,

$$\Delta L = \alpha L_0 \, \Delta T \qquad [10.4]$$

or

$$L - L_0 = \alpha L_0 (T - T_0)$$

where L is the object's final length T is its final temperature, and the proportionality constant α is called the **coefficient of linear expansion** for a given material and has units of $(°C)^{-1}$.

Table 10.1 lists the coefficients of linear expansion for various materials. Note that for these materials α is positive, indicating an increase in length with increasing temperature.

Thermal expansion affects the choice of glassware used in kitchens and laboratories. If hot liquid is poured into a cold container made of ordinary glass, the container may well break due to thermal stress. The inside surface of the glass becomes hot and expands, while the outside surface is at room temperature, and ordinary glass may not withstand the difference in expansion without breaking. Pyrex® glass has a coefficient of linear expansion of about one-third that of ordinary glass, so the thermal stresses are smaller. Kitchen measuring cups and laboratory beakers are often made of Pyrex so they can be used with hot liquids.

TIP 10.1 Coefficients of Expansion are not Constants

The coefficients of expansion can vary somewhat with temperature, so the given coefficients are actually averages.

TABLE 10.1

Average Coefficients of Expansion for Some Materials Near Room Temperature

Material	Average Coefficient of Linear Expansion $[(°C)^{-1}]$	Material	Average Coefficient of Volume Expansion $[(°C)^{-1}]$
Aluminum	24×10^{-6}	Ethyl alcohol	1.12×10^{-4}
Brass and bronze	19×10^{-6}	Benzene	1.24×10^{-4}
Copper	17×10^{-6}	Acetone	1.5×10^{-4}
Glass (ordinary)	9×10^{-6}	Glycerin	4.85×10^{-4}
Glass (Pyrex®)	3.2×10^{-6}	Mercury	1.82×10^{-4}
Lead	29×10^{-6}	Turpentine	9.0×10^{-4}
Steel	11×10^{-6}	Gasoline	9.6×10^{-4}
Invar (Ni-Fe alloy)	0.9×10^{-6}	Air	3.67×10^{-3}
Concrete	12×10^{-6}	Helium	3.665×10^{-3}

EXAMPLE 10.3 Expansion of a Railroad Track

Goal Apply the concept of linear expansion and relate it to stress.

Problem **(a)** A steel railroad track has a length of 30.000 m when the temperature is 0°C. What is its length on a hot day when the temperature is 40.0°C? **(b)** Suppose the track is nailed down so that it can't expand. What stress results in the track due to the temperature change?

Strategy (a) Apply the linear expansion equation, using Table 10.1 and Equation 10.4. (b) A track that cannot expand by ΔL due to external constraints is equivalent to compressing the track by ΔL, creating a stress in the track. Using the equation relating tensile stress to tensile strain together with the linear expansion equation, the amount of (compressional) stress can be calculated using Equation 9.3.

AP/Wide World Photos

(Example 10.3) Thermal expansion: The extreme heat of a July day in Asbury Park, New Jersey, caused these railroad tracks to buckle.

Solution

(a) Find the length of the track at 40.0°C.

Substitute given quantities into Equation 10.4, finding the change in length:

$$\Delta L = \alpha L_0 \Delta T = [11 \times 10^{-6}(°\text{C})^{-1}](30.000\text{ m})(40.0°\text{C}) = 0.013\text{ m}$$

Add the change to the original length to find the final length:

$$L = L_0 + \Delta L = 30.013\text{ m}$$

(b) Find the stress if the track cannot expand.

Substitute into Equation 9.3 to find the stress:

$$\frac{F}{A} = Y\frac{\Delta L}{L} = (2.00 \times 10^{11}\text{ Pa})\left(\frac{0.013\text{ m}}{30.0\text{ m}}\right) = 8.67 \times 10^{7}\text{ Pa}$$

Remarks Repeated heating and cooling is an important part of the weathering process that gradually wears things out, weakening structures over time.

Exercise 10.3

What is the length of the same railroad track on a cold winter day when the temperature is 0°F?

Answer 29.994 m

Applying Physics 10.1 Bimetallic Strips and Thermostats

How can different coefficients of expansion for metals be used as a temperature gauge and control electronic devices such as air conditioners?

Explanation When the temperatures of a brass rod and a steel rod of equal length are raised by the same amount from some common initial value, the brass rod expands more than the steel rod because brass has a larger coefficient of expansion than steel. A simple device that uses this principle is a **bimetallic strip**. Such strips can be found in the thermostats of certain home heating systems. The strip is made by securely bonding two different metals together. As the temperature of the strip increases, the two metals expand by different amounts and the strip bends, as in Figure 10.9 (page 330). The change in shape can make or break an electrical connection.

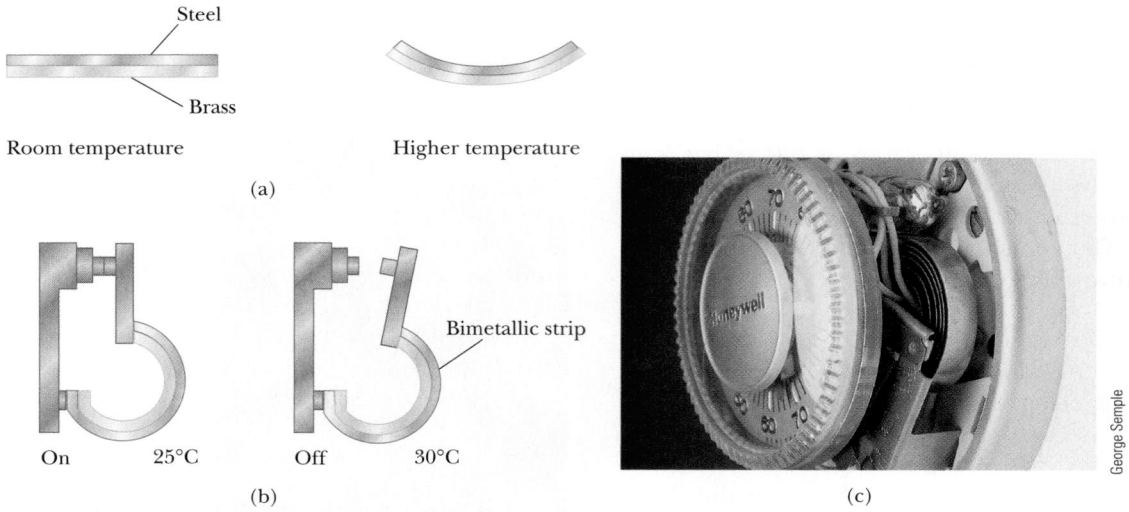

Figure 10.9 (Applying Physics 10.1) (a) A bimetallic strip bends as the temperature changes because the two metals have different coefficients of expansion. (b) A bimetallic strip used in a thermostat to break or make electrical contact. (c) The interior of a thermostat, showing the coiled bimetallic strip. Why do you suppose the strip is coiled?

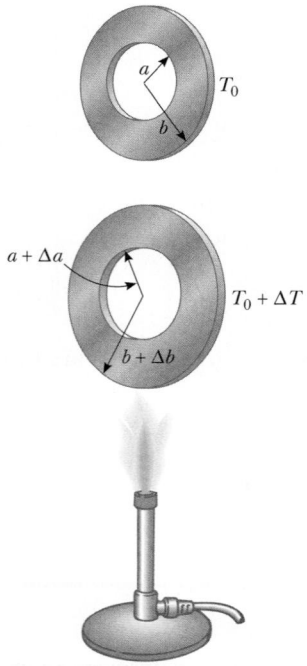

ACTIVE FIGURE 10.10
Thermal expansion of a homogeneous metal washer. As the washer is heated, all dimensions increase. (Note that the expansion is exaggerated in this figure.)

PhysicsNow™
Log into PhysicsNow at **www.cp7e.com**, and go to Active Figure 10.10 to compare expansions for various temperatures of the burner and various materials from which the washer is made.

It may be helpful to picture a thermal expansion as a magnification or a photographic enlargement. For example, as the temperature of a metal washer increases (Active Fig. 10.10), all dimensions, including the radius of the hole, increase according to Equation 10.4.

One practical application of thermal expansion is the common technique of using hot water to loosen a metal lid stuck on a glass jar. This works because the circumference of the lid expands more than the rim of the jar.

Because the linear dimensions of an object change due to variations in temperature, it follows that surface area and volume of the object also change. Consider a square of material having an initial length L_0 on a side and therefore an initial area $A_0 = L_0^2$. As the temperature is increased, the length of each side increases to

$$L = L_0 + \alpha L_0 \Delta T$$

The new area A is

$$A = L^2 = (L_0 + \alpha L_0 \Delta T)(L_0 + \alpha L_0 \Delta T) = L_0^2 + 2\alpha L_0^2 \Delta T + \alpha^2 L_0^2 (\Delta T)^2$$

The last term in this expression contains the quantity $\alpha \Delta T$ raised to the second power. Because $\alpha \Delta T$ is much less than one, squaring it makes it even smaller. Consequently, we can neglect this term to get a simpler expression:

$$A = L_0^2 + 2\alpha L_0^2 \Delta T$$

$$A = A_0 + 2\alpha A_0 \Delta T$$

so that

$$\Delta A = A - A_0 = \gamma A_0 \Delta T \qquad [10.5]$$

where $\gamma = 2\alpha$. The quantity γ (Greek letter gamma) is called the **coefficient of area expansion**.

EXAMPLE 10.4 Rings and Rods

Goal Apply the equation of area expansion.

Problem **(a)** A circular copper ring at 20.0°C has a hole with an area of 9.98 cm^2. What minimum temperature must it have so that it can be slipped onto a steel metal rod having a cross-sectional area of 10.0 cm^2? **(b)** Suppose the ring and the rod are heated simultaneously. What change in temperature of both will allow the ring to be slipped onto the end of the rod? (Assume no significant change in the coefficients of linear expansion over this temperature range.)

Strategy In part (a), finding the necessary temperature change is just a matter of substituting given values into Equation 10.5, the equation of area expansion. Remember that $\gamma = 2\alpha$. Part (b) is a little harder, because now the rod is also expanding. If the ring is to slip onto the rod, however, the final cross-sectional areas of both ring and rod must be equal. Write this condition in mathematical terms, using Equation 10.5 on both sides of the equation, and solve for ΔT.

Solution

(a) Find the temperature of the ring that will allow it to slip onto the rod.

Substitute the desired change in area into Equation 10.5, finding the necessary change in temperature:

$$\Delta A = \gamma A_0 \Delta T = [34 \times 10^{-6}\ (°C)^{-1}](9.98\ cm^2)(\Delta T)$$
$$= 0.02\ cm^2$$

Solve for ΔT, then add this change to the initial temperature to get the final temperature:

$$\Delta T = 58.9°C$$
$$T = T_0 + \Delta T = 20.0°C + 58.9°C = 78.9°C$$

(b) Increase the temperature of both ring and rod to remove the ring.

Set the final areas of the copper ring and steel rod equal to each other:

$$A_C + \Delta A_C = A_S + \Delta A_S$$

Substitute for each change in area, ΔA:

$$A_C + \gamma_C A_C \Delta T = A_S + \gamma_S A_S \Delta T$$

Rearrange terms to get ΔT on one side only, factor it out and solve:

$$\gamma_C A_C \Delta T - \gamma_S A_S \Delta T = A_S - A_C$$
$$(\gamma_C A_C - \gamma_S A_S)\ \Delta T = A_S - A_C$$
$$\Delta T = \frac{A_S - A_C}{\gamma_C A_C - \gamma_S A_S}$$
$$= \frac{10.0\ cm^2 - 9.98\ cm^2}{(34 \times 10^{-6}\ °C^{-1})(9.98\ cm^2) - (22 \times 10^{-6}\ °C^{-1})(10.0\ cm^2)}$$
$$\Delta T = 168°C$$

Exercise 10.4

A steel ring with a hole having area of 3.99 cm^2 is to be placed on an aluminum rod with cross-sectional area of 4.00 cm^2. Both rod and ring are initially at a temperature of 35.0°C. At what common temperature can the steel ring be slipped onto one end of the aluminum rod?

Answer −61°C

We can also show that the *increase in volume* of an object accompanying a change in temperature is

$$\Delta V = \beta V_0 \Delta T \qquad [10.6]$$

where β, the **coefficient of volume expansion**, is equal to 3α. (Note that $\gamma = 2\alpha$ and $\beta = 3\alpha$ only if the coefficient of linear expansion of the object is the same in all directions.) The proof of Equation 10.6 is similar to the proof of Equation 10.5.

As Table 10.1 indicates, each substance has its own characteristic coefficients of expansion.

APPLICATION
Rising Sea Levels

The thermal expansion of water has a profound influence on rising ocean levels. At current rates of global warming, scientists predict that about one-half of the expected rise in sea level will be caused by thermal expansion; the remainder will be due to the melting of polar ice.

Quick Quiz 10.2

If you quickly plunge a room-temperature mercury thermometer into very hot water, the mercury level will (a) go up briefly before reaching a final reading, (b) go down briefly before reaching a final reading, or (c) not change.

Quick Quiz 10.3

If you are asked to make a very sensitive glass thermometer, which of the following working fluids would you choose? (a) mercury (b) alcohol (c) gasoline (d) glycerin

Quick Quiz 10.4

Two spheres are made of the same metal and have the same radius, but one is hollow and the other is solid. The spheres are taken through the same temperature increase. Which sphere expands more? (a) solid sphere, (b) hollow sphere, (c) they expand by the same amount, or (d) not enough information to say.

EXAMPLE 10.5 Global Warming and Coastal Flooding

Goal Apply the volume expansion equation together with linear expansion.

Problem **(a)** Estimate the fractional change in the volume of Earth's oceans due to an average temperature change of 1°C. **(b)** Use the fact that the average depth of the ocean is 4.00×10^3 m to estimate the change in depth. Note that $\beta_{\text{water}} = 2.07 \times 10^{-4}(°\text{C})^{-1}$

Strategy In part (a), solve the volume expansion expression, Equation 10.6, for $\Delta V/V$. For part (b), use linear expansion to estimate the increase in depth. Neglect the expansion of landmasses, which would reduce the rise in sea level only slightly.

Solution

(a) Find the fractional change in volume.

Divide the volume expansion equation by V_0 and substitute:

$$\Delta V = \beta V_0 \, \Delta T$$

$$\left(\frac{\Delta V}{V_0}\right) = \beta \Delta T = (2.07 \times 10^{-4}(°\text{C})^{-1} \cdot (1°\text{C}) = \boxed{2 \times 10^{-4}}$$

(b) Find the approximate increase in depth.

Use the linear expansion equation. Divide the volume expansion coefficient of water by three to get the equivalent linear expansion coefficient:

$$\Delta L = \alpha L_0 \, \Delta T = \left(\frac{\beta}{3}\right) L_0 \, \Delta T$$

$$\Delta L = (6.90 \times 10^{-5}(°\text{C})^{-1}(4\,000 \text{ m})(1°\text{C}) \approx \boxed{0.3 \text{ m}}$$

Remarks Three-tenths of a meter may not seem significant, but combined with increased melting of the polar ice caps, some coastal areas could experience flooding. An increase of several degrees increases the value of ΔL several times and could significantly reduce the value of waterfront property.

Exercise 10.5

A 1.00-liter aluminum cylinder at 5.00°C is filled to the brim with gasoline at the same temperature. If the aluminum and gasoline are warmed to 65.0°C, how much of the gasoline spills out? [*Hint:* Be sure to account for the expansion of the container. Also, ignore the possibility of evaporation, and assume the volume coefficients are good to three digits.]

Answer The volume spilled is 53.3 cm^3. Forgetting to take into account the expansion of the cylinder results in a (wrong) answer of 57.6 cm^3.

The Unusual Behavior of Water

Liquids generally increase in volume with increasing temperature and have volume expansion coefficients about ten times greater than those of solids. Over a small temperature range, water is an exception to this rule, as we can see from its density-versus-temperature curve in Figure 10.11. As the temperature increases from 0°C to 4°C, water contracts, so its density increases. Above 4°C, water exhibits the expected expansion with increasing temperature. The density of water reaches its maximum value of 1 000 kg/m^3 at 4°C.

We can use this unusual thermal expansion behavior of water to explain why a pond freezes slowly from the top down. When the atmospheric temperature drops from 7°C to 6°C, say, the water at the surface of the pond also cools and consequently decreases in volume. This means the surface water is more dense than the water below it, which has not yet cooled nor decreased in volume. As a result, the surface water sinks and warmer water from below is forced to the surface to be cooled, a process called *upwelling*. When the atmospheric temperature is between 4°C and 0°C, however, the surface water expands as it cools, becoming less dense than the water below it. The sinking process stops, and eventually the surface water freezes. As the water freezes, the ice remains on the surface because ice is less dense than water. The ice continues to build up on the surface, and water near the bottom of the pool remains at 4°C. Further, the ice forms an insulating layer that slows heat loss from the underlying water, offering thermal protection for marine life.

Without buoyancy and the expansion of water upon freezing, life on Earth may not have been possible. If ice had been more dense than water, it would have sunk to the bottom of the ocean and built up over time. This could have led to a freezing of the oceans, turning the Earth into an icebound world similar to Hoth in the Star Wars epic *The Empire Strikes Back*.

APPLICATION

Bursting Pipes in Winter

The same peculiar thermal expansion properties of water sometimes cause pipes to burst in winter. As energy leaves the water through the pipe by heat and is transferred to the outside cold air, the outer layers of water in the pipe freeze first.

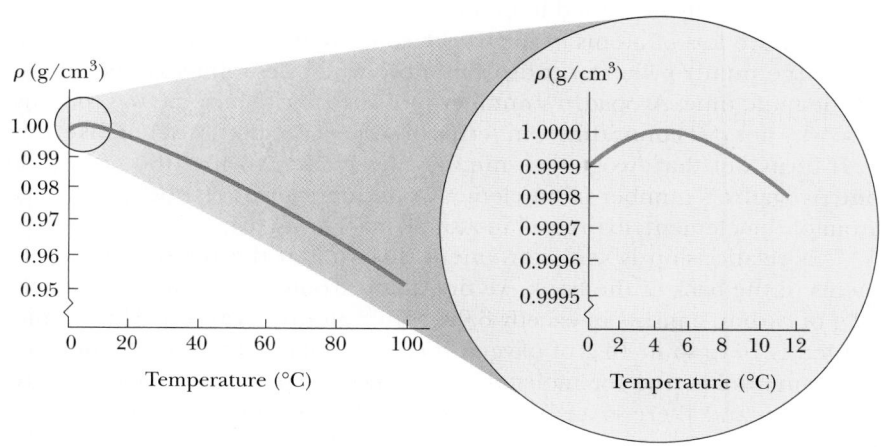

Figure 10.11 The density of water as a function of temperature. The inset at the right shows that the maximum density of water occurs at 4°C.

The continuing energy transfer causes ice to form ever closer to the center of the pipe. As long as there is still an opening through the ice, the water can expand as its temperature approaches 0°C or as it freezes into more ice, pushing itself into another part of the pipe. Eventually, however, the ice will freeze to the center somewhere along the pipe's length, forming a plug of ice at that point. If there is still liquid water between this plug and some other obstruction, such as another ice plug or a spigot, then no additional volume is available for further expansion and freezing. The pressure in the pipe builds and can rupture the pipe.

10.4 MACROSCOPIC DESCRIPTION OF AN IDEAL GAS

The properties of gases are important in a number of thermodynamic processes. Our weather is a good example of the types of processes that depend on the behavior of gases.

If we introduce a gas into a container, it expands to fill the container uniformly, with its pressure depending on the size of the container, the temperature, and the amount of gas. A larger container results in a lower pressure, while higher temperatures or larger amounts of gas result in a higher pressure. The pressure P, volume V, temperature T, and amount n of gas in a container are related to each other by an *equation of state.*

The equation of state can be very complicated, but is found experimentally to be relatively simple if the gas is maintained at a low pressure (or a low density). Such a low-density gas approximates what is called an **ideal gas**. Most gases at room temperature and atmospheric pressure behave approximately as ideal gases. **An ideal gas is a collection of atoms or molecules that move randomly and exert no long-range forces on each other. Each particle of the ideal gas is individually point-like, occupying a negligible volume.**

A gas usually consists of a very large number of particles, so it's convenient to express the amount of gas in a given volume in terms of the number of **moles**, n. A mole is a number. The same number of particles is found in a mole of helium as in a mole of iron or aluminum. This number is known as *Avogadro's number* and is given by

Avogadro's number ▶

$$N_A = 6.02 \times 10^{23} \text{ particles/mole}$$

Avogadro's number and the definition of a mole are fundamental to chemistry and related branches of physics. The number of moles of a substance is related to its mass m by the expression

$$n = \frac{m}{\text{molar mass}} \qquad \textbf{[10.7]}$$

where the molar mass of the substance is defined as the mass of one mole of that substance, usually expressed in grams per mole.

There are lots of atoms in the world, so it's natural and convenient to choose a very large number like Avogadro's number when describing collections of atoms. At the same time, Avogadro's number must be special in some way, because otherwise why not just count things in terms of some large power of ten, like 10^{24}?

It turns out that Avogadro's number was chosen so that the mass in grams of one Avogadro's number of an element is numerically the same as the mass of one atom of the element, expressed in atomic mass units (u).

This relationship is very convenient. Looking at the periodic table of the elements in the back of the book, we find that carbon has an atomic mass of 12 u, so 12 g of carbon consists of exactly 6.02×10^{23} atoms of carbon. The atomic mass of oxygen is 16 u, so in 16 g of oxygen there are again 6.02×10^{23} atoms of oxygen. The same holds true for molecules: The molecular mass of molecular hydrogen, H_2, is 2 u, and there is an Avogadro's number of molecules in 2 g of molecular hydrogen.

The technical definition of a mole is as follows: **One mole (mol) of any substance is that amount of the substance that contains as many particles (atoms, molecules, or other particles) as there are atoms in 12 g of the isotope carbon-12.**

Taking carbon-12 as a test case, let's find the mass of an Avogadro's number of carbon-12 atoms. A carbon-12 atom has an atomic mass of 12 u, or 12 atomic mass units. One atomic mass unit is equal to 1.66×10^{-24} g, about the same as the mass of a neutron or proton—particles that make up atomic nuclei. The mass m of an Avogadro's number of carbon-12 atoms is then given by

$$m = N_A(12\text{ u}) = 6.02 \times 10^{23}(12\text{ u})\left(\frac{1.66 \times 10^{-24}\text{ g}}{\text{u}}\right) = 12.0\text{ g}$$

So we see that Avogadro's number is deliberately chosen to be the inverse of the number of grams in an atomic mass unit. In this way, the atomic mass of an atom expressed in atomic mass units is numerically the same as the mass of an Avogadro's number of that kind of atom expressed in grams. Because there are 6.02×10^{23} particles in one mole of *any* element, the mass per atom for a given element is

$$m_{\text{atom}} = \frac{\text{molar mass}}{N_A}$$

For example, the mass of a helium atom is

$$m_{\text{He}} = \frac{4.00\text{ g/mol}}{6.02 \times 10^{23}\text{ atoms/mol}} = 6.64 \times 10^{-24}\text{ g/atom}$$

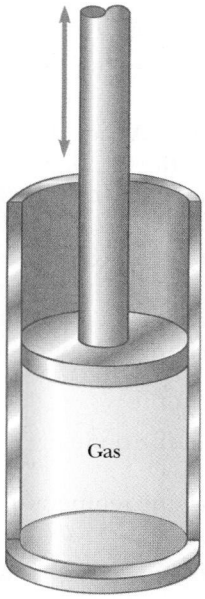

ACTIVE FIGURE 10.12
A gas confined to a cylinder whose volume can be varied with a movable piston.

PhysicsNow™
Log into PhysicsNow at **www.cp7e.com**, and go to Active Figure 10.12, where you can choose to keep either the temperature or the pressure constant and verify the ideal gas law.

Now suppose an ideal gas is confined to a cylindrical container with a volume that can be changed by moving a piston, as in Active Figure 10.12. Assume that the cylinder doesn't leak, so the number of moles remains constant. Experiments yield the following observations: First, when the gas is kept at a constant temperature, its pressure is inversely proportional to its volume (Boyle's law). Second, when the pressure of the gas is kept constant, the volume of the gas is directly proportional to the temperature (Charles's law.) Third, when the volume of the gas is held constant, the pressure is directly proportional to the temperature (Gay-Lussac's law). These observations can be summarized by the following equation of state, known as the **ideal gas law**:

$$PV = nRT \qquad [10.8]$$

◀ Equation of state for an ideal gas

In this equation, R is a constant for a specific gas that must be determined from experiments, while T is the temperature in kelvins. Each point on a P versus V diagram would represent a different state of the system. Experiments on several gases show that, as the pressure approaches zero, the quantity PV/nT approaches the same value of R for all gases. For this reason, R is called the **universal gas constant**. In SI units, where pressure is expressed in pascals and volume in cubic meters,

$$R = 8.31\text{ J/mol}\cdot\text{K} \qquad [10.9]$$

◀ The universal gas constant

TIP 10.2 Only Kelvin Works!
Temperatures used in the ideal gas law must always be in kelvins.

If the pressure is expressed in atmospheres and the volume is given in liters (recall that $1\text{ L} = 10^3\text{ cm}^3 = 10^{-3}\text{ m}^3$), then

$$R = 0.0821\text{ L}\cdot\text{atm/mol}\cdot\text{K}$$

Using this value of R and Equation 10.8, the volume occupied by 1 mol of any ideal gas at atmospheric pressure and at 0°C (273 K) is 22.4 L.

TIP 10.3 Standard Temperature and Pressure
Chemists often define standard temperature and pressure (STP) to be 20°C and 1.0 atm. We choose STP to be 0°C and 1.0 atm. (See Table 9.3.)

Quick Quiz 10.5

A helium-filled balloon is released into the atmosphere. Assuming constant temperature, as the balloon rises, it (a) expands, (b) contracts, or (c) remains unchanged in size.

EXAMPLE 10.6 An Expanding Gas

Goal Use the ideal gas law to analyze a system of gas.

Problem An ideal gas at 20.0°C and a pressure of 1.50×10^5 Pa is in a container having a volume of 1.00 L. **(a)** Determine the number of moles of gas in the container. **(b)** The gas pushes against a piston, expanding to twice its original volume, while the pressure falls to atmospheric pressure. Find the final temperature.

Strategy (a) Solve the ideal gas equation of state for the number of moles, n, and substitute the known quantities. Be sure to convert the temperature from Celsius to Kelvin! (b) When comparing two states of a gas, it's often most convenient to divide the ideal gas equation of the final state by the equation of the initial state. Then quantities that don't change can immediately be cancelled, simplifying the algebra.

Solution

(a) Find the number of moles of gas.

Convert the temperature to kelvins:

$$T = T_C + 273 = 20.0 + 273 = 293 \text{ K}$$

Solve the ideal gas law for n and substitute:

$$PV = nRT$$

$$n = \frac{PV}{RT} = \frac{(1.50 \times 10^5 \text{ Pa})(1.00 \times 10^{-3} \text{ m}^3)}{(8.31 \text{ J/mol}\cdot\text{K})(293 \text{ K})}$$

$$= 6.16 \times 10^{-2} \text{ mol}$$

(b) Find the temperature after the gas expands to 2.00 L.

Divide the ideal gas law for the final state by the ideal gas law for the initial state:

$$\frac{P_f V_f}{P_i V_i} = \frac{nRT_f}{nRT_i}$$

Cancel the number of moles n and the gas constant R, and solve for T_f:

$$\frac{P_f V_f}{P_i V_i} = \frac{T_f}{T_i}$$

$$T_f = \frac{P_f V_f}{P_i V_i} T_i = \frac{(1.01 \times 10^5 \text{ Pa})(2.00 \text{ L})}{(1.50 \times 10^5 \text{ Pa})(1.00 \text{ L})}(293 \text{ K})$$

$$= 395 \text{ K}$$

Remark Remember the trick used in part (b), it's often useful in ideal gas problems. Notice that it wasn't necessary to convert units from liters to cubic meters, since the units were going to cancel anyway.

Exercise 10.6

Suppose the temperature of 4.50 L of ideal gas drops from 375 K to 275 K. (a) If the volume remains constant and the initial pressure is atmospheric pressure, find the final pressure. (b) Find the number of moles of gas.

Answer (a) 7.41×10^4 Pa (b) 0.146 mol

EXAMPLE 10.7 Message in a Bottle

Goal Apply the ideal gas law in tandem with Newton's second law.

Problem A beachcomber finds a corked bottle containing a message. The air in the bottle is at atmospheric pressure and a temperature of 30.0°C. The cork has a cross-sectional area of 2.30 cm^2. The beachcomber places the bottle over a fire, figuring the increased pressure will push out the cork. At a temperature of 99°C the cork is ejected from the bottle. **(a)** What was the pressure in the bottle just before the cork left it? **(b)** What force of friction held the cork in place? Neglect any change in volume of the bottle.

Strategy (a) The number of moles of air in the bottle remains the same as it warms over the fire. Take the ideal gas equation for the final state and divide by the ideal gas equation for the initial state. Solve for the final pressure. (b) There are three forces acting on the cork: a friction force, the exterior force of the atmosphere pushing in, and the force of the air inside the bottle pushing out. Apply Newton's second law. Just before the cork begins to move, the three forces are in equilibrium and the static friction force has its maximum value.

Solution

(a) Find the final pressure.

Divide the ideal gas law at the final point by the ideal gas law at the initial point:

$$\frac{P_f V_f}{P_i V_i} = \frac{nRT_f}{nRT_i} \tag{1}$$

Cancel n, R, and V, which don't change, and solve for P_f:

$$\frac{P_f}{P_i} = \frac{T_f}{T_i} \quad \rightarrow \quad P_f = P_i \frac{T_f}{T_i}$$

Substitute known values, obtaining the final pressure:

$$P_f = (1.01 \times 10^5 \text{ Pa}) \frac{372 \text{ K}}{303 \text{ K}} = 1.24 \times 10^5 \text{ Pa}$$

(b) Find the magnitude of the friction force acting on the cork.

Apply Newton's second law to the cork just before it leaves the bottle. P_{in} is the pressure inside the bottle, P_{out} the pressure outside.

$$\Sigma F = 0 \quad \rightarrow \quad P_{in}A - P_{out}A - F_{friction} = 0$$

$$\begin{aligned} F_{friction} &= P_{in}A - P_{out}A = (P_{in} - P_{out})A \\ &= (1.24 \times 10^5 \text{ Pa} - 1.01 \times 10^5 \text{ Pa}) \\ &\quad \times (2.30 \times 10^{-4} \text{ m}^2) \end{aligned}$$

$$F_{friction} = 5.29 \text{ N}$$

Remark Notice the use, once again, of the ideal gas law in Equation (1). Whenever comparing the state of a gas at two different points, this is the best way to do the math. One other point: heating the gas blasted the cork out of the bottle, which meant the gas did work on the cork. The work done by an expanding gas—driving pistons and generators—is one of the foundations of modern technology and will be studied extensively in Chapter 12.

Exercise 10.7

A tire contains air at a gauge pressure of 5.00×10^4 Pa at a temperature of 30.0°C. After nightfall, the temperature drops to −10.0°C. Find the new gauge pressure in the tire. (Recall that gauge pressure is absolute pressure minus atmospheric pressure. Assume constant volume.)

Answer 3.01×10^4 Pa

EXAMPLE 10.8 Submerging a Balloon

Goal Combine the ideal gas law with the equation of hydrostatic equilibrium and buoyancy.

Problem A sturdy balloon with volume 0.500 m^3 is attached to a 2.50×10^2-kg iron weight and tossed overboard into a freshwater lake. The balloon is made of a light material of negligible mass and elasticity (though it can be compressed). The air in the balloon is initially at atmospheric pressure. The system fails to sink and there are no more weights, so a skin diver decides to drag it deep enough so that the balloon will remain submerged. **(a)** Find the volume of the balloon at the point where the system will remain submerged, in equilibrium. **(b)** What's the balloon's pressure at that point? **(c)** Assuming constant temperature, to what minimum depth must the balloon be dragged?

Strategy As the balloon and weight are dragged deeper into the lake, the air in the balloon is compressed and the volume is reduced along with the buoyancy. At some depth h the total buoyant force acting on the balloon and weight, $B_{bal} + B_{Fe}$, will equal the total weight, $w_{bal} + w_{Fe}$, and the balloon will remain at that depth. Substitute these forces into Newton's second law and solve for the unknown volume of the balloon, answering part (a). Then use the ideal gas law to find the pressure, and the equation of hydrostatic equilibrium to find the depth.

Solution

(a) Find the volume of the balloon at the equilibrium point.

Find the volume of the iron, V_{Fe}:

$$V_{Fe} = \frac{m_{Fe}}{\rho_{Fe}} = \frac{2.50 \times 10^2 \text{ kg}}{7.86 \times 10^3 \text{ kg/m}^3} = 0.0318 \text{ m}^3$$

Find the mass of the balloon, which is equal to the mass of the air if we neglect the mass of the balloon's material:

$$m_{bal} = \rho_{air} V_{bal} = (1.29 \text{ kg/m}^3)(0.500 \text{ m}^3) = 0.65 \text{ kg}$$

Apply Newton's second law to the system when it's in equilibrium:

$$B_{Fe} - w_{Fe} + B_{bal} - w_{bal} = 0$$

Substitute the appropriate expression for each term:

$$\rho_{wat} V_{Fe} g - m_{Fe} g + \rho_{wat} V_{bal} g - m_{bal} g = 0$$

Cancel the g's and solve for the volume of the balloon, V_{bal}:

$$V_{bal} = \frac{m_{bal} + m_{Fe} - \rho_{wat} V_{Fe}}{\rho_{wat}}$$

$$= \frac{0.65 \text{ kg} + 2.50 \times 10^2 \text{ kg} - (1.00 \times 10^3 \text{ kg/m}^3)(0.0318 \text{ m}^3)}{1.00 \times 10^3 \text{ kg/m}^3}$$

$$V_{bal} = 0.219 \text{ m}^3$$

(b) What's the balloon's pressure at the equilibrium point?

Now use the ideal gas law to find the pressure, assuming constant temperature, so that $T_i = T_f$.

$$\frac{P_f V_f}{P_i V_i} = \frac{nRT_f}{nRT_i} = 1$$

$$P_f = \frac{V_i}{V_f} P_i = \frac{0.500 \text{ m}^3}{0.219 \text{ m}^3} (1.01 \times 10^5 \text{ Pa})$$

$$= 2.31 \times 10^5 \text{ Pa}$$

(c) To what minimum depth must the balloon be dragged?

Use the equation of hydrostatic equilibrium to find the depth:

$$P_f = P_{atm} + \rho g h$$

$$h = \frac{P_f - P_{atm}}{\rho g} = \frac{2.31 \times 10^5 \text{ Pa} - 1.01 \times 10^5 \text{ Pa}}{(1.00 \times 10^3 \text{ kg/m}^3)(9.80 \text{ m/s}^2)}$$

$$= 13.3 \text{ m}$$

Remark Once again, the ideal gas law was used to good effect. This problem shows how even answering a fairly simple question can require the application of several different physical concepts: density, buoyancy, the ideal gas law, and hydrostatic equilibrium.

Exercise 10.8

A boy takes a 30.0-cm^3 balloon holding air at 1.00 atm at the surface of a freshwater lake down to a depth of 4.00 m. Find the volume of the balloon at this depth. Assume the balloon is made of light material of little elasticity (though it can be compressed), and that the temperature of the trapped air remains constant.

Answer 21.6 cm^3

As previously stated, the number of molecules contained in one mole of any gas is Avogadro's number, $N_A = 6.02 \times 10^{23}$ particles/mol, so

$$n = \frac{N}{N_A} \qquad [10.10]$$

where n is the number of moles and N is the number of molecules in the gas. With Equation 10.10, we can rewrite the ideal gas law in terms of the total number of molecules as

$$PV = nRT = \frac{N}{N_A} RT$$

or

$$PV = Nk_BT \qquad [10.11]$$

◀ Ideal gas law

where

$$k_B = \frac{R}{N_A} = 1.38 \times 10^{-23}\ \text{J/K} \qquad [10.12]$$

◀ Boltzmann's constant

is **Boltzmann's constant**. This reformulation of the ideal gas law will be used in the next section to relate the temperature of a gas to the average kinetic energy of particles in the gas.

10.5 THE KINETIC THEORY OF GASES

In Section 10.4, we discussed the macroscopic properties of an ideal gas, including pressure, volume, number of moles, and temperature. In this section we consider the ideal gas model from the microscopic point of view. We will show that the macroscopic properties can be understood on the basis of what is happening on the atomic scale. In addition, we reexamine the ideal gas law in terms of the behavior of the individual molecules that make up the gas.

Using the model of an ideal gas, we will describe the **kinetic theory of gases**. With this theory we can interpret the pressure and temperature of an ideal gas in terms of microscopic variables. The kinetic theory of gases model makes the following assumptions:

◀ Assumptions of kinetic theory for an ideal gas

1. **The number of molecules in the gas is large, and the average separation between them is large compared with their dimensions**. The fact that the number of molecules is large allows us to analyze their behavior statistically. The large separation between molecules means that the molecules occupy a negligible volume in the container. This assumption is consistent with the ideal gas model, in which we imagine the molecules to be pointlike.
2. **The molecules obey Newton's laws of motion, but as a whole they move randomly**. By "randomly" we mean that any molecule can move in any direction with equal probability, with a wide distribution of speeds.
3. **The molecules interact only through short-range forces during elastic collisions**. This assumption is consistent with the ideal gas model, in which the molecules exert no long-range forces on each other.
4. **The molecules make elastic collisions with the walls**.
5. **All molecules in the gas are identical**.

Although we often picture an ideal gas as consisting of single atoms, *molecular* gases exhibit ideal behavior at low pressures. On average, effects associated with molecular structure have no effect on the motions considered, so we can apply the results of the following development to molecular gases as well as to monatomic gases.

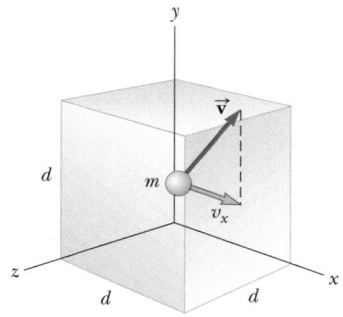

Figure 10.13 A cubical box with sides of length d containing an ideal gas. The molecule shown moves with velocity $\vec{v}$.

Molecular Model for the Pressure of an Ideal Gas

As a first application of kinetic theory, we derive an expression for the pressure of an ideal gas in a container in terms of microscopic quantities. The pressure of the gas is the result of collisions between the gas molecules and the walls of the container. During these collisions, the gas molecules undergo a change of momentum as a result of the force exerted on them by the walls.

We now derive an expression for the pressure of an ideal gas consisting of N molecules in a container of volume V. In this section, we use m to represent the mass of one molecule. The container is a cube with edges of length d (Fig. 10.13). Consider the collision of one molecule moving with a velocity $-v_x$ toward the left-hand face of the box (Fig. 10.14). After colliding elastically with the wall, the molecule moves in the positive x-direction with a velocity $+v_x$. Because the momentum of the molecule is $-mv_x$ before the collision and $+mv_x$ afterward, the change in its momentum is

$$\Delta p_x = mv_x - (-mv_x) = 2mv_x$$

If F_1 is the magnitude of the average force exerted by a molecule on the *wall* in the time Δt, then applying Newton's second law to the wall gives

$$F_1 = \frac{\Delta p_x}{\Delta t} = \frac{2mv_x}{\Delta t}$$

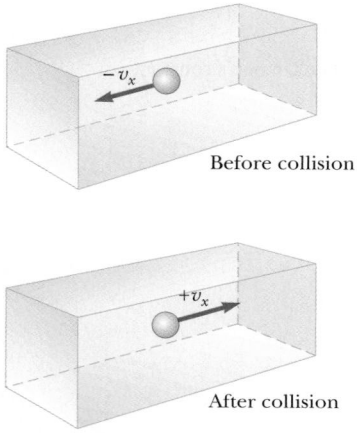

Figure 10.14 A molecule moving along the x-axis makes elastic collisions with the walls of the container. In colliding with a wall, the molecule's momentum is reversed, and the molecule exerts a force on the wall.

In order for the molecule to make two collisions with the same wall, it must travel a distance $2d$ along the x-direction in a time Δt. Therefore, the time interval between two collisions with the same wall is $\Delta t = 2d/v_x$, and the force imparted to the wall by a single molecule is

$$F_1 = \frac{2mv_x}{\Delta t} = \frac{2mv_x}{2d/v_x} = \frac{mv_x^2}{d}$$

The total force F exerted by all the molecules on the wall is found by adding the forces exerted by the individual molecules:

$$F = \frac{m}{d}(v_{1x}^2 + v_{2x}^2 + \cdots)$$

In this equation, v_{1x} is the x-component of velocity of molecule 1, v_{2x} is the x-component of velocity of molecule 2, and so on. The summation terminates when we reach N molecules because there are N molecules in the container.

Note that the average value of the square of the velocity in the x-direction for N molecules is

$$\overline{v_x^2} = \frac{v_{1x}^2 + v_{2x}^2 + \cdots + v_{Nx}^2}{N}$$

where $\overline{v_x^2}$ is the average value of v_x^2. The total force on the wall can then be written

$$F = \frac{Nm}{d}\overline{v_x^2}$$

© R. Folwell/Science Photo Library/Photo Researchers, Inc.

The glass vessel contains dry ice (solid carbon dioxide). The white cloud is carbon dioxide vapor, which is denser than air and hence falls from the vessel as shown.

Now we focus on one molecule in the container traveling in some arbitrary direction with velocity $\vec{v}$ and having components v_x, v_y, and v_z. In this case, we must express the total force on the wall in terms of the speed of the molecules rather than just a single component. The Pythagorean theorem relates the square of the speed to the square of these components according to the expression $v^2 = v_x^2 + v_y^2 + v_z^2$. Hence, the average value of v^2 for all the molecules in the container is related to the average values $\overline{v_x^2}$, $\overline{v_y^2}$, and $\overline{v_z^2}$ according to the expression $\overline{v^2} = \overline{v_x^2} + \overline{v_y^2} + \overline{v_z^2}$. Because the motion is completely random, the average values $\overline{v_x^2}$, $\overline{v_y^2}$, and $\overline{v_z^2}$ are equal to each other. Using this fact and the earlier equation for $\overline{v_x^2}$, we find that

$$\overline{v_x^2} = \tfrac{1}{3}\overline{v^2}$$

The total force on the wall, then, is

$$F = \frac{N}{3}\left(\frac{m\overline{v^2}}{d}\right)$$

This expression allows us to find the total pressure exerted on the wall by dividing by the force by the area:

$$P = \frac{F}{A} = \frac{F}{d^2} = \tfrac{1}{3}\left(\frac{N}{d^3}\, m\overline{v^2}\right) = \tfrac{1}{3}\left(\frac{N}{V}\right)m\overline{v^2}$$

$$P = \tfrac{2}{3}\left(\frac{N}{V}\right)\left(\tfrac{1}{2}m\overline{v^2}\right) \qquad \textbf{[10.13]}$$

◀ Pressure of an ideal gas

Equation 10.13 says that **the pressure is proportional to the number of molecules per unit volume and to the average translational kinetic energy of a molecule, $\frac{1}{2}m\overline{v^2}$.** With this simplified model of an ideal gas, we have arrived at an important result that relates the large-scale quantity of pressure to an atomic quantity—the average value of the square of the molecular speed. This relationship provides a key link between the atomic world and the large-scale world.

Equation 10.13 captures some familiar features of pressure. One way to increase the pressure inside a container is to increase the number of molecules per unit volume in the container. You do this when you add air to a tire. The pressure in the tire can also be increased by increasing the average translational kinetic energy of the molecules in the tire. As we will see shortly, this can be accomplished by increasing the temperature of the gas inside the tire. That's why the pressure inside a tire increases as the tire warms up during long trips. The continuous flexing of the tires as they move along the road transfers energy to the air inside them, increasing the air's temperature, which in turn raises the pressure.

Molecular Interpretation of Temperature

Having related the pressure of a gas to the average kinetic energy of the gas molecules, we now relate temperature to a microscopic description of the gas. We can obtain some insight into the meaning of temperature by multiplying Equation 10.13 by the volume:

$$PV = \tfrac{2}{3}N\left(\tfrac{1}{2}\, m\overline{v^2}\right)$$

Comparing this equation with the equation of state for an ideal gas in the form of Equation 10.11, $PV = Nk_BT$, we note that the left-hand sides of the two equations are identical. Equating the right-hand sides, we obtain

$$T = \frac{2}{3k_B}\left(\tfrac{1}{2}\, m\overline{v^2}\right) \qquad \textbf{[10.14]}$$

◀ Temperature is proportional to average kinetic energy

This means that **the temperature of a gas is a direct measure of the average molecular kinetic energy of the gas.** As the temperature of a gas increases, the molecules move with higher average kinetic energy.

Rearranging Equation 10.14, we can relate the translational molecular kinetic energy to the temperature:

$$\tfrac{1}{2}\, m\overline{v^2} = \tfrac{3}{2}\, k_BT \qquad \textbf{[10.15]}$$

◀ Average kinetic energy per molecule

So the average translational kinetic energy per molecule is $\frac{3}{2}k_BT$. The total translational kinetic energy of N molecules of gas is simply N times the average energy per molecule,

$$KE_{\text{total}} = N\left(\tfrac{1}{2}\, m\overline{v^2}\right) = \tfrac{3}{2}\, Nk_BT = \tfrac{3}{2}\, nRT \qquad \textbf{[10.16]}$$

◀ Total kinetic energy of N molecules

where we have used $k_B = R/N_A$ for Boltzmann's constant and $n = N/N_A$ for the number of moles of gas. From this result, we see that **the total translational kinetic energy of a system of molecules is proportional to the absolute temperature of the system.**

For a monatomic gas, translational kinetic energy is the only type of energy the molecules can have, so Equation 10.16 gives the **internal energy U for a monatomic gas**:

$$U = \tfrac{3}{2}nRT \quad \text{(monatomic gas)} \qquad \textbf{[10.17]}$$

For diatomic and polyatomic molecules, additional possibilities for energy storage are available in the vibration and rotation of the molecule.

The square root of $\overline{v^2}$ is called the **root-mean-square (rms) speed** of the molecules. From Equation 10.15, we get, for the rms speed,

Root-mean-square speed ▶

$$v_{\text{rms}} = \sqrt{\overline{v^2}} = \sqrt{\frac{3k_BT}{m}} = \sqrt{\frac{3RT}{M}} \qquad \textbf{[10.18]}$$

where M is the molar mass in *kilograms per mole*, if R is given in SI units. Equation 10.18 shows that, at a given temperature, lighter molecules tend to move faster than heavier molecules. For example, if gas in a vessel consists of a mixture of hydrogen and oxygen, the hydrogen (H_2) molecules, with a molar mass of 2.0×10^{-3} kg/mol, move four times faster than the oxygen (O_2) molecules, with molar mass 32×10^{-3} kg/mol. If we calculate the rms speed for hydrogen at room temperature (~ 300 K), we find

$$v_{\text{rms}} = \sqrt{\frac{3RT}{M}} = \sqrt{\frac{3(8.31\ \text{J/mol}\cdot\text{K})(300\ \text{K})}{2.0 \times 10^{-3}\ \text{kg/mol}}} = 1.9 \times 10^3\ \text{m/s}$$

TIP 10.4 Kilograms, not Grams Per Mole

In the equation for the rms speed, the units of molar mass M must be consistent with the units of the gas constant R. In particular, if R is in SI units, M must be expressed in kilograms per mole, not grams per mole.

This speed is about 17% of the escape speed for Earth, as calculated in Chapter 7. Because it is an average speed, a large number of molecules have much higher speeds and can therefore escape from Earth's atmosphere. This is why Earth's atmosphere doesn't currently contain hydrogen—it has all bled off into space.

Table 10.2 lists the rms speeds for various molecules at 20°C. A system of gas at a given temperature will exhibit a variety of speeds. This distribution of speeds is known as the *Maxwell velocity distribution.* An example of such a distribution for nitrogen gas at two different temperatures is given in Active Figure 10.15. The horizontal axis is speed, and the vertical axis is the number of molecules per unit speed. Notice that three speeds are of special interest: the most probable speed, corresponding to the peak in the graph; the average speed, which is found by averaging over all the possible speeds; and the rms speed. For every gas, $v_{\text{mp}} < v_{\text{av}} < v_{\text{rms}}$. As the temperature rises, these three speeds shift to the right.

TABLE 10.2

Some rms Speeds

Gas	Molar Mass (kg/mol)	v_{rms} at 20°C (m/s)
H_2	2.02×10^{-3}	1 902
He	4.0×10^{-3}	1 352
H_2O	18×10^{-3}	637
Ne	20.2×10^{-3}	602
N_2 and CO	28.0×10^{-3}	511
NO	30.0×10^{-3}	494
O_2	32.0×10^{-3}	478
CO_2	44.0×10^{-3}	408
SO_2	64.1×10^{-3}	338

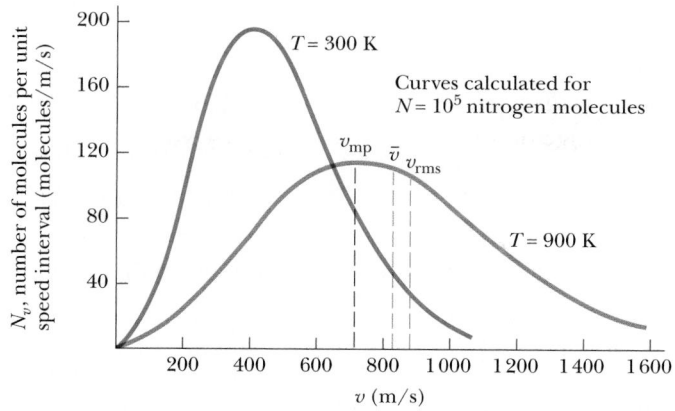

ACTIVE FIGURE 10.15
The Maxwell speed distribution for 10^5 nitrogen molecules at 300 K and 900 K. The total area under either curve equals the total number of molecules. The most probable speed v_{mp}, the average speed v_{av}, and the root-mean-square speed v_{rms} are indicated for the 900-K curve.

Physics Now™
Log into PhysicsNow at **www.cp7e.com**, and go to Active Figure 10.15 to set the desired temperature and see the effect on the distribution curve.

Quick Quiz 10.6

One container is filled with argon gas and another with helium gas. Both containers are at the same temperature. Which atoms have the higher rms speed? (a) argon, (b) helium, (c) they have the same speed, or (d) not enough information to say.

Applying Physics 10.2 Expansion and Temperature

Imagine a gas in an insulated cylinder with a movable piston. The piston has been pushed inward, compressing the gas, and is now released. As the molecules of the gas strike the piston, they move it outward. Explain, from the point of view of the kinetic theory, how the expansion of this gas causes its temperature to drop.

Explanation From the point of view of kinetic theory, a molecule colliding with the piston causes the piston to move with some velocity. According to the conservation of momentum, the molecule must rebound with less speed than it had before the collision. As these collisions occur, therefore, the average speed of the collection of molecules is reduced. Because temperature is related to the average speed of the molecules, the temperature of the gas drops.

EXAMPLE 10.9 A Cylinder of Helium

Goal Calculate the internal energy of a system and the average kinetic energy per molecule.

Problem A cylinder contains 2.00 mol of helium gas at 20.0°C. Assume that the helium behaves like an ideal gas. **(a)** Find the total internal energy of the system. **(b)** What is the average kinetic energy per molecule? **(c)** How much energy would have to be added to the system to double the rms speed? The molar mass of helium is 4.00×10^{-3} kg/mol.

Strategy This problem requires substitution of given information into the appropriate equations: Equation 10.17 for part (a) and Equation 10.15 for part (b). In part (c), use the equations for the rms speed and internal energy together. A *change* in the internal energy must be computed.

Solution

(a) Find the total internal energy of the system.

Substitute values into Equation 10.17 with $n = 2.00$ and $T = 293$ K:

$$U = \tfrac{3}{2}(2.00 \text{ mol})(8.31 \text{ J/mol}\cdot\text{K})(293 \text{ K}) = \boxed{7.30 \times 10^3 \text{ J}}$$

(b) What is the average kinetic energy per molecule?

Substitute given values into Equation 10.15:

$$\tfrac{1}{2}m\overline{v^2} = \tfrac{3}{2}k_BT = \tfrac{3}{2}(1.38 \times 10^{-23} \text{ J/K})(293 \text{ K})$$
$$= \boxed{6.07 \times 10^{-21} \text{ J}}$$

(c) How much energy must be added to double the rms speed?

From Equation 10.18, doubling the rms speed requires quadrupling T. Calculate the required change of internal energy, which is the energy that must be put into the system:

$$\Delta U = U_f - U_i = \tfrac{3}{2}nRT_f - \tfrac{3}{2}nRT_i = \tfrac{3}{2}nR(T_f - T_i)$$

$$\Delta U = \tfrac{3}{2}(2.00 \text{ mol})(8.31 \text{ J/mol}\cdot\text{K})((4.00 \times 293 \text{ K}) - 293 \text{ K}) = 2.19 \times 10^4 \text{ J}$$

Remark Computing changes in internal energy will be important in understanding engine cycles in Chapter 12.

Exercise 10.9

The temperature of 5.00 moles of argon gas is lowered from 3.00×10^2 K to 2.40×10^2 K. (a) Find the change in the internal energy, ΔU, of the gas. (b) Find the change in the average kinetic energy per atom.

Answer (a) $\Delta U = -3.74 \times 10^3$ J (b) -1.24×10^{-21} J

SUMMARY

PhysicsNow™ Take a practice test by logging into PhysicsNow at **www.cp7e.com** and clicking on the Pre-Test link for this chapter.

10.1 Temperature and the Zeroth Law of Thermodynamics

Two systems are in **thermal contact** if energy can be exchanged between them, and in **thermal equilibrium** if they're in contact and there is no net exchange of energy. The exchange of energy between two objects because of differences in their temperatures is called **heat**.

The **zeroth law of thermodynamics** states that if two objects A and B are separately in thermal equilibrium with a third object, then A and B are in thermal equilibrium with each other. Equivalently, if the third object is a **thermometer**, then the **temperature** it measures for A and B will be the same. Two objects in thermal equilibrium are at the same temperature.

10.2 Thermometers and Temperature Scales

Thermometers measure temperature and are based on physical properties, such as the temperature-dependent expansion or contraction of a solid, liquid, or gas. These changes in volume are related to a linear scale, the most common being the **Fahrenheit**, **Celsius**, and **Kelvin scales**. The Kelvin temperature scale takes its zero point as **absolute zero** (0 K = −273.15°C), the point at which, by extrapolation, the pressure of all gases falls to zero.

The relationship between the Celsius temperature T_C and the Kelvin (absolute) temperature T is

$$T_C = T - 273.15 \qquad \textbf{[10.1]}$$

The relationship between the Fahrenheit and Celsius temperatures is

$$T_F = \tfrac{9}{5}T_C + 32 \qquad \textbf{[10.2a]}$$

10.3 Thermal Expansion of Solids and Liquids

Ordinarily a substance expands when heated. If an object has an initial length L_0 at some temperature and undergoes a change in temperature ΔT, its linear dimension changes by the amount ΔL, which is proportional to the object's initial length and the temperature change:

$$\Delta L = \alpha L_0 \Delta T \qquad \textbf{[10.4]}$$

The parameter α is called the **coefficient of linear expansion**. The change in area of a substance with change in temperature is given by

$$\Delta A = \gamma A_0 \Delta T \qquad \textbf{[10.5]}$$

where $\gamma = 2\alpha$ is the **coefficient of area expansion.** Similarly, the change in volume with temperature of most substances is proportional to the initial volume V_0 and the temperature change ΔT;

$$\Delta V = \beta V_0 \Delta T \qquad \textbf{[10.6]}$$

where $\beta = 3\alpha$ is the **coefficient of volume expansion**.

The expansion and contraction of material due to changes in temperature creates stresses and strains, sometimes sufficient to cause fracturing.

10.4 Macroscopic Description of an Ideal Gas

Avogadro's number is $N_A = 6.02 \times 10^{23}$ particles/mol. A mole of anything, by definition, consists of an Avogadro's number of particles. The number is defined so that one mole of carbon-12 atoms has a mass of exactly 12 g. The mass of one mole of a pure substance in grams is the same, numerically, as that substance's atomic (or molecular) mass.

An **ideal gas** obeys the equation

$$PV = nRT \qquad \textbf{[10.8]}$$

where P is the pressure of the gas, V is its volume, n is the number of moles of gas, R is the universal gas constant

(8.31 J/mol·K), and T is the absolute temperature in kelvins. A real gas at very low pressures behaves approximately as an ideal gas.

Solving problems usually entails comparing two different states of the same system of gas, dividing the ideal gas equation for the final state by the ideal gas equation for the initial state, canceling factors that don't change and solving for the unknown quantity.

10.5 The Kinetic Theory of Gases

The **pressure** of N molecules of an ideal gas contained in a volume V is given by

$$P = \frac{2}{3}\left(\frac{N}{V}\right)\left(\frac{1}{2}m\overline{v^2}\right) \qquad [10.13]$$

where $\frac{1}{2}m\overline{v^2}$ is the **average kinetic energy per molecule**.

The average kinetic energy of the molecules of a gas is directly proportional to the absolute temperature of the gas:

$$\frac{1}{2}m\overline{v^2} = \frac{3}{2}k_B T \qquad [10.15]$$

The quantity k_B is **Boltzmann's constant** (1.38×10^{-23} J/K).

The internal energy of n moles of a monatomic ideal gas is

$$U = \frac{3}{2}nRT \qquad [10.17]$$

The **root-mean-square (rms) speed** of the molecules of a gas is

$$v_{\text{rms}} = \sqrt{\frac{3k_B T}{m}} = \sqrt{\frac{3RT}{M}} \qquad [10.18]$$

CONCEPTUAL QUESTIONS

1. Why does an ordinary glass dish usually break when placed on a hot stove? Dishes made of Pyrex glass don't break so easily. What characteristic of Pyrex prevents breakage?
2. The balance wheel of a mechanical watch governs the frequency with which the watch ticks. A wheel cut from a single piece of metal will expand upon heating, thus increasing its moment of inertia. Will this expansion cause the watch to speed up or slow down?
3. In an astronomy class, the temperature at the core of a star is given by the teacher as 1.5×10^7 degrees. A student asks if this is Kelvins or degrees Celsius. How would you respond?
4. When a car engine overheats, you are warned not to remove the radiator cap to add cold water until there is time for the engine to cool down. Is this good advice? Why or why not?
5. Common thermometers are made of a mercury column in a glass tube. Based on the operation of these common thermometers, which has the larger coefficient of linear expansion—glass or mercury? (Don't answer this question by looking in a table.)
6. A steel wheel bearing is 1 mm smaller in diameter than an axle. How can the bearing be fit onto the axle without removing any material from the axle?
7. Objects deep beneath the surface of the ocean are subjected to extremely high pressures, as we saw in Chapter 9. Some bacteria in these environments have adapted to pressures as much as a thousand times atmospheric pressure. How might such bacteria be affected if they were rapidly moved to the surface of the ocean?
8. Why is a power line more likely to break in winter than in summer, even if it is loaded with the same weight?
9. Although the average speed of gas molecules in thermal equilibrium at some temperature is greater than zero, the average velocity is zero. Explain.
10. After food is cooked in a pressure cooker, why is it very important to cool the container with cold water before attempting to remove the lid?
11. Some picnickers stop at a convenience store to buy food, including bags of potato chips. They then drive up into the mountains to their picnic site. When they unload the food, they notice that the bags of chips are puffed up like balloons. Why did this happen?
12. Markings to indicate length are placed on a steel tape in a room that is at a temperature of 22°C. Measurements are then made with the same tape on a day when the temperature is 27°C. Are the measurements too long, too short, or accurate?
13. Why do vapor bubbles in a pot of boiling water get larger as they approach the surface?
14. Why do small planets tend to have little or no atmosphere?

PROBLEMS

1, 2, 3 = straightforward, intermediate, challenging □ = full solution available in *Student Solutions Manual/Study Guide*
PhysicsNow™ = coached problem with hints available at **www.cp7e.com** = biomedical application

Section 10.1 Temperature and the Zeroth Law of Thermodynamics

Section 10.2 Thermometers and Temperature Scales

1. For each of the following temperatures, find the equivalent temperature on the indicated scale: (a) $-273.15°C$ on the Fahrenheit scale, (b) 98.6°F on the Celsius scale, and (c) 100 K on the Fahrenheit scale.

2. The pressure in a constant-volume gas thermometer is 0.700 atm at 100°C and 0.512 atm at 0°C. (a) What is the temperature when the pressure is 0.0400 atm? (b) What is the pressure at 450°C?

3. Convert the following temperatures to their values on the Fahrenheit and Kelvin scales: (a) the boiling point of liquid hydrogen, $-252.87°C$; (b) the temperature of a room at 20°C.

4. Death Valley holds the record for the highest recorded temperature in the United States. On July 10, 1913, at a place called Furnace Creek Ranch, the temperature rose to 134°F. The lowest U.S. temperature ever recorded occurred at Prospect Creek Camp in Alaska on January 23, 1971, when the temperature plummeted to $-79.8°$ F. Convert these temperatures to the Celsius scale.

5. Show that the temperature $-40°$ is unique in that it has the same numerical value on the Celsius and Fahrenheit scales.

6. A constant-volume gas thermometer is calibrated in dry ice ($-80.0°C$) and in boiling ethyl alcohol (78.0°C). The respective pressures are 0.900 atm and 1.635 atm. (a) What value of absolute zero does the calibration yield? (b) What pressures would be found at the freezing and boiling points of water? (Note that we have the linear relationship $P = A + BT$, where A and B are constants.)

7. Show that if the temperature on the Celsius scale changes by ΔT_C, the Fahrenheit temperature changes by $\Delta T_F = (9/5)\Delta T_C$.

8. The temperature difference between the inside and the outside of an automobile engine is 450°C. Express this difference on (a) the Fahrenheit scale and (b) the Kelvin scale.

9. The melting point of gold is 1 064°C, and the boiling point is 2 660°C. (a) Express these temperatures in Kelvins. (b) Compute the difference of the two temperatures in Celsius degrees and in Kelvins.

Section 10.3 Thermal Expansion of Solids and Liquids

10. A cylindrical brass sleeve is to be shrink-fitted over a brass shaft whose diameter is 3.212 cm at 0°C. The diameter of the sleeve is 3.196 cm at 0°C. (a) To what temperature must the sleeve be heated before it will slip over the shaft? (b) Alternatively, to what temperature must the shaft be cooled before it will slip into the sleeve?

11. The New River Gorge bridge in West Virginia is a 518-m-long steel arch. How much will its length change between temperature extremes of $-20°C$ and 35°C?

12. A grandfather clock is controlled by a swinging brass pendulum that is 1.3 m long at a temperature of 20°C. (a) What is the length of the pendulum rod when the temperature drops to 0.0°C? (b) If a pendulum's period is given by $T = 2\pi\sqrt{L/g}$, where L is its length, does the change in length of the rod cause the clock to run fast or slow?

13. A pair of eyeglass frames are made of epoxy plastic (coefficient of linear expansion $= 1.30 \times 10^{-4}\ °C^{-1}$). At room temperature (20.0°C), the frames have circular lens holes 2.20 cm in radius. To what temperature must the frames be heated if lenses 2.21 cm in radius are to be inserted into them?

14. A cube of solid aluminum has a volume of $1.00\ m^3$ at 20°C. What temperature change is required to produce a $100\text{-}cm^3$ increase in the volume of the cube?

15. **Physics Now™** A brass ring of diameter 10.00 cm at 20.0°C is heated and slipped over an aluminum rod of diameter 10.01 cm at 20.0°C. Assuming the average coefficients of linear expansion are constant, (a) to what temperature must the combination be cooled to separate the two metals? Is that temperature attainable? (b) What if the aluminum rod were 10.02 cm in diameter?

16. Show that the coefficient of volume expansion, β, is related to the coefficient of linear expansion, α, through the expression $\beta = 3\alpha$.

17. A gold ring has an inner diameter of 2.168 cm at a temperature of 15.0°C. Determine its inner diameter at 100°C ($\alpha_{gold} = 1.42 \times 10^{-5}\ °C^{-1}$).

18. A construction worker uses a steel tape to measure the length of an aluminum support column. If the measured length is 18.700 m when the temperature is 21.2°C, what is the measured length when the temperature rises to 29.4°C? (*Note:* Don't neglect the expansion of the tape.)

19. The band in Figure P10.19 is stainless steel (coefficient of linear expansion $= 17.3 \times 10^{-6}\ °C^{-1}$; Young's modulus $= 18 \times 10^{10}\ N/m^2$). It is essentially circular with an initial mean radius of 5.0 mm, a height of 4.0 mm, and a thickness of 0.50 mm. If the band just fits snugly over the tooth when heated to a temperature of 80°C, what is the tension in the band when it cools to a temperature of 37°C?

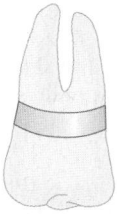

Figure P10.19

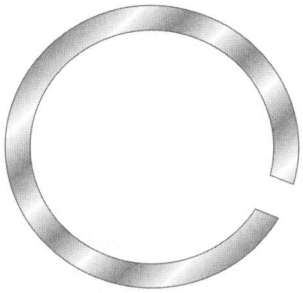

Figure P10.25

20. The Trans-Alaskan pipeline is 1 300 km long, reaching from Prudhoe Bay to the port of Valdez, and is subject to temperatures ranging from $-73°C$ to $+35°C$. How much does the steel pipeline expand due to the difference in temperature? How can this expansion be compensated for?

21. An automobile fuel tank is filled to the brim with 45 L (12 gal) of gasoline at 10°C. Immediately afterward, the vehicle is parked in the sunlight, where the temperature is 35°C. How much gasoline overflows from the tank as a result of the expansion? (Neglect the expansion of the tank.)

22. When the hot water in a certain upstairs bathroom is turned on, a series of 18 "ticks" is heard as the copper hot-water pipe slowly heats up and increases in length. The pipe runs vertically from the hot-water heater in the basement, through a hole in the floor 5.0 m above the water heater. The "ticks" are caused by the pipe sticking in the hole in the floor until the tension in the expanding pipe is great enough to unstick the pipe, enabling it to jump a short distance through the hole. If the hot-water temperature is 46°C and room temperature is 20°C, determine (a) the distance the pipe moves with each "tick" and (b) the force required to unstick the pipe if the cross-sectional area of the copper in the pipe is $3.55 \times 10^{-5}\ m^2$.

23. The average coefficient of volume expansion for carbon tetrachloride is $5.81 \times 10^{-4}\ (°C)^{-1}$. If a 50.0-gal steel container is filled completely with carbon tetrachloride when the temperature is 10.0°C, how much will spill over when the temperature rises to 30.0°C?

24. On a day when the temperature is 20.0°C, a concrete walk is poured in such a way that its ends are unable to move. (a) What is the stress in the cement when its temperature is 50.0°C on a hot, sunny day? (b) Does the concrete fracture? Take Young's modulus for concrete to be $7.00 \times 10^9\ N/m^2$ and the compressive strength to be $2.00 \times 10^7\ N/m^2$.

25. Figure P10.25 shows a circular steel casting with a gap. If the casting is heated, (a) does the width of the gap increase or decrease? (b) The gap width is 1.600 cm when the temperature is 30.0°C. Determine the gap width when the temperature is 190°C.

26. A hollow aluminum cylinder 20.0 cm deep has an internal capacity of 2.000 L at 20.0°C. It is completely filled with turpentine and then warmed to 80.0°C. (a) How much turpentine overflows? (b) If it is then cooled back to 20.0°C, how far below the surface of the cylinder's rim is the turpentine surface?

Section 10.4 Macroscopic Description of an Ideal Gas

27. Physics Now™ One mole of oxygen gas is at a pressure of 6.00 atm and a temperature of 27.0°C. (a) If the gas is heated at constant volume until the pressure triples, what is the final temperature? (b) If the gas is heated so that both the pressure and volume are doubled, what is the final temperature?

28. Gas is contained in an 8.0-L vessel at a temperature of 20°C and a pressure of 9.0 atm. (a) Determine the number of moles of gas in the vessel. (b) How many molecules are in the vessel?

29. (a) An ideal gas occupies a volume of $1.0\ cm^3$ at 20°C and atmospheric pressure. Determine the number of molecules of gas in the container. (b) If the pressure of the $1.0\text{-}cm^3$ volume is reduced to 1.0×10^{-11} Pa (an extremely good vacuum) while the temperature remains constant, how many moles of gas remain in the container?

30. A tank having a volume of $0.100\ m^3$ contains helium gas at 150 atm. How many balloons can the tank blow up if each filled balloon is a sphere 0.300 m in diameter at an absolute pressure of 1.20 atm?

31. A cylinder with a movable piston contains gas at a temperature of 27.0°C, a volume of $1.50\ m^3$, and an absolute pressure of 0.200×10^5 Pa. What will be its final temperature if the gas is compressed to $0.700\ m^3$ and the absolute pressure increases to 0.800×10^5 Pa?

32. The density of helium gas at $T = 0°C$ is $\rho_0 = 0.179\ kg/m^3$. The temperature is then raised to $T = 100°C$, but the pressure is kept constant. Assuming that the helium is an ideal gas, calculate the new density ρ_f of the gas.

33. A weather balloon is designed to expand to a maximum radius of 20 m at its working altitude, where the air pressure is 0.030 atm and the temperature is 200 K. If the balloon is filled at atmospheric pressure and 300 K, what is its radius at liftoff?

34. A cylindrical diving bell 3.00 m in diameter and 4.00 m tall with an open bottom is submerged to a depth of 220 m in the ocean. The surface temperature is 25.0°C, and the temperature 220 m down is 5.00°C. The density of seawater is 1 025 kg/m^3. How high does the seawater rise in the bell when it is submerged?

35. An air bubble has a volume of 1.50 cm^3 when it is released by a submarine 100 m below the surface of a lake. What is the volume of the bubble when it reaches the surface? Assume that the temperature and the number of air molecules in the bubble remain constant during its ascent.

Section 10.5 The Kinetic Theory of Gases

36. A sealed cubical container 20.0 cm on a side contains three times Avogadro's number of molecules at a temperature of 20.0°C. Find the force exerted by the gas on one of the walls of the container.

37. What is the average kinetic energy of a molecule of oxygen at a temperature of 300 K?

38. (a) What is the total random kinetic energy of all the molecules in 1 mole of hydrogen at a temperature of 300 K? (b) With what speed would a mole of hydrogen have to move so that the kinetic energy of the mass as a whole would be equal to the total random kinetic energy of its molecules?

39. Use Avogadro's number to find the mass of a helium atom.

40. The temperature near the top of the atmosphere on Venus is 240 K. (a) Find the rms speed of hydrogen (H_2) at that point in Venus's atmosphere. (b) Repeat for carbon dioxide (CO_2). (c) It has been found that if the rms speed exceeds one-sixth of the planet's escape velocity, the gas eventually leaks out of the atmosphere and into outer space. If the escape velocity on Venus is 10.3 km/s, does hydrogen escape? Does carbon dioxide?

41. A cylinder contains a mixture of helium and argon gas in equilibrium at a temperature of 150°C. (a) What is the average kinetic energy of each type of molecule? (b) What is the rms speed of each type of molecule?

42. Three moles of nitrogen gas, N_2, at 27.0°C are contained in a 22.4-L cylinder. Find the pressure the gas exerts on the cylinder walls.

43. PhysicsNow™ Superman leaps in front of Lois Lane to save her from a volley of bullets. In a 1-minute interval, an automatic weapon fires 150 bullets, each of mass 8.0 g, at 400 m/s. The bullets strike his mighty chest, which has an area of 0.75 m^2. Find the average force exerted on Superman's chest if the bullets bounce back after an elastic, head-on collision.

44. In a period of 1.0 s, 5.0×10^{23} nitrogen molecules strike a wall of area 8.0 cm^2. If the molecules move at 300 m/s and strike the wall head on in a perfectly elastic collision, find the pressure exerted on the wall. (The mass of one N_2 molecule is 4.68×10^{-26} kg.)

ADDITIONAL PROBLEMS

45. Inside the wall of a house, an L-shaped section of hot-water pipe consists of a straight horizontal piece 28.0 cm long, an elbow, and a straight vertical piece 134 cm long (Fig. P10.45). A stud and a second-story floorboard hold the ends of this section of copper pipe stationary. Find the magnitude and direction of the displacement of the pipe elbow when the water flow is turned on, raising the temperature of the pipe from 18.0°C to 46.5°C.

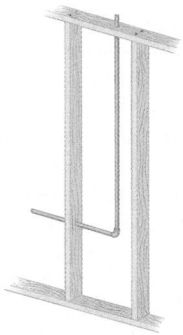

Figure P10.45

46. The active element of a certain laser is an ordinary glass rod 20 cm long and 1.0 cm in diameter. If the temperature of the rod increases by 75°C, find its increases in (a) length, (b) diameter, and (c) volume.

47. A popular brand of cola contains 6.50 g of carbon dioxide dissolved in 1.00 L of soft drink. If the evaporating carbon dioxide is trapped in a cylinder at 1.00 atm and 20.0°C, what volume does the gas occupy?

48. A 1.5-m-long glass tube that is closed at one end is weighted and lowered to the bottom of a freshwater lake. When the tube is recovered, an indicator mark shows that water rose to within 0.40 m of the closed end. Determine the depth of the lake. Assume constant temperature.

49. Long-term space missions require reclamation of the oxygen in the carbon dioxide exhaled by the crew. In one method of reclamation, 1.00 mol of carbon dioxide produces 1.00 mol of oxygen, with 1.00 mol of methane as a by-product. The methane is stored in a tank under pressure and is available to control the attitude of the spacecraft by controlled venting. A single astronaut exhales 1.09 kg of carbon dioxide each day. If the methane generated in the recycling of three astronauts' respiration during one week of flight is stored in an originally empty 150-L tank at $-45.0°C$, what is the final pressure in the tank?

50. A vertical cylinder of cross-sectional area $0.050\ m^2$ is fitted with a tight-fitting, frictionless piston of mass 5.0 kg (Fig. P10.50). If there are 3.0 mol of an ideal gas in the cylinder at 500 K, determine the height h at which the piston will be in equilibrium under its own weight.

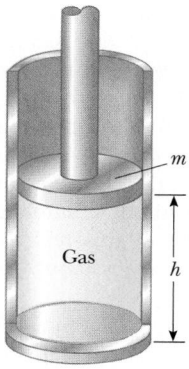

Figure P10.50

51. A liquid with a coefficient of volume expansion of β just fills a spherical flask of volume V_0 at temperature T (Fig. P10.51). The flask is made of a material that has a coefficient of linear expansion of α. The liquid is free to expand into a capillary of cross-sectional area A at the top. (a) Show that if the temperature increases by ΔT, the liquid rises in the capillary by the amount $\Delta h = (V_0/A)(\beta - 3\alpha)\Delta T$. (b) For a typical system, such as a mercury thermometer, why is it a good approximation to neglect the expansion of the flask?

52. A hollow aluminum cylinder is to be fitted over a steel piston. At 20°C, the inside diameter of the cylinder is 99% of the outside diameter of the piston. To what common temperature should the two pieces be heated in order that the cylinder just fit over the piston?

53. A steel measuring tape was designed to read correctly at 20°C. A parent uses the tape to measure the height of a 1.1-m-tall child. If the measurement is made on a day when the temperature is 25°C, is the tape reading longer or shorter than the actual height, and by how much?

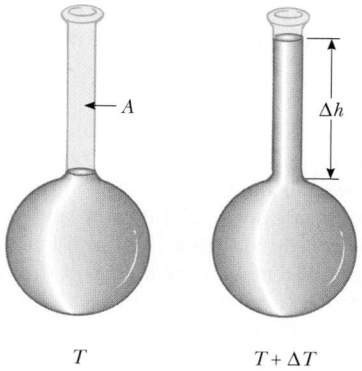

Figure P10.51

54. Before beginning a long trip on a hot day, a driver inflates an automobile tire to a gauge pressure of 1.80 atm at 300 K. At the end of the trip, the gauge pressure has increased to 2.20 atm. (a) Assuming that the volume has remained constant, what is the temperature of the air inside the tire? (b) What percentage of the original mass of air in the tire should be released so the pressure returns to its original value? Assume that the temperature remains at the value found in (a) and the volume of the tire remains constant as air is released.

55. Two concrete spans of a 250-m-long bridge are placed end to end so that no room is allowed for expansion (Fig. P10.55a). If the temperature increases by 20.0°C, what is the height y to which the spans rise when they buckle (Fig. P10.55b)?

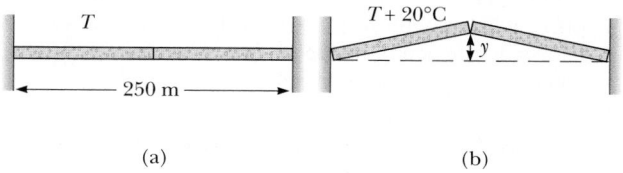

Figure P10.55

56. A copper rod and a steel rod are heated. At 0°C, the copper rod has a length L_C and the steel one has a length L_S. When the rods are being heated or cooled, a difference of 5.00 cm is maintained between their lengths. Determine the values of L_C and L_S.

57. If 9.00 g of water is placed in a 2.00-L pressure cooker and heated to 500°C, what is the pressure inside the container?

58. An expandable cylinder has its top connected to a spring with force constant 2.00×10^3 N/m. (See Fig. P10.58.) The cylinder is filled with 5.00 L of gas with the spring relaxed at a pressure of 1.00 atm and a temperature of 20.0°C. (a) If the lid has a cross-sectional area of 0.010 0 m^2 and negligible mass, how high will the lid rise when the temperature is raised to 250°C? (b) What is the pressure of the gas at 250°C?

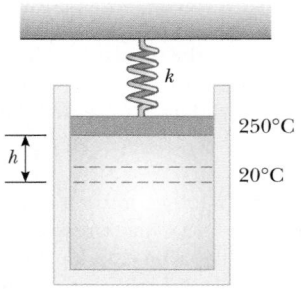

Figure P10.58

59. A swimmer has 0.820 L of dry air in his lungs when he dives into a lake. Assuming the pressure of the dry air is 95% of the external pressure at all times, what is the volume of the dry air at a depth of 10.0 m? Assume that atmospheric pressure at the surface is 1.013×10^5 Pa.

60. Two small containers, each with a volume of 100 cm^3, contain helium gas at 0°C and 1.00 atm pressure. The two containers are joined by a small open tube of negligible volume, allowing gas to flow from one container to the other. What common pressure will exist in the two containers if the temperature of one container is raised to 100°C while the other container is kept at 0°C?

61. **Physics Now™** A bimetallic bar is made of two thin strips of dissimilar metals bonded together. As they are heated, the one with the larger average coefficient of expansion expands more than the other, forcing the bar into an arc, with the outer strip having both a larger radius and a larger circumference. (See Fig. P10.61.) (a) Derive an expression for the angle of bending, θ, as a function of the initial length of the strips, their average coefficients of linear expansion, the change in temperature, and the separation of the centers of the strips ($\Delta r = r_2 - r_1$). (b) Show that the angle of bending goes to zero when ΔT goes to zero or when the two coefficients of expansion become equal. (c) What happens if the bar is cooled?

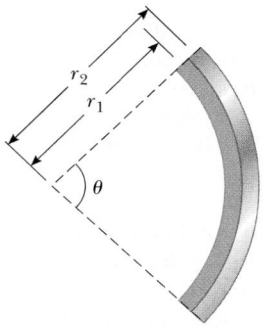

Figure P10.61

62. A 250-m-long bridge is improperly designed so that it cannot expand with temperature. It is made of concrete with $\alpha = 12 \times 10^{-6}$ °C^{-1}. (a) Assuming that the maximum change in temperature at the site is expected to be 20°C, find the change in length the span would undergo if it were free to expand. (b) Show that the stress on an object with Young's modulus Y when raised by ΔT with its ends firmly fixed is given by $\alpha Y \Delta T$. (c) If the maximum stress the bridge can withstand without crumbling is 2.0×10^7 Pa, will it crumble because of this temperature increase? Young's modulus for concrete is about 2.0×10^{10} Pa.

63. The density of gasoline is 730 kg/m^3 at 0°C. Its volume expansion coefficient is 9.6×10^{-4} °C^{-1}. If 1.00 gal of gasoline occupies 0.003 8 m^3, how many extra kilograms of gasoline are obtained when 10 gallons of gasoline are bought at 0°C rather than at 20°C?

ACTIVITIES

A.1. Fill one basin with hot tap water (not to exceed about 100°F). Fill another basin with cold tap water, and add ice until about one-third of the mixture is ice. Fill the third basin with an equal mixture of hot and cold tap water. Place your left hand in the hot water and your right hand in the cold water for about 15 s. Then place both hands in the basin of lukewarm water for 15 s. Describe whether the water feels hot or cold to either of your hands and why this effect occurs.

A.2. Tape two plastic straws tightly together along their entire length, but with a 2-cm offset. Hold them in a stream of very hot water from a faucet so that the water pours through one straw but not the other. Quickly hold the straws up and sight along their length. You should be able to see a very slight curvature in the tape. The effect is small, so look closely. Running cold water through the same straw and again sighting along the length will help you see the small change in shape more clearly. Explain these observations.

A.3. You can study the thermal expansion of air with a Florence flask and a balloon. To do so, boil a small amount of water in the flask. Then quickly remove the flask from the heat source and place the mouth of the balloon over the mouth of the flask. Observe what happens to the balloon when you place the flask in a basin of cold water. Explain your observation. Reheat the water in the flask with the balloon still in place, and explain your observations.

Glacier fragments fall into the sea. Global warming could melt enough ice to swell the oceans and threaten coastal cities around the world.

Steve Bly/Getty Images

CHAPTER

11

OUTLINE

11.1 Heat and Internal Energy
11.2 Specific Heat
11.3 Calorimetry
11.4 Latent Heat and Phase Change
11.5 Energy Transfer
11.6 Global Warming and Greenhouse Gases

Energy in Thermal Processes

When two objects with different temperatures are placed in thermal contact, the temperature of the warmer object decreases while the temperature of the cooler object increases. With time, they reach a common equilibrium temperature somewhere in between their initial temperatures. During this process, we say that energy is transferred from the warmer object to the cooler one.

Until about 1850, the subjects of thermodynamics and mechanics were considered two distinct branches of science, and the principle of conservation of energy seemed to describe only certain kinds of mechanical systems. Experiments performed by the English physicist James Joule (1818–1889) and others showed that the decrease in mechanical energy (kinetic plus potential) of an isolated system was equal to the increase in internal energy of the system. Today, internal energy is treated as a form of energy that can be transformed into mechanical energy and vice versa. Once the concept of energy was broadened to include internal energy, the law of conservation of energy emerged as a universal law of nature.

This chapter focuses on some of the processes of energy transfer between a system and its surroundings.

11.1 HEAT AND INTERNAL ENERGY

A major distinction must be made between internal energy and heat. These terms are not interchangeable—heat involves a *transfer* of internal energy from one location to another. The following formal definitions will make the distinction precise.

Internal energy U is the energy associated with the microscopic components of a system—the atoms and molecules of the system. The internal energy includes kinetic and potential energy associated with the random translational, rotational, and vibrational motion of the particles that make up the system, and any potential energy bonding the particles together.

◀ Internal energy

In Chapter 10 we showed that the internal energy of a monatomic ideal gas is associated with the translational motion of its atoms. In this special case, the internal energy is the total translational kinetic energy of the atoms; the higher the temperature of the gas, the greater the kinetic energy of the atoms and the greater the internal energy of the gas. For more complicated diatomic and polyatomic gases, internal energy includes other forms of molecular energy, such as rotational kinetic energy and the kinetic and potential energy associated with molecular vibrations. Internal energy is also associated with the intermolecular potential energy ("bond energy") between molecules in a liquid or solid.

Heat was introduced in Chapter 5 as one possible method of transferring energy between a system and its environment, and we provide a formal definition here:

By kind permission of the President and Council of the Royal Society

JAMES PRESCOTT JOULE,
British physicist (1818–1889)

Joule received some formal education in mathematics, philosophy, and chemistry from John Dalton, but was in large part self-educated. Joule's most active research period, from 1837 through 1847, led to the establishment of the principle of conservation of energy and the relationship between heat and other forms of energy transfer. His study of the quantitative relationship among electrical, mechanical, and chemical effects of heat culminated in his announcement in 1843 of the amount of work required to produce a unit of internal energy.

Heat is the transfer of energy between a system and its environment due to a temperature difference between them.

The symbol Q is used to represent the amount of energy transferred by heat between a system and its environment. For brevity, we will often use the phrase "the energy Q transferred to a system . . ." rather than "the energy Q transferred by heat to a system . . ."

If a pan of water is heated on the burner of a stove, it's incorrect to say more heat is in the water. Heat is the *transfer* of thermal energy, just as work is the transfer of mechanical energy. When an object is pushed, it doesn't have more work; rather, it has more mechanical energy transferred *by* work. Similarly, the pan of water has more thermal energy transferred by heat.

Units of Heat

Early in the development of thermodynamics, before scientists realized the connection between thermodynamics and mechanics, heat was defined in terms of the temperature changes it produced in an object, and a separate unit of energy, the **calorie**, was used for heat. The calorie (cal) is defined as **the energy necessary to raise the temperature of 1 g of water from 14.5° to 15.5°C.** (The "Calorie," with a capital "C," used in describing the energy content of foods, is actually a kilocalorie.) Likewise, the unit of heat in the U.S. customary system, the **British thermal unit** (Btu), was defined as **the energy required to raise the temperature of 1 lb of water from 63°F to 64°F.**

◀ Definition of the calorie

In 1948, scientists agreed that because heat (like work) is a measure of the transfer of energy, its SI unit should be the joule. The calorie is now defined to be exactly 4.186 J:

$$1 \text{ cal} \equiv 4.186 \text{ J} \qquad [11.1]$$

◀ The mechanical equivalent of heat

This definition makes no reference to raising the temperature of water. The calorie is a general energy unit, introduced here for historical reasons, though we will make little use of it. The definition in Equation 11.1 is known, from the historical background we have discussed, as the **mechanical equivalent of heat**.

EXAMPLE 11.1 Working Off Breakfast

Goal Relate caloric energy to mechanical energy.

Problem A student eats a breakfast consisting of two bowls of cereal and milk, containing a total of 3.20×10^2 Calories of energy. He wishes to do an equivalent amount of work in the gymnasium by doing curls with a 25.0-kg barbell (Fig. 11.1). How many times must he raise the weight to expend that much energy? Assume that he raises it through a vertical displacement of 0.400 m each time, the distance from his lap to his upper chest.

Figure 11.1 (Example 11.1)

Strategy Convert the energy in Calories to joules, then equate that energy to the work necessary to do n repetitions of the barbell exercise. The work he does lifting the barbell can be found from the work–energy theorem and the change in potential energy of the barbell. He does negative work on the barbell going down, to keep it from speeding up. The net work on the barbell during one repetition is zero, but his muscles expend the same energy both in raising and lowering.

Solution

Convert his breakfast Calories, E, to joules:

$$E = (3.20 \times 10^2\ \text{Cal})\left(\frac{1.00 \times 10^3\ \text{cal}}{1.00\ \text{Cal}}\right)\left(\frac{4.186\ \text{J}}{\text{cal}}\right)$$

$$= 1.34 \times 10^6\ \text{J}$$

Use the work–energy theorem to find the work necessary to lift the barbell up to its maximum height.

$$W = \Delta KE + \Delta PE = (0 - 0) + (mgh - 0) = mgh$$

The student must expend the same amount of energy lowering the barbell, making $2mgh$ per repetition. Multiply this amount by n repetitions and set it equal to the food energy E:

$$n(2mgh) = E$$

Solve for n, substituting the food energy for E:

$$n = \frac{E}{2mgh} = \frac{1.34 \times 10^6\ \text{J}}{2(25.0\ \text{kg})(9.80\ \text{m/s}^2)(0.400\ \text{m})}$$

$$= 6.84 \times 10^3\ \text{times}$$

Remarks If the student does one repetition every five seconds, it will take him 9.5 hours to work off his breakfast! In exercising, a large fraction of energy is lost through heat, however, due to the inefficiency of the body in doing work. This transfer of energy dramatically reduces the exercise requirement by at least three-quarters, a little over two hours. All the same, it might be best to forego that second bowl of cereal!

Exercise 11.1

How many sprints from rest to a speed of 5.0 m/s would a 65-kg woman have to complete in order to burn off 5.0×10^2 Calories? (Assume 100% efficiency in converting food energy to mechanical energy).

Answer 2.6×10^3 sprints

APPLICATION

Physiology of Exercise

Getting proper exercise is an important part of staying healthy and keeping weight under control. As seen in the preceding example, the body expends energy when doing mechanical work, and these losses are augmented by the inefficiency of converting the body's internal stores of energy into useful work, with three-quarters or more leaving the body through heat. In addition, exercise tends to elevate the body's general metabolic rate, which persists even after the exercise is over. The increase in metabolic rate due to exercise, more so than the exercise itself, is helpful in weight reduction.

11.2 SPECIFIC HEAT

The historical definition of the calorie is the amount of energy necessary to raise the temperature of one gram of a specific substance—water—by one degree. That amount is 4.186 J. Raising the temperature of one kilogram of water by 1° requires 4 186 J of energy. The amount of energy required to raise the temperature of one kilogram of an arbitrary substance by 1° varies with the substance. For example, the energy required to raise the temperature of one kilogram of copper by 1.0°C is 387 J. Every substance requires a unique amount of energy per unit mass to change the temperature of that substance by 1.0°C.

If a quantity of energy Q is transferred to a substance of mass m, changing its temperature by $\Delta T = T_f - T_i$, the **specific heat** c of the substance is defined by

$$c \equiv \frac{Q}{m\Delta T} \qquad [11.2]$$

SI unit: Joule per kilogram-degree Celsius (J/kg·°C)

Table 11.1 lists specific heats for several substances. From the definition of the calorie, the specific heat of water is 4 186 J/kg · °C.

From the definition of specific heat, we can express the energy Q needed to raise the temperature of a system of mass m by ΔT as

$$Q = mc\Delta T \qquad [11.3]$$

The energy required to raise the temperature of 0.500 kg of water by 3.00°C, for example, is $Q = (0.500\text{ kg})(4\,186\text{ J/kg}\cdot\text{°C})(3.00\text{°C}) = 6.28 \times 10^3$ J. Note that when the temperature increases, ΔT and Q are *positive*, corresponding to energy flowing *into* the system. When the temperature decreases, ΔT and Q are *negative*, and energy flows *out* of the system.

Table 11.1 shows that water has the highest specific heat relative to most other common substances. This high specific heat is responsible for the moderate temperatures found in regions near large bodies of water. As the temperature of a body of water decreases during winter, the water transfers energy to the air, which carries the energy landward when prevailing winds are toward the land. Off the western coast of the United States, the energy liberated by the Pacific Ocean is carried to the east, keeping coastal areas much warmer than they would otherwise be. Winters are generally colder in the eastern coastal states, because the prevailing winds tend to carry the energy away from land.

The fact that the specific heat of water is higher than the specific heat of sand is responsible for the pattern of airflow at a beach. During the day, the Sun adds roughly equal amounts of energy to the beach and the water, but the lower specific heat of sand causes the beach to reach a higher temperature than the water. As a result, the air above the land reaches a higher temperature than the air above the water. The denser cold air pushes the less dense hot air upward (due to Archimedes's principle), resulting in a breeze from ocean to land during the day. Because the hot air gradually cools as it rises, it subsequently sinks, setting up the circulation pattern shown in Figure 11.2.

A similar effect produces rising layers of air called *thermals* that can help eagles soar higher and hang gliders stay in flight longer. A thermal is created when a portion of the Earth reaches a higher temperature than neighboring regions. This often happens to plowed fields, which are warmed by the Sun to higher temperatures

TABLE 11.1

Specific Heats of Some Materials at Atmospheric Pressure

Substance	J/kg · °C	cal/g · °C
Aluminum	900	0.215
Beryllium	1 820	0.436
Cadmium	230	0.055
Copper	387	0.0924
Germanium	322	0.077
Glass	837	0.200
Gold	129	0.0308
Ice	2 090	0.500
Iron	448	0.107
Lead	128	0.0305
Mercury	138	0.033
Silicon	703	0.168
Silver	234	0.056
Steam	2 010	0.480
Water	4 186	1.00

TIP 11.1 Finding ΔT

In Equation 11.3, be sure to remember that ΔT is *always* the final temperature minus the initial temperature: $\Delta T = T_f - T_i$.

Figure 11.2 Circulation of air at the beach. On a hot day, the air above the sand warms faster than the air above the cooler water. The warmer air floats upward due to Archimedes's principle, resulting in the movement of cooler air toward the beach.

APPLICATION

Sea Breezes and Thermals

than nearby fields shaded by vegetation. The cooler, denser air over the vegetation-covered fields pushes the expanding air over the plowed field upwards, and a thermal is formed.

Quick Quiz 11.1

Suppose you have 1 kg each of iron, glass, and water, and all three samples are at 10°C. (a) Rank the samples from lowest to highest temperature after 100 J of energy is added to each by heat. (b) Rank them from least to greatest amount of energy transferred by heat if enough energy is transferred so that each increases in temperature by 20°C.

EXAMPLE 11.2 Stressing a Strut

Goal Use the energy transfer equation in the context of linear expansion and compressional stress.

Problem A steel strut near a ship's furnace is 2.00 m long, with a mass of 1.57 kg and cross-sectional area of $1.00 \times 10^{-4}\ \text{m}^2$. During operation of the furnace, the strut absorbs thermal energy in a net amount of 2.50×10^5 J. **(a)** Find the change in temperature of the strut. **(b)** Find the increase in length of the strut. **(c)** If the strut is not allowed to expand because it's bolted at each end, find the compressional stress developed in the strut.

Strategy This problem can be solved by substituting given quantities into three different equations. In part (a), the change in temperature can be computed by substituting into Equation 11.3, which relates temperature change to the energy transferred by heat. In part (b), substituting the result of part (a) into the linear expansion equation yields the change in length. If that change of length is thwarted by poor design, as in part (c), the result is compressional stress, found with the compressional stress–strain equation.

Solution

(a) Find the change in temperature.

Solve Equation 11.3 for the change in temperature and substitute:

$$Q = m_s c_s \Delta T \quad \rightarrow \quad \Delta T = \frac{Q}{m_s c_s}$$

$$\Delta T = \frac{(2.50 \times 10^5\ \text{J})}{(1.57\ \text{kg})(448\ \text{J/kg}\cdot°\text{C})} = 355°\text{C}$$

(b) Find the change in length of the strut if it's allowed to expand.

Substitute into the linear expansion equation:

$$\Delta L = \alpha L_0 \Delta T = (11 \times 10^{-6}\ °\text{C}^{-1})(2.00\ \text{m})(355°\text{C})$$
$$= 7.8 \times 10^{-3}\ \text{m}$$

(c) Find the compressional stress in the strut if it is not allowed to expand.

Substitute into the compressional stress–strain equation:

$$\frac{F}{A} = Y\frac{\Delta L}{L_0} = (2.00 \times 10^{11}\ \text{Pa})\frac{7.8 \times 10^{-3}\ \text{m}}{2.01\ \text{m}}$$
$$= 7.8 \times 10^8\ \text{Pa}$$

Remarks Notice the use of 2.01 m in the denominator of the last calculation, rather than 2.00 m. This is because, in effect, the strut was compressed back to the original length from the length to which it would have expanded. (The difference is negligible, however.) The answer exceeds the ultimate compressive strength of steel and underscores the importance of allowing for thermal expansion. Of course, it's likely the strut would bend, relieving some of the stress (creating some shear stress in the process). Finally, if the strut is attached at both ends by bolts, thermal expansion and contraction would exert sheer stresses on the bolts, possibly weakening or loosening them over time.

Exercise 11.2

Suppose a steel strut with cross-sectional area $5.00 \times 10^{-4}\ \text{m}^2$ and length 2.50 m is bolted between two rigid bulkheads in the engine room of a submarine. (a) Calculate the change in temperature of the strut if it absorbs an energy of 3.00×10^5 J. (b) Calculate the compressional stress in the strut.

Answers (a) 68.2°C (b) 1.50×10^8 Pa

11.3 CALORIMETRY

One technique for measuring the specific heat of a solid or liquid is to raise the temperature of the substance to some value, place it into a vessel containing cold water of known mass and temperature, and measure the temperature of the combination after equilibrium is reached. Define the system as the substance and the water. If the vessel is assumed to be a good insulator, so that energy doesn't leave the system, then we can assume the system is isolated. Vessels having this property are called **calorimeters**, and analysis performed using such vessels is called **calorimetry**.

The principle of conservation of energy for this isolated system requires that the net result of all energy transfers is zero. If one part of the system loses energy, another part has to gain the energy, because the system is isolated and the energy has nowhere else to go. When a warm object is placed in the cooler water of a calorimeter, the warm object becomes cooler while the water becomes warmer. This principle can be written

$$Q_{\text{cold}} = -Q_{\text{hot}} \qquad \textbf{[11.4]}$$

Q_{cold} is positive because energy is flowing into cooler objects, and Q_{hot} is negative because energy is leaving the hot object. The negative sign on the right-hand side of Equation 11.4 ensures that the right-hand side is a positive number, consistent with the left-hand side. The equation is valid only when the system it describes is isolated.

Calorimetry problems involve solving Equation 11.4 for an unknown quantity, usually either a specific heat or a temperature.

EXAMPLE 11.3 Finding a Specific Heat

Goal Solve a calorimetry problem involving only two substances.

Problem A 125-g block of an unknown substance with a temperature of 90.0°C is placed in a Styrofoam cup containing 0.326 kg of water at 20.0°C. The system reaches an equilibrium temperature of 22.4°C. What is the specific heat, c_x, of the unknown substance if the heat capacity of the cup is neglected?

Strategy The water gains thermal energy Q_{cold}, while the block loses thermal energy Q_{hot}. Using Equation 11.3, substitute expressions into Equation 11.4 and solve for the unknown specific heat, c_x.

Solution

Let T be the final temperature, and let T_w and T_x be the initial temperatures of the water and block, respectively. Apply Equations 11.3 and 11.4:

$$Q_{\text{cold}} = -Q_{\text{hot}}$$

$$m_w c_w (T - T_w) = -m_x c_x (T - T_x)$$

Solve for c_x and substitute numerical values:

$$c_x = \frac{m_w c_w (T - T_w)}{m_x (T_x - T)}$$

$$= \frac{(0.326\ \text{kg})(4\,190\ \text{J/kg}\cdot°\text{C})(22.4°\text{C} - 20.0°\text{C})}{(0.125\ \text{kg})(90.0°\text{C} - 22.4°\text{C})}$$

$$c_x = 388\ \text{J/kg}\cdot°\text{C}$$

Remarks Comparing our results to values given in Table 11.1, the unknown substance is probably copper.

Exercise 11.3

A 255-g block of gold at 85.0°C is immersed in 155 g of water at 25.0°C. Find the equilibrium temperature, assuming the system is isolated and the heat capacity of the cup can be neglected.

Answer 27.9°C

TIP 11.2 Celsius versus Kelvin

In equations in which T appears, such as the ideal gas law, the Kelvin temperature *must* be used. In equations involving ΔT, such as calorimetry equations, it's possible to use either Celsius or Kelvin temperatures, because a change in temperature is the same on both scales. When in doubt, use Kelvin.

As long as there are no more than two substances involved, Equation 11.4 can be used to solve elementary calorimetry problems. Sometimes, however, there may be three (or more) substances exchanging thermal energy, each at a different temperature. If the problem requires finding the final temperature, it may not be clear whether the substance with the middle temperature gains or loses thermal energy. In such cases, Equation 11.4 can't be used reliably.

For example, suppose we want to calculate the final temperature of a system consisting initially of a glass beaker at 25°C, hot water at 40°C, and a block of aluminum at 37°C. We know that after the three are combined, the glass beaker warms up and the hot water cools, but we don't know for sure whether the aluminum block gains or loses energy because the final temperature is unknown.

Fortunately, we can still solve such a problem as long as it's set up correctly. With an unknown final temperature T_f, the expression $Q = mc(T_f - T_i)$ will be positive if $T_f > T_i$ and negative if $T_f < T_i$. Equation 11.4 can be written as

$$\Sigma Q_k = 0 \qquad \textbf{[11.5]}$$

where Q_k is the energy change in the kth object. Equation 11.5 says that the sum of all the gains and losses of thermal energy must add up to zero, as required by the conservation of energy for an isolated system. Each term in Equation 11.5 will have the correct sign automatically. Applying Equation 11.5 to the water, aluminum, and glass problem, we get

$$Q_w + Q_{\text{al}} + Q_g = 0$$

There's no need to decide in advance whether a substance in the system is gaining or losing energy. This equation is similar in style to the conservation of mechanical energy equation, where the gains and losses of kinetic and potential energies sum to zero for an isolated system: $\Delta K + \Delta PE = 0$. As will be seen, changes in thermal energy can be included on the left-hand side of this equation.

When more than two substances exchange thermal energy, it's easy to make errors substituting numbers, so it's a good idea to construct a table to organize and assemble all the data. This strategy is illustrated in the next example.

EXAMPLE 11.4 Calculate an Equilibrium Temperature

Goal Solve a calorimetry problem involving three substances at three different temperatures.

Problem Suppose 0.400 kg of water initially at 40.0°C is poured into a 0.300-kg glass beaker having a temperature of 25.0°C. A 0.500-kg block of aluminum at 37.0°C is placed in the water, and the system insulated. Calculate the final equilibrium temperature of the system.

Strategy The energy transfer for the water, aluminum, and glass will be designated Q_w, Q_{al}, and Q_g, respectively. The sum of these transfers must equal zero, by conservation of energy. Construct a table, assemble the three terms from the given data, and solve for the final equilibrium temperature, T.

Solution

Apply Equation 11.5 to the system:

$$Q_w + Q_{al} + Q_g = 0$$

$$m_w c_w (T - T_w) + m_{al} c_{al} (T - T_{al}) + m_g c_g (T - T_g) = 0 \quad \textbf{(1)}$$

Construct a data table:

Q (J)	m (kg)	c (J/kg °C)	T_f	T_i
Q_w	0.400	4 190	T	40.0°C
Q_{al}	0.500	9.00×10^2	T	37.0°C
Q_g	0.300	837	T	25.0°C

Using the table, substitute into Equation 1:

$$(1.68 \times 10^3 \text{ J/°C})(T - 40.0\text{°C}) + (4.50 \times 10^2 \text{ J/°C})(T - 37.0\text{°C}) + (2.51 \times 10^2 \text{ J/°C})(T - 25.0\text{°C}) = 0$$

$$(1.68 \times 10^3 \text{ J/°C} + 4.50 \times 10^2 \text{ J/°C} + 2.51 \times 10^2 \text{ J/°C})\,T = 9.01 \times 10^4 \text{ J}$$

$$T = 37.9\text{°C}$$

Remarks The answer turned out to be very close to the aluminum's initial temperature, so it would have been impossible to guess in advance whether the aluminum would lose or gain energy. Notice the way the table was organized, mirroring the order of factors in the different terms. This kind of organization helps prevent substitution errors, which are common in these problems.

Exercise 11.4

A 20.0-kg gold bar at 35.0°C is placed in a large, insulated 0.800-kg glass container at 15.0°C and 2.00 kg of water at 25.0°C. Calculate the final equilibrium temperature.

Answer 26.6°C

11.4 LATENT HEAT AND PHASE CHANGE

A substance usually undergoes a change in temperature when energy is transferred between the substance and its environment. In some cases, however, the transfer of energy doesn't result in a change in temperature. This can occur when the physical characteristics of the substance change from one form to another, commonly referred to as a **phase change**. Some common phase changes are solid to liquid (melting), liquid to gas (boiling), and a change in the crystalline structure of a solid. Any such phase change involves a change in the internal energy, but *no change in the temperature.*

◀ Latent heat

The energy Q needed to change the phase of a given pure substance is

$$Q = \pm mL \quad \textbf{[11.6]}$$

where L, called the **latent heat** of the substance, depends on the nature of the phase change as well as on the substance.

The unit of latent heat is the joule per kilogram (J/kg). The word *latent* means "lying hidden within a person or thing." The positive sign in Equation 11.6 is chosen when energy is absorbed by a substance, as when ice is melting. The negative sign is chosen when energy is removed from a substance, as when steam condenses to water.

TIP 11.3 Signs are Critical

For phase changes, use the correct explicit sign in Equation 11.6, positive if you are adding energy to the substance, negative if you're taking it away.

The **latent heat of fusion** L_f is used when a phase change occurs during melting or freezing, while the **latent heat of vaporization** L_v is used when a phase change occurs during boiling or condensing.[1] For example, at atmospheric pressure the

[1] When a gas cools, it eventually returns to the liquid phase, or *condenses.* The energy per unit mass given up during the process is called the *heat of condensation,* and it equals the heat of vaporization. When a liquid cools, it eventually solidifies, and the *heat of solidification* equals the heat of fusion.

TABLE 11.2

Latent Heats of Fusion and Vaporization

Substance	Melting Point (°C)	Latent Heat of Fusion (J/kg)	(cal/g)	Boiling Point (°C)	Latent Heat of Vaporization (J/kg)	(cal/g)
Helium	−269.65	5.23×10^3	(1.25)	−268.93	2.09×10^4	(4.99)
Nitrogen	−209.97	2.55×10^4	(6.09)	−195.81	2.01×10^5	(48.0)
Oxygen	−218.79	1.38×10^4	(3.30)	−182.97	2.13×10^5	(50.9)
Ethyl alcohol	−114	1.04×10^5	(24.9)	78	8.54×10^5	(204)
Water	0.00	3.33×10^5	(79.7)	100.00	2.26×10^6	(540)
Sulfur	119	3.81×10^4	(9.10)	444.60	3.26×10^5	(77.9)
Lead	327.3	2.45×10^4	(5.85)	1 750	8.70×10^5	(208)
Aluminum	660	3.97×10^5	(94.8)	2 450	1.14×10^7	(2 720)
Silver	960.80	8.82×10^4	(21.1)	2 193	2.33×10^6	(558)
Gold	1 063.00	6.44×10^4	(15.4)	2 660	1.58×10^6	(377)
Copper	1 083	1.34×10^5	(32.0)	1 187	5.06×10^6	(1 210)

latent heat of fusion for water is 3.33×10^5 J/kg, and the latent heat of vaporization for water is 2.26×10^6 J/kg. The latent heats of different substances vary considerably, as can be seen in Table 11.2.

Another process, sublimation, is the passage from the solid to the gaseous phase without going through a liquid phase. The fuming of dry ice (frozen carbon dioxide) illustrates this process, which has its own latent heat associated with it—the heat of sublimation.

EXAMPLE 11.5 Boiling Liquid Helium

Goal Apply the concept of latent heat of vaporization to liquid helium.

Problem Liquid helium has a very low boiling point, 4.2 K, as well as a low latent heat of vaporization, 2.09×10^4 J/kg. If energy is transferred to a container of liquid helium at the boiling point from an immersed electric heater at a rate of 10.0 W, how long does it take to boil away 2.00 kg of the liquid?

Strategy Because $L_v = 2.09 \times 10^4$ J/kg, boiling away each kilogram of liquid helium requires 2.09×10^4 J of energy. Joules divided by watts is time, so find the total energy needed and divide by the power to find the time.

Solution

Find the energy needed to vaporize 2.00 kg of liquid helium at its boiling point:

$$Q = mL_v = (2.00 \text{ kg})(2.09 \times 10^4 \text{ J/kg}) = 4.18 \times 10^4 \text{ J}$$

Divide this result by the power to find the time:

$$\Delta t = \frac{Q}{\mathcal{P}} = \frac{mL_v}{\mathcal{P}} = \frac{4.18 \times 10^4 \text{ J}}{10.0 \text{ W}}$$

$$\Delta t = 4.18 \times 10^3 \text{ s} = \boxed{69.7 \text{ min}}$$

Remark Notice that no change of temperature was involved. During such processes, the transferred energy goes into changing the state of the substance involved.

Exercise 11.5

If 10.0 W of power is supplied to 2.00 kg of water at 1.00×10^2°C, how long will it take for the water to completely boil away?

Answer 126 h

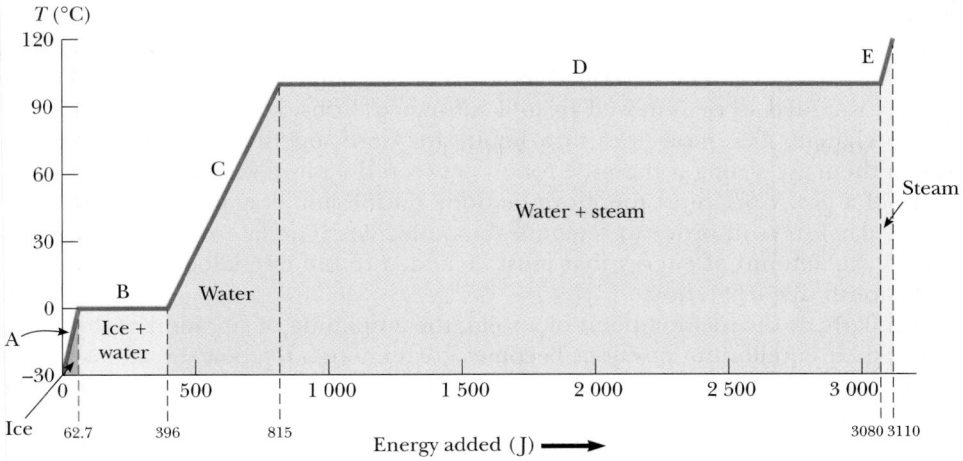

Figure 11.3 A plot of temperature versus energy added when 1.00 g of ice, initially at −30.0°C, is converted to steam at 120°C.

To better understand the physics of phase changes, consider the addition of energy to a 1.00-g cube of ice at − 30.0°C in a container held at constant pressure. Suppose this input of energy turns the ice to steam (water vapor) at 120.0°C. Figure 11.3 is a plot of the experimental measurement of temperature as energy is added to the system. We examine each portion of the curve separately.

Part A During this portion of the curve, the temperature of the ice changes from − 30.0°C to 0.0°C. Because the specific heat of ice is 2 090 J/kg · °C, we can calculate the amount of energy added from Equation 11.3:

$$Q = mc_{\text{ice}}\,\Delta T = (1.00 \times 10^{-3}\ \text{kg})(2\,090\ \text{J/kg}\cdot{}^\circ\text{C})(30.0^\circ\text{C}) = 62.7\ \text{J}$$

Part B When the ice reaches 0°C, the ice–water mixture remains at that temperature—even though energy is being added—until all the ice melts to become water at 0°C. According to Equation 11.6, the energy required to melt 1.00 g of ice at 0°C is

$$Q = mL_f = (1.00 \times 10^{-3}\ \text{kg})(3.33 \times 10^5\ \text{J/kg}) = 333\ \text{J}$$

Part C Between 0°C and 100°C, no phase change occurs. The energy added to the water is used to increase its temperature, as in part A. The amount of energy necessary to increase the temperature from 0°C to 100°C is

$$Q = mc_{\text{water}}\,\Delta T = (1.00 \times 10^{-3}\ \text{kg})(4.19 \times 10^3\ \text{J/kg}\cdot{}^\circ\text{C})(1.00 \times 10^2\ {}^\circ\text{C})$$

$$Q = 4.19 \times 10^2\ \text{J}$$

Part D At 100°C, another phase change occurs as the water changes to steam at 100°C. As in Part B, the water–steam mixture remains at constant temperature, this time at 100°C—even though energy is being added—until all the liquid has been converted to steam. The energy required to convert 1.00 g of water at 100°C to steam at 100°C is

$$Q = mL_v = (1.00 \times 10^{-3}\ \text{kg})(2.26 \times 10^6\ \text{J/kg}) = 2.26 \times 10^3\ \text{J}$$

Part E During this portion of the curve, as in parts A and C, no phase change occurs, so all the added energy goes into increasing the temperature of the steam. The energy that must be added to raise the temperature of the steam to 120.0°C is

$$Q = mc_{\text{steam}}\,\Delta T = (1.00 \times 10^{-3}\ \text{kg})(2.01 \times 10^3\ \text{J/kg}\cdot{}^\circ\text{C})(20.0^\circ\text{C}) = 40.2\ \text{J}$$

The total amount of energy that must be added to change 1.00 g of ice at − 30.0°C to steam at 120.0°C is the sum of the results from all five parts of the curve—3.11×10^3 J. Conversely, to cool 1.00 g of steam at 120.0°C down to the point at which it becomes ice at −30.0°C, 3.11×10^3 J of energy must be removed.

Phase changes can be described in terms of rearrangements of molecules when energy is added to or removed from a substance. Consider first the liquid-to-gas phase change. The molecules in a liquid are close together, and the forces between them are stronger than the forces between the more widely separated molecules of a gas. Work must therefore be done on the liquid against these attractive molecular forces in order to separate the molecules. The latent heat of vaporization is the amount of energy that must be added to the one kilogram of liquid to accomplish this separation.

Similarly, at the melting point of a solid, the amplitude of vibration of the atoms about their equilibrium positions becomes great enough to allow the atoms to pass the barriers of adjacent atoms and move to their new positions. On average, these new positions are less symmetrical than the old ones and therefore have higher energy. The latent heat of fusion is equal to the work required at the molecular level to transform the mass from the ordered solid phase to the disordered liquid phase.

The average distance between atoms is much greater in the gas phase than in either the liquid or the solid phase. Each atom or molecule is removed from its neighbors, overcoming the attractive forces of nearby neighbors. Therefore, more work is required at the molecular level to vaporize a given mass of a substance than to melt it, so in general the latent heat of vaporization is much greater than the latent heat of fusion (Table 11.2).

Quick Quiz 11.2

Calculate the slopes for the A, C, and E portions of Figure 11.3. Rank the slopes from least to greatest and explain what your ranking means. (a) A, C, E (b) C, A, E (c) E, A, C (d) E, C, A

Problem-Solving Strategy
Calorimetry with Phase Changes

1. **Make a table for all data.** Include separate rows for different phases and for any transition between phases. Include columns for each quantity used and a final column for the combination of the quantities. Transfers of thermal energy in this last column are given by $Q = mc\,\Delta T$, while phase changes are given by $Q = \pm mL_f$ for changes between liquid and solid, and by $Q = \pm mL_v$ for changes between liquid and gas.
2. **Apply conservation of energy.** If the system is isolated, use $\Sigma Q_k = 0$ (Eq. 11.5). For a nonisolated system, the net energy change should replace the zero on the right-hand side of that equation. ΣQ_k is just the sum of all the terms in the last column of the table.
3. **Solve** for the unknown quantity.

EXAMPLE 11.6 Ice Water

Goal Solve a problem involving heat transfer and a phase change from solid to liquid.

Problem At a party, 6.00 kg of ice at −5.00°C is added to a cooler holding 30 liters of water at 20.0°C. What is the temperature of the water when it comes to equilibrium?

Strategy In this problem, it's best to make a table. With the addition of thermal energy Q_{ice}, the ice will warm to 0°C; then melt at 0°C with the addition of energy Q_{melt}. Next, the melted ice will warm to some final temperature T by absorbing energy $Q_{\text{ice-water}}$, obtained from the energy change of the original liquid water, Q_{water}. By conservation of energy, these quantities must sum to zero.

Solution

Calculate the mass of liquid water:

$$m_{\text{water}} = \rho_{\text{water}} V$$
$$= (1.00 \times 10^3\ \text{kg/m}^3)(30.0\ \text{L})\frac{1.00\ \text{m}^3}{1.00 \times 10^3\ \text{L}}$$
$$= 30.0\ \text{kg}$$

Write the equation of thermal equilibrium:

$$Q_{\text{ice}} + Q_{\text{melt}} + Q_{\text{ice-water}} + Q_{\text{water}} = 0 \qquad \textbf{(1)}$$

Construct a comprehensive table:

Q	m (kg)	c (J/kg · °C)	L (J/kg)	T_f (°C)	T_i (°C)	Expression
Q_{ice}	6.00	2 090		0	−5.00	$m_{\text{ice}}c_{\text{ice}}(T_f - T_i)$
Q_{melt}	6.00		3.33×10^5	0	0	$m_{\text{ice}}L_f$
$Q_{\text{ice-water}}$	6.00	4 190		T	0	$m_{\text{ice}}c_{\text{wat}}(T_f - T_i)$
Q_{water}	30.0	4 190		T	20.0	$m_{\text{wat}}c_{\text{wat}}(T_f - T_i)$

Substitute all quantities in the second through sixth columns into the last column and sum (which is the evaluation of Equation 1), and solve for T:

$$6.27 \times 10^4\ \text{J} + 2.00 \times 10^6\ \text{J} + (2.51 \times 10^4\ \text{J/°C})(T - 0\text{°C}) + (1.26 \times 10^5\ \text{J/°C})(T - 20.0\text{°C}) = 0$$

$$T = 3.03\text{°C}$$

Remarks Making a table is optional. However, simple substitution errors are extremely common, and the table makes such errors less likely.

Exercise 11.6

What mass of ice at −10.0°C is needed to cool a whale's water tank, holding $1.20 \times 10^3\ \text{m}^3$ of water, from 20.0°C down to a more comfortable 10.0°C?

Answer 1.27×10^5 kg

EXAMPLE 11.7 Partial Melting

Goal Understand how to handle an incomplete phase change.

Problem A 5.00-kg block of ice at 0°C is added to an insulated container partially filled with 10.0 kg of water at 15.0°C. **(a)** Find the final temperature, neglecting the heat capacity of the container. **(b)** Find the mass of the ice that was melted.

Strategy Part (a) is tricky, because the ice does not entirely melt in this example. When there is any doubt concerning whether there will be a complete phase change, some preliminary calculations are necessary. First, find the total energy required to melt the ice, Q_{melt}, and then find Q_{water}, the maximum energy that can be delivered by the water above 0°C. If the energy delivered by the water is high enough, all the ice melts. If not, there will usually be a final mixture of ice and water at 0°C, unless the ice starts at a temperature far below 0°C, in which case all the liquid water freezes.

Solution

(a) Find the equilibrium temperature.

First, compute the amount of energy necessary to completely melt the ice:

$$Q_{\text{melt}} = m_{\text{ice}}L_f = (5.00\ \text{kg})(3.33 \times 10^5\ \text{J/kg})$$
$$= 1.67 \times 10^6\ \text{J}$$

Next, calculate the maximum energy that can be lost by the initial mass of liquid water without freezing it:

$$Q_{\text{water}} = m_{\text{water}}c\Delta T$$
$$= (10.0\ \text{kg})(4\,190\ \text{J/kg}\cdot\text{°C})(0\text{°C} - 15.0\text{°C})$$
$$= -6.29 \times 10^5\ \text{J}$$

This is less than half the energy necessary to melt all the ice, so the final state of the system is a mixture of water and ice at the freezing point:

$$T = 0°\text{C}$$

(b) Compute the mass of ice melted.

Set the total available energy equal to the heat of fusion of m grams of ice, mL_f:

$$6.29 \times 10^5\,\text{J} = mL_f = m(3.33 \times 10^5\,\text{J/kg})$$

$$m = 1.89\text{ kg}$$

Remarks If this problem is solved assuming (wrongly) that all the ice melts, a final temperature of $T = -16.5°\text{C}$ is obtained. The only way that could happen is if the system were not isolated, contrary to the statement of the problem. In the following exercise, you must also compute the thermal energy needed to warm the ice to its melting point.

Exercise 11.7

If 8.00 kg of ice at $-5.00°$ is added to 12.0 kg of water at 20.0°, compute the final temperature. How much ice remains, if any?

Answer $T = 0°\text{C}$, 5.22 kg

Sometimes problems involve changes in mechanical energy. During a collision, for example, some kinetic energy can be transformed to the internal energy of the colliding objects. This kind of transformation is illustrated in Example 11.8, involving a possible impact of a comet on Earth. In this example, a number of liberties will be taken in order to estimate the magnitude of the destructive power of such a catastrophic event. The specific heats depend on temperature and pressure, for example, but that will be ignored. Also, the ideal gas law doesn't apply at the temperatures and pressures attained, and the result of the collision wouldn't be superheated steam, but a plasma of charged particles. Despite all these simplifications, the example yields good order-of-magnitude results.

EXAMPLE 11.8 Armageddon!

Goal Link mechanical energy to thermal energy, phase changes, and the ideal gas law to create an estimate.

Problem A comet half a kilometer in radius consisting of ice at 273 K hits Earth at a speed of 4.00×10^4 m/s. For simplicity, assume that all the kinetic energy converts to thermal energy on impact and that all the thermal energy goes into warming the comet. **(a)** Calculate the volume and mass of the ice. **(b)** Use conservation of energy to find the final temperature of the comet material. Assume, contrary to fact, that the result is superheated steam and that the usual specific heats are valid, though in fact they depend on both temperature and pressure. **(c)** Assuming the steam retains a spherical shape and has the same initial volume as the comet, calculate the pressure of the steam using the ideal gas law. This law actually doesn't apply to a system at such high pressure and temperature, but can be used to get an estimate.

Strategy Part (a) requires the volume formula for a sphere and the definition of density. In part (b), conservation of energy can be applied. There are four processes involved: (1) melting the ice, (2) warming the ice water to the boiling point, (3) converting the boiling water to steam, and (4) warming the steam. The energy needed for these processes will be designated Q_{melt}, Q_{water}, Q_{vapor}, and Q_{steam}, respectively. These quantities plus the change in kinetic energy ΔK sum to zero because they are assumed to be internal to the system. In this case, the first three Q's can be neglected compared to the (extremely large) kinetic energy term. Solve for the unknown temperature, and substitute it into the ideal gas law in part (c).

Solution

(a) Find the volume and mass of the ice.

Apply the volume formula for a sphere:

$$V = \frac{4}{3}\pi r^3 = \frac{4}{3}(3.14)(5.00 \times 10^2\,\text{m})^3$$

$$= 5.23 \times 10^8\,\text{m}^3$$

Apply the density formula to find the mass of the ice:

$$m = \rho V = (917\ \text{kg/m}^3)(5.23 \times 10^8\ \text{m}^3)$$
$$= 4.80 \times 10^{11}\ \text{kg}$$

(b) Find the final temperature of the cometary material.

Use conservation of energy:

$$Q_{\text{melt}} + Q_{\text{water}} + Q_{\text{vapor}} + Q_{\text{steam}} + \Delta K = 0 \quad \textbf{(1)}$$
$$mL_f + mc_{\text{water}}\,\Delta T_{\text{water}} + mL_v + mc_{\text{steam}}\,\Delta T_{\text{steam}} + (0 - \tfrac{1}{2}mv^2) = 0 \quad \textbf{(2)}$$

The first three terms are negligible compared to the kinetic energy. The steam term involves the unknown final temperature, so retain only it and the kinetic energy, canceling the mass and solving for T:

$$mc_{\text{steam}}\,(T - 373\ \text{K}) - \tfrac{1}{2}mv^2 = 0$$
$$T = \frac{\frac{1}{2}v^2}{c_{\text{steam}}} + 373\ \text{K} = \frac{\frac{1}{2}(4.00 \times 10^4\text{m/s})^2}{2\,010\ \text{J/kg}\cdot\text{K}} + 373\ \text{K}$$
$$T = 3.98 \times 10^5\ \text{K}$$

(c) Estimate the pressure of the gas, using the ideal gas law.

First, compute the number of moles of steam:

$$n = (4.80 \times 10^{11}\ \text{kg})\left(\frac{1\ \text{mol}}{0.018\ \text{kg}}\right) = 2.67 \times 10^{13}\ \text{mol}$$

Solve for the pressure, using $PV = nRT$:

$$P = \frac{nRT}{V}$$
$$= \frac{(2.67 \times 10^{13}\ \text{mol})(8.31\ \text{J/mol}\cdot\text{K})(3.98 \times 10^5\ \text{K})}{5.23 \times 10^8\ \text{m}^3}$$
$$P = 1.69 \times 10^{11}\ \text{Pa}$$

Remarks The estimated pressure is several hundred times greater than the ultimate shear stress of steel! This high-pressure region would expand rapidly, destroying everything within a very large radius. Fires would ignite across a continent-sized region, and tidal waves would wrap around the world, wiping out coastal regions everywhere. The sun would be obscured for at least a decade, and numerous species, possibly including *Homo sapiens*, would become extinct. Such extinction events are rare, but in the long run represent a significant threat to life on Earth.

Exercise 11.8

Suppose a lead bullet with mass 5.00 g and an initial temperature of 65.0°C hits a wall and completely liquifies. What minimum speed did it have before impact? (*Hint:* The minimum speed corresponds to the case where all the kinetic energy becomes internal energy of the lead and the final temperature of the lead is at its melting point. Don't neglect any terms here!)

Answer 341 m/s

11.5 ENERGY TRANSFER

For some applications it's necessary to know the rate at which energy is transferred between a system and its surroundings and the mechanisms responsible for the transfer. This is particularly important in weatherproofing buildings or in medical applications, such as human survival time when exposed to the elements.

Earlier in this chapter we defined heat as a transfer of energy between a system and its surroundings due to a temperature difference between them. In this section, we take a closer look at heat as a means of energy transfer and consider the processes of thermal conduction, convection, and radiation.

George Semple

Figure 11.4 Conduction makes the metal handle of a cooking pan hot.

Thermal Conduction

The energy transfer process most closely associated with a temperature difference is called **thermal conduction** or simply **conduction**. In this process, the transfer can be viewed on an atomic scale as an exchange of kinetic energy between microscopic particles—molecules, atoms, and electrons—with less energetic particles gaining energy as they collide with more energetic particles. An inexpensive pot, as in Figure 11.4, may have a metal handle with no surrounding insulation. As the pot is warmed, the temperature of the metal handle increases, and the cook must hold it with a cloth potholder to avoid being burned.

The way the handle warms up can be understood by looking at what happens to the microscopic particles in the metal. Before the pot is placed on the stove, the particles are vibrating about their equilibrium positions. As the stove coil warms up, those particles in contact with it begin to vibrate with larger amplitudes. These particles collide with their neighbors and transfer some of their energy in the collisions. Metal atoms and electrons farther and farther from the flame gradually increase the amplitude of their vibrations, until eventually those in the metal near your hand are affected. This increased vibration represents an increase in temperature of the metal (and possibly a burned hand!).

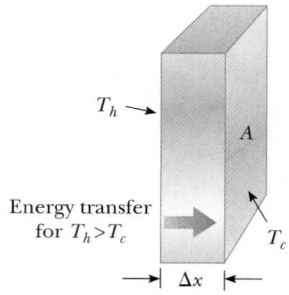

Figure 11.5 Energy transfer through a conducting slab of cross-sectional area A and thickness L. The opposite faces are at different temperatures T_c and T_h.

Although the transfer of energy through a substance can be partly explained by atomic vibrations, the rate of conduction depends on the properties of the substance. For example, it's possible to hold a piece of asbestos in a flame indefinitely. This fact implies that very little energy is conducted through the asbestos. In general, metals are good thermal conductors because they contain large numbers of electrons that are relatively free to move through the metal and can transport energy from one region to another. In a good conductor such as copper, conduction takes place via the vibration of atoms and the motion of free electrons. Materials such as asbestos, cork, paper, and fiberglass are poor thermal conductors. Gases are also poor thermal conductors because of the large distance between their molecules.

Conduction occurs only if there is a difference in temperature between two parts of the conducting medium. The temperature difference drives the flow of energy. Consider a slab of material of thickness Δx and cross-sectional area A with its opposite faces at different temperatures T_c and T_h, where $T_h > T_c$ (Fig. 11.5). The slab allows energy to transfer from the region of higher temperature to the region of lower temperature by thermal conduction. The rate of energy transfer, $\mathcal{P} = Q/\Delta t$, is proportional to the cross-sectional area of the slab and the temperature difference and is inversely proportional to the thickness of the slab:

$$\mathcal{P} = \frac{Q}{\Delta t} \propto A\frac{\Delta T}{\Delta x}$$

Note that $\mathcal{P}$ has units of watts when Q is in joules and Δt is in seconds.

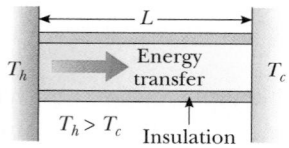

Figure 11.6 Conduction of energy through a uniform, insulated rod of length L. The opposite ends are in thermal contact with energy reservoirs at different temperatures.

Suppose a substance is in the shape of a long, uniform rod of length L, as in Figure 11.6. We assume the rod is insulated, so thermal energy can't escape by conduction from its surface except at the ends. One end is in thermal contact with an energy reservoir at temperature T_c and the other end is in thermal contact with a reservoir at temperature $T_h > T_c$. When a steady state is reached, the temperature at each point along the rod is constant in time. In this case, $\Delta T = T_h - T_c$ and $\Delta x = L$, so

$$\frac{\Delta T}{\Delta x} = \frac{T_h - T_c}{L}$$

The rate of energy transfer by conduction through the rod is given by

$$\mathcal{P} = kA\frac{(T_h - T_c)}{L} \qquad [11.7]$$

TIP 11.4 Blankets and Coats in Cold Weather

When you sleep under a blanket in the winter or wear a warm coat outside, the blanket or coat serves as a layer of material with low thermal conductivity in order to reduce the transfer of energy away from your body by heat. The primary insulating medium is the air trapped in small pockets within the material.

where k, a proportionality constant that depends on the material, is called the **thermal conductivity**. Substances that are good conductors have large thermal conductivities, whereas good insulators have low thermal conductivities. Table 11.3 lists the thermal conductivities for various substances.

TABLE 11.3

Thermal Conductivities

Substance	Thermal Conductivity ($J/s \cdot m \cdot °C$)
Metals (at 25°C)	
Aluminum	238
Copper	397
Gold	314
Iron	79.5
Lead	34.7
Silver	427
Gases (at 20°C)	
Air	0.023 4
Helium	0.138
Hydrogen	0.172
Nitrogen	0.023 4
Oxygen	0.023 8
Nonmetals	
Asbestos	0.25
Concrete	1.3
Glass	0.84
Ice	1.6
Rubber	0.2
Water	0.60
Wood	0.10

Quick Quiz 11.3

Will an ice cube wrapped in a wool blanket remain frozen for (a) less time, (b) the same length of time, or (c) a longer time than an identical ice cube exposed to air at room temperature?

Quick Quiz 11.4

Two rods of the same length and diameter are made from different materials. The rods are to connect two regions of different temperature so that energy will transfer through the rods by heat. They can be connected in series, as in Figure 11.7a, or in parallel, as in Figure 11.7b. In which case is the rate of energy transfer by heat larger? (a) When the rods are in series (b) When the rods are in parallel (c) The rate is the same in both cases.

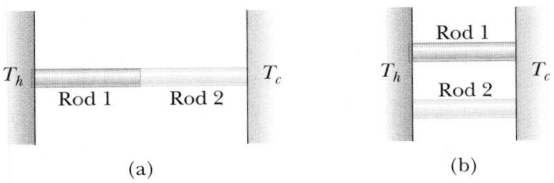

Figure 11.7 (Quick Quiz 11.4) In which case is the rate of energy transfer larger?

EXAMPLE 11.9 Energy Transfer through a Concrete Wall

Goal Apply the equation of heat conduction.

Problem Find the energy transferred in 1.00 h by conduction through a concrete wall 2.0 m high, 3.65 m long, and 0.20 m thick if one side of the wall is held at 20°C and the other side is at 5°C.

Strategy Equation 11.7 gives the rate of energy transfer by conduction in joules per second. Multiply by the time and substitute given values to find the total thermal energy transferred.

Solution

Multiply Equation 11.7 by Δt to find an expression for the total energy Q transferred through the wall:

$$Q = \mathcal{P}\,\Delta t = kA\left(\frac{T_h - T_c}{L}\right)\Delta t$$

Substitute the numerical values to obtain Q, consulting Table 11.3 for k:

$$Q = (1.3\ \mathrm{J/s \cdot m \cdot °C})(7.3\ \mathrm{m}^2)\left(\frac{15°\mathrm{C}}{0.20\ \mathrm{m}}\right)(3\,600\ \mathrm{s})$$

$$= \boxed{2.6 \times 10^6\ \mathrm{J}}$$

Remarks Early houses were insulated with thick masonry walls, which restrict energy loss by conduction because k is relatively low. The large thickness L also decreases energy loss by conduction, as shown by Equation 11.7. There are much better insulating materials, however, and layering is also helpful. Despite the low thermal conductivity of masonry, the amount of energy lost is still rather large—enough to raise the temperature of 600 kg of water by more than 1°C. There are better insulating materials than masonry.

Exercise 11.9

A wooden shelter has walls constructed of wooden planks 1.00 cm thick. If the exterior temperature is $-20.0°C$ and the interior is $5.00°C$, find the rate of energy loss through a wall that has dimensions 2.00 m by 2.00 m.

Answer 1.00×10^3 W

Home Insulation

To determine whether to add insulation to a ceiling or some other part of a building, the preceding discussion of conduction must be extended, for two reasons:

1. The insulating properties of materials used in buildings are usually expressed in engineering (U.S. customary) rather than SI units. Measurements stamped on a package of fiberglass insulating board will be in units such as British thermal units, feet, and degrees Fahrenheit.
2. In dealing with the insulation of a building, conduction through a compound slab must be considered, with each portion of the slab having a certain thickness and a specific thermal conductivity. A typical wall in a house consists of an array of materials, such as wood paneling, drywall, insulation, sheathing, and wood siding.

The rate of energy transfer by conduction through a compound slab is

$$\frac{Q}{\Delta t} = \frac{A(T_h - T_c)}{\sum_i L_i/k_i} \qquad [11.8]$$

where T_h and T_c are the temperatures of the *outer extremities* of the slab and the summation is over all portions of the slab. This formula can be derived algebraically, using the facts that the temperature at the interface between two insulating materials must be the same and that the rate of energy transfer through one insulator must be the same as through all the other insulators. If the slab consists of three different materials, the denominator is the sum of three terms. In engineering practice, the term L/k for a particular substance is referred to as the

TABLE 11.4

***R* Values for Some Common Building Materials**

Material	*R* value ($ft^2 \cdot °F \cdot h/Btu$)
Hardwood siding (1.0 in. thick)	0.91
Wood shingles (lapped)	0.87
Brick (4.0 in. thick)	4.00
Concrete block (filled cores)	1.93
Styrofoam (1.0 in. thick)	5.0
Fiber glass batting (3.5 in. thick)	10.90
Fiber glass batting (6.0 in. thick)	18.80
Fiber glass board (1.0 in. thick)	4.35
Cellulose fiber (1.0 in. thick)	3.70
Flat glass (0.125 in. thick)	0.89
Insulating glass (0.25-in. space)	1.54
Vertical air space (3.5 in. thick)	1.01
Stagnant layer of air	0.17
Dry wall (0.50 in. thick)	0.45
Sheathing (0.50 in. thick)	1.32

***R* value** of the material, so Equation 11.8 reduces to

$$\frac{Q}{\Delta t} = \frac{A(T_h - T_c)}{\sum_i R_i} \qquad [11.9]$$

The R values for a few common building materials are listed in Table 11.4. Note the unit of R and the fact that the R values are defined for specific thicknesses.

Next to any vertical outside surface is a very thin, stagnant layer of air that must be considered when the total R value for a wall is computed. The thickness of this stagnant layer depends on the speed of the wind. As a result, energy loss by conduction from a house on a day when the wind is blowing is greater than energy loss on a day when the wind speed is zero. A representative R value for a stagnant air layer is given in Table 11.4.

INTERACTIVE EXAMPLE 11.10 The *R* Value of a Typical Wall

Goal Calculate the R value of a wall consisting of several layers of insulating material.

Problem Calculate the total R value for a wall constructed as shown in Figure 11.8a. Starting outside the house (to the left in the figure) and moving inward, the wall consists of 4.0 in. brick, 0.50 in. sheathing, an air space 3.5 in. thick, and 0.50 in. drywall.

Strategy Add all the R values together, remembering the stagnant air layers inside and outside the house.

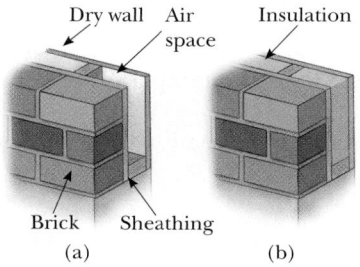

Figure 11.8 (Example 11.10) A cross-sectional view of an exterior wall containing (a) an air space and (b) insulation.

Solution

Refer to Table 11.4, and sum. All quantities are in units of $\text{ft}^2 \cdot {}^\circ\text{F} \cdot \text{h/Btu}$.

$$R_{\text{total}} = R_{\text{outside air layer}} + R_{\text{brick}} + R_{\text{sheath}} + R_{\text{air space}} + R_{\text{drywall}} + R_{\text{inside air layer}} = (0.17 + 4.00 + 1.32 + 1.01 + 0.45 + 0.17)\,\text{ft}^2 \cdot {}^\circ\text{F} \cdot \text{h/Btu}$$

$$R_{\text{total}} = 7.12\ \text{ft}^2 \cdot {}^\circ\text{F} \cdot \text{h/Btu}$$

Exercise 11.10

If a layer of fiber glass insulation 3.5 in. thick is placed inside the wall to replace the air space, as in Figure 11.8b, what is the new total R value? By what factor is the energy loss reduced?

Answer $R = 17\ \text{ft}^2 \cdot {}^\circ\text{F} \cdot \text{h/Btu}$; 2.4

PhysicsNow™ Study the R values of various types of common building materials by logging into PhysicsNow at **www.cp7e.com** and going to Interactive Example 11.10.

EXAMPLE 11.11 Staying Warm in the Arctic

Goal Combine two layers of insulation.

Problem An arctic explorer builds a wooden shelter out of wooden planks that are 1.0 cm thick. To improve the insulation, he covers the shelter with a layer of ice 3.2 cm thick. **(a)** Compute the R factors for the wooden planks and the ice. **(b)** If the temperature outside the shelter is $-20.0°\text{C}$ and the temperature inside is $5.00°\text{C}$, find the rate of energy loss through one of the walls, if the wall has dimensions 2.00 m by 2.00 m. **(c)** Find the temperature at the interface between the wood and the ice.

Strategy After finding the R values, substitute into Equation 11.9 to get the rate of energy transfer. To answer part (c), use Equation 11.7 for one of the layers, setting it equal to the rate found in part (b), solving for the temperature.

Solution

(a) Compute the R values using the data in Table 11.3.

Find the R value for the wooden wall:

$$R_{\text{wood}} = \frac{L_{\text{wood}}}{k_{\text{wood}}} = \frac{0.01\ \text{m}}{0.10\ \text{J/s}\cdot\text{m}\cdot{}^\circ\text{C}} = 0.10\ \text{m}^2\cdot\text{s}\cdot{}^\circ\text{C/J}$$

Find the R-value for the ice layer:

$$R_{\text{ice}} = \frac{L_{\text{ice}}}{k_{\text{ice}}} = \frac{0.032\ \text{m}}{1.6\ \text{J/s}\cdot\text{m}\cdot{}^\circ\text{C}} = 0.020\ \text{m}^2\cdot\text{s}\cdot{}^\circ\text{C/J}$$

(b) Find the rate of heat loss.

Apply Equation 11.9:

$$\mathcal{P} = \frac{Q}{\Delta t} = \frac{A(T_h - T_c)}{\sum_i R_i}$$

$$= \frac{(4.00\ \text{m}^2)(5.00^\circ\text{C} - (-20.0^\circ\text{C}))}{0.12\ \text{m}^2\cdot\text{s}\cdot{}^\circ\text{C/J}}$$

$$\mathcal{P} = 830\ \text{W}$$

(c) Find the temperature in between the ice and wood.

Apply the equation of heat conduction to the wood:

$$\frac{k_{\text{wood}}A(T_h - T_c)}{L} = \mathcal{P}$$

$$\frac{(0.10\ \text{J/s}\cdot\text{m}\cdot{}^\circ\text{C})(4.00\ \text{m}^2)(5.00^\circ\text{C} - T)}{0.010\ \text{m}} = 830\ \text{W}$$

Solve for the unknown temperature:

$$T = -16^\circ\text{C}$$

Remarks The outer side of the wooden wall and the inner surface of the ice must have the same temperature, and the rate of energy transfer through the ice must be the same as through the wooden wall. Using Equation 11.7 for ice instead of wood gives the same answer. This rate of energy transfer is only a modest improvement over the thousand-watt rate in Exercise 11.9. The choice of insulating material is important!

Exercise 11.11

Rather than use ice to cover the wooden shelter, the explorer glues pressed cork with thickness 0.500 cm to the outside of his wooden shelter. Find the new rate of energy loss through the same wall. (Note that $k_{\text{cork}} = 0.046\ \text{J/s}\cdot\text{m}\cdot{}^\circ\text{C}$.)

Answer 480 W

Figure 11.9 Warming a hand by convection.

Convection

When you warm your hands over an open flame, as illustrated in Figure 11.9, the air directly above the flame, being warmed, expands. As a result, the density of this air decreases and the air rises, warming your hands as it flows by. **The transfer of energy by the movement of a substance is called convection.** When the movement results from differences in density, as with air around a fire, it's referred to as *natural convection.* Airflow at a beach is an example of natural convection, as is the mixing that occurs as surface water in a lake cools and sinks. When the substance is forced to move by a fan or pump, as in some hot air and hot water heating systems, the process is called *forced convection.*

Convection currents assist in the boiling of water. In a teakettle on a hot stovetop, the lower layers of water are warmed first. The warmed water has a lower density and rises to the top, while the denser, cool water at the surface sinks to the bottom of the kettle and is warmed.

The same process occurs when a radiator raises the temperature of a room. The hot radiator warms the air in the lower regions of the room. The warm air expands and, because of its lower density, rises to the ceiling. The denser cooler air from above sinks, setting up the continuous air current pattern shown in Figure 11.10.

An automobile engine is maintained at a safe operating temperature by a combination of conduction and forced convection. Water (actually, a mixture of water and antifreeze) circulates in the interior of the engine. As the metal of the engine block increases in temperature, energy passes from the hot metal to the cooler water by thermal conduction. The water pump forces water out of the engine and into the radiator, carrying energy along with it (by forced convection). In the radiator, the hot water passes through metal pipes that are in contact with the cooler outside air, and energy passes into the air by conduction. The cooled water is then returned to the engine by the water pump to absorb more energy. The process of air being pulled past the radiator by the fan is also forced convection.

The algal blooms often seen in temperate lakes and ponds during the spring or fall are caused by convection currents in the water. To understand this process, consider Figure 11.11. During the summer, bodies of water develop temperature gradients, with an upper, warm layer of water separated from a lower, cold layer by a buffer zone called a thermocline. In the spring or fall, temperature changes in the water break down this thermocline, setting up convection currents that mix the water. The mixing process transports nutrients from the bottom to the surface. The nutrient-rich water forming at the surface can cause a rapid, temporary increase in the algae population.

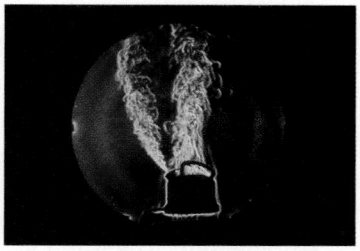

Gary Settles/Science Source/Photo Researchers, Inc.

Photograph of a teakettle, showing steam and turbulent convection air currents.

APPLICATION

Cooling Automobile Engines

APPLICATION

Algal Blooms in Ponds and Lakes

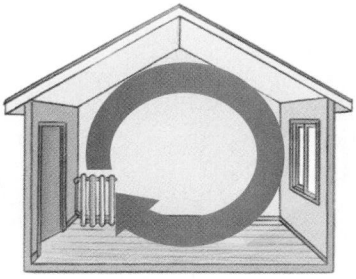

Figure 11.10 Convection currents are set up in a room warmed by a radiator.

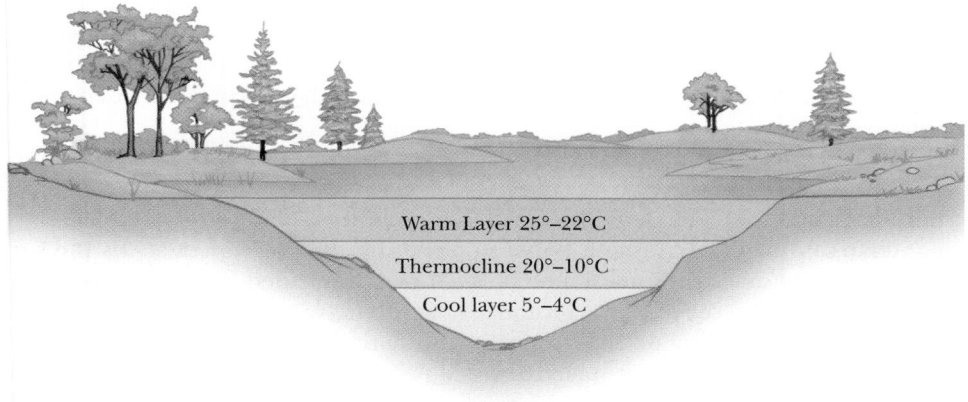

(a) Summer layering of water

(b) Fall and spring upwelling

Figure 11.11 (a) During the summer, a warm upper layer of water is separated from a cooler lower layer by a thermocline. (b) Convection currents during the spring or fall mix the water and can cause algal blooms.

Applying Physics 11.1 Body Temperature

The body temperature of mammals ranges from about 35°C to 38°C, while that of birds ranges from about 40°C to 43°C. How can these narrow ranges of body temperature be maintained in cold weather?

Explanation A natural method of maintaining body temperature is via layers of fat beneath the skin. Fat protects against both conduction and convection because of its low thermal conductivity and because there are few blood vessels in fat to carry blood to the surface, where energy losses by convection can occur. Birds ruffle their feathers in cold weather in order to trap a layer of air with a low thermal conductivity between the feathers and the skin. Bristling the fur produces the same effect in fur-bearing animals.

Humans keep warm with wool sweaters and down jackets that trap the warmer air in regions close to their bodies, reducing energy loss by convection and conduction.

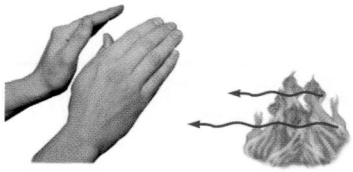

Figure 11.12 Warming hands by radiation.

Radiation

Another process of transferring energy is through **radiation**. Figure 11.12 shows how your hands can be warmed at an open flame through radiation. Because your hands aren't in physical contact with the flame and the conductivity of air is very low, conduction can't account for the energy transfer. Nor can convection be responsible for any transfer of energy, because your hands aren't above the flame in the path of convection currents. The warmth felt in your hands, therefore, must come from the transfer of energy by radiation.

All objects radiate energy continuously in the form of electromagnetic waves due to thermal vibrations of their molecules. These vibrations create the orange glow of an electric stove burner, an electric space heater, and the coils of a toaster.

The rate at which an object radiates energy is proportional to the fourth power of its absolute temperature. This is known as **Stefan's law**, expressed in equation form as

Stefan's law ▶

$$\mathcal{P} = \sigma A e T^4 \qquad [11.10]$$

where $\mathcal{P}$ is the power in watts (or joules per second) radiated by the object, σ is the Stefan–Boltzmann constant, equal to $5.669\,6 \times 10^{-8}\ \mathrm{W/m^2 \cdot K^4}$, A is the surface area of the object in square meters, e is a constant called the **emissivity** of the object, and T is the object's Kelvin temperature. The value of e can vary between zero and one, depending on the properties of the object's surface.

Approximately 1 340 J of electromagnetic radiation from the Sun passes through each square meter at the top of the Earth's atmosphere every second. This radiation is primarily visible light, accompanied by significant amounts of infrared and ultraviolet. We will study these types of radiation in detail in Chapter 21. Some of this energy is reflected back into space, and some is absorbed by the atmosphere, but enough arrives at the surface of the Earth each day to supply all our energy needs hundreds of times over—if it could be captured and used efficiently. The growth in the number of solar houses in the United States is one example of an attempt to make use of this abundant energy. Radiant energy from the Sun affects our day-to-day existence in a number of ways, influencing Earth's average temperature, ocean currents, agriculture, and rain patterns. It can also affect behavior.

As another example of the effects of energy transfer by radiation, consider what happens to the atmospheric temperature at night. If there is a cloud cover above Earth, the water vapor in the clouds absorbs part of the infrared radiation emitted by Earth and re-emits it back to the surface. Consequently, the temperature at the surface remains at moderate levels. In the absence of cloud cover, there is nothing to prevent the radiation from escaping into space, so the temperature drops more on a clear night than when it's cloudy.

As an object radiates energy at a rate given by Equation 11.10, it also absorbs radiation. If it didn't, the object would eventually radiate all its energy and its temperature would reach absolute zero. The energy an object absorbs comes from its environment, which consists of other bodies that radiate energy. If an object is at a temperature T, and its surroundings are at a temperature T_0, the net energy gained or lost each second by the object as a result of radiation is

$$\mathcal{P}_{\text{net}} = \sigma A e(T^4 - T_0^4) \qquad [11.11]$$

When an object is in equilibrium with its surroundings, it radiates and absorbs energy at the same rate, so its temperature remains constant. When an object is hotter than its surroundings, it radiates more energy than it absorbs and so cools.

An **ideal absorber** is an object that absorbs all the light radiation incident on it, including invisible infrared and ultraviolet light. Such an object is called a **black body** because a room-temperature black body would look black. Since a black body doesn't reflect radiation at any wavelength, any light coming from it is due to atomic and molecular vibrations alone. A perfect black body has emissivity $e = 1$. An ideal absorber is also an ideal radiator of energy. The Sun, for example, is nearly a perfect black body. This statement may seem contradictory, because the Sun is bright, not dark; however, the light that comes from the Sun is emitted, not reflected. Black bodies are perfect absorbers that look black at room temperature because they don't reflect any light. All black bodies, except those at absolute zero, emit light that has a characteristic spectrum, to be discussed in Chapter 27. In contrast to black bodies, an object for which $e = 0$ absorbs none of the energy incident on it, reflecting it all. Such a body is an **ideal reflector**.

APPLICATION
Light-Colored Summer Clothing

White clothing is more comfortable to wear in the summer than black clothing. Black fabric acts as a good absorber of incoming sunlight and as a good emitter of this absorbed energy. About half of the emitted energy, however, travels toward the body, causing the person wearing the garment to feel uncomfortably warm. White or light-colored clothing reflects away much of the incoming energy.

APPLICATION
Thermography

The amount of energy radiated by an object can be measured with temperature-sensitive recording equipment via a technique called **thermography**. An image of the pattern formed by varying radiation levels, called a **thermogram**, is brightest in the warmest areas. Figure 11.13 reproduces a thermogram of a house. More energy escapes in the lighter regions, such as the door and windows. The owners of this house could conserve energy and reduce their heating costs by adding insulation to the attic area and by installing thermal draperies over the windows. Thermograms have also been used to image injured or diseased tissue in medicine, since such areas are often at a different temperature than surrounding healthy tissue, though many radiologists consider thermograms inadequate as a diagnostic tool.

APPLICATION
Radiation Thermometers for Measuring Body Temperature

Figure 11.14 shows a recently developed radiation thermometer that has removed most of the risk of taking the temperature of young children or the aged with a rectal thermometer—risks such as bowel perforation or bacterial contamination. The instrument measures the intensity of the infrared radiation leaving

Daedalus Enterprises, Inc./Peter Arnold, Inc.

Figure 11.13 This thermogram of a house, made during cold weather, shows colors ranging from white and yellow (areas of greatest energy loss) to blue and purple (areas of least energy loss).

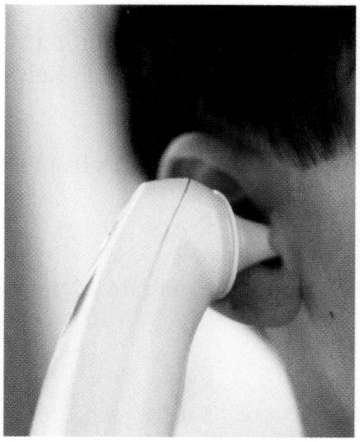

Photodisc/Getty Images

Figure 11.14 A radiation thermometer measures a patient's temperature by monitoring the intensity of infrared radiation leaving the ear.

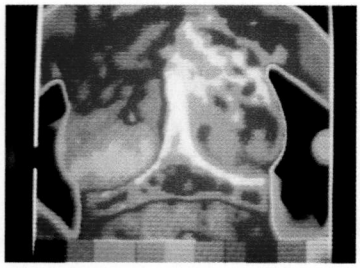

SPL/Photo Researchers, Inc.

Thermogram of a woman's breasts. Her left breast is diseased (red and orange) and her right breast (blue) is healthy.

the eardrum and surrounding tissues and converts this information to a standard numerical reading. The eardrum is a particularly good location to measure body temperature because it's near the hypothalamus—the body's temperature control center.

Quick Quiz 11.5

Stars A and B have the same temperature, but star A has twice the radius of star B. (a) What is the ratio of star A's power output to star B's output due to electromagnetic radiation? The emissivity of both stars can assumed to be 1. (b) Repeat the question if the stars have the same radius, but star A has twice the absolute temperature of star B. (c) What's the ratio if star A has both twice the radius and twice the absolute temperature of star B?

Applying Physics 11.2 Thermal Radiation and Night Vision

How can thermal radiation be used to see objects in near total darkness?

Explanation There are two methods of night vision, one enhancing a combination of very faint visible light and infrared light, and another using infrared light only. The latter is valuable for creating images in absolute darkness. Because all objects above absolute zero emit thermal radiation due to the vibrations of their atoms, the infrared (invisible) light can be focused by a special lens and scanned by an array of infrared detector elements. These elements create a thermogram. The information from thousands of separate points in the field of view is converted to electrical impulses and translated by a microchip into a form suitable for display. Different temperature areas are assigned different colors, which can then be easily discerned on the display.

EXAMPLE 11.12 Polar Bear Club

Goal Apply Stefan's law.

Problem A member of the Polar Bear Club, dressed only in bathing trunks of negligible size, prepares to plunge into the Baltic Sea from the beach in St. Petersburg, Russia. The air is calm, with a temperature of 5°C. If the swimmer's surface body temperature is 25°C, compute the net rate of energy loss from his skin due to radiation. How much energy is lost in 10.0 min? Assume his emissivity is 0.900, and his surface area is $1.50\ \text{m}^2$.

Strategy Use Equation 11.11, the thermal radiation equation, substituting the given information. Remember to convert temperatures to Kelvin by adding 273 to each value in degrees Celsius!

Solution

Convert temperatures from Celsius to Kelvin:

$$T_{5°\text{C}} = T_C + 273 = 5 + 273 = 278\ \text{K}$$

$$T_{25°\text{C}} = T_C + 273 = 25 + 273 = 298\ \text{K}$$

Compute the net rate of energy loss, using Equation 11.11:

$$\begin{aligned}\mathcal{P}_{\text{net}} &= \sigma A e (T^4 - T_0^4)\\ &= (5.67 \times 10^{-8}\ \text{W/m}^2\cdot\text{K}^4)(1.50\ \text{m}^2)\\ &\quad \times (0.90)[(298\ \text{K})^4 - (278\ \text{K})^4]\\ \mathcal{P}_{\text{net}} &= \boxed{146\ \text{W}}\end{aligned}$$

Multiply the preceding result by the time, 10 minutes, to get the energy lost in that time due to radiation:

$$Q = \mathcal{P}_{\text{net}} \times \Delta t = (146)(6.00 \times 10^2\ \text{s}) = \boxed{8.76 \times 10^4\ \text{J}}$$

Remarks Energy is also lost from the body through convection and conduction. Clothing traps layers of air next to the skin, which are warmed by radiation and conduction. In still air these warm layers are more readily retained. Even a Polar Bear Club member enjoys some benefit from the still air, better retaining a stagnant air layer next to the surface of his skin.

Exercise 11.12

Repeat the calculation when the man is standing in his bedroom, with an ambient temperature of 20.0°C. Assume his body surface temperature is 27.0°C, with emissivity of 0.900.

Answer 55.9 W, 3.35×10^4 J

The Dewar Flask

The Thermos bottle, also called a **Dewar flask** (after its inventor), is designed to minimize energy transfer by conduction, convection, and radiation. The thermos can store either cold or hot liquids for long periods. The standard vessel (Fig. 11.15) is a double-walled Pyrex glass with silvered walls. The space between the walls is evacuated to minimize energy transfer by conduction and convection. The silvered surface minimizes energy transfer by radiation because silver is a very good reflector and has very low emissivity. A further reduction in energy loss is achieved by reducing the size of the neck. Dewar flasks are commonly used to store liquid nitrogen (boiling point 77 K) and liquid oxygen (boiling point 90 K).

To confine liquid helium (boiling point 4.2 K), which has a very low heat of vaporization, it's often necessary to use a double Dewar system in which the Dewar flask containing the liquid is surrounded by a second Dewar flask. The space between the two flasks is filled with liquid nitrogen.

Some of the principles of the Thermos bottle are used in the protection of sensitive electronic instruments in orbiting space satellites. In half of its orbit around the Earth a satellite is exposed to intense radiation from the Sun, and in the other half it lies in the Earth's cold shadow. Without protection, its interior would be subjected to tremendous extremes of temperature. The interior of the satellite is wrapped with blankets of highly reflective aluminum foil. The foil's shiny surface reflects away much of the Sun's radiation while the satellite is in the unshaded part of the orbit and helps retain interior energy while the satellite is in the Earth's shadow.

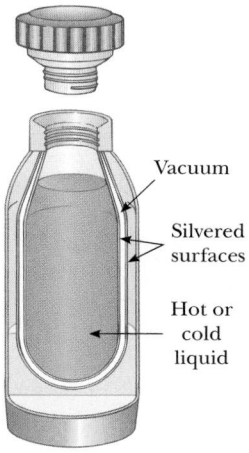

Figure 11.15 A cross-sectional view of a Thermos bottle designed to store hot or cold liquids.

APPLICATION

Thermos Bottles

11.6 GLOBAL WARMING AND GREENHOUSE GASES

Many of the principles of energy transfer, and opposition to it, can be understood by studying the operation of a glass greenhouse. During the day, sunlight passes into the greenhouse and is absorbed by the walls, soil, plants, and so on. This absorbed visible light is subsequently reradiated as infrared radiation, causing the temperature of the interior to rise.

In addition, convection currents are inhibited in a greenhouse. As a result, warmed air can't rapidly pass over the surfaces of the greenhouse that are exposed to the outside air and thereby cause an energy loss by conduction through those surfaces. Most experts now consider this restriction to be a more important warming effect than the trapping of infrared radiation. In fact, experiments have shown that when the glass over a greenhouse is replaced by a special glass known to transmit infrared light, the temperature inside is lowered only slightly. On the basis of this evidence, the primary mechanism that raises the temperature of a greenhouse is not the trapping of infrared radiation, but the inhibition of airflow that occurs under any roof (in an attic, for example).

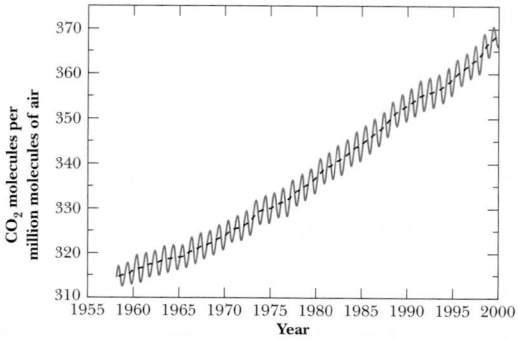

Figure 11.16 The concentration of atmospheric carbon dioxide in parts per million (ppm) of dry air as a function of time during the latter part of the 20th century. These data were recorded at Mauna Loa Observatory in Hawaii. The yearly variations (red curve) coincide with growing seasons, because vegetation absorbs carbon dioxide from the air. The steady increase (black curve) is of concern to scientists.

A phenomenon commonly known as the **greenhouse effect** can also play a major role in determining the Earth's temperature. First, note that the Earth's atmosphere is a good transmitter (and hence a poor absorber) of visible radiation and a good absorber of infrared radiation. The visible light that reaches the Earth's surface is absorbed and reradiated as infrared light, which in turn is absorbed (trapped) by the Earth's atmosphere. An extreme case is the warmest planet, Venus, which has a carbon dioxide (CO_2) atmosphere and temperatures approaching 850°F.

As fossil fuels (coal, oil, and natural gas) are burned, large amounts of carbon dioxide are released into the atmosphere, causing it to retain more energy. This is of great concern to scientists and governments throughout the world. Many scientists are convinced that the 10% increase in the amount of atmospheric carbon dioxide in the past 30 years could lead to drastic changes in world climate. The increase in concentration of atmospheric carbon dioxide in the latter part of the 20th century is shown in Figure 11.16. According to one estimate, doubling the carbon dioxide content in the atmosphere will cause temperatures to increase by 2°C. In temperate regions, such as Europe and the United States, a 2°C temperature rise would save billions of dollars per year in fuel costs. Unfortunately, it would also melt a large amount of ice from the polar ice caps, which could cause flooding and destroy many coastal areas. A 2°C rise would also increase the frequency of droughts, and consequently decrease already low crop yields in tropical and subtropical countries. Even slightly higher average temperatures might make it impossible for certain plants and animals to survive in their customary ranges.

At present, about 3.5×10^{11} tons of CO_2 are released into the atmosphere each year. Most of this gas results from human activities such as the burning of fossil fuels, the cutting of forests, and manufacturing processes. Another greenhouse gas is methane (CH_4), which is released in the digestive process of cows and other ruminants. This gas originates from that part of the animal's stomach called the *rumen*, where cellulose is digested. Termites are also major producers of this gas. Finally, greenhouse gases such as nitrous oxide (N_2O) and sulfur dioxide (SO_2) are increasing due to automobile and industrial pollution.

Whether the increasing greenhouse gases are responsible or not, there is convincing evidence that global warming is underway. The evidence comes from the melting of ice in Antarctica and the retreat of glaciers at widely scattered sites throughout the world (see Fig. 11.17). For example, satellite images of Antarctica show James Ross Island completely surrounded by water for the first time since maps were made, about 100 years ago. Previously, the island was connected to the mainland by an ice bridge. In addition, at various places across the continent, ice shelves are retreating, some at a rapid rate.

Perhaps at no place in the world are glaciers monitored with greater interest than in Switzerland. There, it is found that the Alps have lost about 50% of their glacial ice compared to 130 years ago. The retreat of glaciers on high-altitude peaks in the tropics is even more severe than in Switzerland. The Lewis glacier on Mount Kenya and the snows of Kilimanjaro are two examples. However, in certain regions of the planet where glaciers are near large bodies of water and are fed by

Tom Bean

(a) (b)

Figure 11.17 Glacier melting in Alaska. (a) Hiker on a ridge above the Muir Glacier in Glacier Bay National Park, August 26, 1978. (b) Hiker on the same ridge, June 27, 1993.

large and frequent snows, glaciers continue to advance, so the overall picture of a catastrophic global-warming scenario may be premature. In about 50 years, however, the amount of carbon dioxide in the atmosphere is expected to be about twice what it was in the preindustrial era. Because of the possible catastrophic consequences, most scientists voice the concern that reductions in greenhouse gas emissions need to be made now.

SUMMARY

Physics Now™ Take a practice test by logging into PhysicsNow at **www.cp7e.com** and clicking on the Pre-Test link for this chapter.

11.1 Heat and Internal Energy

Internal energy is associated with a system's microscopic components. Internal energy includes the kinetic energy of translation, rotation, and vibration of molecules, as well as potential energy.

Heat is the transfer of energy across the boundary of a system resulting from a temperature difference between the system and its surroundings. The symbol Q represents the amount of energy transferred.

The **calorie** is the amount of energy necessary to raise the temperature of 1 g of water from 14.5°C to 15.5°C. The **mechanical equivalent of heat** is 4.186 J/cal.

11.2 Specific Heat

11.3 Calorimetry

The energy required to change the temperature of a substance of mass m by an amount ΔT is

$$Q = mc\,\Delta T \qquad [11.3]$$

where c is the **specific heat** of the substance. In calorimetry problems, the specific heat of a substance can be determined by placing it in water of known temperature, isolating the system, and measuring the temperature at equilibrium. The sum of all energy gains and losses for all the objects in an isolated system is given by

$$\Sigma Q_k = 0 \qquad [11.5]$$

where Q_k is the energy change in the kth object in the system. This equation can be solved for the unknown specific heat, or used to determine an equilibrium temperature.

11.4 Latent Heat and Phase Change

The energy required to change the phase of a pure substance of mass m is

$$Q = \pm mL \qquad [11.6]$$

where L is the **latent heat** of the substance. The latent heat of fusion, L_f, describes an energy transfer during a change from a solid phase to a liquid phase (or vice-versa), while the latent heat of vaporizaion, L_v, describes an energy transfer during a change from a liquid phase to a gaseous phase (or vice-versa). Calorimetry problems involving phase changes are handled with Equation 11.5, with latent heat terms added to the specific heat terms.

11.5 Energy Transfer

Energy can be transferred by several different processes, including work, discussed in Chapter 5, and by conduction, convection, and radiation. **Conduction** can be viewed as an exchange of kinetic energy between colliding molecules or electrons. The rate at which energy transfers by conduction through a slab of area A and thickness L is

$$\mathcal{P} = kA\frac{(T_h - T_c)}{L} \qquad [11.7]$$

where k is the **thermal conductivity** of the material making up the slab.

Energy is transferred by **convection** as a substance moves from one place to another.

All objects emit **radiation** from their surfaces in the form of electromagnetic waves at a net rate of

$$\mathcal{P}_{\text{net}} = \sigma Ae(T^4 - T_0{}^4) \qquad [11.11]$$

where T is the temperature of the object and T_0 is the temperature of the surroundings. An object that is hotter than its surroundings radiates more energy than it absorbs, whereas a body that is cooler than its surroundings absorbs more energy than it radiates.

CONCEPTUAL QUESTIONS

1. Rub the palm of your hand on a metal surface for 30–45 seconds. Place the palm of your other hand on an unrubbed portion of the surface and then the rubbed portion. The rubbed portion will feel warmer. Now repeat this process on a wooden surface. Why does the temperature difference between the rubbed and unrubbed portions of the wood surface seem larger than for the metal surface?

2. Pioneers stored fruits and vegetables in underground cellars. Discuss fully this choice for a storage site.

3. In usually warm climates that experience an occasional hard freeze, fruit growers will spray the fruit trees with water, hoping that a layer of ice will form on the fruit. Why would such a layer be advantageous?

4. In winter, why did the pioneers (mentioned in Question 2) store an open barrel of water alongside their produce?

5. Cups of water for coffee or tea can be warmed with a coil that is immersed in the water and raised to a high temperature by means of electricity. Why do the instructions warn users not to operate the coils in the absence of water? Can the immersion coil be used to warm up a cup of stew?

6. The U.S. penny is now made of copper-coated zinc. Can a calorimetric experiment be devised to test for the metal content in a collection of pennies? If so, describe the procedure.

7. On a clear, cold night, why does frost tend to form on the tops, rather than the sides, of mailboxes and cars?

8. A warning sign often seen on highways just before a bridge is "Caution—Bridge Surface Freezes before Road Surface." Of the three energy transfer processes discussed in Sections 11.5 to 11.7, which is most important in causing a bridge surface to freeze before the road surface on very cold days?

9. A tile floor may feel uncomfortably cold to your bare feet, but a carpeted floor in an adjoining room at the same temperature feels warm. Why?

10. On a very hot day, it's possible to cook an egg on the hood of a car. Would you select a black car or a white car on which to cook your egg? Why?

11. Concrete has a higher specific heat than does soil. Use this fact to explain (partially) why a city has a higher average temperature than the surrounding countryside. Would you expect evening breezes to blow from city to country or from country to city? Explain.

12. You need to pick up a very hot cooking pot in your kitchen. You have a pair of hot pads. Should you soak them in cold water or keep them dry in order to pick up the pot most comfortably?

13. In a daring demonstration, a professor dips her wetted fingers into molten lead (327°C) and withdraws them quickly without getting burned. How is this possible?

14. The air temperature above coastal areas is profoundly influenced by the large specific heat of water. One reason is that the energy released when 1 cubic meter of water cools by 1.0°C will raise the temperature of an enormously larger volume of air by 1.0°C. Estimate that volume of air. The specific heat of air is approximately 1.0 kJ/kg · °C. Take the density of air to be 1.3 kg/m^3.

15. Ethyl alcohol has about one-half the specific heat of water. Compare the temperature increases of equal masses of alcohol and water in separate beakers that are supplied with the same amount of energy.

16. Energy is added to ice, raising its temperature from −10°C to −5°C. A larger amount of energy is added to the same mass of liquid water, raising its temperature from 15°C to 20°C. From these results, we can conclude that (a) overcoming the latent heat of fusion of ice requires an input of energy (b) the latent heat of fusion of ice delivers some energy to the system (c) the specific heat of ice is less than that of water (d) the specific heat of ice is greater than that of water.

17. The specific heat of substance A is greater than the specific heat of substance B. Both A and B are at the same initial temperature when equal amounts of energy are added to them. Assuming no melting, freezing, or evaporation occurs, which of the following can be concluded about the final temperature T_A of substance A and the final temperature T_B of substance B? (a) $T_A > T_B$ (b) $T_A < T_B$ (c) $T_A = T_B$ (d) more information is needed.

PROBLEMS

1, 2, 3 = straightforward, intermediate, challenging □ = full solution available in *Student Solutions Manual/Study Guide*
PhysicsNow™ = coached problem with hints available at **www.cp7e.com** = biomedical application

Section 11.1 Heat and Internal Energy
Section 11.2 Specific Heat

1. Water at the top of Niagara Falls has a temperature of 10.0°C. If it falls a distance of 50.0 m and all of its potential energy goes into heating the water, calculate the temperature of the water at the bottom of the falls.

2. A 50.0-g piece of cadmium is at 20°C. If 400 cal of energy is transferred to the cadmium, what is its final temperature?

3. Lake Erie contains roughly 4.00×10^{11} m^3 of water. (a) How much energy is required to raise the temperature of that volume of water from 11.0°C to 12.0°C? (b) How many years would it take to supply this amount of energy by

using the 1 000-MW exhaust energy of an electric power plant?

4. An aluminum rod is 20.0 cm long at 20°C and has a mass of 350 g. If 10 000 J of energy is added to the rod by heat, what is the change in length of the rod?

5. How many joules of energy are required to raise the temperature of 100 g of gold from 20.0°C to 100°C?

6. As part of an exercise routine, a 50.0-kg person climbs 10.0 meters up a vertical rope. How many (food) Calories are expended in a single climb up the rope? (1 food Calorie = 10^3 calories)

7. **Physics Now™** A 75.0-kg weight watcher wishes to climb a mountain to work off the equivalent of a large piece of chocolate cake rated at 500 (food) Calories. How high must the person climb? (1 food Calorie = 10^3 calories)

8. The apparatus shown in Figure P11.8 was used by Joule to measure the mechanical equivalent of heat. Work is done on the water by a rotating paddle wheel, which is driven by two blocks falling at a constant speed. The temperature of the stirred water increases due to the friction between the water and the paddles. If the energy lost in the bearings and through the walls is neglected, then the loss in potential energy associated with the blocks equals the work done by the paddle wheel on the water. If each block has a mass of 1.50 kg and the insulated tank is filled with 200 g of water, what is the increase in temperature of the water after the blocks fall through a distance of 3.00 m?

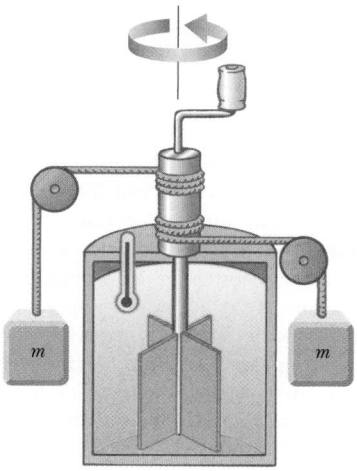

Figure P11.8 The falling weights rotate the paddles, causing the temperature of the water to increase.

9. A 5.00-g lead bullet traveling at 300 m/s is stopped by a large tree. If half the kinetic energy of the bullet is transformed into internal energy and remains with the bullet while the other half is transmitted to the tree, what is the increase in temperature of the bullet?

10. A 1.5-kg copper block is given an initial speed of 3.0 m/s on a rough horizontal surface. Because of friction, the block finally comes to rest. (a) If the block absorbs 85% of its initial kinetic energy as internal energy, calculate its increase in temperature. (b) What happens to the remaining energy?

11. A 200-g aluminum cup contains 800 g of water in thermal equilibrium with the cup at 80°C. The combination of cup and water is cooled uniformly so that the temperature decreases by 1.5°C per minute. At what rate is energy being removed? Express your answer in watts.

Section 11.3 Calorimetry

12. Lead pellets, each of mass 1.00 g, are heated to 200°C. How many pellets must be added to 500 g of water that is initially at 20.0°C to make the equilibrium temperature 25.0°C? Neglect any energy transfer to or from the container.

13. What mass of water at 25.0°C must be allowed to come to thermal equilibrium with a 3.00-kg gold bar at 100°C in order to lower the temperature of the bar to 50.0°C?

14. In a showdown on the streets of Laredo, the good guy drops a 5.0-g silver bullet at a temperature of 20°C into a 100-cm^3 cup of water at 90°C. Simultaneously, the bad guy drops a 5.0-g copper bullet at the same initial temperature into an identical cup of water. Which one ends the showdown with the coolest cup of water in the west? Neglect any energy transfer into or away from the container.

15. **Physics Now™** An aluminum cup contains 225 g of water and a 40-g copper stirrer, all at 27°C. A 400-g sample of silver at an initial temperature of 87°C is placed in the water. The stirrer is used to stir the mixture until it reaches its final equilibrium temperature of 32°C. Calculate the mass of the aluminum cup.

16. It is desired to cool iron parts from 500°F to 100°F by dropping them into water that is initially at 75°F. Assuming that all the heat from the iron is transferred to the water and that none of the water evaporates, how many kilograms of water are needed per kilogram of iron?

17. A 100-g aluminum calorimeter contains 250 g of water. The two substances are in thermal equilibrium at 10°C. Two metallic blocks are placed in the water. One is a 50-g piece of copper at 80°C. The other sample has a mass of 70 g and is originally at a temperature of 100°C.

The entire system stabilizes at a final temperature of 20°C. Determine the specific heat of the unknown second sample.

18. When a driver brakes an automobile, the friction between the brake drums and the brake shoes converts the car's kinetic energy to thermal energy. If a 1 500-kg automobile traveling at 30 m/s comes to a halt, how much does the temperature rise in each of the four 8.0-kg iron brake drums? (The specific heat of iron is 448 J/kg · °C.)

19. A student drops two metallic objects into a 120-g steel container holding 150 g of water at 25°C. One object is a 200-g cube of copper that is initially at 85°C, and the other is a chunk of aluminum that is initially at 5.0°C. To the surprise of the student, the water reaches a final temperature of 25°C, precisely where it started. What is the mass of the aluminum chunk?

Section 11.4 Latent Heat and Phase Change

20. A 50-g ice cube at 0°C is heated until 45 g has become water at 100°C and 5.0 g has become steam at 100°C. How much energy was added to accomplish the transformation?

21. A 100-g cube of ice at 0°C is dropped into 1.0 kg of water that was originally at 80°C. What is the final temperature of the water after the ice has melted?

22. How much energy is required to change a 40-g ice cube from ice at −10°C to steam at 110°C?

23. What mass of steam that is initially at 120°C is needed to warm 350 g of water and its 300-g aluminum container from 20°C to 50°C?

24. A resting adult of average size converts chemical energy in food into internal energy at the rate of 120 W, called her *basal metabolic rate.* To stay at a constant temperature, energy must be transferred out of the body at the same rate. Several processes exhaust energy from your body. Usually the most important is thermal conduction into the air in contact with your exposed skin. If you are not wearing a hat, a convection current of warm air rises vertically from your head like a plume from a smokestack. Your body also loses energy by electromagnetic radiation, by your exhaling warm air, and by the evaporation of perspiration. Now consider still another pathway for energy loss: moisture in exhaled breath. Suppose you breathe out 22.0 breaths per minute, each with a volume of 0.600 L. Suppose also that you inhale dry air and exhale air at 37°C containing water vapor with a vapor pressure of 3.20 kPa. The vapor comes from the evaporation of liquid water in your body. Model the water vapor as an ideal gas. Assume its latent heat of evaporation at 37°C is the same as its heat of vaporization at 100°C. Calculate the rate at which you lose energy by exhaling humid air.

25. A 75-kg cross-country skier glides over snow as in Figure P11.25. The coefficient of friction between skis and snow is 0.20. Assume all the snow beneath his skis is at 0°C and that all the internal energy generated by friction is added to snow, which sticks to his skis until it melts. How far would he have to ski to melt 1.0 kg of snow?

Nathan Bilow/Leo de Wys, Inc.

Figure P11.25 A cross-country skier.

26. When you jog, most of the food energy you burn above your basal metabolic rate (BMR) ends up as internal energy that would raise your body temperature if it were not eliminated. The evaporation of perspiration is the primary mechanism for eliminating this energy. Determine the amount of water you lose to evaporation when running for 30 minutes at a rate that uses 400 kcal/h above your BMR. (That amount is often considered to be the "maximum fat-burning" energy output.) The metabolism of 1 gram of fat generates approximately 9.0 kcal of energy and produces approximately 1 gram of water. (The hydrogen atoms in the fat molecule are transferred to oxygen to form water.) What fraction of your need for water will be provided by fat metabolism? (The latent heat of vaporization of water at room temperature is 2.5×10^6 J/kg).

27. A 40-g block of ice is cooled to −78°C and is then added to 560 g of water in an 80-g copper calorimeter at a temperature of 25°C. Determine the final temperature of the system consisting of the ice, water, and calorimeter. (If not all the ice melts, determine how much ice is left.) Remember that the ice must first warm to 0°C, melt, and then continue warming as water. The specific heat of ice is 0.500 cal/g · °C = 2090 J/kg · °C.

28. A 60.0-kg runner expends 300 W of power while running a marathon. Assuming that 10.0% of the energy is delivered to the muscle tissue and that the excess energy is

removed from the body primarily by sweating, determine the volume of bodily fluid (assume it is water) lost per hour. (At 37.0°C, the latent heat of vaporization of water is 2.41×10^6 J/kg.)

© Jim Bourg/Reuters/Corbis

Figure P11.28 Timothy Cherigat of Kenya, winner of the Boston Marathon in 2004.

29. A high-end gas stove usually has at least one burner rated at 14 000 Btu/h. If you place a 0.25-kg aluminum pot containing 2.0 liters of water at 20°C on this burner, how long will it take to bring the water to a boil, assuming all of the heat from the burner goes into the pot? How long will it take to boil all of the water out of the pot?

30. A beaker of water sits in the sun until it reaches an equilibrium temperature of 30°C. The beaker is made of 100 g of aluminum and contains 180 g of water. In an attempt to cool this system, 100 g of ice at 0°C is added to the water. (a) Determine the final temperature of the system. If $T_f = 0°C$, determine how much ice remains. (b) Repeat your calculations for 50 g of ice.

31. Steam at 100°C is added to ice at 0°C. (a) Find the amount of ice melted and the final temperature when the mass of steam is 10 g and the mass of ice is 50 g. (b) Repeat with steam of mass 1.0 g and ice of mass 50 g.

Section 11.5 Energy Transfer

32. The average thermal conductivity of the walls (including windows) and roof of a house in Figure P11.32 is 4.8×10^{-4} kW/m · °C, and their average thickness is 21.0 cm. The house is heated with natural gas, with a heat of combustion (energy released per cubic meter of gas burned) of 9300 kcal/m^3. How many cubic meters of gas must be burned each day to maintain an inside temperature of 25.0°C if the outside temperature is 0.0°C? Disregard radiation and energy loss by heat through the ground.

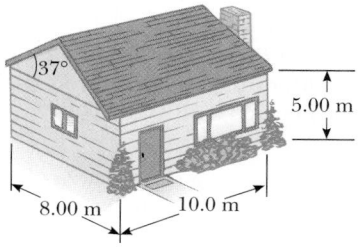

Figure P11.32

33. (a) Find the rate of energy flow through a copper block of cross-sectional area 15 cm^2 and length 8.0 cm when a temperature difference of 30°C is established across the block. Repeat the calculation, assuming that the material is (b) a block of stagnant air with the given dimensions; (c) a block of wood with the given dimensions.

34. A window has a glass surface area of 1.6×10^3 cm^2 and a thickness of 3.0 mm. (a) Find the rate of energy transfer by conduction through the window when the temperature of the inside surface of the glass is 70°F and the outside temperature is 90°F. (b) Repeat for the same inside temperature and an outside temperature of 0°F.

35. **Physics Now™** A steam pipe is covered with 1.50-cm-thick insulating material of thermal conductivity 0.200 cal/cm · °C · s. How much energy is lost every second when the steam is at 200°C and the surrounding air is at 20.0°C? The pipe has a circumference of 800 cm and a length of 50.0 m. Neglect losses through the ends of the pipe.

36. A box with a total surface area of 1.20 m^2 and a wall thickness of 4.00 cm is made of an insulating material. A 10.0-W electric heater inside the box maintains the inside temperature at 15.0°C above the outside temperature. Find the thermal conductivity k of the insulating material.

37. Determine the R value for a wall constructed as follows: The outside of the house consists of lapped wood shingles placed over 0.50-in.-thick sheathing, over 3.0 in. of cellulose fiber, over 0.50 in. of drywall.

38. A thermopane window consists of two glass panes, each 0.50 cm thick, with a 1.0-cm-thick sealed layer of air in between. If the inside surface temperature is 23°C and the outside surface temperature is 0.0°C, determine the rate

of energy transfer through 1.0 m^2 of the window. Compare your answer with the rate of energy transfer through 1.0 m^2 of a single 1.0-cm-thick pane of glass.

39. A copper rod and an aluminum rod of equal diameter are joined end to end in good thermal contact. The temperature of the free end of the copper rod is held constant at 100°C, and that of the far end of the aluminum rod is held at 0°C. If the copper rod is 0.15 m long, what must be the length of the aluminum rod so that the temperature at the junction is 50°C?

40. A Styrofoam box has a surface area of 0.80 m^2 and a wall thickness of 2.0 cm. The temperature of the inner surface is 5.0°C, and the outside temperature is 25°C. If it takes 8.0 h for 5.0 kg of ice to melt in the container, determine the thermal conductivity of the Styrofoam.

41. A sphere that is a perfect blackbody radiator has a radius of 0.060 m and is at 200°C in a room where the temperature is 22°C. Calculate the net rate at which the sphere radiates energy.

42. The surface temperature of the Sun is about 5 800 K. Taking the radius of the Sun to be 6.96×10^8 m, calculate the total energy radiated by the Sun each second. (Assume $e = 0.965$.)

43. A large, hot pizza 70 cm in diameter and 2.0 cm thick, at a temperature of 100°C, floats in outer space. Assume its emissivity is 0.8. What is the order of magnitude of its rate of energy loss?

44. Calculate the temperature at which a tungsten filament that has an emissivity of 0.90 and a surface area of 2.5×10^{-5} m^2 will radiate energy at the rate of 25 W in a room where the temperature is 22°C.

45. Measurements on two stars indicate that Star X has a surface temperature of 5 727°C and Star Y has a surface temperature of 11 727°C. If both stars have the same radius, what is the ratio of the luminosity (total power output) of Star Y to the luminosity of Star X? Both stars can be considered to have an emissivity of 1.0.

46. At high noon, the Sun delivers 1.00 kW to each square meter of a blacktop road. If the hot asphalt loses energy only by radiation, what is its equilibrium temperature?

ADDITIONAL PROBLEMS

47. The bottom of a copper kettle has a 10-cm radius and is 2.0 mm thick. The temperature of the outside surface is 102°C, and the water inside the kettle is boiling at 1 atm of pressure. Find the rate at which energy is being transferred through the bottom of the kettle.

48. A family comes home from a long vacation with laundry to do and showers to take. The water heater has been turned off during the vacation. If the heater has a capacity of 50.0 gallons and a 4 800-W heating element, how much time is required to raise the temperature of the water from 20.0°C to 60.0°C? Assume that the heater is well insulated and no water is withdrawn from the tank during that time.

49. Solar energy can be the primary source of winter space heating for a typical house (with floor area 130 m^2 = 1 400 ft^2) in the north central United States. If the house has very good insulation, you may model it as losing energy by heat steadily at the rate of 1 000 W during the winter, when the average exterior temperature is −5°C. The passive solar-energy collector can consist simply of large windows facing south. Sunlight shining in during the daytime is absorbed by the floor, interior walls, and objects in the house, raising their temperature to 30°C. As the sun goes down, insulating draperies or shutters are closed over the windows. During the period between 4 PM and 8 AM, the temperature of the house will drop, and a sufficiently large "thermal mass" is required to keep it from dropping too far. The thermal mass can be a large quantity of stone (with specific heat 800 J/kg · °C) in the floor and the interior walls exposed to sunlight. What mass of stone is required if the temperature is not to drop below 18°C overnight?

50. A water heater is operated by solar power. If the solar collector has an area of 6.00 m^2, and the intensity delivered by sunlight is 550 W/m^2, how long does it take to increase the temperature of 1.00 m^3 of water from 20.0°C to 60.0°C?

51. A 40-g ice cube floats in 200 g of water in a 100-g copper cup; all are at a temperature of 0°C. A piece of lead at 98°C is dropped into the cup, and the final equilibrium temperature is 12°C. What is the mass of the lead?

52. The evaporation of perspiration is the primary mechanism for cooling the human body. Estimate the amount of water you will lose when you bake in the sun on the beach for an hour. Use a value of 1 000 W/m^2 for the intensity of sunlight, and note that the energy required to evaporate a liquid at a particular temperature is approximately equal to the sum of the energy required to raise its temperature to the boiling point and the latent heat of vaporization (determined at the boiling point).

53. A 200-g block of copper at a temperature of 90°C is dropped into 400 g of water at 27°C. The water is contained in a 300-g glass container. What is the final temperature of the mixture?

54. A class of 10 students taking an exam has a power output per student of about 200 W. Assume that the initial temperature of the room is 20°C and that its dimensions are

6.0 m by 15.0 m by 3.0 m. What is the temperature of the room at the end of 1.0 h if all the energy remains in the air in the room and none is added by an outside source? The specific heat of air is 837 J/kg · °C, and its density is about 1.3×10^{-3} g/cm^3.

55. The human body must maintain its core temperature inside a rather narrow range around 37°C. Metabolic processes (notably, muscular exertion) convert chemical energy into internal energy deep in the interior. From the interior, energy must flow out to the skin or lungs, to be lost by heat to the environment. During moderate exercise, an 80-kg man can metabolize food energy at the rate of 300 kcal/h, do 60 kcal/h of mechanical work, and put out the remaining 240 kcal/h of energy by heat. Most of the energy is carried from the interior of the body out to the skin by "forced convection" (as a plumber would say): Blood is warmed in the interior and then cooled at the skin, which is a few degrees cooler than the body core. Without blood flow, living tissue is a good thermal insulator, with a thermal conductivity about 0.210 W/m · °C. Show that blood flow is essential to keeping the body cool by calculating the rate of energy conduction, in kcal/h, through the tissue layer under the skin. Assume that its area is 1.40 m^2, its thickness is 2.50 cm, and it is maintained at 37.0°C on one side and at 34.0°C on the other side.

56. An aluminum rod and an iron rod are joined end to end in good thermal contact. The two rods have equal lengths and radii. The free end of the aluminum rod is maintained at a temperature of 100°C, and the free end of the iron rod is maintained at 0°C. (a) Determine the temperature of the interface between the two rods. (b) If each rod is 15 cm long and each has a cross-sectional area of 5.0 cm^2, what quantity of energy is conducted across the combination in 30 min?

57. Water is being boiled in an open kettle that has a 0.500-cm-thick circular aluminum bottom with a radius of 12.0 cm. If the water boils away at a rate of 0.500 kg/min, what is the temperature of the lower surface of the bottom of the kettle? Assume that the top surface of the bottom of the kettle is at 100°C.

58. A 3.00-g copper penny at 25.0°C drops 50.0 m to the ground. (a) If 60.0% of the initial potential energy associated with the penny goes into increasing its internal energy, determine the final temperature of the penny. (b) Does the result depend on the mass of the coin? Explain.

59. A bar of gold (Au) is in thermal contact with a bar of silver (Ag) of the same length and area (Fig. P11.59). One end of the compound bar is maintained at 80.0°C, while the opposite end is at 30.0°C. Find the temperature at the junction when the energy flow reaches a steady state.

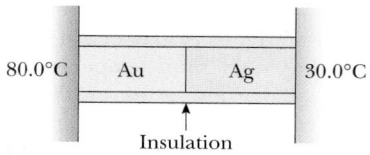

Figure P11.59

60. An iron plate is held against an iron wheel so that a sliding frictional force of 50 N acts between the two pieces of metal. The relative speed at which the two surfaces slide over each other is 40 m/s. (a) Calculate the rate at which mechanical energy is converted to internal energy. (b) The plate and the wheel have masses of 5.0 kg each, and each receives 50% of the internal energy. If the system is run as described for 10 s, and each object is then allowed to reach a uniform internal temperature, what is the resultant temperature increase?

61. An automobile has a mass of 1 500 kg, and its aluminum brakes have an overall mass of 6.0 kg. (a) Assuming that all of the internal energy transformed by friction when the car stops is deposited in the brakes, and neglecting energy transfer, how many times could the car be braked to rest starting from 25 m/s (56 mi/h) before the brakes would begin to melt? (Assume an initial temperature of 20°C.) (b) Identify some effects that are neglected in part (a), but are likely to be important in a more realistic assessment of the temperature increase of the brakes.

62. A 1.0-m-long aluminum rod of cross-sectional area 2.0 cm^2 is inserted vertically into a thermally insulated vessel containing liquid helium at 4.2 K. The rod is initially at 300 K. If half of the rod is inserted into the helium, how many liters of helium boil off in the very short time while the inserted half cools to 4.2 K? The density of liquid helium at 4.2 K is 122 kg/m^3.

63. **PhysicsNow™** A *flow calorimeter* is an apparatus used to measure the specific heat of a liquid. The technique is to measure the temperature difference between the input and output points of a flowing stream of the liquid while adding energy at a known rate. (a) Start with the equations $Q = mc(\Delta T)$ and $m = \rho V$, and show that the rate at which energy is added to the liquid is given by the expression $\Delta Q/\Delta t = \rho c(\Delta T)(\Delta V/\Delta t)$. (b) In a particular experiment, a liquid of density 0.72 g/cm^3 flows through the calorimeter at the rate of 3.5 cm^3/s. At steady state, a temperature difference of 5.8°C is established between the input and output points when energy is supplied at the rate of 40 J/s. What is the specific heat of the liquid?

64. Three liquids are at temperatures of 10°C, 20°C, and 30°C, respectively. Equal masses of the first two liquids are mixed, and the equilibrium temperature is 17°C. Equal masses of the second and third are then mixed, and the equilibrium temperature is 28°C. Find the equilibrium temperature when equal masses of the first and third are mixed.

65. At time $t = 0$, a vessel contains a mixture of 10 kg of water and an unknown mass of ice in equilibrium at 0°C. The temperature of the mixture is measured over a period of an hour, with the following results: During the first 50 min, the mixture remains at 0°C; from 50 min to 60 min, the temperature increases steadily from 0°C to 2°C. Neglecting the heat capacity of the vessel, determine the mass of ice that was initially placed in it. Assume a constant power input to the container.

66. A wood stove is used to heat a single room. The stove is cylindrical in shape, with a diameter of 40.0 cm and a length of 50.0 cm and operates at a temperature of 400°F. (a) If the temperature of the room is 70.0°F determine the amount of radiant energy delivered to the room by the stove each second if the emissivity is 0.920. (b) If the room is a square with walls that are 8.00 ft high and 25.0 ft wide, determine the R value needed in the walls and ceiling to maintain the inside temperature at 70.0°F if the outside temperature is 32.0°F. Note that we are ignoring any heat conveyed by the stove via convection and any energy lost through the walls (and windows!) via convection or radiation.

67. A "solar cooker" consists of a curved reflecting mirror that focuses sunlight onto the object to be heated (Fig. P11.67). The solar power per unit area reaching the Earth at the location of a 0.50-m-diameter solar cooker is 600 W/m^2. Assuming that 50% of the incident energy is converted to thermal energy, how long would it take to boil away 1.0 L of water initially at 20°C? (Neglect the specific heat of the container.)

Figure P11.67

68. For bacteriological testing of water supplies and in medical clinics, samples must routinely be incubated for 24 h at 37°C. A standard constant temperature bath with electric heating and thermostatic control is not suitable in developing nations without continuously operating electric power lines. Peace Corps volunteer and MIT engineer Amy Smith invented a low cost, low maintenance incubator to fill the need. The device consists of a foam-insulated box containing several packets of a waxy material that melts at 37.0°C, interspersed among tubes, dishes, or bottles containing the test samples and growth medium (food for bacteria). Outside the box, the waxy material is first melted by a stove or solar energy collector. Then it is put into the box to keep the test samples warm as it solidifies. The heat of fusion of the phase-change material is 205 kJ/kg. Model the insulation as a panel with surface area 0.490 m^2, thickness 9.50 cm, and conductivity 0.012 0 W/m°C. Assume the exterior temperature is 23.0°C for 12.0 h and 16.0°C for 12.0 h. (a) What mass of the waxy material is required to conduct the bacteriological test? (b) Explain why your calculation can be done without knowing the mass of the test samples or of the insulation.

69. What mass of steam initially at 130°C is needed to warm 200 g of water in a 100-g glass container from 20.0°C to 50.0°C?

ACTIVITIES

1. A plot of the decreasing temperature of a substance over time is called a cooling curve and has the same shape and basic explanation as the curve shown in Figure 11.3. You can plot such a curve by observing some water in a container in the freezer compartment of a refrigerator. Place a thermometer in the liquid, and record the reading of the thermometer every minute until about five minutes after the liquid has frozen completely. Explain your observations. A material that is a little easier to work with is naphthalene (mothballs). You can plot the cooling curve in this case without a freezer. Melt a small amount of naphthalene in a container, and plot a graph of temperature versus time as before. Again, explain your observations.

2. You have probably heard someone say that hot water freezes faster than cold water. Is this an urban legend or is it true? To test this hypothesis, fill one container with hot water, at about 200°F, and another with cooler water, at about 70°F. Place the two containers in the freezer compartment of a refrigerator, and find out for yourself. There are a number of variables that you need to attempt to control in such an experiment: (1) The two containers need to be placed at similar locations in the freezer compartment. That is, one should not be near the door while the other is in the back of the compartment. (2) The two containers should not be placed close together, or an un-

wanted exchange of energy will take place between them. (3) If the freezer is one of the old types that forms frost on its walls, the hot container should not be allowed to melt through the frost and make intimate contact with the cold walls of the freezer. Can you list any more variables that you need to control?

3. You may have heard that you can greatly reduce the baking time for potatoes in a *conventional* oven by inserting a nail through each potato. Are there any scientific reasons for believing that this hypothesis is true? Test it with a couple of similar-sized potatoes—but don't bake them in a microwave oven!

Lance Armstrong is an engine: he requires fuel and oxygen to burn it, and the result is work that drives him up the mountainside as his excess, waste energy is expelled in his evaporating sweat.

© Tim De Waele/Corbis

CHAPTER

12

OUTLINE

12.1 Work in Thermodynamic Processes
12.2 The First Law of Thermodynamics
12.3 Heat Engines and the Second Law of Thermodynamics
12.4 Entropy
12.5 Human Metabolism

The Laws of Thermodynamics

According to the first law of thermodynamics, the internal energy of a system can be increased either by adding energy to the system or by doing work on it. This means the internal energy of a system, which is just the sum of the molecular kinetic and potential energies, can change as a result of two separate types of energy transfer across the boundary of the system. Although the first law imposes conservation of energy for both energy added by heat and work done on a system, it doesn't predict which of several possible energy-conserving processes actually occur in nature.

The second law of thermodynamics constrains the first law by establishing which processes allowed by the first law actually occur. For example, the second law tells us that energy never flows by heat spontaneously from a cold object to a hot object. One important application of this law is in the study of heat engines (such as the internal combustion engine) and the principles that limit their efficiency.

12.1 WORK IN THERMODYNAMIC PROCESSES

Energy can be transferred to a system by heat and by work done on the system. In most cases of interest treated here, the system is a volume of gas, which is important in understanding engines. All such systems of gas will be assumed to be in thermodynamic equilibrium, so that every part of the gas is at the same temperature and pressure. If that were not the case, the ideal gas law wouldn't apply and most of the results presented here wouldn't be valid. Consider a gas contained by a cylinder fitted with a movable piston (Active Fig. 12.1a) and in equilibrium. The gas occupies a volume V and exerts a uniform pressure P on the cylinder walls and the piston. The gas is compressed slowly enough so the system remains essentially in thermodynamic equilibrium at all times. As the piston is pushed downward by an external force F through a distance Δy, the work done on the gas is

$$W = -F\Delta y = -PA\,\Delta y$$

where we have set the magnitude F of the external force equal to PA, possible because the pressure is the same everywhere in the system (by the assumption of equilibrium). Note that if the piston is pushed downward, $\Delta y = y_f - y_i$ is negative, so we need an explicit negative sign in the expression for W to make the work positive. The change in volume of the gas is $\Delta V = A\,\Delta y$, which leads to the following definition:

The **work W done on a gas** at constant pressure is given by

$$W = -P\Delta V \qquad [12.1]$$

where P is the pressure throughout the gas and ΔV is the change in volume of the gas during the process.

If the gas is compressed as in Active Figure 12.1b, ΔV is negative and the work done on the gas is positive. If the gas expands, ΔV is positive and the work done on the gas is negative. The work done by the gas on its environment, W_{env}, is simply the negative of the work done on the gas. In the absence of a change in volume, the work is zero.

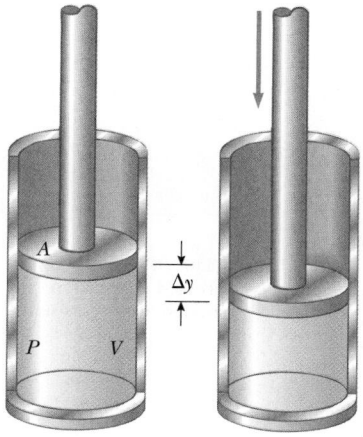

ACTIVE FIGURE 12.1
(a) A gas in a cylinder occupying a volume V at a pressure P. (b) Pushing the piston down compresses the gas.

Physics Now™
Log into PhysicsNow at **www.cp7e.com**, and go to Active Figure 12.1 to move the piston and see the resulting work done on the gas.

EXAMPLE 12.1 Work Done by an Expanding Gas

Goal Apply the definition of work at constant pressure.

Problem In a system similar to that shown in Active Figure 12.1, the gas in the cylinder is at a pressure of 1.01×10^5 Pa and the piston has an area of $0.100\ \text{m}^2$. As energy is slowly added to the gas by heat, the piston is pushed up a distance of 4.00 cm. Calculate the work done by the expanding gas on the surroundings, W_{env}, assuming the pressure remains constant.

Strategy The work done on the environment is the negative of the work done on the gas given in Equation 12.1. Compute the change in volume and multiply by the pressure.

Solution

Find the change in volume of the gas, ΔV, which is the cross-sectional area times the displacement:

$$\Delta V = A\,\Delta y = (0.100\ \text{m}^2)(4.00 \times 10^{-2}\ \text{m}) = 4.00 \times 10^{-3}\ \text{m}^3$$

Multiply this result by the pressure, getting the work the gas does on the environment, W_{env}:

$$W_{\text{env}} = P\Delta V = (1.01 \times 10^5\ \text{Pa})(4.00 \times 10^{-3}\ \text{m}^3) = 404\ \text{J}$$

Remark The volume of the gas increases, so the work done on the environment is positive. The work done on the system during this process is $W = -404$ J. The energy required to perform positive work on the environment must come from the energy of the gas. (See the next section for more details.)

Exercise 12.1

Gas in a cylinder similar to Figure 12.1 moves a piston with area $0.20\ \text{m}^2$ as energy is slowly added to the system. If 2.00×10^3 J of work is done on the environment and the pressure of the gas in the cylinder remains constant at 1.01×10^5 Pa, find the displacement of the piston.

Answer 9.90×10^{-2} m

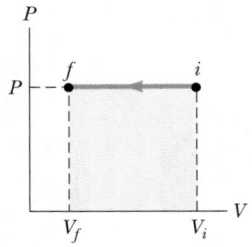

Figure 12.2 The *PV* diagram for a gas being compressed at constant pressure. The shaded area represents the work done on the gas.

Equation 12.1 can be used to calculate the work done on the system *only* when the pressure of the gas remains constant during the expansion or compression. A process in which the pressure remains constant is called an **isobaric process**. The pressure vs. volume graph, or ***PV* diagram**, of an isobaric process is shown in Figure 12.2. The curve on such a graph is called the *path* taken between the initial and final states, with the arrow indicating the direction the process is going, in this case from smaller to larger volume. The area under the graph is

$$\text{Area} = P(V_f - V_i) = P\Delta V$$

The area under the graph in a *PV* diagram is equal in magnitude to the work done on the gas.

This is true in general, whether or not the process proceeds at constant pressure. Just draw the *PV* diagram of the process, find the area underneath the graph (and above the horizontal axis), and that area will be the equal to the magnitude of the work done on the gas. If the arrow on the graph points toward larger volumes, the work done on the gas is negative. If the arrow on the graph points toward smaller volumes, the work done on the gas is positive.

Whenever negative work is done on a system, positive work is done by the system on its environment. The negative work done on the system represents a loss of energy from the system—the cost of doing positive work on the environment.

Quick Quiz 12.1

By visual inspection, order the *PV* diagrams shown in Figure 12.3 from the most negative work done on the system to the most positive work done on the system.
(a) a,b,c,d (b) a,c,b,d (c) d,b,c,a (d) d,a,c,b

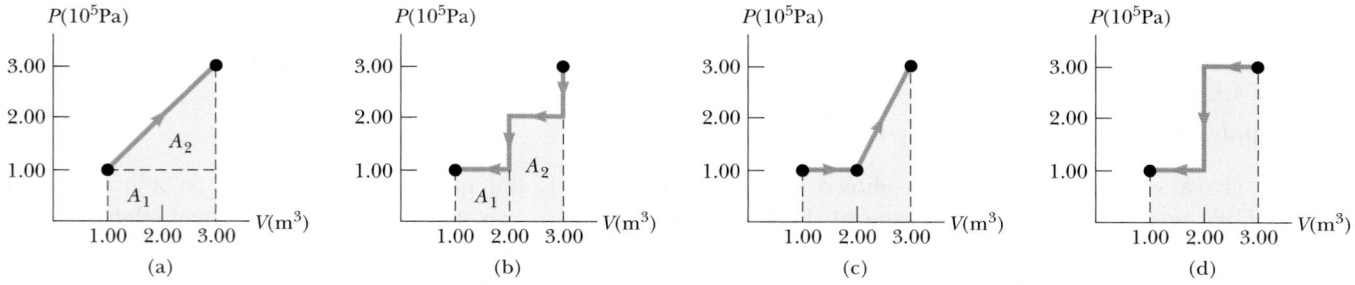

Figure 12.3 (Quick Quiz 12.1 and Example 12.2)

Notice that the graphs in Figure 12.3 all have the same endpoints, but the areas beneath the curves are different. The work done on a system depends on the path taken in the *PV* diagram.

EXAMPLE 12.2 Work and *PV* Diagrams

Goal Calculate work from a *PV* diagram.

Problem Find the numeric value of the work done on the gas in **(a)** Figure 12.3a and **(b)** Figure 12.3b.

Strategy The regions in question are composed of rectangles and triangles. Use basic geometric formulas to find the area underneath each curve. Check the direction of the arrow to determine signs.

Solution

(a) Find the work done on the gas in Figure 12.3a.

Compute the areas A_1 and A_2 in Figure 12.3a. A_1 is a rectangle and A_2 is a triangle.

$$A_1 = \text{height} \times \text{width} = (1.00 \times 10^5\ \text{Pa})(2.00\ \text{m}^3)$$
$$= 2.00 \times 10^5\ \text{J}$$

$$A_2 = \text{one-half base} \times \text{height}$$
$$= \tfrac{1}{2}(2.00\ \text{m}^3)(2.00 \times 10^5\ \text{Pa}) = 2.00 \times 10^5\ \text{J}$$

Sum the areas (the arrows point to increasing volume, so the work done on the gas is negative):

$$\text{Area} = A_1 + A_2 = 4.00 \times 10^5\,\text{J} \rightarrow$$
$$W = -4.00 \times 10^5\,\text{J}$$

(b) Find the work done on the gas in Figure 12.3b.

Compute the areas of the two rectangular regions:

$$A_1 = \text{height} \times \text{width} = (1.00 \times 10^5\,\text{Pa})(1.00\,\text{m}^3)$$
$$= 1.00 \times 10^5\,\text{J}$$
$$A_2 = \text{height} \times \text{width} = (2.00 \times 10^5\,\text{Pa})(1.00\,\text{m}^3)$$
$$= 2.00 \times 10^5\,\text{J}$$

Sum the areas (the arrows point to decreasing volume, so the work done on the gas is positive):

$$\text{Area} = A_1 + A_2 = 3.00 \times 10^5\,\text{J} \rightarrow$$
$$W = +3.00 \times 10^5\,\text{J}$$

Remarks Notice that in both cases the paths in the *PV* diagrams start and end at the same points, but the answers are different.

Exercise 12.2

Compute the work done on the system in Figures 12.3c and 12.3d.

Answers $-3.00 \times 10^5\,\text{J}$, $+4.00 \times 10^5\,\text{J}$

12.2 THE FIRST LAW OF THERMODYNAMICS

The **first law of thermodynamics** is another energy conservation law that relates changes in internal energy—the energy associated with the position and jiggling of all the molecules of a system—to energy transfers due to heat and work. The first law is universally valid, applicable to all kinds of processes, providing a connection between the microscopic and macroscopic worlds.

There are two ways energy can be transferred between a system and its surroundings: by doing work, which requires a macroscopic displacement of an object through the application of a force; and by heat, which occurs through random molecular collisions. Both mechanisms result in a *change in internal energy*, ΔU, of the system and therefore in measurable changes in the macroscopic variables of the system, such as the pressure, temperature, and volume. This change in the internal energy can be summarized in the **first law of thermodynamics**:

◀ First law of thermodynamics

If a system undergoes a change from an initial state to a final state, where Q is the energy transferred to the system by heat and W is the work done on the system, the change in the internal energy of the system, ΔU, is given by

$$\Delta U = U_f - U_i = Q + W \qquad \textbf{[12.2]}$$

The quantity Q is positive when energy is transferred into the system by heat and negative when energy is transferred out of the system by heat. The quantity W is positive when work is done on the system and negative when the system does work on its environment. All quantities in the first law, Equation 12.2, must have the same energy units. Any change in the internal energy of a system—the positions and vibrations of the molecules—is due to the transfer of energy by heat or work (or both).

From Equation 12.2, we also see that the internal energy of any isolated system must remain constant, so that $\Delta U = 0$. Even when a system isn't isolated, the change in internal energy will be zero if the system goes through a cyclic process in which all the thermodynamic variables—pressure, volume, temperature, and moles of gas—return to their original values.

It's important to remember that the quantities in Equation 12.2 concern a *system*, not the effect on the system's environment through work. If the system is hot

TIP 12.1 Dual Sign Conventions

Many physics and engineering textbooks present the first law as $\Delta U = Q - W$, with a minus sign between the heat and the work. The reason is that work is defined in these treatments as the work done *by* the gas rather than *on* the gas, as in our treatment. This form of the first law represents the original interest in applying it to steam engines, where the primary concern is the work extracted from the engine.

steam expanding against a piston, for example, the system work W is *negative*, because the piston can only expand at the expense of the internal energy of the gas. The work W_{env} done by the hot steam on the *environment*—in this case, moving a piston which moves the train—is positive, but that's not the work W in Equation 12.2. This way of defining work in the first law makes it consistent with the concept of work defined in Chapter 5. There, positive work done on a system (for example, a block) increased its mechanical energy, while negative work decreased its energy. In this chapter, positive work done on a system (typically, a volume of gas) increases its internal energy, and negative work decreases that internal energy. In both the mechanical and thermal cases, the effect on the system is the same: positive work increases the system's energy, and negative work decreases the system's energy.

Some textbooks identify W as the work done by the gas on its environment. This is an equivalent formulation, but it means that W must carry a minus sign in the first law. That convention isn't consistent with previous discussions of the energy of a system, because when W is positive the system *loses* energy, whereas in Chapter 5 positive W means the system *gains* energy. For that reason, the old convention is not used in this book.

EXAMPLE 12.3 Heating a Gas

Goal Combine the first law of thermodynamics with work done during a constant pressure process.

Problem An ideal gas absorbs 5.00×10^3 J of energy while doing 2.00×10^3 J of work on the environment during a constant pressure process. **(a)** Compute the change in the internal energy of the gas. **(b)** If the internal energy now drops by 4.50×10^3 J and 2.00×10^3 J is expelled from the system, find the change in volume, assuming a constant pressure process at 1.01×10^5 Pa.

Strategy Part (a) requires substitution of the given information into the first law, Equation 12.2. Notice, however, that the given work is done on the *environment*. The negative of this amount is the work done on the *system*, representing a loss of internal energy. Part (b) is a matter of substituting the equation for work at constant pressure into the first law and solving for the change in volume.

Solution

(a) Compute the change in internal energy.

Substitute values into the first law, noting that the work done on the gas is negative:

$$\Delta U = Q + W = 5.00 \times 10^3\,\text{J} - 2.00 \times 10^3\,\text{J}$$
$$= 3.00 \times 10^3\,\text{J}$$

(b) Find the change in volume, noting that ΔU and Q are both negative in this case.

Substitute the equation for work done at constant pressure into the first law:

$$\Delta U = Q + W = Q - P\Delta V$$
$$-4.50 \times 10^3\,\text{J} = -2.00 \times 10^3\,\text{J} - (1.01 \times 10^5\,\text{J})\Delta V$$

Solve for the change in volume, ΔV:

$$\Delta V = 2.48 \times 10^{-2}\,\text{m}^3$$

Remarks The change in volume is positive, so the system expands, doing positive work on the environment, while the work W on the system is negative.

Exercise 12.3

Suppose the internal energy of an ideal gas drops by 3.00×10^3 J at a constant pressure of 1.00×10^5 Pa, while the system gains 5.00×10^2 J of energy by heat. Find the change in volume of the system.

Answer $3.50 \times 10^{-2}\,\text{m}^3$

Recall that an expression for the internal energy of an ideal gas is

$$U = \tfrac{3}{2}nRT \qquad \textbf{[12.3a]}$$

This expression is valid only for a *monatomic* ideal gas, which means the particles of the gas consist of single atoms. The change in the internal energy, ΔU, for such a gas is given by

$$\Delta U = \tfrac{3}{2}nR\Delta T \qquad \textbf{[12.3b]}$$

The **molar specific heat at constant volume** of a monatomic ideal gas, C_v, is defined by

$$C_v \equiv \tfrac{3}{2}R \qquad \textbf{[12.4]}$$

The change in internal energy of an ideal gas can then be written

$$\Delta U = nC_v\Delta T \qquad \textbf{[12.5]}$$

For ideal gases, this expression is always valid, even when the volume isn't constant. The value of the molar specific heat, however, depends on the gas and can vary under different conditions of temperature and pressure.

A gas with a larger molar specific heat requires more energy to realize a given temperature change. The size of the molar specific heat depends on the structure of the gas molecule and how many different ways it can store energy. A monatomic gas such as helium can store energy as motion in three different directions. A gas such as hydrogen, on the other hand, is diatomic in normal temperature ranges, and aside from moving in three directions, it can also tumble, rotating in two different directions. So hydrogen molecules can store energy in the form of translational motion, and in addition can store energy through tumbling. Further, molecules can also store energy in the vibrations of their constituent atoms. A gas composed of molecules with more ways to store energy will have a larger molar specific heat.

Each different way a gas molecule can store energy is called a *degree of freedom.* Each degree of freedom contributes $\frac{1}{2}R$ to the molar specific heat. Because an atomic ideal gas can move in three directions, it has a molar specific heat capacity $C_v = 3(\frac{1}{2}R) = \frac{3}{2}R$. A diatomic gas like molecular oxygen, O_2, can also tumble in two different directions. This adds $2 \times \frac{1}{2}R = R$ to the molar heat specific heat, so $C_v = \frac{5}{2}R$ for diatomic gases. The spinning about the long axis connecting the two atoms is generally negligible. Vibration of the atoms in a molecule can also contribute to the heat capacity. A full analysis of a given system is often complex, so in general, molar specific heats must be determined by experiment. Some representative values of C_v can be found in Table 12.1 (page 392).

There are four basic types of thermal processes, which will be studied and illustrated by their effect on an ideal gas.

Isobaric Processes

Recall from Section 12.1 that in an isobaric process the pressure remains constant as the gas expands or is compressed. An expanding gas does work on its environment, given by $W_{\text{env}} = P\Delta V$. The PV diagram of an isobaric expansion is given in Figure 12.2. As previously discussed, the magnitude of the work done on the gas is just the area under the path in its PV diagram: height times length, or $P\Delta V$. The negative of this quantity, $W = -P\Delta V$, is the energy lost by the gas because the gas does work as it expands. This is the quantity that should be substituted into the first law.

The work done by the gas on its environment must come at the expense of the change in its internal energy, ΔU. Because the change in the internal energy of an ideal gas is given by $\Delta U = nC_v\Delta T$, the temperature of an expanding gas must decrease as the internal energy decreases. Expanding volume and decreasing temperature means the pressure must also decrease, in conformity with the ideal gas law, $PV = nRT$. Consequently, the only way such a process can remain at constant

TABLE 12.1

Molar Specific Heats of Various Gases

Gas	Molar Specific Heat (J/mol · K)[a]			
	C_P	C_V	$C_P - C_V$	$\gamma = C_P/C_V$
Monatomic Gases				
He	20.8	12.5	8.33	1.67
Ar	20.8	12.5	8.33	1.67
Ne	20.8	12.7	8.12	1.64
Kr	20.8	12.3	8.49	1.69
Diatomic Gases				
H_2	28.8	20.4	8.33	1.41
N_2	29.1	20.8	8.33	1.40
O_2	29.4	21.1	8.33	1.40
CO	29.3	21.0	8.33	1.40
Cl_2	34.7	25.7	8.96	1.35
Polyatomic Gases				
CO_2	37.0	28.5	8.50	1.30
SO_2	40.4	31.4	9.00	1.29
H_2O	35.4	27.0	8.37	1.30
CH_4	35.5	27.1	8.41	1.31

[a]All values except that for water were obtained at 300 K.

pressure is if thermal energy Q is transferred into the gas by heat. Rearranging the first law, we obtain

$$Q = \Delta U - W = \Delta U + P\Delta V$$

Now we can substitute the expression in Equation 12.3b for ΔU and use the ideal gas law to substitute $P\Delta V = nR\Delta T$:

$$Q = \tfrac{3}{2}nR\Delta T + nR\Delta T = \tfrac{5}{2}nR\Delta T$$

Another way to express this transfer by heat is

$$Q = nC_p\Delta T \qquad \textbf{[12.6]}$$

where $C_p = \frac{5}{2}R$. For ideal gases, the molar heat capacity at constant pressure, C_p, is the sum of the molar heat capacity at constant volume, C_v, and the gas constant R:

$$C_p = C_v + R \qquad \textbf{[12.7]}$$

This can be seen in the fourth column of Table 12.1, where $C_p - C_v$ is calculated for a number of different gases. The difference works out to be approximately R in virtually every case.

EXAMPLE 12.4 Expanding Gas

Goal Use molar specific heats and the first law in a constant pressure process.

Problem Suppose a system of monatomic ideal gas at 2.00×10^5 Pa and an initial temperature of 293 K slowly expands at constant pressure from a volume of 1.00 L to 2.50 L. **(a)** Find the work done on the environment. **(b)** Find the change in internal energy of the gas. **(c)** Use the first law of thermodynamics to obtain the thermal energy absorbed by the gas during the process. **(d)** Use the molar heat capacity at constant pressure to find the thermal energy absorbed. **(e)** How would the answers change for a diatomic ideal gas?

Strategy This problem mainly involves substituting into the appropriate equations. Substitute into the equation for work at constant pressure to obtain the answer to part (a). In part (b), use the ideal gas law twice, to find the temperature when $V = 2.00$ L and to find the number of moles of the gas. These quantities can then be used to obtain the change in internal energy, ΔU. Part (c) can then be solved by substituting into the first law, yielding Q, the answer checked in part (d) with Equation 12.6. Repeat these steps for part (e) after increasing the molar specific heats by R because of the extra two degrees of freedom associated with a diatomic gas.

Solution

(a) Find the work done on the environment.

Apply the definition of work at constant pressure:

$$W_{\text{env}} = P\Delta V = (2.00 \times 10^5 \text{ Pa})(2.50 \times 10^{-3} \text{ m}^3 - 1.00 \times 10^{-3} \text{ m}^3)$$

$$W_{\text{env}} = 3.00 \times 10^2 \text{ J}$$

(b) Find the change in the internal energy of the gas.

First, obtain the final temperature, using the ideal gas law, noting that $P_i = P_f$:

$$\frac{P_f V_f}{P_i V_i} = \frac{T_f}{T_i} \rightarrow T_f = T_i \frac{V_f}{V_i} = (293 \text{ K}) \frac{(2.50 \times 10^{-3} \text{ m}^3)}{(1.00 \times 10^{-3} \text{ m}^3)}$$

$$T_f = 733 \text{ K}$$

Again using the ideal gas law, obtain the number of moles of gas:

$$n = \frac{P_i V_i}{RT_i} = \frac{(2.00 \times 10^5 \text{ Pa})(1.00 \times 10^{-3} \text{ m}^3)}{(8.31 \text{ J/K} \cdot \text{mol})(293 \text{ K})}$$

$$= 8.21 \times 10^{-2} \text{ mol}$$

Use these results and given quantities to calculate the change in internal energy, ΔU:

$$\Delta U = nC_v \Delta T = \tfrac{3}{2} nR\Delta T$$

$$= \tfrac{3}{2}(8.21 \times 10^{-2} \text{ mol})(8.31 \text{ J/K} \cdot \text{mol})(733 \text{ K} - 293 \text{ K})$$

$$\Delta U = 4.50 \times 10^2 \text{ J}$$

(c) Use the first law to obtain the energy transferred by heat.

Solve the first law for Q, and substitute ΔU and $W = -W_{\text{env}} = -3.00 \times 10^2$ J:

$$\Delta U = Q + W \rightarrow Q = \Delta U - W$$

$$Q = 4.50 \times 10^2 \text{ J} - (-3.00 \times 10^2 \text{ J}) = 7.50 \times 10^2 \text{ J}$$

(d) Use the molar heat capacity at constant pressure to obtain Q:

Substitute values into Equation 12.6:

$$Q = nC_p \Delta T = \tfrac{5}{2} nR\Delta T$$

$$= \tfrac{5}{2}(8.21 \times 10^{-2} \text{ mol})(8.31 \text{ J/K} \cdot \text{mol})(733 \text{ K} - 293 \text{ K})$$

$$= 7.50 \times 10^2 \text{ J}$$

(e) How would the answers change for a diatomic gas?

Obtain the new change in internal energy, ΔU, noting that $C_v = \frac{5}{2}R$ for a diatomic gas:

$$\Delta U = nC_v \Delta T = (\tfrac{3}{2} + 1) nR\Delta T$$

$$= \tfrac{5}{2}(8.21 \times 10^{-2} \text{ mol})(8.31 \text{ J/K} \cdot \text{mol})(733 \text{ K} - 293 \text{ K})$$

$$\Delta U = 7.50 \times 10^2 \text{ J}$$

Obtain the new energy transferred by heat, Q:

$$Q = nC_p \Delta T = (\tfrac{5}{2} + 1) nR\Delta T$$

$$= \tfrac{7}{2}(8.21 \times 10^{-2} \text{ mol})(8.31 \text{ J/K} \cdot \text{mol})(733 \text{ K} - 293 \text{ K})$$

$$Q = 1.05 \times 10^3 \text{ J}$$

Remarks Notice that problems involving diatomic gases are no harder than those with monatomic gases. It's just a matter of adjusting the molar specific heats.

Exercise 12.4

Suppose an ideal monatomic gas at an initial temperature of 475 K is compressed from 3.00 L to 2.00 L while its pressure remains constant at 1.00×10^5 Pa. Find (a) the work done on the gas, (b) the change in internal energy, and (c) the energy lost by heat, Q.

Answers (a) 1.00×10^2 J (b) -150 J (c) -250 J

Adiabatic Processes

In an adiabatic process, no energy enters or leaves the system by heat. Such a system is insulated—thermally isolated from its environment. In general, however, the system isn't mechanically isolated, so it can still do work. A sufficiently rapid process may be considered approximately adiabatic because there isn't time for any significant transfer of energy by heat.

For adiabatic processes $Q = 0$, so the first law becomes

$$\Delta U = W$$

The work done during an adiabatic process can be calculated by finding the change in the internal energy. Alternately, the work can be computed from a PV diagram. For an ideal gas undergoing an adiabatic process, it can be shown that

$$PV^{\gamma} = \text{constant} \qquad \textbf{[12.8a]}$$

where

$$\gamma = \frac{C_p}{C_v} \qquad \textbf{[12.8b]}$$

is called the *adiabatic index* of the gas. Values of the adiabatic index for several different gases are given in Table 12.1. After computing the constant on the right-hand side of Equation 12.8a and solving for the pressure P, the area under the curve in the PV diagram can be found by counting boxes, yielding the work.

If a hot gas is allowed to expand so quickly that there is no time for energy to enter or leave the system by heat, the work done on the gas is negative and the internal energy decreases. This decrease occurs because kinetic energy is transferred from the gas molecules to the moving piston. Such an adiabatic expansion is of practical importance and is nearly realized in an internal combustion engine when a gasoline–air mixture is ignited and expands rapidly against a piston. The following example illustrates this process.

EXAMPLE 12.5 Work and an Engine Cylinder

Goal Use the first law to find the work done in an adiabatic expansion.

Problem In a car engine operating at 1.80×10^3 rev/min, the expansion of hot, high-pressure gas against a piston occurs in about 10 ms. Because energy transfer by heat typically takes a time on the order of minutes or hours, it's safe to assume that little energy leaves the hot gas during the expansion. Estimate the work done by the gas on the piston during this adiabatic expansion by assuming the engine cylinder contains 0.100 moles of an ideal monatomic gas which goes from 1.200×10^3 K to 4.00×10^2 K, typical engine temperatures, during the expansion.

Strategy Find the change in internal energy using the given temperatures. For an adiabatic process, this equals the work done on the gas, which is the negative of the work done on the environment—in this case, the piston.

Solution

Start with the first law, taking $Q = 0$.

$$W = \Delta U - Q = \Delta U - 0 = \Delta U$$

Find ΔU from the expression for the internal energy of an ideal monatomic gas.

$$\Delta U = U_f - U_i = \tfrac{3}{2}nR(T_f - T_i)$$
$$= \tfrac{3}{2}(0.100\text{ mol})(8.31\text{ J/mol}\cdot\text{K})(4.00 \times 10^2\text{ K} - 1.20 \times 10^3\text{ K})$$
$$\Delta U = -9.97 \times 10^2\text{ J}$$

The change in internal energy equals the work done on the system, which is the negative of the work done on the piston.

$$W_{\text{piston}} = -W = -\Delta U = \boxed{9.97 \times 10^2\text{ J}}$$

Remarks The work done on the piston comes at the expense of the internal energy of the gas. In an ideal adiabatic expansion, the loss of internal energy is completely converted into useful work. In a real engine, there are always losses.

Exercise 12.5

A monatomic ideal gas with volume 0.200 L is rapidly compressed, so the process can be considered adiabatic. If the gas is initially at 1.01×10^5 Pa and 3.00×10^2 K and the final temperature is 477 K, find the work done by the gas on the environment, W_{env}.

Answer 17.9 J

EXAMPLE 12.6 An Adiabatic Expansion

Goal Use the adiabatic pressure vs. volume relation to find a change in pressure and the work done on a gas.

Problem A monatomic ideal gas at a pressure 1.01×10^5 Pa expands adiabatically from an initial volume of 1.50 m^3, doubling its volume. **(a)** Find the new pressure. **(b)** Sketch the *PV* diagram and estimate the work done on the gas.

Strategy There isn't enough information to solve this problem with the ideal gas law. Instead, use Equation 12.8 and the given information to find the adiabatic index and the constant *C* for the process. For part (b), sketch the *PV* diagram and count boxes to estimate the area under the graph, which gives the work.

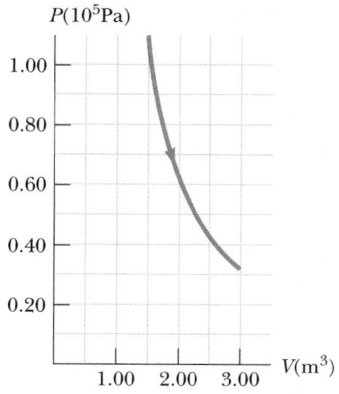

Figure 12.4 (Example 12.6) The *PV* diagram of an adiabatic expansion: the graph of $P = CV^{-\gamma}$, where *C* is a constant and $\gamma = C_p/C_v$.

Solution

(a) Find the new pressure.

First, calculate the adiabatic index:

$$\gamma = \frac{C_p}{C_v} = \frac{\frac{5}{2}R}{\frac{3}{2}R} = \frac{5}{3}$$

Use Equation 12.8a to find the constant *C*:

$$C = P_1V_1^{\gamma} = (1.01 \times 10^5 \text{ Pa})(1.50 \text{ m}^3)^{5/3}$$
$$= 1.99 \times 10^5 \text{ Pa}\cdot\text{m}^5$$

The constant *C* is fixed for the entire process and can be used to find P_2:

$$C = P_2V_2^{\gamma} = P_2(3.00 \text{ m}^3)^{5/3}$$
$$1.99 \times 10^5 \text{ Pa}\cdot\text{m}^5 = P_2\,(6.24 \text{ m}^5)$$
$$P_2 = 3.19 \times 10^4 \text{ Pa}$$

(b) Estimate the work done on the gas from a *PV* diagram.

Count the boxes between $V_1 = 1.50$ m^3 and $V_2 = 3.00$ m^3 in the graph of $P = (1.99 \times 10^5 \text{ Pa}\cdot\text{m}^5)\,V^{-5/3}$ in the *PV* diagram shown in Figure 12.4:

number of boxes ≈ 17

Each box has 'area' 5.00×10^3 J.

$$W \approx -17 \cdot 5.00 \times 10^3 \text{ J} = -8.5 \times 10^4 \text{ J}$$

Remarks The exact answer, obtained with calculus, is -8.43×10^4 J, so our result is a very good estimate. The answer is negative because the gas is expanding, doing positive work on the environment, thereby reducing its own internal energy.

Exercise 12.6

Repeat the preceding calculations for an ideal diatomic gas expanding adiabatically from an initial volume of 0.500 m^3 to a final volume of 1.25 m^3, starting at a pressure of $P_1 = 1.01 \times 10^5$ Pa. (You must sketch the curve to find the work.)

Answers $P_2 = 2.80 \times 10^4$ Pa, $W \approx -4 \times 10^4$ J

Isovolumetric Processes

An **isovolumetric process**, sometimes called an *isochoric* process (which is harder to remember), proceeds at constant volume, corresponding to vertical lines in a *PV* diagram. If the volume doesn't change, no work is done on or by the system, so $W = 0$, and the first law of thermodynamics reads

$$\Delta U = Q \quad \text{(isovolumetric process)}$$

This result tells us that **in an isovolumetric process, the change in internal energy of a system equals the energy transferred to the system by heat**. From Equation 12.5, the energy transferred by heat in constant volume processes is given by

$$Q = nC_v\Delta T \qquad \textbf{[12.9]}$$

EXAMPLE 12.7 An Isovolumetric Process

Goal Apply the first law to a constant-volume process.

Problem How much thermal energy must be added to 5.00 moles of monatomic ideal gas at 3.00×10^2 K and with a constant volume of 1.50 L in order to raise the temperature of the gas by 3.80×10^2 K?

Strategy The energy transferred by heat is equal to the change in the internal energy of the gas, which can be calculated by substitution into Equation 12.9.

Solution

Apply Equation 12.9, using the fact that $C_v = 3R/2$ for an ideal monatomic gas:

$$Q = \Delta U = nC_v\Delta T = \tfrac{3}{2}nR\Delta T$$
$$= \tfrac{3}{2}(5.00 \text{ mol})(8.31 \text{ J/K}\cdot\text{mol})(80.0^\circ \text{ K})$$
$$Q = 4.99 \times 10^3 \text{ J}$$

Remark Constant volume processes are the simplest to handle, and include such processes as heating a solid or liquid, in which the work of expansion is negligible.

Exercise 12.7

Find the change in temperature of 22.0 mol of a monatomic ideal gas if it absorbs 9 750 J at constant volume.

Answer 35.6 K

Isothermal Processes

During an isothermal process, the temperature of a system doesn't change. In an ideal gas the internal energy U depends only on the temperature, so it follows that $\Delta U = 0$ because $\Delta T = 0$. In this case, the first law of thermodynamics gives

$$W = -Q \quad \text{(isothermal process)}$$

We see that if the system is an ideal gas undergoing an isothermal process, the work done on the system is equal to the negative of the thermal energy transferred to the system. Such a process can be visualized in Figure 12.5. A cylinder filled with gas is in contact with a large energy reservoir that can exchange energy with the gas without changing its temperature. For a constant temperature ideal gas,

$$P = \frac{nRT}{V}$$

where the numerator on the right-hand side is constant. The *PV* diagram of a typical isothermal process is graphed in Figure 12.6, contrasted with an adiabatic process. When the process is adiabatic, the pressure falls off more rapidly.

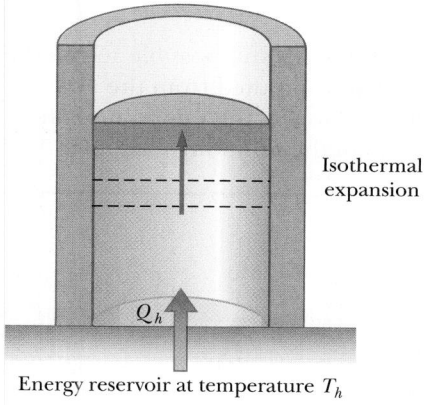

Figure 12.5 The gas in the cylinder expands isothermally while in contact with a reservoir at temperature T_h.

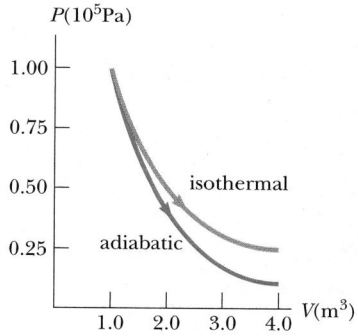

Figure 12.6 The PV diagram of an isothermal expansion, graph of $P = CV^{-1}$, where C is a constant. Contrasted with an adiabatic expansion, $P = C_A V^{-\gamma}$. C_A is a constant equal in magnitude to C in this case, but carrying different units.

Using methods of calculus, it can be shown that the work done on the environment during an isothermal process is given by

$$W_{\text{env}} = nRT \ln\left(\frac{V_f}{V_i}\right) \qquad \textbf{[12.10]}$$

The symbol "ln" in Equation 12.10 is an abbreviation for the natural logarithm, discussed in Appendix A. The work W done on the gas is just the negative of W_{env}.

EXAMPLE 12.8 An Isothermally Expanding Balloon

Goal Find the work done during an isothermal expansion.

Problem A balloon contains 5.00 moles of a monatomic ideal gas. As energy is added to the system by heat (say, by absorption from the Sun), the volume increases by 25% at a constant temperature of 27.0°C. Find the work W_{env} done by the gas in expanding the balloon, the thermal energy Q transferred to the gas, and the work W done on the gas.

Strategy Be sure to convert temperatures to kelvins. Use the equation for isothermal work to find the work done on the balloon, which is the work done on the environment. The latter is equal to the thermal energy Q transferred to the gas, and the negative of this quantity is the work done on the gas.

Solution

Substitute into Equation 12.10, finding the work done during the isothermal expansion. Note that $T = 27.0°\text{C} = 3.00 \times 10^2$ K.

$$W_{\text{env}} = nRT \ln\left(\frac{V_f}{V_i}\right)$$

$$= (5.00 \text{ mol})(8.31 \text{ J/K}\cdot\text{mol})(3.00 \times 10^2 \text{ K}) \times \ln\left(\frac{1.25 V_0}{V_0}\right)$$

$$W_{\text{env}} = 2.78 \times 10^3 \text{ J}$$

$$Q = W_{\text{env}} = 2.78 \times 10^3 \text{ J}$$

The negative of this amount is the work done on the gas:

$$W = -W_{\text{env}} = -2.78 \times 10^3 \text{ J}$$

Remarks Notice the relationship between the work done on the gas, the work done on the environment, and the energy transferred. These relationships are true of all isothermal processes.

Exercise 12.8

Suppose that subsequent to this heating, 1.50×10^4 J of thermal energy is removed from the gas isothermally. Find the final volume in terms of the initial volume of the example, V_0. (*Hint:* Follow the same steps as in the example, but in reverse. Also note that the initial volume in this exercise is $1.25 V_0$.)

Answer $0.375 V_0$

General Case

When a process follows none of the four given models, it's still possible to use the first law to get information about it. The work can be computed from the area under the curve of the *PV* diagram, and if the temperatures at the endpoints can be found, ΔU follows from Equation 12.5, as illustrated in the following example.

EXAMPLE 12.9 A General Process

Goal Find thermodynamic quantities for a process that doesn't fall into any of the four previously discussed categories.

Problem A quantity of 4.00 moles of a monatomic ideal gas expands from an initial volume of 0.100 m^3 to a final volume of 0.300 m^3 and pressure of 2.5×10^5 Pa (Fig. 12.7a). Compute **(a)** the work done on the gas, **(b)** the change in internal energy of the gas, and **(c)** the thermal energy transferred to the gas.

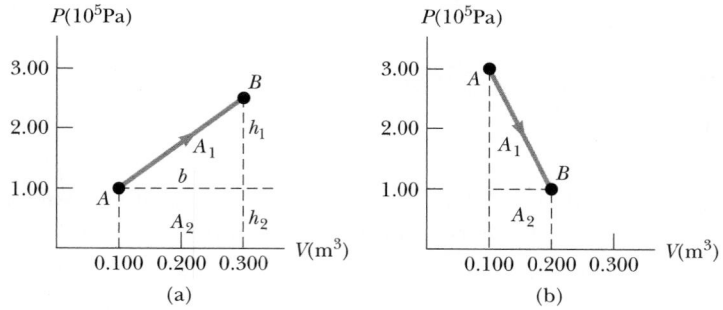

Figure 12.7 (a) (Example 12.9) (b) (Exercise 12.9)

Strategy The work done on the gas is just equal to the negative of the area under the curve in the *PV* diagram. Use the ideal gas law to get the temperature change and, subsequently, the change in internal energy. Finally, the first law gives the thermal energy transferred by heat.

Solution

(a) Find the work done on the gas by computing the area under the curve in Figure 12.7a.

Find A_1, the area of the triangle:

$$A_1 = \tfrac{1}{2}bh_1 = \tfrac{1}{2}(0.200\ \text{m}^3)(1.50 \times 10^5\ \text{Pa}) = 1.50 \times 10^4\ \text{J}$$

Find A_2, the area of the rectangle:

$$A_2 = bh_2 = (0.200\ \text{m}^3)(1.00 \times 10^5\ \text{Pa}) = 2.00 \times 10^4\ \text{J}$$

Sum the two areas (the gas is expanding, so the work done on the gas is negative and a minus sign must be supplied):

$$W = -(A_1 + A_2) = -3.50 \times 10^4\ \text{J}$$

(b) Find the change in the internal energy during the process.

Compute the temperature at points *A* and *B* with the ideal gas law:

$$T_A = \frac{P_A V_A}{nR} = \frac{(1.00 \times 10^5\ \text{Pa})(0.100\ \text{m}^3)}{(4.00\ \text{mol})(8.31\ \text{J/K}\cdot\text{mol})} = 301\ \text{K}$$

$$T_B = \frac{P_B V_B}{nR} = \frac{(2.50 \times 10^5\ \text{Pa})(0.300\ \text{m}^3)}{(4.00\ \text{mol})(8.31\ \text{J/K}\cdot\text{mol})} = 2.26 \times 10^3\ \text{K}$$

Compute the change in internal energy:

$$\Delta U = \tfrac{3}{2}nR\Delta T$$
$$= \tfrac{3}{2}(4.00\ \text{mol})(8.31\ \text{J/K}\cdot\text{mol})(2.26 \times 10^3\ \text{K} - 301\ \text{K})$$
$$\Delta U = 9.77 \times 10^4\ \text{J}$$

(c) Compute *Q* with the first law:

$$Q = \Delta U - W = 9.77 \times 10^4\ \text{J} - (-3.50 \times 10^4\ \text{J})$$
$$= 1.33 \times 10^5\ \text{J}$$

Remarks As long as it's possible to compute the work, cycles involving these more exotic processes can be completely analyzed. Usually, however, it's necessary to use calculus.

Exercise 12.9

Figure 12.7b represents a process involving 3.00 moles of a monatomic ideal gas expanding from 0.100 m^3 to 0.200 m^3. Find the work done on the system, the change in the internal energy of the system, and the thermal energy transferred in the process.

Answers $W = -2.00 \times 10^4$ J, $\Delta U = -1.50 \times 10^4$ J, $Q = 5.00 \times 10^3$ J

TABLE 12.2

The First Law and Thermodynamic Processes (Ideal Gases)

Process	ΔU	Q	W
Isobaric	$nC_v\Delta T$	$nC_p\Delta T$	$-P\Delta V$
Adiabatic	$nC_v\Delta T$	0	ΔU
Isovolumetric	$nC_v\Delta T$	ΔU	0
Isothermal	0	$-W$	$-nRT\ln\left(\frac{V_f}{V_i}\right)$
General	$nC_v\Delta T$	$\Delta U - W$	(PV Area)

Given all the different processes and formulae, it's easy to become confused when approaching one of these ideal gas problems, though most of the time only substitution into the correct formula is required. The essential facts and formulas are compiled in Table 12.2, both for easy reference and also to display the similarities and differences between the processes.

Quick Quiz 12.2

Identify the paths A, B, C, and D in Figure 12.8 as isobaric, isothermal, isovolumetric, or adiabatic. For path B, $Q = 0$.

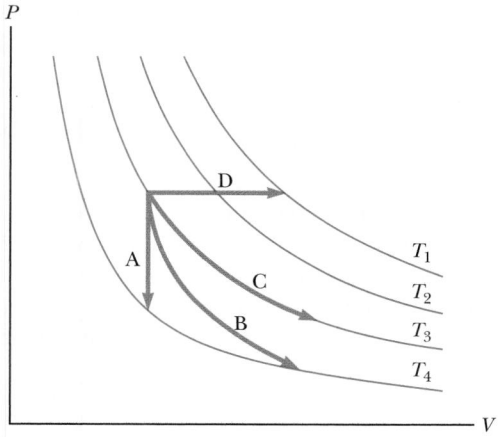

Figure 12.8 (Quick Quiz 12.2) Identify the nature of paths A, B, C, and D.

12.3 HEAT ENGINES AND THE SECOND LAW OF THERMODYNAMICS

A **heat engine** takes in energy by heat and partially converts it to other forms, such as electrical and mechanical energy. In a typical process for producing electricity in a power plant, for instance, coal or some other fuel is burned, and the resulting internal energy is used to convert water to steam. The steam is then directed at the blades of a turbine, setting it rotating. Finally, the mechanical energy associated with this rotation is used to drive an electric generator. In another heat engine—the internal combustion engine in an automobile—energy enters the engine as fuel is injected into the cylinder and combusted, and a fraction of this energy is converted to mechanical energy.

In general, a heat engine carries some working substance through a **cyclic process**[1] during which (1) energy is transferred by heat from a source at a high

◀ Cyclic process

[1]Strictly speaking, the internal combustion engine is not a heat engine according to the description of the cyclic process, because the air–fuel mixture undergoes only one cycle and is then expelled through the exhaust system.

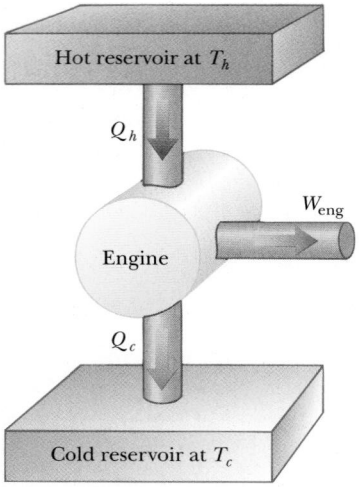

ACTIVE FIGURE 12.9
A schematic representation of a heat engine. The engine receives energy Q_h from the hot reservoir, expels energy Q_c to the cold reservoir, and does work W.

PhysicsNow™
Log into PhysicsNow at **www.cp7e.com** and go to Active Figure 12.9 to select the efficiency of the engine and observe the transfer of energy.

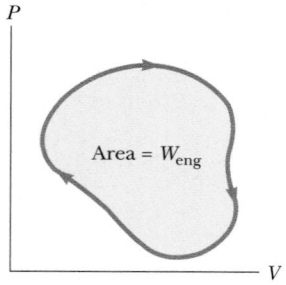

Figure 12.10 The PV diagram for an arbitrary cyclic process. The area enclosed by the curve equals the net work done.

temperature, (2) work is done by the engine, and (3) energy is expelled by the engine by heat to a source at lower temperature. As an example, consider the operation of a steam engine in which the working substance is water. The water in the engine is carried through a cycle in which it first evaporates into steam in a boiler and then expands against a piston. After the steam is condensed with cooling water, it returns to the boiler, and the process is repeated.

It's useful to draw a heat engine schematically, as in Active Figure 12.9. The engine absorbs energy Q_h from the hot reservoir, does work W_{eng}, then gives up energy Q_c to the cold reservoir. (Note that *negative* work is done *on* the engine, so that $W = -W_{eng}$.) Because the working substance goes through a cycle, always returning to its initial thermodynamic state, its initial and final internal energies are equal, so $\Delta U = 0$. From the first law of thermodynamics, therefore,

$$\Delta U = 0 = Q + W \quad \rightarrow \quad Q_{net} = -W = W_{eng}$$

The last equation shows that **the work W_{eng} done by a heat engine equals the net energy absorbed by the engine**. As we can see from Active Figure 12.9, $Q_{net} = |Q_h| - |Q_c|$. Therefore,

$$W_{eng} = |Q_h| - |Q_c| \qquad \textbf{[12.11]}$$

Ordinarily, a transfer of thermal energy Q can be either positive or negative, so the use of absolute value signs makes the signs of Q_h and Q_c explicit.

If the working substance is a gas, then **the work done by the engine for a cyclic process is the area enclosed by the curve representing the process on a *PV* diagram**. This area is shown for an arbitrary cyclic process in Figure 12.10.

The **thermal efficiency** e of a heat engine is defined as the work done by the engine, W_{eng}, divided by the energy absorbed during one cycle:

$$e \equiv \frac{W_{eng}}{|Q_h|} = \frac{|Q_h| - |Q_c|}{|Q_h|} = 1 - \frac{|Q_c|}{|Q_h|} \qquad \textbf{[12.12]}$$

We can think of thermal efficiency as the ratio of the benefit received (work) to the cost incurred (energy transfer at the higher temperature). Equation 12.12 shows that a heat engine has 100% efficiency ($e = 1$) only if $Q_c = 0$—meaning no energy is expelled to the cold reservoir. In other words, a heat engine with perfect efficiency would have to expel all the input energy by doing mechanical work. This isn't possible, as will be seen in Section 12.4.

EXAMPLE 12.10 The Efficiency of an Engine

Goal Apply the efficiency formula to a heat engine.

Problem During one cycle, an engine extracts 2.00×10^3 J of energy from a hot reservoir and transfers 1.50×10^3 J to a cold reservoir. **(a)** Find the thermal efficiency of the engine. **(b)** How much work does this engine do in one cycle? **(c)** How much power does the engine generate if it goes through four cycles in 2.50 s?

Strategy Apply Equation 12.12 to obtain the thermal efficiency, then use the first law, adapted to engines (Equation 12.11), to find the work done in one cycle. To obtain the power generated, just divide the work done in four cycles by the time it takes to run those cycles.

Solution

(a) Find the engine's thermal efficiency.

Substitute Q_c and Q_h into Equation 12.12:

$$e = 1 - \frac{|Q_c|}{|Q_h|} = 1 - \frac{1.50 \times 10^3\,\text{J}}{2.00 \times 10^3\,\text{J}} = 0.250, \text{ or } 25.0\%$$

(b) How much work does this engine do in one cycle?

Apply the first law in the form of Equation 12.11 to find the work done by the engine:

$$W_{\text{eng}} = |Q_h| - |Q_c| = 2.00 \times 10^3\,\text{J} - 1.50 \times 10^3\,\text{J}$$
$$= 5.00 \times 10^2\,\text{J}$$

(c) Find the power output of the engine.

Multiply the answer of part (b) by four and divide by time:

$$\mathcal{P} = \frac{W}{\Delta t} = \frac{4.00 \times (5.00 \times 10^2\,\text{J})}{2.50\,\text{s}} = 8.00 \times 10^2\,\text{W}$$

Remark Problems like this usually reduce to solving two equations and two unknowns, as here, where the two equations are the efficiency equation and the first law and the unknowns are the efficiency and the work done by the engine.

Exercise 12.10

The energy absorbed by an engine is three times as great as the work it performs. (a) What is its thermal efficiency? (b) What fraction of the energy absorbed is expelled to the cold reservoir? (c) What is the power output of the engine if the energy input is 1 650 J each cycle and it goes through two cycles every 3 seconds?

Answer (a) 1/3 (b) 2/3 (c) 367 W

EXAMPLE 12.11 Analyzing an Engine Cycle

Goal Combine several concepts to analyze an engine cycle.

Problem A heat engine contains an ideal monatomic gas confined to a cylinder by a movable piston. The gas starts at A, where $T = 3.00 \times 10^2$ K. (See Fig. 12.11a.) $B \rightarrow C$ is an isothermal expansion. **(a)** Find the number n of moles of gas and the temperature at B. **(b)** Find ΔU, Q, and W for the isovolumetric process $A \rightarrow B$. **(c)** Repeat for the isothermal process $B \rightarrow C$. **(d)** Repeat for the isobaric process $C \rightarrow A$. **(e)** Find the net change in the internal energy for the complete cycle. **(f)** Find the thermal energy Q_h transferred into the system, the thermal energy rejected, Q_c, the thermal efficiency, and net work on the environment performed by the engine.

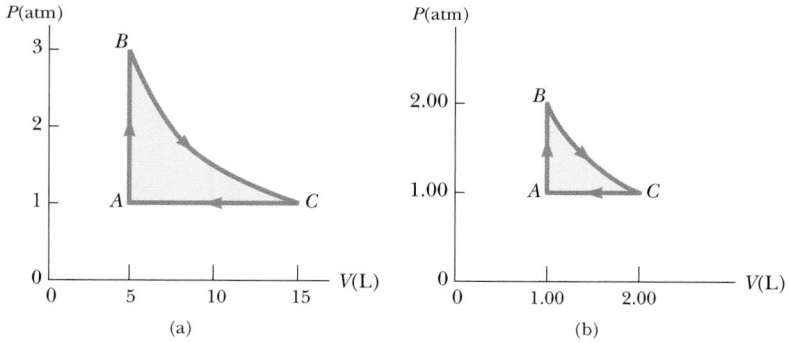

Figure 12.11 (a) (Example 12.11) (b) (Exercise 12.11)

Strategy In part (a) n, T, and V can be found from the ideal gas law, which connects the equilibrium values of P, V, and T. Once the temperature T is known at the points A, B, and C, the change in internal energy, ΔU, can be computed from the formula in Table 12.2 for each process. Q and W can be similarly computed, or deduced from the first law, using the techniques applied in the single process examples.

Solution

(a) Find n and T_B with the ideal gas law:

$$n = \frac{P_A V_A}{R T_A} = \frac{(1.00\,\text{atm})(5.00\,\text{L})}{(0.0821\,\text{L}\cdot\text{atm/mol}\cdot\text{K})(3.00 \times 10^2\,\text{K})}$$
$$= 0.203\,\text{mol}$$

$$T_B = \frac{P_B V_B}{nR} = \frac{(3.00\,\text{atm})(5.00\,\text{L})}{(0.203\,\text{mol})(0.0821\,\text{L}\cdot\text{atm/mol}\cdot\text{K})}$$
$$= 9.00 \times 10^2\,\text{K}$$

(b) Find ΔU_{AB}, Q_{AB}, and W_{AB} for the constant volume process $A \rightarrow B$.

Compute ΔU_{AB}, noting that $C_v = \frac{3}{2}R = 12.5\ \text{J/mol}\cdot\text{K}$:

$$\Delta U_{AB} = nC_v\Delta T = (0.203\ \text{mol})(12.5\ \text{J/mol}\cdot\text{K}) \times (9.00 \times 10^2\ \text{K} - 3.00 \times 10^2\ \text{K})$$

$$\Delta U_{AB} = 1.52 \times 10^3\ \text{J}$$

$\Delta V = 0$ for isovolumetric processes, so no work is done:

$$W_{AB} = 0$$

We can find Q_{AB} from the first law:

$$Q_{AB} = \Delta U_{AB} = 1.52 \times 10^3\ \text{J}$$

(c) Find ΔU_{BC}, Q_{BC}, and W_{BC} for the isothermal process $B \rightarrow C$.

This process is isothermal, so the temperature doesn't change, and the change in internal energy is zero:

$$\Delta U_{BC} = nC_v\Delta T = 0$$

Compute the work done on the system, using the negative of Equation 12.10:

$$W_{BC} = -nRT\ \ln\left(\frac{V_C}{V_B}\right) = -(0.203\ \text{mol})(8.31\ \text{J/mol}\cdot\text{K})(9.00 \times 10^2\ \text{K}) \times \ln\left(\frac{1.50 \times 10^{-2}\ \text{m}^3}{5.00 \times 10^{-3}\ \text{m}^3}\right)$$

$$W_{BC} = -1.67 \times 10^3\ \text{Pa}$$

Compute Q_{BC} from the first law:

$$0 = Q_{BC} + W_{BC} \quad \rightarrow \quad Q_{BC} = -W_{BC} = 1.67 \times 10^3\ \text{J}$$

(d) Find ΔU_{CA}, Q_{CA}, and W_{CA} for the isobaric process $C \rightarrow A$.

Compute the work on the system, with pressure constant:

$$W_{CA} = -P\Delta V = -(1.01 \times 10^5\ \text{Pa})(5.00 \times 10^{-3}\ \text{m}^3 - 1.50 \times 10^{-2}\ \text{m}^3)$$

$$W_{CA} = 1.01 \times 10^3\ \text{J}$$

Find the change in internal energy, ΔU_{CA}:

$$\Delta U_{CA} = \frac{3}{2}nRT = \frac{3}{2}(0.203\ \text{mol})(8.31\ \text{J/K}\cdot\text{mol}) \times (3.00 \times 10^2\ \text{K} - 9.00 \times 10^2\ \text{K})$$

$$\Delta U_{CA} = -1.52 \times 10^3\ \text{J}$$

Compute the thermal energy, Q_{CA}, from the first law:

$$Q_{CA} = \Delta U_{CA} - W_{CA} = -1.52 \times 10^3\ \text{J} - 1.01 \times 10^3\ \text{J} = -2.53 \times 10^3\ \text{J}$$

(e) Find the net change in internal energy, ΔU_{net}, for the cycle:

$$\Delta U_{\text{net}} = \Delta U_{AB} + \Delta U_{BC} + \Delta U_{CA} = 1.52 \times 10^3\ \text{J} + 0 - 1.52 \times 10^3\ \text{J} = 0$$

(f) Find the energy input, Q_h; the energy rejected, Q_c; the thermal efficiency; and the net work performed by the engine:

Sum all the positive contributions to find Q_h:

$$Q_h = Q_{AB} + Q_{BC} = 1.52 \times 10^3\ \text{J} + 1.67 \times 10^3\ \text{J} = 3.19 \times 10^3\ \text{J}$$

Sum any negative contributions (in this case, there is only one):

$$Q_c = -2.53 \times 10^3\ \text{J}$$

Find the engine efficiency and the net work done by the engine:

$$e = 1 - \frac{|Q_c|}{|Q_h|} = 1 - \frac{2.53 \times 10^3\,\text{J}}{3.19 \times 10^3\,\text{J}} = 0.207$$

$$\begin{aligned} W_{\text{eng}} &= -(W_{AB} + W_{BC} + W_{CA}) \\ &= -(0 - 1.67 \times 10^3\,\text{J} + 1.01 \times 10^3\,\text{J}) \\ &= 6.60 \times 10^2\,\text{J} \end{aligned}$$

Remarks Cyclic problems are rather lengthy; however, the individual steps are often short substitutions. Notice that the change in internal energy for the cycle is zero and that the net work done on the environment is identical to the net thermal energy transferred, both as they should be.

Exercise 12.11

4.05×10^{-2} mol of monatomic ideal gas goes through the process shown in Figure 12.11b. The temperature at point A is 3.00×10^2 K and is 6.00×10^2 K during the isothermal process $B \rightarrow C$. (a) Find Q, ΔU, and W for the constant volume process $A \rightarrow B$. (b) Do the same for the isothermal process $B \rightarrow C$. (c) Repeat, for the constant pressure process $C \rightarrow A$. (d) Find Q_h, Q_c, and the efficiency. (e) Find W_{eng}.

Answers (a) $Q_{AB} = \Delta U_{AB} = 151\,\text{J}$, $W_{AB} = 0$ (b) $\Delta U_{BC} = 0$, $Q_{BC} = -W_{BC} = 1.40 \times 10^2\,\text{J}$ (c) $Q_{CA} = -252\,\text{J}$, $\Delta U_{CA} = -151\,\text{J}$, $W_{CA} = 101\,\text{J}$ (d) $Q_h = 291\,\text{J}$, $Q_c = -252\,\text{J}$, $e = 0.134$ (e) $W_{\text{eng}} = 39.0\,\text{J}$

Refrigerators and Heat Pumps

Heat engines can operate in reverse. In this case, energy is injected into the engine, modeled as work W in Active Figure 12.12, resulting in energy being extracted from the cold reservoir and transferred to the hot reservoir. The system now operates as a heat pump, a common example being a refrigerator (Fig. 12.13). Energy Q_c is extracted from the interior of the refrigerator and delivered as energy Q_h to the warmer air in the kitchen. The work is done in the compressor unit of the refrigerator, compressing a refrigerant such as freon, causing its temperature to increase.

A household air conditioner is another example of a heat pump. Some homes are both heated and cooled by heat pumps. In the winter, the heat pump extracts energy Q_c from the cool outside air and delivers energy Q_h to the warmer air inside. In summer, energy Q_c is removed from the cool inside air, while energy Q_h is ejected to the warm air outside.

For a refrigerator or an air conditioner—a heat pump operating in cooling mode—work W is what you pay for, in terms of electrical energy running the compressor, while Q_c is the desired benefit. The most efficient refrigerator or air conditioner is one that removes the greatest amount of energy from the cold reservoir in exchange for the least amount of work.

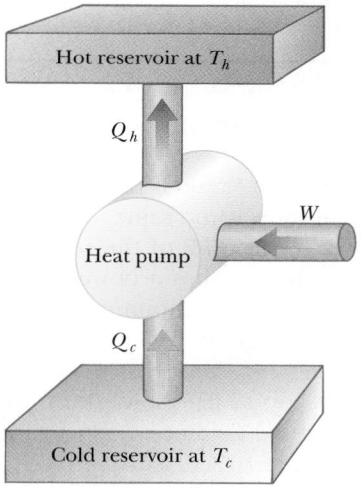

ACTIVE FIGURE 12.12
Schematic diagram of a heat pump, which takes in energy $Q_c > 0$ from a cold reservoir and expels energy $Q_h < 0$ to a hot reservoir. Work W is done *on* the heat pump. A refrigerator works the same way.

PhysicsNow™
Log into PhysicsNow at **www.cp7e.com**, and go to Active Figure 12.12 to select the coefficient of performance (COP) of the heat pump and observe the transfer of energy.

The coefficient of performance for a refrigerator or an air conditioner is the magnitude of the energy extracted from the cold reservoir, $|Q_c|$, divided by the work W performed by the device:

$$\text{COP(cooling mode)} = \frac{|Q_c|}{W} \qquad [12.13]$$

SI unit: dimensionless

The larger this ratio, the better the performance, since more energy is being removed for a given amount of work. A good refrigerator or air conditioner will have a COP of 5 or 6.

Charles D. Winters

Figure 12.13 The coils on the back of a refrigerator transfer energy by heat to the air.

A heat pump operating in heating mode warms the inside of a house in winter by extracting energy from the colder outdoor air. This may seem paradoxical, but recall that this is equivalent to a refrigerator removing energy from its interior and ejecting it into the kitchen.

The coefficient of performance of a heat pump operating in the heating mode is the magnitude of the energy rejected to the hot reservoir, $|Q_h|$, divided by the work W done by the pump:

$$\text{COP(heating mode)} = \frac{|Q_h|}{W} \qquad [12.14]$$

SI unit: dimensionless

In effect, the COP of a heat pump in the heating mode is the ratio of what you gain (energy delivered to the interior of your home) to what you give (work input). Typical values for this COP are greater than one, because $|Q_h|$ is usually greater than W.

In a groundwater heat pump, energy is extracted in the winter from water deep in the ground rather than from the outside air, while energy is delivered to that water in the summer. This strategy increases the year-round efficiency of the heating and cooling unit, because the groundwater is at a higher temperature than the air in winter and at a cooler temperature than the air in summer.

EXAMPLE 12.12 Cooling the Leftovers

Goal Apply the coefficient of performance of a refrigerator.

Problem 2.00 L of leftover soup at a temperature of 323 K is placed in a refrigerator. Assume the specific heat of the soup is the same as that of water and the density is 1.25×10^3 kg/m^3. The refrigerator cools the soup to 283 K. **(a)** If the COP of the refrigerator is 5.00, find the energy needed, in the form of work, to cool the soup. **(b)** If the compressor has a power rating of 0.250 hp, for what minimum length of time must it operate to cool the soup to 283 K? (The minimum time assumes the soup cools at the same rate that the heat pump ejects thermal energy from the refrigerator.)

Strategy The solution to this problem requires three steps. First, find the total mass m of the soup. Second, using $Q = mc\Delta T$, where $Q = Q_c$, find the energy transfer required to cool the soup. Third, substitute Q_c and the COP into Equation 12.13, solving for W. Divide the work by the power to get an estimate of the time required to cool the soup.

Solution

(a) Find the work needed to cool the soup.

Calculate the mass of the soup:

$$m = \rho V = (1.25 \times 10^3 \text{ kg/m}^3)(2.00 \times 10^{-3} \text{ m}^3) = 2.50 \text{ kg}$$

Find the energy transfer required to cool the soup:

$$\begin{aligned} Q_c = Q &= mc\Delta T \\ &= (2.50 \text{ kg})(4\,190 \text{ J/kg}\cdot\text{K})(283 \text{ K} - 323 \text{ K}) \\ &= -4.19 \times 10^5 \text{ J} \end{aligned}$$

Substitute Q_c and the COP into Equation 12.13:

$$\text{COP} = \frac{|Q_c|}{W} = \frac{4.19 \times 10^5 \text{ J}}{W} = 5.00$$

$$W = 8.38 \times 10^4 \text{ J}$$

(b) Find the time needed to cool the food.

Convert horsepower to watts:

$$\mathcal{P} = (0.250 \text{ hp})(746 \text{ W}/1 \text{ hp}) = 187 \text{ W}$$

Divide the work by the power to find the elapsed time: $\Delta t = \frac{W}{\mathcal{P}} = \frac{8.38 \times 10^4 \text{ J}}{187 \text{ W}} = 448 \text{ s}$

Remarks This example illustrates how cooling different substances requires differing amounts of work, due to differences in specific heats. The problem doesn't take into account the insulating properties of the soup container and of the soup itself, which retard the cooling process.

Exercise 12.12

(a) How much work must a heat pump with a COP of 2.50 do in order to extract 1.00 MJ of thermal energy from the outdoors (the cold reservoir)? (b) If the unit operates at 0.500 hp, how long will the process take? (Be sure to use the correct COP!)

Answers (a) 6.67×10^5 J (b) 1.79×10^3 s

The Second Law of Thermodynamics

There are limits to the efficiency of heat engines. The ideal engine would convert all input energy into useful work, but it turns out that such an engine is impossible to construct. The Kelvin–Planck formulation of the **second law of thermodynamics** can be stated as follows:

> No heat engine operating in a cycle can absorb energy from a reservoir and use it entirely for the performance of an equal amount of work.

This form of the second law means that the efficiency $e = W_{\text{eng}}/|Q_h|$ of engines must always be less than one. Some energy Q_c must always be lost to the environment. In other words, it's theoretically impossible to construct a heat engine with an efficiency of 100%.

To summarize, the first law says **we can't get a greater amount of energy out of a cyclic process than we put in**, and the second law says **we can't break even**.

J-L. Charmet/SPL/Photo Researchers, Inc.

LORD KELVIN, British Physicist and Mathematician (1824–1907)

Born William Thomson in Belfast, Kelvin was the first to propose the use of an absolute scale of temperature. His study of Carnot's theory led to the idea that energy cannot pass spontaneously from a colder object to a hotter object; this principle is known as the second law of thermodynamics.

Reversible and Irreversible Processes

No engine can operate with 100% efficiency, but different designs yield different efficiencies, and it turns out one design in particular delivers the maximum possible efficiency. This design is the Carnot cycle, discussed in the next subsection. Understanding it requires the concepts of reversible and irreversible processes. In a **reversible** process, every state along the path is an equilibrium state, so the system can return to its initial conditions by going along the same path in the reverse direction. A process that doesn't satisfy this requirement is **irreversible**.

Most natural processes are known to be irreversible—the reversible process is an idealization. Although real processes are always irreversible, some are *almost* reversible. If a real process occurs so slowly that the system is virtually always in equilibrium, the process can be considered reversible. Imagine compressing a gas very slowly by dropping grains of sand onto a frictionless piston, as in Figure 12.14. The temperature can be kept constant by placing the gas in thermal contact with an energy reservoir. The pressure, volume, and temperature of the gas are well defined during this isothermal compression. Each added grain of sand represents a change to a new equilibrium state. The process can be reversed by slowly removing grains of sand from the piston.

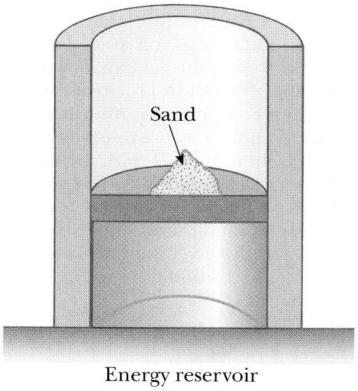

Figure 12.14 A gas in thermal contact with an energy reservoir is compressed slowly by grains of sand dropped onto a piston. The compression is isothermal and reversible.

The Carnot Engine

In 1824, in an effort to understand the efficiency of real engines, a French engineer named Sadi Carnot (1796–1832) described a theoretical engine now called a *Carnot engine* that is of great importance from both a practical and a theoretical

J-L. Charmet/SPL/Photo Researchers, Inc.

SADI CARNOT, French Engineer (1796–1832)

Carnot is considered to be the founder of the science of thermodynamics. Some of his notes found after his death indicate that he was the first to recognize the relationship between work and heat.

viewpoint. He showed that a heat engine operating in an ideal, reversible cycle—now called a **Carnot cycle**—between two energy reservoirs is the most efficient engine possible. Such an engine establishes an upper limit on the efficiencies of all real engines. **Carnot's theorem** can be stated as follows:

> No real engine operating between two energy reservoirs can be more efficient than a Carnot engine operating between the same two reservoirs.

In a Carnot cycle, an ideal gas is contained in a cylinder with a movable piston at one end. The temperature of the gas varies between T_c and T_h. The cylinder walls and the piston are thermally nonconducting. Active Figure 12.15 shows the four stages of the Carnot cycle, and Active Figure 12.16 is the *PV* diagram for the cycle. The cycle consists of two adiabatic and two isothermal processes, all reversible:

1. The process $A \rightarrow B$ is an isothermal expansion at temperature T_h in which the gas is placed in thermal contact with a hot reservoir (a large oven, for example) at temperature T_h (Active Fig. 12.15a). During the process, the gas absorbs energy Q_h from the reservoir and does work W_{AB} in raising the piston.
2. In the process $B \rightarrow C$, the base of the cylinder is replaced by a thermally nonconducting wall and the gas expands adiabatically, so no energy enters or

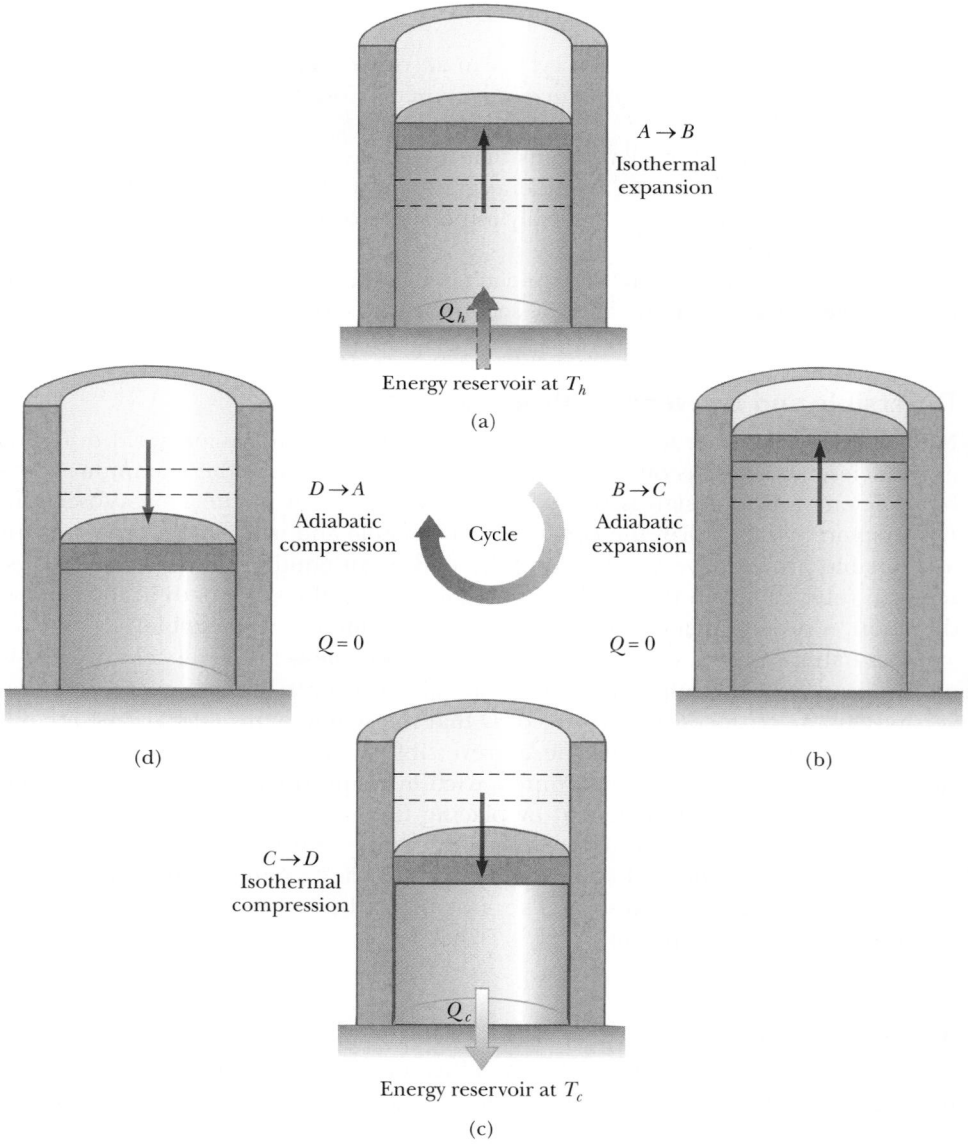

ACTIVE FIGURE 12.15
The Carnot cycle. In process $A \rightarrow B$, the gas expands isothermally while in contact with a reservoir at T_h. In process $B \rightarrow C$, the gas expands adiabatically ($Q = 0$). In process $C \rightarrow D$, the gas is compressed isothermally while in contact with a reservoir at $T_c < T_h$. In process $D \rightarrow A$, the gas is compressed adiabatically. The upward arrows on the piston indicate the removal of sand during the expansions, and the downward arrows indicate the addition of sand during the compressions.

PhysicsNow™
Log into PhysicsNow at **www.cp7e.com**, and go to Active Figure 12.15 to observe the motion of the piston in the Carnot cycle while you also observe the cycle on the *PV* diagram of Active Figure 12.16.

leaves the system by heat (Active Fig. 12.15b). During the process, the temperature falls from T_h to T_c and the gas does work W_{BC} in raising the piston.

3. In the process $C \rightarrow D$, the gas is placed in thermal contact with a cold reservoir at temperature T_c (Active Fig. 12.15c) and is compressed isothermally at temperature T_c. During this time, the gas expels energy Q_c to the reservoir and the work done on the gas is W_{CD}.
4. In the final process, $D \rightarrow A$, the base of the cylinder is again replaced by a thermally nonconducting wall (Active Fig. 12.15d) and the gas is compressed adiabatically. The temperature of the gas increases to T_h, and the work done on the gas is W_{DA}.

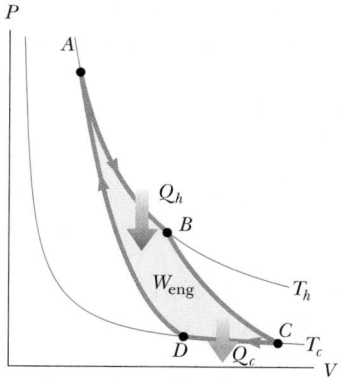

ACTIVE FIGURE 12.16
The *PV* diagram for the Carnot cycle. The net work done, W_{eng}, equals the net energy received by heat in one cycle, $|Q_h| - |Q_c|$.

Physics Now™
Log into PhysicsNow at **www.cp7e.com**, and go to Active Figure 12.15 to observe the Carnot cycle on the *PV* diagram while you also observe the motion of the piston in Active Figure 12.15.

For a Carnot engine, the following relationship between the thermal energy transfers and the absolute temperatures can be derived:

$$\frac{|Q_c|}{|Q_h|} = \frac{T_c}{T_h} \qquad [12.15]$$

Substituting this expression into Equation 12.12, we find that the thermal efficiency of a Carnot engine is

$$e_C = 1 - \frac{T_c}{T_h} \qquad [12.16]$$

where T must be in kelvins. From this result, we see that **all Carnot engines operating reversibly between the same two temperatures have the same efficiency**.

Equation 12.16 can be applied to any working substance operating in a Carnot cycle between two energy reservoirs. According to that equation, the efficiency is zero if $T_c = T_h$. The efficiency increases as T_c is lowered and as T_h is increased. The efficiency can be one (100%), however, only if $T_c = 0$ K. According to the **third law of thermodynamics**, it's impossible to lower the temperature of a system to absolute zero in a finite number of steps, so such reservoirs are not available, and the maximum efficiency is always less than one. In most practical cases, the cold reservoir is near room temperature, about 300 K, so increasing the efficiency requires raising the temperature of the hot reservoir. **All real engines operate irreversibly, due to friction and the brevity of their cycles, and are therefore *less* efficient than the Carnot engine.**

TIP 12.2 Don't Shop for a Carnot Engine
The Carnot engine is only an idealization. If a Carnot engine were developed in an effort to maximize efficiency, it would have zero power output, because, in order for all of the processes to be reversible, the engine would have to run infinitely slowly.

Quick Quiz 12.3

Three engines operate between reservoirs separated in temperature by 300 K. The reservoir temperatures are as follows:

Engine A: $T_h = 1\,000$ K, $T_c = 700$ K

Engine B: $T_h = 800$ K, $T_c = 500$ K

Engine C: $T_h = 600$ K, $T_c = 300$ K

Rank the engines in order of their theoretically possible efficiency, from highest to lowest. (a) A, B, C (b) B, C, A (c) C, B, A (d) C, A, B

EXAMPLE 12.13 The Steam Engine

Goal Apply the equations of an ideal (Carnot) engine.

Problem A steam engine has a boiler that operates at 5.00×10^2 K. The energy from the boiler changes water to steam, which drives the piston. The temperature of the exhaust is that of the outside air, 3.00×10^2 K. **(a)** What is the engine's efficiency if it's an ideal engine? **(b)** If the 3.50×10^3 J of energy is supplied from the boiler, find the work done by the engine on its environment.

Strategy This problem requires substitution into Equations 12.15 and 12.16, both applicable to a Carnot engine. The first equation relates the ratio Q_c/Q_h to the ratio T_c/T_h, and the second gives the Carnot engine efficiency.

Solution

(a) Find the engine's efficiency, assuming it's ideal.

Substitute into Equation 12.16, the equation for the efficiency of a Carnot engine:

$$e_C = 1 - \frac{T_c}{T_h} = 1 - \frac{3.00 \times 10^2\ \text{K}}{5.00 \times 10^2\ \text{K}} = 0.400, \text{ or } 40\%$$

(b) Find the work done on the environment if 3.50×10^3 J is delivered to the engine during one cycle.

The ratio of energies equals the ratio of temperatures:

$$\frac{|Q_c|}{|Q_h|} = \frac{T_c}{T_h} \quad \rightarrow \quad |Q_c| = |Q_h|\frac{T_c}{T_h}$$

$$|Q_c| = (3.50 \times 10^3\ \text{J})\left(\frac{3.00 \times 10^2\ \text{K}}{5.00 \times 10^2\ \text{K}}\right) = 2.10 \times 10^3\ \text{J}$$

Use Equation 12.11 to find W:

$$W_{\text{eng}} = |Q_h| - |Q_c| = 3.50 \times 10^3\ \text{J} - 2.10 \times 10^3\ \text{J}$$
$$= 1.40 \times 10^3\ \text{J}$$

Remarks This problem differs from the earlier examples on work and efficiency because we used the special Carnot relationships, Equations 12.15 and 12.16. Remember that these equations can only be used when the cycle is identified as ideal or a Carnot.

Exercise 12.13

The highest theoretical efficiency of a gasoline engine based on the Carnot cycle, is 0.300, or 30.0%. (a) If this engine expels its gases into the atmosphere, which has a temperature of 3.00×10^2 K, what is the temperature in the cylinder immediately after combustion? (b) If the heat engine absorbs 837 J of energy from the hot reservoir during each cycle, how much work can it perform in each cycle?

Answers (a) 429 K (b) 251 J

12.4 ENTROPY

Temperature and internal energy, associated with the zeroth and first laws of thermodynamics, respectively, are both state variables, meaning they can be used to describe the thermodynamic state of a system. A state variable called the **entropy** S is related to the second law of thermodynamics. We define entropy on a macroscopic scale as the German physicist Rudolf Clausius (1822–1888) first expressed it in 1865:

Entropy ▶

Let Q_r be the energy absorbed or expelled during a reversible, constant temperature process between two equilibrium states. Then the change in entropy during any constant temperature process connecting the two equilibrium states is defined as

$$\Delta S \equiv \frac{Q_r}{T} \qquad [12.17]$$

SI unit: joules/kelvin (J/K)

TIP 12.3 Entropy ≠ Energy

Don't confuse energy and entropy—though the names sound similar, they are different concepts.

A similar formula holds when the temperature isn't constant, but its derivation entails calculus and won't be considered here. Calculating the change in entropy, ΔS, during a transition between two equilibrium states requires finding a reversible path that connects the states. The entropy change calculated on that reversible path is taken to be ΔS for the *actual* path. This approach is necessary because quantities such as the temperature of a system can be defined only for systems in equilibrium, and a reversible path consists of a sequence of equilibrium states. The subscript r on the term Q_r emphasizes that the path chosen for the calculation must be reversible. The change in entropy ΔS, like changes in internal energy ΔU and changes in potential energy, depends only on the endpoints, and not on the path connecting them.

The concept of entropy gained wide acceptance in part because it provided another variable to describe the state of a system, along with pressure, volume, and temperature. Its significance was enhanced when it was found that **the entropy of the Universe increases in all natural processes**. This is yet another way of stating the second law of thermodynamics.

Although the entropy of the *Universe* increases in all natural processes, the entropy of a *system* can decrease. For example, if system A transfers energy Q to system B by heat, the entropy of system A decreases. This transfer, however, can only occur if the temperature of system B is less than the temperature of system A. Because temperature appears in the denominator in the definition of entropy, system B's increase in entropy will be greater than system A's decrease, so taken together, the entropy of the Universe increases.

For centuries, individuals have attempted to build perpetual motion machines that operate continuously without any input of energy or increase in entropy. The laws of thermodynamics preclude the invention of any such machines.

The concept of entropy is satisfying because it enables us to present the second law of thermodynamics in the form of a mathematical statement. In the next section, we find that entropy can also be interpreted in terms of probabilities, a relationship that has profound implications.

AIP Niels Bohr Library, Lande Collection

RUDOLF CLAUSIUS, German Physicist (1822–1888)

Born with the name Rudolf Gottlieb, he adopted the classical name of Clausius, which was a popular thing to do in his time. "I propose . . . to call S the entropy of a body, after the Greek word 'transformation.' I have designedly coined the word 'entropy' to be similar to energy, for these two quantities are so analogous in their physical significance, that an analogy of denominations seems to be helpful."

Quick Quiz 12.4

Which of the following is true for the entropy change of a system that undergoes a reversible, adiabatic process? (a) $\Delta S < 0$ (b) $\Delta S = 0$ (c) $\Delta S > 0$

EXAMPLE 12.14 Melting a Piece of Lead

Goal Calculate the change in entropy due to a phase change.

Problem **(a)** Find the change in entropy of 3.00×10^2 g of lead when it melts at 327°C. Lead has a latent heat of fusion of 2.45×10^4 J/kg. **(b)** Suppose the same amount of energy is used to melt part of a piece of silver, which is already at its melting point of 961°C. Find the change in the entropy of the silver.

Strategy This problem can be solved by substitution into Equation 12.17. Be sure to use the Kelvin temperature scale.

Solution

(a) Find the entropy change of the lead.

Find the energy necessary to melt the lead:

$$Q = mL_f = (0.300\text{ kg})(2.45 \times 10^4\text{ J/kg}) = 7.35 \times 10^3\text{ J}$$

Convert the temperature in degrees Celsius to kelvins:

$$T = T_C + 273 = 327 + 273 = 6.00 \times 10^2\text{ K}$$

Substitute the quantities found into the entropy equation:

$$\Delta S = \frac{Q}{T} = \frac{7.35 \times 10^3\text{ J}}{6.00 \times 10^2\text{ K}} = \boxed{12.3\text{ J/K}}$$

(b) Find the entropy change of the silver.

The added energy is the same as in part (a), by supposition. Substitute into the entropy equation, after first converting the melting point of silver to kelvins:

$$T = T_C + 273 = 961 + 273 = 1.234 \times 10^2\text{ K}$$

$$\Delta S = \frac{Q}{T} = \frac{7.35 \times 10^3\text{ J}}{1.234 \times 10^3\text{ K}} = \boxed{5.96\text{ J/K}}$$

Remarks This example shows that adding a given amount of energy to a system increases its disorder, but adding the same amount of energy to another system at higher temperature results in a smaller increase in disorder. This is because the change in entropy is inversely proportional to the temperature.

Exercise 12.14
Find the change in entropy of a 2.00-kg block of gold at 1 063°C when it melts to become liquid gold at 1 063°C.

Answer 96.4 J/K

EXAMPLE 12.15 Ice, Steam, and the Entropy of the Universe

Goal Calculate the change in entropy for a system and its environment.

Problem A block of ice at 273 K is put in thermal contact with a container of steam at 373 K, converting 25.0 g of ice to water at 273 K while condensing some of the steam to water at 373 K. **(a)** Find the change in entropy of the ice. **(b)** Find the change in entropy of the steam. **(c)** Find the change in entropy of the Universe.

Strategy First, calculate the energy transfer necessary to melt the ice. The amount of energy gained by the ice is lost by the steam. Compute the entropy change for each process, and sum to get the entropy change of the universe.

Solution
(a) Find the change in entropy of the ice.

Use the latent heat of fusion, L_f, to compute the thermal energy needed to melt 25.0 g of ice:

$$Q_{\text{ice}} = mL_f = (0.025\text{ kg})(3.33 \times 10^5\text{ J}) = 8.33 \times 10^3\text{ J}$$

Calculate the change in entropy of the ice:

$$\Delta S_{\text{ice}} = \frac{Q_{\text{ice}}}{T_{\text{ice}}} = \frac{8.33 \times 10^3\text{ J}}{273\text{ K}} = 30.5\text{ J/K}$$

(b) Find the change in entropy of the steam.

By supposition, the thermal energy lost by the steam is equal to the thermal energy gained by the ice:

$$\Delta S_{\text{steam}} = \frac{Q_{\text{steam}}}{T_{\text{steam}}} = \frac{-8.33 \times 10^3\text{ J}}{373\text{ K}} = -22.3\text{ J/K}$$

(c) Find the change in entropy of the Universe.

Sum the two changes in entropy:

$$\Delta S_{\text{universe}} = \Delta S_{\text{ice}} + \Delta S_{\text{steam}} = 30.5\text{ J/k} - 22.3\text{ J/K}$$
$$= +8.2\text{ J/K}$$

Remark Notice that the entropy of the Universe increases, as it must in all natural processes.

Exercise 12.15
A 4.00-kg block of ice at 273 K encased in a thin plastic shell of negligible mass melts in a large lake at 293 K. At the instant the ice has completely melted in the shell and is still at 273 K, calculate the change in entropy of (a) the ice (b) the lake (which essentially remains at 293 K), and (c) the universe.

Answers (a) 4.88×10^3 J/K (b) -4.55×10^3 J/K (c) $+3.30 \times 10^2$ J/K

EXAMPLE 12.16 A Falling Boulder

Goal Combine mechanical energy and entropy.

Problem A chunk of rock of mass 1.00×10^3 kg at 293 K falls from a cliff of height 125 m into a large lake, also at 293 K. Find the change in entropy of the lake, assuming that all of the rock's kinetic energy upon entering the lake converts to thermal energy absorbed by the lake.

Strategy Gravitational potential energy when the rock is at the top of the cliff converts to kinetic energy of the rock before it enters the lake, and then is transferred to the lake as thermal energy. The change in the lake's temperature is negligible (due to its mass). Divide the mechanical energy of the rock by the temperature of the lake to estimate the lake's change in entropy.

Solution

Calculate the gravitational potential energy associated with the rock at the top of the cliff:

$$PE = mgh = (1.00 \times 10^3 \text{ kg})(9.80 \text{ m/s}^2)(125 \text{ m}) = 1.23 \times 10^6 \text{ J}$$

This energy is transferred to the lake as thermal energy, resulting in an entropy increase of the lake:

$$\Delta S = \frac{Q}{T} = \frac{1.23 \times 10^6 \text{ J}}{293 \text{ K}} = 4.20 \times 10^3 \text{ J/K}$$

Remarks This example shows how even simple mechanical processes can bring about increases in the Universe's entropy.

Exercise 12.16

Estimate the change in entropy of a tree trunk at 15.0°C when a bullet of mass 5.00 g traveling at 1.00×10^3 m/s embeds itself in it. (Assume the kinetic energy of the bullet transforms to thermal energy, all of which is absorbed by the tree.)

Answer 8.68 J/K

Entropy and Disorder

A large element of chance is inherent in natural processes. The spacing between trees in a natural forest, for example, is random; if you discovered a forest where all the trees were equally spaced, you would conclude that it had been planted. Likewise, leaves fall to the ground with random arrangements. It would be highly unlikely to find the leaves laid out in perfectly straight rows. We can express the results of such observations by saying that **a disorderly arrangement is much more probable than an orderly one if the laws of nature are allowed to act without interference**.

Entropy originally found its place in thermodynamics, but its importance grew tremendously as the field of statistical mechanics developed. This analytical approach employs an alternate interpretation of entropy. In statistical mechanics, the behavior of a substance is described by the statistical behavior of the atoms and molecules contained in it. One of the main conclusions of the statistical mechanical approach is that **isolated systems tend toward greater disorder, and entropy is a measure of that disorder**.

In light of this new view of entropy, Boltzmann found another method for calculating entropy through use of the relation

$$S = k_B \ln W \qquad [12.18]$$

where $k_B = 1.38 \times 10^{-23}$ J/K is Boltzmann's constant and W is a number proportional to the probability that the system has a particular configuration. The symbol "ln" again stands for natural logarithm, discussed in Appendix A.

TIP 12.4 Don't Confuse the *W*'s

The symbol W used here is a *probability*, not to be confused with the same symbol used for work.

Equation 12.18 could be applied to a bag of marbles. Imagine that you have 100 marbles—50 red and 50 green—stored in a bag. You are allowed to draw four marbles from the bag according to the following rules: Draw one marble, record its color, return it to the bag, and draw again. Continue this process until four marbles have been drawn. Note that because each marble is returned to the bag before the next one is drawn, the probability of drawing a red marble is always the same as the probability of drawing a green one.

The results of all possible drawing sequences are shown in Table 12.3. For example, the result RRGR means that a red marble was drawn first, a red one second, a green one third, and a red one fourth. The table indicates that there is only one possible way to draw four red marbles. There are four possible sequences that produce one green and three red marbles, six sequences that produce two green

TABLE 12.3

Possible Results of Drawing Four Marbles from a Bag

End Result	Possible Draws	Total Number of Same Results
All R	RRRR	1
1G, 3R	RRRG, RRGR, RGRR, GRRR	4
2G, 2R	RRGG, RGRG, GRRG, RGGR, GRGR, GGRR	6
3G, 1R	GGGR, GGRG, GRGG, RGGG	4
All G	GGGG	1

and two red, four sequences that produce three green and one red, and one sequence that produces all green. From Equation 12.18, we see that the state with the greatest disorder (two red and two green marbles) has the highest entropy, because it is most probable. In contrast, the most ordered states (all red marbles and all green marbles) are least likely to occur and are states of lowest entropy.

The outcome of the draw can range between these highly ordered (lowest-entropy) and highly disordered (highest-entropy) states. Entropy can be regarded as an index of how far a system has progressed from an ordered to a disordered state.

The second law of thermodynamics is really a statement of what is most probable rather than of what must be. Imagine placing an ice cube in contact with a hot piece of pizza. There is nothing in nature that absolutely forbids the transfer of energy by heat from the ice to the much warmer pizza. Statistically, it's possible for a slow-moving molecule in the ice to collide with a faster-moving molecule in the pizza so that the slow one transfers some of its energy to the faster one. However, when the great number of molecules present in the ice and pizza are considered, the odds are overwhelmingly in favor of the transfer of energy from the faster-moving molecules to the slower-moving molecules. Furthermore, this example demonstrates that a system naturally tends to move from a state of order to a state of disorder. The initial state, in which all the pizza molecules have high kinetic energy and all the ice molecules have lower kinetic energy, is much more ordered than the final state after energy transfer has taken place and the ice has melted.

APPLICATION

The Direction of Time

Even more generally, the second law of thermodynamics defines the direction of time for all events as the direction in which the entropy of the universe increases. Although conservation of energy isn't violated if energy flows spontaneously from a cold object (the ice cube) to a hot object (the pizza slice), that event violates the second law because it represents a spontaneous increase in order of course, such an event also violates everyday experience. If the melting ice cube is filmed and the film speeded up, the difference between running the film in forward and reverse directions would be obvious to an audience. The same would be true of filming any event involving a large number of particles, such as a dish dropping to the floor and shattering.

As another example, suppose you were able to measure the velocities of all the air molecules in a room at some instant. It's very unlikely that you would find all molecules moving in the same direction with the same speed—that would be a highly ordered state, indeed. The most probable situation is a system of molecules moving haphazardly in all directions with a wide distribution of speeds—a highly disordered state. This physical situation can be compared to the drawing of marbles from a bag: If a container held 10^{23} molecules of a gas, the probability of finding all of the molecules moving in the same direction with the same speed at some instant would be similar to that of drawing a marble from the bag 10^{23} times and getting a red marble on every draw—clearly an unlikely set of events.

The tendency of nature to move toward a state of disorder affects the ability of a system to do work. Consider a ball thrown toward a wall. The ball has kinetic energy, and its state is an ordered one, which means all of the atoms and molecules of the ball move in unison at the same speed and in the same direction (apart from their random internal motions). When the ball hits the wall, however, part of the ball's kinetic energy is transformed into the random, disordered, internal motion of the molecules in the ball and the wall, and the temperatures of the ball and the wall both increase slightly. Before the collision, the ball was capable of doing work. It could drive a nail into the wall, for example. With the transformation of part of the ordered energy into disordered internal energy, this capability of doing work is reduced. The ball rebounds with less kinetic energy than it originally had, because the collision is inelastic.

Various forms of energy can be converted to internal energy, as in the collision between the ball and the wall, but the reverse transformation is never complete. In general, given two kinds of energy, A and B, if A can be completely converted to B and vice versa, we say that A and B are of the *same grade*. However, if A can be completely converted to B and the reverse is never complete, then A is of a *higher grade* of energy than B. In the case of a ball hitting a wall, the kinetic energy of the ball is of a higher grade than the internal energy contained in the ball and the wall after the collision. When high-grade energy is converted to internal energy, it can never be fully recovered as high-grade energy.

This conversion of high-grade energy to internal energy is referred to as the **degradation of energy**. The energy is said to be degraded because it takes on a form that is less useful for doing work. In other words, **in all real processes, the energy available for doing work decreases**.

Finally, note once again that the statement that entropy must increase in all natural processes is true only for isolated systems. There are instances in which the entropy of some system decreases, but with a corresponding net increase in entropy for some other system. When all systems are taken together to form the Universe, **the entropy of the Universe always increases**.

Ultimately, the entropy of the Universe should reach a maximum. When it does, the Universe will be in a state of uniform temperature and density. All physical, chemical, and biological processes will cease, because a state of perfect disorder implies no available energy for doing work. This gloomy state of affairs is sometimes referred to as the ultimate "heat death" of the Universe.

George Semple

(a)

George Semple

(b)

A full house is a very good hand in the game of poker. Can you calculate the probability of being dealt a full house (a pair and three of a kind) from a standard deck of 52 cards? (a) A royal flush is a highly ordered poker hand with a low probability of occurrence. (b) A disordered and worthless poker hand. The probability of this *particular* hand occurring is the same as that of the royal flush. There are so many worthless hands, however, that the probability of being dealt a worthless hand is much higher than that of being dealt a royal flush.

Quick Quiz 12.5

Suppose you are throwing two dice in a friendly game of craps. For any given throw, the two numbers that are face up can have a sum of 2, 3, 4, 5, 6, 7, 8, 9, 10, 11, or 12. Which outcome is most probable? Which is least probable?

12.5 HUMAN METABOLISM

Animals do work and give off energy by heat, and this lead us to believe the first law of thermodynamics can be applied to living organisms to describe them in a general way. The internal energy stored in humans goes into other forms needed for maintaining and repairing the major body organs and is transferred out of the body by work as a person walks or lifts a heavy object, and by heat when the body is warmer than its surroundings. Because the rates of change of internal energy, energy loss by heat, and energy loss by work vary widely with the intensity and duration of human activity, it's best to measure the time rates of change of ΔU, Q, and W. Rewriting the first law, these time rates of change are related by

$$\frac{\Delta U}{\Delta t} = \frac{Q}{\Delta t} + \frac{W}{\Delta t} \qquad [12.19]$$

On average, energy Q flows *out* of the body, and work is done *by* the body on its surroundings, so both $Q/\Delta t$ and $W/\Delta t$ are negative. This means that $\Delta U/\Delta t$ would be negative and the internal energy and body temperature would decrease with time if a human were a closed system with no way of ingesting matter or replenishing internal energy stores. Because all animals are actually open systems, they acquire internal energy (chemical potential energy) by eating and breathing, so their internal energy and temperature are kept constant. Overall, the energy from the oxidation of food ultimately supplies the work done by the body and energy lost from the body by heat, and this is the interpretation we give Equation 12.19. That is, $\Delta U/\Delta t$ is the rate at which internal energy is added to our bodies by food, and this term just balances the rate of energy loss by heat, $Q/\Delta t$, and by work, $W/\Delta t$. Finally, if we have a way of measuring $\Delta U/\Delta t$ and $W/\Delta t$ for a human, we can calculate $Q/\Delta t$ from Equation 12.19 and gain useful information on the efficiency of the body as a machine.

Measuring the Metabolic Rate $\Delta U/\Delta t$

The value of $W/\Delta t$, the work done by a person per unit time, can easily be determined by measuring the power output supplied by the person (in pedaling a bike, for example). **The metabolic rate $\Delta U/\Delta t$ is the rate at which chemical potential energy in food and oxygen are transformed into internal energy to just balance the body losses of internal energy by work and heat.** Although the mechanisms of food oxidation and energy release in the body are complicated, involving many intermediate reactions and enzymes (organic compounds that speed up the chemical reactions taking place at "low" body temperatures), an amazingly simple rule summarizes these processes: **The metabolic rate is directly proportional to the rate of oxygen consumption by volume.** It is found that for an average diet, the consumption of one liter of oxygen releases 4.8 kcal, or 20 kJ, of energy. We may write this important summary rule as

$$\frac{\Delta U}{\Delta t} = 4.8\,\frac{\Delta V_{O_2}}{\Delta t} \qquad \textbf{[12.20]}$$

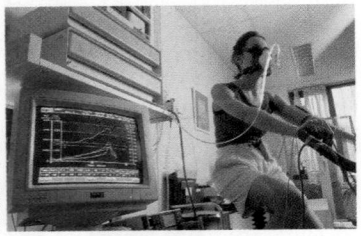

© BSIP/Laurent Science Source/Photo Researchers, Inc.

Figure 12.17 This bike rider is being monitored for oxygen consumption.

where the metabolic rate $\Delta U/\Delta t$ is measured in kcal/s and $\Delta V_{O_2}/\Delta t$, the volume rate of oxygen consumption, is in L/s. Measuring the rate of oxygen consumption during various activities ranging from sleep to intense bicycle racing effectively measures the variation of metabolic rate or the variation in the total power the body generates. A simultaneous measurement of the work per unit time done by a person along with the metabolic rate allows the efficiency of the body as a machine to be determined. Figure 12.17 shows a person monitored for oxygen consumption while riding a bike attached to a dynamometer, a device for measuring power output.

Metabolic Rate, Activity, and Weight Gain

Table 12.4 shows the measured rate of oxygen consumption in milliliters per minute per kilogram of body mass and the calculated metabolic rate for a 65-kg male engaged in various activities. A sleeping person uses about 80 W of power, the **basal metabolic rate**, just to maintain and run different body organs—heart, lungs, liver, kidneys, brain, and skeletal muscles. More intense activity increases the metabolic rate to a maximum of about 1 600 W for a superb racing cyclist, although such a high rate can only be maintained for periods of a few seconds.

TABLE 12.4

Oxygen Consumption and Metabolic Rates for Various Activities for a 65-kg Male[a]

Activity	O_2 Use Rate (mL/min · kg)	Metabolic Rate (kcal/h)	Metabolic Rate (W)
Sleeping	3.5	70	80
Light activity (dressing, walking slowly, desk work)	10	200	230
Moderate activity (walking briskly)	20	400	465
Heavy activity basketball, swimming a fast breaststroke)	30	600	700
Extreme activity (bicycle racing)	70	1400	1600

[a]*Source: A Companion to Medical Studies*, 2/e, R. Passmore, Philadelphia, F. A. Davis, 1968.

When we sit watching a riveting film, we give off about as much energy by heat as a bright (250-W) lightbulb.

Regardless of level of activity, the daily food intake should just balance the loss in internal energy if a person is not to gain weight. Further, exercise is a poor substitute for dieting as a method of losing weight. For example, the loss of one pound of body fat requires the muscles to expend 4 100 kcal of energy. If the goal is to lose a pound of fat in 35 days, a jogger could run an extra mile a day, because a 65-kg jogger uses about 120 kcal to jog one mile (35 days × 120 kcal/day = 4 200 kcal). An easier way to lose the pound of fat would be to diet and eat two fewer slices of bread per day for 35 days, because bread has a calorie content of 60 kcal/slice (35 days × 2 slices/day × 60 kcal/slice = 4 200 kcal).

EXAMPLE 12.17 Fighting Fat

Goal Estimate human energy usage during a typical day.

Problem In the course of 24 hours, a 65-kg person spends 8 h at a desk, 2 h puttering around the house, 1 h jogging 5 miles, 5 h in moderate activity, and 8 h sleeping. What is the change in her internal energy during this period?

Strategy The time rate of energy usage—or power—multiplied by time gives the amount of energy used during a given activity. Use Table 12.4 to find the power $\mathcal{P}_i$ needed for each activity, multiply each by the time, and sum them all up.

Solution

$$\Delta U = -\Sigma \mathcal{P}_i \Delta t_i = -(\mathcal{P}_1 \Delta t_1 + \mathcal{P}_2 \Delta t_2 + \ldots + \mathcal{P}_n \Delta t_n)$$
$$= -(200\ \text{kcal/h})(10\ \text{h}) - (5\ \text{mi/h})(120\ \text{kcal/mi})(1\ \text{h}) - (400\ \text{kcal/h})(5\ \text{h}) - (70\ \text{kcal/h})(8\ \text{h})$$

$$\Delta U = -5\ 000\ \text{kcal}$$

Remarks If this is a typical day in the woman's life, she will have to consume less than 5 000 kilocalories on a daily basis in order to lose weight. A complication lies in the fact that human metabolism tends to drop when food intake is reduced.

Exercise 12.17

If a 60.0-kg woman ingests 3 000 kcal a day and spends 6 h sleeping, 4 h walking briskly, 8 h sitting at a desk job, 1 h swimming a fast breaststroke, and 5 h watching action movies on TV, about how much weight will the woman gain or lose every day? (*Note:* Recall that using about 4 100 kcal of energy will burn off a pound of fat.)

Answer She'll lose a little over half a pound of fat a day.

Physical Fitness and Efficiency of the Human Body as a Machine

One measure of a person's physical fitness is his or her maximum capacity to use or consume oxygen. This "aerobic" fitness can be increased and maintained with regular exercise, but falls when training stops. Typical maximum rates of oxygen consumption and corresponding fitness levels are shown in Table 12.5; we see that the maximum oxygen consumption rate varies from 28 mL/min · kg of body mass for poorly conditioned subjects to 70 mL/min · kg for superb athletes.

We have already pointed out that the first law of thermodynamics can be rewritten to relate the metabolic rate $\Delta U/\Delta t$ to the rate at which energy leaves the body by work and by heat:

$$\frac{\Delta U}{\Delta t} = \frac{Q}{\Delta t} + \frac{W}{\Delta t}$$

Now consider the body as a machine capable of supplying mechanical power to the outside world, and ask for its efficiency. The body's efficiency e is defined as

TABLE 12.5

Physical Fitness and Maximum Oxygen Consumption Rate[a]

Fitness Level	Maximum Oxygen Consumption Rate (mL/min · kg)
Very poor	28
Poor	34
Fair	42
Good	52
Excellent	70

[a]*Source: Aerobics*, K. H. Cooper, Bantam Books, New York, 1968.

TABLE 12.6

Metabolic Rate, Power Output, and Efficiency for Different Activities[a]

Activity	Metabolic Rate $\frac{\Delta U}{\Delta t}$ (watts)	Power Output $\frac{W}{\Delta t}$ (watts)	Efficiency e
Cycling	505	96	0.19
Pushing loaded coal cars in a mine	525	90	0.17
Shoveling	570	17.5	0.03

[a]*Source:* "Inter- and Intra-Individual Differences in Energy Expenditure and Mechanical Efficiency," C. H. Wyndham et al., *Ergonomics* **9**, 17 (1966).

the ratio of the mechanical power supplied by a human to the metabolic rate or the total power input to the body:

$$e = \text{body's efficiency} = \frac{\left|\frac{W}{\Delta t}\right|}{\left|\frac{\Delta U}{\Delta t}\right|} \qquad [12.21]$$

In this definition, absolute-value signs are used to show that e is a positive number and to avoid explicitly using minus signs required by our definitions of W and Q in the first law. Table 12.6 shows the efficiency of workers engaged in different activities for several hours. These values were obtained by measuring the power output and simultaneous oxygen consumption of mine workers and calculating the metabolic rate from their oxygen consumption. The table shows that a person can steadily supply mechanical power for several hours at about 100 W with an efficiency of about 17%. It also shows the dependence of efficiency on activity, and that e can drop to values as low as 3% for highly inefficient activities like shoveling, which involves many starts and stops. Finally, it is interesting in comparison to the average results of Table 12.6 that a superbly-conditioned athlete efficiently coupled to a mechanical device for extracting power (a bike!) can supply a power of around 300 W for about 30 minutes at a peak efficiency of 22%.

SUMMARY

PhysicsNow™ Take a practice test by logging into PhysicsNow at **www.cp7e.com** and clicking on the Pre-Test link for this chapter.

12.1 Work in Thermodynamic Processes

The work done on a gas at a constant pressure is

$$W = -P\Delta V \qquad [12.1]$$

The work done on the gas is positive if the gas is compressed (ΔV is negative) and negative if the gas expands (ΔV is positive). In general, the work done on a gas that takes it from some initial state to some final state is the negative of the area under the curve on a PV diagram.

12.2 The First Law of Thermodynamics

According to the first law of thermodynamics, when a system undergoes a change from one state to another, the **change in its internal energy** ΔU is

$$\Delta U = U_f - U_i = Q + W \qquad [12.2]$$

where Q is the energy transferred into the system by heat and W is the work done on the system. Q is positive when energy enters the system by heat and negative when the system loses energy. W is positive when work is done on the system (for example, by compression) and negative when the system does positive work on its environment.

The change of the internal energy, ΔU, of an ideal gas is given by

$$\Delta U = nC_v\Delta T \qquad [12.5]$$

where C_v is the molar specific heat at constant volume.

An **isobaric process** is one that occurs at constant pressure. The work done on the system in such a process is $-P\,\Delta V$, while the thermal energy transferred by heat is given by

$$Q = nC_p\Delta T \qquad [12.6]$$

with the molar heat capacity at constant pressure given by $C_p = C_v + R$.

In an **adiabatic process** no energy is transferred by heat between the system and its surroundings ($Q = 0$). In this case, the first law gives $\Delta U = W$, which means the internal energy changes solely as a consequence of work being done on the system. The pressure and volume in adiabatic processes are related by

$$PV^{\gamma} = \text{constant} \qquad [12.8a]$$

where $\gamma = C_p/C_v$ is the adiabatic index.

In an **isovolumetric process** the volume doesn't change and no work is done. For such processes, the first law gives $\Delta U = Q$.

An **isothermal process** occurs at constant temperature. The work done by an ideal gas on the environment is

$$W_{\text{env}} = nRT \ln\left(\frac{V_f}{V_i}\right) \qquad [12.10]$$

12.3 Heat Engines and the Second Law of Thermodynamics

In a **cyclic process** (in which the system returns to its initial state), $\Delta U = 0$ and therefore $Q = W_{\text{eng}}$, meaning the energy transferred into the system by heat equals the work done on the system during the cycle.

A **heat engine** takes in energy by heat and partially converts it to other forms of energy, such as mechanical and electrical energy. The work W_{eng} done by a heat engine in carrying a working substance through a cyclic process ($\Delta U = 0$) is

$$W_{\text{eng}} = |Q_h| - |Q_c| \qquad [12.11]$$

where Q_h is the energy absorbed from a hot reservoir and Q_c is the energy expelled to a cold reservoir.

The **thermal efficiency** of a heat engine is defined as the ratio of the work done by the engine to the energy transferred into the engine per cycle:

$$e = \frac{W_{\text{eng}}}{|Q_h|} = 1 - \frac{|Q_c|}{|Q_h|} \qquad [12.12]$$

Heat pumps are heat engines in reverse. In a refrigerator, the heat pump removes thermal energy from inside the refrigerator. Heat pumps operating in cooling mode have coefficient of performance given by

$$\text{COP(cooling mode)} = \frac{|Q_c|}{W} \qquad [12.13]$$

A heat pump in heating mode has coefficient of performance

$$\text{COP(heating mode)} = \frac{|Q_h|}{W} \qquad [12.14]$$

Real processes proceed in an order governed by the **second law of thermodynamics**, which can be stated in two ways:

1. Energy will not flow spontaneously by heat from a cold object to a hot object.
2. No heat engine operating in a cycle can absorb energy from a reservoir and perform an equal amount of work.

No real heat engine operating between the Kelvin temperatures T_h and T_c can exceed the efficiency of an engine operating between the same two temperatures in a **Carnot cycle**, given by

$$e_C = 1 - \frac{T_c}{T_h} \qquad [12.16]$$

Perfect efficiency of a Carnot engine requires a cold reservoir of 0 K, absolute zero. According to the **third law of thermodynamics**, however, it is impossible to lower the temperature of a system to absolute zero in a finite number of steps.

12.4 Entropy

The second law can also be stated in terms of a quantity called **entropy** (S). The **change in entropy** of a system is equal to the energy flowing by heat into (or out of) the system as the system changes from one state (A) to another (B) by a reversible process, divided by the absolute temperature:

$$\Delta S \equiv \frac{Q_r}{T} \qquad [12.17]$$

One of the primary findings of statistical mechanics is that systems tend toward disorder and that entropy is a measure of that disorder. An alternate statement of the second law is that the entropy of the Universe increases in all natural processes.

CONCEPTUAL QUESTIONS

1. What are some factors that affect the efficiency of automobile engines?
2. "Energy is the mistress of the Universe and entropy is her shadow." Writing for an audience of general readers, argue for the truth of this statement with examples. Alternately, argue for the view that entropy is like a decisive hands-on executive instantly determining what will happen, while energy is like a wretched back-office bookkeeper telling us how little we can afford.
3. For an ideal gas in an isothermal process, there is no change in internal energy. Suppose the gas does work W during such a process. How much energy was transferred by heat?
4. If you shake a jar full of jelly beans of different sizes, the larger beans tend to appear near the top and the smaller ones tend to fall to the bottom. Why does this occur? Does this process violate the second law of thermodynamics?
5. Consider the human body performing a strenuous exercise, such as lifting weights or riding a bicycle. Work is being done by the body, and energy is leaving by conduction from the skin into the surrounding air. According to the first law of thermodynamics, the temperature of the body should be steadily decreasing during the exercise. This isn't what happens, however. Is the first law invalid for this situation? Explain.
6. Clearly distinguish among temperature, heat, and internal energy.
7. What is wrong with the following statement? "Given any two objects, the one with the higher temperature contains more heat."

8. A steam-driven turbine is one major component of an electric power plant. Why is it advantageous to increase the temperature of the steam as much as possible?
9. When a sealed Thermos bottle full of hot coffee is shaken, what changes, if any, take place in (a) the temperature of the coffee and (b) its internal energy?
10. In solar ponds constructed in Israel, the Sun's energy is concentrated near the bottom of a salty pond. With the proper layering of salt in the water, convection is prevented, and temperatures of 100°C may be reached. Can you guess the maximum efficiency with which useful mechanical work can be extracted from the pond?
11. Is it possible to construct a heat engine that creates no thermal pollution?
12. Suppose your roommate is "Mr. Clean" and tidies up your messy room after a big party. Because more order is being created by your roommate, does this tidying up represent a violation of the second law of thermodynamics?
13. A thermodynamic process occurs in which the entropy of a system changes by $-8.0\ \text{J/K}$. According to the second law of thermodynamics, what can you conclude about the entropy change of the environment?
14. If a supersaturated sugar solution is allowed to evaporate slowly, sugar crystals form in the container. Hence, sugar molecules go from a disordered form (in solution) to a highly ordered, crystalline form. Does this process violate the second law of thermodynamics? Explain.
15. The first law of thermodynamics says we can't get more out of a process than we put in, but the second law says that we can't break even. Explain this statement.
16. Give some examples of irreversible processes that occur in nature. Give an example of a process in nature that is nearly reversible.
17. Imagine a gas in an insulated cylinder with a movable piston. The piston has been pushed inward, compressing the gas, and is now released. As the molecules of the gas strike the piston, they move it outward. From the point of view of energy principles, explain how this expansion causes the temperature of the gas to drop.

PROBLEMS

1, 2, 3 = straightforward, intermediate, challenging ☐ = full solution available in *Student Solutions Manual/Study Guide*
PhysicsNow™ = coached problem with hints available at **www.cp7e.com** = biomedical application

Section 12.1 Work in Thermodynamic Processes

1. PhysicsNow™ The only form of energy possessed by molecules of a monatomic ideal gas is translational kinetic energy. Using the results from the discussion of kinetic theory in Section 10.5, show that the internal energy of a monatomic ideal gas at pressure P and occupying volume V may be written as $U = \frac{3}{2}PV$.
2. Sketch a PV diagram and find the work done by the gas during the following stages: (a) A gas is expanded from a volume of 1.0 L to 3.0 L at a constant pressure of 3.0 atm. (b) The gas is then cooled at constant volume until the pressure falls to 2.0 atm. (c) The gas is then compressed at a constant pressure of 2.0 atm from a volume of 3.0 L to 1.0 L. (*Note:* Be careful of signs.) (d) The gas is heated until its pressure increases from 2.0 atm to 3.0 atm at a constant volume. (e) Find the net work done during the complete cycle.
3. A container of volume $0.40\ \text{m}^3$ contains 3.0 mol of argon gas at 30°C. Assuming argon behaves as an ideal gas, find the total internal energy of the gas. (*Hint:* See Problem 1.)
4. A 40.0-g projectile is launched by the expansion of hot gas in an arrangement shown in Figure P12.4a. The cross-sectional area of the launch tube is $1.0\ \text{cm}^2$, and the length that the projectile travels down the tube after starting from rest is 32 cm. As the gas expands, the pressure varies as shown in Figure P12.4b. The values for the initial pressure and volume are $P_i = 11 \times 10^5$ Pa and $V_i = 8.0\ \text{cm}^3$ while the final values are $P_f = 1.0 \times 10^5$ Pa and $V_f = 40.0\ \text{cm}^3$. Friction between the projectile and the launch tube is negligible. (a) If the projectile is launched into a vacuum, what is the speed of the projectile as it leaves the launch tube? (b) If instead the projectile is launched into air at a pressure of 1.0×10^5 Pa, what fraction of the work done by the expanding gas in the tube is spent by the projectile pushing air out of the way as it proceeds down the tube?

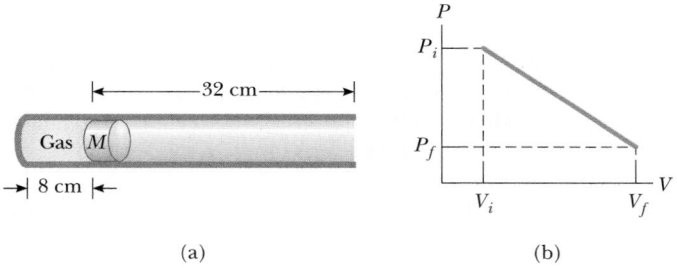

Figure P12.4

5. A gas expands from I to F along the three paths indicated in Figure P12.5. Calculate the work done *on* the gas along paths (a) IAF, (b) IF, and (c) IBF.

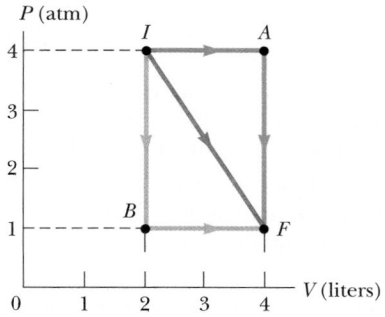

Figure P12.5 (Problems 5 and 15)

6. Sketch a PV diagram of the following processes: (a) A gas expands at constant pressure P_1 from volume V_1 to volume V_2. It is then kept at constant volume while the pressure is reduced to P_2. (b) A gas is reduced in pressure from P_1 to P_2 while its volume is held constant at V_1. It is then expanded at constant pressure P_2 to a final volume V_2. (c) In which of the processes is more work done *by* the gas? Why?

7. Gas in a container is at a pressure of 1.5 atm and a volume of 4.0 m^3. What is the work done *on* the gas (a) if it expands at constant pressure to twice its initial volume? (b) if it is compressed at constant pressure to one-quarter its initial volume?

8. A movable piston having a mass of 8.00 kg and a cross-sectional area of 5.00 cm^2 traps 0.200 moles of an ideal gas in a vertical cylinder. If the piston slides without friction in the cylinder, how much work is done *on* the gas when its temperature is increased from 20°C to 300°C?

9. One mole of an ideal gas initially at a temperature of $T_i = 0°C$ undergoes an expansion at a constant pressure of 1.00 atm to four times its original volume. (a) Calculate the new temperature T_f of the gas. (b) Calculate the work done *on* the gas during the expansion.

10. (a) Determine the work done *on* a fluid that expands from i to f as indicated in Figure P12.10. (b) How much work is done *on* the fluid if it is compressed from f to i along the same path?

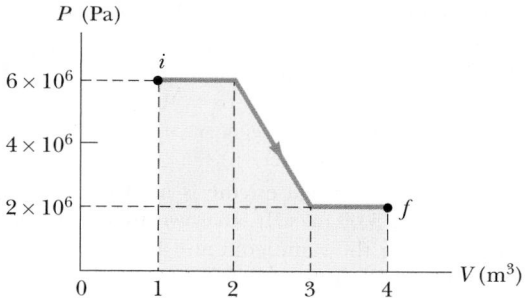

Figure P12.10

Section 12.2 The First Law of Thermodynamics

11. A container is placed in a water bath and held at constant volume as a mixture of fuel and oxygen is burned inside it. The temperature of the water is observed to rise during the burning. (The water is also held at constant volume. (a) Consider the burning mixture to be the system. What are the signs of Q, ΔU, and W? (b) What are the signs of these quantities if the water bath is considered to be the system?

12. A quantity of a monatomic ideal gas undergoes a process in which both its pressure and volume are doubled as shown in Figure P12.12. What is the energy absorbed by heat into the gas during this process? (*Hint:* See Problem 1.)

13. A gas is compressed at a constant pressure of 0.800 atm from 9.00 L to 2.00 L. In the process, 400 J of energy leaves the gas by heat. (a) What is the work done *on* the gas? (b) What is the change in its internal energy?

14. A monatomic ideal gas undergoes the thermodynamic process shown in the PV diagram of Figure P12.14. Determine whether each of the values ΔU, Q, and W for the gas is positive, negative, or zero. (*Hint:* See Problem 1.)

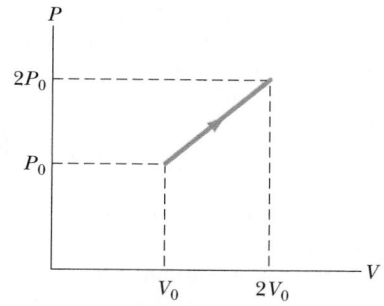

Figure P12.12

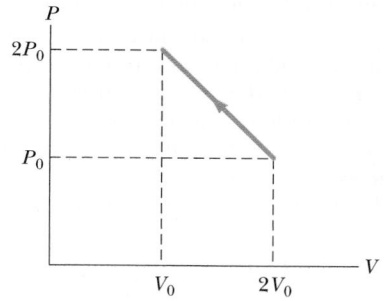

Figure P12.14

15. Physics Now™ A gas expands from I to F in Figure P12.5. The energy added to the gas by heat is 418 J when the gas goes from I to F along the diagonal path. (a) What is the change in internal energy of the gas? (b) How much energy must be added to the gas by heat for the indirect path IAF to give the same change in internal energy?

16. A gas is taken through the cyclic process described by Figure P12.16. (a) Find the net energy transferred to the system by heat during one complete cycle. (b) If the cycle is reversed—that is, the process follows the path $ACBA$—what is the net energy transferred by heat per cycle?

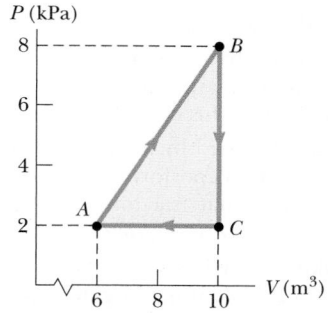

Figure P12.16 (Problems 16 and 18)

17. A gas is enclosed in a container fitted with a piston of cross-sectional area 0.150 m^2. The pressure of the gas is maintained at 6 000 Pa as the piston moves inward 20.0 cm. (a) Calculate the work done by the gas. (b) If the internal energy of the gas decreases by 8.00 J, find the amount of heat removed from the system by heat during the compression.

18. Consider the cyclic process described by Figure P12.16. If Q is negative for the process BC and ΔU is negative for the process CA, determine the signs of Q, W, and ΔU associated with each process.

19. One gram of water changes to ice at a constant pressure of 1.00 atm and a constant temperature of 0°C. In the process, the volume changes from 1.00 cm^3 to 1.09 cm^3. (a) Find the work done *on* the water and (b) the change in the internal energy of the water.

20. A thermodynamic system undergoes a process in which its internal energy decreases by 500 J. If at the same time 220 J of work is done on the system, find the energy transferred to or from it by heat.

21. A 5.0-kg block of aluminum is heated from 20°C to 90°C at atmospheric pressure. Find (a) the work done by the aluminum, (b) the amount of energy transferred to it by heat, and (c) the increase in its internal energy.

22. One mole of gas initially at a pressure of 2.00 atm and a volume of 0.300 L has an internal energy equal to 91.0 J. In its final state, the gas is at a pressure of 1.50 atm and a volume of 0.800 L, and its internal energy equals 180 J. For the paths *IAF*, *IBF*, and *IF* in Figure P12.22, calculate (a) the work done *on* the gas and (b) the net energy transferred to the gas by heat in the process.

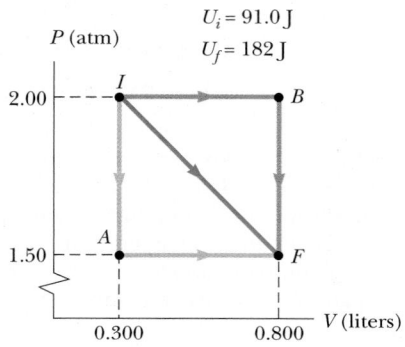

Figure P12.22

Section 12.3 Heat Engines and the Second Law of Thermodynamics

23. A heat engine operates between two reservoirs at temperatures of 20°C and 300°C. What is the maximum efficiency possible for this engine?

24. A steam engine has a boiler that operates at 300°F, and the temperature of the exhaust is 150°F. Find the maximum efficiency of this engine.

25. The energy absorbed by an engine is three times greater than the work it performs. (a) What is its thermal efficiency? (b) What fraction of the energy absorbed is expelled to the cold reservoir?

26. A particular engine has a power output of 5.00 kW and an efficiency of 25.0%. If the engine expels 8 000 J of energy in each cycle, find (a) the energy absorbed in each cycle and (b) the time required to complete each cycle.

27. One of the most efficient engines ever built is a coal-fired steam turbine engine in the Ohio River valley, driving an electric generator as it operates between 1 870°C and 430°C. (a) What is its maximum theoretical efficiency? (b) Its actual efficiency is 42.0%. How much mechanical power does the engine deliver if it absorbs 1.40×10^5 J of energy each second from the hot reservoir?

28. A gun is a heat engine. In particular, it is an internal combustion piston engine that does not operate in a cycle, but comes apart during its adiabatic expansion process. A certain gun consists of 1.80 kg of iron. It fires one 2.40-g bullet at 320 m/s with an energy efficiency of 1.10%. Assume that the body of the gun absorbs all of the energy exhaust and increases uniformly in temperature for a short time before it loses any energy by heat into the environment. Find its temperature increase.

29. An engine absorbs 1 700 J from a hot reservoir and expels 1 200 J to a cold reservoir in each cycle. (a) What is the engine's efficiency? (b) How much work is done in each cycle? (c) What is the power output of the engine if each cycle lasts for 0.300 s?

30. A power plant has been proposed that would make use of the temperature gradient in the ocean. The system is to operate between 20.0°C (surface water temperature) and 5.00°C (water temperature at a depth of about 1 km). (a) What is the maximum efficiency of such a system? (b) If the useful power output of the plant is 75.0 MW, how much energy is absorbed per hour? (c) In view of your answer to (a), do you think such a system is worthwhile (considering that there is no charge for fuel)?

31. In one cycle, a heat engine absorbs 500 J from a high-temperature reservoir and expels 300 J to a low-temperature reservoir. If the efficiency of this engine is 60% of the efficiency of a Carnot engine, what is the ratio of the low temperature to the high temperature in the Carnot engine?

32. A heat engine operates in a Carnot cycle between 80.0°C and 350°C. It absorbs 21 000 J of energy per cycle from the hot reservoir. The duration of each cycle is 1.00 s. (a) What is the mechanical power output of this engine? (b) How much energy does it expel in each cycle by heat?

33. A nuclear power plant has an electrical power output of 1 000 MW and operates with an efficiency of 33%. If excess energy is carried away from the plant by a river with a flow rate of 1.0×10^6 kg/s, what is the rise in temperature of the flowing water?

34. A 20.0%-efficient real engine is used to speed up a train from rest to 5.00 m/s. It is known that an ideal (Carnot) engine using the same cold and hot reservoirs would accelerate the same train from rest to a speed of 6.50 m/s using the same amount of fuel. If the engines use air at 300 K as a cold reservoir, find the temperature of the steam serving as the hot reservoir.

Section 12.4 Entropy

35. PhysicsNow™ A freezer is used to freeze 1.0 L of water completely into ice. The water and the freezer remain at a constant temperature of $T = 0$°C. Determine (a) the change in the entropy of the water and (b) the change in the entropy of the freezer.

36. What is the change in entropy of 1.00 kg of liquid water at 100°C as it changes to steam at 100°C?

37. A 70-kg log falls from a height of 25 m into a lake. If the log, the lake, and the air are all at 300 K, find the change in entropy of the Universe during this process.

38. Two 2 000-kg cars, both traveling at 20 m/s, undergo a head-on collision and stick together. Find the change in entropy of the Universe resulting from the collision if the temperature is 23°C.

39. The surface of the Sun is approximately at 5 700 K, and the temperature of the Earth's surface is approximately 290 K. What entropy change occurs when 1 000 J of energy is transferred by heat from the Sun to the Earth?

40. Repeat the procedure used to construct Table 12.3 (a) for the case in which you draw three marbles rather than four from your bag and (b) for the case in which you draw five rather than four.

41. Prepare a table like Table 12.3 for the following occurrence: You toss four coins into the air simultaneously and record all the possible results of the toss in terms of the numbers of heads and tails that can result. (For example, HHTH and HTHH are two possible ways in which three heads and one tail can be achieved.) (a) On the basis of your table, what is the most probable result of a toss? In terms of entropy, (b) what is the most ordered state, and (c) what is the most disordered?

42. Consider a standard deck of 52 playing cards that has been thoroughly shuffled. (a) What is the probability of drawing the ace of spades in one draw? (b) What is the probability of drawing any ace? (c) What is the probability of drawing any spade?

ADDITIONAL PROBLEMS

43. A student claims that she has constructed a heat engine that operates between the temperatures of 200 K and 100 K with 60% efficiency. The professor does not give her credit for the project. Why not?

44. A Carnot engine operates between the temperatures $T_h = 100°C$ and $T_c = 20°C$. By what factor does the theoretical efficiency increase if the temperature of the hot reservoir is increased to 550°C?

45. A Carnot heat engine extracts energy Q_h from a hot reservoir at constant temperature T_h and rejects energy Q_c to a cold reservoir at constant temperature T_c. Find the entropy changes of (a) the hot reservoir, (b) the cold reservoir, (c) the engine, and (d) the complete system.

46. One end of a copper rod is in thermal contact with a hot reservoir at $T = 500$ K, and the other end is in thermal contact with a cooler reservoir at $T = 300$ K. If 8 000 J of energy is transferred from one end of the rod to the other, with no change in the temperature distribution, find the entropy change of each reservoir and the total entropy change of the Universe.

47. Find the change in temperature of a river due to the exhausted energy from a nuclear power plant. Assume that the input power to the boiler in the plant is 25×10^8 W, the efficiency of use of this power is 30%, and the river flow rate is 9.0×10^6 kg/min.

48. A Carnot engine operates between 100°C and 20°C. How much ice can the engine melt from its exhaust after it has done 5.0×10^4 J of work?

49. A 1500-kW heat engine operates at 25% efficiency. The heat energy expelled at the low temperature is absorbed by a stream of water that enters the cooling coils at 20°C. If 60 L flows across the coils per second, determine the increase in temperature of the water.

50. When a gas follows path 123 on the *PV* diagram in Figure P12.50, 418 J of energy flows into the system by heat and −167 J of work is done *on the gas*. (a) What is the change in the internal energy of the system? (b) How much energy Q flows into the system if the gas follows path 143? The work done on the gas along this path is −63.0 J. What net work would be done on or by the system if the system followed (c) path 12341? (d) path 14321? (e) What is the change in internal energy of the system in the processes described in parts (c) and (d)?

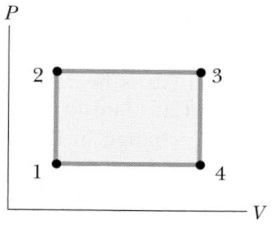

Figure P12.50

51. A substance undergoes the cyclic process shown in Figure P12.51. Work output occurs along path *AB* while work input is required along path *BC*, and no work is involved in the constant volume process *CA*. Energy transfers by heat occur during each process involved in the cycle. (a) What is the work output during process *AB*? (b) How much work input is required during process *BC*? (c) What is the net energy input Q during this cycle?

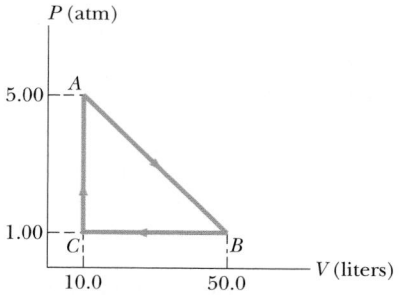

Figure P12.51

52. A power plant having a Carnot efficiency produces 1 000 MW of electrical power from turbines that take in steam at 500 K and eject water at 300 K into a flowing river. The water downstream is 6.00 K warmer due to the output of the plant. Determine the flow rate of the river.

53. A 100-kg steel support rod in a building has a length of 2.0 m at a temperature of 20°C. The rod supports a hanging load of 6 000 kg. Find (a) the work done *on* the rod as the temperature increases to 40°C, (b) the energy Q added to the rod (assume the specific heat of steel is the same as that for iron), and (c) the change in internal energy of the rod.

54. An ideal gas initially at pressure P_0, volume V_0, and temperature T_0 is taken through the cycle described in Figure P12.54. (a) Find the net work done *by* the gas per cycle in

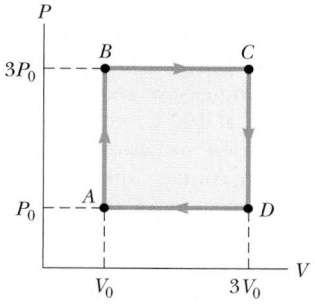

Figure P12.54

terms of P_0 and V_0. (b) What is the net energy Q added to the system per cycle? (c) Obtain a numerical value for the net work done per cycle for 1.00 mol of gas initially at 0°C. (*Hint:* Recall that the work done by the system equals the area under a PV curve.)

55. One mole of neon gas is heated from 300 K to 420 K at constant pressure. Calculate (a) the energy Q transferred to the gas, (b) the change in the internal energy of the gas, and (c) the work done *on* the gas. Note that neon has a molar specific heat of $c = 20.79\ \text{J/mol}\cdot\text{K}$ for a constant-pressure process.

56. A 1.0-kg block of aluminum is heated at atmospheric pressure so that its temperature increases from 22°C to 40°C. Find (a) the work done *on* the aluminum, (b) the energy Q added to the aluminum, and (c) the change in internal energy of the aluminum.

57. **Physics Now™** Suppose a heat engine is connected to two energy reservoirs, one a pool of molten aluminum at 660°C and the other a block of solid mercury at −38.9°C. The engine runs by freezing 1.00 g of aluminum and melting 15.0 g of mercury during each cycle. The latent heat of fusion of aluminum is 3.97×10^5 J/kg, and that of mercury is 1.18×10^4 J/kg. (a) What is the efficiency of this engine? (b) How does the efficiency compare with that of a Carnot engine?

58. One mole of an ideal gas is taken through the cycle shown in Figure P12.58. At point A, the pressure, volume, and temperature are P_0, V_0, and T_0. In terms of R and T_0, find (a) the total energy entering the system by heat per cycle, (b) the total energy leaving the system by heat per cycle, (c) the efficiency of an engine operating in this cycle, and (d) the efficiency of an engine operating in a Carnot cycle between the temperature extremes for this process. (*Hint:* Recall that work done on the gas is the negative of the area under a PV curve.)

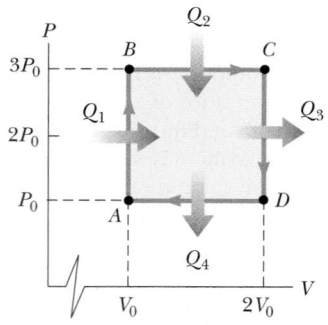

Figure P12.58

59. An electrical power plant has an overall efficiency of 15%. The plant is to deliver 150 MW of electrical power to a city, and its turbines use coal as fuel. The burning coal produces steam at 190°C, which drives the turbines. The steam is condensed into water at 25°C by passing through coils that are in contact with river water. (a) How many metric tons of coal does the plant consume each day (1 metric ton = 1×10^3 kg)? (b) What is the total cost of the fuel per year if the delivery price is \$8 per metric ton? (c) If the river water is delivered at 20°C, at what minimum rate must it flow over the cooling coils in order that its temperature not exceed 25°C? (*Note:* The heat of combustion of coal is 7.8×10^3 cal/g.)

60. At atmospheric pressure (1.013×10^5 Pa) and 20.0°C, 1.00 g of water occupies a volume of 1.00 cm^3. (a) Find the change in internal energy when the water is heated to the boiling point. (b) When the water is boiled, it becomes 1 671 cm^3 of steam. Calculate the change in internal energy for this process. Assume the steam vapor doesn't mix with the surrounding air and that it expands at atmospheric pressure.

61. A gas is enclosed in a container fitted with a piston of cross-sectional area 0.10 m^2. The pressure of the gas is maintained at 8 000 Pa while energy is slowly added by heat; as a result, the piston is pushed up a distance of 4.0 cm. (Recall that any process in which the pressure remains constant is an isobaric process.) (a) If 42 J of energy is added to the system by heat during the expansion, what is the change in internal energy of the system? (b) If 42 J of energy is added by heat to the system with the piston clamped in a *fixed* position, what is the work done by the gas? What is the change in its internal energy?

62. Hydrothermal vents deep on the ocean floor spout water at temperatures as high as 570°C. This temperature is below the boiling point of water because of the immense pressure at that depth. Since the surrounding ocean temperature is at 4.0°C, an organism could use the temperature gradient as a source of energy. (a) Assuming the specific heat of water under these conditions is 1.0 cal/g · °C, how much energy is released when 1.0 liter of water is cooled from 570°C to 4.0°C? (b) What is the maximum useable energy an organism can extract from this energy source? (Assume that the organism has some internal type of heat engine acting between the two temperature extremes.) (c) Water from these vents contains hydrogen sulfide (H_2S) at a concentration of 0.90 mmole/liter. Oxidation of one mole of H_2S produces 310 kJ of energy. How much energy is available through H_2S oxidation of 1.0 L of water?

63. Suppose you spend 30.0 minutes on a stair-climbing machine, climbing at a rate of 90.0 steps per minute, with each step 8.00 inches high. If you weigh 150 lb and the machine reports that 600 kcal have been burned at the end of the workout, what efficiency is the machine using in obtaining this result? If your actual efficiency is 0.18, how many kcal did you really burn?

ACTIVITIES

1. Bend a paper clip back and forth several times and touch it to your lip. Why does it warm up? Repeat the process with a rubber band after it is stretched several times. Notice that it also warms up after stretching. Now take the stretched rubber band from your lip and wait for it to come to equilibrium with the surrounding air. Touch the stretched band to your lip, let it relax to its unstretched length, and note that it becomes cooler. Explain how the second law of thermodynamics helps you understand these results.

2. You should be able to locate a helium tank, a nitrogen tank, or a CO_2 tank in the chemistry or physics department of your university. Ask to use one of these for a short period for the following observation: Learn to open and close the valve or valves correctly. Touch the tanks and

nozzle to verify that they are at room temperature. Open a tank for 2–3 seconds and then close it. Touch the nozzle again, and you will find that it is very cold. Why? (*Hint:* The expansion of the gas is nearly adiabatic.)

3. You can study order and disorder and verify the answer to Quick Quiz 12.5 by rolling a pair of dice 100 times and recording the number of spots appearing on the dice for each throw. Which total comes up most frequently?

4. Another way to study order and disorder is to toss pennies. Predict in advance which is the more likely occurrence when tossing three pennies: having the pennies ordered with all the same orientation (all heads or all tails) or having them disordered (not all the same). Show that the ordered result includes two out of eight possibilities, while the disordered result includes six out of eight. Toss the three pennies 100 times to see how close you come to this prediction. Repeat the process for four pennies (The all-the-same result should be expected two times in sixteen throws.)

Ocean waves combine properties of both transverse and longitudinal waves. With proper balance and timing, a surfer can capture some of the wave's energy and take it for a ride.

© Rick Doyle/Corbis

CHAPTER

13

OUTLINE

13.1 Hooke's Law
13.2 Elastic Potential Energy
13.3 Comparing Simple Harmonic Motion with Uniform Circular Motion
13.4 Position, Velocity, and Acceleration as a Function of Time
13.5 Motion of a Pendulum
13.6 Damped Oscillations
13.7 Waves
13.8 Frequency, Amplitude, and Wavelength
13.9 The Speed of Waves on Strings
13.10 Interference of Waves
13.11 Reflection of Waves

Vibrations and Waves

Periodic motion, from masses on springs to vibrations of atoms, is one of the most important kinds of physical behavior. In this chapter we take a more detailed look at Hooke's law, where the force is proportional to the displacement, tending to restore objects to some equilibrium position. A large number of physical systems can be successfully modeled with this simple idea, including the vibrations of strings, the swinging of a pendulum, and the propagation of waves of all kinds. All these physical phenomena involve periodic motion.

Periodic vibrations can cause disturbances that move through a medium in the form of waves. Many kinds of waves occur in nature, such as sound waves, water waves, seismic waves, and electromagnetic waves. These very different physical phenomena are described by common terms and concepts introduced here.

13.1 HOOKE'S LAW

One of the simplest types of vibrational motion is that of an object attached to a spring, previously discussed in the context of energy in Chapter 5. We assume that the object moves on a frictionless horizontal surface. If the spring is stretched or compressed a small distance x from its unstretched or equilibrium position and then released, it exerts a force on the object as shown in Active Figure 13.1. From experiment this spring force is found to obey the equation

Hooke's law ▶

$$F_s = -kx \qquad [13.1]$$

where x is the displacement of the object from its equilibrium position ($x = 0$) and k is a positive constant called the **spring constant**. This force law for springs was discovered by Robert Hooke in 1678 and is known as **Hooke's law**. The value of k is a measure of the stiffness of the spring. Stiff springs have large k values, and soft springs have small k values.

The negative sign in Equation 13.1 means that the force exerted by the spring is always directed *opposite* the displacement of the object. When the object is to the right of the equilibrium position, as in Active Figure 13.1a, x is positive and F_s is negative. This means that the force is in the negative direction, to the left. When the object is to the left of the equilibrium position, as in Active Figure 13.1c, x is negative and F_s is positive, indicating that the direction of the force is to the right. Of course, when $x = 0$, as in Active Figure 13.1b, the spring is unstretched and $F_s = 0$. Because the spring force always acts toward the equilibrium position, it is sometimes called a restoring force. **A restoring force always pushes or pulls the object toward the equilibrium position.**

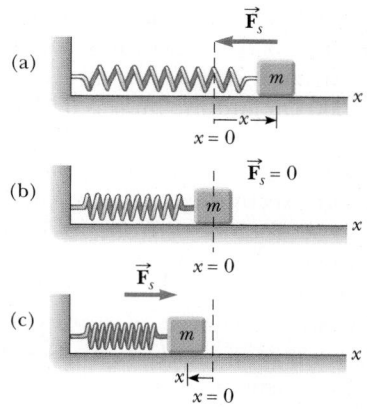

ACTIVE FIGURE 13.1
The force exerted by a spring on an object varies with the displacement of the object from the equilibrium position, $x = 0$. (a) When x is positive (the spring is stretched), the spring force is to the left. (b) When x is zero (the spring is unstretched), the spring force is zero. (c) When x is negative (the spring is compressed), the spring force is to the right.

Physics Now™
Log into PhysicsNow at **www.cp7e.com**, and go to Active Figure 13.1 to choose the spring constant and the initial position and velocity of the block and see the resulting simple harmonic motion.

Suppose the object is initially pulled a distance A to the right and released from rest. The force exerted by the spring on the object pulls it back toward the equilibrium position. As the object moves toward $x = 0$, the magnitude of the force decreases (because x decreases) and reaches zero at $x = 0$. However, the object gains speed as it moves toward the equilibrium position, reaching its maximum speed when $x = 0$. The momentum gained by the object causes it to overshoot the equilibrium position and compress the spring. As the object moves to the left of the equilibrium position (negative x-values), the spring force acts on it to the right, steadily increasing in strength, and the speed of the object decreases. The object finally comes briefly to rest at $x = -A$ before accelerating back towards $x = 0$ and ultimately returning to the original position at $x = A$. The process is then repeated, and the object continues to oscillate back and forth over the same path. This type of motion is called **simple harmonic motion. Simple harmonic motion occurs when the net force along the direction of motion obeys Hooke's law—when the net force is proportional to the displacement from the equilibrium point and is always directed toward the equilibrium point.**

Not all periodic motions over the same path can be classified as simple harmonic motion. A ball being tossed back and forth between a parent and a child moves repetitively, but the motion isn't simple harmonic motion, because the force acting on the ball doesn't take the form of Hooke's law, Equation 13.1.

The motion of an object suspended from a vertical spring is also simple harmonic. In this case, the force of gravity acting on the attached object stretches the spring until equilibrium is reached and the object is suspended at rest. By definition the equilibrium position of the object is $x = 0$. When the object is moved away from equilibrium by a distance x and released, a net force acts toward the equilibrium position. Because the net force is proportional to x, the motion is simple harmonic.

The following three concepts are important in discussing any kind of periodic motion:

- The **amplitude** A is the maximum distance of the object from its equilibrium position. In the absence of friction, an object in simple harmonic motion oscillates between the positions $x = -A$ and $x = +A$.
- The **period** T is the time it takes the object to move through one complete cycle of motion, from $x = A$ to $x = -A$ and back to $x = A$.
- The **frequency** f is the number of complete cycles or vibrations per unit of time, and is the reciprocal of the period ($f = 1/T$).

EXAMPLE 13.1 Measuring the Spring Constant

Goal Use Newton's second law together with Hooke's law to calculate a spring constant.

Problem A common technique used to evaluate a spring constant is illustrated in Figure 13.2. A spring is hung vertically (Fig. 13.2a), and an object of mass m is attached to the lower end of the spring and slowly lowered a distance d to the equilibrium point (Fig. 13.2b). Find the value of the spring constant if the spring is displaced by 2.00 cm and the mass is 0.550 kg.

Strategy This is an application of Newton's second law. The spring is stretched by a distance d from its initial position under the action of the load mg. The spring force is upward, balancing the downward force of gravity mg *when the system is in equilibrium.* (See Fig. 13.2c.) The suspended mass is in equilibrium, so set the sum of the forces equal to zero.

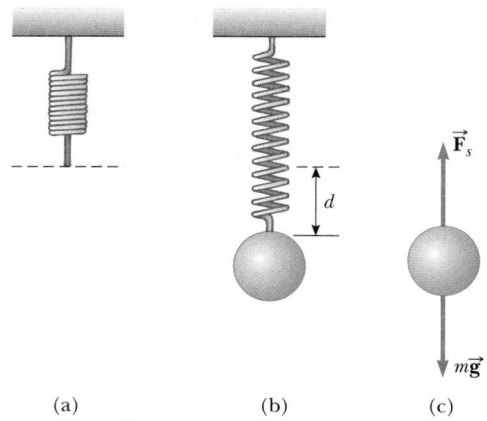

Figure 13.2 (Example 13.1) Determining the spring constant. The elongation d of the spring is due to the suspended weight mg. Because the upward spring force balances the weight when the system is in equilibrium, it follows that $k = mg/d$.

Solution

Apply the second law (with $a = 0$) and solve for the spring constant k:

$$\Sigma F = F_g + F_s = -mg + kd = 0$$

$$k = \frac{mg}{d} = \frac{(0.550\text{ kg})(9.80\text{ m/s}^2)}{2.00 \times 10^{-2}\text{ m}} = 2.70 \times 10^2\text{ N/m}$$

Remarks In this case the spring force is positive, because it's directed upward. Once the mass is pulled down from the equilibrium position and released, it oscillates around the equilibrium position, just like the horizontal spring.

Exercise 13.1

A spring with constant $k = 475$ N/m stretches 4.50 cm when an object of mass 25.0 kg is attached to the end of the spring. Find the acceleration of gravity in this location.

Answer 0.855 m/s^2 (The location is evidently an asteroid or small moon.)

The acceleration of an object moving with simple harmonic motion can be found by using Hooke's law in the equation for Newton's second law, $F = ma$. This gives

$$ma = F = -kx$$

Acceleration in simple harmonic motion ▶

$$a = -\frac{k}{m}x \qquad [13.2]$$

Equation 13.2, an example of a *harmonic oscillator equation,* gives the acceleration as a function of position. Because the maximum value of x is defined to be the amplitude A, the acceleration ranges over the values $-kA/m$ to $+kA/m$. In the next section we will find equations for velocity as a function of position and for position as a function of time.

TIP 13.1 Constant-Acceleration Equations Don't Apply

The acceleration a of a particle in simple harmonic motion is *not* constant; it changes, varying with x, so we can't apply the constant acceleration kinematic equations of Chapter 2.

Quick Quiz 13.1

A block on the end of a spring is pulled to position $x = A$ and released. Through what total distance does it travel in one full cycle of its motion? (a) $A/2$ (b) A (c) $2A$ (d) $4A$

Quick Quiz 13.2

For a simple harmonic oscillator, which of the following pairs of vector quantities can't both point in the same direction? (The position vector is the displacement from equilibrium.) (a) position and velocity (b) velocity and acceleration (c) position and acceleration

EXAMPLE 13.2 Simple Harmonic Motion on a Frictionless Surface

Goal Calculate forces and accelerations for a horizontal spring system.

Problem A 0.350-kg object attached to a spring of force constant 1.30×10^2 N/m is free to move on a frictionless horizontal surface, as in Active Figure 13.1. If the object is released from rest at $x = 0.100$ m, find the force on it and its acceleration at $x = 0.100$ m, $x = 0.050\,0$ m, $x = 0$ m, $x = -0.050\,0$ m, and $x = -0.100$ m.

Strategy Substitute given quantities into Hooke's law to find the forces, then calculate the accelerations with Newton's second law. The amplitude A is the same as the point of release from rest, $x = 0.100$ m.

Solution

Write Hooke's force law:

$$F_s = -kx$$

Substitute the value for k, and take $x = A = 0.100$ m, finding the force at that point:

$$F_{max} = -kA = -(1.30 \times 10^2 \text{ N/m})(0.100 \text{ m}) = -13.0 \text{ N}$$

Solve Newton's second law for a and substitute to find the acceleration at $x = A$:

$$ma = F_{max}$$

$$a = \frac{F_{max}}{m} = \frac{-13.0 \text{ N}}{0.350 \text{ kg}} = -37.1 \text{ m/s}^2$$

Repeat the same process for the other four points, assembling a table:

Position (m)	Force (N)	Acceleration (m/s^2)
0.100	−13.0	−37.1
0.050	−6.50	−18.6
0	0	0
−0.050	+6.50	+18.6
−0.100	+13.0	+37.1

Remarks Notice that when the initial position is halved, the force and acceleration are also halved. Further, positive values of x give negative values of the force and acceleration, while negative values of x give positive values of the force and acceleration. As the object moves to the left and passes the equilibrium point, the spring force becomes positive (for negative values of x), slowing the object down.

Exercise 13.2

For the same spring and mass system, find the force exerted by the spring and the position x when the object's acceleration is $+9.00$ m/s^2.

Answers 3.15 N, −2.42 cm

13.2 ELASTIC POTENTIAL ENERGY

In this section we review the material covered in Section 4 of Chapter 5.

A system of interacting objects has potential energy associated with the configuration of the system. A compressed spring has potential energy that, when allowed to expand, can do work on an object, transforming spring potential energy into the object's kinetic energy. As an example, Figure 13.3 (page 428) shows a ball being projected from a spring-loaded toy gun, where the spring is compressed a distance x. As the gun is fired, the compressed spring does work on the ball and imparts kinetic energy to it.

Recall that the energy stored in a stretched or compressed spring or some other elastic material is called elastic potential energy, PE_s, given by

$$PE_s \equiv \tfrac{1}{2}kx^2 \qquad [13.3]$$

◀ Elastic potential energy

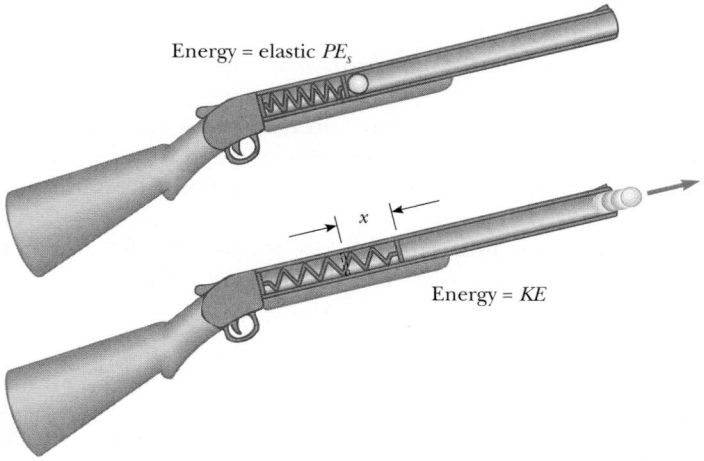

Figure 13.3 A ball projected from a spring-loaded gun. The elastic potential energy stored in the spring is transformed into the kinetic energy of the ball.

Recall also that the law of conservation of energy, including both gravitational and spring potential energy, is given by

$$(KE + PE_g + PE_s)_i = (KE + PE_g + PE_s)_f \qquad \textbf{[13.4]}$$

If nonconservative forces such as friction are present, then the change in mechanical energy must equal the work done by the nonconservative forces:

$$W_{nc} = (KE + PE_g + PE_s)_f - (KE + PE_g + PE_s)_i \qquad \textbf{[13.5]}$$

Rotational kinetic energy must be included in both Equation 13.4 and Equation 13.5 for systems involving torques.

As an example of the energy conversions that take place when a spring is included in a system, consider Figure 13.4. A block of mass m slides on a frictionless horizontal surface with constant velocity $\vec{\mathbf{v}}_i$ and collides with a coiled spring. The description that follows is greatly simplified by assuming that the spring is very light and therefore has negligible kinetic energy. As the spring is compressed, it exerts a force to the left on the block. At maximum compression, the block comes to rest for just an instant (Fig. 13.4c). The initial total energy in the system (block plus spring) before the collision is the kinetic energy of the block. After the block

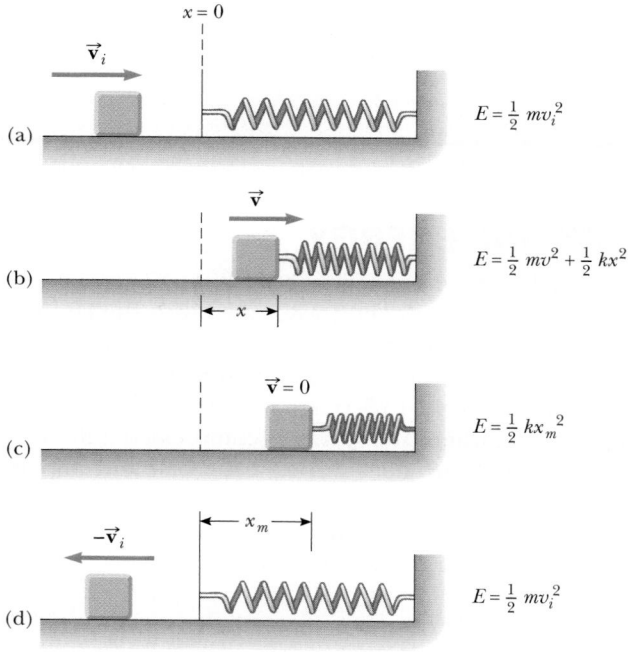

Figure 13.4 A block sliding on a frictionless horizontal surface collides with a light spring. (a) Initially, the mechanical energy is entirely the kinetic energy of the block. (b) The mechanical energy at some arbitrary position is the sum of the kinetic energy of the block and the elastic potential energy stored in the spring. (c) When the block comes to rest, the mechanical energy is entirely elastic potential energy stored in the compressed spring. (d) When the block leaves the spring, the mechanical energy is equal to the block's kinetic energy. The total energy remains constant.

collides with the spring and the spring is partially compressed, as in Figure 13.4b, the block has kinetic energy $\frac{1}{2}mv^2$ (where $v < v_i$) and the spring has potential energy $\frac{1}{2}kx^2$. When the block stops for an instant at the point of maximum compression, the kinetic energy is zero. Because the spring force is conservative and because there are no external forces that can do work on the system, **the total mechanical energy of the system consisting of the block and spring remains constant.** Energy is transformed from the kinetic energy of the block to the potential energy stored in the spring. As the spring expands, the block moves in the opposite direction and regains all of its initial kinetic energy, as in Figure 13.4d.

When an archer pulls back on a bowstring, elastic potential energy is stored in both the bent bow and stretched bowstring (Fig. 13.5). When the arrow is released, the potential energy stored in the system is transformed into the kinetic energy of the arrow. Devices such as crossbows and slingshots work the same way.

Eric Lars Baleke/Black Star

Figure 13.5 Elastic potential energy is stored in this drawn bow.

APPLICATION

Archery

Quick Quiz 13.3

When an object moving in simple harmonic motion is at its maximum displacement from equilibrium, which of the following is at a maximum? (a) velocity, (b) acceleration, or (c) kinetic energy.

EXAMPLE 13.3 Stop That Car!

Goal Apply conservation of energy and the work–energy theorem with spring and gravitational potential energy.

Problem A 13 000-N car starts at rest and rolls down a hill from a height of 10.0 m (Fig. 13.6). It then moves across a level surface and collides with a light spring-loaded guardrail. **(a)** Neglecting any losses due to friction, find the maximum distance the spring is compressed. Assume a spring constant of 1.0×10^6 N/m. **(b)** Calculate the maximum acceleration of the car after contact with the spring, assuming no frictional losses. **(c)** If the spring is compressed by only 0.30 m, find the energy lost through friction.

Figure 13.6 (Example 13.3) A car starts from rest on a hill at the position shown. When the car reaches the bottom of the hill, it collides with a spring-loaded guardrail.

Strategy Because friction losses are neglected, use conservation of energy in the form of Equation 13.4 to solve for the spring displacement in part (a). The initial and final values of the car's kinetic energy are zero, so the initial potential energy of the car–spring–Earth system is completely converted to elastic potential energy in the spring at the end of the ride. In part (b), apply Newton's second law, substituting the answer to part (a) for x because the maximum compression will give the maximum acceleration. In part (c) friction is no longer neglected, so use the work–energy theorem, Equation 13.5. The change in mechanical energy must equal the mechanical energy lost due to friction.

Solution

(a) Find the maximum spring compression, assuming no energy losses due to friction.

Apply conservation of mechanical energy. Initially, there is only gravitational potential energy, and at maximum compression of the guardrail, there is only spring potential energy.

$$(KE + PE_g + PE_s)_i = (KE + PE_g + PE_s)_f$$

$$0 + mgh + 0 = 0 + 0 + \tfrac{1}{2}kx^2$$

Solve for x:

$$x = \sqrt{\frac{2mgh}{k}} = \sqrt{\frac{2(13\,000\text{ N})(10.0\text{ m})}{1.0 \times 10^6\text{ N/m}}} = \boxed{0.51\text{ m}}$$

(b) Calculate the maximum acceleration of the car by the spring, neglecting friction.

Apply Newton's second law:

$$ma = -kx \rightarrow a = -\frac{kx}{m} = -\frac{kxg}{mg} = -\frac{kxg}{w}$$

Substitute values:

$$a = -\frac{(1.0 \times 10^6 \text{ N/m})(0.51 \text{ m})(9.8 \text{ m/s}^2)}{13\,000 \text{ N}}$$

$$= -380 \text{ m/s}^2$$

(c) If the compression of the guardrail is only 0.30 m, find the mechanical energy lost due to friction.

Use the work–energy theorem:

$$W_{nc} = (KE + PE_g + PE_s)_f - (KE + PE_g + PE_s)_i$$
$$= (0 + 0 + \tfrac{1}{2}kx^2) - (0 + mgh + 0)$$
$$= \tfrac{1}{2}(1.0 \times 10^6 \text{ N/m})(0.30)^2 - (13\,000 \text{ N})(10.0 \text{ m})$$
$$W_{nc} = -8.5 \times 10^4 \text{ J}$$

Remarks The answer to part (b) is about 40 times greater than the acceleration of gravity, so we'd better be wearing our seat belts. Note that the solution didn't require calculation of the velocity of the car.

Exercise 13.3

A spring-loaded gun fires a 0.100-kg puck along a tabletop. The puck slides up a curved ramp and flies straight up into the air. If the spring is displaced 12.0 cm from equilibrium and the spring constant is 875 N/m, how high does the puck rise, neglecting friction? (b) If instead it only rises to a height of 5.00 m because of friction, what is the change in mechanical energy?

Answer (a) 6.43 m (b) −1.40 J

In addition to studying the preceding example, it's a good idea to review those given in Section 5.4.

Velocity as a Function of Position

Conservation of energy provides a simple method of deriving an expression for the velocity of an object undergoing periodic motion as a function of position. The object in question is initially at its maximum extension A (Fig. 13.7a) and is then released from rest. The initial energy of the system is entirely elastic potential energy stored in the spring, $\frac{1}{2}kA^2$. As the object moves toward the origin to some new position x (Fig. 13.7b), part of this energy is transformed into kinetic energy, and the potential energy stored in the spring is reduced to $\frac{1}{2}kx^2$. Because the total energy of the system is equal to $\frac{1}{2}kA^2$ (the initial energy stored in the spring), we can equate this quantity to the sum of the kinetic and potential energies at the position x:

$$\tfrac{1}{2}kA^2 = \tfrac{1}{2}mv^2 + \tfrac{1}{2}kx^2$$

Solving for v, we get

$$v = \pm\sqrt{\frac{k}{m}(A^2 - x^2)} \qquad [13.6]$$

This expression shows that the object's speed is a maximum at $x = 0$ and is zero at the extreme positions $x = \pm A$.

The right side of Equation 13.6 is preceded by the $\pm$ sign because the square root of a number can be either positive or negative. If the object in Figure 13.7 is moving to the right, v is positive; if the object is moving to the left, v is negative.

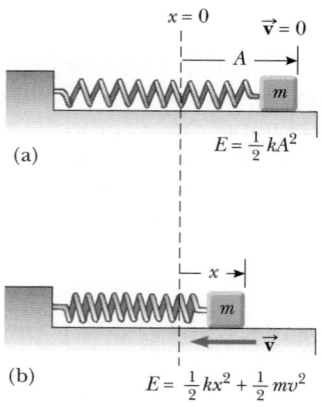

Figure 13.7 (a) An object attached to a spring on a frictionless surface is released from rest with the spring extended a distance A. Just before the object is released, the total energy is the elastic potential energy $kA^2/2$. (b) When the object reaches position x, it has kinetic energy $mv^2/2$ and the elastic potential energy has decreased to $kx^2/2$.

EXAMPLE 13.4 The Object–Spring System Revisited

Goal Apply the time-independent velocity expression, Equation 13.6.

Problem A 0.500-kg object connected to a light spring with a spring constant of 20.0 N/m oscillates on a frictionless horizontal surface. **(a)** Calculate the total energy of the system and the maximum speed of the object if the amplitude of the motion is 3.00 cm. **(b)** What is the velocity of the object when the displacement is 2.00 cm? **(c)** Compute the kinetic and potential energies of the system when the displacement is 2.00 cm.

Strategy The total energy of the system can be found most easily at the amplitude $x = A$, where the kinetic energy is zero. There, the potential energy alone is equal to the total energy. Conservation of energy then yields the speed at $x = 0$. For part (b), obtain the velocity by substituting the given value of x into the time-independent velocity equation. Using this result, the kinetic energy asked for in part (c) can be found by substitution, and the potential energy from substitution into Equation 13.3.

Solution

(a) Calculate the total energy and maximum speed if the amplitude is 3.00 cm.

Substitute $x = A = 3.00$ cm and $k = 20.0$ N/m into the equation for the total mechanical energy E:

$$\begin{aligned} E &= KE + PE_g + PE_s \\ &= 0 + 0 + \tfrac{1}{2}kA^2 = \tfrac{1}{2}(20.0\ \text{N/m})(3.00 \times 10^{-2}\ \text{m})^2 \\ &= 9.00 \times 10^{-3}\ \text{J} \end{aligned}$$

Use conservation of energy with $x_i = A$ and $x_f = 0$ to compute the speed of the object at the origin:

$$\begin{aligned} (KE + PE_g + PE_s)_i &= (KE + PE_g + PE_s)_f \\ 0 + 0 + \tfrac{1}{2}kA^2 &= \tfrac{1}{2}mv_{\text{max}}^2 + 0 + 0 \\ \tfrac{1}{2}mv_{\text{max}}^2 &= 9.00 \times 10^{-3}\ \text{J} \\ v_{\text{max}} &= \sqrt{\frac{18.0 \times 10^{-3}\ \text{J}}{0.500\ \text{kg}}} = \boxed{0.190\ \text{m/s}} \end{aligned}$$

(b) Compute the velocity of the object when the displacement is 2.00 cm.

Substitute known values directly into Equation 13.6:

$$\begin{aligned} v &= \pm\sqrt{\frac{k}{m}(A^2 - x^2)} \\ &= \pm\sqrt{\frac{20.0\ \text{N/m}}{0.500\ \text{kg}}((0.030\ 0\ \text{m})^2 - (0.020\ 0\ \text{m})^2)} \\ &= \boxed{\pm 0.141\ \text{m/s}} \end{aligned}$$

(c) Compute the kinetic and potential energies when the displacement is 2.00 cm.

Substitute into the equation for kinetic energy:

$$KE = \tfrac{1}{2}mv^2 = \tfrac{1}{2}(0.500\ \text{kg})(0.141\ \text{m/s})^2 = \boxed{4.97 \times 10^{-3}\ \text{J}}$$

Substitute into the equation for spring potential energy:

$$\begin{aligned} PE_s &= \tfrac{1}{2}kx^2 = \tfrac{1}{2}(20.0\ \text{N/m})(2.00 \times 10^{-2}\ \text{m})^2 \\ &= \boxed{4.00 \times 10^{-3}\ \text{J}} \end{aligned}$$

Remark With the given information, it is impossible to choose between the positive and negative solutions in part (b). Notice that the sum $KE + PE_s$ in part (c) equals the total energy E found in part (a), as it should (except for a small discrepancy due to rounding).

Exercise 13.4

For what values of x is the speed of the object 0.10 m/s?

Answer ± 2.55 cm

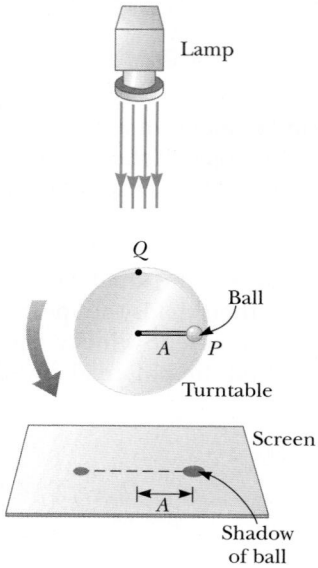

ACTIVE FIGURE 13.8
An experimental setup for demonstrating the connection between simple harmonic motion and uniform circular motion. As the ball rotates on the turntable with constant angular speed, its shadow on the screen moves back and forth with simple harmonic motion.

PhysicsNow™
Log into PhysicsNow at **www.cp7e.com**, and go to Active Figure 13.8 to adjust the frequency and radial position of the ball and see the resulting simple harmonic motion of the shadow.

13.3 COMPARING SIMPLE HARMONIC MOTION WITH UNIFORM CIRCULAR MOTION

We can better understand and visualize many aspects of simple harmonic motion along a straight line by looking at its relationship to uniform circular motion. Active Figure 13.8 is a top view of an experimental arrangement that is useful for this purpose. A ball is attached to the rim of a turntable of radius A, illuminated from the side by a lamp. We find that **as the turntable rotates with constant angular speed, the shadow of the ball moves back and forth with simple harmonic motion.**

This fact can be understood from Equation 13.6, which says that the velocity of an object moving with simple harmonic motion is related to the displacement by

$$v = C\sqrt{A^2 - x^2}$$

where C is a constant. To see that the shadow also obeys this relation, consider Figure 13.9, which shows the ball moving with a constant speed v_0 in a direction tangent to the circular path. At this instant, the velocity of the ball in the x-direction is given by $v = v_0 \sin\theta$, or

$$\sin\theta = \frac{v}{v_0}$$

From the larger triangle in the figure we can obtain a second expression for $\sin\theta$:

$$\sin\theta = \frac{\sqrt{A^2 - x^2}}{A}$$

Equating the right-hand sides of the two expressions for $\sin\theta$, we find the following relationship between the velocity v and the displacement x:

$$\frac{v}{v_0} = \frac{\sqrt{A^2 - x^2}}{A}$$

or

$$v = \frac{v_0}{A}\sqrt{A^2 - x^2} = C\sqrt{A^2 - x^2}$$

The velocity of the ball in the x-direction is related to the displacement x in exactly the same way as the velocity of an object undergoing simple harmonic motion. The shadow therefore moves with simple harmonic motion.

A valuable example of the relationship between simple harmonic motion and circular motion can be seen in vehicles and machines that use the back-and-forth motion of a piston to create rotational motion in a wheel. Consider the drive wheel of a locomotive. In Figure 13.10, the curved housing at the left contains a piston that moves back and forth in simple harmonic motion. The piston is connected to an arrangement of rods that transforms its back-and-forth motion into rotational motion of the wheels. A similar mechanism in an automobile engine transforms the back-and-forth motion of the pistons to rotational motion of the crankshaft.

APPLICATION
Pistons and Drive Wheels

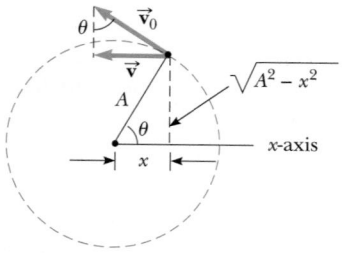

Figure 13.9 The ball rotates with constant speed v_0. The x-component of the ball's velocity equals the projection of $\vec{v}_0$ on the x-axis.

Period and Frequency

The period T of the shadow in Active Figure 13.8, which represents the time required for one complete trip back and forth, is also the time it takes the ball to make one complete circular trip on the turntable. Because the ball moves through the distance $2\pi A$ (the circumference of the circle) in the time T, the speed v_0 of the ball around the circular path is

$$v_0 = \frac{2\pi A}{T}$$

and the period is

$$T = \frac{2\pi A}{v_0} \qquad [13.7]$$

© Link/Visuals Unlimited

Figure 13.10 The drive wheel mechanism of an old locomotive.

Imagine that the ball moves from P to Q, a quarter of a revolution, in Active Figure 13.8. The motion of the shadow is equivalent to the horizontal motion of an object on the end of a spring. For this reason, the radius A of the circular motion is the same as the amplitude A of the simple harmonic motion of the shadow. During the quarter of a cycle shown, the shadow moves from a point where the energy of the system (ball and spring) is solely elastic potential energy to a point where the energy is solely kinetic energy. By conservation of energy, we have

$$\tfrac{1}{2}kA^2 = \tfrac{1}{2}mv_0{}^2$$

which can be solved for A/v_0:

$$\frac{A}{v_0} = \sqrt{\frac{m}{k}}$$

Substituting this expression for A/v_0 in Equation 13.7, we find that the period is

$$T = 2\pi\sqrt{\frac{m}{k}} \qquad [13.8]$$

◀ The period of an object–spring system moving with simple harmonic motion

Equation 13.8 represents the time required for an object of mass m attached to a spring with spring constant k to complete one cycle of its motion. The square root of the mass is in the numerator, so a large mass will mean a large period, in agreement with intuition. The square root of the spring constant k is in the denominator, so a large spring constant will yield a small period, again agreeing with intuition. It's also interesting that the period doesn't depend on the amplitude A.

The inverse of the period is the frequency of the motion:

$$f = \frac{1}{T} \qquad [13.9]$$

Therefore, the frequency of the periodic motion of a mass on a spring is

$$f = \frac{1}{2\pi}\sqrt{\frac{k}{m}} \qquad [13.10]$$

◀ Frequency of an object–spring system

The units of frequency are cycles per second (s^{-1}), or **hertz** (Hz). The **angular frequency** ω is

$$\omega = 2\pi f = \sqrt{\frac{k}{m}} \qquad [13.11]$$

◀ Angular frequency of an object–spring system

The frequency and angular frequency are actually closely related concepts. The unit of frequency is cycles per second, where a cycle may be thought of as a unit of angular measure corresponding to 2π radians, or 360°. Viewed in this way, angular frequency is just a unit conversion of frequency. Radian measure is used for angles mainly because it provides a convenient and natural link between linear and angular quantities.

TIP 13.2 Twin Frequencies

The *frequency* gives the number of cycles per second, while the *angular frequency* gives the number of radians per second. These two physical concepts are nearly identical—linked by the conversion factor 2π rad/cycle.

Although an ideal mass–spring system has a period proportional to the square root of the object's mass m, experiments show that a graph of T^2 versus m doesn't pass through the origin. This is because the spring itself has a mass. The coils of the spring oscillate just like the object, except the amplitudes are smaller for all coils but the last. For a cylindrical spring, energy arguments can be used to show that the *effective* additional mass of a light spring is one-third the mass of the spring. The square of the period is proportional to the total oscillating mass, so a graph of T^2 versus total mass (the mass hung on the spring plus the effective oscillating mass of the spring) would pass through the origin.

Quick Quiz 13.4

An object of mass m is attached to a horizontal spring, stretched to a displacement A from equilibrium and released, undergoing harmonic oscillations on a frictionless surface with period T_0. The experiment is then repeated with a mass of $4m$. What's the new period of oscillation? (a) $2T_0$ (b) T_0 (c) $T_0/2$ (d) $T_0/4$.

Quick Quiz 13.5

Consider the situation in Quick Quiz 13.4. The subsequent total mechanical energy of the object with mass $4m$ is (a) greater than, (b) less than, or (c) equal to the original total mechanical energy.

Applying Physics 13.1 Bungee Jumping

A bungee cord can be modeled as a spring. If you go bungee jumping, you will bounce up and down at the end of the elastic cord after your dive off a bridge (Fig. 13.11). Suppose you perform a dive and measure the frequency of your bouncing. You then move to another bridge, but find that the bungee cord is too long for dives off this bridge. What possible solutions might be applied? In terms of the original frequency, what is the frequency of vibration associated with the solution?

Telegraph Colour Library/FPG International/Getty Images

Figure 13.11 (Applying Physics 13.1) Bungee jumping from a bridge.

Explanation There are two possible solutions: Make the bungee cord smaller or fold it in half. The latter would be the safer of the two choices, as we'll see. The force exerted by the bungee cord, modeled as a spring, is proportional to the separation of the coils as the spring is extended. First, we extend the spring by a given distance and measure the distance between coils. We then cut the spring in half. If one of the half-springs is now extended by the same distance, the coils will be twice as far apart as they were for the complete spring. Therefore, it takes twice as much force to stretch the half-spring through the same displacement, so the half-spring has a spring constant twice that of the complete spring. The folded bungee cord can then be modeled as two half-springs in parallel. Each half has a spring constant that is twice the original spring constant of the bungee cord. In addition, an object hanging on the folded bungee cord will experience two forces—one from each half-spring. As a result, the required force for a given extension will be four times as much as for the original bungee cord. The effective spring constant of the folded bungee cord is therefore four times as large as the original spring constant. Because the frequency of oscillation is proportional to the square root of the spring constant, your bouncing frequency on the folded cord will be twice what it was on the original cord.

This discussion neglects the fact that the coils of a spring have an initial separation. It's also important to remember that a shorter coil may lose elasticity more readily, possibly even going beyond the elastic limit for the material, with disastrous results. Bungee jumping is dangerous—discretion is advised!

EXAMPLE 13.5 That Car Needs Shock Absorbers!

Goal Understand the relationships between period, frequency, and angular frequency.

Problem A 1.30×10^3-kg car is constructed on a frame supported by four springs. Each spring has a spring constant of 2.00×10^4 N/m. If two people riding in the car have a combined mass of 1.60×10^2 kg, find the frequency of vibration of the car when it is driven over a pothole in the road. Find also the period and the angular frequency. Assume the weight is evenly distributed.

Strategy Because the weight is evenly distributed, each spring supports one-fourth of the mass. Substitute this value and the spring constant into Equation 13.10 to get the frequency. The reciprocal is the period, and multiplying the frequency by 2π gives the angular frequency.

Solution

Compute one-quarter of the total mass:

$$m = \tfrac{1}{4}(m_{\text{car}} + m_{\text{pass}}) = \tfrac{1}{4}(1.30 \times 10^3\ \text{kg} + 1.60 \times 10^2\ \text{kg}) = 365\ \text{kg}$$

Substitute into Equation 13.10 to find the frequency:

$$f = \frac{1}{2\pi}\sqrt{\frac{k}{m}} = \frac{1}{2\pi}\sqrt{\frac{2.00 \times 10^4\ \text{N/m}}{365\ \text{kg}}} = 1.18\ \text{Hz}$$

Invert the frequency to get the period:

$$T = \frac{1}{f} = \frac{1}{1.18\ \text{Hz}} = 0.847\ \text{s}$$

Multiply the frequency by 2π to get the angular frequency:

$$\omega = 2\pi f = 2\pi(1.18\ \text{Hz}) = 7.41\ \text{rad/s}$$

Remark Solving this problem didn't require any knowledge of the size of the pothole, because the frequency doesn't depend on the amplitude of the motion.

Exercise 13.5

A 45.0-kg boy jumps on a 5.00-kg pogo stick with spring constant 3 650 N/m. Find (a) the angular frequency, (b) the frequency, and (c) the period of the boy's motion.

Answers (a) 8.54 rad/s (b) 1.36 Hz (c) 0.735 s

13.4 POSITION, VELOCITY, AND ACCELERATION AS A FUNCTION OF TIME

We can obtain an expression for the position of an object moving with simple harmonic motion as a function of time by returning to the relationship between simple harmonic motion and uniform circular motion. Again, consider a ball on the rim of a rotating turntable of radius A, as in Active Figure 13.12. We refer to the circle made by the ball as the *reference circle* for the motion. We assume that the turntable revolves at a *constant* angular speed ω. As the ball rotates on the reference circle, the angle θ made by the line OP with the x-axis changes with time. Meanwhile, the projection of P on the x-axis, labeled point Q, moves back and forth along the axis with simple harmonic motion.

From the right triangle OPQ, we see that $\cos\theta = x/A$. Therefore, the x-coordinate of the ball is

$$x = A\cos\theta$$

Because the ball rotates with constant angular speed, it follows that $\theta = \omega t$ (see Chapter 7), so we have

$$x = A\cos(\omega t) \qquad [13.12]$$

In one complete revolution, the ball rotates through an angle of 2π rad in a time equal to the period T. In other words, the motion repeats itself every T seconds. Therefore,

$$\omega = \frac{\Delta\theta}{\Delta t} = \frac{2\pi}{T} = 2\pi f \qquad [13.13]$$

where f is the frequency of the motion. The angular speed of the ball as it moves around the reference circle is the same as the angular frequency of the projected simple harmonic motion. Consequently, Equation 13.12 can be written

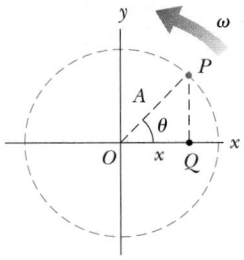

ACTIVE FIGURE 13.12
A reference circle. As the ball at P rotates in a circle with uniform angular speed, its projection Q along the x-axis moves with simple harmonic motion.

PhysicsNow™
Log into PhysicsNow at **www.cp7e.com**, and go to Active Figure 13.12 to compare the oscillations of two blocks starting from different initial positions and to verify that the frequency is independent of the amplitude.

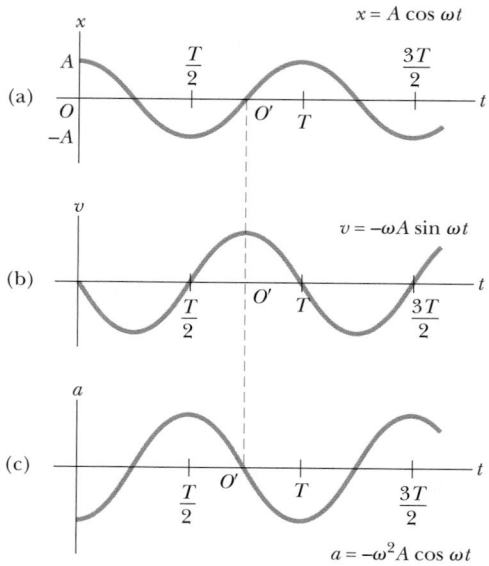

ACTIVE FIGURE 13.13
(a) Displacement, (b) velocity, and (c) acceleration versus time for an object moving with simple harmonic motion under the initial conditions $x_0 = A$ and $v_0 = 0$ at $t = 0$.

PhysicsNow™
Log into PhysicsNow at **www.cp7e.com**, and go to Active Figure 13.13 to adjust the graphical representation and see the resulting simple harmonic motion of the block.

$$x = A\cos(2\pi ft) \qquad \textbf{[13.14a]}$$

This cosine function represents the position of an object moving with simple harmonic motion as a function of time, and is graphed in Active Figure 13.13a. Because the cosine function varies between 1 and -1, x varies between A and $-A$. The shape of the graph is called *sinusoidal*.

Active Figures 13.13b and 13.13c represent curves for velocity and acceleration as a function of time. To find the equation for the velocity, use Equations 13.6 and 13.14a together with the identity $\cos^2\theta + \sin^2\theta = 1$, obtaining

$$v = -A\omega\sin(2\pi ft) \qquad \textbf{[13.14b]}$$

where we have used the fact that $\omega = \sqrt{k/m}$. The $\pm$ sign is no longer needed, because sine can take both positive and negative values. Deriving an expression for the acceleration involves substituting Equation 13.14a into Equation 13.2, Newton's second law for springs:

$$a = -A\omega^2\cos(2\pi ft) \qquad \textbf{[13.14c]}$$

The detailed steps of these derivations are left as an exercise for the student. Notice that when the displacement x is at a maximum, at $x = A$ or $x = -A$, the velocity is zero, and when x is zero, the magnitude of the velocity is a maximum. Further, when $x = +A$, its most positive value, the acceleration is a maximum but in the negative x-direction, and when x is at its most negative position, $x = -A$, the acceleration has its maximum value in the positive x-direction. These facts are consistent with our earlier discussion of the points at which v and a reach their maximum, minimum, and zero values.

The maximum values of the position, velocity, and acceleration are always equal to the magnitude of the expression in front of the trigonometric function in each equation, because the largest value of either cosine or sine is 1.

Figure 13.14 illustrates one experimental arrangement that demonstrates the sinusoidal nature of simple harmonic motion. An object connected to a spring has a marking pen attached to it. While the object vibrates vertically, a sheet of paper is moved horizontally with constant speed. The pen traces out a sinusoidal pattern.

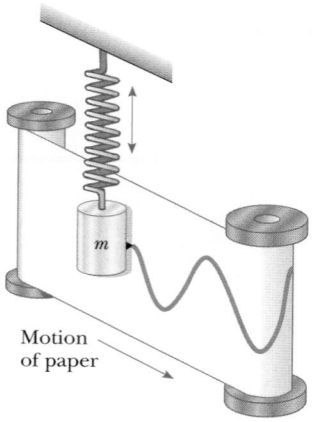

Figure 13.14
An experimental apparatus for demonstrating simple harmonic motion. A pen attached to the oscillating object traces out a sinusoidal wave on the moving chart paper.

Quick Quiz 13.6

If the amplitude of a system moving in simple harmonic motion is doubled, which of the following quantities *doesn't* change? (a) total energy, (b) maximum speed, (c) maximum acceleration, (d) period.

EXAMPLE 13.6 The Vibrating Object–Spring System

Goal Identify the physical parameters of a harmonic oscillator from its mathematical description.

Problem (a) Find the amplitude, frequency, and period of motion for an object vibrating at the end of a horizontal spring if the equation for its position as a function of time is

$$x = (0.250\text{ m})\cos\left(\frac{\pi}{8.00}t\right)$$

(b) Find the maximum magnitude of the velocity and acceleration. **(c)** What is the position, velocity, and acceleration of the object after 1.00 s has elapsed?

Strategy In part (a), the amplitude and frequency can be found by comparing the given equation with the standard form in Equation 13.14a, matching up the numerical values with the corresponding terms in the standard form. (b) The maximum speed will occur when the sine function in Equation 13.14b equals 1 or -1, the extreme values of the sine function (and similarly for the acceleration and the cosine function). In each case, find the magnitude of the expression in front of the trigonometric function. Part (c) is just a matter of substituting values into Equations 13.14a, b, and c.

Solution

(a) Find the amplitude, frequency, and period.

Write the standard form given by Equation 13.14a, and underneath it write the given equation:

$$x = A\cos(2\pi ft) \quad \textbf{(1)}$$

$$x = (0.250\text{ m})\cos\left(\frac{\pi}{8.00}t\right) \quad \textbf{(2)}$$

Equate the factors in front of the cosine functions to find the amplitude:

$$A = 0.250\text{ m}$$

The angular frequency ω is the factor in front of t in equations (1) and (2). Equate these factors:

$$\omega = 2\pi f = \frac{\pi}{8.00}\text{ rad/s} = 0.393\text{ rad/s}$$

Divide ω by 2π to get the frequency f:

$$f = \frac{\omega}{2\pi} = 0.062\,5\text{ Hz}$$

The period T is the reciprocal of the frequency:

$$T = \frac{1}{f} = 16.0\text{ s}$$

(b) Find the maximum magnitudes of the velocity and the acceleration.

Calculate the maximum speed from the factor in front of the sine function in Equation 13.14b:

$$v_{\max} = A\omega = (0.250\text{ m})(0.393\text{ rad/s}) = 0.098\,3\text{ m/s}$$

Calculate the maximum acceleration from the factor in front of the cosine function in Equation 13.14c:

$$a_{\max} = A\omega^2 = (0.250\text{ m})(0.393\text{ rad/s})^2 = 0.038\,6\text{ m/s}^2$$

(c) Find the position, velocity, and acceleration of the object after 1.00 s.

Substitute $t = 1.00$ s in the given equation:

$$x = (0.250\text{ m})\cos(0.393\text{ rad}) = 0.231\text{ m}$$

Substitute values into the velocity equation:

$$v = -A\omega\sin(\omega t)$$
$$= -(0.250\text{ m})(0.393\text{ rad/s})\sin(0.393\text{ rad/s}\cdot 1.00\text{ s})$$
$$v = -0.037\,6\text{ m/s}$$

Substitute values into the acceleration equation:

$$a = -A\omega^2\cos(\omega t)$$
$$= -(0.250\text{ m})(0.393\text{ rad/s}^2)^2\cos(0.393\text{ rad/s}\cdot 1.00\text{ s})$$
$$a = -0.035\,7\text{ m/s}^2$$

Remarks In evaluating the cosine function, the angle is in radians, so you should either set your calculator to evaluate trigonometric functions based on radian measure or convert from radians to degrees.

Exercise 13.6

If the object–spring system is described by $x = (0.330\text{ m})\cos(1.50t)$, find (a) the amplitude, the angular frequency, the frequency, and the period, (b) the maximum magnitudes of the velocity and acceleration, and (c) the position, velocity, and acceleration when $t = 0.250$ s.

Answers (a) $A = 0.330$ m, $\omega = 1.50$ rad/s, $f = 0.239$ Hz, $T = 4.19$ s; (b) $v_{\max} = 0.495$ m/s, $a_{\max} = 0.743$ m/s^2; (c) $x = 0.307$ m, $v = -0.181$ m/s, $a = -0.691$ m/s^2

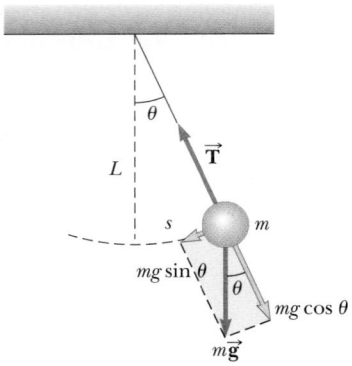

ACTIVE FIGURE 13.15
A simple pendulum consists of a bob of mass m suspended by a light string of length L. (L is the distance from the pivot to the center of mass of the bob.) The restoring force that causes the pendulum to undergo simple harmonic motion is the component of gravitational force tangent to the path of motion, $mg \sin \theta$.

PhysicsNow™
Log into PhysicsNow at **www.cp7e.com**, and go to Active Figure 13.15 to adjust the mass of the bob, the length of the string, and the initial angle and see the resulting oscillation of the pendulum. The period is slightly larger for larger initial angles.

13.5 MOTION OF A PENDULUM

A simple pendulum is another mechanical system that exhibits periodic motion. It consists of a small bob of mass m suspended by a light string of length L fixed at its upper end, as in Active Figure 13.15. (By a light string, we mean that the string's mass is assumed to be very small compared with the mass of the bob and hence can be ignored.) When released, the bob swings to and fro over the same path; but is its motion simple harmonic?

Answering this question requires examining the restoring force—the force of gravity—that acts on the pendulum. The pendulum bob moves along a circular arc, rather than back and forth in a straight line. When the oscillations are small, however, the motion of the bob is nearly straight, so Hooke's law may apply approximately.

In Active Figure 13.15, s is the displacement of the bob from equilibrium along the arc. Hooke's law is $F = -kx$, so we are looking for a similar expression involving s, $F_t = -ks$, where F_t is the force acting in a direction tangent to the circular arc. From the figure, the restoring force is

$$F_t = -mg \sin \theta$$

Since $s = L\theta$, the equation for F_t can be written as

$$F_t = -mg \sin\left(\frac{s}{L}\right)$$

This expression isn't of the form $F_t = -ks$, so in general, the motion of a pendulum is *not* simple harmonic. For small angles less than about 15 degrees, however, the angle θ measured in radians and the sine of the angle are approximately equal. For example, $\theta = 10.0° = 0.175$ rad, and $\sin(10.0°) = 0.174$. Therefore, if we restrict the motion to *small* angles, the approximation $\sin \theta \approx \theta$ is valid, and the restoring force can be written

$$F_t = -mg \sin \theta \approx -mg\,\theta$$

Substituting $\theta = s/L$, we obtain

$$F_t = -\left(\frac{mg}{L}\right) s$$

This equation follows the general form of Hooke's force law $F_t = -ks$, with $k = mg/L$. We are justified in saying that a pendulum undergoes simple harmonic motion only when it swings back and forth at small amplitudes (or, in this case, small values of θ, so that $\sin \theta \cong \theta$).

TIP 13.3 Pendulum Motion is not Harmonic
Remember that the pendulum *does not* exhibit true simple harmonic motion for *any* angle. If the angle is less than about 15°, the motion can be *modeled* as approximately simple harmonic.

Recall that for the object–spring system, the angular frequency is given by Equation 13.11:

$$\omega = 2\pi f = \sqrt{\frac{k}{m}}$$

Substituting the expression of k for a pendulum, we obtain

$$\omega = \sqrt{\frac{mg/L}{m}} = \sqrt{\frac{g}{L}}$$

This angular frequency can be substituted into Equation 13.12, which then mathematically describes the motion of a pendulum. The frequency is just the angular frequency divided by 2π, while the period is the reciprocal of the frequency, or

The period of a simple pendulum depends only on L and g ▶

$$T = 2\pi\sqrt{\frac{L}{g}} \qquad [13.15]$$

This equation reveals the somewhat surprising result that the period of a simple pendulum doesn't depend on the mass, but only on the pendulum's length and on the free-fall acceleration. Furthermore, the amplitude of the motion isn't a

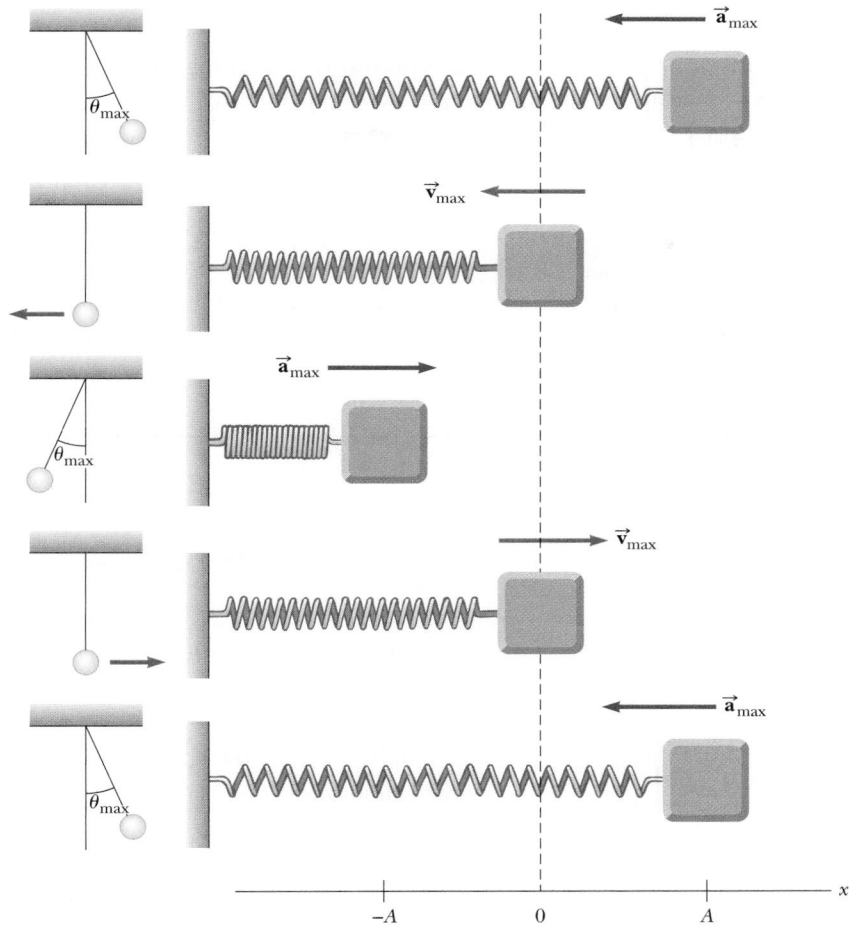

ACTIVE FIGURE 13.16
Simple harmonic motion for an object–spring system, and its analogy, the motion of a simple pendulum.

PhysicsNow™
Log into PhysicsNow at **www.cp7e.com**, and go to Active Figure 13.16 to set the initial position of the block and see the block–spring motion and the analogous pendulum motion.

factor as long as it's relatively small. The analogy between the motion of a simple pendulum and the object–spring system is illustrated in Active Figure 13.16.

Galileo first noted that the period of a pendulum was independent of its amplitude. He supposedly observed this while attending church services at the cathedral in Pisa. The pendulum he studied was a swinging chandelier that was set in motion when someone bumped it while lighting candles. Galileo was able to measure its period by timing the swings with his pulse.

APPLICATION
Pendulum Clocks

The dependence of the period of a pendulum on its length and on the free-fall acceleration allows us to use a pendulum as a timekeeper for a clock. A number of clock designs employ a pendulum, with the length adjusted so that its period serves as the basis for the rate at which the clock's hands turn. Of course, these clocks are used at different locations on the Earth, so there will be some variation of the free-fall acceleration. To compensate for this variation, the pendulum of a clock should have some movable mass so that the effective length can be adjusted.

APPLICATION
Use of Pendulum in Prospecting

Geologists often make use of the simple pendulum and Equation 13.15 when prospecting for oil or minerals. Deposits beneath the Earth's surface can produce irregularities in the free-fall acceleration over the region being studied. A specially designed pendulum of known length is used to measure the period, which in turn is used to calculate g. Although such a measurement in itself is inconclusive, it's an important tool for geological surveys.

Quick Quiz 13.7

A simple pendulum is suspended from the ceiling of a stationary elevator, and the period is measured. If the elevator moves with constant velocity, does the period (a) increase, (b) decrease, or (c) remain the same? If the elevator accelerates upward, does the period (a) increase, (b) decrease, or (c) remain the same?

Quick Quiz 13.8

A pendulum clock depends on the period of a pendulum to keep correct time. Suppose a pendulum clock is keeping correct time and then Dennis the Menace slides the bob of the pendulum downward on the oscillating rod. Does the clock run (a) slow, (b) fast, or (c) correctly?

Quick Quiz 13.9

The period of a simple pendulum is measured to be T on Earth. If the same pendulum were set in motion on the Moon, would its period be (a) less than T, (b) greater than T, or (c) equal to T?

EXAMPLE 13.7 Measuring the Value of g

Goal Determine g from pendulum motion.

Problem Using a small pendulum of length 0.171 m, a geologist counts 72.0 complete swings in a time of 60.0 s. What is the value of g in this location?

Strategy First calculate the period of the pendulum by dividing the total time by the number of complete swings. Solve Equation 13.15 for g and substitute values.

Solution

Calculate the period by dividing the total elapsed time by the number of complete oscillations:

$$T = \frac{\text{time}}{\text{\# of oscillations}} = \frac{60.0\text{ s}}{72.0} = 0.833\text{ s}$$

Solve Equation 13.15 for g and substitute values:

$$T = 2\pi\sqrt{\frac{L}{g}} \quad \rightarrow \quad T^2 = 4\pi^2\frac{L}{g}$$

$$g = \frac{4\pi^2 L}{T^2} = \frac{(39.5)(0.171\text{ m})}{(0.833\text{ s})^2} = 9.73\text{ m/s}^2$$

Exercise 13.7

What would be the period of the same pendulum on the Moon, where the acceleration of gravity is 1.62 m/s^2?

Answer 2.04 s

The Physical Pendulum

The simple pendulum discussed thus far consists of a mass attached to a string. A pendulum, however, can be made from an object of any shape. The general case is called the *physical pendulum.*

In Figure 13.17, a rigid object is pivoted at point O, which is a distance L from the object's center of mass. The center of mass oscillates along a circular arc, just like the simple pendulum. The period of a physical pendulum is given by

$$T = 2\pi\sqrt{\frac{I}{mgL}} \qquad \textbf{[13.16]}$$

where I is the object's moment of inertia and m is the object's mass. As a check, notice that in the special case of a simple pendulum with an arm of length L and negligible mass, the moment of inertia is $I = mL^2$. Substituting into Equation 13.16 results in

$$T = 2\pi\sqrt{\frac{mL^2}{mgL}} = 2\pi\sqrt{\frac{L}{g}}$$

which is the correct period for a simple pendulum.

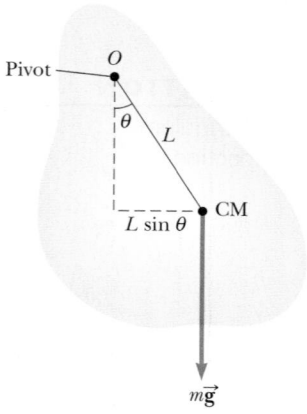

Figure 13.17 A physical pendulum pivoted at O.

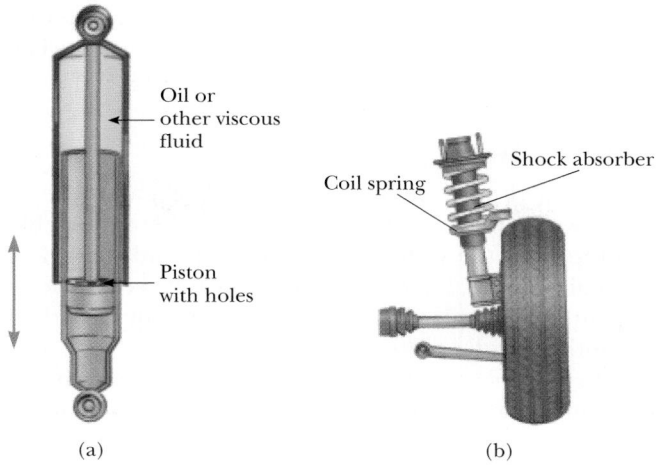

Figure 13.18 (a) A shock absorber consists of a piston oscillating in a chamber filled with oil. As the piston oscillates, the oil is squeezed through holes between the piston and the chamber, causing a damping of the piston's oscillations. (b) One type of automotive suspension system, in which a shock absorber is placed inside a coil spring at each wheel.

13.6 DAMPED OSCILLATIONS

The vibrating motions we have discussed so far have taken place in ideal systems that *oscillate indefinitely* under the action of a linear restoring force. In all real mechanical systems, forces of friction retard the motion, so the systems don't oscillate indefinitely. The friction reduces the mechanical energy of the system as time passes, and the motion is said to be **damped**.

Shock absorbers in automobiles (Fig. 13.18) are one practical application of damped motion. A shock absorber consists of a piston moving through a liquid such as oil. The upper part of the shock absorber is firmly attached to the body of the car. When the car travels over a bump in the road, holes in the piston allow it to move up and down in the fluid in a damped fashion.

Damped motion varies with the fluid used. For example, if the fluid has a relatively low viscosity, the vibrating motion is preserved but the amplitude of vibration decreases in time and the motion ultimately ceases. This is known as *underdamped* oscillation. The position vs. time curve for an object undergoing such oscillation appears in Active Figure 13.19. Figure 13.20 compares three types of damped motion, with curve (a) representing underdamped oscillation. If the fluid viscosity is increased, the object returns rapidly to equilibrium after it's released and doesn't oscillate. In this case, the system is said to be *critically damped*, and is shown as curve (b) in Figure 13.20. The piston returns to the equilibrium position in the shortest time possible without once overshooting the equilibrium position. If the viscosity is made greater still, the system is said to be *overdamped*. In this case, the piston returns to equilibrium without ever passing through the equilibrium point, but the time required to reach equilibrium is greater than in critical damping, as illustrated by curve (c) in Figure 13.20.

To make automobiles more comfortable to ride in, shock absorbers are designed to be slightly underdamped. This can be demonstrated by a sharp downward push on the hood of a car: After the applied force is removed, the body of the car oscillates a few times about the equilibrium position before returning to its fixed position.

APPLICATION

Shock Absorbers

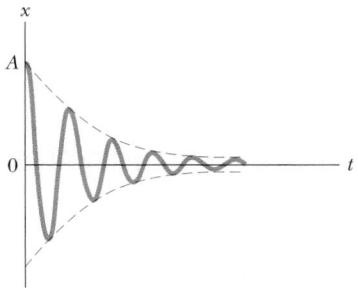

ACTIVE FIGURE 13.19
A graph of displacement versus time for an underdamped oscillator. Note the decrease in amplitude with time.

PhysicsNow™
Log into PhysicsNow at **www.cp7e.com**, and go to Active Figure 13.19 to adjust the spring constant, the mass of the object, and the damping constant and see the resulting damped oscillation of the object.

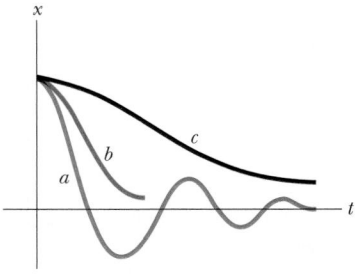

Figure 13.20 Plots of displacement versus time for (a) an underdamped oscillator, (b) a critically damped oscillator, and (c) an overdamped oscillator.

13.7 WAVES

The world is full of waves: sound waves, waves on a string, seismic waves, and electromagnetic waves, such as visible light, radio waves, television signals, and x-rays. All of these waves have as their source a vibrating object, so we can apply the concepts of simple harmonic motion in describing them.

In the case of sound waves, the vibrations that produce waves arise from sources such as a person's vocal chords or a plucked guitar string. The vibrations of electrons in an antenna produce radio or television waves, and the simple up-and-down motion of a hand can produce a wave on a string. Certain concepts are common to all waves, regardless of their nature. In the remainder of this chapter, we focus our attention on the general properties of waves. In later chapters we will study specific types of waves, such as sound waves and electromagnetic waves.

What Is a Wave?

When you drop a pebble into a pool of water, the disturbance produces water waves, which move away from the point where the pebble entered the water. A leaf floating near the disturbance moves up and down and back and forth about its original position, but doesn't undergo any net displacement attributable to the disturbance. This means that the water wave (or disturbance) moves from one place to another, *but the water isn't carried with it.*

When we observe a water wave, we see a rearrangement of the water's surface. Without the water, there wouldn't be a wave. Similarly, a wave traveling on a string wouldn't exist without the string. Sound waves travel through air as a result of pressure variations from point to point. Therefore, we can consider a wave to be *the motion of a disturbance.* In a later chapter we will discuss electromagnetic waves, which don't require a medium.

The mechanical waves discussed in this chapter require (1) some source of disturbance, (2) a medium that can be disturbed, and (3) some physical connection or mechanism through which adjacent portions of the medium can influence each other. All waves carry energy and momentum. The amount of energy transmitted through a medium and the mechanism responsible for the transport of energy differ from case to case. The energy carried by ocean waves during a storm, for example, is much greater than the energy carried by a sound wave generated by a single human voice.

Applying Physics 13.2 Burying Bond

At one point in *On Her Majesty's Secret Service,* a James Bond film from the 1960s, Bond was escaping on skis. He had a good lead and was a hard-to-hit moving target. There was no point in wasting bullets shooting at him, so why did the bad guys open fire?

Explanation These misguided gentlemen had a good understanding of the physics of waves. An impulsive sound, like a gunshot, can cause an acoustical disturbance that propagates through the air. If it impacts a ledge of snow that is ready to break free, an avalanche can result. Such a disaster occurred in 1916 during World War I when Austrian soldiers in the Alps were smothered by an avalanche caused by cannon fire. So the bad guys, who have never been able to hit Bond with a bullet, decided to use the sound of gunfire to start an avalanche.

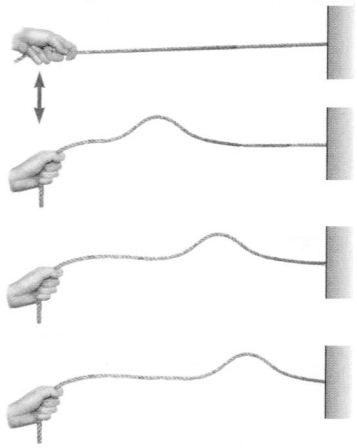

Figure 13.21 A wave pulse traveling along a stretched rope. The shape of the pulse is approximately unchanged as it travels.

Types of Waves

One of the simplest ways to demonstrate wave motion is to flip one end of a long rope that is under tension and has its opposite end fixed, as in Figure 13.21. The bump (called a pulse) travels to the right with a definite speed. A disturbance of this type is called a **traveling wave**. The figure shows the shape of the rope at three closely spaced times.

As such a wave pulse travels along the rope, **each segment of the rope that is disturbed moves in a direction perpendicular to the wave motion**. Figure 13.22 illustrates this point for a particular tiny segment *P*. The rope never moves in the direction of the wave. A traveling wave in which the particles of the disturbed medium move in a direction perpendicular to the wave velocity is called a **transverse wave**. Figure 13.23a illustrates the formation of transverse waves on a long spring.

In another class of waves, called **longitudinal waves, the elements of the medium undergo displacements parallel to the direction of wave motion**. Sound waves in air are longitudinal. Their disturbance corresponds to a series of high- and low-pressure regions that may travel through air or through any material medium with a certain speed. A longitudinal pulse can easily be produced in a stretched spring, as in Figure 13.23b. The free end is pumped back and forth along the length of the spring. This action produces compressed and stretched regions of the coil that travel along the spring, parallel to the wave motion.

Waves need not be purely transverse or purely longitudinal: ocean waves exhibit a superposition of both types. When an ocean wave encounters a cork, the cork executes a circular motion, going up and down while going forward and back.

Another type of wave, called a **soliton**, consists of a solitary wave front that propagates in isolation. Ordinary water waves generally spread out and dissipate, but solitons tend to maintain their form. The study of solitons began in 1849, when the Scottish engineer John Scott Russell noticed a solitary wave leaving the turbulence in front of a barge and propagating forward all on its own. The wave maintained its shape and traveled down a canal at about 10 mi/h. Russell chased the wave two miles on horseback before losing it. Only in the 1960s did scientists take solitons seriously; they are now widely used to model physical phenomena, from elementary particles to the Giant Red Spot of Jupiter.

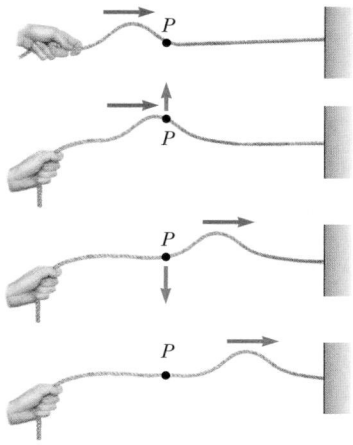

Figure 13.22 A pulse traveling on a stretched rope is a transverse wave. Any element P on the rope moves (blue arrows) in a direction perpendicular to the direction of propagation of the wave motion (red arrows).

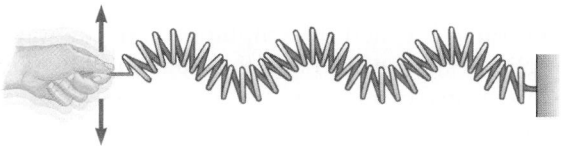

(a) Transverse wave

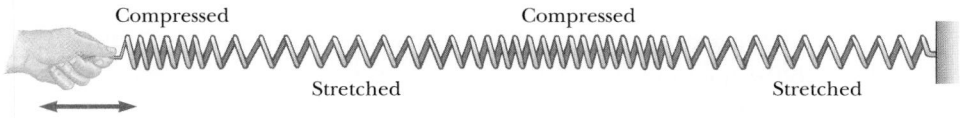

(b) Longitudinal wave

Figure 13.23 (a) A transverse wave is set up in a spring by moving one end of the spring perpendicular to its length. (b) A longitudinal pulse along a stretched spring. The displacement of the coils is in the direction of the wave motion. For the starting motion described in the text, the compressed region is followed by a stretched region.

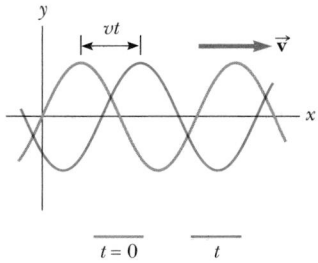

ACTIVE FIGURE 13.24
A one-dimensional sinusoidal wave traveling to the right with a speed v. The brown curve is a snapshot of the wave at $t = 0$, and the blue curve is another snapshot at some later time t.

Physics Now™
Log into PhysicsNow at **www.cp7e.com**, and go to Active Figure 13.24 to watch the wave move and to take snapshots of it at various times.

Picture of a Wave

Active Figure 13.24 shows the curved shape of a vibrating string. This pattern is a sinusoidal curve, the same as in simple harmonic motion. The brown curve can be thought of as a snapshot of a traveling wave taken at some instant of time, say, $t = 0$; the blue curve is a snapshot of the same traveling wave at a later time. This picture can also be used to represent a wave on water. In such a case, a high point would correspond to the *crest* of the wave and a low point to the *trough* of the wave.

The same waveform can be used to describe a longitudinal wave, even though no up-and-down motion is taking place. Consider a longitudinal wave traveling on a spring. Figure 13.25a is a snapshot of this wave at some instant, and Figure 13.25b shows the sinusoidal curve that represents the wave. Points where the coils of the spring are compressed correspond to the crests of the waveform, and stretched regions correspond to troughs.

The type of wave represented by the curve in Figure 13.25b is often called a *density wave* or *pressure wave*, because the crests, where the spring coils are compressed, are regions of high density, and the troughs, where the coils are stretched, are regions of low density. Sound waves are longitudinal waves, propagating as a series of high- and low-density regions.

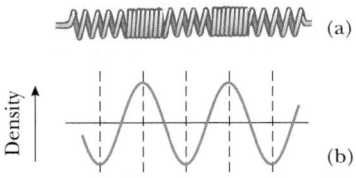

Figure 13.25 (a) A longitudinal wave on a spring. (b) The crests of the waveform correspond to compressed regions of the spring, and the troughs correspond to stretched regions of the spring.

13.8 FREQUENCY, AMPLITUDE, AND WAVELENGTH

Active Figure 13.26 illustrates a method of producing a continuous wave or a steady stream of pulses on a very long string. One end of the string is connected to a blade that is set vibrating. As the blade oscillates vertically with simple harmonic motion, a traveling wave moving to the right is set up in the string. Active Figure 13.26 consists of views of the wave at intervals of one-quarter of a period. Note that **each small segment of the string, such as *P*, oscillates vertically in the *y*-direction with simple harmonic motion**. This must be the case, because each segment follows the simple harmonic motion of the blade. Every segment of the string can therefore be treated as a simple harmonic oscillator vibrating with the same frequency as the blade that drives the string.

The frequencies of the waves studied in this course will range from rather low values for waves on strings and waves on water, to values for sound waves between 20 Hz and 20 000 Hz (recall that $1\ \text{Hz} = 1\ \text{s}^{-1}$), to much higher frequencies for electromagnetic waves. These waves have different physical sources, but can be described with the same concepts.

The horizontal dashed line in Active Figure 13.26 represents the position of the string when no wave is present. The maximum distance the string moves above or below this equilibrium value is called the **amplitude** A of the wave. For the waves we work with, the amplitudes at the crest and the trough will be identical.

Active Figure 13.26b illustrates another characteristic of a wave. The horizontal arrows show the distance between two successive points that behave identically. This distance is called the **wavelength** λ (the Greek letter lambda).

We can use these definitions to derive an expression for the speed of a wave. We start with the defining equation for the **wave speed** v**:**

$$v = \frac{\Delta x}{\Delta t}$$

The wave speed is the speed at which a particular part of the wave—say, a crest—moves through the medium.

A wave advances a distance of one wavelength in a time interval equal to one period of the vibration. Taking $\Delta x = \lambda$ and $\Delta t = T$, we see that

$$v = \frac{\lambda}{T}$$

Because the frequency is the reciprocal of the period, we have

Wave speed ▶

$$v = f\lambda \qquad [13.17]$$

This important general equation applies to many different types of waves, such as sound waves and electromagnetic waves.

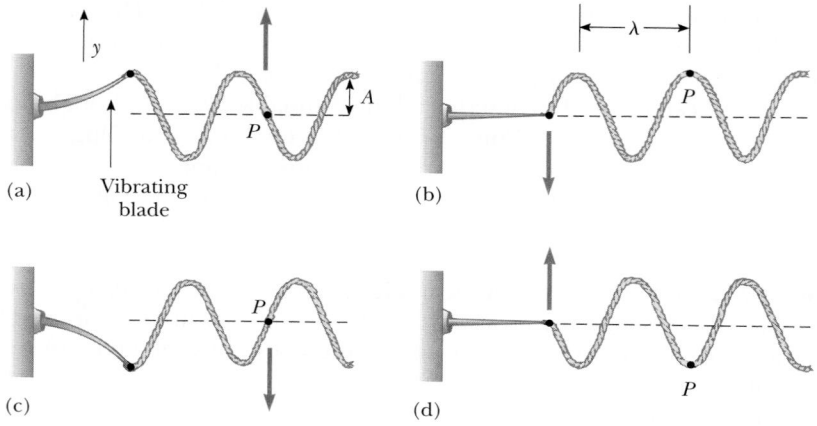

ACTIVE FIGURE 13.26
One method for producing traveling waves on a continuous string. The left end of the string is connected to a blade that is set vibrating. Every part of the string, such as point *P*, oscillates vertically with simple harmonic motion.

PhysicsNow™
Log into PhysicsNow at **www.cp7e.com**, and go to Active Figure 13.26 to adjust the frequency of the blade.

EXAMPLE 13.8 A Traveling Wave

Goal Obtain information about a wave directly from its graph.

Problem A wave traveling in the positive x-direction is pictured in Figure 13.27a. Find the amplitude, wavelength, speed, and period of the wave if it has a frequency of 8.00 Hz. In Figure 13.27a, $\Delta x = 40.0$ cm and $\Delta y = 15.0$ cm.

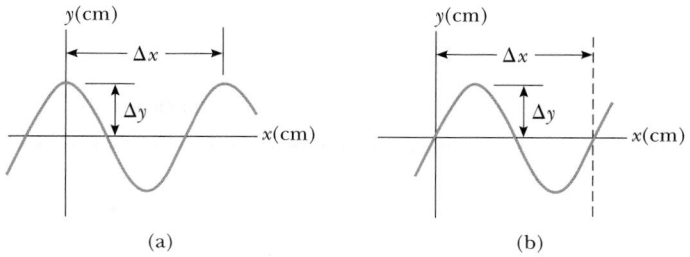

Figure 13.27 (a) (Example 13.8) (b) (Exercise 13.8)

Strategy The amplitude and wavelength can be read directly from the figure: The maximum vertical displacement is the amplitude, and the distance from one crest to the next is the wavelength. Multiplying the wavelength by the frequency gives the speed, while the period is just the reciprocal of the frequency.

Solution

The maximum wave displacement is the amplitude A:

$$A = \Delta y = 15.0 \text{ cm} = 0.150 \text{ m}$$

The distance from crest to crest is the wavelength:

$$\lambda = \Delta x = 40.0 \text{ cm} = 0.400 \text{ m}$$

Multiply the wavelength by the frequency to get the speed of the wave.

$$v = f\lambda = (8.00 \text{ Hz})(0.400 \text{ m}) = 3.20 \text{ m/s}$$

Take the reciprocal of the frequency to get the period:

$$T = \frac{1}{f} = \frac{1}{8.00} \text{ s} = 0.125 \text{ s}$$

Exercise 13.8

A wave traveling in the positive x-direction is pictured in Figure 13.27b. Find the amplitude, wavelength, speed, and period of the wave if it has a frequency of 15.0 Hz. In the figure, $\Delta x = 72.0$ cm and $\Delta y = 25.0$ cm.

Answers $A = 0.25$ m, $\lambda = 0.720$ m, $v = 10.8$ m/s, $T = 0.066\,7$ s

EXAMPLE 13.9 Sound and Light

Goal Perform elementary calculations using speed, wavelength, and frequency.

Problem A wave has a wavelength of 3.00 m. Calculate the frequency of the wave if it is **(a)** a sound wave and **(b)** a light wave. Take the speed of sound as 343 m/s and the speed of light as 3.00×10^8 m/s.

Solution

(a) Find the frequency of a sound wave with $\lambda = 3.00$ m.

Solve Equation 3.17 for the frequency and substitute:

$$f = \frac{v}{\lambda} = \frac{343 \text{ m/s}}{3.00 \text{ m}} = 114 \text{ Hz} \qquad (1)$$

(b) Find the frequency of a light wave with $\lambda = 3.00$ m.

Substitute into Equation (1), using the speed of light for c:

$$f = \frac{c}{\lambda} = \frac{3.00 \times 10^8 \text{ m/s}}{3.00 \text{ m}} = 1.00 \times 10^8 \text{ Hz}$$

Remark The same equation can be used to find the frequency in each case, despite the great difference between the physical phenomena. Notice how much larger frequencies of light waves are than frequencies of sound waves.

Exercise 13.9

(a) Find the wavelength of an electromagnetic wave with frequency 9.00 GHz = 9.00×10^9 Hz (G = giga = 10^9), which is in the microwave range. (b) Find the speed of a sound wave in an unknown fluid medium if a frequency of 567 Hz has a wavelength of 2.50 m.

Answers (a) 0.0333 m (b) 1.42×10^3 m/s

13.9 THE SPEED OF WAVES ON STRINGS

In this section we focus our attention on the speed of a transverse wave on a stretched string.

For a vibrating string, there are two speeds to consider. One is the speed of the physical string that vibrates up and down, transverse to the string, in the y-direction. The other is the *wave speed*, which is the rate at which the disturbance propagates along the length of the string in the x-direction. We wish to find an expression for the wave speed.

If a horizontal string under tension is pulled vertically and released, it starts at its maximum displacement, $y = A$, and takes a certain amount of time to go to $y = -A$ and back to A again. This amount of time is the period of the wave, and is the same as the time needed for the wave to advance *horizontally* by one wavelength. Dividing the wavelength by the period of one transverse oscillation gives the wave speed.

For a fixed wavelength, a string under greater tension F has a greater wave speed because the period of vibration is shorter, and the wave advances one wavelength during one period. It also makes sense that a string with greater mass per unit length, μ, vibrates more slowly, leading to a longer period and a slower wave speed. The wave speed is given by

$$v = \sqrt{\frac{F}{\mu}} \qquad \textbf{[13.18]}$$

where F is the tension in the string and μ is the mass of the string per unit length, called the *linear density*. From Equation 13.18, it's clear that a larger tension F results in a larger wave speed, while a larger linear density μ gives a slower wave speed, as expected.

According to Equation 13.18, the propagation speed of a mechanical wave, such as a wave on a string, depends only on the properties of the string through which the disturbance travels. It doesn't depend on the amplitude of the vibration. This turns out to be generally true of waves in various media.

A proof of Equation 13.18 requires calculus, but dimensional analysis can easily verify that the expression is dimensionally correct. The dimensions of F are ML/T^2, and the dimensions of μ are M/L. The dimensions of F/μ are therefore L^2/T^2, so those of $\sqrt{F/\mu}$ are L/T, giving the dimensions of speed. No other combination of F and μ is dimensionally correct, so in the case where the tension and mass density are the only relevant physical factors, we have verified Equation 13.18 up to an overall constant.

APPLICATION

Bass Guitar Strings

According to Equation 13.18, we can increase the speed of a wave on a stretched string by increasing the tension in the string. Increasing the mass per unit length, on the other hand, decreases the wave speed. These physical facts lie behind the metallic windings on the bass strings of pianos and guitars. The windings increase the mass per unit length, μ, decreasing the wave speed and hence the frequency, resulting in a lower tone. Tuning a string to a desired frequency is a simple matter of changing the tension in the string.

INTERACTIVE EXAMPLE 13.10 A Pulse Traveling on a String

Goal Calculate the speed of a wave on a string.

Problem A uniform string has a mass M of 0.030 0 kg and a length L of 6.00 m. Tension is maintained in the string by suspending a block of mass $m = 2.00$ kg from one end (Fig. 13.28). **(a)** Find the speed of a transverse wave pulse on this string. **(b)** Find the time it takes the pulse to travel from the wall to the pulley.

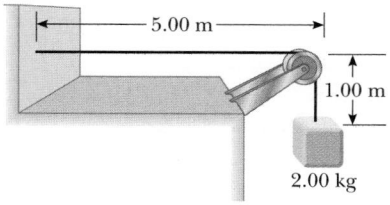

Figure 13.28 (Interactive Example 13.10) The tension F in the string is maintained by the suspended block. The wave speed is given by the expression $v = \sqrt{F/\mu}$.

Strategy The tension F can be obtained from Newton's second law for equilibrium applied to the block, and the mass per unit length of the string is $\mu = M/L$. With these quantities, the speed of the transverse pulse can be found by substitution into Equation 13.18. Part (b) requires the formula $d = vt$.

Solution

(a) Find the speed of the wave pulse.

Apply the second law to the block: the tension F is equal and opposite to the force of gravity.

$$\Sigma F = F - mg = 0 \quad \rightarrow \quad F = mg$$

Substitute expressions for F and μ into Equation 13.18:

$$v = \sqrt{\frac{F}{\mu}} = \sqrt{\frac{mg}{M/L}}$$

$$= \sqrt{\frac{(2.00\text{ kg})(9.80\text{ m/s}^2)}{(0.030\,0\text{ kg})/(6.00\text{ m})}} = \sqrt{\frac{19.6\text{ N}}{0.005\,00\text{ kg/m}}}$$

$$= 62.6\text{ m/s}$$

(b) Find the time it takes the pulse to travel from the wall to the pulley.

Solve the distance formula for time:

$$t = \frac{d}{v} = \frac{5.00\text{ m}}{62.6\text{ m/s}} = 0.0799\text{ s}$$

Exercise 13.10

To what tension must a string with mass 0.010 0 kg and length 2.50 m be tightened so that waves will travel on it at a speed of 125 m/s?

Answer 62.5 N

PhysicsNow™ Investigate the transmission of such pulses by logging into PhysicsNow at **www.cp7e.com** and going to Interactive Example 13.10.

13.10 INTERFERENCE OF WAVES

Many interesting wave phenomena in nature require two or more waves passing through the same region of space at the same time. **Two traveling waves can meet and pass through each other without being destroyed or even altered.** For instance, when two pebbles are thrown into a pond, the expanding circular waves don't destroy each other. In fact, the ripples pass through each other. Likewise, when sound waves from two sources move through air, they pass through each other. In the region of overlap, the resultant wave is found by adding the displacements of the individual waves. For such analyses, the **superposition principle** applies:

> When two or more traveling waves encounter each other while moving through a medium, the resultant wave is found by adding together the displacements of the individual waves point by point.

◀ Superposition principle

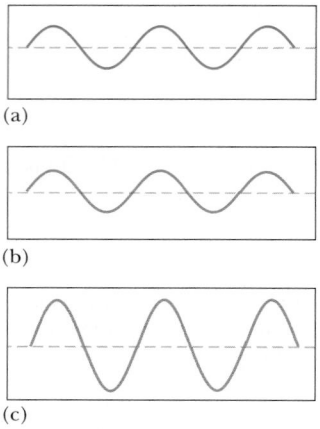

Figure 13.29 Constructive interference. If two waves having the same frequency and amplitude are in phase, as in (a) and (b), the resultant wave when they combine (c) has the same frequency as the individual waves, but twice their amplitude.

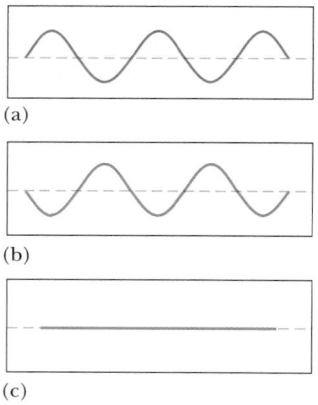

Figure 13.30 Destructive interference. When two waves with the same frequency and amplitude are 180° out of phase, as in (a) and (b), the result when they combine (c) is complete cancellation.

Experiments show that the superposition principle is valid only when the individual waves have small amplitudes of displacement—an assumption we make in all our examples.

Figures 13.29a and 13.29b show two waves of the same amplitude and frequency. If at some instant of time these two waves were traveling through the same region of space, the resultant wave at that instant would have a shape like that shown in Figure 13.29c. For example, suppose the waves are water waves of amplitude 1 m. At the instant they overlap so that crest meets crest and trough meets trough, the resultant wave has an amplitude of 2 m. Waves coming together like this are said to be *in phase* and to exhibit **constructive interference**.

Figures 13.30a and 13.30b show two similar waves. In this case, however, the crest of one coincides with the trough of the other, so one wave is *inverted* relative to the other. The resultant wave, shown in Figure 13.30c, is seen to be a state of complete cancellation. If these were water waves coming together, one of the waves would exert an upward force on an individual drop of water at the same instant the other wave was exerting a downward force. The result would be no motion of the water at all. In such a situation, the two waves are said to be 180° out of phase and to exhibit **destructive interference**. Figure 13.31 illustrates the interference of water waves produced by drops of water falling into a pond.

Active Figure 13.32 shows constructive interference in two pulses moving toward each other along a stretched string; Active Figure 13.33 shows destructive interference in two pulses. Notice in both figures that when the two pulses separate, their shapes are unchanged, as if they had never met!

13.11 REFLECTION OF WAVES

In our discussion so far, we have assumed that waves could travel indefinitely without striking anything. Often, such conditions are not realized in practice. Whenever a traveling wave reaches a boundary, part or all of the wave is reflected. For example, consider a pulse traveling on a string that is fixed at one end (Active Fig. 13.34). When the pulse reaches the wall, it is reflected.

Note that the reflected pulse is inverted. This can be explained as follows: When the pulse meets the wall, the string exerts an upward force on the wall. According to Newton's third law, the wall must exert an equal and opposite (downward) reaction force on the string. This downward force causes the pulse to invert on reflection.

Now suppose the pulse arrives at the string's end, and the end is attached to a ring of negligible mass that is free to slide along the post without friction (Active Fig. 13.35). Again the pulse is reflected, but this time it is not inverted. On reaching

Martin Dohrn/SPL/Photo Researchers, Inc.

Figure 13.31 Interference patterns produced by outward-spreading waves from many drops of liquid falling into a body of water.

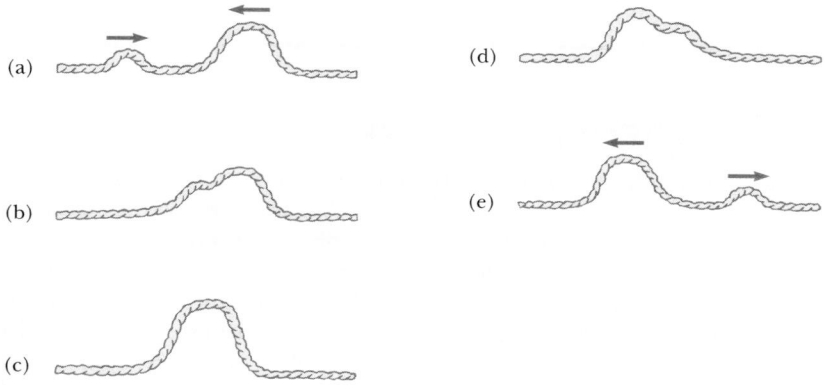

ACTIVE FIGURE 13.32
Two wave pulses traveling on a stretched string in opposite directions pass through each other. When the pulses overlap, as in (b), (c), and (d), the net displacement of the string equals the sum of the displacements produced by each pulse.

PhysicsNow™
Log into PhysicsNow at **www.cp7e.com**, and go to Active Figure 13.32 to choose the amplitude and orientation of each of the pulses and study the interference between them as they pass each other.

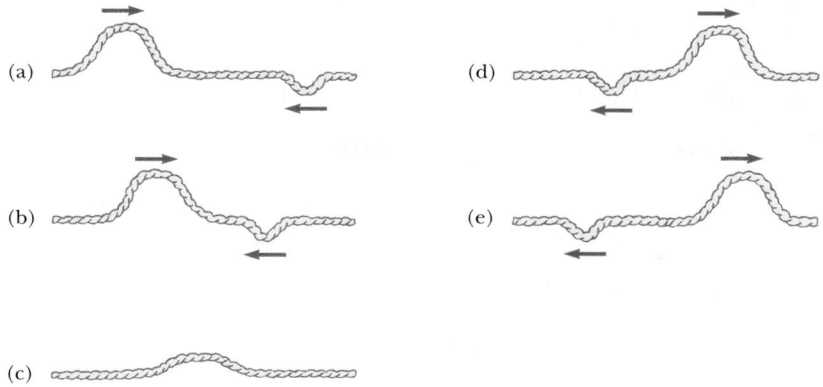

ACTIVE FIGURE 13.33
Two wave pulses traveling in opposite directions with displacements that are inverted relative to each other. When the two overlap, as in (c), their displacements subtract from each other.

Physics Now™
Log into PhysicsNow at **www.cp7e.com**, and go to Active Figure 13.33 to choose the amplitude and orientation of each of the pulses and study the interference between them as they pass each other.

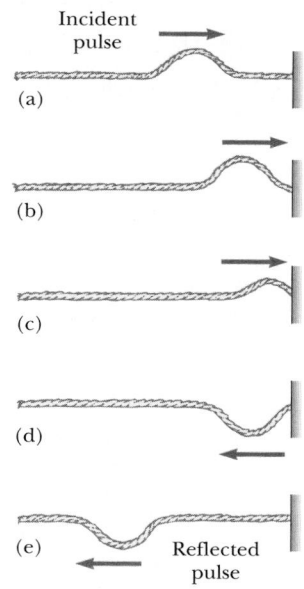

ACTIVE FIGURE 13.34
The reflection of a traveling wave at the fixed end of a stretched string. Note that the reflected pulse is inverted, but its shape remains the same.

Physics Now™
Log into PhysicsNow at **www.cp7e.com**, and go to Active Figure 13.34 to adjust the linear mass density of the string and the transverse direction of the initial pulse.

the post, the pulse exerts a force on the ring, causing it to accelerate upward. The ring is then returned to its original position by the downward component of the tension in the string.

An alternate method of showing that a pulse is reflected without inversion when it strikes a free end of a string is to send the pulse down a string hanging vertically. When the pulse hits the free end, it's reflected without inversion, just as is the pulse in Active Figure 13.35.

Finally, when a pulse reaches a boundary, it's partly reflected and partly transmitted past the boundary into the new medium. This effect is easy to observe in the case of two ropes of different density joined at some boundary.

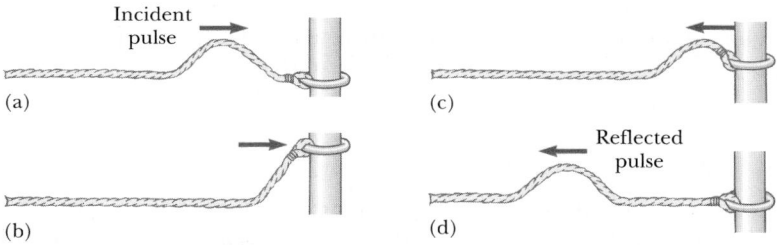

ACTIVE FIGURE 13.35
The reflection of a traveling wave at the free end of a stretched string. In this case, the reflected pulse is not inverted.

Physics Now™
Log into PhysicsNow at **www.cp7e.com**, and go to Active Figure 13.35 to adjust the linear mass density of the string and the transverse direction of the initial pulse.

SUMMARY

Physics Now™ Take a practice test by logging into PhysicsNow at **www.cp7e.com** and clicking on the Pre-Test link for this chapter.

13.1 Hooke's Law

Simple harmonic motion occurs when the net force on an object along the direction of motion is proportional to the object's displacement and in the opposite direction:

$$F_s = -kx \qquad [13.1]$$

This is called Hooke's law. The time required for one complete vibration is called the **period** of the motion. The reciprocal of the period is the **frequency** of the motion, which is the number of oscillations per second.

When an object moves with simple harmonic motion, its **acceleration** as a function of position is

$$a = -\frac{k}{m}x \qquad [13.2]$$

13.2 Elastic Potential Energy

The energy stored in a stretched or compressed spring or in some other elastic material is called **elastic potential energy**:

$$PE_s \equiv \tfrac{1}{2}kx^2 \qquad [13.3]$$

The **velocity** of an object as a function of position, when the object is moving with simple harmonic motion, is

$$v = \pm\sqrt{\frac{k}{m}(A^2 - x^2)} \qquad [13.6]$$

13.4 Position, Velocity, and Acceleration as a Function of Time

The **period** of an object of mass m moving with simple harmonic motion while attached to a spring of spring constant k is

$$T = 2\pi\sqrt{\frac{m}{k}} \qquad [13.8]$$

where T is independent of the amplitude A.

The **frequency** of an object–spring system is $f = 1/T$. The **angular frequency** ω of the system in rad/s is

$$\omega = 2\pi f = \sqrt{\frac{k}{m}} \qquad [13.11]$$

When an object is moving with simple harmonic motion, the **position**, **velocity**, and **acceleration** of the object as a function of time are given by

$$x = A\cos(2\pi ft) \qquad [13.14a]$$

$$v = -A\omega\sin(2\pi ft) \qquad [13.14b]$$

$$a = -A\omega^2\cos(2\pi ft) \qquad [13.14c]$$

13.5 Motion of a Pendulum

A **simple pendulum** of length L moves with simple harmonic motion for small angular displacements from the vertical, with a period of

$$T = 2\pi\sqrt{\frac{L}{g}} \qquad [13.15]$$

13.7 Waves

In a **transverse wave** the elements of the medium move in a direction perpendicular to the direction of the wave. An example is a wave on a stretched string.

In a **longitudinal wave** the elements of the medium move parallel to the direction of the wave velocity. An example is a sound wave.

13.8 Frequency, Amplitude, and Wavelength

The relationship of the speed, wavelength, and frequency of a wave is

$$v = f\lambda \qquad [13.17]$$

This relationship holds for a wide variety of different waves.

13.9 The Speed of Waves on Strings

The speed of a wave traveling on a stretched string of mass per unit length μ and under tension F is

$$v = \sqrt{\frac{F}{\mu}} \qquad [13.18]$$

13.10 Interference of Waves

The **superposition principle** states that if two or more traveling waves are moving through a medium, the resultant wave is found by adding the individual waves together point by point. When waves meet crest to crest and trough to trough, they undergo **constructive interference**. When crest meets trough, the waves undergo **destructive interference**.

13.11 Reflection of Waves

When a wave pulse reflects from a rigid boundary, the pulse is inverted. When the boundary is free, the reflected pulse is not inverted.

CONCEPTUAL QUESTIONS

1. If one end of a heavy rope is attached to one end of a light rope, the speed of a wave will change as the wave goes from the heavy rope to the light one. Will the speed increase or decrease? What happens to the frequency? To the wavelength?
2. If a spring is cut in half, what happens to its spring constant?
3. An object–spring system undergoes simple harmonic motion with an amplitude A. Does the total energy change if the mass is doubled but the amplitude isn't changed? Are the kinetic and potential energies at a given point in its motion affected by the change in mass? Explain.
4. The speed of sound in air as given in this chapter (343 m/s) is an enormous speed compared to the speed of common objects. Yet the speed of sound is of the same order of magnitude as the rms speed of air molecules at 1 atmosphere and 20°C as given by the kinetic theory of gases. Is this just a remarkable coincidence? Explain.
5. An object is hung on a spring, and the frequency of oscillation of the system, f, is measured. The object, a second identical object, and the spring are carried to space in the Space Shuttle. The two objects are attached to the ends of the spring, and the system is taken out into space on a space walk. The spring is extended, and the system is released to oscillate while floating in space. The coils of the spring don't bump into one another. What is the frequency of oscillation for this system, in terms of f?
6. If an object–spring system is hung vertically and set into oscillation, why does the motion eventually stop?
7. Is a bouncing ball an example of simple harmonic motion? Is the daily movement of a student from home to school and back simple harmonic motion?
8. If a pendulum clock keeps perfect time at the base of a mountain, will it also keep perfect time when it is moved to the top of the mountain? Explain.
9. A pendulum bob is made from a sphere filled with water. What would happen to the frequency of vibration of this pendulum if the sphere had a hole in it that allowed the water to leak out slowly?
10. If a grandfather clock were running slow, how could we adjust the length of the pendulum to correct the time?
11. A grandfather clock depends on the period of a pendulum to keep correct time. Suppose such a clock is calibrated

correctly and then the temperature of the room in which it resides increases. Does the clock run slow, fast, or correctly? [*Hint:* A material expands when its temperature increases.]

12. If you stretch a rubber hose and pluck it, you can observe a pulse traveling up and down the hose. What happens to the speed of the pulse if you stretch the hose more tightly? What happens to the speed if you fill the hose with water?

13. In a long line of people waiting to buy tickets at a movie theater, when the first person leaves, a pulse of motion occurs as people step forward to fill in the gap. The gap moves through the line of people. What determines the speed of this pulse? Is it transverse or longitudinal? How about the "wave" at a baseball game, where people in the stands stand up and shout as the wave arrives at their location (Fig. Q13.13) and the pulse moves around the stadium—what determines the speed of this pulse? Is it transverse or longitudinal?

Gregg Adams/Stone/Getty Images

Figure Q13.13

14. As part of a physics open house, a department sets up a bungee jump from the top of the physics building. Assume that one end of the elastic band will be firmly attached to the top of the building and the other to the waist of a courageous participant. The participant will step off the edge of the building, to be slowed down and brought back up by the elastic band before hitting the ground. Estimate the length and spring constant of the elastic you would recommend. (Question 14 is courtesy of Edward F. Redish. For more questions of this type, see http://www.physics.umd.edu/perg/.)

15. In mechanics, massless strings are often assumed. Why is this not a good assumption when discussing waves on strings?

16. What happens to the wavelength of a wave on a string when the frequency is doubled? Assume that the tension in the string remains the same.

17. Explain why the kinetic and potential energies of an object–spring system can never be negative.

18. What happens to the speed of a wave on a string when the frequency is doubled? Assume that the tension in the string remains the same.

19. By what factor would you have to multiply the tension in a stretched spring in order to double the wave speed?

20. The left end of a spring is attached to a wall, and its right end is attached to a cart lying on a frictionless horizontal surface. An experimenter pulls the cart away from the wall and holds it there.
(a) What forces are acting on the spring? What is the total force on the spring?
(b) What forces are acting on the cart? What is the total force on the cart?
(c) If the cart is released, describe its motion.

PROBLEMS

1, 2, 3 = straightforward, intermediate, challenging □ = full solution available in *Student Solutions Manual/Study Guide*
PhysicsNow™ = coached problem with hints available at **www.cp7e.com** = biomedical application

Section 13.1 Hooke's Law

1. A 0.40-kg object is attached to a spring with force constant 160 N/m so that the object is allowed to move on a horizontal frictionless surface. The object is released from rest when the spring is compressed 0.15 m. Find (a) the force on the object and (b) its acceleration at that instant.

2. A load of 50 N attached to a spring hanging vertically stretches the spring 5.0 cm. The spring is now placed horizontally on a table and stretched 11 cm. (a) What force is required to stretch the spring by that amount? (b) Plot a graph of force (on the y-axis) versus spring displacement from the equilibrium position along the x-axis.

3. **PhysicsNow™** A ball dropped from a height of 4.00 m makes a perfectly elastic collision with the ground. Assuming that no mechanical energy is lost due to air resistance, (a) show that the motion is periodic and (b) determine the period of the motion. (c) Is the motion simple harmonic? Explain.

4. A small ball is set in horizontal motion by rolling it with a speed of 3.00 m/s across a room 12.0 m long between two walls. Assume that the collisions made with each wall are perfectly elastic and that the motion is perpendicular to the two walls. (a) Show that the motion is periodic and determine its period. (b) Is the motion simple harmonic? Explain.

5. A spring is hung from a ceiling, and an object attached to its lower end stretches the spring by a distance of 5.00 cm from its unstretched position when the system is in equilibrium. If the spring constant is 47.5 N/m, determine the mass of the object.

6. An archer must exert a force of 375 N on the bowstring shown in Figure P13.6a (page 452) such that the string makes an angle of $\theta = 35.0°$ with the vertical. (a) Determine the tension in the bowstring. (b) If the applied force is replaced by a stretched spring as in Figure P13.6b, and the spring is stretched 30.0 cm from its unstretched length, what is the spring constant?

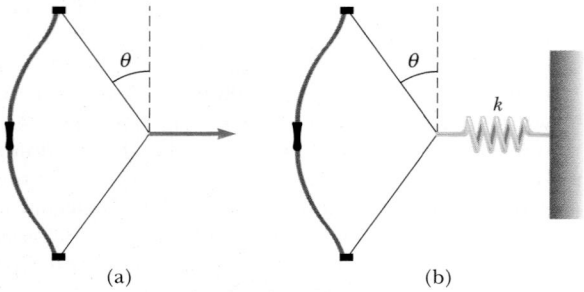

Figure P13.6

Section 13.2 Elastic Potential Energy

7. A slingshot consists of a light leather cup containing a stone. The cup is pulled back against two parallel rubber bands. It takes a force of 15 N to stretch either one of these bands 1.0 cm. (a) What is the potential energy stored in the two bands together when a 50-g stone is placed in the cup and pulled back 0.20 m from the equilibrium position? (b) With what speed does the stone leave the slingshot?

8. An archer pulls her bowstring back 0.400 m by exerting a force that increases uniformly from zero to 230 N. (a) What is the equivalent spring constant of the bow? (b) How much work is done in pulling the bow?

9. A child's toy consists of a piece of plastic attached to a spring (Fig. P13.9). The spring is compressed against the floor a distance of 2.00 cm, and the toy is released. If the toy has a mass of 100 g and rises to a maximum height of 60.0 cm, estimate the force constant of the spring.

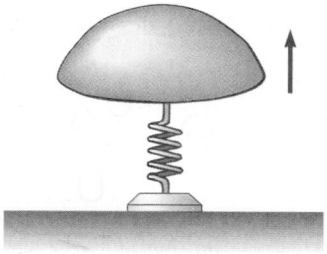

Figure P13.9

10. An automobile having a mass of 1 000 kg is driven into a brick wall in a safety test. The bumper behaves like a spring with constant 5.00×10^6 N/m and is compressed 3.16 cm as the car is brought to rest. What was the speed of the car before impact, assuming that no energy is lost in the collision with the wall?

11. A simple harmonic oscillator has a total energy E. (a) Determine the kinetic and potential energies when the displacement is one-half the amplitude. (b) For what value of the displacement does the kinetic energy equal the potential energy?

12. A 1.50-kg block at rest on a tabletop is attached to a horizontal spring having constant 19.6 N/m, as in Figure P13.12. The spring is initially unstretched. A constant 20.0-N horizontal force is applied to the object, causing the spring to stretch. (a) Determine the speed of the block after it has moved 0.300 m from equilibrium if the surface between the block and tabletop is frictionless. (b) Answer part (a) if the coefficient of kinetic friction between block and tabletop is 0.200.

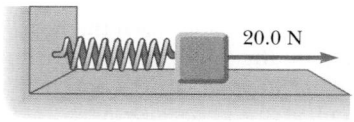

Figure P13.12

13. A 10.0-g bullet is fired into, and embeds itself in, a 2.00-kg block attached to a spring with a force constant of 19.6 N/m and whose mass is negligible. How far is the spring compressed if the bullet has a speed of 300 m/s just before it strikes the block and the block slides on a frictionless surface? [*Note:* You must use conservation of momentum in this problem. Why?]

14. A 1.5-kg block is attached to a spring with a spring constant of 2 000 N/m. The spring is then stretched a distance of 0.30 cm and the block is released from rest. (a) Calculate the speed of the block as it passes through the equilibrium position if no friction is present. (b) Calculate the speed of the block as it passes through the equilibrium position if a constant frictional force of 2.0 N retards its motion. (c) What would be the strength of the frictional force if the block reached the equilibrium position the first time with zero velocity?

Section 13.3 Comparing Simple Harmonic Motion with Uniform Circular Motion

Section 13.4 Position, Velocity, and Acceleration as a Function of Time

15. A 0.40-kg object connected to a light spring with a force constant of 19.6 N/m oscillates on a frictionless horizontal surface. If the spring is compressed 4.0 cm and released from rest, determine (a) the maximum speed of the object, (b) the speed of the object when the spring is compressed 1.5 cm, and (c) the speed of the object when the spring is stretched 1.5 cm. (d) For what value of x does the speed equal one-half the maximum speed?

16. An object–spring system oscillates with an amplitude of 3.5 cm. If the spring constant is 250 N/m and the object has a mass of 0.50 kg, determine (a) the mechanical energy of the system, (b) the maximum speed of the object, and (c) the maximum acceleration of the object.

17. At an outdoor market, a bunch of bananas is set into oscillatory motion with an amplitude of 20.0 cm on a spring with a force constant of 16.0 N/m. It is observed that the maximum speed of the bunch of bananas is 40.0 cm/s. What is the weight of the bananas in newtons?

18. A 50.0-g object is attached to a horizontal spring with a force constant of 10.0 N/m and released from rest with an amplitude of 25.0 cm. What is the velocity of the object when it is halfway to the equilibrium position if the surface is frictionless?

19. While riding behind a car traveling at 3.00 m/s, you notice that one of the car's tires has a small hemispherical bump on its rim, as in Figure P13.19. (a) Explain why the bump, from your viewpoint behind the car, executes simple harmonic motion. (b) If the radius of the car's tires is 0.30 m, what is the bump's period of oscillation?

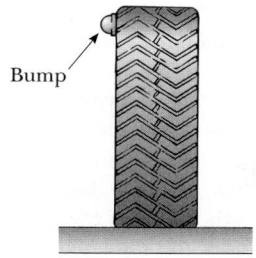

Figure P13.19

20. An object moves uniformly around a circular path of radius 20.0 cm, making one complete revolution every 2.00 s. What are (a) the translational speed of the object, (b) the frequency of motion in hertz, and (c) the angular speed of the object?

21. Consider the simplified single-piston engine in Figure P13.21. If the wheel rotates at a constant angular speed ω, explain why the piston rod oscillates in simple harmonic motion.

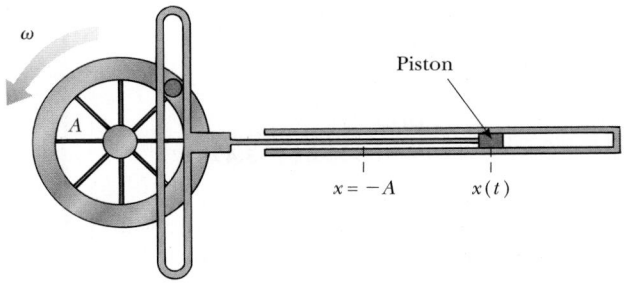

Figure P13.21

22. The frequency of vibration of an object–spring system is 5.00 Hz when a 4.00-g mass is attached to the spring. What is the force constant of the spring?

23. A spring stretches 3.9 cm when a 10-g object is hung from it. The object is replaced with a block of mass 25 g that oscillates in simple harmonic motion. Calculate the period of motion.

24. When four people with a combined mass of 320 kg sit down in a car, they find that the car drops 0.80 cm lower on its springs. Then they get out of the car and bounce it up and down. What is the frequency of the car's vibration if its mass (when it is empty) is 2.0×10^3 kg?

25. A cart of mass 250 g is placed on a frictionless horizontal air track. A spring having a spring constant of 9.5 N/m is attached between the cart and the left end of the track. When in equilibrium, the cart is located 12 cm from the left end of the track. If the cart is displaced 4.5 cm from its equilibrium position, find (a) the period at which it oscillates, (b) its maximum speed, and (c) its speed when it is 14 cm from the left end of the track.

26. The motion of an object is described by the equation

$$x = (0.30 \text{ m}) \cos\left(\frac{\pi t}{3}\right)$$

Find (a) the position of the object at $t = 0$ and $t = 0.60$ s, (b) the amplitude of the motion, (c) the frequency of the motion, and (d) the period of the motion.

27. A 2.00-kg object on a frictionless horizontal track is attached to the end of a horizontal spring whose force constant is 5.00 N/m. The object is displaced 3.00 m to the right from its equilibrium position and then released, initiating simple harmonic motion. (a) What is the force (magnitude and direction) acting on the object 3.50 s after it is released? (b) How many times does the object oscillate in 3.50 s?

28. A spring of negligible mass stretches 3.00 cm from its relaxed length when a force of 7.50 N is applied. A 0.500-kg particle rests on a frictionless horizontal surface and is attached to the free end of the spring. The particle is pulled horizontally so that it stretches the spring 5.00 cm and is then released from rest at $t = 0$. (a) What is the force constant of the spring? (b) What are the angular frequency ω, the frequency, and the period of the motion? (c) What is the total energy of the system? (d) What is the amplitude of the motion? (e) What are the maximum velocity and the maximum acceleration of the particle? (f) Determine the displacement x of the particle from the equilibrium position at $t = 0.500$ s.

29. **Physics Now™** Given that $x = A \cos(\omega t)$ is a sinusoidal function of time, show that v (velocity) and a (acceleration) are also sinusoidal functions of time. [*Hint:* Use Equations 13.6 and 13.2.]

Section 13.5 Motion of a Pendulum

30. A man enters a tall tower, needing to know its height. He notes that a long pendulum extends from the ceiling almost to the floor and that its period is 15.5 s. (a) How tall is the tower? (b) If this pendulum is taken to the Moon, where the free-fall acceleration is 1.67 m/s^2, what is the period there?

31. A simple 2.00-m-long pendulum oscillates at a location where $g = 9.80$ m/s^2. How many complete oscillations does it make in 5.00 min?

32. An aluminum clock pendulum having a period of 1.00 s keeps perfect time at 20.0°C. (a) When placed in a room at a temperature of −5.0°C, will it gain time or lose time? (b) How much time will it gain or lose every hour? [*Hint:* See Chapter 10.]

33. A pendulum clock that works perfectly on Earth is taken to the Moon. (a) Does it run fast or slow there? (b) If the clock is started at 12:00 midnight, what will it read after one Earth day (24.0 h)? Assume that the free-fall acceleration on the Moon is 1.63 m/s^2.

34. A simple pendulum is 5.00 m long. (a) What is the period of simple harmonic motion for this pendulum if it is located in an elevator accelerating upward at 5.00 m/s^2? (b) What is its period if the elevator is accelerating downward at 5.00 m/s^2? (c) What is the period of simple harmonic motion for the pendulum if it is placed in a truck that is accelerating horizontally at 5.00 m/s^2?

35. The free-fall acceleration on Mars is 3.7 m/s^2. (a) What length of pendulum has a period of 1 s on Earth? What length of pendulum would have a 1-s period on Mars? (b) An object is suspended from a spring with force constant 10 N/m. Find the mass suspended from this spring that would result in a period of 1 s on Earth and on Mars.

Section 13.6 Damped Oscillations

Section 13.7 Waves

Section 13.8 Frequency, Amplitude, and Wavelength

36. A cork on the surface of a pond bobs up and down two times per second on ripples having a wavelength of

8.50 cm. If the cork is 10.0 m from shore, how long does it take a ripple passing the cork to reach the shore?

37. A wave traveling in the positive x-direction has a frequency of 25.0 Hz, as in Figure P13.37. Find the (a) amplitude, (b) wavelength, (c) period, and (d) speed of the wave.

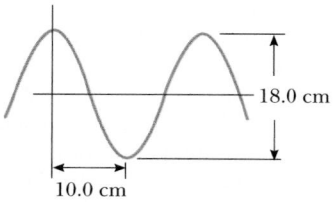

Figure P13.37

38. A bat can detect small objects, such as an insect, whose size is approximately equal to one wavelength of the sound the bat makes. If bats emit a chirp at a frequency of 60.0 kHz, and if the speed of sound in air is 340 m/s, what is the smallest insect a bat can detect?

39. If the frequency of oscillation of the wave emitted by an FM radio station is 88.0 MHz, determine (a) the wave's period of vibration and (b) its wavelength. (Radio waves travel at the speed of light, 3.00×10^8 m/s.)

40. The distance between two successive maxima of a transverse wave is 1.20 m. Eight crests, or maxima, pass a given point along the direction of travel every 12.0 s. Calculate the wave speed.

41. A harmonic wave is traveling along a rope. It is observed that the oscillator that generates the wave completes 40.0 vibrations in 30.0 s. Also, a given maximum travels 425 cm along the rope in 10.0 s. What is the wavelength?

42. Ocean waves are traveling to the east at 4.0 m/s with a distance of 20 m between crests. With what frequency do the waves hit the front of a boat (a) when the boat is at anchor and (b) when the boat is moving westward at 1.0 m/s?

Section 13.9 The Speed of Waves on Strings

43. A phone cord is 4.00 m long and has a mass of 0.200 kg. A transverse wave pulse is produced by plucking one end of the taut cord. The pulse makes four trips down and back along the cord in 0.800 s. What is the tension in the cord?

44. A circus performer stretches a tightrope between two towers. He strikes one end of the rope and sends a wave along it toward the other tower. He notes that it takes the wave 0.800 s to reach the opposite tower, 20.0 m away. If a 1-m length of the rope has a mass of 0.350 kg, find the tension in the tightrope.

45. Transverse waves with a speed of 50.0 m/s are to be produced on a stretched string. A 5.00-m length of string with a total mass of 0.060 0 kg is used. (a) What is the required tension in the string? (b) Calculate the wave speed in the string if the tension is 8.00 N.

46. An astronaut on the Moon wishes to measure the local value of g by timing pulses traveling down a wire that has a large object suspended from it. Assume a wire of mass 4.00 g is 1.60 m long and has a 3.00-kg object suspended from it. A pulse requires 36.1 ms to traverse the length of the wire. Calculate g_{Moon} from these data. (You may neglect the mass of the wire when calculating the tension in it.)

47. Physics Now™ A simple pendulum consists of a ball of mass 5.00 kg hanging from a uniform string of mass 0.0600 kg and length L. If the period of oscillation of the pendulum is 2.00 s, determine the speed of a transverse wave in the string when the pendulum hangs vertically.

48. A string is 50.0 cm long and has a mass of 3.00 g. A wave travels at 5.00 m/s along this string. A second string has the same length, but half the mass of the first. If the two strings are under the same tension, what is the speed of a wave along the second string?

49. Tension is maintained in a string as in Figure P13.49. The observed wave speed is 24 m/s when the suspended mass is 3.0 kg. (a) What is the mass per unit length of the string? (b) What is the wave speed when the suspended mass is 2.0 kg?

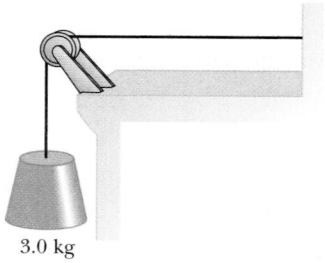

Figure P13.49

50. The elastic limit of a piece of steel wire is 2.70×10^9 Pa. What is the maximum speed at which transverse wave pulses can propagate along the wire without exceeding its elastic limit? (The density of steel is 7.86×10^3 kg/m^3.)

51. Transverse waves travel at 20.0 m/s on a string that is under a tension of 6.00 N. What tension is required for a wave speed of 30.0 m/s in the string?

Section 13.10 Interference of Waves
Section 13.11 Reflection of Waves

52. A series of pulses of amplitude 0.15 m is sent down a string that is attached to a post at one end. The pulses are reflected at the post and travel back along the string without loss of amplitude. What is the amplitude at a point on the string where two pulses are crossing (a) if the string is rigidly attached to the post? (b) if the end at which reflection occurs is free to slide up and down?

53. A wave of amplitude 0.30 m interferes with a second wave of amplitude 0.20 m traveling in the same direction. What are (a) the largest and (b) the smallest resultant amplitudes that can occur, and under what conditions will these maxima and minima arise?

ADDITIONAL PROBLEMS

54. The position of a 0.30-kg object attached to a spring is described by

$$x = (0.25 \text{ m}) \cos(0.4\pi t)$$

Find (a) the amplitude of the motion, (b) the spring constant, (c) the position of the object at $t = 0.30$ s, and (d) the object's speed at $t = 0.30$ s.

55. A large block P executes horizontal simple harmonic motion as it slides across a frictionless surface with a frequency $f = 1.50$ Hz. Block B rests on it, as shown in Figure P13.55, and the coefficient of static friction between the two is $\mu_s = 0.600$. What maximum amplitude of oscillation can the system have if block B is not to slip?

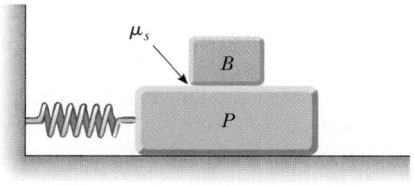

Figure P13.55

56. A 500-g block is released from rest and slides down a frictionless track that begins 2.00 m above the horizontal, as shown in Figure P13.56. At the bottom of the track, where the surface is horizontal, the block strikes and sticks to a light spring with a spring constant of 20.0 N/m. Find the maximum distance the spring is compressed.

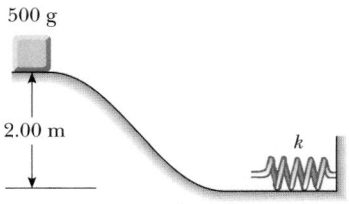

Figure P13.56

57. A 3.00-kg object is fastened to a light spring, with the intervening cord passing over a pulley (Fig. P13.57). The pulley is frictionless, and its inertia may be neglected. The object is released from rest when the spring is unstretched. If the object drops 10.0 cm before stopping, find (a) the spring constant of the spring and (b) the speed of the object when it is 5.00 cm below its starting point.

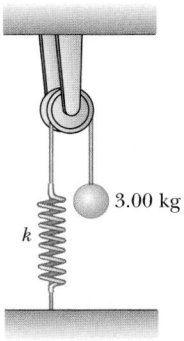

Figure P13.57

58. A 5.00-g bullet moving with an initial speed of 400 m/s is fired into and passes through a 1.00-kg block, as in Figure P13.58. The block, initially at rest on a frictionless horizontal surface, is connected to a spring with a spring constant of 900 N/m. If the block moves 5.00 cm to the right after impact, find (a) the speed at which the bullet emerges from the block and (b) the mechanical energy lost in the collision.

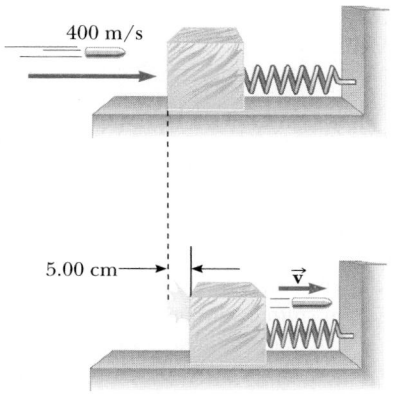

Figure P13.58

59. A 25-kg block is connected to a 30-kg block by a light string that passes over a frictionless pulley. The 30-kg block is connected to a light spring of force constant 200 N/m, as in Figure P13.59. The spring is unstretched when the system is as shown in the figure, and the incline is smooth. The 25-kg block is pulled 20 cm down the incline (so that the 30-kg block is 40 cm above the floor) and is released from rest. Find the speed of each block when the 30-kg block is 20 cm above the floor (that is, when the spring is unstretched).

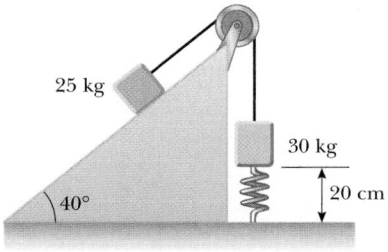

Figure P13.59

60. A spring in a toy gun has a spring constant of 9.80 N/m and can be compressed 20.0 cm beyond the equilibrium position. A 1.00-g pellet resting against the spring is propelled forward when the spring is released. (a) Find the muzzle speed of the pellet. (b) If the pellet is fired horizontally from a height of 1.00 m above the floor, what is its range?

61. A 2.00-kg block hangs without vibrating at the end of a spring ($k = 500$ N/m) that is attached to the ceiling of an elevator car. The car is rising with an upward acceleration of $g/3$ when the acceleration suddenly ceases (at $t = 0$). (a) What is the angular frequency of oscillation of the block after the acceleration ceases? (b) By what amount is the spring stretched during the time that the elevator car is accelerating? This distance will be the amplitude of the ensuing oscillation of the block.

62. An object of mass m is connected to two rubber bands of length L, each under tension F, as in Figure P13.62. The object is displaced vertically by a small distance y. Assuming the tension does not change, show that (a) the restoring force is $-(2F/L)y$ and (b) the system exhibits simple harmonic motion with an angular frequency $\omega = \sqrt{2F/mL}$.

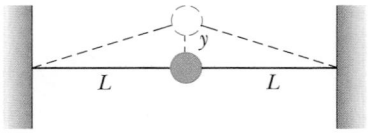

Figure P13.62

63. A light balloon filled with helium of density 0.180 kg/m^3 is tied to a light string of length $L = 3.00$ m. The string is tied to the ground, forming an "inverted" simple pendulum (Fig. P13.63a). If the balloon is displaced slightly from equilibrium, as in Figure P13.63b, show that the motion is simple harmonic, and determine the period of the motion. Take the density of air to be 1.29 kg/m^3. [*Hint:* Use an analogy with the simple pendulum discussed in the text, and see Chapter 9.]

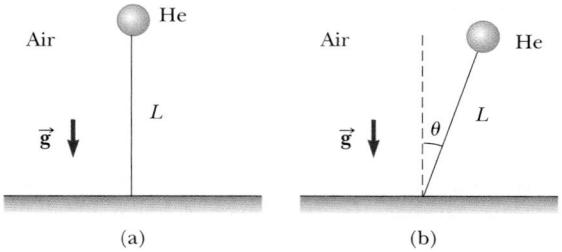

Figure P13.63

64. A light string of mass 10.0 g and length $L = 3.00$ m has its ends tied to two walls that are separated by the distance $D = 2.00$ m. Two objects, each of mass $M = 2.00$ kg, are suspended from the string as in Figure P13.64. If a wave pulse is sent from point A, how long does it take to travel to point B?

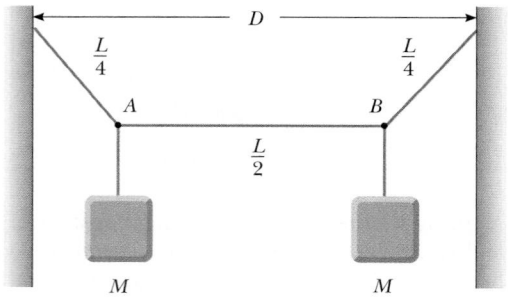

Figure P13.64

65. Assume that a hole is drilled through the center of the Earth. It can be shown that an object of mass m at a distance r from the center of the Earth is pulled toward the center only by the material in the shaded portion of Figure P13.65.

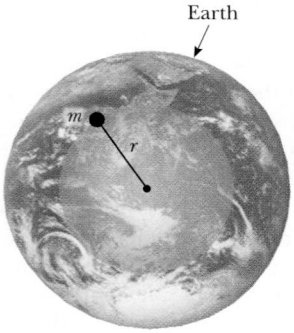

Figure P13.65

Assume Earth has a uniform density ρ. Write down Newton's law of gravitation for an object at a distance r from the center of the Earth, and show that the force on it is of the form of Hooke's law, $F = -kr$, with an effective force constant of $k = (\frac{4}{3})\pi\rho Gm$, where G is the gravitational constant.

66. A 60.0-kg firefighter slides down a pole while a constant frictional force of 300 N retards his motion. A horizontal 20.0-kg platform is supported by a spring at the bottom of the pole to cushion the fall. The firefighter starts from rest 5.00 m above the platform, and the spring constant is 2500 N/m. Find (a) the firefighter's speed just before he collides with the platform and (b) the maximum distance the spring is compressed. Assume that the frictional force acts during the entire motion. [*Hint:* The collision between the firefighter and the platform is perfectly inelastic.]

67. PhysicsNow™ An object of mass $m_1 = 9.0$ kg is in equilibrium while connected to a light spring of constant $k = 100$ N/m that is fastened to a wall, as in Figure P13.67a. A second object, of mass $m_2 = 7.0$ kg, is slowly pushed up against m_1, compressing the spring by the amount $A = 0.20$ m, as shown in Figure P13.67b. The system is then released, causing both objects to start moving to the right on the frictionless surface. (a) When m_1 reaches the equilibrium point, m_2 loses contact with m_1 (Fig. P13.67c) and moves to the right with velocity $\vec{\mathbf{v}}$. Determine the magnitude of $\vec{\mathbf{v}}$. (b) How far apart are the objects when the spring is fully stretched for the first time (Fig. P13.67d)? [*Hint:* First determine the period of oscillation and the amplitude of the m_1–spring system after m_2 loses contact with m_1.]

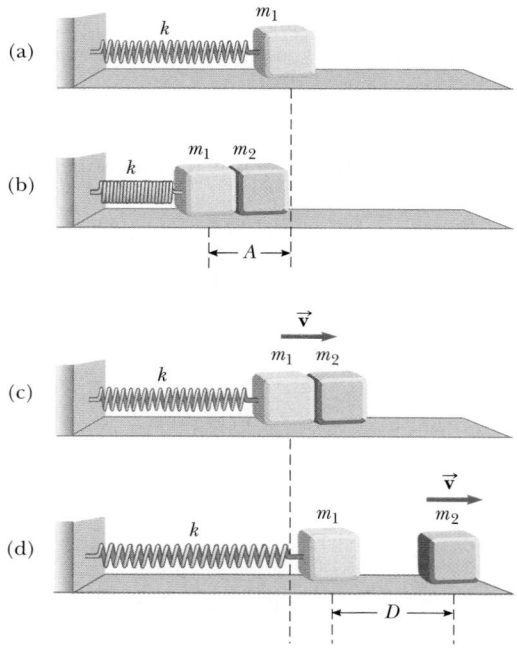

Figure P13.67

68. An 8.00-kg block travels on a rough horizontal surface and collides with a spring. The speed of the block just before the collision is 4.00 m/s. As it rebounds to the left with the spring uncompressed, the block travels at 3.00 m/s. If the coefficient of kinetic friction between the block and the surface is 0.400, determine (a) the loss in

mechanical energy due to friction while the block is in contact with the spring and (b) the maximum distance the spring is compressed.

69. Two points, A and B, on Earth are at the same longitude and 60.0° apart in latitude. An earthquake at point A sends two waves toward B. A transverse wave travels along the surface of Earth at 4.50 km/s, and a longitudinal wave travels through Earth at 7.80 km/s. (a) Which wave arrives at B first? (b) What is the time difference between the arrivals of the two waves at B? Take the radius of Earth to be 6.37×10^6 m.

70. Figure P13.70 shows a crude model of an insect wing. The mass m represents the entire mass of the wing, which pivots about the fulcrum F. The spring represents the surrounding connective tissue. Motion of the wing corresponds to vibration of the spring. Suppose the mass of the wing is 0.30 g and the effective spring constant of the tissue is 4.7×10^{-4} N/m. If the mass m moves up and down a distance of 2.0 mm from its position of equilibrium, what is the maximum speed of the outer tip of the wing?

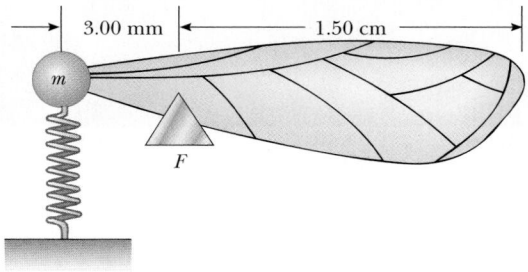

Figure P13.70

71. A 1.6-kg block on a horizontal surface is attached to a spring with a force constant of 1.0×10^3 N/m, as in Active Figure 13.1. The spring is compressed a distance of 2.0 cm, and the block is released from rest. (a) Calculate the speed of the block as it passes through the equilibrium position, $x = 0$, if the surface is frictionless. (b) Calculate the speed of the block as it passes through the equilibrium position if a constant frictional force of 4.0 N retards its motion. (c) How far does the block travel before coming to rest in part (b)?

ACTIVITIES

A.1. Construct a simple pendulum by tying a metal bolt to one end of a string and taping the other end to the top of a doorframe. Adjust the length of the pendulum to be about 1.0 m, and measure it precisely. Obtain the period by timing 25 complete oscillations, making sure the string always makes small angles with the vertical. Repeat the measurements for precisely measured pendulum lengths ranging from 0.4 m to 1.6 m in increments of 0.2 m. Plot the square of the period versus the length of the pendulum and measure the slope of the line best fitting your data points. Does your slope agree with that predicted by $T^2 = (4\pi^2/g)L$? What value of g do you obtain from your data?

While you have your pendulum in position, use a procedure similar to the preceding to verify that the period is also independent of the amplitude for small angles and that the period is also independent of the mass.

A.2. Attach one end of a rope (or a spring such as a Slinky™) to a wall, stretch it taut, and use the system to study the following aspects of wave motion:

(a) Send a pulse down the rope by striking it sharply from the side. An observer watching from the side can measure the time elapsed during 3–5 trips of the pulse from one end to the other. Dividing the total distance traveled by the elapsed time yields the wave speed. To get reliable results, take a number of readings and average them.

(b) Use the same setup as in part (a) to test whether the initial amplitude of the pulse changes the wave speed.

(c) Have two people hold opposite ends of the rope (or spring). Then, at the same instant, have both people hit the rope sharply from the side. Observe what happens when the two pulses meet. The superposition lasts only for the short time that the pulses overlap, and you must look carefully to see the effect.

(d) Using the same setup, devise a way to check whether pulses traveling toward one another pass through when they collide or reflect off each other.

(e) Tie one end of a rope to a doorknob and send a pulse down it. Observe what happens when the pulse reflects from the door. Does it return on the same side of the rope, or does it invert?

The characteristic sound of any instrument is referred to as the quality of that sound. What is it about the sound from the tuba that allows us to distinguish between it and the sound from a flute?

Patrick Ward/CORBIS

CHAPTER

14

OUTLINE

14.1 Producing a Sound Wave
14.2 Characteristics of Sound Waves
14.3 The Speed of Sound
14.4 Energy and Intensity of Sound Waves
14.5 Spherical and Plane Waves
14.6 The Doppler Effect
14.7 Interference of Sound Waves
14.8 Standing Waves
14.9 Forced Vibrations and Resonance
14.10 Standing Waves in Air Columns
14.11 Beats
14.12 Quality of Sound
14.13 The Ear

Sound

Sound waves are the most important example of longitudinal waves. In this chapter we discuss the characteristics of sound waves: how they are produced, what they are, and how they travel through matter. We then investigate what happens when sound waves interfere with each other. The insights gained in this chapter will help you understand how we hear.

14.1 PRODUCING A SOUND WAVE

Whether it conveys the shrill whine of a jet engine or the soft melodies of a crooner, any sound wave has its source in a vibrating object. Musical instruments produce sounds in a variety of ways. The sound of a clarinet is produced by a vibrating reed, the sound of a drum by the vibration of the taut drumhead, the sound of a piano by vibrating strings, and the sound from a singer by vibrating vocal cords.

Sound waves are longitudinal waves traveling through a medium, such as air. In order to investigate how sound waves are produced, we focus our attention on the tuning fork, a common device for producing pure musical notes. A tuning fork consists of two metal prongs, or tines, that vibrate when struck. Their vibration disturbs the air near them, as shown in Figure 14.1. (The amplitude of vibration of the tine shown in the figure has been greatly exaggerated for clarity.) When a tine swings to the right, as in Figure 14.1a, the molecules in an element of air in front of its movement are forced closer together than normal. Such a region of high molecular density and high air pressure is called a **compression**. This compression moves away from the fork like a ripple on a pond. When the tine swings to the left, as in Figure 14.1b, the molecules in an element of air to the right of the tine spread apart, and the density and air pressure in this region are then lower than normal. Such a region of reduced density is called a **rarefaction** (pronounced "rare a fak' shun"). Molecules to the right of the rarefaction in the figure move to the left. The rarefaction itself therefore moves to the right, following the previously produced compression.

As the tuning fork continues to vibrate, a succession of compressions and rarefactions forms and spreads out from it. The resultant pattern in the air is somewhat like that pictured in Figure 14.2a. We can use a sinusoidal curve to represent a sound wave, as in Figure 14.2b. Notice that there are crests in the sinusoidal wave at the points where the sound wave has compressions and troughs where the sound wave has rarefactions. The compressions and rarefactions of the sound waves are superposed on the random thermal motion of the atoms and molecules of the air (discussed in Chapter 10), so sound waves in gases travel at about the molecular rms speed.

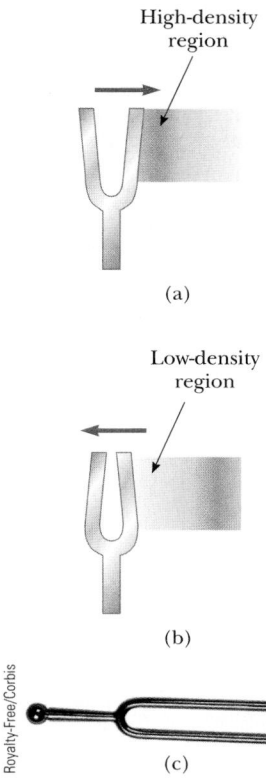

Royalty-Free/Corbis

Figure 14.1 A vibrating tuning fork. (a) As the right tine of the fork moves to the right, a high-density region (compression) of air is formed in front of its movement. (b) As the right tine moves to the left, a low-density region (rarefaction) of air is formed behind it. (c) A tuning fork.

14.2 CHARACTERISTICS OF SOUND WAVES

As already noted, the general motion of elements of air near a vibrating object is back and forth between regions of compression and rarefaction. This back-and-forth motion of elements of the medium in the direction of the disturbance is characteristic of a longitudinal wave. **The motion of the elements of the medium in a longitudinal sound wave is back and forth along the direction in which the wave travels**. By contrast, **in a transverse wave, the vibrations of the elements of the medium are at right angles to the direction of travel of the wave**.

Categories of Sound Waves

Sound waves fall into three categories covering different ranges of frequencies. **Audible waves** are longitudinal waves that lie within the range of sensitivity of the human ear, approximately 20 to 20 000 Hz. **Infrasonic waves** are longitudinal waves with frequencies below the audible range. Earthquake waves are an example. **Ultrasonic waves** are longitudinal waves with frequencies above the audible range for humans and are produced by certain types of whistles. Animals such as dogs can hear the waves emitted by these whistles.

Applications of Ultrasound

APPLICATION

Medical Uses of Ultrasound

Ultrasonic waves are sound waves with frequencies greater than 20 kHz. Because of their high frequency and corresponding short wavelengths, ultrasonic waves can be used to produce images of small objects and are currently in wide use in medical applications, both as a diagnostic tool and in certain treatments. Internal organs can be examined via the images produced by the reflection and absorption of ultrasonic waves. Although ultrasonic waves are far safer than x-rays, their images don't always have as much detail. Certain organs, however, such as the liver and the spleen, are invisible to x-rays but can be imaged with ultrasonic waves.

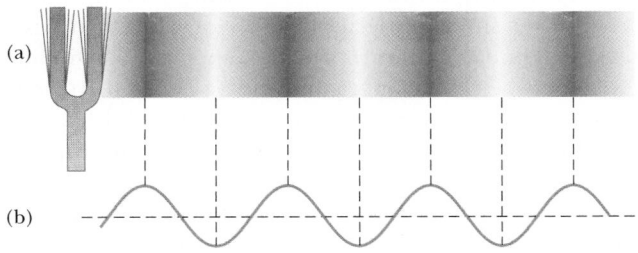

Figure 14.2 (a) As the tuning fork vibrates, a series of compressions and rarefactions moves outward, away from the fork. (b) The crests of the wave correspond to compressions, the troughs to rarefactions.

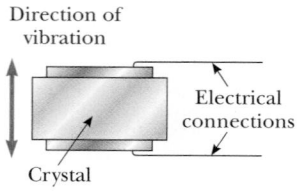

Figure 14.3 An alternating voltage applied to the faces of a piezoelectric crystal causes the crystal to vibrate.

Medical workers can measure the speed of the blood flow in the body with a device called an ultrasonic flow meter, which makes use of the Doppler effect (discussed in Section 14.6). The flow speed is found by comparing the frequency of the waves scattered by the flowing blood with the incident frequency.

Figure 14.3 illustrates the technique that produces ultrasonic waves for clinical use. Electrical contacts are made to the opposite faces of a crystal, such as quartz or strontium titanate. If an alternating voltage of high frequency is applied to these contacts, the crystal vibrates at the same frequency as the applied voltage, emitting a beam of ultrasonic waves. At one time, a technique like this was used to produce sound in nearly all headphones. This method of transforming electrical energy into mechanical energy, called the **piezoelectric effect**, is reversible: If some external source causes the crystal to vibrate, an alternating voltage is produced across it. A single crystal can therefore be used to both generate and receive ultrasonic waves.

The primary physical principle that makes ultrasound imaging possible is the fact that a sound wave is partially reflected whenever it is incident on a boundary between two materials having different densities. If a sound wave is traveling in a material of density ρ_i and strikes a material of density ρ_t, the percentage of the incident sound wave intensity reflected, PR, is given by

$$PR = \left(\frac{\rho_i - \rho_t}{\rho_i + \rho_t}\right)^2 \times 100$$

This equation assumes that the direction of the incident sound wave is perpendicular to the boundary and that the speed of sound is approximately the same in the two materials. The latter assumption holds very well for the human body because the speed of sound doesn't vary much in the organs of the body.

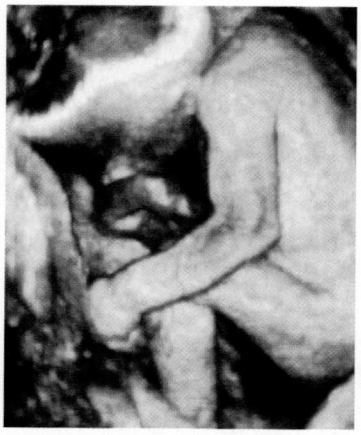

Bernard Benoit/Photo Researchers, Inc.

An ultrasound image of a human fetus in the womb.

Physicians commonly use ultrasonic waves to observe fetuses. This technique presents far less risk than do x-rays, which deposit more energy in cells and can produce birth defects. First the abdomen of the mother is coated with a liquid, such as mineral oil. If this were not done, most of the incident ultrasonic waves from the piezoelectric source would be reflected at the boundary between the air and the mother's skin. Mineral oil has a density similar to that of skin, and a very small fraction of the incident ultrasonic wave is reflected when $\rho_i \approx \rho_t$. The ultrasound energy is emitted in pulses rather than as a continuous wave, so the same crystal can be used as a detector as well as a transmitter. An image of the fetus is obtained by using an array of transducers placed on the abdomen. The reflected sound waves picked up by the transducers are converted to an electric signal, which is used to form an image on a fluorescent screen. Difficulties such as the likelihood of spontaneous abortion or of breech birth are easily detected with this technique. Fetal abnormalities such as spina bifida and water on the brain are also readily observed.

A relatively new medical application of ultrasonics is the *cavitron ultrasonic surgical aspirator* (CUSA). This device has made it possible to surgically remove brain tumors that were previously inoperable. The probe of the CUSA emits ultrasonic waves (at about 23 kHz) at its tip. When the tip touches a tumor, the part of the tumor near the probe is shattered and the residue can be sucked up (aspirated) through the hollow probe. Using this technique, neurosurgeons are able to remove brain tumors without causing serious damage to healthy surrounding tissue.

Ultrasound is also used to break up kidney stones that are otherwise too large to pass. Previously, invasive surgery was more often required.

APPLICATION

Ultrasonic Ranging Unit for Cameras

Another interesting application of ultrasound is the ultrasonic ranging unit used in some cameras to provide an almost instantaneous measurement of the distance between the camera and the object to be photographed. The principal component of this device is a crystal that acts as both a loudspeaker and a microphone. A pulse of ultrasonic waves is transmitted from the transducer to the object, which then reflects part of the signal, producing an echo that is detected by the device. The time interval between the outgoing pulse and the detected echo is electronically converted to a distance, because the speed of sound is a known quantity.

14.3 THE SPEED OF SOUND

The speed of a sound wave in a fluid depends on the fluid's compressibility and inertia. If the fluid has a bulk modulus B and an equilibrium density ρ, the speed of sound in it is

$$v = \sqrt{\frac{B}{\rho}} \qquad [14.1]$$

◀ Speed of sound in a fluid

Equation 14.1 also holds true for a gas. Recall from Chapter 9 that the bulk modulus is defined as the ratio of the change in pressure, ΔP, to the resulting fractional change in volume, $\Delta V/V$:

$$B \equiv -\frac{\Delta P}{\Delta V/V} \qquad [14.2]$$

B is always positive because an increase in pressure (positive ΔP) results in a decrease in volume. Hence, the ratio $\Delta P/\Delta V$ is always negative.

It's interesting to compare Equation 14.1 with Equation 13.18 for the speed of transverse waves on a string, $v = \sqrt{F/\mu}$, discussed in Chapter 13. In both cases, the wave speed depends on an elastic property of the medium (B or F) and on an inertial property of the medium (ρ or μ). In fact, the speed of all mechanical waves follows an expression of the general form

$$v = \sqrt{\frac{\text{elastic property}}{\text{inertial property}}}$$

Another example of this general form is the **speed of a longitudinal wave in a solid rod**, which is

$$v = \sqrt{\frac{Y}{\rho}} \qquad [14.3]$$

where Y is the Young's modulus of the solid (see Eqn. 9.3), and ρ is its density. This expression is valid only for a thin, solid rod.

Table 14.1 lists the speeds of sound in various media. The speed of sound is much higher in solids than in gases, because the molecules in a solid interact more strongly with each other than do molecules in a gas. Striking a long steel rail with a hammer, for example, produces two sound waves, one moving through the rail and a slower wave moving through the air. A student with an ear pressed against the rail first hears the faster sound moving through the rail, then the sound moving through air. In general, sound travels faster through solids than liquids and faster through liquids than gases, although there are exceptions.

The speed of sound also depends on the temperature of the medium. For sound traveling through air, the relationship between the speed of sound and temperature is

$$v = (331 \text{ m/s}) \sqrt{\frac{T}{273 \text{ K}}} \qquad [14.4]$$

where 331 m/s is the speed of sound in air at 0°C and T is the absolute (Kelvin) temperature. Using this equation the speed of sound in air at 293 K (a typical room temperature) is approximately 343 m/s.

TABLE 14.1

Speeds of Sound in Various Media

Medium	v (m/s)
Gases	
Air (0°C)	331
Air (100°C)	386
Hydrogen (0°C)	1 290
Oxygen (0°C)	317
Helium (0°C)	972
Liquids at 25°C	
Water	1 490
Methyl alcohol	1 140
Sea water	1 530
Solids	
Aluminum	5 100
Copper	3 560
Iron	5 130
Lead	1 320
Vulcanized rubber	54

Quick Quiz 14.1

Which of the following actions will increase the speed of sound in air? (a) decreasing the air temperature (b) increasing the frequency of the sound (c) increasing the air temperature (d) increasing the amplitude of the sound wave (e) reducing the pressure of the air.

Applying Physics 14.1 The Sounds Heard During a Storm

How does lightning produce thunder, and what causes the extended rumble?

Explanation Assume that you're at ground level, and neglect ground reflections. When lightning strikes, a channel of ionized air carries a large electric current from a cloud to the ground. This results in a rapid temperature increase of the air in the channel as the current moves through it, causing a similarly rapid expansion of the air. The expansion is so sudden and so intense that a tremendous disturbance is produced in the air—thunder. The entire length of the channel produces the sound at essentially the same instant of time. Sound produced at the bottom of the channel reaches you first, because that's the point closest to you. Sounds from progressively higher portions of the channel reach you at later times, resulting in an extended roar. If the lightning channel were a perfectly straight line, the roar might be steady, but the zigzag shape of the path results in the rumbling variation in loudness, with different quantities of sound energy from different segments arriving at any given instant.

INTERACTIVE EXAMPLE 14.1 Sound Waves in Various Media

Goal Calculate and compare the speeds of sound in different media.

Problem **(a)** If a solid bar of aluminum 1.00 m long is struck at one end with a hammer, a longitudinal pulse propagates down the bar. Find the speed of sound in the bar, which has a Young's modulus of 7.0×10^{10} Pa and a density of 2.7×10^3 kg/m^3. **(b)** Calculate the speed of sound in ethyl alcohol, which has a density of 806 kg/m^3 and bulk modulus of 1.0×10^9 Pa. **(c)** Compute the speed of sound in air at 35.0°C.

Strategy Substitute the given values into the appropriate equations.

Solution

(a) Compute the speed of sound in an aluminum bar.

Substitute values into Equation 14.3:

$$v_{Al} = \sqrt{\frac{Y}{\rho}} = \sqrt{\frac{7.0 \times 10^{10}\ \text{Pa}}{2.7 \times 10^3\ \text{kg/m}^3}}$$

$$= 5\,100\ \text{m/s, or about } 11\,000\ \text{mi/h !}$$

(b) Compute the speed of sound in ethyl alcohol.

Substitute values into Equation 14.1:

$$v = \sqrt{\frac{B}{\rho}} = \sqrt{\frac{1.0 \times 10^9\ \text{Pa}}{806\ \text{kg/m}^3}} = 1.1 \times 10^3\ \text{m/s}$$

(c) Compute the speed of sound in air at 35.0°C.

Substitute values into Equation 14.4:

$$v = (331\ \text{m/s})\sqrt{\frac{(273\ \text{K} + 35.0\ \text{K})}{273\ \text{K}}} = 352\ \text{m/s}$$

Remark The speed of sound in aluminum is dramatically higher than in either liquid alcohol or air.

Exercise 14.1

Compute the speed of sound in the following substances at 273 K: (a) lead ($Y = 1.6 \times 10^{10}$ Pa), (b) mercury ($B = 2.8 \times 10^{10}$ Pa), and (c) air at −15.0°C.

Answers (a) 1.2×10^3 m/s (b) 1.4×10^3 m/s (c) 322 m/s

PhysicsNow™

You can compare the speeds of sound through various media by logging into PhysicsNow at **www.cp7e.com** and going to Interactive Example 14.1.

14.4 ENERGY AND INTENSITY OF SOUND WAVES

As the tines of a tuning fork move back and forth through the air, they exert a force on a layer of air and cause it to move. In other words, the tines do work on the layer of air. The fact that the fork pours sound energy into the air is one of the reasons the vibration of the fork slowly dies out. (Other factors, such as the energy lost to friction as the tines bend, are also responsible for the lessening of movement.)

The average **intensity** I of a wave on a given surface is defined as the rate at which energy flows through the surface, $\Delta E/\Delta t$, divided by the surface area A:

$$I \equiv \frac{1}{A}\frac{\Delta E}{\Delta t} \qquad [14.5]$$

where the direction of energy flow is perpendicular to the surface at every point.

SI unit: watt per meter squared (W/m^2)

A rate of energy transfer is power, so Equation 14.5 can be written in the alternate form

$$I \equiv \frac{\text{power}}{\text{area}} = \frac{\mathcal{P}}{A} \qquad [14.6]$$

◀ Intensity of a wave

where $\mathcal{P}$ is the sound power passing through the surface, measured in watts, and the intensity again has units of watts per square meter.

The faintest sounds the human ear can detect at a frequency of 1 000 Hz have an intensity of about 1×10^{-12} W/m^2. This intensity is called the **threshold of hearing**. The loudest sounds the ear can tolerate have an intensity of about 1 W/m^2 (the **threshold of pain**). At the threshold of hearing, the increase in pressure in the ear is approximately 3×10^{-5} Pa over normal atmospheric pressure. Because atmospheric pressure is about 1×10^5 Pa, this means the ear can detect pressure fluctuations as small as about 3 parts in 10^{10}! The maximum displacement of an air molecule at the threshold of hearing is about 1×10^{-11} m—a remarkably small number! If we compare this displacement with the diameter of a molecule (about 10^{-10} m), we see that the ear is an extremely sensitive detector of sound waves.

The loudest sounds the human ear can tolerate at 1 kHz correspond to a pressure variation of about 29 Pa away from normal atmospheric pressure, with a maximum displacement of air molecules of 1×10^{-5} m.

Intensity Level in Decibels

The loudest tolerable sounds have intensities about 1.0×10^{12} times greater than the faintest detectable sounds. The most intense sound, however, isn't perceived as being 1.0×10^{12} times louder than the faintest sound, because the sensation of loudness is approximately logarithmic in the human ear. (For a review of logarithms, see Section A.3, heading G, in Appendix A.) The relative intensity of a sound is called the **intensity level** or **decibel level**, defined by

$$\beta \equiv 10 \log\left(\frac{I}{I_0}\right) \qquad [14.7]$$

◀ Intensity level

The constant $I_0 = 1.0 \times 10^{-12}$ W/m^2 is the reference intensity, the sound intensity at the threshold of hearing—I is the intensity, and β is the corresponding intensity

TIP 14.1 Intensity Versus Intensity Level

Don't confuse intensity with intensity level. Intensity is a physical quantity with units of watts per meter squared; intensity level, or decibel level, is a convenient mathematical transformation of intensity to a logarithmic scale.

level measured in decibels (dB). (The word *decibel*, which is one-tenth of a *bel*, comes from the name of the inventor of the telephone, Alexander Graham Bell (1847–1922).

To get a feel for various decibel levels, we can substitute a few representative numbers into Equation 14.7, starting with $I = 1.0 \times 10^{-12}\ \text{W/m}^2$:

$$\beta = 10 \log\left(\frac{1.0 \times 10^{-12}\ \text{W/m}^2}{1.0 \times 10^{-12}\ \text{W/m}^2}\right) = 10 \log(1) = 0\ \text{dB}$$

From this result, we see that the lower threshold of human hearing has been chosen to be zero on the decibel scale. Progressing upward by powers of ten yields

$$\beta = 10 \log\left(\frac{1.0 \times 10^{-11}\ \text{W/m}^2}{1.0 \times 10^{-12}\ \text{W/m}^2}\right) = 10 \log(10) = 10\ \text{dB}$$

$$\beta = 10 \log\left(\frac{1.0 \times 10^{-10}\ \text{W/m}^2}{1.0 \times 10^{-12}\ \text{W/m}^2}\right) = 10 \log(100) = 20\ \text{dB}$$

Notice the pattern: *Multiplying* a given intensity by ten *adds* 10 db to the intensity level. This pattern holds throughout the decibel scale. For example, a 50-dB sound is 10 times as intense as a 40-dB sound, while a 60-dB sound is 100 times as intense as a 40-dB sound.

On this scale, the threshold of pain ($I = 1.0\ \text{W/m}^2$) corresponds to an intensity level of $\beta = 10 \log(1/1 \times 10^{-12}) = 10 \log(10^{12}) = 120$ dB. Nearby jet airplanes can create intensity levels of 150 dB, and subways and riveting machines have levels of 90 to 100 dB. The electronically amplified sound heard at rock concerts can attain levels of up to 120 dB, the threshold of pain. Exposure to such high intensity levels can seriously damage the ear. Earplugs are recommended whenever prolonged intensity levels exceed 90 dB. Recent evidence suggests that noise pollution, which is common in most large cities and in some industrial environments, may be a contributing factor to high blood pressure, anxiety, and nervousness. Table 14.2 gives the approximate intensity levels of various sounds.

TABLE 14.2

Intensity Levels in Decibels for Different Sources

Source of Sound	β(dB)
Nearby jet airplane	150
Jackhammer, machine gun	130
Siren, rock concert	120
Subway, power mower	100
Busy traffic	80
Vacuum cleaner	70
Normal conversation	50
Mosquito buzzing	40
Whisper	30
Rustling leaves	10
Threshold of hearing	0

EXAMPLE 14.2 A Noisy Grinding Machine

Goal Work with watts and decibels.

Problem A noisy grinding machine in a factory produces a sound intensity of $1.00 \times 10^{-5}\ \text{W/m}^2$. Calculate **(a)** the decibel level of this machine, and **(b)** the new intensity level when a second, identical machine is added to the factory. **(c)** A certain number of additional such machines are put into operation alongside these two. When all the machines are running at the same time the decibel level is 77.0 dB. Find the sound intensity.

Strategy Parts (a) and (b) require substituting into the decibel formula, Equation 14.7, with the intensity in part (b) twice the intensity in part (a). In part (c), the intensity level in decibels is given, and it's necessary to work backwards, using the inverse of the logarithm function, to get the intensity in watts per meter squared.

Solution

(a) Calculate the intensity level of the single grinder.

Substitute the intensity into the decibel formula:

$$\beta = 10 \log\left(\frac{1.00 \times 10^{-5}\ \text{W/m}^2}{1.00 \times 10^{-12}\ \text{W/m}^2}\right) = 10 \log(10^7)$$

$$= 70.0\ \text{dB}$$

(b) Calculate the new intensity level when an additional machine is added.

Substitute twice the intensity of part (a) into the decibel formula:

$$\beta = 10 \log\left(\frac{2.00 \times 10^{-5}\ \text{W/m}^2}{1.00 \times 10^{-12}\ \text{W/m}^2}\right) = 73.0\ \text{dB}$$

(c) Find the intensity corresponding to an intensity level of 77.0 dB.

Substitute 77.0 dB into the decibel formula and divide both sides by 10:

$$\beta = 77.0 \text{ dB} = 10 \log \left(\frac{I}{I_0} \right)$$

$$7.70 = \log \left(\frac{I}{10^{-12} \text{ W/m}^2} \right)$$

Make each side the exponent of 10. On the right-hand side, $10^{\log u} = u$, by definition of base ten logarithms.

$$10^{7.70} = 5.01 \times 10^7 = \frac{I}{1.00 \times 10^{-12} \text{ W/m}^2}$$

$$I = 5.01 \times 10^{-5} \text{ W/m}^2$$

Remark The answer is five times the intensity of the single grinder, so in part (c) there are five such machines operating simultaneously. Because of the logarithmic definition of intensity level, large changes in intensity correspond to small changes in intensity level.

Exercise 14.2

Suppose a manufacturing plant has an average sound intensity level of 97.0 dB created by 25 identical machines. (a) Find the total intensity created by all the machines. (b) Find the sound intensity created by one such machine. (c) What's the sound intensity level if five such machines are running?

Answers (a) 5.01×10^{-3} W/m^2 (b) 2.00×10^{-4} W/m^2 (c) 90.0 dB

APPLICATION

OSHA Noise-Level Regulations

Federal regulations now demand that no office or factory worker be exposed to noise levels that average more than 90 dB over an 8-h day. From a management point of view, here's the good news: one machine in the factory may produce a noise level of 70 dB, but a second machine, while doubling the total intensity, increases the noise level by only 3 dB. Because of the logarithmic nature of intensity levels, doubling the intensity doesn't double the intensity level; in fact, it alters it by a surprisingly small amount. This means that equipment can be added to the factory without appreciably altering the intensity level of the environment.

Now here's the bad news: as you remove noisy machinery, the intensity level isn't lowered appreciably. In Exercise 14.2, reducing the intensity level by 7 dB would require the removal of 20 of the 25 machines! To lower the level another 7 dB would require removing 80% of the remaining machines, in which case only one machine would remain.

14.5 SPHERICAL AND PLANE WAVES

If a small spherical object oscillates so that its radius changes periodically with time, a spherical sound wave is produced (Fig. 14.4, page 466). The wave moves outward from the source at a constant speed.

Because all points on the vibrating sphere behave in the same way, we conclude that the energy in a spherical wave propagates equally in all directions. This means that no one direction is preferred over any other. If $\mathcal{P}_{av}$ is the average power emitted by the source, then at any distance r from the source, this power must be distributed over a spherical surface of area $4\pi r^2$, assuming no absorption in the medium. (Recall that $4\pi r^2$ is the surface area of a sphere.) Hence, the **intensity** of the sound at a distance r from the source is

$$I = \frac{\text{average power}}{\text{area}} = \frac{\mathcal{P}_{av}}{A} = \frac{\mathcal{P}_{av}}{4\pi r^2} \qquad [14.8]$$

This equation shows that the intensity of a wave decreases with increasing distance from its source, as you might expect. The fact that I varies as $1/r^2$ is a result of the

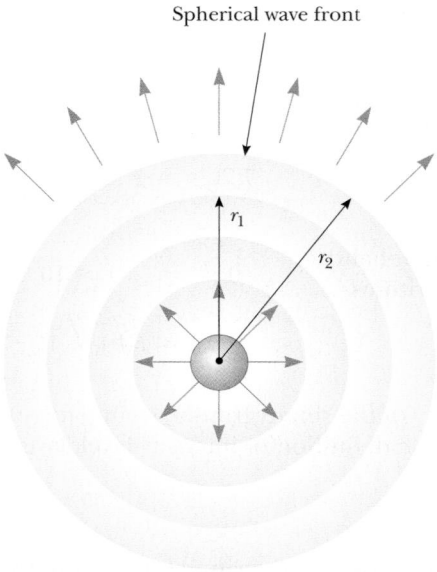

Figure 14.4 A spherical wave propagating radially outward from an oscillating sphere. The intensity of the wave varies as $1/r^2$.

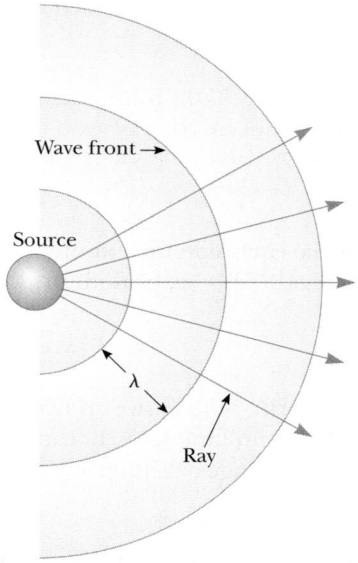

Figure 14.5 Spherical waves emitted by a point source. The circular arcs represent the spherical wave fronts concentric with the source. The rays are radial lines pointing outward from the source, perpendicular to the wavefronts.

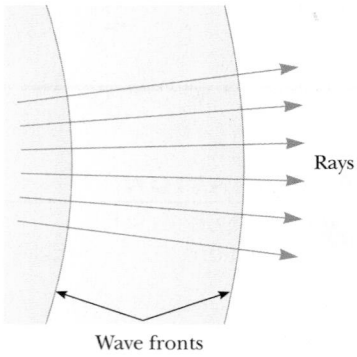

Figure 14.6 Far away from a point source, the wave fronts are nearly parallel planes and the rays are nearly parallel lines perpendicular to the planes. Hence, a small segment of a spherical wavefront is approximately a plane wave.

assumption that the small source (sometimes called a **point source**) emits a spherical wave. (In fact, light waves also obey this so-called inverse-square relationship.) Because the average power is the same through any spherical surface centered at the source, we see that the intensities at distances r_1 and r_2 (Fig. 14.4) from the center of the source are

$$I_1 = \frac{\mathcal{P}_{av}}{4\pi r_1^2} \qquad I_2 = \frac{\mathcal{P}_{av}}{4\pi r_2^2}$$

The ratio of the intensities at these two spherical surfaces is

$$\frac{I_1}{I_2} = \frac{r_2^2}{r_1^2} \qquad \textbf{[14.9]}$$

It's useful to represent spherical waves graphically with a series of circular arcs (lines of maximum intensity) concentric with the source representing part of a spherical surface, as in Figure 14.5. We call such an arc a **wave front**. The distance between adjacent wave fronts equals the wavelength λ. The radial lines pointing outward from the source and perpendicular to the arcs are called **rays**.

Now consider a small portion of a wave front that is at a *great* distance (relative to λ) from the source, as in Figure 14.6. In this case, the rays are nearly parallel to each other and the wave fronts are very close to being planes. At distances from the source that are great relative to the wavelength, therefore, we can approximate the wave front with parallel planes, called **plane waves**. Any small portion of a spherical wave that is far from the source can be considered a plane wave. Figure 14.7 illustrates a plane wave propagating along the x-axis. If the positive x-direction is taken to be the direction of the wave motion (or ray) in this figure, then the wave fronts are parallel to the plane containing the y- and z-axes.

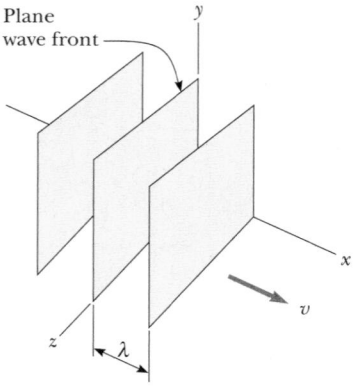

Figure 14.7 A representation of a plane wave moving in the positive x-direction with a speed v. The wavefronts are planes parallel to the yz-plane.

EXAMPLE 14.3 Intensity Variations of a Point Source

Goal Relate sound intensities and their distances from a point source.

Problem A small source emits sound waves with a power output of 80.0 W. **(a)** Find the intensity 3.00 m from the source. **(b)** At what distance would the intensity be one-fourth as much as it is at $r = 3.00$ m? **(c)** Find the distance at which the sound level is 40.0 dB.

Strategy The source is small, so the emitted waves are spherical and the intensity in part (a) can be found by substituting values into Equation 14.8. Part (b) involves solving for r in Equation 14.8 followed by substitution (though Equation 14.9 can be used instead). In part (c), convert from the sound intensity level to the intensity in W/m^2, using Equation 14.7. Then substitute into Equation 14.9 (though 14.8 could be used, instead) and solve for r_2.

Solution

(a) Find the intensity 3.00 m from the source.

Substitute $\mathcal{P}_{av} = 80.0$ W and $r = 3.00$ m into Equation 14.8:

$$I = \frac{\mathcal{P}_{av}}{4\pi r^2} = \frac{80.0\text{ W}}{4\pi(3.00\text{ m})^2} = 0.707\text{ W/m}^2$$

(b) At what distance would the intensity be one-fourth as much as it is at $r = 3.00$ m?

Take $I = (0.707\text{ W/m}^2)/4$, and solve for r in Equation 14.8:

$$r = \left(\frac{\mathcal{P}_{av}}{4\pi I}\right)^{1/2} = \left(\frac{80.0\text{ W}}{4\pi(0.707\text{ W/m}^2)/4.0}\right)^{1/2} = 6.00\text{ m}$$

(c) Find the distance at which the sound level is 40.0 dB.

Convert the intensity level of 40.0 dB to an intensity in W/m^2 by solving Equation 14.7 for I:

$$40.0 = 10\log\left(\frac{I}{I_0}\right) \rightarrow 4.00 = \log\left(\frac{I}{I_0}\right)$$

$$10^{4.00} = \frac{I}{I_0} \rightarrow I = 10^{4.00}I_0 = 1.00 \times 10^{-8}\text{ W/m}^2$$

Solve Equation 14.9 for r_2^2, substitute the intensity and the result of part (a), and take the square root:

$$\frac{I_1}{I_2} = \frac{r_2^2}{r_1^2} \rightarrow r_2^2 = r_1^2\frac{I_1}{I_2}$$

$$r_2^2 = (3.00\text{ m})^2\left(\frac{0.707\text{ W/m}^2}{1.00 \times 10^{-8}\text{ W/m}^2}\right)$$

$$r_2 = 2.52 \times 10^4\text{ m}$$

Remarks Once the intensity is known at one position a certain distance away from the source, it's easier to use Equation 14.9 rather than Equation 14.8 to find the intensity at any other location. This is particularly true for part (b), where, using Equation 14.9, we can see right away that doubling the distance reduces the intensity to one-quarter its previous value.

Exercise 14.3

Suppose a certain jet plane creates an intensity level of 125 dB at a distance of 5.00 m. What intensity level does it create on the ground directly underneath it when flying at an altitude of 2.00 km?

Answer 73.0 dB

14.6 THE DOPPLER EFFECT

If a car or truck is moving while its horn is blowing, the frequency of the sound you hear is higher as the vehicle approaches you and lower as it moves away from you. This is one example of the *Doppler effect*, named for the Austrian physicist Christian Doppler (1803–1853), who discovered it. The same effect is heard if you're on a motorcycle and the horn is stationary: the frequency is higher as you approach the source and lower as you move away.

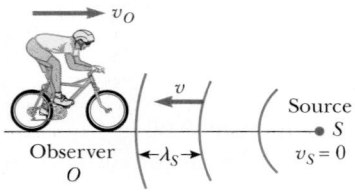

ACTIVE FIGURE 14.8
An observer moving with a speed v_O *toward* a stationary point source (S) hears a frequency f_O that is *greater* than the source frequency f_S.

PhysicsNow™
Log into PhysicsNow at **www.cp7e.com** and go to Active Figure 14.8 to adjust the speed of the observer.

Although the Doppler effect is most often associated with sound, it's common to all waves, including light.

In deriving the Doppler effect, we assume that the air is stationary and that all speed measurements are made relative to this stationary medium. The speed v_O is the speed of the observer, v_S is the speed of the source, and v is the speed of sound.

Case 1: The Observer Is Moving Relative to a Stationary Source

In Active Figure 14.8 an observer is moving with a speed of v_O toward the source (considered a point source), which is at rest ($v_S = 0$).

We take the frequency of the source to be f_S, the wavelength of the source to be λ_S, and the speed of sound in air to be v. If both observer and source are stationary, the observer detects f_S wave fronts per second. (That is, when $v_O = 0$ and $v_S = 0$, the observed frequency f_O equals the source frequency f_S.) When moving toward the source, the observer moves a distance of $v_O t$ in t seconds. During this interval, **the observer detects an additional number of wave fronts**. The number of extra wave fronts is equal to the distance traveled, $v_O t$, divided by the wavelength λ_S:

$$\text{Additional wave fronts detected} = \frac{v_O t}{\lambda_S}$$

Divide this equation by the time t to get the number of additional wave fronts detected *per second*, v_O/λ_S. Hence, the frequency heard by the observer is *increased* to

$$f_O = f_S + \frac{v_O}{\lambda_S}$$

Substituting $\lambda_S = v/f_S$ into this expression for f_O we obtain

$$f_O = f_S\left(\frac{v + v_O}{v}\right) \qquad \textbf{[14.10]}$$

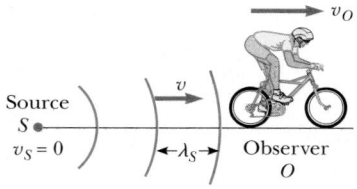

Figure 14.9 An observer moving with a speed of v_O *away* from a stationary source hears a frequency f_O that is *lower* than the source frequency f_S.

When the observer is *moving away* from a stationary source (Fig. 14.9), the observed frequency decreases. A derivation yields the same result as Equation 14.10, but with $v - v_O$ in the numerator. Therefore, when the observer is moving away from the source, substitute $-v_O$ for v_O in Equation 14.10.

TIP 14.2 Doppler Effect Doesn't Depend on Distance
The sound from a source approaching at constant speed will increase in intensity, but the observed (elevated) frequency will remain unchanged. The Doppler effect doesn't depend on distance.

Case 2: The Source Is Moving Relative to a Stationary Observer

Now consider a source moving toward an observer at rest, as in Active Figure 14.10. Here, the wave fronts passing observer A are closer together because the source is moving in the direction of the outgoing wave. As a result, the wavelength λ_O measured by observer A is shorter than the wavelength λ_S of the source at rest. During each vibration, which lasts for an interval T (the period), the source moves a distance $v_S T = v_S/f_S$ and **the wavelength is shortened by that amount**. The observed wavelength is therefore given by

$$\lambda_O = \lambda_S - \frac{v_S}{f_S}$$

Because $\lambda_S = v/f_S$, the frequency observed by A is

$$f_O = \frac{v}{\lambda_O} = \frac{v}{\lambda_S - \dfrac{v_S}{f_S}} = \frac{v}{\dfrac{v}{f_S} - \dfrac{v_S}{f_S}}$$

or

$$f_O = f_S\left(\frac{v}{v - v_S}\right) \qquad \textbf{[14.11]}$$

As expected, **the observed frequency increases when the source is moving toward the observer**. When the source is *moving away* from an observer at rest, the minus sign in the denominator must be replaced with a plus sign, so the factor becomes $(v + v_S)$.

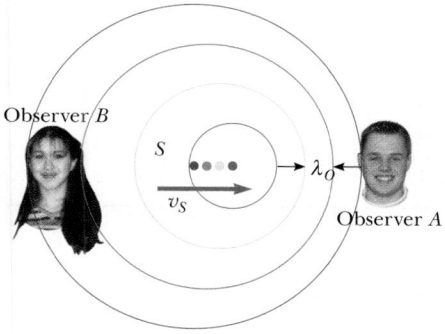

(a)

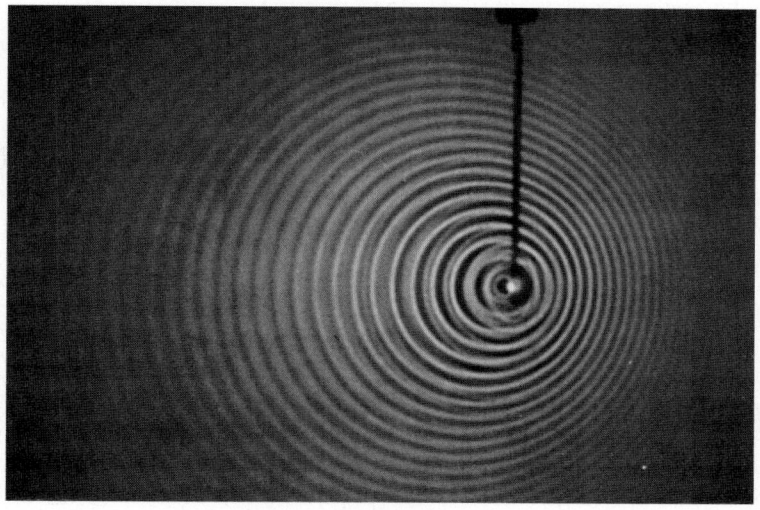

(b)

Courtesy of Educational Development Center, Newton, MA.

ACTIVE FIGURE 14.10
(a) A source S moving with speed v_S toward stationary observer A and away from stationary observer B. Observer A hears an *increased* frequency, and observer B hears a *decreased* frequency. (b) The Doppler effect in water, observed in a ripple tank. The source producing the water waves is moving to the right.

PhysicsNow™
Log into PhysicsNow at **www.cp7e.com**, and go to Active Figure 14.10 to adjust the speed of the source.

General Case

When both the source and the observer are in motion relative to Earth, Equations 14.10 and 14.11 can be combined to give

$$f_O = f_S \left(\frac{v + v_O}{v - v_S} \right) \qquad \textbf{[14.12]}$$

◀ Doppler shift equation—observer and source in motion

In this expression, the signs for the values substituted for v_O and v_S depend on the direction of the velocity. When the observer moves *toward* the source, a *positive* speed is substituted for v_O; when the observer moves *away from* the source, a *negative* speed is substituted for v_O. Similarly, a *positive* speed is substituted for v_S when the source moves *toward* the observer, a *negative* speed when the source moves away from the observer.

Choosing incorrect signs is the most common mistake made in working a Doppler effect problem. The following rules may be helpful: The word *toward* is associated with an *increase* in the observed frequency; the words *away from* are associated with a *decrease* in the observed frequency.

These two rules derive from the physical insight that when the observer is moving toward the source (or the source toward the observer), there is a smaller observed period between wave crests, hence a larger frequency, with the reverse holding—a smaller observed frequency—when the observer is moving away from the source (or the source away from the observer). Keep the physical insight in mind whenever you're in doubt about the signs in Equation 14.12: Adjust them as necessary to get the correct physical result.

The second most common mistake made in applying Equation 14.12 is to accidentally reverse numerator and denominator. Some find it helpful to remember the equation in the following form:

$$\frac{f_O}{v + v_O} = \frac{f_S}{v - v_S}$$

The advantage of this form is its symmetry: both sides are very nearly the same, with O's on the left and S's on the right. Forgetting which side has the plus sign and which has the minus sign is not a serious problem, as long as physical insight is used to check the answer and make adjustments as necessary.

Quick Quiz 14.2

Suppose you're on a hot air balloon ride, carrying a buzzer that emits a sound of frequency f. If you accidentally drop the buzzer over the side while the balloon is rising at constant speed, what can you conclude about the sound you hear as the buzzer falls toward the ground? (a) the frequency and intensity increase, (b) the frequency decreases and the intensity increases, (c) the frequency decreases and the intensity decreases, or (d) the frequency remains the same, but the intensity decreases.

Applying Physics 14.2 Out-of-Tune Speakers

Suppose you place your stereo speakers far apart and run past them from right to left or left to right. If you run rapidly enough and have excellent pitch discrimination, you may notice that the music playing seems to be out of tune when you're between the speakers. Why?

Explanation When you are between the speakers, you are running away from one of them and toward the other, so there is a Doppler shift downward for the sound from the speaker behind you and a Doppler shift upward for the sound from the speaker ahead of you. As a result, the sound from the two speakers will not be in tune. A calculation shows that a world-class sprinter could run fast enough to generate about a semitone difference in the sound from the two speakers.

EXAMPLE 14.4 Listen, but Don't Stand on the Track

Goal Solve a Doppler shift problem when only the source is moving.

Problem A train moving at a speed of 40.0 m/s sounds its whistle, which has a frequency of 5.00×10^2 Hz. Determine the frequency heard by a stationary observer as the train *approaches* the observer. The ambient temperature is 24.0°C.

Strategy Use Equation 14.4 to get the speed of sound at the ambient temperature, then substitute values into Equation 14.12 for the Doppler shift. Because the train approaches the observer, the observed frequency will be larger. Choose the sign of v_S to reflect this fact.

Solution

Use Equation 14.4 to calculate the speed of sound in air at $T = 24.0°\text{C}$:

$$v = (331 \text{ m/s}) \sqrt{\frac{T}{273 \text{ K}}}$$

$$= (331 \text{ m/s}) \sqrt{\frac{(273 + 24.0) \text{ K}}{273 \text{ K}}} = 345 \text{ m/s}$$

The observer is stationary, so $v_O = 0$. The train is moving *toward* the observer, so $v_S = +40.0$ m/s (*positive*). Substitute these values and the speed of sound into the Doppler shift equation:

$$f_O = f_S \left(\frac{v + v_O}{v - v_S} \right)$$

$$= (5.00 \times 10^2 \text{ Hz}) \left(\frac{345 \text{ m/s}}{345 \text{ m/s} - 40.0 \text{ m/s}} \right)$$

$$= 566 \text{ Hz}$$

Remark If the train were going away from the observer, $v_S = -40.0$ m/s would have been chosen instead.

Exercise 14.4

Determine the frequency heard by the stationary observer as the train *recedes* from the observer.

Answer 448 Hz

INTERACTIVE EXAMPLE 14.5 The Noisy Siren

Goal Solve a Doppler shift problem when both the source and observer are moving.

Problem An ambulance travels down a highway at a speed of 75.0 mi/h, its siren emitting sound at a frequency of 4.00×10^2 Hz. What frequency is heard by a passenger in a car traveling at 55.0 mi/h in the opposite direction as the car and ambulance **(a)** *approach* each other and **(b)** pass and *move away* from each other? Take the speed of sound in air to be $v = 345$ m/s.

Strategy Aside from converting from mi/h to m/s, this problem only requires substitution into the Doppler formula, but two signs must be chosen correctly in each part. In part (a), the observer moves toward the source and the source moves toward the observer, so both v_O and v_S should be chosen to be positive. Switch signs after they pass each other.

Solution

Convert the speeds from mi/h to m/s:

$$v_S = (75.0 \text{ mi/h})\left(\frac{0.447 \text{ m/s}}{1.00 \text{ mi/h}}\right) = 33.5 \text{ m/s}$$

$$v_O = (55.0 \text{ mi/h})\left(\frac{0.447 \text{ m/s}}{1.00 \text{ mi/h}}\right) = 24.6 \text{ m/s}$$

(a) Compute the observed frequency as the ambulance and car approach each other.

Each vehicle goes toward the other, so substitute $v_O = +24.6$ m/s and $v_S = +33.5$ m/s into the Doppler shift formula:

$$f_O = f_S\left(\frac{v + v_O}{v - v_S}\right)$$

$$= (4.00 \times 10^2 \text{ Hz})\left(\frac{345 \text{ m/s} + 24.6 \text{ m/s}}{345 \text{ m/s} - 33.5 \text{ m/s}}\right) = \boxed{475 \text{ Hz}}$$

(b) Compute the observed frequency as the ambulance and car recede from each other.

Each vehicle goes away from the other, so substitute $v_O = -24.6$ m/s and $v_S = -33.5$ m/s into the Doppler shift formula:

$$f_O = f_S\left(\frac{v + v_O}{v - v_S}\right)$$

$$= (4.00 \times 10^2 \text{ Hz})\left(\frac{345 \text{ m/s} + (-24.6 \text{ m/s})}{345 \text{ m/s} - (-33.5 \text{ m/s})}\right)$$

$$= \boxed{339 \text{ Hz}}$$

Remarks Notice how the signs were handled: In part (b), the negative signs were required on the speeds because both observer and source were moving away from each other. Sometimes, of course, one of the speeds is negative and the other is positive.

Exercise 14.5

Repeat this problem, but assume the ambulance and car are going the same direction, with the ambulance initially behind the car. The speeds and the frequency of the siren are the same as in the example. Find the frequency heard by the observer in the car (a) before and (b) after the ambulance passes the car. [*Note:* The highway patrol subsequently gives the driver of the car a ticket for not pulling over for an emergency vehicle!]

Answers (a) 411 Hz (b) 391 Hz

PhysicsNow™

You can alter the relative speeds of two submarines and observe the Doppler-shifted frequency by logging into PhysicsNow at **www.cp7e.com** and going to Interactive Example 14.5.

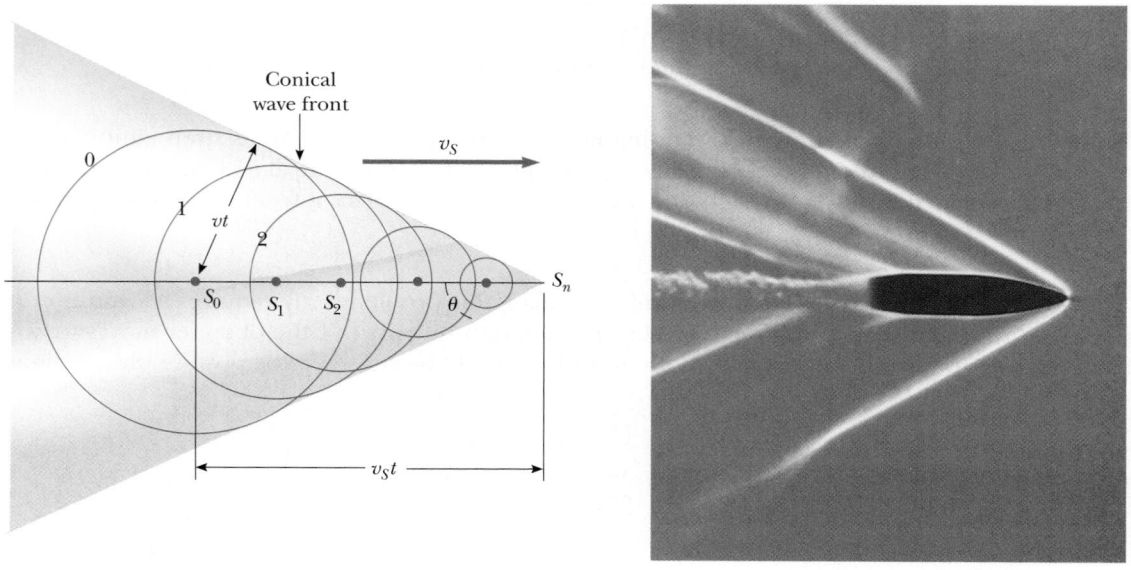

(a) (b)

Figure 14.11 (a) A representation of a shock wave, produced when a source moves from S_0 to S_n with a speed v_s that is *greater* than the wave speed v in that medium. The envelope of the wave fronts forms a cone with half-angle of $\sin\theta = v/v_s$. (b) A stroboscopic photograph of a bullet moving at supersonic speed through the hot air above a candle.

Shock Waves

What happens when the source speed v_S *exceeds* the wave velocity v? Figure 14.11a describes this situation graphically. The circles represent spherical wave fronts emitted by the source at various times during its motion. At $t = 0$, the source is at point S_0, and at some later time t, the source is at point S_n. In the interval t, the wave front centered at S_0 reaches a radius of vt. In this same interval, the source travels to S_n, a distance of $v_s t$. At the instant the source is at S_n, the waves just beginning to be generated at this point have wave fronts of zero radius. The line drawn from S_n to the wave front centered on S_0 is tangent to all other wave fronts generated at intermediate times. All such tangent lines lie on the surface of a cone. The angle θ between one of these tangent lines and the direction of travel is given by

$$\sin\theta = \frac{v}{v_s}$$

Figure 14.12 The V-shaped bow wave of a boat is formed because the boat travels at a speed greater than the speed of the water waves. A bow wave is analogous to a shock wave formed by an airplane traveling faster than sound.

The ratio v_s/v is called the **Mach number**. The conical wave front produced when $v_s > v$ (supersonic speeds) is known as a **shock wave**. Figure 14.11b is a photograph of a bullet traveling at supersonic speed through the hot air rising above a candle. Notice the shock waves in the vicinity of the bullet. Another interesting example of a shock wave is the V-shaped wave front produced by a boat (the bow wave) when the boat's speed exceeds the speed of the water waves (Fig. 14.12).

Jet aircraft and space shuttles traveling at supersonic speeds produce shock waves that are responsible for the loud explosion, or sonic boom, heard on the ground. A shock wave carries a great deal of energy concentrated on the surface of the cone, with correspondingly great pressure variations. Shock waves are unpleasant to hear and can damage buildings when aircraft fly supersonically at low altitudes. In fact, an airplane flying at supersonic speeds produces a double boom, because two shock waves are formed—one from the nose of the plane and one from the tail (Fig. 14.13).

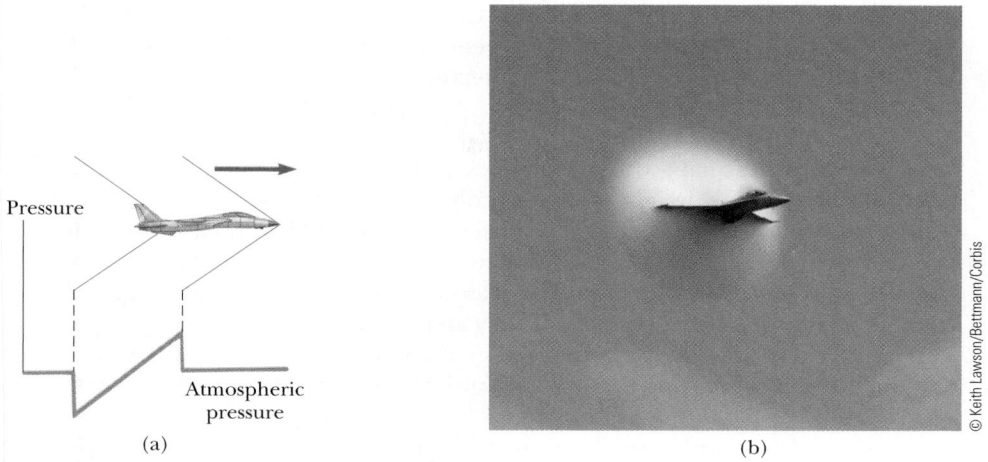

Figure 14.13 (a) The two shock waves produced by the nose and tail of a jet airplane traveling at supersonic speed. (b) A shock wave due to a jet traveling at the speed of sound is made visible as a fog of water vapor. The large pressure variation in the shock wave causes the water in the air to condense into water droplets.

Quick Quiz 14.3

As an airplane flying with constant velocity moves from a cold air mass into a warm air mass, does the Mach number (a) increase, (b) decrease, or (c) remain the same?

14.7 INTERFERENCE OF SOUND WAVES

Sound waves can be made to interfere with each other, a phenomenon that can be demonstrated with the device shown in Figure 14.14. Sound from a loudspeaker at S is sent into a tube at *P*, where there is a T-shaped junction. The sound splits and follows two separate pathways, indicated by the red arrows. Half of the sound travels upward, half downward. Finally, the two sounds merge at an opening where a listener places her ear. If the two paths r_1 and r_2 have the same length, waves that enter the junction will separate into two halves, travel the two paths, and then combine again at the ear. This reuniting of the two waves produces *constructive interference*, so the listener hears a loud sound. If the upper path is adjusted to be one full wavelength longer than the lower path, constructive interference of the two waves occurs again, and a loud sound is detected at the receiver. We have the following result: **If the path difference $r_2 - r_1$ is zero or some integer multiple of wavelengths, then constructive interference occurs and**

$$r_2 - r_1 = n\lambda \quad (n = 0, 1, 2, \ldots) \qquad [14.13]$$

◀ Condition for constructive interference

Suppose, however, that one of the path lengths, r_2, is adjusted so that the upper path is half a wavelength *longer* than the lower path r_1. In this case, an entering sound wave splits and travels the two paths as before, but now the wave along the upper path must travel a distance equivalent to half a wavelength farther than the wave traveling along the lower path. As a result, the crest of one wave meets the trough of the other when they merge at the receiver, causing the two waves to

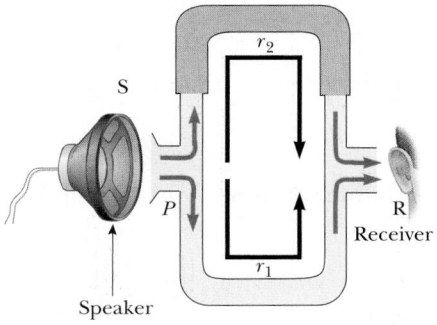

Figure 14.14 An acoustical system for demonstrating interference of sound waves. Sound from the speaker enters the tube and splits into two parts at *P*. The two waves combine at the opposite side and are detected at *R*. The upper path length is varied by the sliding section.

cancel each other. This phenomenon is called *totally destructive interference*, and no sound is detected at the receiver. In general, **if the path difference $r_2 - r_1$ is $\frac{1}{2}$, $1\frac{1}{2}$, $2\frac{1}{2}$ . . . wavelengths, destructive interference occurs** and

Condition for destructive interference ▶

$$r_2 - r_1 = (n + \tfrac{1}{2})\lambda \ (n = 0, 1, 2, \ldots) \qquad \textbf{[14.14]}$$

Nature provides many other examples of interference phenomena, most notably in connection with light waves, described in Chapter 24.

APPLICATION

Connecting Your Stereo Speakers

In connecting the wires between your stereo system and loudspeakers, you may notice that the wires are usually color coded and that the speakers have positive and negative signs on the connections. The reason for this is that the speakers need to be connected with the same "polarity." If they aren't, then the same electrical signal fed to both speakers will result in one speaker cone moving outward at the same time that the other speaker cone is moving inward. In this case, the sound leaving the two speakers will be 180° out of phase with each other. If you are sitting midway between the speakers, the sounds from both speakers travel the same distance and preserve the phase difference they had when they left. In an ideal situation, for a 180° phase difference, you would get complete destructive interference and no sound! In reality, the cancellation is not complete and is much more significant for bass notes (which have a long wavelength) than for the shorter wavelength treble notes. Nevertheless, to avoid a significant reduction in the intensity of bass notes, the color-coded wires and the signs on the speaker connections should be carefully noted.

TIP 14.3 Do Waves Really Interfere?

In popular usage, to *interfere* means "to come into conflict with" or "to intervene to affect an outcome." This differs from its use in physics, where waves pass through each other and interfere, but don't affect each other in any way.

EXAMPLE 14.6 Two Speakers Driven by the Same Source

Goal Use the concept of interference to compute a frequency.

Problem Two speakers placed 3.00 m apart are driven by the same oscillator (Fig. 14.15). A listener is originally at point O, which is located 8.00 m from the center of the line connecting the two speakers. The listener then walks to point P, which is a perpendicular distance 0.350 m from O, before reaching the *first minimum* in sound intensity. What is the frequency of the oscillator? Take the speed of sound in air to be $v_s = 343$ m/s.

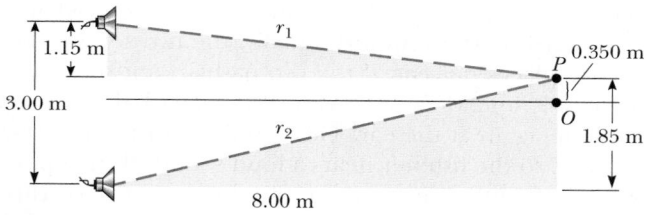

Figure 14.15 (Example 14.6) Two loudspeakers driven by the same source can produce interference.

Strategy The position of the first minimum in sound intensity is given, which is a point of destructive interference. We can find the path lengths r_1 and r_2 with the Pythagorean theorem and then use Equation 14.14 for destructive interference to find the wavelength λ. Using $v = f\lambda$ then yields the frequency.

Solution

Use the Pythagorean theorem to find the path lengths r_1 and r_2:

$$r_1 = \sqrt{(8.00\text{ m})^2 + (1.15\text{ m})^2} = 8.08\text{ m}$$

$$r_2 = \sqrt{(8.00\text{ m})^2 + (1.85\text{ m})^2} = 8.21\text{ m}$$

Substitute these values and $n = 0$ into Equation 14.14, solving for the wavelength:

$$r_2 - r_1 = (n + \tfrac{1}{2})\lambda$$

$$8.21\text{ m} - 8.08\text{ m} = 0.13\text{ m} = \lambda/2 \quad \rightarrow \quad \lambda = 0.26\text{ m}$$

Solve $v = \lambda f$ for the frequency f and substitute the speed of sound and the wavelength:

$$f = \frac{v}{\lambda} = \frac{343\text{ m/s}}{0.26\text{ m}} = \boxed{1.3\text{ kHz}}$$

Remark For problems involving constructive interference, the only difference is that Equation 14.13, $r_2 - r_1 = n\lambda$, would be used instead of Equation 14.14.

Exercise 14.6

If the oscillator frequency is adjusted so that the location of the first minimum is at a distance of 0.750 m from O, what is the new frequency?

Answer 0.642 kHz

14.8 STANDING WAVES

Standing waves can be set up in a stretched string by connecting one end of the string to a stationary clamp and connecting the other end to a vibrating object, such as the end of a tuning fork, or by shaking the hand holding the string up and down at a steady rate (Fig. 14.16). Traveling waves then reflect from the ends and move in both directions on the string. The incident and reflected waves combine according to the **superposition principle**. (See Section 13.10.) If the string vibrates at exactly the right frequency, the wave appears to stand still—hence its name, **standing wave**. A **node** occurs where the two traveling waves always have the same magnitude of displacement but the opposite sign, so the net displacement is zero at that point. There is no motion in the string at the nodes, but midway between two adjacent nodes, at an **antinode**, the string vibrates with the largest amplitude.

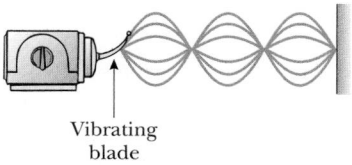

Figure 14.16 Standing waves can be set up in a stretched string by connecting one end of the string to a vibrating blade. When the blade vibrates at one of the natural frequencies of the string, large-amplitude standing waves are created.

Figure 14.17 shows snapshots of the oscillation of a standing wave during half of a cycle. The pink arrows indicate the direction of motion of different parts of the string. Notice that **all points on the string oscillate together vertically with the same frequency, but different points have different amplitudes of motion.** The points of attachment to the wall and all other stationary points on the string are called nodes, labeled N in Figure 14.17a. From the figure, observe that the distance between adjacent nodes is one-half the wavelength of the wave:

$$d_{NN} = \tfrac{1}{2}\lambda$$

Consider a string of length L that is fixed at both ends, as in Active Figure 14.18. For a string, we can set up standing-wave patterns at many frequencies—the more loops, the higher the frequency. Three such patterns are shown in Active Figures 14.18b, 14.18c, and 14.18d. Each has a characteristic frequency, which we will now calculate.

First, **the ends of the string must be nodes, because these points are fixed.** If the string is displaced at its midpoint and released, the vibration shown in Active Figure 14.18b can be produced, in which case the center of the string is an antinode, labeled A. Note that from end to end, the pattern is N–A–N. The distance from a node to its adjacent antinode, N–A, is always equal to a quarter wavelength, $\lambda_1/4$. There are two such segments, N–A and A–N, so $L = 2(\lambda_1/4) = \lambda_1/2$, and $\lambda_1 = 2L$. The frequency of this vibration is therefore

$$f_1 = \frac{v}{\lambda_1} = \frac{v}{2L} \qquad \textbf{[14.15]}$$

Recall that the speed of a wave on a string is $v = \sqrt{F/\mu}$, where F is the tension in the string and μ is its mass per unit length (Chapter 13). Substituting into Equation 14.15, we obtain

$$f_1 = \frac{1}{2L}\sqrt{\frac{F}{\mu}} \qquad \textbf{[14.16]}$$

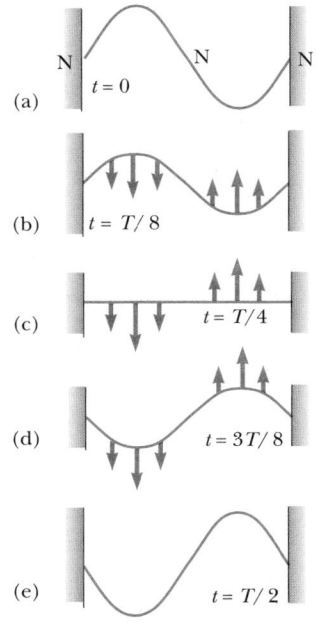

Figure 14.17 A standing-wave pattern in a stretched string, shown by snapshots of the string during one-half of a cycle. In part (a), N denotes a node.

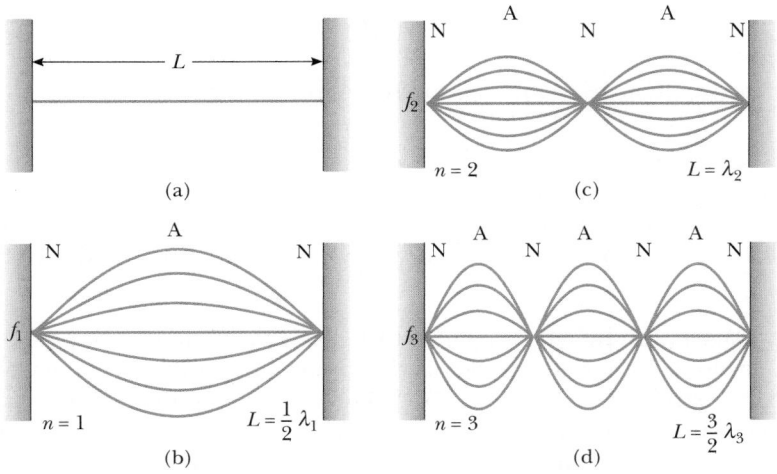

ACTIVE FIGURE 14.18
(a) Standing waves in a stretched string of length L fixed at both ends. The characteristic frequencies of vibration form a harmonic series: (b) the fundamental frequency, or first harmonic; (c) the second harmonic; and (d) the third harmonic. Note that N denotes a node, A an antinode.

PhysicsNow™
Log into PhysicsNow at **www.cp7e.com** and go to Active Figure 14.18 to choose the mode number and see the corresponding standing wave.

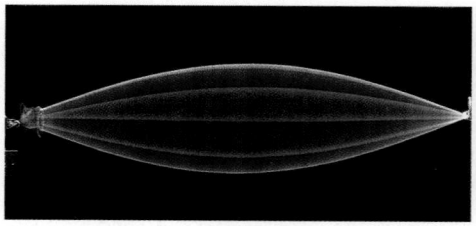

(a)

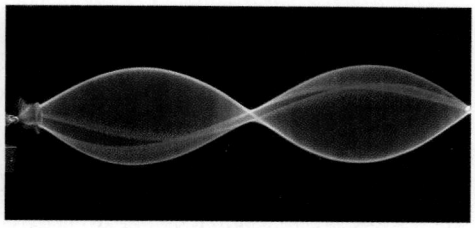

(b)

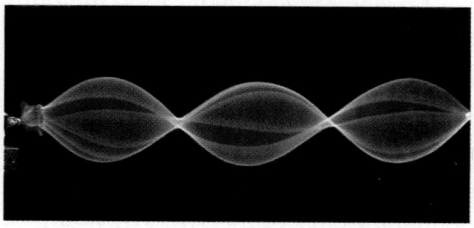

(c)

© Richard Megna, Fundamental Photographs

Multiflash photographs of standing-wave patterns in a cord driven by a vibrator at the left end. The single-loop pattern in (a) represents the fundamental frequency ($n = 1$), the two-loop pattern in (b) the second harmonic ($n = 2$), and the three-loop pattern in (c) the third harmonic ($n = 3$).

This lowest frequency of vibration is called the **fundamental frequency** of the vibrating string, or the **first harmonic**.

The first harmonic has nodes only at the ends—the points of attachment, with node–antinode pattern of N–A–N. The next harmonic, called the **second harmonic** (also called the **first overtone**) can be constructed by inserting an additional node–antinode segment between the endpoints. This makes the pattern N–A–N–A–N, as in Active Figure 14.18c. We count the node–antinode pairs: N–A, A–N, N–A, and A–N, four segments in all, each representing a quarter wavelength. We then have $L = 4(\lambda_2/4) = \lambda_2$, and the second harmonic (first overtone) is

$$f_2 = \frac{v}{\lambda_2} = \frac{v}{L} = 2\left(\frac{v}{2L}\right) = 2f_1$$

This frequency is equal to *twice* the fundamental frequency. The **third harmonic** (**second overtone**) is constructed similarly. Inserting one more N–A segment, we obtain Active Figure 14.18c, the pattern of nodes reading N–A–N–A–N–A–N. There are six node–antinode segments, so $L = 6(\lambda_3/4) = 3(\lambda_3/2)$, which means that $\lambda_3 = 2L/3$, giving

$$f_3 = \frac{v}{\lambda_3} = \frac{3v}{2L} = 3f_1$$

All the higher harmonics, it turns out, are positive integer multiples of the fundamental:

$$f_n = nf_1 = \frac{n}{2L}\sqrt{\frac{F}{\mu}} \qquad n = 1, 2, 3 \ldots \qquad \textbf{[14.17]}$$

The frequencies f_1, $2f_1$, $3f_1$, and so on form a **harmonic series**.

Quick Quiz 14.4

Which of the following frequencies are higher harmonics of a string with fundamental frequency of 150 Hz? (a) 200 Hz (b) 300 Hz (c) 400 Hz (d) 500 Hz (e) 600 Hz.

When a stretched string is distorted to a shape that corresponds to any one of its harmonics, after being released it vibrates only at the frequency of that harmonic. If the string is struck or bowed, however, the resulting vibration includes different amounts of various harmonics, including the fundamental frequency. Waves not in the harmonic series are quickly damped out on a string fixed at both ends. In effect, when disturbed, the string "selects" the standing-wave frequencies. As we'll see later, the presence of several harmonics on a string gives stringed instruments their characteristic sound, which enables us to distinguish one from another even when they are producing identical fundamental frequencies.

APPLICATION
Tuning a Musical Instrument

The frequency of a string on a musical instrument can be changed by varying either the tension or the length. The tension in guitar and violin strings is varied by turning pegs on the neck of the instrument. As the tension is increased, the frequency of the harmonic series increases according to Equation 14.17. Once the instrument is tuned, the musician varies the frequency by pressing the strings against the neck at a variety of positions, thereby changing the effective lengths of the vibrating portions of the strings. As the length is reduced, the frequency again increases, as follows from Equation 14.17.

Finally, Equation 14.17 shows that a string of fixed length can be made to vibrate at a lower fundamental frequency by increasing its mass per unit length. This is achieved in the bass strings of guitars and pianos by wrapping them with metal windings.

INTERACTIVE EXAMPLE 14.7 Guitar Fundamentals

Goal Apply standing-wave concepts to a stringed instrument.

Problem The high E string on a certain guitar measures 64.0 cm in length and has a fundamental frequency of 329 Hz. When a guitarist presses down so that the string is in contact with the first fret (Fig. 14.19a), the string is shortened so that it plays an F note that has a frequency of 349 Hz. **(a)** How far is the fret from the nut? **(b)** Overtones can be produced on a guitar string by gently placing the index finger in the location of a node of a higher harmonic. The string should be touched, but not depressed against a fret. (Given the width of a finger, pressing too hard will damp out higher harmonics as well.) The fundamental frequency is thereby suppressed, making it possible to hear overtones. Where on the guitar string relative to the nut should the finger be lightly placed so as to hear the second harmonic? The fourth harmonic? (This is equivalent to finding the location of the nodes in each case.)

Strategy For part (a) use Equation 14.15, corresponding to the fundamental frequency, to find the speed of waves on the string. Shortening the string by playing a higher note doesn't affect the wave speed, which depends only on the tension and linear density of the string (which are unchanged). Solve Equation 14.15 for the new length L, using the new fundamental frequency, and subtract this length from the original length to find the distance from the nut to the first fret. In part (b), remember that the distance from node to node is half a wavelength. Calculate the wavelength, divide it in two, and locate the nodes, which are integral numbers of half-wavelengths from the nut. [*Note:* The nut is a small piece of wood or ebony at the top of the fret board. The distance from the nut to the bridge (below the sound hole) is the length of the string. (See Fig. 14.19b.)]

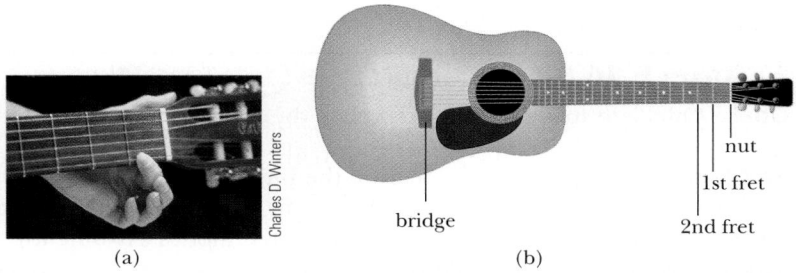

Figure 14.19 (Example 14.7) (a) Playing an F note on a guitar. (b) Some parts of a guitar.

Solution

(a) Find the distance from the nut to the first fret.

Substitute $L_0 = 0.640$ m and $f_1 = 329$ Hz into Equation 14.15, finding the wave speed on the string:

$$f_1 = \frac{v}{2L_0}$$

$$v = 2L_0 f_1 = 2(0.640\text{ m})(329\text{ Hz}) = 421\text{ m/s}$$

Solve Equation 14.15 for the length L, and substitute the wave speed and the frequency of F.

$$L = \frac{v}{2f} = \frac{421\text{ m/s}}{2(349\text{ Hz})} = 0.603\text{ m} = 60.3\text{ cm}$$

Subtract this length from the original length L_0 to find the distance from the nut to the first fret:

$$\Delta x = L_0 - L = 64.0\text{ cm} - 60.3\text{ cm} = 3.7\text{ cm}$$

(b) Find the locations of nodes for the second and fourth harmonics.

The second harmonic has a wavelength $\lambda_2 = L_0 = 64.0$ cm. The distance from nut to node corresponds to half a wavelength.

$$\Delta x = \tfrac{1}{2}\lambda_2 = \tfrac{1}{2}L_0 = 32.0 \text{ cm}$$

The fourth harmonic, of wavelength $\lambda_4 = \frac{1}{2}L_0 = 32.0$ cm, has three nodes between the endpoints:

$$\Delta x = \tfrac{1}{2}\lambda_4 = \boxed{16.0 \text{ cm}},\ \Delta x = 2(\lambda_4/2) = \boxed{32.0 \text{ cm}},$$
$$\Delta x = 3(\lambda_4/2) = \boxed{48.0 \text{ cm}}$$

Remarks Placing a finger at the position $\Delta x = 32.0$ cm damps out the fundamental and odd harmonics, but not all the higher even harmonics. The second harmonic dominates, however, because the rest of the string is free to vibrate. Placing the finger at $\Delta x = 16.0$ cm or 48.0 cm damps out the first through third harmonics, allowing the fourth harmonic to be heard.

Exercise 14.7

Pressing the E-string down on the fret board just above the second fret pinches the string firmly against the fret, giving an F sharp, which has frequency 3.70×10^2 Hz. (a) Where should the second fret be located? (b) Find two locations where you could touch the open E-string and hear the third harmonic.

Answer (a) 7.1 cm from the nut and 3.4 cm from the first fret. Note that the distance from the first to the second fret isn't the same as from the nut to the first fret. (b) 21.3 cm and 42.7 cm from the nut.

PhysicsNow™

Explore this situation by logging into PhysicsNow at **www.cp7e.com** and going to Interactive Example 14.7.

EXAMPLE 14.8 Harmonics of a Stretched Wire

Goal Calculate string harmonics, relate them to sound and combine them with tensile stress.

Problem **(a)** Find the frequencies of the fundamental, second, and third harmonics of a steel wire 1.00 m long with a mass per unit length of 2.00×10^{-3} kg/m and under a tension of 80.0 N. **(b)** Find the wavelengths of the sound waves created by the vibrating wire for all three modes. Assume the speed of sound in air is 345 m/s. **(c)** Suppose the wire is carbon steel with a density of 7.80×10^3 kg/m^3, a cross-sectional area $A = 2.56 \times 10^{-7}$ m^2, and an elastic limit of 2.80×10^8 Pa. Find the fundamental frequency if the wire is tightened to the elastic limit. Neglect any stretching of the wire (which would slightly reduce the mass per unit length).

Strategy (a) It's easiest to find the speed of waves on the wire then substitute into Equation 14.15 to find the first harmonic. The next two are multiples of the first, given by Equation 14.17. (b) The frequencies of the sound waves are the same as the frequencies of the vibrating wire, but the wavelengths are different. Use $v_s = f\lambda$, where v_s is the speed of sound in air, to find the wavelengths in air. (c) Find the force corresponding to the elastic limit, and substitute it into Equation 14.16.

Solution

(a) Find the first three harmonics at the given tension.

Use Equation 13.17 to calculate the speed of the wave on the wire:

$$v = \sqrt{\frac{F}{\mu}} = \sqrt{\frac{80.0 \text{ N}}{2.00 \times 10^{-3} \text{ kg/m}}} = 2.00 \times 10^2 \text{ m/s}$$

Find the wire's fundamental frequency from Equation 14.15:

$$f_1 = \frac{v}{2L} = \frac{2.00 \times 10^2 \text{ m/s}}{2(1.00 \text{ m})} = \boxed{1.00 \times 10^2 \text{ Hz}}$$

Find the next two harmonics by multiplication:

$$f_2 = 2f_1 = \boxed{2.00 \times 10^2 \text{ Hz}},\ f_3 = 3f_1 = \boxed{3.00 \times 10^2 \text{ Hz}}$$

(b) Find the wavelength of the sound waves produced.

Solve $v_s = f\lambda$ for the wavelength and substitute the frequencies.

$$\lambda_1 = v_s/f_1 = (345\text{ m/s})/(1.00 \times 10^2\text{ Hz}) = 3.45\text{ m}$$

$$\lambda_2 = v_s/f_2 = (345\text{ m/s})/(2.00 \times 10^2\text{ Hz}) = 1.73\text{ m}$$

$$\lambda_3 = v_s/f_3 = (345\text{ m/s})/(3.00 \times 10^2\text{ Hz}) = 1.15\text{ m}$$

(c) Find the fundamental frequency corresponding to the elastic limit.

Calculate the tension in the wire from the elastic limit:

$$\frac{F}{A} = \text{elastic limit} \quad \rightarrow \quad F = (\text{elastic limit})A$$

$$F = (2.80 \times 10^8\text{ Pa})(2.56 \times 10^{-7}\text{ m}^2) = 71.7\text{ N}$$

Substitute the values of F, μ, and L into Equation 14.16:

$$f_1 = \frac{1}{2L}\sqrt{\frac{F}{\mu}}$$

$$f_1 = \frac{1}{2(1.00\text{ m})}\sqrt{\frac{71.7\text{ N}}{2.00 \times 10^{-3}\text{ kg/m}}} = 94.7\text{ Hz}$$

Remarks From the answer to part (c), it appears we need to choose a thicker wire or use a better grade of steel with a higher elastic limit. The frequency corresponding to the elastic limit is smaller than the fundamental!

Exercise 14.8

(a) Find the fundamental frequency and second harmonic if the tension in the wire is increased to 115 N. (Assume the wire doesn't stretch or break.) (b) Using a sound speed of 345 m/s, find the wavelengths of the sound waves produced.

Answer (a) 1.20×10^2 Hz, 2.40×10^2 Hz (b) 2.88 m, 1.44 m

14.9 FORCED VIBRATIONS AND RESONANCE

In Chapter 13 we learned that the energy of a damped oscillator decreases over time because of friction. It's possible to compensate for this energy loss by applying an external force that does positive work on the system.

For example, suppose an object–spring system having some natural frequency of vibration f_0 is pushed back and forth by a periodic force with frequency f. The system vibrates at the frequency f of the driving force. This type of motion is referred to as a **forced vibration**. Its amplitude reaches a maximum when the frequency of the driving force equals the natural frequency of the system f_0, called the **resonant frequency** of the system. Under this condition, the system is said to be in **resonance**.

In Section 14.8 we learned that a stretched string can vibrate in one or more of its natural modes. Here again, if a periodic force is applied to the string, the amplitude of vibration increases as the frequency of the applied force approaches one of the string's natural frequencies of vibration.

Resonance vibrations occur in a wide variety of circumstances. Figure 14.20 illustrates one experiment that demonstrates a resonance condition. Several pendulums of different lengths are suspended from a flexible beam. If one of them, such as A, is set in motion, the others begin to oscillate because of vibrations in the flexible beam. Pendulum C, the same length as A, oscillates with the greatest amplitude because its natural frequency matches that of pendulum A (the driving force).

Another simple example of resonance is a child being pushed on a swing, which is essentially a pendulum with a natural frequency that depends on its length. The

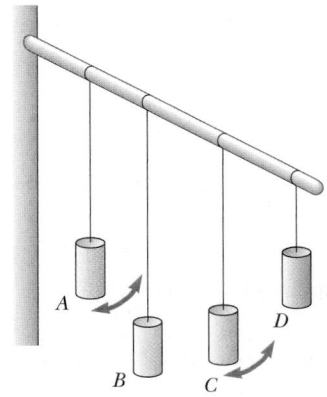

Figure 14.20 Resonance. If pendulum A is set in oscillation, only pendulum C, with a length matching that of A, will eventually oscillate with a large amplitude, or resonate. The arrows indicate motion perpendicular to the page.

APPLICATION

Shattering Goblets with the Voice

swing is kept in motion by a series of appropriately timed pushes. For its amplitude to increase, the swing must be pushed each time it returns to the person's hands. This corresponds to a frequency equal to the natural frequency of the swing. If the energy put into the system per cycle of motion equals the energy lost due to friction, the amplitude remains constant.

Opera singers have been known to set crystal goblets in audible vibration with their powerful voices, as shown in Figure 14.21. This is yet another example of resonance: The sound waves emitted by the singer can set up large-amplitude vibrations in the glass. If a highly amplified sound wave has the right frequency, the amplitude of forced vibrations in the glass increases to the point where the glass becomes heavily strained and shatters.

The classic example of structural resonance occurred in 1940, when the Tacoma Narrows bridge in the state of Washington was set in oscillation by the wind (Fig. 14.22). The amplitude of the oscillations increased rapidly and reached a high value until the bridge ultimately collapsed (probably because of metal fatigue). In recent years, however, a number of researchers have called this explanation into question. Gusts of wind, in general, don't provide the periodic force necessary for a sustained resonance condition, and the bridge exhibited large twisting oscillations, rather than the simple up-and-down oscillations expected of resonance.

A more recent example of destruction by structural resonance occurred during the Loma Prieta earthquake near Oakland, California, in 1989. In a mile-long section of the double-decker Nimitz Freeway, the upper deck collapsed onto the lower deck, killing several people. The collapse occurred because that particular section was built on mud fill while other parts were built on bedrock. As seismic waves pass through mud fill or other loose soil, their speed decreases and their amplitude increases. The section of the freeway that collapsed oscillated at the same frequency as other sections, but at a much larger amplitude.

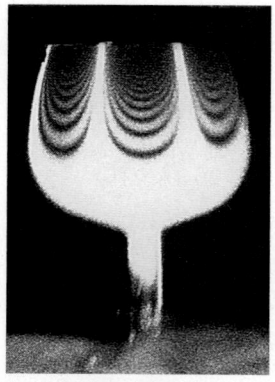

Courtesy of Professor Thomas D. Rossing/Getty Images
Northern Illinois University

Ben Rose/The IMAGE Bank/Getty Images

Figure 14.21 (*Top*) Standing-wave pattern in a vibrating wineglass. The glass will shatter if the amplitude of vibration becomes too large. (*Bottom*) A wineglass shattered by the amplified sound of a human voice.

14.10 STANDING WAVES IN AIR COLUMNS

Standing longitudinal waves can be set up in a tube of air, such as an organ pipe, as the result of interference between sound waves traveling in opposite directions. The relationship between the incident wave and the reflected wave depends on whether the reflecting end of the tube is open or closed. A portion of the sound wave is reflected back into the tube even at an open end. **If one end is closed, a node must exist at that end because the movement of air is restricted. If the end is open, the elements of air have complete freedom of motion, and an antinode exists.**

Figure 14.23a shows the first three modes of vibration of a pipe open at both ends. When air is directed against an edge at the left, longitudinal standing waves are formed and the pipe vibrates at its natural frequencies. Note that, from end to end, the pattern is A–N–A, the same pattern as in the vibrating string, except node and antinode have exchanged positions. As before, an antinode and its adjacent node, A–N, represent a quarter-wavelength, and there are two, A–N and N–A, so $L = 2(\lambda_1/4) = \lambda_1/2$ and $\lambda_1 = 2L$. The fundamental frequency of the pipe open at both ends is then $f_1 = v/\lambda_1 = v/2L$. The next harmonic has an addi-

United Press International Photo

Figure 14.22 The collapse of the Tacoma Narrows suspension bridge in 1940 has been cited as a demonstration of mechanical resonance. High winds set up standing waves in the bridge, causing it to oscillate at one of its natural frequencies. Once established, the resonance may have led to the bridge's collapse (although this interpretation is currently being challenged by mathematicians and physical scientists).

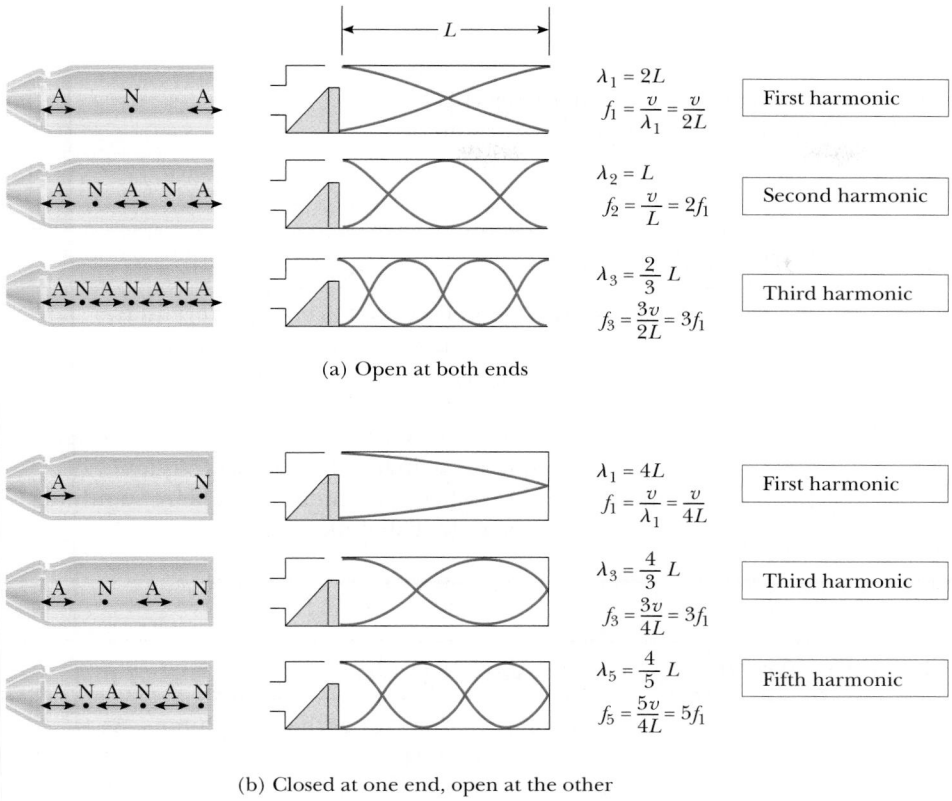

Figure 14.23 (a) Standing longitudinal waves in an organ pipe open at both ends. The natural frequencies f_1, $2f_1$, $3f_1$ . . . form a harmonic series. (b) Standing longitudinal waves in an organ pipe closed at one end. Only *odd* harmonics are present, and the natural frequencies are f_1, $3f_1$, $5f_1$, and so on.

TIP 14.4 Sound Waves Are Not Transverse

The standing longitudinal waves in Figure 14.23 are drawn as transverse waves only because it's difficult to draw longitudinal displacements—they're in the same direction as the wave propagation. In the figure, the vertical axis represents either pressure or horizontal displacement of the elements of the medium.

tional node and antinode between the ends, creating the pattern A–N–A–N–A. We count the pairs: A–N, N–A, A–N, and N–A, making four segments, each with length $\lambda_2/4$. We have $L = 4(\lambda_2/4) = \lambda_2$, and the second harmonic (first overtone) is $f_2 = v/\lambda_2 = v/L = 2(v/2L) = 2f_1$. All higher harmonics, it turns out, are positive integer multiples of the fundamental:

$$f_n = n\frac{v}{2L} = nf_1 \quad n = 1, 2, 3, \ldots \qquad \textbf{[14.18]}$$

◀ Pipe open at both ends; all harmonics are present

where v is the speed of sound in air. Notice the similarity to Equation 14.17, which also involves multiples of the fundamental.

If a pipe is open at one end and closed at the other, the open end is an antinode while the closed end is a node (Fig. 14.23b). In such a pipe, the fundamental frequency consists of a single antinode–node pair, A–N, so $L = \lambda_1/4$ and $\lambda_1 = 4L$. The fundamental harmonic for a pipe closed at one end is then $f_1 = v/\lambda_1 = v/4L$. The first overtone has another node and antinode between the open end and closed end, making the pattern A–N–A–N. There are three antinode–node segments in this pattern (A–N, N–A, and A–N), so $L = 3(\lambda_3/4)$ and $\lambda_3 = 4L/3$. The first overtone, therefore, has frequency $f_3 = v/\lambda_3 = 3v/4L = 3f_1$. Similarly, $f_5 = 5f_1$. In contrast to the pipe open at both ends, **there are no even multiples of the fundamental harmonic**. The odd harmonics for a pipe open at one end only are given by

$$f_n = n\frac{v}{4L} = nf_1 \quad n = 1, 3, 5, \ldots \qquad \textbf{[14.19]}$$

◀ Pipe closed at one end; only odd harmonics are present

Quick Quiz 14.5

A pipe open at both ends resonates at a fundamental frequency f_{open}. When one end is covered and the pipe is again made to resonate, the fundamental frequency is f_{closed}. Which of the following expressions describes how these two resonant frequencies compare? (a) $f_{\text{closed}} = f_{\text{open}}$, (b) $f_{\text{closed}} = \frac{3}{2}f_{\text{open}}$, (c) $f_{\text{closed}} = 2\,f_{\text{open}}$, (d) $f_{\text{closed}} = \frac{1}{2}f_{\text{open}}$, (e) none of these.

Quick Quiz 14.6

Balboa Park in San Diego has an outdoor organ. When the air temperature increases, the fundamental frequency of one of the organ pipes (a) increases (b) decreases (c) stays the same (d) impossible to determine. (The thermal expansion of the pipe is negligible.)

Applying Physics 14.3 Oscillations in a Harbor

Why do passing ocean waves sometimes cause the water in a harbor to undergo very large oscillations, called a *seiche* (pronounced *saysh*)?

Explanation Water in a harbor is enclosed and possesses a natural frequency based on the size of the harbor. This is similar to the natural frequency of the enclosed air in a bottle, which can be excited by blowing across the edge of the opening. Ocean waves pass by the opening of the harbor at a certain frequency. If this frequency matches that of the enclosed harbor, then a large standing wave can be set up in the water by resonance. This situation can be simulated by carrying a fish tank filled with water. If your walking frequency matches the natural frequency of the water as it sloshes back and forth, a large standing wave develops in the fish tank.

Applying Physics 14.4 Why Are Instruments Warmed Up?

Why do the strings go flat and the wind instruments go sharp during a performance if an orchestra doesn't warm up beforehand?

Explanation Without warming up, all the instruments will be at room temperature at the beginning of the concert. As the wind instruments are played, they fill with warm air from the player's exhalation. The increase in temperature of the air in the instruments causes an increase in the speed of sound, which raises the resonance frequencies of the air columns. As a result, the instruments go sharp. The strings on the stringed instruments also increase in temperature due to the friction of rubbing with the bow. This results in thermal expansion, which causes a decrease in tension in the strings. With the decrease in tension, the wave speed on the strings drops, and the fundamental frequencies decrease, so the stringed instruments go flat.

Applying Physics 14.5 How Do Bugles Work?

A bugle has no valves, keys, slides, or finger holes. How can it be used to play a song?

Explanation Songs for the bugle are limited to harmonics of the fundamental frequency, because there is no control over frequencies without valves, keys, slides, or finger holes. The player obtains different notes by changing the tension in the lips as the bugle is played, in order to excite different harmonics. The normal playing range of a bugle is among the third, fourth, fifth, and sixth harmonics of the fundamental. "Reveille," for example, is played with just the three notes G, C, and F. And "Taps" is played with these three notes and the G one octave above the lower G.

EXAMPLE 14.9 Harmonics of a Pipe

Goal Find frequencies of open and closed pipes.

Problem A pipe is 2.46 m long. **(a)** Determine the frequencies of the first three harmonics if the pipe is open at both ends. Take 343 m/s as the speed of sound in air. **(b)** How many harmonic frequencies of this pipe lie in the audible range, from 20 Hz to 20 000 Hz? **(c)** What are the three lowest possible frequencies if the pipe is closed at one end and open at the other?

Strategy Substitute into Equation 14.18 for part (a) and Equation 14.19 for part (c). All harmonics, $n = 1, 2, 3 \ldots$ are available for the pipe open at both ends, but only the harmonics with $n = 1, 3, 5, \ldots$ for the pipe closed at one end. For part (b), set the frequency in Equation 14.18 equal to 2.00×10^4 Hz.

Solution

(a) Find the frequencies if the pipe is open at both ends.

Substitute into Equation 14.18, with $n = 1$:

$$f_1 = \frac{v}{2L} = \frac{343 \text{ m/s}}{2(2.46 \text{ m})} = \boxed{69.7 \text{ Hz}}$$

Multiply to find the second and third harmonics:

$$f_2 = 2f_1 = \boxed{139 \text{ Hz}} \qquad f_3 = 3f_1 = \boxed{209 \text{ Hz}}$$

(b) How many harmonics lie between 20 Hz and 20 000 Hz for this pipe?

Set the frequency in Equation 14.18 equal to 2.00×10^4 and solve for n:

$$f_n = n\frac{v}{2L} = n\frac{343 \text{ m/s}}{2 \cdot 2.46 \text{ m}} = 2.00 \times 10^4 \text{ Hz}$$

This works out to $n = 286.88$, which must be truncated down ($n = 287$ gives a frequency over 2.00×10^4 Hz).

$$\boxed{n = 286}$$

(c) Find the frequencies for the pipe closed at one end.

Apply Equation 14.19 with $n = 1$:

$$f_1 = \frac{v}{4L} = \frac{343 \text{ m/s}}{4(2.46 \text{ m})} = \boxed{34.9 \text{ Hz}}$$

The next two harmonics are odd multiples of the first:

$$f_3 = 3f_1 = \boxed{105 \text{ Hz}} \qquad f_5 = 5f_1 = \boxed{175 \text{ Hz}}$$

Exercise 14.9

(a) What length pipe open at both ends has a fundamental frequency of 3.70×10^2 Hz? Find the first overtone. (b) If the one end of this pipe is now closed, what is the new fundamental frequency? Find the first overtone. (c) If the pipe is open at one end only, how many harmonics are possible in the normal hearing range from 20 to 20 000 Hz?

Answer (a) 0.464 m, 7.40×10^2 Hz (b) 185 Hz, 555 Hz (c) 54

EXAMPLE 14.10 Resonance in a Tube of Variable Length

Goal Understand resonance in tubes and perform elementary calculations.

Problem Figure 14.24a shows a simple apparatus for demonstrating resonance in a tube. A long tube open at both ends is partially submerged in a beaker of water, and a vibrating tuning fork of unknown frequency is placed near the top of the tube. The length of the air column, L, is adjusted by moving the tube vertically. The sound waves generated by the fork are reinforced when the length of the air column corresponds to one of the resonant frequencies of the tube. Suppose the smallest value of L for which a peak occurs in the sound intensity is 9.00 cm. **(a)** With this measurement, determine the frequency of the tuning fork. **(b)** Find the wavelength and the next two air-column lengths giving resonance. Take the speed of sound to be 345 m/s.

Strategy Once the tube is in the water, the setup is the same as a pipe closed at one end. For part (a), substitute values for v and L into Equation 14.19 with $n = 1$ and find the frequency of the tuning fork. (b) The next resonance maximum occurs when the water level is low enough to allow a second node, which is another half-wavelength in distance. The third resonance occurs when the third node is reached, requiring yet another half-wavelength of distance. The frequency in each case is the same, because it's generated by the tuning fork.

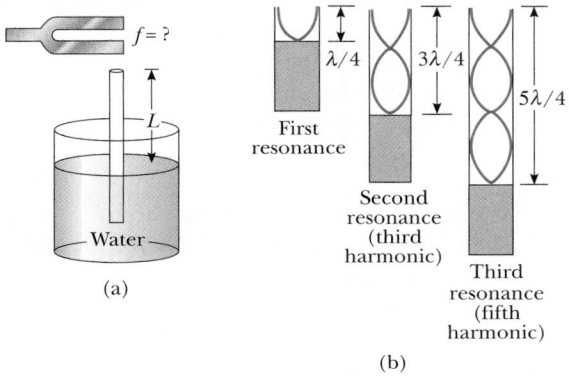

Figure 14.24 (Example 14.10) (a) Apparatus for demonstrating the resonance of sound waves in a tube closed at one end. The length L of the air column is varied by moving the tube vertically while it is partially submerged in water. (b) The first three resonances of the system.

Solution

(a) Find the frequency of the tuning fork.

Substitute $n = 1$, $v = 345$ m/s, and $L_1 = 9.00 \times 10^{-2}$ m into Equation 14.19:

$$f_1 = \frac{v}{4L_1} = \frac{345 \text{ m/s}}{4(9.00 \times 10^{-2} \text{ m})} = \boxed{958 \text{ Hz}}$$

(b) Find the wavelength and the next two water levels giving resonance.

Calculate the wavelength, using the fact that, for a tube open at one end, $\lambda = 4L$ for the fundamental.

$$\lambda = 4L_1 = 4(9.00 \times 10^{-2} \text{ m}) = \boxed{0.360 \text{ m}}$$

Add a half-wavelength of distance to L_1 to get the next resonance position:

$$L_2 = L_1 + \lambda/2 = 0.0900 \text{ m} + 0.180 \text{ m} = \boxed{0.270 \text{ m}}$$

Add another half-wavelength to L_2 to obtain the third resonance position:

$$L_3 = L_2 + \lambda/2 = 0.270 \text{ m} + 0.180 \text{ m} = \boxed{0.450 \text{ m}}$$

Remark This experimental arrangement is often used to measure the speed of sound, in which case the frequency of the tuning fork must be known in advance.

Exercise 14.10

An unknown gas is introduced into the aforementioned apparatus using the same tuning fork, and the first resonance occurs when the air column is 5.84 cm long. Find the speed of sound in the gas.

Answer 224 m/s

14.11 BEATS

The interference phenomena we have been discussing so far have involved the superposition of two or more waves with the same frequency, traveling in opposite directions. Another type of interference effect results from the superposition of two waves with slightly different frequencies. In such a situation, the waves at some fixed point are periodically in and out of phase, corresponding to an alternation in time between constructive and destructive interference. In order to understand this phenomenon, consider Active Figure 14.25. The two waves shown in Active Figure 14.25a were emitted by two tuning forks having slightly different frequencies; Active Figure 14.25b shows the superposition of these waves. At some time t_a the waves are in phase and constructive interference occurs, as demonstrated by the resultant curve in Active Figure 14.25b. At some later time, however, the vibrations of the two forks move out of step with each other. At time t_b, one fork emits a compression while the other emits a rarefaction, and destructive interference occurs, as demonstrated by the curve shown. As time passes, the vibrations of the two forks move out of phase, then into phase again, and so on. As a consequence, a listener at some fixed point hears an alternation in loudness, known as **beats**. The number of beats per second, or the *beat frequency*, equals the difference in frequency between the two sources:

$$f_b = |f_2 - f_1| \qquad \textbf{[14.20]}$$

where f_b is the beat frequency and f_1 and f_2 are the two frequencies. The absolute value is used because the beat frequency is a positive quantity and will occur regardless of the order of subtraction.

APPLICATION

Using Beats to Tune a Musical Instrument

A stringed instrument such as a piano can be tuned by beating a note on the instrument against a note of known frequency. The string can then be tuned to the desired frequency by adjusting the tension until no beats are heard.

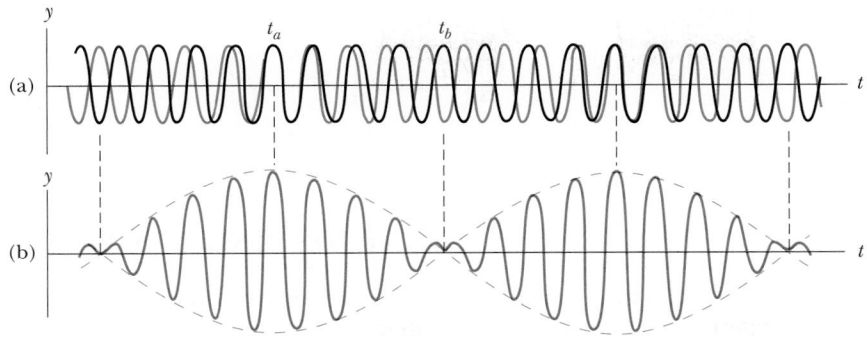

ACTIVE FIGURE 14.25
Beats are formed by the combination of two waves of slightly different frequencies traveling in the same direction. (a) The individual waves heard by an observer at a fixed point in space. (b) The combined wave has an amplitude (dashed line) that oscillates in time.

PhysicsNow™
Log into PhysicsNow at **www.cp7e.com**, and go to Active Figure 14.25 to choose two frequencies and see the corresponding beats.

Quick Quiz 14.7

You are tuning a guitar by comparing the sound of the string with that of a standard tuning fork. You notice a beat frequency of 5 Hz when both sounds are present. As you tighten the guitar string, the beat frequency rises steadily to 8 Hz. In order to tune the string exactly to the tuning fork, you should (a) continue to tighten the string (b) loosen the string (c) impossible to determine from the given information.

EXAMPLE 14.11 Sour Notes

Goal Apply the beat frequency concept.

Problem A certain piano string is supposed to vibrate at a frequency of 4.40×10^2 Hz. In order to check its frequency, a tuning fork known to vibrate at a frequency of 4.40×10^2 Hz is sounded at the same time the piano key is struck, and a beat frequency of 4 beats per second is heard. **(a)** Find the two possible frequencies at which the string could be vibrating. **(b)** Suppose the piano tuner runs toward the piano, holding the vibrating tuning fork while his assistant plays the note, which is at 436 Hz. At his maximum speed, the piano tuner notices the beat frequency drops from 4 Hz to 2 Hz (without going through a beat frequency of zero). How fast is he moving? Use a sound speed of 343 m/s. **(c)** While the piano tuner is running, what beat frequency is observed by the assistant? [*Note:* Assume all numbers are accurate to two decimal places, necessary for this last calculation.]

Strategy (a) The beat frequency is equal to the absolute value of the difference in frequency between the two sources of sound and occurs if the piano string is tuned either too high or too low. Solve Equation 14.20 for these two possible frequencies. (b) Moving toward the piano raises the observed piano string frequency. Solve the Doppler shift formula, Equation 14.12, for the speed of the observer. (c) The assistant observes a Doppler shift for the tuning fork. Apply Equation 14.12.

Solution

(a) Find the two possible frequencies.

Case 1: $f_2 - f_1$ is already positive, so just drop the absolute-value signs:

$$f_b = f_2 - f_1 \quad \rightarrow \quad 4\text{ Hz} = f_2 - 4.40 \times 10^2\text{ Hz}$$

$$f_2 = 444\text{ Hz}$$

Case 2: $f_2 - f_1$ is negative, so drop the absolute-value signs, but apply an overall negative sign:

$$f_b = -(f_2 - f_1) \quad \rightarrow \quad 4\text{ Hz} = -(f_2 - 4.40 \times 10^2\text{ Hz})$$

$$f_2 = 436\text{ Hz}$$

(b) Find the speed of the observer if running toward the piano results in a beat frequency of 2 Hz.

Apply the Doppler shift to the case where frequency of the piano string heard by the running observer is $f_O = 438$ Hz:

$$f_O = f_S\left(\frac{v + v_O}{v - v_S}\right)$$

$$438\text{ Hz} = (436\text{ Hz})\left(\frac{343\text{ m/s} + v_O}{343\text{ m/s}}\right)$$

$$v_O = \left(\frac{438\text{ Hz} - 436\text{ Hz}}{436\text{ Hz}}\right)(343\text{ m/s}) = 1.57\text{ m/s}$$

(c) What beat frequency does the assistant observe?

Apply Equation 14.12. Now the source is the tuning fork, so $f_S = 4.40 \times 10^2$ Hz.

$$f_O = f_S\left(\frac{v + v_O}{v - v_S}\right)$$

$$= (4.40 \times 10^2 \text{ Hz})\left(\frac{343 \text{ m/s}}{343 \text{ m/s} - 1.57 \text{ m/s}}\right) = 442 \text{ Hz}$$

Compute the beat frequency.

$$f_b = f_2 - f_1 = 442 \text{ Hz} - 436 \text{ Hz} = \boxed{6 \text{ Hz}}$$

Remarks The assistant on the piano bench and the tuner running with the fork observe different beat frequencies. Many physical observations depend on the state of motion of the observer, a subject discussed more fully in Chapter 26, on relativity.

Exercise 14.11

The assistant adjusts the tension in the same piano string, and a beat frequency of 2.00 Hz is heard when the note and the tuning fork are struck at the same time. (a) Find the two possible frequencies of the string. (b) Assume the actual string frequency is the higher frequency. If the piano tuner runs away from the piano at 4.00 m/s while holding the vibrating tuning fork, what beat frequency does he hear? (c) What beat frequency does the assistant on the bench hear? Use 343 m/s for the speed of sound.

Answers (a) 438 Hz, 442 Hz (b) 3 Hz (c) 7 Hz

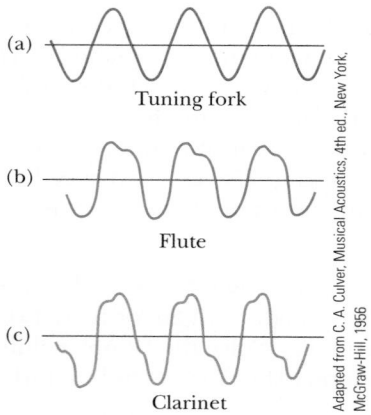

Figure 14.26 Waveforms produced by (a) a tuning fork, (b) a flute, and (c) a clarinet, all at approximately the same frequency. Pressure is plotted vertically, time horizontally.

14.12 QUALITY OF SOUND

The sound-wave patterns produced by most musical instruments are complex. Figure 14.26 shows characteristic waveforms (pressure is plotted on the vertical axis, time on the horizontal axis) produced by a tuning fork, a flute, and a clarinet, each playing the same steady note. Although each instrument has its own characteristic pattern, the figure reveals that each of the waveforms is periodic. Note that the tuning fork produces only one harmonic (the fundamental frequency), but the two instruments emit mixtures of harmonics. Figure 14.27 graphs the harmonics of the waveforms of Figure 14.26. When the note is played on the flute (Fig. 14.26b), part of the sound consists of a vibration at the fundamental frequency, an even higher intensity is contributed by the second harmonic, the fourth harmonic produces about the same intensity as the fundamental, and so on. These sounds add together according to the principle of superposition to give the complex waveform shown. The clarinet emits a certain intensity at a frequency of the first harmonic, about half as much intensity at the frequency of the second harmonic, and so forth. The resultant superposition of these frequencies produces the pattern shown in Figure 14.26c. The tuning fork (Figs. 14.26a and 14.27a) emits sound only at the frequency of the first harmonic.

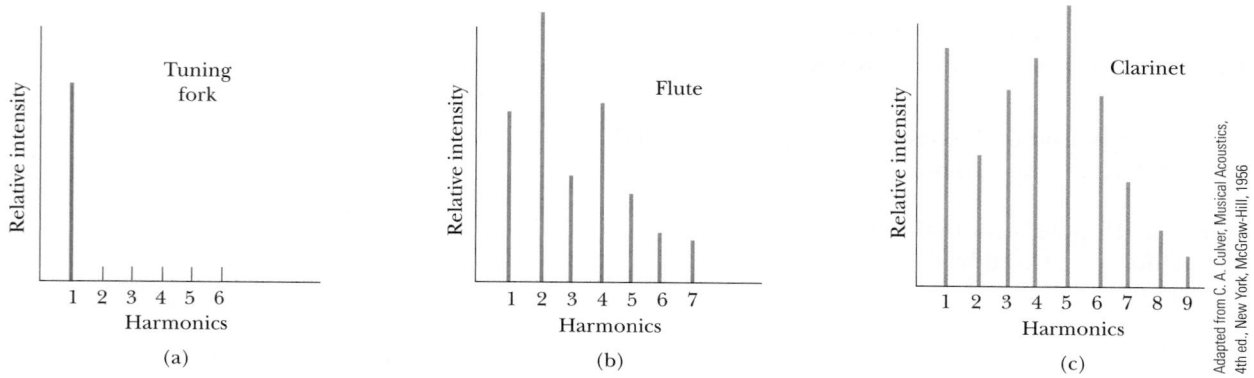

Figure 14.27 Harmonics of the waveforms in Figure 14.26. Note their variation in intensity.

In music, the characteristic sound of any instrument is referred to as the *quality*, or *timbre*, of the sound. The quality depends on the mixture of harmonics in the sound. We say that the note C on a flute differs in quality from the same C on a clarinet. Instruments such as the bugle, trumpet, violin, and tuba are rich in harmonics. A musician playing a wind instrument can emphasize one or another of these harmonics by changing the configuration of the lips, thereby playing different musical notes with the same valve openings.

TIP 14.5 Pitch is Not The Same as Frequency

Although pitch is related mostly (but not completely) to frequency, they are not the same. A phrase such as "the pitch of the sound" is incorrect, because pitch is not a physical property of the sound. Frequency is the physical measurement of the number of oscillations per second of the sound. Pitch is a psychological reaction to sound that enables a human being to place the sound on a scale from high to low or from treble to bass. Frequency is the stimulus and pitch is the response.

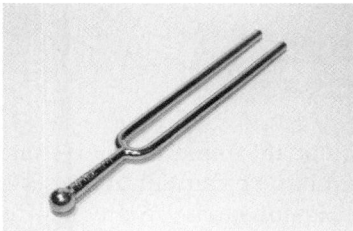

(a)

(b)

(c)

a–c, Royalty-Free/Corbis

Each musical instrument has its own characteristic sound and mixture of harmonics. (See Figures 14.26 and 14.27.) Instruments shown are (a) the tuning fork, (b) the flute, and (c) the clarinet.

Applying Physics 14.6 Why Does the Professor Sound Like Donald Duck?

A professor performs a demonstration in which he breathes helium and then speaks with a comical voice. One student explains, "The velocity of sound in helium is higher than in air, so the fundamental frequency of the standing waves in the mouth is increased." Another student says, "No, the fundamental frequency is determined by the vocal folds and cannot be changed. Only the quality of the voice has changed." Which student is correct?

Explanation The second student is correct. The fundamental frequency of the complex tone from the voice is determined by the vibration of the vocal folds and is not changed by substituting a different gas in the mouth. The introduction of the helium into the mouth results in harmonics of higher frequencies being excited more than in the normal voice, but the fundamental frequency of the voice is the same—only the quality has changed. The unusual inclusion of the higher frequency harmonics results in a common description of this effect as a "high-pitched" voice, but that description is incorrect. (It is really a "quacky" timbre.)

14.13 THE EAR

The human ear is divided into three regions: the outer ear, the middle ear, and the inner ear (Fig. 14.28, page 488). The *outer ear* consists of the ear canal (which is open to the atmosphere), terminating at the eardrum (tympanum). Sound waves travel down the ear canal to the eardrum, which vibrates in and out in phase with the pushes and pulls caused by the alternating high and low pressures of the waves. Behind the eardrum are three small bones of the *middle ear*, called the hammer, the anvil, and the stirrup because of their shapes. These bones transmit the vibration to the *inner ear*, which contains the cochlea, a snail-shaped tube about 2 cm long. The cochlea makes contact with the stirrup at the oval window and is divided along its length by the basilar membrane, which consists of small hairs (cilia) and nerve fibers. This membrane varies in mass per unit length and in tension along its length, and different portions of it resonate at different frequencies. (Recall that the natural frequency of a string depends on its mass per unit length

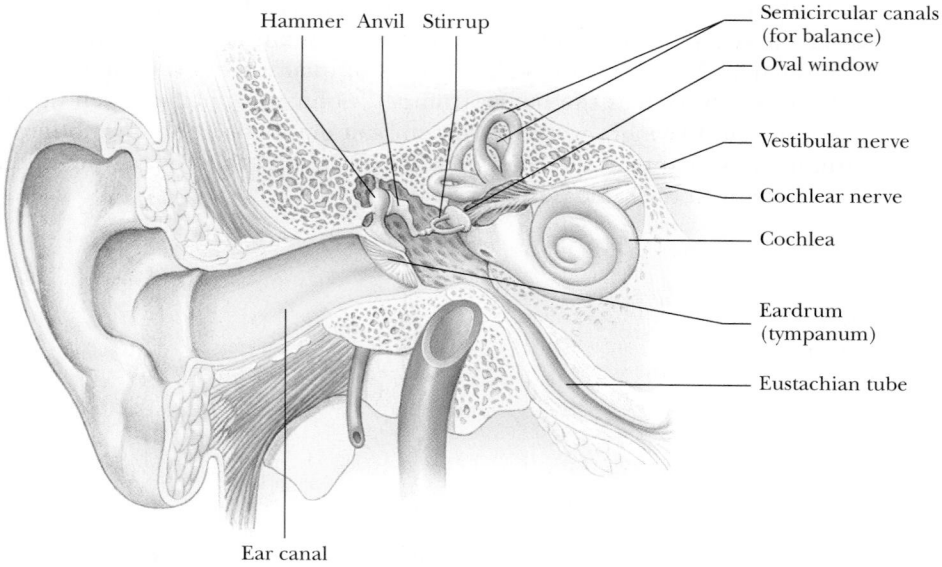

Figure 14.28 The structure of the human ear. The three tiny bones (ossicles) that connect the eardrum to the window of the cochlea act as a double-lever system to decrease the amplitude of vibration and hence increase the pressure on the fluid in the cochlea.

and on the tension in it.) Along the basilar membrane are numerous nerve endings, which sense the vibration of the membrane and in turn transmit impulses to the brain. The brain interprets the impulses as sounds of varying frequency, depending on the locations along the basilar membrane of the impulse-transmitting nerves and on the rates at which the impulses are transmitted.

Figure 14.29 shows the frequency response curves of an average human ear for sounds of equal loudness, ranging from 0 to 120 dB. To interpret this series of graphs, take the bottom curve as the threshold of hearing. Compare the intensity level on the vertical axis for the two frequencies 100 Hz and 1 000 Hz. The vertical axis shows that the 100-Hz sound must be about 38 dB greater than the 1 000-Hz sound to be at the threshold of hearing, which means that the threshold of hearing is very strongly dependent on frequency. The easiest frequencies to hear are around 3 300 Hz; those above 12 000 Hz or below about 50 Hz must be relatively intense to be heard.

Now consider the curve labeled 80. This curve uses a 1 000-Hz tone at an intensity level of 80 dB as its reference. The curve shows that a tone of frequency 100 Hz would have to be about 4 dB louder than the 80-dB, 1 000-Hz tone in order to sound as loud. Notice that the curves flatten out as the intensities levels of the sounds increase, so when sounds are loud, all frequencies can be heard equally well.

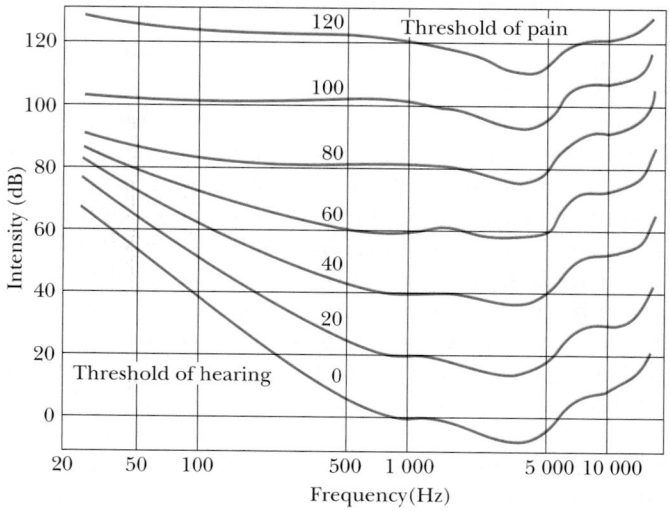

Figure 14.29 Curves of intensity level versus frequency for sounds that are perceived to be of equal loudness. Note that the ear is most sensitive at a frequency of about 3 300 Hz. The lowest curve corresponds to the threshold of hearing for only about 1% of the population.

The small bones in the middle ear represent an intricate lever system that increases the force on the oval window. The pressure is greatly magnified because the surface area of the eardrum is about 20 times that of the oval window (in analogy with a hydraulic press). The middle ear, together with the eardrum and oval window, in effect acts as a matching network between the air in the outer ear and the liquid in the inner ear. The overall energy transfer between the outer ear and the inner ear is highly efficient, with pressure amplification factors of several thousand. In other words, pressure variations in the inner ear are much greater than those in the outer ear.

The ear has its own built-in protection against loud sounds. The muscles connecting the three middle-ear bones to the walls control the volume of the sound by changing the tension on the bones as sound builds up, thus hindering their ability to transmit vibrations. In addition, the eardrum becomes stiffer as the sound intensity increases. These two events make the ear less sensitive to loud incoming sounds. There is a time delay between the onset of a loud sound and the ear's protective reaction, however, so a very sudden loud sound can still damage the ear.

The complex structure of the human ear is believed to be related to the fact that mammals evolved from seagoing creatures. In comparison, insect ears are considerably simpler in design, because insects have always been land residents. A typical insect ear consists of an eardrum exposed directly to the air on one side and to an air-filled cavity on the other side. Nerve cells communicate directly with the cavity and the brain, without the need for the complex intermediary of an inner and middle ear. This simple design allows the ear to be placed virtually anywhere on the body. For example, a grasshopper has its ears on its legs. One advantage of the simple insect ear is that the distance and orientation of the ears can be varied so that it is easier to locate sources of sound, such as other insects.

One of the most amazing medical advances in recent decades is the cochlear implant, allowing the deaf to hear. Deafness can occur when the hairlike sensors (cilia) in the cochlea break off over a lifetime or sometimes because of prolonged exposure to loud sounds. Because the cilia don't grow back, the ear loses sensitivity to certain frequencies of sound. The cochlear implant stimulates the nerves in the ear electronically to restore hearing loss that is due to damaged or absent cilia.

SUMMARY

PhysicsNow™ Take a practice test by logging into PhysicsNow at **www.cp7e.com** and clicking on the Pre-Test link for this chapter.

14.2 Characteristics of Sound Waves

Sound waves are longitudinal waves. **Audible waves** are sound waves with frequencies between 20 and 20 000 Hz. **Infrasonic waves** have frequencies below the audible range, and **ultrasonic waves** have frequencies above the audible range.

14.3 The Speed of Sound

The speed of sound in a medium of bulk modulus B and density ρ is

$$v = \sqrt{\frac{B}{\rho}} \qquad [14.1]$$

The speed of sound also depends on the temperature of the medium. The relationship between temperature and the speed of sound in air is

$$v = (331 \text{ m/s}) \sqrt{\frac{T}{273 \text{ K}}} \qquad [14.4]$$

where T is the absolute (Kelvin) temperature and 331 m/s is the speed of sound in air at 0°C.

14.4 Energy and Intensity of Sound Waves

The **average intensity** of sound incident on a surface is defined by

$$I \equiv \frac{\text{power}}{\text{area}} = \frac{\mathcal{P}}{A} \qquad [14.6]$$

where the power $\mathcal{P}$ is the energy per unit time flowing through the surface, which has area A. The **intensity level** of a sound wave is given by

$$\beta \equiv 10 \log \left(\frac{I}{I_0} \right) \qquad [14.7]$$

The constant $I_0 = 1.0 \times 10^{-12}$ W/m^2 is a reference intensity, usually taken to be at the threshold of hearing, and I is the intensity at level β, measured in **decibels** (dB).

14.5 Spherical and Plane Waves

The **intensity** of a *spherical wave* produced by a point source is proportional to the average power emitted and inversely proportional to the square of the distance from the source:

$$I = \frac{\mathcal{P}_{\text{av}}}{4\pi r^2} \qquad [14.8]$$

14.6 The Doppler Effect

The change in frequency heard by an observer whenever there is relative motion between a source of sound and the observer is called the **Doppler effect**. If the observer is moving with speed v_O and the source is moving with speed v_S, the observed frequency is

$$f_O = f_S\left(\frac{v + v_O}{v - v_S}\right) \qquad [14.12]$$

where v is the speed of sound. A positive speed is substituted for v_O when the observer moves toward the source, a negative speed when the observer moves away from the source. Similarly, a positive speed is substituted for v_S when the sources moves toward the observer, a negative speed when the source moves away.

14.7 Interference of Sound Waves

When waves interfere, the resultant wave is found by adding the individual waves together point by point. When crest meets crest and trough meets trough, the waves undergo **constructive interference**, with path length difference

$$r_2 - r_1 = n\lambda \quad (n = 0, 1, 2, \ldots) \qquad [14.13]$$

When crest meets trough, **destructive interference** occurs, with path length difference

$$r_2 - r_1 = (n + \tfrac{1}{2})\lambda \quad (n = 0, 1, 2, \ldots) \qquad [14.14]$$

14.8 Standing Waves

Standing waves are formed when two waves having the same frequency, amplitude, and wavelength travel in opposite directions through a medium. The natural frequencies of vibration of a stretched string of length L, fixed at both ends, are

$$f_n = nf_1 = \frac{n}{2L}\sqrt{\frac{F}{\mu}} \quad n = 1, 2, 3, \ldots \qquad [14.17]$$

where F is the tension in the string and μ is its mass per unit length.

14.9 Forced Vibrations and Resonance

A system capable of oscillating is said to be in **resonance** with some driving force whenever the frequency of the driving force matches one of the natural frequencies of the system. When the system is resonating, it oscillates with maximum amplitude.

14.10 Standing Waves in Air Columns

Standing waves can be produced in a tube of air. If the reflecting end of the tube is *open*, all harmonics are present and the natural frequencies of vibration are

$$f_n = n\frac{v}{2L} = nf_1 \quad n = 1, 2, 3, \ldots \qquad [14.18]$$

If the tube is *closed* at the reflecting end, only the *odd* harmonics are present and the natural frequencies of vibration are

$$f_n = n\frac{v}{4L} = nf_1 \quad n = 1, 3, 5, \ldots \qquad [14.19]$$

14.11 Beats

The phenomenon of **beats** is an interference effect that occurs when two waves with slightly different frequencies combine at a fixed point in space. For sound waves, the intensity of the resultant sound changes periodically with time. The *beat frequency* is

$$f_b = |f_2 - f_1| \qquad [14.20]$$

where f_2 and f_1 are the two source frequencies.

CONCEPTUAL QUESTIONS

1. (a) You are driving down the highway in your car when a police car sounding its siren overtakes you and passes you. If its frequency at rest is f_0, is the frequency you hear while the car is catching up to you higher or lower than f_0? (b) What about the frequency you hear after the car has passed you?
2. A crude model of the human throat is that of a pipe open at both ends with a vibrating source to introduce the sound into the pipe at one end. Assuming the vibrating source produces a range of frequencies, discuss the effect of changing the pipe's length.
3. An autofocus camera sends out a pulse of sound and measures the time taken for the pulse to reach an object, reflect off of it, and return to be detected. Can the temperature affect the camera's focus?
4. To keep animals away from their cars, some people mount short, thin pipes on the fenders. The pipes give out a high-pitched wail when the cars are moving. How do they create the sound?
5. Secret agents in the movies always want to get to a secure phone with a voice scrambler. How do these devices work?
6. When a bell is rung, standing waves are set up around its circumference. What boundary conditions must be satisfied by the resonant wavelengths? How does a crack in the bell, such as in the Liberty Bell, affect the satisfying of the boundary conditions and the sound emanating from the bell?
7. How does air temperature affect the tuning of a wind instrument?
8. Explain how the distance to a lightning bolt can be determined by counting the seconds between the flash and the sound of thunder.
9. You are driving toward a cliff and you honk your horn. Is there a Doppler shift of the sound when you hear the echo? Is it like a moving source or moving observer? What if the reflection occurs not from a cliff, but from the forward edge of a huge alien spacecraft moving toward you as you drive?

10. Of the following sounds, state which is most likely to have an intensity level of 60 dB: a rock concert, the turning of a page in this text, a normal conversation, a cheering crowd at a football game, and background noise at a church?

11. Guitarists sometimes play a "harmonic" by lightly touching a string at its exact center and plucking the string. The result is a clear note one octave higher than the fundamental frequency of the string, even though the string is not pressed to the fingerboard. Why does this happen?

12. Will two separate 50-dB sounds together constitute a 100-dB sound? Explain.

13. An archer shoots an arrow from a bow. Does the string of the bow exhibit standing waves after the arrow leaves? If so, and if the bow is perfectly symmetric so that the arrow leaves from the center of the string, what harmonics are excited?

14. The radar systems used by police to detect speeders are sensitive to the Doppler shift of a pulse of radio waves. Discuss how this sensitivity can be used to measure the speed of a car.

15. As oppositely moving pulses of the same shape (one upward, one downward) on a string pass through each other, there is one instant at which the string shows no displacement from the equilibrium position at any point. Has the energy carried by the pulses disappeared at this instant of time? If not, where is it?

16. A soft drink bottle resonates as air is blown across its top. What happens to the resonant frequency as the level of fluid in the bottle decreases?

17. A blowing whistle is attached to the roof of a car that moves around a circular race track. Assuming you're standing near the outside of the track, explain the nature of the sound you hear as the whistle comes by each time.

18. Despite a reasonably steady hand, a person often spills his coffee when carrying it to his seat. Discuss resonance as a possible cause of this difficulty, and devise a means for solving the problem.

19. An airplane mechanic notices that the sound from a twin-engine aircraft varies rapidly in loudness when both engines are running. What could be causing this variation from loud to soft?

20. Why does a vibrating guitar string sound louder when placed on the instrument than it would if allowed to vibrate in the air while off the instrument?

PROBLEMS

1, 2, 3 = straightforward, intermediate, challenging □ = full solution available in *Student Solutions Manual/Study Guide*
PhysicsNow™ = coached problem with hints available at **www.cp7e.com** = biomedical application

Section 14.2 Characteristics of Sound Waves
Section 14.3 The Speed of Sound

Unless otherwise stated, use 345 m/s as the speed of sound in air.

1. Suppose that you hear a clap of thunder 16.2 s after seeing the associated lightning stroke. The speed of sound waves in air is 343 m/s and the speed of light in air is 3.00×10^8 m/s. How far are you from the lightning stroke?

2. A dolphin located in sea water at a temperature of 25°C emits a sound directed toward the bottom of the ocean 150 m below. How much time passes before it hears an echo?

3. A sound wave has a frequency of 700 Hz in air and a wavelength of 0.50 m. What is the temperature of the air?

4. The range of human hearing extends from approximately 20 Hz to 20 000 Hz. Find the wavelengths of these extremes at a temperature of 27°C.

5. A group of hikers hears an echo 3.00 s after shouting. If the temperature is 22.0°C, how far away is the mountain that reflected the sound wave?

6. A stone is dropped from rest into a well. The sound of the splash is heard exactly 2.00 s later. Find the depth of the well if the air temperature is 10.0°C.

7. □ You are watching a pier being constructed on the far shore of a saltwater inlet when some blasting occurs. You hear the sound in the water 4.50 s before it reaches you through the air. How wide is the inlet? [*Hint:* See Table 14.1. Assume the air temperature is 20°C.]

8. The speed of sound in a column of air is measured to be 356 m/s. What is the temperature of the air?

Section 14.4 Energy and Intensity of Sound Waves

9. The toadfish makes use of resonance in a closed tube to produce very loud sounds. The tube is its swim bladder, used as an amplifier. The sound level of this creature has been measured as high as 100 dB. (a) Calculate the intensity of the sound wave emitted. (b) What is the intensity level if three of these fish try to imitate three frogs by saying "Budweiser" at the same time.

10. The area of a typical eardrum is about 5.0×10^{-5} m^2. Calculate the sound power (the energy per second) incident on an eardrum at (a) the threshold of hearing and (b) the threshold of pain.

11. There is evidence that elephants communicate via infrasound, generating rumbling vocalizations as low as 14 hz that can travel up to 10 km. The intensity level of these sounds can reach 103 dB, measured a distance of 5.0 m from the source. Determine the intensity level of the infrasound 10 km from the source, assuming the sound energy radiates uniformly in all directions.

12. Two sounds have measured intensities of $I_1 = 100$ W/m^2 and $I_2 = 200$ W/m^2. By how many decibels is the level of sound 1 lower than that of sound 2?

13. □ A noisy machine in a factory produces sound with a level of 80 dB. How many identical machines could you add to the factory without exceeding the 90-dB limit?

14. A family ice show is held at an enclosed arena. The skaters perform to music playing at a level of 80.0 dB. This intensity level is too loud for your baby, who yells at 75.0 dB. (a) What total sound intensity engulfs you? (b) What is the combined sound level?

15. Calculate the sound level in decibels of a sound wave that has an intensity of 4.00 μW/m^2.

Section 14.5 Spherical and Plane Waves

16. An outside loudspeaker (considered a small source) emits sound waves with a power output of 100 W. (a) Find the intensity 10.0 m from the source. (b) Find the intensity level in decibels at that distance. (c) At what distance would you experience the sound at the threshold of pain, 120 dB?

17. Physics Now™ A train sounds its horn as it approaches an intersection. The horn can just be heard at a level of 50 dB by an observer 10 km away. (a) What is the average power generated by the horn? (b) What intensity level of the horn's sound is observed by someone waiting at an intersection 50 m from the train? Treat the horn as a point source and neglect any absorption of sound by the air.

18. A skyrocket explodes 100 m above the ground (Fig. P14.18). Three observers are spaced 100 m apart, with the first (A) directly under the explosion. (a) What is the ratio of the sound intensity heard by observer A to that heard by observer B? (b) What is the ratio of the intensity heard by observer A to that heard by observer C?

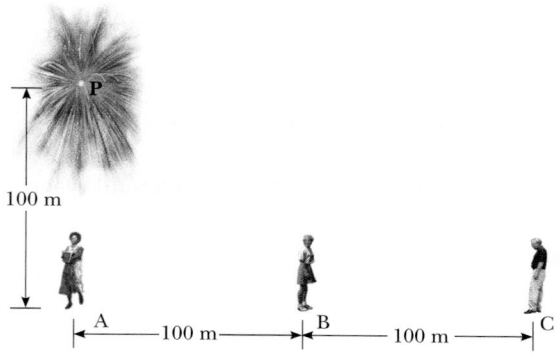

Figure P14.18

19. Show that the difference in decibel levels β_1 and β_2 of a sound source is related to the ratio of its distances r_1 and r_2 from the receivers by the formula

$$\beta_2 - \beta_1 = 20 \log \left(\frac{r_1}{r_2} \right)$$

Section 14.6 The Doppler Effect

20. An airplane traveling at half the speed of sound (v = 172 m/s) emits a sound of frequency 5.00 kHz. At what frequency does a stationary listener hear the sound (a) as the plane approaches? (b) after it passes?

21. A commuter train passes a passenger platform at a constant speed of 40.0 m/s. The train horn is sounded at its characteristic frequency of 320 Hz. (a) What overall change in frequency is detected by a person on the platform as the train moves from approaching to receding? (b) What wavelength is detected by a person on the platform as the train approaches?

22. At rest, a car's horn sounds the note A (440 Hz). The horn is sounded while the car is moving down the street. A bicyclist moving in the same direction with one-third the car's speed hears a frequency of 415 Hz. What is the speed of the car? Is the cyclist ahead of or behind the car?

23. Two trains on separate tracks move towards one another. Train 1 has a speed of 130 km/h, train 2 a speed of 90.0 km/h. Train 2 blows its horn, emitting a frequency of 500 Hz. What is the frequency heard by the engineer on train 1?

24. A bat flying at 5.0 m/s emits a chirp at 40 kHz. If this sound pulse is reflected by a wall, what is the frequency of the echo received by the bat?

25. An alert physics student stands beside the tracks as a train rolls slowly past. He notes that the frequency of the train whistle is 442 Hz when the train is approaching him and 441 Hz when the train is receding from him. Using these frequencies, he calculates the speed of the train. What value does he find?

26. Expectant parents are thrilled to hear their unborn baby's heartbeat, revealed by an ultrasonic motion detector. Suppose the fetus's ventricular wall moves in simple harmonic motion with amplitude 1.80 mm and frequency 115 per minute. (a) Find the maximum linear speed of the heart wall. Suppose the motion detector in contact with the mother's abdomen produces sound at precisely 2 MHz, which travels through tissue at 1.50 km/s. (b) Find the maximum frequency at which sound arrives at the wall of the baby's heart. (c) Find the maximum frequency at which reflected sound is received by the motion detector. (By electronically "listening" for echoes at a frequency different from the broadcast frequency, the motion detector can produce beeps of audible sound in synchrony with the fetal heartbeat.)

27. A tuning fork vibrating at 512 Hz falls from rest and accelerates at 9.80 m/s^2. How far below the point of release is the tuning fork when waves of frequency 485 Hz reach the release point? Take the speed of sound in air to be 340 m/s.

28. A supersonic jet traveling at Mach 3 at an altitude of 20 000 m is directly overhead at time $t = 0$, as in Figure P14.28. (a) How long will it be before the ground observer encounters the shock wave? (b) Where will the plane be when it is finally heard? (Assume an average value of 330 m/s for the speed of sound in air.)

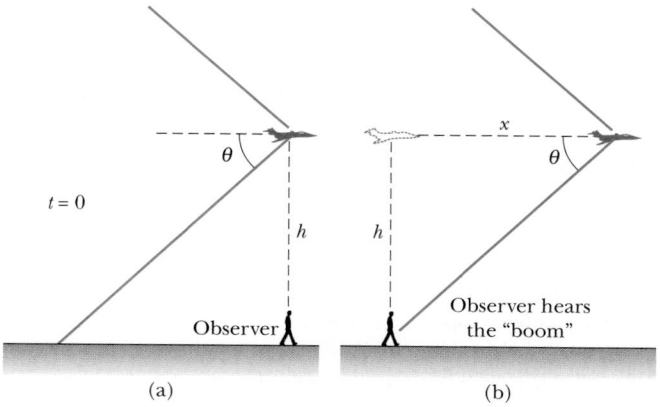

Figure P14.28

29. The now-discontinued *Concorde* flew at Mach 1.5, which meant the speed of the plane was 1.5 times the speed of sound in air. What was the angle between the direction of propagation of the shock wave and the direction of the plane's velocity?

Section 14.7 Interference of Sound Waves

30. The acoustical system shown in Figure 14.14 is driven by a speaker emitting a 400-Hz note. If *destructive* interference occurs at a particular instant, how much must the path length in the U-shaped tube be increased in order to hear (a) constructive interference and (b) destructive interference once again?

31. The ship in Figure P14.31 travels along a straight line parallel to the shore and 600 m from it. The ship's radio receives simultaneous signals of the same frequency from antennas *A* and *B*. The signals interfere constructively at point *C*, which is equidistant from *A* and *B*. The signal goes through the first minimum at point *D*. Determine the wavelength of the radio waves.

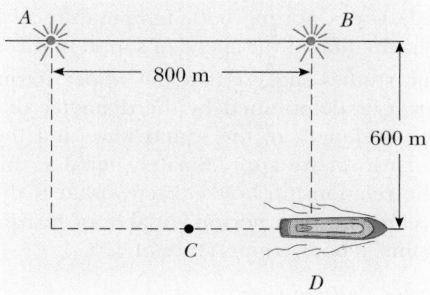

Figure P14.31

32. Two loudspeakers are placed above and below one another, as in Figure 14.15, and are driven by the same source at a frequency of 500 Hz. (a) What minimum distance should the top speaker be moved back in order to create destructive interference between the speakers? (b) If the top speaker is moved back twice the distance calculated in part (a), will there be constructive or destructive interference?

33. A pair of speakers separated by 0.700 m are driven by the same oscillator at a frequency of 690 Hz. An observer originally positioned at one of the speakers begins to walk along a line perpendicular to the line joining the speakers. (a) How far must the observer walk before reaching a relative maximum in intensity? (b) How far will the observer be from the speaker when the first relative minimum is detected in the intensity?

Section 14.8 Standing Waves

34. A steel wire in a piano has a length of 0.700 0 m and a mass of 4.300×10^{-3} kg. To what tension must this wire be stretched in order that the fundamental vibration correspond to middle C ($f_C = 261.6$ Hz on the chromatic musical scale)?

35. A stretched string fixed at each end has a mass of 40.0 g and a length of 8.00 m. The tension in the string is 49.0 N. (a) Determine the positions of the nodes and antinodes for the third harmonic. (b) What is the vibration frequency for this harmonic?

36. Resonance of sound waves can be produced within an aluminum rod by holding the rod at its midpoint and stroking it with an alcohol-saturated paper towel. In this resonance mode, the middle of the rod is a node while the ends are antinodes; no other nodes or antinodes are present. What is the frequency of the resonance if the rod is 1.00 m long?

37. Physics Now™ Two speakers are driven by a common oscillator at 800 Hz and face each other at a distance of 1.25 m. Locate the points along a line joining the speakers where relative minima of the amplitude of the pressure would be expected. (Use $v = 343$ m/s.)

38. Two pieces of steel wire with identical cross sections have lengths of L and $2L$. The wires are each fixed at both ends and stretched so that the tension in the longer wire is four times greater than in the shorter wire. If the fundamental frequency in the shorter wire is 60 Hz, what is the frequency of the second harmonic in the longer wire?

39. A 12-kg object hangs in equilibrium from a string of total length $L = 5.0$ m and linear mass density $\mu = 0.001\,0$ kg/m. The string is wrapped around two light, frictionless pulleys that are separated by the distance $d = 2.0$ m (Fig. P14.39a). (a) Determine the tension in the string. (b) At what frequency must the string between the pulleys vibrate in order to form the standing-wave pattern shown in Figure P14.39b?

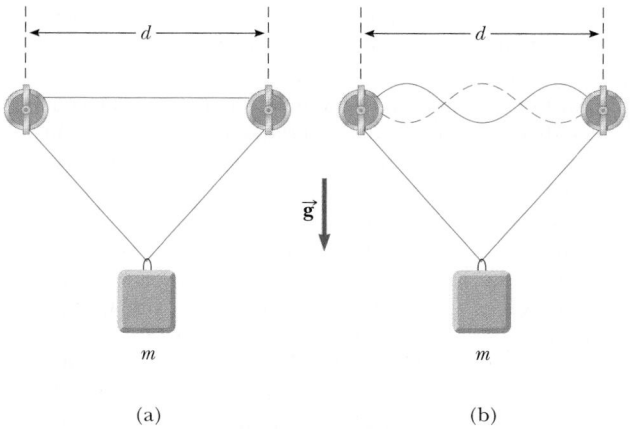

Figure P14.39

40. In the arrangement shown in Figure P14.40, an object of mass $m = 5.0$ kg hangs from a cord around a light pulley. The length of the cord between point *P* and the pulley is $L = 2.0$ m. (a) When the vibrator is set to a frequency of 150 Hz, a standing wave with six loops is formed. What must be the linear mass density of the cord? (b) How many loops (if any) will result if m is changed to 45 kg? (c) How many loops (if any) will result if m is changed to 10 kg?

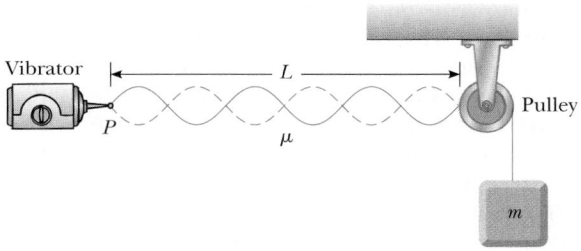

Figure P14.40

41. A 60.00-cm guitar string under a tension of 50.000 N has a mass per unit length of 0.100 00 g/cm. What is the highest resonant frequency that can be heard by a person capable of hearing frequencies up to 20 000 Hz?

Section 14.9 Forced Vibrations and Resonance

42. Standing-wave vibrations are set up in a crystal goblet with four nodes and four antinodes equally spaced around the 20.0-cm circumference of its rim. If transverse waves move around the glass at 900 m/s, an opera singer would have to produce a high harmonic with what frequency in order to shatter the glass with a resonant vibration?

Section 14.10 Standing Waves in Air Columns

43. The windpipe of a typical whooping crane is about 5.0 feet long. What is the lowest resonant frequency of this pipe, assuming that it is closed at one end? Assume a temperature of 37°C.

44. The overall length of a piccolo is 32.0 cm. The resonating air column vibrates as in a pipe that is open at both ends. (a) Find the frequency of the lowest note a piccolo can play, assuming the speed of sound in air is 340 m/s. (b) Opening holes in the side effectively shortens the length of the resonant column. If the highest note a piccolo can sound is 4 000 Hz, find the distance between adjacent antinodes for this mode of vibration.

45. The human ear canal is about 2.8 cm long. If it is regarded as a tube that is open at one end and closed at the eardrum, what is the fundamental frequency around which we would expect hearing to be most sensitive? Take the speed of sound to be 340 m/s.

46. A shower stall measures 86.0 cm × 86.0 cm × 210 cm. When you sing in the shower, which frequencies will sound the richest (because of resonance)? Assume the stall acts as a pipe closed at both ends, with nodes at opposite sides. Assume also that the voices of various singers range from 130 Hz to 2 000 Hz. Let the speed of sound in the hot shower stall be 355 m/s.

47. PhysicsNow™ A pipe open at both ends has a fundamental frequency of 300 Hz when the temperature is 0°C. (a) What is the length of the pipe? (b) What is the fundamental frequency at a temperature of 30°C?

48. A 2.00-m-long air column is open at both ends. The frequency of a certain harmonic is 410 Hz, and the frequency of the next-higher harmonic is 492 Hz. Determine the speed of sound in the air column.

Section 14.11 Beats

49. Two identical mandolin strings under 200 N of tension are sounding tones with frequencies of 523 Hz. The peg of one string slips slightly, and the tension in it drops to 196 N. How many beats per second are heard?

50. The G string on a violin has a fundamental frequency of 196 Hz. It is 30.0 cm long and has a mass of 0.500 g. While this string is sounding, a nearby violinist effectively shortens the G string on her identical violin (by sliding her finger down the string) until a beat frequency of 2.00 Hz is heard between the two strings. When this occurs, what is the effective length of her string?

51. Two train whistles have identical frequencies of 180 Hz. When one train is at rest in the station, sounding its whistle, a beat frequency of 2 Hz is heard from a moving train. What two possible speeds and directions can the moving train have?

52. Two pipes of equal length are each open at one end. Each has a fundamental frequency of 480 Hz at 300 K. In one pipe, the air temperature is increased to 305 K. If the two pipes are sounded together, what beat frequency results?

53. A student holds a tuning fork oscillating at 256 Hz. He walks toward a wall at a constant speed of 1.33 m/s. (a) What beat frequency does he observe between the tuning fork and its echo? (b) How fast must he walk away from the wall to observe a beat frequency of 5.00 Hz?

Section 14.13 The Ear

54. If a human ear canal can be thought of as resembling an organ pipe, closed at one end, that resonates at a fundamental frequency of 3 000 Hz, what is the length of the canal? Use a normal body temperature of 37°C for your determination of the speed of sound in the canal.

55. Some studies suggest that the upper frequency limit of hearing is determined by the diameter of the eardrum. The wavelength of the sound wave and the diameter of the eardrum are approximately equal at this upper limit. If the relationship holds exactly, what is the diameter of the eardrum of a person capable of hearing 20 000 Hz? (Assume a body temperature of 37°C.)

ADDITIONAL PROBLEMS

56. A commuter train blows its horn as it passes a passenger platform at a constant speed of 40.0 m/s. The horn sounds at a frequency of 320 Hz when the train is at rest. What is the frequency observed by a person on the platform (a) as the train approaches and (b) as the train recedes from him? (c) What wavelength does the observer find in each case?

57. A quartz watch contains a crystal oscillator in the form of a block of quartz that vibrates by contracting and expanding. Two opposite faces of the block, 7.05 mm apart, are antinodes, moving alternately towards and away from each other. The plane halfway between these two faces is a node of the vibration. The speed of sound in quartz is 3.70 km/s. Find the frequency of the vibration. An oscillating electric voltage accompanies the mechanical oscillation, so the quartz is described as *piezoelectric*. An electric circuit feeds in energy to maintain the oscillation and also counts the voltage pulses to keep time.

58. A flowerpot is knocked off a balcony 20.0 m above the sidewalk and falls toward an unsuspecting 1.75-m-tall man who is standing below. How close to the sidewalk can the flowerpot fall before it is too late for a warning shouted from the balcony to reach the man in time? Assume that the man below requires 0.300 s to respond to the warning.

59. PhysicsNow™ On a workday, the average decibel level of a busy street is 70 dB, with 100 cars passing a given point every minute. If the number of cars is reduced to 25 every minute on a weekend, what is the decibel level of the street?

60. A variable-length air column is placed just below a vibrating wire that is fixed at both ends. The length of the column, open at one end, is gradually increased from zero until the first position of resonance is observed at

$L = 34.0$ cm. The wire is 120 cm long and is vibrating in its third harmonic. If the speed of sound in air is 340 m/s, what is the speed of transverse waves in the wire?

61. A block with a speaker bolted to it is connected to a spring having spring constant $k = 20.0$ N/m, as shown in Figure P14.61. The total mass of the block and speaker is 5.00 kg, and the amplitude of the unit's motion is 0.500 m. If the speaker emits sound waves of frequency 440 Hz, determine the lowest and highest frequencies heard by the person to the right of the speaker.

Figure P14.61

62. A flute is designed so that it plays a frequency of 261.6 Hz, middle C, when all the holes are covered and the temperature is 20.0°C. (a) Consider the flute to be a pipe open at both ends, and find its length, assuming that the middle-C frequency is the fundamental frequency. (b) A second player, nearby in a colder room, also attempts to play middle C on an identical flute. A beat frequency of 3.00 beats/s is heard. What is the temperature of the room?

63. When at rest, two trains have sirens that emit a frequency of 300 Hz. The trains travel toward one another and toward an observer stationed between them. One of the trains moves at 30.0 m/s, and the observer hears a beat frequency of 3.0 beats per second. What is the speed of the second train, which travels faster than 30.0 m/s?

64. Many artists sing very high notes in *ad lib* ornaments and cadenzas. The highest note written for a singer in a published score was F-sharp above high C, 1.480 kHz, sung by Zerbinetta in the original version of Richard Strauss's opera *Ariadne auf Naxos.* (a) Find the wavelength of this sound in air. (b) In response to complaints, Strauss later transposed the note down to F above high C, 1.397 kHz. By what increment did the wavelength change?

65. A speaker at the front of a room and an identical speaker at the rear of the room are being driven at 456 Hz by the same sound source. A student walks at a uniform rate of 1.50 m/s away from one speaker and towards the other. How many beats does the student hear per second?

66. Two identical speakers separated by 10.0 m are driven by the same oscillator with a frequency of $f = 21.5$ Hz (Fig. P14.66). Explain why a receiver at A records a minimum in sound intensity from the two speakers. (b) If the receiver is moved in the plane of the speakers, what path should it take so that the intensity remains at a minimum? That is, determine the relationship between x and y (the coordinates of the receiver) such that the receiver will record a minimum in sound intensity. Take the speed of sound to be 344 m/s.

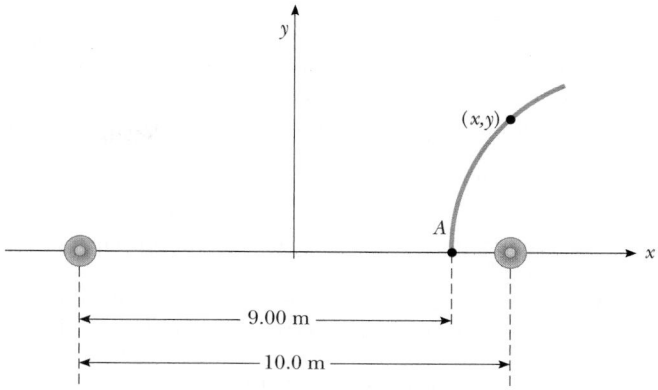

Figure P14.66

67. By proper excitation, it is possible to produce both longitudinal and transverse waves in a long metal rod. In a particular case, the rod is 150 cm long and 0.200 cm in radius and has a mass of 50.9 g. Young's modulus for the material is 6.80×10^{10} Pa. Determine the required tension in the rod so that the ratio of the speed of longitudinal waves to the speed of transverse waves is 8.

68. A student stands several meters in front of a smooth reflecting wall, holding a board on which a wire is fixed at each end. The wire, vibrating in its third harmonic, is 75.0 cm long, has a mass of 2.25 g, and is under a tension of 400 N. A second student, moving towards the wall, hears 8.30 beats per second. What is the speed of the student approaching the wall? Use 340 m/s as the speed of sound in air.

69. Two ships are moving along a line due east. The trailing vessel has a speed of 64.0 km/h relative to a land-based observation point, and the leading ship has a speed of 45.0 km/h relative to the same station. The trailing ship transmits a sonar signal at a frequency of 1 200 Hz. What frequency is monitored by the leading ship? (Use 1 520 m/s as the speed of sound in ocean water.)

70. The Doppler equation presented in the text is valid when the motion between the observer and the source occurs on a straight line, so that the source and observer are moving either directly toward or directly away from each other. If this restriction is relaxed, one must use the more general Doppler equation

$$f_O = \left[\frac{v + v_O \cos(\theta_O)}{v - v_S \cos(\theta_S)}\right] f_S$$

where θ_O and θ_S are defined in Figure P14.70a. (a) If both observer and source are moving away from each other along a straight line, show that the preceding equation yields the same result as Equation 14.12 in the text.

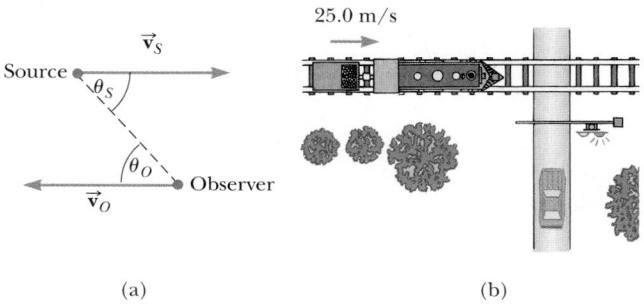

Figure P14.70

(b) Use the preceding equation to solve the following problem: A train moves at a constant speed of 25.0 m/s toward the intersection shown in Figure P14.70b. A car is stopped near the intersection, 30.0 m from the tracks. If the train's horn emits a frequency of 500 Hz, what is the frequency heard by the passengers in the car when the train is 40.0 m to the left of the intersection? Take the speed of sound to be 343 m/s.

71. A rescue plane flies horizontally at a constant speed, searching for a disabled boat. When the plane is directly above the boat, the boat's crew blows a loud horn. By the time the plane's sound detector perceives the horn's sound, the plane has traveled a distance equal to one-half its altitude above the ocean. If the sound takes 2.00 s to reach the plane, determine (a) the plane's altitude and (b) its speed.

72. In order to determine her speed, a skydiver carries a tone generator. A friend on the ground at the landing site has equipment for receiving and analyzing sound waves. While the skydiver is falling at terminal speed, her tone generator emits a steady tone of 1.80 kHz. (Assume that the air is calm, that the speed of sound is 343 m/s, independent of altitude.) (a) If her friend on the ground (directly beneath the skydiver) receives waves of frequency 2.15 kHz, what is the skydiver's speed of descent? (b) If the skydiver were also carrying sound-receiving equipment sensitive enough to detect waves reflected from the ground, what frequency of waves would she receive?

ACTIVITIES

A.1. Use an empty 1-liter soft-drink container, blow over the open end, and listen to the sound that is produced. Add some water to the container to change the height of the air column, and repeat the procedure. How does the frequency that you hear change with the height of the air column?

If you want to investigate this phenomenon in more detail, construct a musical instrument made up of several soft-drink bottles with different amounts of water in each. You can play your instrument as a wind instrument by blowing over the mouths of the bottles.

A.2. Beats can easily be heard on a guitar. When a finger is placed at the fifth fret of the second string, the note produced when the string is plucked should be identical to the note from the first string when it is played without fingering. With your finger in position on the second string, pluck the two strings simultaneously. If one of the strings is slightly out of tune, a very pronounced beat frequency will be heard. What happens to the beat frequency as the string tension is changed in small increments from too low for the intended tuning to too high?

A.3. Attach a rope to a door and shake the other end to see how many of the standing-wave patterns in Figure 14.18 you can produce. When a pattern is formed, note that the amplitude of the rope's vibration is much larger than the movement of your hand.

A.4. Snip off the corners of one end of a paper straw so that the end tapers to a point, as shown in Figure A14.4. Chew on this end to flatten it, and you have created a double-reed instrument. Put your lips around the tapered end of the straw, press them together slightly, and blow through the straw. When you hear a steady tone, slowly snip off a piece of the straw at the other end. Be careful to keep about the same amount of pressure with your lips. How does the frequency of the sound change as the straw becomes shorter? Why does this change occur? You may be able to produce more than one tone for any given length of the straw. Why?

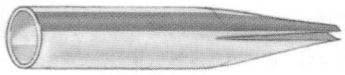

Figure A14.4

A.5. Inflate a balloon just enough to form a small sphere. Measure its diameter. Use a marker to color in a 1-cm square on its surface. Now continue inflating the balloon until it reaches twice the original diameter. Measure the size of the square now. Note how the color of the marked area has changed. Use the information in Section 14.5 to explain these results.

"I love hearing that lonesome wail of the train whistle as the magnitude of the frequency of the wave changes due to the Doppler effect."

APPENDIX A
Mathematical Review

A.1 MATHEMATICAL NOTATION

Many mathematical symbols are used throughout this book. You are no doubt familiar with some, such as the symbol $=$ to denote the equality of two quantities.

The symbol $\propto$ denotes a proportionality. For example, $y \propto x^2$ means that y is proportional to the square of x.

The symbol $<$ means *is less than,* and $>$ means *is greater than.* For example, $x > y$ means x is greater than y.

The symbol $\ll$ means *is much less than,* and $\gg$ means *is much greater than.*

The symbol $\approx$ indicates that two quantities are *approximately equal* to each other.

The symbol $\equiv$ means *is defined as.* This is a stronger statement than a simple $=$.

It is convenient to use the notation Δx (read as "delta x") to indicate the *change in the quantity* x. (Note that Δx does not mean "the product of Δ and x.") For example, suppose that a person out for a morning stroll starts measuring her distance away from home when she is 10 m from her doorway. She then moves along a straight-line path and stops strolling 50 m from the door. Her change in position during the walk is $\Delta x = 50\text{ m} - 10\text{ m} = 40\text{ m}$ or, in symbolic form,

$$\Delta x = x_f - x_i$$

In this equation x_f is the *final position* and x_i is the *initial position.*

We often have occasion to add several quantities. A useful abbreviation for representing such a sum is the Greek letter Σ (capital sigma). Suppose we wish to add a set of five numbers represented by x_1, x_2, x_3, x_4, and x_5. In the abbreviated notation, we would write the sum as

$$x_1 + x_2 + x_3 + x_4 + x_5 = \sum_{i=1}^{5} x_i$$

where the subscript i on x represents any one of the numbers in the set. For example, if there are five masses in a system, m_1, m_2, m_3, m_4, and m_5, the total mass of the system $M = m_1 + m_2 + m_3 + m_4 + m_5$ could be expressed as

$$M = \sum_{i=1}^{5} m_i$$

Finally, the magnitude of a quantity x, written $|x|$, is simply the absolute value of that quantity. The sign of $|x|$ is always positive, regardless of the sign of x. For example, if $x = -5$, $|x| = 5$; if $x = 8$, $|x| = 8$.

A.2 SCIENTIFIC NOTATION

Many quantities that scientists deal with often have very large or very small values. For example, the speed of light is about 300 000 000 m/s and the ink required to make the dot over an i in this textbook has a mass of about 0.000 000 001 kg. Obviously, it is cumbersome to read, write, and keep track of numbers such as these. We avoid this problem by using a method dealing with powers of the number 10:

$$10^0 = 1$$
$$10^1 = 10$$
$$10^2 = 10 \times 10 = 100$$
$$10^3 = 10 \times 10 \times 10 = 1\,000$$
$$10^4 = 10 \times 10 \times 10 \times 10 = 10\,000$$
$$10^5 = 10 \times 10 \times 10 \times 10 \times 10 = 100\,000$$

and so on. The number of zeros corresponds to the power to which 10 is raised, called the **exponent** of 10. For example, the speed of light, 300 000 000 m/s, can be expressed as 3×10^8 m/s.

For numbers less than one, we note the following:

$$10^{-1} = \frac{1}{10} = 0.1$$

$$10^{-2} = \frac{1}{10 \times 10} = 0.01$$

$$10^{-3} = \frac{1}{10 \times 10 \times 10} = 0.001$$

$$10^{-4} = \frac{1}{10 \times 10 \times 10 \times 10} = 0.000\,1$$

$$10^{-5} = \frac{1}{10 \times 10 \times 10 \times 10 \times 10} = 0.000\,01$$

In these cases, the number of places the decimal point is to the left of the digit 1 equals the value of the (negative) exponent. Numbers that are expressed as some power of 10 multiplied by another number between 1 and 10 are said to be in **scientific notation.** For example, the scientific notation for 5 943 000 000 is 5.943×10^9 and that for 0.000 083 2 is 8.32×10^{-5}.

When numbers expressed in scientific notation are being multiplied, the following general rule is very useful:

$$10^n \times 10^m = 10^{n+m} \qquad \textbf{[A.1]}$$

where n and m can be *any* numbers (not necessarily integers). For example, $10^2 \times 10^5 = 10^7$. The rule also applies if one of the exponents is negative. For example, $10^3 \times 10^{-8} = 10^{-5}$.

When dividing numbers expressed in scientific notation, note that

$$\frac{10^n}{10^m} = 10^n \times 10^{-m} = 10^{n-m} \qquad \textbf{[A.2]}$$

EXERCISES

With help from the above rules, verify the answers to the following:

1. $86\,400 = 8.64 \times 10^4$
2. $9\,816\,762.5 = 9.816\,762\,5 \times 10^6$
3. $0.000\,000\,039\,8 = 3.98 \times 10^{-8}$
4. $(4.0 \times 10^8)(9.0 \times 10^9) = 3.6 \times 10^{18}$
5. $(3.0 \times 10^7)(6.0 \times 10^{-12}) = 1.8 \times 10^{-4}$
6. $\dfrac{75 \times 10^{-11}}{5.0 \times 10^{-3}} = 1.5 \times 10^{-7}$
7. $\dfrac{(3 \times 10^6)(8 \times 10^{-2})}{(2 \times 10^{17})(6 \times 10^5)} = 2 \times 10^{-18}$

A.3 ALGEBRA

A. Some Basic Rules

When algebraic operations are performed, the laws of arithmetic apply. Symbols such as x, y, and z are frequently used to represent quantities that are not specified, what are called the **unknowns.**

First, consider the equation

$$8x = 32$$

If we wish to solve for x, we can divide (or multiply) each side of the equation by the same factor without destroying the equality. In this case, if we divide both sides by 8, we have

$$\frac{8x}{8} = \frac{32}{8}$$
$$x = 4$$

Next consider the equation

$$x + 2 = 8$$

In this type of expression, we can add or subtract the same quantity from each side. If we subtract 2 from each side, we get

$$x + 2 - 2 = 8 - 2$$
$$x = 6$$

In general, if $x + a = b$, then $x = b - a$.

Now consider the equation

$$\frac{x}{5} = 9$$

If we multiply each side by 5, we are left with x on the left by itself and 45 on the right:

$$\left(\frac{x}{5}\right)(5) = 9 \times 5$$
$$x = 45$$

In all cases, **whatever operation is performed on the left side of the equality must also be performed on the right side.**

The following rules for multiplying, dividing, adding, and subtracting fractions should be recalled, where a, b, and c are three numbers:

	Rule	Example
Multiplying	$\left(\frac{a}{b}\right)\left(\frac{c}{d}\right) = \frac{ac}{bd}$	$\left(\frac{2}{3}\right)\left(\frac{4}{5}\right) = \frac{8}{15}$
Dividing	$\frac{(a/b)}{(c/d)} = \frac{ad}{bc}$	$\frac{2/3}{4/5} = \frac{(2)(5)}{(4)(3)} = \frac{10}{12}$
Adding	$\frac{a}{b} \pm \frac{c}{d} = \frac{ad \pm bc}{bd}$	$\frac{2}{3} - \frac{4}{5} = \frac{(2)(5) - (4)(3)}{(3)(5)} = -\frac{2}{15}$

EXERCISES

In the following exercises, solve for x:

	ANSWERS
1. $a = \frac{1}{1 + x}$	$x = \frac{1 - a}{a}$
2. $3x - 5 = 13$	$x = 6$
3. $ax - 5 = bx + 2$	$x = \frac{7}{a - b}$
4. $\frac{5}{2x + 6} = \frac{3}{4x + 8}$	$x = -\frac{11}{7}$

B. Powers

When powers of a given quantity x are multiplied, the following rule applies:

$$x^n x^m = x^{n+m} \qquad \textbf{[A.3]}$$

For example, $x^2x^4 = x^{2+4} = x^6$.

When dividing the powers of a given quantity, note that

$$\frac{x^n}{x^m} = x^{n-m} \qquad \textbf{[A.4]}$$

For example, $x^8/x^2 = x^{8-2} = x^6$.

A power that is a fraction, such as $\frac{1}{3}$, corresponds to a root as follows:

$$x^{1/n} = \sqrt[n]{x} \qquad \textbf{[A.5]}$$

For example, $4^{1/3} = \sqrt[3]{4} = 1.587\,4$. (A scientific calculator is useful for such calculations.)

Finally, any quantity x^n that is raised to the mth power is

$$(x^n)^m = x^{nm} \qquad \textbf{[A.6]}$$

Table A.1 summarizes the rules of exponents.

TABLE A.1

Rules of Exponents
$x^0 = 1$
$x^1 = x$
$x^n x^m = x^{n+m}$
$x^n/x^m = x^{n-m}$
$x^{1/n} = \sqrt[n]{x}$
$(x^n)^m = x^{nm}$

EXERCISES

Verify the following:

1. $3^2 \times 3^3 = 243$
2. $x^5 x^{-8} = x^{-3}$
3. $x^{10}/x^{-5} = x^{15}$
4. $5^{1/3} = 1.709\,975$ (Use your calculator.)
5. $60^{1/4} = 2.783\,158$ (Use your calculator.)
6. $(x^4)^3 = x^{12}$

C. Factoring

Some useful formulas for factoring an equation are

$ax + ay + az = a(x + y + z)$	common factor
$a^2 + 2ab + b^2 = (a + b)^2$	perfect square
$a^2 - b^2 = (a + b)(a - b)$	differences of squares

D. Quadratic Equations

The general form of a quadratic equation is

$$ax^2 + bx + c = 0 \qquad \textbf{[A.7]}$$

where x is the unknown quantity and a, b, and c are numerical factors referred to as **coefficients** of the equation. This equation has two roots, given by

$$x = \frac{-b \pm \sqrt{b^2 - 4ac}}{2a} \qquad \textbf{[A.8]}$$

If $b^2 \geq 4ac$, the roots will be real.

EXAMPLE

The equation $x^2 + 5x + 4 = 0$ has the following roots corresponding to the two signs of the square root term:

$$x = \frac{-5 \pm \sqrt{5^2 - (4)(1)(4)}}{2(1)} = \frac{-5 \pm \sqrt{9}}{2} = \frac{-5 \pm 3}{2}$$

that is,

$$x_+ = \frac{-5 + 3}{2} = -1 \qquad x_- = \frac{-5 - 3}{2} = -4$$

where x_+ refers to the root corresponding to the positive sign and x_- refers to the root corresponding to the negative sign.

EXERCISES

Solve the following quadratic equations:

	ANSWERS	
1. $x^2 + 2x - 3 = 0$	$x_+ = 1$	$x_- = -3$
2. $2x^2 - 5x + 2 = 0$	$x_+ = 2$	$x_- = 1/2$
3. $2x^2 - 4x - 9 = 0$	$x_+ = 1 + \sqrt{22}/2$	$x_- = 1 - \sqrt{22}/2$

E. Linear Equations

A linear equation has the general form

$$y = ax + b \qquad \textbf{[A.9]}$$

where a and b are constants. This equation is referred to as being linear because the graph of y versus x is a straight line, as shown in Figure A.1. The constant b, called the **intercept,** represents the value of y at which the straight line intersects the y axis. The constant a is equal to the **slope** of the straight line. If any two points on the straight line are specified by the coordinates (x_1, y_1) and (x_2, y_2), as in Figure A.1, then the slope of the straight line can be expressed

$$\text{Slope} = \frac{y_2 - y_1}{x_2 - x_1} = \frac{\Delta y}{\Delta x} \qquad \textbf{[A.10]}$$

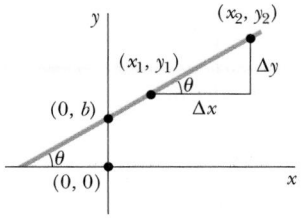

Figure A.1

Note that a and b can have either positive or negative values. If $a > 0$, the straight line has a positive slope, as in Figure A.1. If $a < 0$, the straight line has a *negative* slope. In Figure A.1, both a and b are positive. Three other possible situations are shown in Figure A.2: $a > 0$, $b < 0$; $a < 0$, $b > 0$; and $a < 0$, $b < 0$.

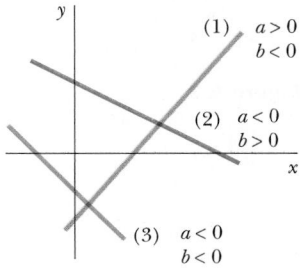

Figure A.2

EXERCISES

1. Draw graphs of the following straight lines:
   (a) $y = 5x + 3$ **(b)** $y = -2x + 4$ **(c)** $y = -3x - 6$
2. Find the slopes of the straight lines described in Exercise 1.
   Answers: **(a)** 5 **(b)** -2 **(c)** -3
3. Find the slopes of the straight lines that pass through the following sets of points: **(a)** $(0, -4)$ and $(4, 2)$, **(b)** $(0, 0)$ and $(2, -5)$, and **(c)** $(-5, 2)$ and $(4, -2)$
   Answers: **(a)** $\frac{3}{2}$ **(b)** $-\frac{5}{2}$ **(c)** $-\frac{4}{9}$

F. Solving Simultaneous Linear Equations

Consider an equation such as $3x + 5y = 15$, which has two unknowns, x and y. Such an equation does not have a unique solution. That is, $(x = 0,\ y = 3)$, $(x = 5, y = 0)$ and $(x = 2, y = \frac{9}{5})$ are all solutions to this equation.

If a problem has two unknowns, a unique solution is possible only if we have *two* independent equations. In general, if a problem has n unknowns, its solution requires n independent equations. In order to solve two simultaneous equations involving two unknowns, x and y, we solve one of the equations for x in terms of y and substitute this expression into the other equation.

EXAMPLE

Solve the following two simultaneous equations:

$$\textbf{(1)}\quad 5x + y = -8 \qquad \textbf{(2)}\quad 2x - 2y = 4$$

Solution From (2), we find that $x = y + 2$. Substitution of this into (1) gives

$$5(y + 2) + y = -8$$
$$6y = -18$$
$$y = -3$$
$$x = y + 2 = -1$$

Alternate Solution Multiply each term in (1) by the factor 2 and add the result to (2):

$$10x + 2y = -16$$
$$2x - 2y = 4$$
$$12x = -12$$
$$x = -1$$
$$y = x - 2 = -3$$

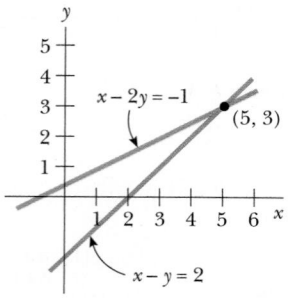

Figure A.3

Two linear equations with two unknowns can also be solved by a graphical method. If the straight lines corresponding to the two equations are plotted in a conventional coordinate system, the intersection of the two lines represents the solution. For example, consider the two equations

$$x - y = 2$$
$$x - 2y = -1$$

These are plotted in Figure A.3. The intersection of the two lines has the coordinates $x = 5$, $y = 3$. This represents the solution to the equations. You should check this solution by the analytical technique discussed above.

EXERCISES

Solve the following pairs of simultaneous equations involving two unknowns:

	ANSWERS
1. $x + y = 8$ $x - y = 2$	$x = 5, y = 3$
2. $98 - T = 10a$ $T - 49 = 5a$	$T = 65, a = 3.27$
3. $6x + 2y = 6$ $8x - 4y = 28$	$x = 2, y = -3$

G. Logarithms

Suppose that a quantity x is expressed as a power of some quantity a:

$$x = a^y \qquad \textbf{[A.11]}$$

The number a is called the **base** number. The **logarithm** of x with respect to the base a is equal to the exponent to which the base must be raised in order to satisfy the expression $x = a^y$:

$$y = \log_a x \qquad \textbf{[A.12]}$$

Conversely, the **antilogarithm** of y is the number x:

$$x = \text{antilog}_a\, y \qquad \textbf{[A.13]}$$

In practice, the two bases most often used are base 10, called the *common* logarithm base, and base $e = 2.718\ .\ .\ .$, called the *natural* logarithm base. When common logarithms are used,

$$y = \log_{10} x \qquad (\text{or } x = 10^y) \qquad \textbf{[A.14]}$$

When natural logarithms are used,

$$y = \ln_e x \qquad (\text{or } x = e^y) \qquad \textbf{[A.15]}$$

For example, $\log_{10} 52 = 1.716$, so that $\text{antilog}_{10}\, 1.716 = 10^{1.716} = 52$. Likewise, $\ln_e 52 = 3.951$, so $\text{antiln}_e\, 3.951 = e^{3.951} = 52$.

In general, note that you can convert between base 10 and base e with the equality

$$\ln_e x = (2.302\,585)\log_{10} x \qquad \textbf{[A.16]}$$

Finally, some useful properties of logarithms are

$$\log(ab) = \log a + \log b \qquad \ln e = 1$$

$$\log(a/b) = \log a - \log b \qquad \ln e^a = a$$

$$\log(a^n) = n \log a \qquad \ln\left(\frac{1}{a}\right) = -\ln a$$

A.4 GEOMETRY

Table A.2 gives the areas and volumes for several geometric shapes used throughout this text:

TABLE A.2

Useful Information for Geometry

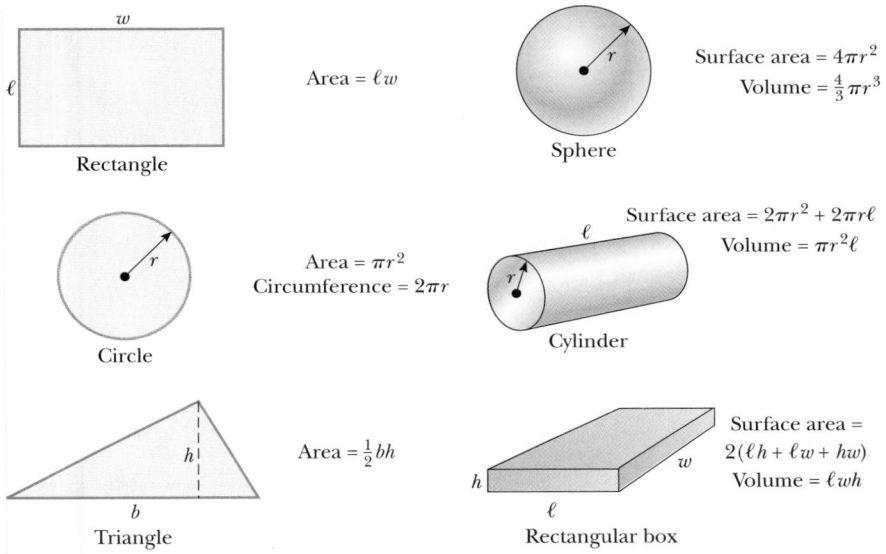

A.5 TRIGONOMETRY

Some of the most basic facts concerning trigonometry are presented in Chapter 1, and we encourage you to study the material presented there if you are having trouble with this branch of mathematics. In addition to the discussion of Chapter 1, certain useful trig identities that can be of value to you follow.

$$\sin^2\theta + \cos^2\theta = 1$$

$$\sin\theta = \cos(90^\circ - \theta)$$

$$\cos\theta = \sin(90^\circ - \theta)$$

$$\sin 2\theta = 2\sin\theta\cos\theta$$

$$\cos 2\theta = \cos^2\theta - \sin^2\theta$$

$$\sin(\theta \pm \phi) = \sin\theta\cos\phi \pm \cos\theta\sin\phi$$

$$\cos(\theta \pm \phi) = \cos\theta\cos\phi \mp \sin\theta\sin\phi$$

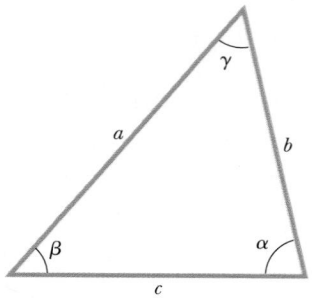

Figure A.4

The following relationships apply to *any* triangle, as shown in Figure A.4:

$$\alpha + \beta + \gamma = 180°$$

Law of cosines

$$a^2 = b^2 + c^2 - 2bc \cos \alpha$$

$$b^2 = a^2 + c^2 - 2ac \cos \beta$$

$$c^2 = a^2 + b^2 - 2ab \cos \gamma$$

Law of sines

$$\frac{a}{\sin \alpha} = \frac{b}{\sin \beta} = \frac{c}{\sin \gamma}$$

APPENDIX B
An Abbreviated Table of Isotopes

Atomic Number Z	Element	Symbol	Chemical Atomic Mass (u)	Mass Number (* Indicates Radioactive) A	Atomic Mass (u)	Percent Abundance	Half-Life (If Radioactive) $T_{1/2}$
0	(Neutron)	n		1*	1.008 665		10.4 min
1	Hydrogen	H	1.007 94	1	1.007 825	99.988 5	
	Deuterium	D		2	2.014 102	0.011 5	
	Tritium	T		3*	3.016 049		12.33 yr
2	Helium	He	4.002 602	3	3.016 029	0.000 137	
				4	4.002 603	99.999 863	
3	Lithium	Li	6.941	6	6.015 122	7.5	
				7	7.016 004	92.5	
4	Beryllium	Be	9.012 182	7*	7.016 929		53.3 days
				9	9.012 182	100	
5	Boron	B	10.811	10	10.012 937	19.9	
				11	11.009 306	80.1	
6	Carbon	C	12.010 7	10*	10.016 853		19.3 s
				11*	11.011 434		20.4 min
				12	12.000 000	98.93	
				13	13.003 355	1.07	
				14*	14.003 242		5 730 yr
7	Nitrogen	N	14.006 7	13*	13.005 739		9.96 min
				14	14.003 074	99.632	
				15	15.000 109	0.368	
8	Oxygen	O	15.999 4	15*	15.003 065		122 s
				16	15.994 915	99.757	
				18	17.999 160	0.205	
9	Fluorine	F	18.998 403 2	19	18.998 403	100	
10	Neon	Ne	20.179 7	20	19.992 440	90.48	
				22	21.991 385	9.25	
11	Sodium	Na	22.989 77	22*	21.994 437		2.61 yr
				23	22.989 770	100	
				24*	23.990 963		14.96 h
12	Magnesium	Mg	24.305 0	24	23.985 042	78.99	
				25	24.985 837	10.00	
				26	25.982 593	11.01	
13	Aluminum	Al	26.981 538	27	26.981 539	100	
14	Silicon	Si	28.085 5	28	27.976 926	92.229 7	
15	Phosphorus	P	30.973 761	31	30.973 762	100	
				32*	31.973 907		14.26 days
16	Sulfur	S	32.066	32	31.972 071	94.93	
				35*	34.969 032		87.5 days
17	Chlorine	Cl	35.452 7	35	34.968 853	75.78	
				37	36.965 903	24.22	
18	Argon	Ar	39.948	40	39.962 383	99.600 3	
19	Potassium	K	39.098 3	39	38.963 707	93.258 1	
				40*	39.963 999	0.011 7	1.28×10^9 yr

(Continued)

Atomic Number Z	Element	Symbol	Chemical Atomic Mass (u)	Mass Number (* Indicates Radioactive) A	Atomic Mass (u)	Percent Abundance	Half-Life (If Radioactive) $T_{1/2}$
20	Calcium	Ca	40.078	40	39.962 591	96.941	
21	Scandium	Sc	44.955 910	45	44.955 910	100	
22	Titanium	Ti	47.867	48	47.947 947	73.72	
23	Vanadium	V	50.941 5	51	50.943 964	99.750	
24	Chromium	Cr	51.996 1	52	51.940 512	83.789	
25	Manganese	Mn	54.938 049	55	54.938 050	100	
26	Iron	Fe	55.845	56	55.934 942	91.754	
27	Cobalt	Co	58.933 200	59	58.933 200	100	
				60*	59.933 822		5.27 yr
28	Nickel	Ni	58.693 4	58	57.935 348	68.076 9	
				60	59.930 790	26.223 1	
29	Copper	Cu	63.546	63	62.929 601	69.17	
				65	64.927 794	30.83	
30	Zinc	Zn	65.39	64	63.929 147	48.63	
				66	65.926 037	27.90	
				68	67.924 848	18.75	
31	Gallium	Ga	69.723	69	68.925 581	60.108	
				71	70.924 705	39.892	
32	Germanium	Ge	72.61	70	69.924 250	20.84	
				72	71.922 076	27.54	
				74	73.921 178	36.28	
33	Arsenic	As	74.921 60	75	74.921 596	100	
34	Selenium	Se	78.96	78	77.917 310	23.77	
				80	79.916 522	49.61	
35	Bromine	Br	79.904	79	78.918 338	50.69	
				81	80.916 291	49.31	
36	Krypton	Kr	83.80	82	81.913 485	11.58	
				83	82.914 136	11.49	
				84	83.911 507	57.00	
				86	85.910 610	17.30	
37	Rubidium	Rb	85.467 8	85	84.911 789	72.17	
				87*	86.909 184	27.83	4.75×10^{10} yr
38	Strontium	Sr	87.62	86	85.909 262	9.86	
				88	87.905 614	82.58	
				90*	89.907 738		29.1 yr
39	Yttrium	Y	88.905 85	89	88.905 848	100	
40	Zirconium	Zr	91.224	90	89.904 704	51.45	
				91	90.905 645	11.22	
				92	91.905 040	17.15	
				94	93.906 316	17.38	
41	Niobium	Nb	92.906 38	93	92.906 378	100	
42	Molybdenum	Mo	95.94	92	91.906 810	14.84	
				95	94.905 842	15.92	
				96	95.904 679	16.68	
				98	97.905 408	24.13	
43	Technetium	Tc		98*	97.907 216		4.2×10^{6} yr
				99*	98.906 255		2.1×10^{5} yr

Atomic Number Z	Element	Symbol	Chemical Atomic Mass (u)	Mass Number (* Indicates Radioactive) A	Atomic Mass (u)	Percent Abundance	Half-Life (If Radioactive) $T_{1/2}$
44	Ruthenium	Ru	101.07	99	98.905 939	12.76	
				100	99.904 220	12.60	
				101	100.905 582	17.06	
				102	101.904 350	31.55	
				104	103.905 430	18.62	
45	Rhodium	Rh	102.905 50	103	102.905 504	100	
46	Palladium	Pd	106.42	104	103.904 035	11.14	
				105	104.905 084	22.33	
				106	105.903 483	27.33	
				108	107.903 894	26.46	
				110	109.905 152	11.72	
47	Silver	Ag	107.868 2	107	106.905 093	51.839	
				109	108.904 756	48.161	
48	Cadmium	Cd	112.411	110	109.903 006	12.49	
				111	110.904 182	12.80	
				112	111.902 757	24.13	
				113*	112.904 401	12.22	9.3×10^{15} yr
				114	113.903 358	28.73	
49	Indium	In	114.818	115*	114.903 878	95.71	4.4×10^{14} yr
50	Tin	Sn	118.710	116	115.901 744	14.54	
				118	117.901 606	24.22	
				120	119.902 197	32.58	
51	Antimony	Sb	121.760	121	120.903 818	57.21	
				123	122.904 216	42.79	
52	Tellurium	Te	127.60	126	125.903 306	18.84	
				128*	127.904 461	31.74	$>8 \times 10^{24}$ yr
				130*	129.906 223	34.08	$\leq 1.25 \times 10^{21}$ yr
53	Iodine	I	126.904 47	127	126.904 468	100	
				129*	128.904 988		1.6×10^{7} yr
54	Xenon	Xe	131.29	129	128.904 780	26.44	
				131	130.905 082	21.18	
				132	131.904 145	26.89	
				134	133.905 394	10.44	
				136*	135.907 220	8.87	$\geq 2.36 \times 10^{21}$ yr
55	Cesium	Cs	132.905 45	133	132.905 447	100	
56	Barium	Ba	137.327	137	136.905 821	11.232	
				138	137.905 241	71.698	
57	Lanthanum	La	138.905 5	139	138.906 349	99.910	
58	Cerium	Ce	140.116	140	139.905 434	88.450	
				142*	141.909 240	11.114	$>5 \times 10^{16}$ yr
59	Praseodymium	Pr	140.907 65	141	140.907 648	100	
60	Neodymium	Nd	144.24	142	141.907 719	27.2	
				144*	143.910 083	23.8	2.3×10^{15} yr
				146	145.913 112	17.2	
61	Promethium	Pm		145*	144.912 744		17.7 yr
62	Samarium	Sm	150.36	147*	146.914 893	14.99	1.06×10^{11} yr
				149*	148.917 180	13.82	$>2 \times 10^{15}$ yr
				152	151.919 728	26.75	
				154	153.922 205	22.75	

(Continued)

Atomic Number Z	Element	Symbol	Chemical Atomic Mass (u)	Mass Number (* Indicates Radioactive) A	Atomic Mass (u)	Percent Abundance	Half-Life (If Radioactive) $T_{1/2}$
63	Europium	Eu	151.964	151	150.919 846	47.81	
				153	152.921 226	52.19	
64	Gadolinium	Gd	157.25	156	155.922 120	20.47	
				158	157.924 100	24.84	
				160	159.927 051	21.86	
65	Terbium	Tb	158.925 34	159	158.925 343	100	
66	Dysprosium	Dy	162.50	162	161.926 796	25.51	
				163	162.928 728	24.90	
				164	163.929 171	28.18	
67	Holmium	Ho	164.930 32	165	164.930 320	100	
68	Erbium	Er	167.6	166	165.930 290	33.61	
				167	166.932 045	22.93	
				168	167.932 368	26.78	
69	Thulium	Tm	168.934 21	169	168.934 211	100	
70	Ytterbium	Yb	173.04	172	171.936 378	21.83	
				173	172.938 207	16.13	
				174	173.938 858	31.83	
71	Lutecium	Lu	174.967	175	174.940 768	97.41	
72	Hafnium	Hf	178.49	177	176.943 220	18.60	
				178	177.943 698	27.28	
				179	178.945 815	13.62	
				180	179.946 549	35.08	
73	Tantalum	Ta	180.947 9	181	180.947 996	99.988	
74	Tungsten	W	183.84	182	181.948 206	26.50	
	(Wolfram)			183	182.950 224	14.31	
				184*	183.950 933	30.64	$>3 \times 10^{17}$ yr
				186	185.954 362	28.43	
75	Rhenium	Re	186.207	185	184.952 956	37.40	
				187*	186.955 751	62.60	4.4×10^{10} yr
76	Osmium	Os	190.23	188	187.955 836	13.24	
				189	188.958 145	16.15	
				190	189.958 445	26.26	
				192	191.961 479	40.78	
77	Iridium	Ir	192.217	191	190.960 591	37.3	
				193	192.962 924	62.7	
78	Platinum	Pt	195.078	194	193.962 664	32.967	
				195	194.964 774	33.832	
				196	195.964 935	25.242	
79	Gold	Au	196.966 55	197	196.966 552	100	
80	Mercury	Hg	200.59	199	198.968 262	16.87	
				200	199.968 309	23.10	
				201	200.970 285	13.18	
				202	201.970 626	29.86	
81	Thallium	Tl	204.383 3	203	202.972 329	29.524	
				205	204.974 412	70.476	
		(Th C″)		208*	207.982 005		3.053 min
		(Ra C″)		210*	209.990 066		1.30 min

Atomic Number Z	Element	Symbol	Chemical Atomic Mass (u)	Mass Number (* Indicates Radioactive) A	Atomic Mass (u)	Percent Abundance	Half-Life (If Radioactive) $T_{1/2}$
82	Lead	Pb	207.2	204*	203.973 029	1.4	$\geq 1.4 \times 10^{17}$ yr
				206	205.974 449	24.1	
				207	206.975 881	22.1	
				208	207.976 636	52.4	
		(Ra D)		210*	209.984 173		22.3 yr
		(Ac B)		211*	210.988 732		36.1 min
		(Th B)		212*	211.991 888		10.64 h
		(Ra B)		214*	213.999 798		26.8 min
83	Bismuth	Bi	208.980 38	209	208.980 383	100	
		(Th C)		211*	210.987 258		2.14 min
84	Polonium	Po					
		(Ra F)		210*	209.982 857		138.38 days
		(Ra C′)		214*	213.995 186		164 μs
85	Astatine	At		218*	218.008 682		1.6 s
86	Radon	Rn		222*	222.017 570		3.823 days
87	Francium	Fr					
		(Ac K)		223*	223.019 731		22 min
88	Radium	Ra		226*	226.025 403		1 600 yr
		(Ms Th_1)		228*	228.031 064		5.75 yr
89	Actinium	Ac		227*	227.027 747		21.77 yr
90	Thorium	Th	232.038 1				
		(Rd Th)		228*	228.028 731		1.913 yr
		(Th)		232*	232.038 050	100	1.40×10^{10} yr
91	Protactinium	Pa	231.035 88	231*	231.035 879		32.760 yr
92	Uranium	U	238.028 9	232*	232.037 146		69 yr
				233*	233.039 628		1.59×10^{5} yr
		(Ac U)		235*	235.043 923	0.720 0	7.04×10^{8} yr
				236*	236.045 562		2.34×10^{7} yr
		(UI)		238*	238.050 783	99.274 5	4.47×10^{9} yr
93	Neptunium	Np		237*	237.048 167		2.14×10^{6} yr
94	Plutonium	Pu		239*	239.052 156		2.412×10^{4} yr
				242*	242.058 737		3.73×10^{6} yr
				244*	244.064 198		8.1×10^{7} yr

[a]Chemical atomic masses are from T. B. Coplen, "Atomic Weights of the Elements 1999," a technical report to the International Union of Pure and Applied Chemistry, and published in *Pure and Applied Chemistry,* 73(4), 667–683, 2001. Atomic masses of the isotopes are from G. Audi and A. H. Wapstra, "The 1995 Update to the Atomic Mass Evaluation," *Nuclear Physics,* A595, vol. 4, 409–480, December 25, 1995. Percent abundance values are from K. J. R. Rosman and P. D. P. Taylor, "Isotopic Compositions of the Elements 1999", a technical report to the International Union of Pure and Applied Chemistry, and published in *Pure and Applied Chemistry,* 70(1), 217–236, 1998.

APPENDIX C
Some Useful Tables

TABLE C.1

Mathematical Symbols Used in the Text and Their Meaning

Symbol	Meaning
$=$	is equal to
$\neq$	is not equal to
$\equiv$	is defined as
$\propto$	is proportional to
$>$	is greater than
$<$	is less than
$\gg$	is much greater than
$\ll$	is much less than
$\approx$	is approximately equal to
$\sim$	is on the order of magnitude of
Δx	change in x or uncertainty in x
Σx_i	sum of all quantities x_i
$\lvert x \rvert$	absolute value of x (always a positive quantity)

TABLE C.2

Standard Symbols for Units

Symbol	Unit	Symbol	Unit
A	ampere	kcal	kilocalorie
Å	angstrom	kg	kilogram
atm	atmosphere	km	kilometer
Bq	bequerel	kmol	kilomole
Btu	British thermal unit	L	liter
C	coulomb	lb	pound
°C	degree Celsius	ly	light year
cal	calorie	m	meter
cm	centimeter	min	minute
Ci	curie	mol	mole
d	day	N	newton
deg	degree (angle)	nm	nanometer
eV	electronvolt	Pa	pascal
°F	degree Fahrenheit	rad	radian
F	farad	rev	revolution
ft	foot	s	second
G	Gauss	T	tesla
g	gram	u	atomic mass unit
H	henry	V	volt
h	hour	W	watt
hp	horsepower	Wb	weber
Hz	hertz	yr	year
in.	inch	μm	micrometer
J	joule	Ω	ohm
K	kelvin		

TABLE C.3

The Greek Alphabet

Alpha	A	α	Nu	N	ν
Beta	B	β	Xi	Ξ	ξ
Gamma	Γ	γ	Omicron	O	o
Delta	Δ	δ	Pi	Π	π
Epsilon	E	ϵ	Rho	P	ρ
Zeta	Z	ζ	Sigma	Σ	σ
Eta	H	η	Tau	T	τ
Theta	Θ	θ	Upsilon	Υ	υ
Iota	I	ι	Phi	Φ	ϕ
Kappa	K	κ	Chi	X	χ
Lambda	Λ	λ	Psi	Ψ	ψ
Mu	M	μ	Omega	Ω	ω

TABLE C.4

Physical Data Often Used[a]

Average Earth-Moon distance	3.84×10^8 m
Average Earth-Sun distance	1.496×10^{11} m
Equatorial radius of Earth	6.38×10^6 m
Density of air (20°C and 1 atm)	1.20 kg/m^3
Density of water (20°C and 1 atm)	1.00×10^3 kg/m^3
Free-fall acceleration	9.80 m/s^2
Mass of Earth	5.98×10^{24} kg
Mass of Moon	7.36×10^{22} kg
Mass of Sun	1.99×10^{30} kg
Standard atmospheric pressure	1.013×10^5 Pa

[a] These are the values of the constants as used in the text.

TABLE C.5

Some Fundamental Constants[a]

Quantity	Symbol	Value[b]
Atomic mass unit	u	$1.660\,540\,2(10) \times 10^{-27}$ kg $931.494\,32(28)$ MeV/c^2
Avogadro's number	N_A	$6.022\,136\,7(36) \times 10^{23}$ (mol)$^{-1}$
Bohr radius	$a_0 = \dfrac{\hbar^2}{m_e e^2 k_e}$	$0.529\,177\,249(24) \times 10^{-10}$ m
Boltzmann's constant	$k_B = R/N_A$	$1.380\,658(12) \times 10^{-23}$ J/K
Compton wavelength	$\lambda_C = \dfrac{h}{m_e c}$	$2.426\,310\,58(22) \times 10^{-12}$ m
Coulomb constant	$k_e = \dfrac{1}{4\pi\epsilon_0}$	$8.987\,551\,787 \times 10^9$ N·m^2/C^2 (exact)
Electron mass	m_e	$9.109\,389\,7(54) \times 10^{-31}$ kg $5.485\,799\,03(13) \times 10^{-4}$ u $0.510\,999\,06(15)$ MeV/c^2
Electron volt	eV	$1.602\,177\,33(49) \times 10^{-19}$ J
Elementary charge	e	$1.602\,177\,33(49) \times 10^{-19}$ C
Gas constant	R	$8.314\,510(70)$ J/K·mol
Gravitational constant	G	$6.672\,59(85) \times 10^{-11}$ N·m^2/kg^2
Hydrogen ionization energy	$-E_1 = \dfrac{m_e e^4 k_e^2}{2\hbar^2} = \dfrac{e^2 k_e}{2a_0}$	$13.605\,698(40)$ eV
Neutron mass	m_n	$1.674\,928\,6(10) \times 10^{-27}$ kg $1.008\,664\,904(14)$ u $939.565\,63(28)$ MeV/c^2
Permeability of free space	μ_0	$4\pi \times 10^{-7}$ T·m/A (exact)
Permittivity of free space	$\epsilon_0 = 1/\mu_0 c^2$	$8.854\,187\,817 \times 10^{-12}$ C^2/N·m^2 (exact)
Planck's constant	h	$6.626\,075(40) \times 10^{-34}$ J·s
	$\hbar = h/2\pi$	$1.054\,572\,66(63) \times 10^{-34}$ J·s
Proton mass	m_p	$1.672\,623(10) \times 10^{-27}$ kg $1.007\,276\,470(12)$ u $938.272\,3(28)$ MeV/c^2
Rydberg constant	R_H	$1.097\,373\,153\,4(13) \times 10^7$ m^{-1}
Speed of light in vacuum	c	$2.997\,924\,58 \times 10^8$ m/s (exact)

[a] These constants are the values recommended in 1986 by CODATA, based on a least-squares adjustment of data from different measurements. For a more complete list, see Cohen, E. Richard, and Barry N. Taylor, *Rev. Mod. Phys.* **59:**1121, 1987.

[b] The numbers in parentheses for the values below represent the uncertainties in the last two digits.

APPENDIX D
SI Units

TABLE D.1

SI Base Units

Base Quantity	SI Base Unit	
	Name	**Symbol**
Length	meter	m
Mass	kilogram	kg
Time	second	s
Electric current	ampere	A
Temperature	kelvin	K
Amount of substance	mole	mol
Luminous intensity	candela	cd

TABLE D.2

Derived SI Units

Quantity	Name	Symbol	Expression in Terms of Base Units	Expression in Terms of Other SI Units
Plane angle	radian	rad	m/m	
Frequency	hertz	Hz	s^{-1}	
Force	newton	N	$kg \cdot m/s^2$	J/m
Pressure	pascal	Pa	$kg/m \cdot s^2$	N/m^2
Energy: work	joule	J	$kg \cdot m^2/s^2$	$N \cdot m$
Power	watt	W	$kg \cdot m^2/s^3$	J/s
Electric charge	coulomb	C	$A \cdot s$	
Electric potential (emf)	volt	V	$kg \cdot m^2/A \cdot s^3$	W/A, J/C
Capacitance	farad	F	$A^2 \cdot s^4/kg \cdot m^2$	C/V
Electric resistance	ohm	Ω	$kg \cdot m^2/A^2 \cdot s^3$	V/A
Magnetic flux	weber	Wb	$kg \cdot m^2/A \cdot s^2$	$V \cdot s$, $T \cdot m^2$
Magnetic field intensity	tesla	T	$kg/A \cdot s^2$	Wb/m^2
Inductance	henry	H	$kg \cdot m^2/A^2 \cdot s^2$	Wb/A

Answers to Quick Quizzes, Odd-Numbered Conceptual Questions and Problems

Chapter 1

CONCEPTUAL QUESTIONS

1. (a) ~ 0.1 m (b) ~ 1 m (c) Between 10 m and 100 m (d) ~ 10 m (e) ~ 100 m

3. (a) $\sim 10^6$ beats (b) $\sim 10^9$ beats

5. $\sim 10^9$ s

7. The length of a hand varies from person to person, so it isn't a useful standard of length.

9. The ark has an approximate volume of 10^4 m^3, whereas a home has an approximate volume of 10^3 m^3.

11. A dimensionally correct equation isn't necessarily true. For example, the equation 2 dogs = 5 dogs is dimensionally correct, but isn't true. However, if an equation is not dimensionally correct, it cannot be correct.

PROBLEMS

1. (b) $A_{\text{cylinder}} = \pi R^2$, $A_{\text{rectangular plate}} = \text{length} \times \text{width}$

5. $\text{m}^3/(\text{kg} \cdot \text{s}^2)$

7. (a) three significant figures (b) four significant figures (c) three significant figures (d) two significant figures

9. (a) 797 (b) 1.1 (c) 17.66

11. 115.9 m

13. 228.8 cm

15. 2×10^8 fathoms

17. 1.39×10^3 m^2

19. 1×10^{17} ft

21. 6.71×10^8 mi/h

23. 3.08×10^4 m^3

25. 9.82 cm

27. (a) 1.0×10^3 kg (b) $m_{\text{cell}} = 5.2 \times 10^{-16}$ kg, $m_{\text{kidney}} = 0.27$ kg, $m_{\text{fly}} = 1.3 \times 10^{-5}$ kg

29. 1 800 balls ($\sim 10^3$ balls) (assumes 81 games per season, nine innings per game, an average of ten hitters per inning, and one ball lost for every four hitters)

31. $\sim 10^7$ rev

33. $\sim 10^6$ balls

35. (2.0 m, 1.4 m)

37. $r = 2.2$ m, $\theta = 27°$

39. (a) 6.71 m (b) 0.894 (c) 0.746

41. 3.41 m

43. (a) 3.00 (b) 3.00 (c) 0.800 (d) 0.800 (e) 1.33

45. 5.00/7.00; the angle itself is 35.5°

47. The customer is incorrect. The cost of the pizza should be proportional to the area of the pizza, so the large one should cost more than the small one by a factor of $9^2/6^2 = 81/36 = 2.25$. Therefore, if the small one costs \$6, the large one should cost \$13.50.

49. The value of k, a dimensionless constant, cannot be found by dimensional analysis.

51. $\sim 10^{-9}$ m

53. $\sim 10^3$ tuners (assumes one tuner per 10 000 residents and a population of 7.5 million)

55. (a) 3.16×10^7 s (b) Between 10^{10} yr and 10^{11} yr

Chapter 2

QUICK QUIZZES

1. (a) 200 yd (b) 0 (c) 0

2. (a) False (b) True (c) True

3. The velocity vs. time graph (a) has a constant slope, indicating a constant acceleration, which is represented by the acceleration vs. time graph (e).

Graph (b) represents an object with increasing speed, but as time progresses, the lines drawn tangent to the curve have increasing slopes. Since the acceleration is equal to the slope of the tangent line, the acceleration must be increasing, and the acceleration vs. time graph that best indicates this behavior is (d).

Graph (c) depicts an object which first has a velocity that increases at a constant rate, which means that the object's acceleration is constant. The velocity then stops changing, which means that the acceleration of the object is zero. This behavior is best matched by graph (f).

4. (b)

5. (a) blue graph (b) red graph (c) green graph

6. (e)

7. (c)

8. (a) and (f)

CONCEPTUAL QUESTIONS

1. Yes. If the velocity of the particle is nonzero, the particle is in motion. If the acceleration is zero, the velocity of the particle is unchanging or is constant.

3. Yes. If this occurs, the acceleration of the car is opposite to the direction of motion, and the car will be slowing down.

5. No. They can be used only when the acceleration is constant. Yes. Zero is a constant.

7. Once the objects leave the hand, both are in free fall, and both exhibit the same downward acceleration equal in magnitude to the free-fall acceleration g.

9. Yes. Yes.

11. (a) At the maximum height, the ball is momentarily at rest. (That is, it has zero velocity.) The acceleration remains constant, with magnitude equal to the free-fall acceleration g and directed downward. Thus, even though the velocity is momentarily zero, it continues to change, and the ball will begin to gain speed in the downward direction. (b) The acceleration of the ball remains constant in magnitude and direction throughout the ball's free flight, from the instant it leaves the hand until the instant just before it strikes the ground. The acceleration is directed downward and has a magnitude equal to the free-fall acceleration g.

13. No. If the velocity of A is greater than that of B, car A is moving faster than B at this instant. However, car B may be picking up speed (that is, accelerating) at a greater rate than A. The driver of B may have stepped very hard on the gas pedal in the recent past, but car B has not yet reached the speed of A.

15. They are the same! You can see this for yourself by solving the kinematic equations for the two cases. The ball that was thrown upward will take longer to reach the ground, but when it arrives, it has the same velocity as the ball that was thrown downward.

17. Ignoring air resistance, in 16 s the pebble would fall a distance $\frac{1}{2}gt^2 = \frac{1}{2}(9.80\ \text{m/s}^2)(16.0\ \text{s})^2 = 1\ 250\ \text{m} = 1.25\ \text{km}$. Air resistance is an important force after the first few seconds, when the pebble has gained high speed. Also, part of the 16-s interval must be occupied by the sound returning up the well. Thus, the depth of the well is less than 1.25 km, but on the order of 10^3 m.

19. Above. Your ball has zero initial speed and a smaller average speed during the time of flight to the passing point.

PROBLEMS

1. (a) 52.9 km/h (b) 90.0 km
3. (a) Boat A wins by 60 km (b) 0
5. (a) 180 km (b) 63.4 km/h
7. (a) 4.0 m/s (b) − 0.50 m/s (c) − 1.0 m/s (d) 0
9. (a) 2.50 m/s (b) − 2.27 m/s (c) 0
11. 2.80 h, 218 km
13. 274 km/h
15. (a) 5.00 m/s (b) − 2.50 m/s (c) 0 (d) 5.00 m/s
17. (a) 4.0 m/s (b) − 4.0 m/s (c) 0 (d) 2.0 m/s
19. (a) $70.0\ \text{mi/h}\cdot\text{s} = 31.3\ \text{m/s}^2 = 3.19g$
(b) 321 ft = 97.8 m
21. 3.7 s
23. (a) $8.0\ \text{m/s}^2$ (b) about $12\ \text{m/s}^2$
25. $2.74 \times 10^5\ \text{m/s}^2$, or 2.79×10^4 times g
27. (a) $1.3\ \text{m/s}^2$ (b) 8.0 s
29. (a) $2.32\ \text{m/s}^2$ (b) 14.4 s
31. (a) 8.94 s (b) 89.4 m/s
33. $-3.6\ \text{m/s}^2$
35. 200 m
37. (a) 1.5 m/s (b) 32 m
39. (a) 8.2 s (b) 1.3×10^2 m
41. 958 m
43. (a) 31.9 m (b) 2.55 s (c) 2.55 s (d) − 25.0 m/s
45. 38.2 m
47. (a) 21.1 m/s (b) 19.6 m (c) 18.1 m/s; 19.6 m
49. (a) 308 m (b) 8.51 s (c) 16.4 s
51. (a) 10.0 m/s upward (b) 4.68 m/s downward
53. 15.0 s
55. (a) $-3.50 \times 10^5\ \text{m/s}^2$ (b) 2.86×10^{-4} s
57. 0.60 s
59. (a) 3.00 s (b) − 24.5 m/s, −24.5 m/s (c) 23.5 m
61. (a) 5.46 s (b) 73.0 m (c) $v_{\text{Kathy}} = 26.7$ m/s, $v_{\text{Stan}} = 22.6$ m/s
63. (a) $t_1 = 5.0$ s, $t_2 = 85$ s (b) 200 ft/s (c) 18 500 ft from starting point (d) 10 s after starting to slow down (total trip time = 100 s)
65. (a) 5.5×10^3 ft (b) 3.7×10^2 ft/s (c) The plane would travel only 0.002 ft in the time it takes the light from the bolt to reach the eye.
67. (a) 7.82 m (b) 0.782 s
69. (a) 3.45 s (b) 10.0 ft
71. 4.2 m/s

Chapter 3

QUICK QUIZZES

1. (c)
2. (b)
3.

Vector	x-component	y-component
$\vec{\mathbf{A}}$	−	+
$\vec{\mathbf{B}}$	+	−
$\vec{\mathbf{A}} + \vec{\mathbf{B}}$	−	−

4. (b)
5. (a)
6. (c)
7. (b)

CONCEPTUAL QUESTIONS

1. The components of $\vec{\mathbf{A}}$ will both be negative when $\vec{\mathbf{A}}$ lies in the third quadrant. The components of $\vec{\mathbf{A}}$ will have opposite signs when $\vec{\mathbf{A}}$ lies in either the second quadrant or the fourth quadrant.

3. Assuming that the ship sails in a straight line, the wrench will strike the deck at the base of the mast. The wrench has the same horizontal velocity as the ship when it is released, and it will maintain that component of velocity as it falls. Thus, the wrench is always above the same spot on the deck and will strike that spot when it hits the deck. To an observer on the ship, the wrench appears to fall straight downward. To a stationary observer on shore, the wrench is seen to follow a parabolic trajectory.

5. Let v_{0x} and v_{0y} be the initial velocity components for the projectile fired on Earth. Because the projectile fired on the Moon is given the same initial velocity, its initial velocity components have the same values v_{0x} and v_{0y} as those for the projectile on Earth. From $v_y^2 = v_{0y}^2 - 2g\Delta y$, the maximum altitude reached by a projectile is found to be $H = v_{0y}^2/2g$. Because the free-fall acceleration g is smaller on the Moon than on Earth, the maximum altitude will be greater for the projectile fired on the Moon. If they start from the same elevation, the projectile on the Moon will have a longer time of flight and therefore a greater range. The Apollo astronauts made several long golf drives on the Moon.

7. A vector has both magnitude and direction and can be added only to other vectors representing the same physical quantity. A scalar has only a magnitude and can be added only to other scalars representing the same physical quantity.

9. (a) At the top of the projectile's flight, its velocity is horizontal and its acceleration is downward. This is the only point at which the velocity and acceleration vectors are perpendicular. (b) If the projectile is thrown straight up or down, then the velocity and acceleration will be parallel throughout the motion. For any other kind of projectile motion, the velocity and acceleration vectors are never parallel.

11. (a) The acceleration is zero, since both the magnitude and direction of the velocity remain constant. (b) The particle has an acceleration, since the direction of $\vec{\mathbf{v}}$ changes.

13. The spacecraft will follow a parabolic path equivalent to that of a projectile thrown off a cliff with a horizontal velocity. As regards the projectile, gravity provides an acceleration that is always perpendicular to the initial velocity, resulting in a parabolic path. As regards the spacecraft, the initial velocity plays the role of the horizontal velocity of the projectile, and the leaking gas plays the role that gravity plays in the case of the projectile. If the orientation of the spacecraft were to change in response to the gas leak (which is by far the more likely result), then the acceleration would change direction and the motion could become very complicated.

15. For angles $\theta < 45°$, the projectile thrown at angle θ will be in the air for a shorter interval. For the smaller angle, the vertical component of the initial velocity is smaller than that for the larger angle. Thus, the projectile thrown at the smaller angle will not go as high into the air and will spend less time in the air before landing.

17. Yes. The projectile is a freely falling body, because gravity is the only force acting on it. The vertical acceleration will be the local free fall acceleration g; the horizontal acceleration will be zero.

19. If a projectile is to have zero speed at the top of its trajectory, it must be thrown straight upward with zero horizontal velocity. To give a projectile a nonzero speed at the top of its trajectory, the projectile should be thrown at any angle other than 90° to the horizontal.

PROBLEMS

1. Approximately 421 ft at 3° below the horizontal
3. Approximately 15 m at 58° S of E
5. Approximately 310 km at 57° S of W
7. (a) Approximately 5.0 units at $-53°$ (b) Approximately 5.0 units at $+53°$
9. 8.07 m at 42.0° S of E
11. (a) 5.00 blocks at 53.1° N of E (b) 13.0 blocks
13. 47.2 units at 122° from the positive x-axis
15. 157 km
17. 245 km at 21.4° W of N
19. 196 cm at 14.7° below the x-axis
21. (a) (10.0 m, 16.0 m)
23. 6.13 m
25. 12 m/s
27. 48.6 m/s
29. 25 m
31. (a) 32.5 m from the base of the cliff (b) 1.78 s
33. (a) 52.0 m/s horizontally (b) 212 m
35. 390 mi/h at 7.37° N of E
37. 1.88 m
39. 249 ft upstream
41. 18.0 s
43. 196 cm at 14.7° below the positive x-axis
45. (a) 57.7 km/h at 60.0° west of vertical (b) 28.9 km/h downward
47. 68.6 km/h
49. (a) 1.52×10^3 m (b) 36.1 s (c) 4.05×10^3 m
51. $R_{\text{moon}} = 18$ m, $R_{\text{Mars}} = 7.9$ m
53. (a) +, + (b) −, + (c) (i) must be in either the first or second quadrant
55. (a) 42 m/s (b) 3.8 s (c) $v_x = 34$ m/s, $v_y = -13$ m/s; $v = 37$ m/s
57. 7.5 m/s in the direction the ball was thrown
61. $R/2$
63. 7.5 min
65. 10.8 m
67. (b) $y = Ax^2$ with $A = \dfrac{g}{(2v_i^2)}$ where v_i is the muzzle velocity (c) 14.5 m/s
69. 227 paces at 165° from the positive x-axis
71. (a) 20.0° above the horizontal (b) 3.05 s
73. (a) 23 m/s (b) 360 m horizontally from the base of cliff

Chapter 4

QUICK QUIZZES

1. (a) True (b) False
2. (a) True (b) True (c) False
3. False
4. (a) (The value of g is smaller on the Moon than on Earth, so, for equal *weights*, the Moon *mass* must be greater.)
5. (c); (d)
6. (c)
7. (c)
8. (b)
9. (b)
10. (b) By exerting an upward force component on the sled, you reduce the normal force on the ground and so reduce the force of kinetic friction.

CONCEPTUAL QUESTIONS

1. (a) Two external forces act on the ball. (i) One is a downward gravitational force exerted by Earth. (ii) The second force on the ball is an upward normal force exerted by the hand. The reactions to these forces are (i) an upward gravitational force exerted by the ball on Earth and (ii) a downward force exerted by the ball on the hand. (b) After the ball leaves the hand, the only external force acting on the ball is the gravitational force exerted by Earth. The reaction is an upward gravitational force exerted by the ball on Earth.

3. No. If the car moves with constant acceleration, then a net force in the direction of the acceleration must be acting on it.

5. The coefficient of static friction is larger than that of kinetic friction. To start the box moving, you must counterbalance the maximum static friction force. This force exceeds the kinetic friction force that you must counterbalance to maintain the constant velocity of the box once it starts moving.

7. The inertia of the suitcase would keep it moving forward as the bus stops. There would be no tendency for the suitcase to be thrown backward toward the passenger. The case should be dismissed.

9. The force causing an automobile to move is the friction between the tires and the roadway as the automobile attempts to push the roadway backward. The force driving a propeller airplane forward is the reaction force exerted by the air on the propeller as the rotating propeller pushes the air backward (the action). In a rowboat, the rower pushes the water backward with the oars (the action). The water pushes forward on the oars and hence the boat (the reaction).

11. When the bus starts moving, Claudette's mass is accelerated by the force exerted by the back of the seat on her body. Clark is standing, however, and the only force acting on him is the friction between his shoes and the floor of the bus. Thus, when the bus starts moving, his feet accelerate forward, but the rest of his body experiences almost no accelerating force (only that due to his being attached to his accelerating feet!). As a consequence, his body tends to stay almost at rest, according to Newton's first law, relative to the ground. Relative to Claudette, however, he is moving toward her and falls into her lap. Both performers won Academy Awards.

13. The tension in the rope is the maximum force that occurs in *both* directions. In this case, then, since both are pulling with a force of magnitude 200 N, the tension is 200 N. If the rope does not move, then the force on each athlete must equal zero. Therefore, each athlete exerts 200 N against the ground.

15. (a) As the man takes the step, the action is the force his foot exerts on Earth; the reaction is the force exerted by Earth on his foot. (b) Here, the action is the force exerted by the snowball on the girl's back; the reaction is the force exerted by the girl's back on the snowball. (c) This action is the force exerted by the glove on the ball; the reaction is the force exerted by the ball on the glove. (d) This action is the force exerted by the air molecules on the window; the reaction is the force exerted by the window on the air molecules. In each case, we could equally well interchange the terms "action" and "reaction."

17. If the brakes lock, the car will travel farther than it would travel if the wheels continued to roll, because the coefficient of kinetic friction is less than that of static friction. Hence, the force of kinetic friction is less than the maximum force of static friction.

19. (a) The crate accelerates because of a friction force exerted by the floor of the truck on the crate. (b) If the driver slams on the brakes, the crate's inertia tends to keep the crate moving forward.

PROBLEMS

1. (a) 12 N (b) 3.0 m/s^2
3. 2×10^4 N
5. 3.71 N, 58.7 N, 2.27 kg
7. 9.6 N
9. 1.4×10^3 N
11. (a) 0.200 m/s^2 (b) 10.0 m (c) 2.00 m/s
13. 1.1×10^4 N
15. 600 N in vertical cable, 997 N in inclined cable, 796 N in horizontal cable
17. 150 N in vertical cable, 75 N in right-side cable, 130 N in left-side cable
19. 236 N (upper rope), 118 N (lower rope)
21. 613 N
23. 64 N
25. (a) 7.50×10^3 N backward (b) 50.0 m
27. (a) 7.0 m/s^2 horizontal and to the right (b) 21 N (c) 14 N horizontal and to the right
29. 7.90 m/s
31. (a) 14.2 m/s (b) 588 m
33. (a) 2.15×10^3 N forward (b) 645 N forward (c) 645 N rearward (d) 1.02×10^4 N at 74.1° below horizontal and rearward
35. $\mu_s = 0.38$, $\mu_k = 0.31$
37. (a) 0.256 (b) 0.509 m/s^2
39. (a) 14.7 m (b) Neither mass is necessary
41. 32.1 N
43. (a) 33 m/s (b) No. The object will speed up to 33 m/s from any lower speed and will slow down to 33 m/s from any higher speed.
45. $\mu_k = 0.287$
47. (a) 1.78 m/s^2 (b) 0.368 (c) 9.37 N (d) 2.67 m/s
49. 3.30 m/s^2
51. (a) 0.404 (b) 45.8 lb
53. (b) 2-kg block: 5.7 m/s^2 to left; 3-kg block: 5.7 m/s^2 to right; 10-kg block: 5.7 m/s^2 downward (c) 17 N, 41 N
55. (a) 84.9 N upward (b) 84.9 N downward
57. 50 m
59. (a) friction between box and truck (b) 2.94 m/s^2
61. (a) 2.22 m (b) 8.74 m/s down the incline
63. (a) 0.232 m/s^2 (b) 9.68 N
65. (a) 1.7 m/s^2, 17 N (b) 0.69 m/s^2, 17 N
67. 100 N, 204 N
69. (a) $T_1 = 78.0$ N, $T_2 = 35.9$ N (b) 0.656
71. (a) 30.7° (b) 0.843 N
73. 5.5×10^2 N
75. (a) 7.1×10^2 N (b) 8.1×10^2 N (c) 7.1×10^2 N (d) 6.5×10^2 N
77. 72.0 N
79. (a) 0.408 m/s^2 upward (b) 83.3 N
81. (b) 514 N, 558 N, 325 N

Chapter 5

QUICK QUIZZES

1. (C)
2. (d)
3. (c)
4. (c)

CONCEPTUAL QUESTIONS

1. Since no motion is taking place, the rope undergoes no displacement and no work is done on it. For the same reason, no work is being done on the pullers or the ground. Work is being done only within the bodies of the pullers. For example, the heart of each puller is applying forces on the blood to move blood through the body.

3. When the slide is frictionless, changing the length or shape of the slide will not make any difference in the final speed of the child, as long as the difference in the heights of the upper and lower ends of the slide is kept constant. If friction must be considered, the path length along which the friction force does negative work will be greater when the slide is made longer or given bumps. Thus, the child will arrive at the lower end with less kinetic energy (and hence less speed).

5. If we ignore any effects due to rolling friction on the tires of the car, we find that the same amount of work would be done in driving up the switchback and in driving straight up the mountain, since the weight of the car is moved upwards against gravity by the same vertical distance in each case. If we include friction, there is more work done in driving the switchback, since the distance over which the friction force acts is much longer. So why do we use switchbacks? The answer lies in the force required, not the work. The force required from the engine to follow a gentle rise is much smaller than that required to drive straight up the hill. To negotiate roadways running straight uphill, engines would have to be redesigned to enable them to apply much larger forces. (It is for much the same reason that ramps are designed to move heavy objects into trucks, as opposed to lifting the objects vertically.)

7. (a) The tension in the supporting cord does no work, because the motion of the pendulum is always perpendicular to the cord and therefore to the tension force. (b) The air resistance does negative work at all times, because the air resistance is always acting in a direction opposite that of the motion. (c) The force of gravity always acts downwards; therefore, the work done by gravity is positive on the downswing and negative on the upswing.

9. Because the periods are the same for both cars, we need only to compare the work done. Since the sports car is moving twice as fast as the older car at the end of the time interval, it has four times the kinetic energy. Thus, according to the work–energy theorem, four times as much work was done, and the engine must have expended four times the power.

11. During the time that the toe is in contact with the ball, the work done by the toe on the ball is given by

$$W_{\text{toe}} = \tfrac{1}{2} m_{\text{ball}} v^2 - 0 = \tfrac{1}{2} m_{\text{ball}} v^2$$

where v is the speed of the ball as it leaves the toe. After the ball loses contact with the toe, only the gravitational force and the retarding force due to air resistance continue to do work on the ball throughout its flight.

13. Yes, the total mechanical energy of the system is conserved because the only forces acting are conservative: the force of gravity and the spring force. There are two forms of potential energy in this case: gravitational potential energy and elastic potential energy stored in the spring.

15. Let's assume you lift the book slowly. In this case, there are two forces on the book that are almost equal in magnitude: the lifting force and the force of gravity. Thus, the positive work done by you and the negative work done by gravity cancel. There is no net work performed and no net change in the kinetic energy, so the work–energy theorem is satisfied.

17. As the satellite moves in a circular orbit about the Earth, its displacement during any small time interval is perpendicular to the gravitational force, which always acts toward the center of the Earth. Therefore, the work done by the gravitational force during any displacement is zero. (Recall that the work done by a force is defined to be $F\Delta x \cos\theta$, where θ is the angle between the force and the displacement. In this case, the angle is 90°, so the work done is zero.) Since the work–energy theorem says that the net work done on an object during any displacement is equal to the change in its kinetic energy, and the work done in this case is zero, the change in the satellite's kinetic energy is zero: hence, its speed remains constant.

19. If a crate is located on the bed of a truck, and the truck accelerates, the friction force exerted on the crate causes it to undergo the same acceleration as the truck, assuming that the crate doesn't slip. Another example is a car that accelerates because of the frictional forces between the road surface and its tires. This force is in the direction of the motion of the car and produces an increase in the car's kinetic energy.

PROBLEMS

1. 700 J
3. 15.0 MJ
5. (a) 61.3 J (b) −46.3 J (c) 0
7. (a) 79.4 N (b) 1.49 kJ (c) −1.49 kJ
9. (a) 2.00 m/s (b) 200 N
11. 0.265 m/s
13. (a) -5.6×10^2 J (b) 1.2 m
15. (a) 90 J (b) 1.8×10^2 N
17. 1.0 m/s
19. 4.1 m
21. (a) 4.1 m (b) 6.4 m/s
23. $h = 6.94$ m
25. $W_{\text{biceps}} = 120$ J, $W_{\text{chin-up}} = 290$ J, additional muscles must be involved
27. 0.459 m
29. 1.53×10^5 N upwards
31. (a) 10.9 m/s (b) 11.6 m/s
33. (a) 544 N/m (b) 19.7 m/s
35. 10.2 m
37. (a) Yes. There are no nonconservative forces acting on the child, so the total mechanical energy is conserved. (b) No. In the expression for conservation of mechanical energy, the mass of the child is included in every term and therefore cancels out. (c) The answer is the same in each case. (d) The expression would have to be modified to include the work done by the force of friction. (e) 15.3 m/s.
39. 2.1 kN
41. 3.8 m/s
43. (a) 5.42 m/s (b) 0.300 (c) 147 J
45. 289 m
47. (a) 24.5 m/s (b) Yes (c) 206 m (d) Unrealistic; the actual retarding force will vary with speed.
49. (a) 1.24 kW (b) 20.9%
51. 8.01 W
53. (a) 2.38×10^4 W = 32.0 hp (b) 4.77×10^4 W = 63.9 hp
55. (a) 24.0 J (b) −3.00 J (c) 21.0 J
57. (a) The graph is a straight line passing through the points (0 m, −16 N), (2 m, 0 N), and (3 m, 8 N). (b) −12.0 J
59. (a) 575 N/m (b) 46.0 J
61. (a) 4.4 m/s (b) 1.5×10^5 N
63. (a) 3.13 m/s (b) 4.43 m/s (c) 1.00 m
65. (a) 0.588 J (b) 0.588 J (c) 2.42 m/s (d) $PE_C = 0.392$ J (e) $KE_C = 0.196$ J
67. (a) 423 mi/gal (b) 776 mi/gal
69. (a) 28.0 m/s (b) 30.0 m (c) 89.0 m beyond the end of the track
71. 1.68 m/s
73. (a) 6.15 m/s (b) 9.87 m/s
75. (a) 101 J (b) 0.410 m (c) 2.84 m/s (d) −9.80 mm (e) 2.85 m/s
77. 914 N/m
79. $W_{\text{net}} = 0$, $W_{\text{grav}} = -2.0 \times 10^4$ J, $W_{\text{normal}} = 0$, $W_{\text{friction}} = 2.0 \times 10^4$ J
81. (a) 10.2 kW (b) 10.6 kW (c) 5.82×10^6 J
83. $v = (8gh/15)^{1/2}$
85. between 25.2 km/h and 27.0 km/h
87. (a) 6.75 W/m^2 (b) 6.64 kW/m^2 (c) A powerful automobile running on sunlight would have to carry on its roof a solar panel that was huge compared with the size of the car.
89. 4.3 m/s

Chapter 6

QUICK QUIZZES

1. (d)
2. (c)
3. (c)

4. (a)
5. (a) Perfectly inelastic (b) Inelastic (c) Inelastic
6. (a)

CONCEPTUAL QUESTIONS

1. (a) No. It cannot carry more kinetic energy than it possesses. That would violate the law of energy conservation. (b) Yes. By bouncing from the object it strikes, it can deliver more momentum in a collision than it possesses in its flight.

3. If all the kinetic energy disappears, there must be no motion of either of the objects after the collision. If neither is moving, the final momentum of the system is zero, and the initial momentum of the system must also have been zero. A situation in which this could be true would be the head-on collision of two objects having momenta of equal magnitude but opposite direction.

5. Initially, the clay has momentum directed toward the wall. When it collides and sticks to the wall, neither the clay nor the wall appears to have any momentum. Thus, it is tempting to (wrongfully) conclude that momentum is not conserved. However, the "lost" momentum is actually imparted to the wall and Earth, causing both to move. Because of Earth's enormous mass, its recoil speed is too small to detect.

7. Before the step the momentum was zero, so afterward the net momentum must also be zero. Obviously, you have some momentum, so something must have momentum in the opposite direction. That something is Earth, the enormous mass of which ensures that its recoil speed will be too small to detect, but if you want to make Earth move, it is as simple as taking a step.

9. As the water is forced out of the holes in the arm, the arm imparts a horizontal impulse to the water. The water then exerts an equal and opposite impulse on the spray arm, causing the spray arm to rotate in the direction opposite that of the spray.

11. Its speed decreases as its mass increases. There are no external horizontal forces acting on the box, so its momentum cannot change as it moves along the horizontal surface. As the box slowly fills with water, its mass increases with time. Because the product mv must be a constant, and because m is increasing, the speed of the box must decrease.

13. It will be easiest to catch the medicine ball when its speed (and kinetic energy) is lowest. The first option—throwing the medicine ball at the same velocity—will be the most difficult, because the speed will not be reduced at all. The second option, throwing the medicine ball with the same momentum, will reduce the velocity by the ratio of the masses. Since $m_t v_t = m_m v_m$, it follows that

$$v_m = v_t\left(\frac{m_t}{m_m}\right)$$

The third option, throwing the medicine ball with the same kinetic energy, will also reduce the velocity, but only by the square root of the ratio of the masses. Since

$$\frac{1}{2}m_t v_t^2 = \frac{1}{2}m_m v_m^2$$

it follows that

$$v_m = v_t\sqrt{\frac{m_t}{m_m}}$$

Thus, the slowest—and easiest—throw will be made when the momentum is held constant. If you wish to check this answer, try substituting in values of $v_t = 1$ m/s, $m_t = 1$ kg, and $m_m = 100$ kg. Then the same-momentum throw will be caught at 1 cm/s, while the same-energy throw will be caught at 10 cm/s.

15. The follow-through keeps the club in contact with the ball as long as possible, maximizing the impulse. Thus, the ball accrues a larger change in momentum than without the follow-through, and it leaves the club with a higher velocity and travels farther. With a short shot to the green, the primary factor is control, not distance. Hence, there is little or no follow-through, allowing the golfer to have a better feel for how hard he or she is striking the ball.

17. It is the product mv that is the same for both the bullet and the gun. The bullet has a large velocity and a small mass, while the gun has a small velocity and a large mass. Furthermore, the bullet carries much more kinetic energy than the gun.

PROBLEMS

1. 1.39 N · s up
3. (a) 8.35×10^{-21} kg · m/s (b) 4.50 kg · m/s (c) 750 kg · m/s (d) 1.78×10^{29} kg · m/s
5. (a) 31.0 m/s (b) the bullet, 3.38×10^3 J versus 69.7 J
7. $\sim 10^3$ N upward
9. 364 kg · m/s forward, 438 N forward
11. (a) 8.0 N · s (b) 5.3 m/s (c) 3.3 m/s
13. (a) 12 N · s (b) 8.0 N · s (c) 8.0 m/s, 5.3 m/s
15. (a) 9.60×10^{-2} s (b) 3.65×10^5 N (c) 26.6g
17. 6.7×10^3 N toward the west
19. 65 m/s
21. (a) 1.15 m/s (b) 0.346 m/s directed opposite to girl's motion
23. $KE_E/KE_b \sim 10^{-25}$
25. 1.67 m/s
27. (a) 1.80 m/s (b) 2.16×10^4 J
29. 57 m
31. 15.6 m/s
33. 273 m/s
35. (a) − 6.67 cm/s, 13.3 cm/s (b) 0.889
37. 17.1 cm/s (25.0-g object), 22.1 cm/s (10.0-g object)
39. 7.94 cm
41. (a) 2.9 m/s at 32° N of E (b) 7.9×10^2 J converted into internal energy
43. 5.59 m/s north
45. (a) 2.50 m/s at − 60° (b) elastic collision
47. (a) 9.0 m/s (b) − 15 m/s
49. 1.78×10^3 N on truck driver, 8.89×10^3 N on car driver
51. (a) 8/3 m/s (incident particle), 32/3 m/s (target particle) (b) − 16/3 m/s (incident particle), 8/3 m/s (target particle) (c) 7.1×10^{-2} J in case (a), and 2.8×10^{-3} J in case (b). The incident particle loses more kinetic energy in case (a), in which the target mass is 1.0 g.
53. 1.1×10^3 N (upward)
55. (a) 1.33 m/s (b) 235 N (c) 0.681 s (d) −160 N · s, 160 N · s (e) 1.82 m (f) 0.454 m (g) − 427 J (h) 107 J (i) Equal friction forces act through different distances on the person and the cart to do different amounts of work on them. This is a perfectly inelastic collision in which the total work on both person and cart together is − 320 J, which becomes + 320 J of internal energy.

57. (a) -2.33 m/s, 4.67 m/s (b) 0.277 m (c) 2.98 m (d) 1.49 m
59. (a) -0.667 m/s (b) 0.952 m
61. (a) 3.54 m/s (b) 1.77 m (c) 3.54×10^4 N (d) No, the normal force exerted by the rail contributes upward momentum to the system.
63. (a) 0.28 or 28% (b) 1.1×10^{-13} J for the neutron, 4.5×10^{-14} J for carbon
65. $\dfrac{2v_0^2}{9\mu_k g} - \dfrac{4d}{9}$
67. (a) No. After colliding, the cars, moving as a unit, would travel northeast, so they couldn't damage property on the southeast corner. (b) x-component: 16.3 km/h, y-component: 9.17 km/h, angle: the final velocity of the car is 18.7 km/h at 29.4° north of east, consistent with part (a).
69. (a) 4.85 m/s (b) 8.41 m
71. 0.312 N to the right
73. (a) 300 m/s, (b) 3.75 m/s, (c) 1.20 m

Chapter 7

QUICK QUIZZES

1. (c)
2. (b)
3. (b)
4. (b)
5. (a)
6. 1. (e) 2. (a) 3. (b)
7. (c)
8. (b), (c)
9. (e)
10. (d)

CONCEPTUAL QUESTIONS

1. (a) The head will tend to lean toward the right shoulder (that is, toward the outside of the curve). (b) When there is no strap, tension in the neck muscles must produce the centripetal acceleration. (c) With a strap, the tension in the strap performs this function, allowing the neck muscles to remain relaxed.
3. An object can move in a circle even if the total force on it is not perpendicular to its velocity, but then its speed will change. Resolve the total force into an inward radial component and a perpendicular tangential component. If the tangential force acts in the forward direction, the object will speed up, and if the tangential force acts backward, the object will slow down.
5. The speedometer will be inaccurate. The speedometer measures the number of tire revolutions per second, so its readings will be too low.
7. The car cannot round a turn at constant velocity, because "constant velocity" means the direction of the velocity is not changing. The statement is correct if the word "velocity" is replaced by the word "speed."
9. The gravitational force exerted on the moon by the planet produces a centripetal acceleration. Mathematically,

$$m_{\text{moon}}\left(\frac{v_t^2}{r}\right) = \frac{GMm_{\text{moon}}}{r^2}$$

or the mass of the planet is $M = \dfrac{(rv_t^2)}{G}$. Both r and v_t can be determined by observing the motion of the moon, and the mass of the planet is then easily computed.
11. Consider an individual standing against the inside wall of the cylinder with her head pointed toward the axis of the cylinder. As the cylinder rotates, the person tends to move in a straight-line path tangent to the circular path followed by the cylinder wall. As a result, the person presses against the wall, and the normal force exerted on her provides the radial force required to keep her moving in a circular path. If the rotational speed is adjusted such that this normal force is equal in magnitude to her weight on Earth, she will not be able to distinguish between the artificial gravity of the colony and ordinary gravity.
13. The tendency of the water is to move in a straight-line path tangent to the circular path followed by the container. As a result, at the top of the circular path, the water presses against the bottom of the pail, and the normal force exerted by the pail on the water provides the radial force required to keep the water moving in its circular path.
15. Any object that moves such that the *direction* of its velocity changes has an acceleration. A car moving in a circular path will always have a centripetal acceleration.
17. (b) decrease. The free-fall acceleration near the surface of the Earth is given by $g = GM_E/R_E^2$, therefore, doubling both the Earth's mass and its radius would reduce the value of g by a factor of 2.

PROBLEMS

1. (a) 3.2×10^8 rad (b) 5.0×10^7 rev
3. 1.99×10^{-7} rad/s, 0.986°/day
5. (a) 821 rad/s^2 (b) 4.21×10^3 rad
7. 3.2 rad
9. Main rotor: 179 m/s $= 0.522v_{\text{sound}}$
Tail rotor: 221 m/s $= 0.644v_{\text{sound}}$
11. $\sim 10^{-2}$ cm
13. 13.7 rad/s^2
15. (a) 3.37×10^{-2} m/s^2 downward (b) 0
17. (a) 0.35 m/s^2 (b) 1.0 m/s (c) 0.35 m/s^2, 0.94 m/s^2, 1.0 m/s^2 at 20° forward with respect to the direction of a_c
19. (a) 1.10 kN (b) 2.04 times her weight
21. 22.6 m/s
23. (a) 18.0 m/s^2 (b) 900 N (c) 1.84; this large coefficient is unrealistic, and she will not be able to stay on the merry-go-round.
25. (a) 9.8 N (b) 9.8 N (c) 6.3 m/s
27. (a) 1.58 m/s^2 (b) 455 N upward (c) 329 N upward (d) 397 N directed inward and 80.8° above horizontal
29. 321 N toward Earth
31. 1.1×10^{-10} N at 72° above the $+x$-axis
33. (a) 2.50×10^{-5} N toward the 500-kg object (b) Between the two objects and 0.245 m from the 500-kg object
35. (a) 9.58×10^6 m (b) 5.57 h
37. 1.90×10^{27} kg
39. (a) 1.48 h (b) 7.79×10^3 m/s (c) 6.43×10^9 J
41. (a) 9.40 rev/s (b) 44.1 rev/s^2; $a_r = 2\,590$ m/s^2; $a_t = 206$ m/s^2 (c) $F_r = 514$ N; $F_t = 40.7$ N
43. (a) 2.51 m/s (b) 7.90 m/s^2 (c) 4.00 m/s
45. (a) 7.76×10^3 m/s (b) 89.3 min

47. (a) $n = m\left(g - \frac{v^2}{r}\right)$ (b) 17.1 m/s

49. (a) $F_{g,\text{ true}} = F_{g,\text{ apparent}} + mR_E\omega^2$
(b) 732 N (equator), 735 N (either pole)

51. 0.131

55. 11.8 km/s

57. 0.75 m

59. (a) $v_0 = \sqrt{g\left(R - \frac{2h}{3}\right)}$ (b) $h' = \frac{R}{2} + \frac{2h}{3}$

61. (a) 15.3 km (b) 1.66×10^{16} kg (c) 1.13×10^4 s

63. 0.835 rev/s = 50.1 rev/min

65. (a) 10.6 kN (b) 14.1 m/s

67. (a) 106 N (b) 0.396

69. (a) 0.605 m/s (b) 17.3 rad/s (c) 5.82 m/s (d) The crank length is unnecessary.

71. (a) 2.0×10^{12} m/s^2 (b) 2.4×10^{11} N (c) 1.4×10^{12} J

Chapter 8

QUICK QUIZZES

1. (d)
2. (b)
3. (b)
4. (a)
5. (c)
6. (c)
7. (a)

CONCEPTUAL QUESTIONS

1. In order for you to remain in equilibrium, your center of gravity must always be over your point of support, the feet. If your heels are against a wall, your center of gravity cannot remain above your feet when you bend forward, so you lose your balance.

3. There are two major differences between torque and work. The primary difference is that the displacement in the expression for work is directed along the force, while the important distance in the torque expression is perpendicular to the force. The second difference involves whether there is motion. In the case of work, work is done only if the force succeeds in causing a displacement of the point of application of the force. By contrast, a force applied at a perpendicular distance from a rotation axis results in a torque regardless of whether there is motion.

As far as units are concerned, the mathematical expressions for both work and torque are in units that are the product of newtons and meters, but this product is called a joule in the case of work and remains a newton-meter in the case of torque.

5. No. For an object to be in equilibrium, the net external force acting on it must be zero. This is not possible when only one force acts on the object, unless that force should have zero magnitude. In that case, there is really no force acting on the object.

7. As the motorcycle leaves the ground, the friction between the tire and the ground suddenly disappears. If the motorcycle driver keeps the throttle open while leaving the ground, the rear tire will increase its angular speed and, hence, its angular momentum. The airborne motorcycle is now an isolated system, and its angular momentum must be conserved. The increase in angular momentum of the tire directed, say, clockwise must be compensated for by an increase in angular momentum of the entire motorcycle counterclockwise. This rotation results in the nose of the motorcycle rising and the tail dropping.

9. In general, you want the rotational kinetic energy of the system to be as small a fraction of the total energy as possible. That is, you want translation, not rotation. You want the wheels to have as little moment of inertia as possible, so that they present the lowest resistance to changes in rotational motion. Disklike wheels would have lower moments of inertia than hooplike wheels, so disks are preferable. The lower the mass of the wheels, the less is the moment of inertia, so light wheels are preferable. The smaller the radius of the wheels, the less is the moment of inertia, so smaller wheels are preferable—within limits: You want the wheels to be large enough to be able to travel relatively smoothly over irregularities in the road.

11. The angular momentum of the gas cloud is conserved. Thus, the product $I\omega$ remains constant. As the cloud shrinks in size, its moment of inertia decreases, so its angular speed ω must increase.

13. We can assume fairly accurately that the driving motor will run at a constant angular speed and at a constant torque. Therefore, as the radius of the take-up reel increases, the tension in the tape will decrease, in accordance with the equation.

$$T = \tau_{\text{const}}/R_{\text{take-up}} \qquad (1)$$

As the radius of the source reel decreases, given a decreasing tension, the torque in the source reel will decrease even faster, as the following equation shows:

$$\tau_{\text{source}} = TR_{\text{source}} = \tau_{\text{const}}R_{\text{source}}/R_{\text{take-up}} \qquad (2)$$

This torque will be partly absorbed by friction in the feed heads (which we assume to be small); some will be absorbed by friction in the source reel. Another small amount of the torque will be absorbed by the increasing angular speed of the source reel. However, in the case of a sudden jerk on the tape, the changing angular speed of the source reel becomes important. If the source reel is full, then the moment of inertia will be large and the tension in the tape will be large. If the source reel is nearly empty, then the angular acceleration will be large instead. Thus, the tape will be more likely to break when the source reel is nearly full. One sees the same effect in the case of paper towels: It is easier to snap a towel free when the roll is new than when it is nearly empty.

15. The initial angular momentum of the system (mouse plus turntable) is zero. As the mouse begins to walk clockwise, its angular momentum increases, so the turntable must rotate in the counterclockwise direction with an angular momentum whose magnitude equals that of the mouse. This conclusion follows from the fact that the final angular momentum of the system must equal the initial angular momentum (zero).

17. When a ladder leans against a wall, both the wall and the floor exert forces of friction on the ladder. If the floor is perfectly smooth, it can exert no frictional force in the horizontal direction to counterbalance the wall's normal force. Therefore, a ladder on a smooth floor cannot stand in equilibrium. However, a smooth wall can still exert a normal force to hold the ladder in equilibrium against horizontal motion. The counterclockwise torque of this force prevents rotation about the foot of the ladder. So you should choose a rough floor.

PROBLEMS

1. 133 N
3. (a) 30 N · m (counterclockwise) (b) 36 N · m (counterclockwise)
5. 5.1 N · m
7. $F_t = 724$ N, $F_s = 716$ N
9. 312 N
11. $x_{cg} = 3.33$ ft, $y_{cg} = 1.67$ ft
13. 1.01 m in Figure P8.13b; 0.015 m towards the head in Figure P8.13c.
15. $T = 2.71$ kN, $R_x = 2.65$ kN
17. (a) 443 N, (b) 222 N (to the right), 216 N (upwards)
19. $R = 107$ N, $T = 157$ N
21. $T_{\text{left wire}} = \frac{1}{3}w$, $T_{\text{right wire}} = \frac{2}{3}w$
23. $T_1 = 501$ N, $T_2 = 672$ N, $T_3 = 384$ N
25. 6.15 m
27. 209 N
29. (a) 99.0 kg · m^2 (b) 44.0 kg · m^2 (c) 143 kg · m^2
31. (a) 24.0 N · m (b) 0.035 6 rad/s^2 (c) 1.07 m/s^2
33. (a) 34 N (b) 33 cm
35. 177 N
37. 0.524
39. (a) 500 J (b) 250 J (c) 750 J
41. 276 J
43. 149 rad/s
45. 17.5 J · s counterclockwise
47. 8.0 rev/s
49. 6.73 rad/s
51. 5.99×10^{-2} J
53. (a) 0.360 rad/s counterclockwise (b) 99.9 J
55. (a) $\omega = \left(\dfrac{I_1}{I_1 + I_2}\right)\omega_0$ (b) $\dfrac{KE_f}{KE_i} = \dfrac{I_1}{I_1 + I_2} < 1$
57. (a) As the child walks to the right end of the boat, the boat moves left (towards the pier). (b) The boat moves 1.45 m closer to the pier, so the child will be 5.55 m from the pier. (c) No. He will be short of reaching the turtle by 0.45 m.
59. (a) 1.1 m/s^2 (b) $T_1 = 22$ N $T_2 = 44$ N
61. 36.9°
63. $a_{\text{sphere}} = \dfrac{g\sin\theta}{1.4}$, $a_{\text{disk}} = \dfrac{g\sin\theta}{1.5}$, $a_{\text{ring}} = \dfrac{g\sin\theta}{2.0}$
Thus, the sphere wins and the ring comes in last.
65. (a) 1.63 m/s (b) 54.2 rad/s
67. 1.09 m
69. (a) Mvd (b) Mv^2 (c) Mvd (d) $2v$ (e) $4Mv^2$ (f) $3Mv^2$
71. 7.5×10^{-11} s
73. (a) $T = \dfrac{Mmg}{M + 4m}$ (b) $a_t = \dfrac{4mg}{M + 4m}$
75. (a) $\sim 10^{-22}$ s^{-2}, (b) $\sim 10^{16}$ N · m, (c) $\sim 10^{13}$ m
77. 24.5 m/s
79. (a) 3.12 m/s^2 (b) $T_1 = 26.7$ N, $T_2 = 9.37$ N
81. (b) 3×10^2 N
83. $6w = 4.5$ kN

Chapter 9

QUICK QUIZZES

1. (c)
2. (a)
3. (c)
4. (b)
5. (c)
6. (b)
7. (a)

CONCEPTUAL QUESTIONS

1. The density of air is lower in the mile-high city of Denver than it is at lower altitudes, so the effect of air drag is less in Denver than it would be in a city such as New York. The reduced air drag means a well-hit ball will go farther, benefiting home-run hitters. On the other hand, curve ball pitchers prefer to throw at lower altitudes where the higher density air produces greater deflecting forces on a spinning ball.
3. She exerts enough pressure on the floor to dent or puncture the floor covering. The large pressure is caused by the fact that her weight is distributed over the very small cross-sectional area of her high heels. If you are the homeowner, you might want to suggest that she remove her high heels and put on some slippers.
5. If you think of the grain stored in the silo as a fluid, the pressure the grain exerts on the walls of the silo increases with increasing depth, just as water pressure in a lake increases with increasing depth. Thus, the spacing between bands is made smaller at the lower portions to counterbalance the larger outward forces on the walls in these regions.
7. The syringe doesn't really draw the blood from the vein. By moving the plunger in the syringe back, the nurse lowers the pressure inside the syringe below the pressure inside the vein. Then the blood pressure within the body forces blood out into the syringe, which "accepts" the blood.
9. In the ocean, the ship floats due to the buoyant force from *salt water*, which is denser than fresh water. As the ship is pulled up the river, the buoyant force from the fresh water in the river is not sufficient to support the weight of the ship, and it sinks.
11. The balance will not be in equilibrium: The side with the lead will be lower. Despite the fact that the weights on both sides of the balance are the same, the Styrofoam, due to its larger volume, will experience a larger buoyant force from the surrounding air. Thus, the net force of the weight and the buoyant force is larger in the downward direction for the lead than for the Styrofoam.
13. The two cans displace the same volume of water and hence are acted upon by buoyant forces of equal magnitude. The total weight of the can of diet cola must be less than this buoyant force, whereas the total weight of the can of regular cola is greater than the buoyant force. This is possible even though the two containers are identical and contain the same volume of liquid. Because of the difference in the quantities and densities of the sweeteners used, the volume V of the diet mixture will have less mass than an equal volume of the regular mixture.

15. As the truck passes, the air between your car and the truck is squeezed into the channel between the two vehicles and moves at a higher speed than when your car is in the open. According to Bernoulli's principle, this high-speed air has a lower pressure than the air on the outer side of your car. The difference in pressure exerts a net force on your car toward the truck.

17. Opening the windows results in a smaller pressure difference between the exterior and interior of the house and, therefore, less tendency for severe damage to the structure due to the Bernoulli effect.

PROBLEMS

1. 1.3 mm
3. 1.8×10^6 Pa
5. 3.5×10^8 Pa
7. 4.4 mm
9. 0.024 mm
11. 1.9 cm
13. (a) 1.88×10^5 Pa (b) 2.65×10^5 Pa
15. 1.4 atm
17. 3.4×10^2 m
19. 0.133 m
21. 271 kN horizontally toward the cellar
23. 1.05×10^5 Pa
25. 2.1 N · m
27. 9.41 kN
29. (b) 5.9 km
31. 2.67×10^3 kg
33. 5.57 N
35. (a) 1.46×10^{-2} m^3 (b) 2.10×10^3 kg/m^3
37. 1.07 m/s^2
39. 17.3 N (upper scale), 31.7 N (lower scale)
41. (a) 80 g/s (b) 0.27 mm/s
43. 12.6 m/s
45. 9.00 cm
47. 1.47 cm
49. (a) 28.0 m/s (b) 28.0 m/s (c) 2.11 MPa
51. (b) For any y less than $y_{max} = P_0/\rho g$
53. 8.3×10^{-2} N/m
55. 5.6×10^{-2} N/m
57. 8.6 N
59. 2.1 MPa
61. 2.8 μm
63. 0.41 mm
65. $RN = 4.3 \times 10^3$; turbulent flow
67. 1.8×10^{-3} kg/m^3
69. 1.4×10^{-5} N · s/m^2
71. (a) The buoyant forces are the same because the two blocks displace equal amounts of water. (b) The spring scale reads largest value for the iron block. (c) $B = 2.0 \times 10^3$ N for both blocks, $T_{iron} = 13 \times 10^3$ N, $T_{aluminum} = 3.3 \times 10^3$ N.
73. 2.5×10^7 capillaries
75. (a) 1.57 kPa, 1.55×10^{-2} atm, 11.8 mm of Hg
(b) The fluid level in the tap should rise.
(c) Blockage of flow of the cerebrospinal fluid
77. 4.14×10^3 m^3
79. 833 kg/m^3
81. 2.25 m above the level of point B
83. 6.4 m
85. 1.3 cm
87. 17.0 cm above the floor
89. 532 cm^3

Chapter 10

QUICK QUIZZES

1. (c)
2. (b)
3. (c)
4. (c)
5. (a)
6. (b)

CONCEPTUAL QUESTIONS

1. An ordinary glass dish will usually break because of stresses that build up as the glass expands when heated. The expansion coefficient for Pyrex glass is much lower than that of ordinary glass. Thus, the Pyrex dish will expand much less than the dish of ordinary glass and does not normally develop sufficient stress to cause breakage.

3. The accurate answer is that it doesn't matter! Temperatures on the Kelvin and Celsius scales differ by only 273 degrees. This difference is insignificant for temperatures on the order of 10^7 degrees. If we imagine that the temperature is given in kelvins and ignore any problems with significant figures, then the Celsius temperature is $1.499\,972\,7 \times 10^7$ °C.

5. Mercury must have the larger coefficient of expansion. As the temperature of a thermometer rises, both the mercury and the glass expand. If they both had the same coefficient of linear expansion, the mercury and the cavity in the glass would expand by the same amount, and there would be no apparent movement of the end of the mercury column relative to the calibration scale on the glass. If the glass expanded more than the mercury, the reading would go down as the temperature went up! Now that we have argued this conceptually, we can look in a table and find that the coefficient for mercury is about 20 times as large as that for glass, so that the expansion of the glass can sometimes be ignored.

7. We can think of each bacterium as being a small bag of liquid containing bubbles of gas at a very high pressure. The ideal gas law indicates that if the bacterium is raised rapidly to the surface, then its volume must increase dramatically. In fact, the increase in volume is sufficient to rupture the bacterium.

9. Velocity is a vector quantity, so direction must be considered. If there are the same number of particles moving to the right along the x-direction as there are to the left along the $-x$-direction, the x-components of the velocity of all the molecules will sum to zero.

11. The bags of chips contain a sealed sample of air. When the bags are taken up the mountain, the external atmospheric pressure on them is reduced. As a result, the difference between the pressure of the air inside the bags and the reduced pressure outside results in a net force pushing the plastic of the bag outward.

13. Additional water vaporizes into the bubble, so that the number of moles n increases.

PROBLEMS

1. (a) −460°C (b) 37.0°C (c) −280°C
3. (a) −423°C, 20.28 K (b) 68°F, 293 K
9. (a) 1 337 K, 2 933 K (b) 1 596°C, 1 596 K
11. 31 cm
13. 55.0°C
15. (a) −179°C (attainable) (b) −376°C (below 0 K, unattainable)

17. 2.171 cm
19. 2.7×10^2 N
21. 1.1 L (0.29 gal)
23. 0.548 gal
25. (a) increases (b) 1.603 cm
27. (a) 627°C (b) 927°C
29. (a) 2.5×10^{19} molecules (b) 4.1×10^{-21} mol
31. 287°C
33. 7.1 m
35. 16.0 cm^3
37. 6.21×10^{-21} J
39. 6.64×10^{-27} kg
41. (a) 8.76×10^{-21} J
(b) $v_{rms,\,He} = 1.62$ km/s, $v_{rms,\,Ar} = 514$ m/s
43. 16 N
45. 0.663 mm at 78.2° below the horizontal
47. 3.55 L
49. 6.57 MPa
51. The expansion of the mercury is almost 20 times that of the flask (assuming Pyrex glass).
53. shorter, by 0.061 mm
55. 2.74 m
57. 1.61 MPa
59. 0.417 L
61. (a) $\theta = \dfrac{(\alpha_2 - \alpha_1)L_0(\Delta T)}{\Delta r}$
(c) The bar bends in the opposite direction.
63. 0.53 kg

Chapter 11

QUICK QUIZZES

1. (a) Water, glass, iron. (b) Iron, glass, water.
2. (b) The slopes are proportional to the reciprocal of the specific heat, so a larger specific heat results in a smaller slope, meaning more energy is required to achieve a given temperature change.
3. (c)
4. (b)
5. (a) 4 (b) 16 (c) 64

CONCEPTUAL QUESTIONS

1. When you rub the surface, you increase the temperature of the rubbed region. With the metal surface, some of this energy is transferred away from the rubbed site by conduction. Consequently, the temperature in the rubbed area is not as high for the metal as it is for the wood, and it feels relatively cooler than the wood.
3. The fruit loses energy into the air by radiation and convection from its surface. Before ice crystals can form inside the fruit to rupture cell walls, all of the liquid water on the skin will have to freeze. The resulting time delay may prevent damage within the fruit throughout a frosty night. Further, a surface film of ice provides some insulation to slow subsequent energy loss by conduction from within the fruit.
5. The operation of an immersion coil depends on the convection of water to maintain a safe temperature. As the water near a coil warms up, the warmed water floats to the top due to Archimedes' principle. The temperature of the coil cannot go higher than the boiling temperature of water, 100°C. If the coil is operated in air, convection is reduced, and the upper limit of 100°C is removed. As a result, the coil can become hot enough to be damaged. If the coil is used in an attempt to warm a thick liquid like stew, convection cannot occur fast enough to carry energy away from the coil, so that it again may become hot enough to be damaged.
7. One of the ways that objects transfer energy is by radiation. The top of the mailbox is oriented toward the clear sky. Radiation emitted by the top of the mailbox goes upward and into space. There is little radiation coming down from space to the top of the mailbox. Radiation leaving the sides of the mailbox is absorbed by the environment. Radiation from the environment (tree, houses, cars, etc.), however, can enter the sides of the mailbox, keeping them warmer than the top. As a result, the top is the coldest portion and frost forms there first.
9. Tile is a better conductor of energy than carpet, so the tile conducts energy away from your feet more rapidly than does the carpeted floor.
11. The large amount of energy stored in the concrete during the day as the sun falls on it is released at night, resulting in an overall higher average temperature in the city than in the countryside. The heated air in a city rises as it's displaced by cooler air moving in from the countryside, so evening breezes tend to blow from country to city.
13. The fingers are wetted to create a layer of steam between the fingers and the molten lead. The steam acts as an insulator and prevents serious burns. This demonstration is dangerous: we don't recommend it.
15. The increase in the temperature of the ethyl alcohol will be about twice that of the water.
17. (d)

PROBLEMS

1. 10.1°C
3. (a) 1.67×10^{18} J (b) 53.1 yr
5. 1.03×10^3 J
7. 2.85 km
9. 176°C
11. 88 W
13. 185 g
15. 80 g
17. 1.8×10^3 J/kg · °C
19. 0.26 kg
21. 65°C
23. 21 g
25. 2.3 km
27. 16°C
29. $t_{boil} = 2.8$ min, $t_{evaporate} = 18$ min
31. (a) all ice melts, $T_f = 40$°C (b) 8.0 g melts, $T_f = 0$°C
33. (a) 0.22 kW (b) 13 mW (c) 56 mW
35. 402 MW
37. 14 ft^2 · °F · h/Btu
39. 9.0 cm
41. 0.11 kW
43. $\sim 10^3$ W
45. 16:1
47. 12 kW
49. 6.0×10^3 kg
51. 2.3 kg
53. 29°C
55. 30.3 kcal/h
57. 109°C
59. 51.2°C

61. (a) 7 stops (b) Assumes that no energy is lost to the surroundings and that all internal energy generated stays with the brakes.
63. (b) 2.7×10^3 J/kg · °C
65. 1.4 kg
67. 12 h
69. 10.9 g

Chapter 12

QUICK QUIZZES

1. (b)
2. A is isovolumetric, B is adiabatic, C is isothermal, D is isobaric.
3. (c)
4. (b)
5. The number 7 is the most probable outcome. The numbers 2 and 12 are the least probable outcomes.

CONCEPTUAL QUESTIONS

1. First, the efficiency of the automobile engine cannot exceed the Carnot efficiency: it is limited by the temperature of the burning fuel and the temperature of the environment into which the exhaust is dumped. Second, the engine block cannot be allowed to exceed a certain temperature. Third, any practical engine has friction, incomplete burning of fuel, and limits set by timing and energy transfer by heat.
3. If there is no change in internal energy, then, according to the first law of thermodynamics, the heat is equal to the negative of the work done on the gas (and thus equal to the work done *by* the gas). Thus, $Q = -W = W_{\text{by gas}}$.
5. The energy that is leaving the body by work and heat is replaced by means of biological processes that transform chemical energy in the food that the individual ate into internal energy. Thus, the temperature of the body can be maintained.
7. The statement shows a misunderstanding of the concept of heat. Heat is energy in the process of being transferred, not a form of energy that is held or contained. If you wish to speak of energy that is contained, you speak of **internal energy**, not heat. Correct statements would be (1) "Given any two objects in thermal contact, the one with the higher temperature will transfer energy by heat to the other" and (2) "Given any two objects of equal mass, the one with the higher product of absolute temperature and specific heat contains more internal energy."
9. Although no energy is transferred into or out of the system by heat, work is done on the system as the result of the agitation. Consequently, both the temperature and the internal energy of the coffee increase.
11. Practically speaking, it isn't possible to create a heat engine that creates no thermal pollution, because there must be both a hot heat source (energy reservoir) and a cold heat sink (low-temperature energy reservoir). The heat engine will warm the cold heat sink and will cool down the heat source. If either of those two events is undesirable, then there will be thermal pollution.

Under some circumstances, the thermal pollution would be negligible. For example, suppose a satellite in space were to run a heat pump between its sunny side and its dark side. The satellite would intercept some of the energy that gathered on one side and would "dump" it to the dark side. Since neither of those effects would be particularly undesirable, it could be said that such a heat pump produced no thermal pollution.
13. The rest of the Universe must have an entropy change of +8.0 J/K or more.
15. The first law is a statement of conservation of energy that says that we cannot devise a cyclic process that produces more energy than we put into it. If the cyclic process takes in energy by heat and puts out work, we call the device a heat engine. In addition to the first law's limitation, the second law says that, during the operation of a heat engine, some energy must be ejected to the environment by heat. As a result, it is theoretically impossible to construct a heat engine that will work with 100% efficiency.
17. From the point of view of energy principles, the molecules strike the piston and move it through a distance, so the molecules do work on the piston. This work represents a transfer of energy out of the gas. As a result, the internal energy of the gas drops. Because the temperature is related to internal energy, the temperature of the gas drops.

PROBLEMS

3. 1.1×10^4 J
5. (a) −810 J (b) −507 J (c) −203 J
7. (a) -6.1×10^5 J (b) 4.6×10^5 J
9. (a) 1.09×10^3 K (b) −6.81 kJ
11. (a) $Q < 0$, $W = 0$, $\Delta U < 0$ (b) $Q > 0$, $W = 0$, $\Delta U > 0$
13. (a) 567 J (b) 167 J
15. (a) −88.5 J (b) 722 J
17. (a) −180 J (b) +188 J
19. (a) -9.12×10^{-3} J (b) −333 J
21. (a) 0.95 J (b) 3.2×10^5 J (c) 3.2×10^5 J
23. 0.489 (or 48.9%)
25. (a) 0.333 (or 33.3%) (b) 2/3
27. (a) 0.672 (or 67.2%) (b) 58.8 kW
29. (a) 0.294 (or 29.4%) (b) 500 J (c) 1.67 kW
31. 1/3
33. 0.49°C
35. (a) −1.2 kJ/K (b) 1.2 kJ/K
37. 57 J/K
39. 3.27 J/K
41. (a)

End Result	Possible Tosses	Total Number of Same Result
All H	HHHH	1
1T, 3H	HHHT, HHTH, HTHH, THHH	4
2T, 2H	HHTT, HTHT, THHT, HTTH, THTH, TTHH	6
3T, 1H	TTTH, TTHT, THTT, HTTT	4
All T	TTTT	1

Most probable result = 2H and 2T.

(b) all H or all T (c) 2H and 2T
43. The maximum efficiency possible with these reservoirs = 50%; the claim is false.
45. (a) $-|Q_h|/T_h$ (b) $|Q_c|/T_c$ (c) $|Q_h|/T_h - |Q_c|/T_c$ (d) 0
47. 2.8°C
49. 18°C
51. (a) 12.2 kJ (b) 4.05 kJ (c) 8.15 kJ

53. (a) 26 J (b) 9.0×10^5 J (c) 9.0×10^5 J
55. (a) 2.49 kJ (b) 1.50 kJ (c) −990 J
57. (a) 0.554 (or 55.4%) (b) The Carnot efficiency is 0.749 (or 74.9%)
59. (a) 2.6×10^3 metric tons/day (b) $\$7.7 \times 10^6$/yr (c) 4.1×10^4 kg/s
61. (a) 10 J (b) No work is done; $\Delta U = 42$ J
63. 0.146; 486 kcal

Chapter 13

QUICK QUIZZES

1. (d)
2. (c)
3. (b)
4. (a)
5. (c)
6. (d)
7. (c), (b)
8. (a)
9. (b)

CONCEPTUAL QUESTIONS

1. It will increase. The speed of the wave varies inversely with the mass per unit length of the rope, so it is higher in the lighter rope. The frequency is unchanged as the wave passes from one rope to the other one. The wavelength is $\lambda = v/f$, so with the frequency constant and the speed increasing, the wavelength must increase.
3. No. Because the total energy is $E = \frac{1}{2}kA^2$, changing the mass of the object while keeping A constant has no effect on the total energy. When the object is at a displacement x from equilibrium, the potential energy is $\frac{1}{2}kx^2$, independent of the mass, and the kinetic energy is $KE = E - \frac{1}{2}kx^2$, also independent of the mass.
5. When the spring with two objects on opposite ends is set into oscillation in space, the coil at the exact center of the spring does not move. Thus, we can imagine clamping the center coil in place without affecting the motion. If we do this, we have two separate oscillating systems, one on each side of the clamp. The half-spring on each side of the clamp has twice the spring constant of the full spring, as shown by the following argument: The force exerted by a spring is proportional to the separation of the coils as the spring is extended. Imagine that we extend a spring by a given distance and measure the distance between coils. We then cut the spring in half. If one of the half-springs is now extended by the same distance, the coils will be twice as far apart as they were in the complete spring. Thus, it takes twice as much force to stretch the half-spring, from which we conclude that the half-spring has a spring constant which is twice that of the complete spring. Hence, our clamped system of objects on two half-springs will vibrate with a frequency that is higher than f by a factor of the square root of two.
7. The bouncing ball is not an example of simple harmonic motion. The ball does not follow a sinusoidal function for its position as a function of time. The daily movement of a student is also not simple harmonic motion, since the student stays at a fixed location—school—for a long time. If this motion were sinusoidal, the student would move more and more slowly as she approached her desk, and as soon as she sat down at the desk, she would start to move back toward home again.
9. We assume that the buoyant force acting on the sphere is negligible in comparison to its weight, even when the sphere is empty. We also assume that the bob is small compared with the pendulum length. Then, the frequency of the pendulum is $f = 1/T = (1/2\pi)\sqrt{g/L}$, which is independent of mass. Thus, the frequency will not change as the water leaks out.
11. As the temperature increases, the length of the pendulum will increase due to thermal expansion, and with a greater length, the period of the pendulum increases. Thus, it takes longer to execute each swing, so that each second according to the clock will take longer than an actual second. Consequently, the clock will run slow.
13. A pulse in a long line of people is longitudinal, since the movement of people is parallel to the direction of propagation of the pulse. The speed is determined by the reaction time of the people and the speed with which they can move once a space opens up. There is also a psychological factor, in that people will not want to fill a space that opens up in front of them too quickly, so as not to intimidate the person in front of them. The "wave" at a stadium is transverse, since the fans stand up vertically as the wave sweeps past them horizontally. The speed of this pulse depends on the limits of the fans' abilities to rise and sit rapidly and on psychological factors associated with the anticipation of seeing the pulse approach the observer's location.
15. A wave on a massless string would have an infinite speed of propagation, because the linear mass density of the string is zero.
17. The kinetic energy is proportional to the square of the speed, and the potential energy is proportional to the square of the displacement. Therefore, both must be positive quantities.
19. From $v = \sqrt{F/\mu}$, we see that increasing the tension by a factor of four doubles the wave speed.

PROBLEMS

1. (a) 24 N toward the equilibrium position (b) 60 m/s^2 toward the equilibrium position.
3. (b) 1.81 s (c) No, the force is not of the form of Hooke's law.
5. 0.242 kg
7. (a) 60 J (b) 49 m/s
9. 2.94×10^3 N/m
11. (a) $PE = E/4$ $KE = 3E/4$ (b) $x = A/\sqrt{2}$
13. 0.478 m
15. (a) 0.28 m/s (b) 0.26 m/s (c) 0.26 m/s (d) 3.5 cm
17. 39.2 N
19. (a) You observe uniform circular motion projected on a plane perpendicular to the motion. (b) 0.628 s
21. The horizontal displacement is described by $x(t) = A \cos \omega t$, where A is the distance from the center of the wheel to the crankpin.
23. 0.63 s
25. (a) 1.0 s (b) 0.28 m/s (c) 0.25 m/s
27. (a) 11.0 N toward the left (b) 0.881 oscillations
29. $v = \pm \omega A \sin \omega t$, $a = -\omega^2 A \cos \omega t$
31. 105 complete oscillations
33. (a) slow (b) 9:47 A.M.
35. (a) $L_{\text{Earth}} = 25$ cm, $L_{\text{Mars}} = 9.4$ cm, (b) $m_{\text{Earth}} = m_{\text{Mars}} = 0.25$ kg

37. (a) 9.00 cm (b) 20.0 cm (c) 40.0 ms (d) 5.00 m/s
39. (a) 11.4 ns (b) 3.41 m
41. 31.9 cm
43. 80.0 N
45. (a) 30.0 N (b) 25.8 m/s
47. 28.5 m/s
49. (a) 0.051 kg/m (b) 20 m/s
51. 13.5 N
53. (a) Constructive interference gives $A = 0.50$ m
(b) Destructive interference gives $A = 0.10$ m
55. 6.62 cm
57. (a) 588 N/m (b) 0.700 m/s
59. 1.1 m/s
61. (a) 15.8 rad/s (b) 5.23 cm
63. $F_{\text{tangential}} = -[(\rho_{\text{air}} - \rho_{\text{He}})Vg/L]s$, $T = 1.40$ s
67. (a) 0.50 m/s (b) 8.6 cm
69. (a) the longitudinal wave (b) 11.1 min
71. (a) 0.50 m/s (b) 0.39 m/s (c) 5.0 cm

Chapter 14

QUICK QUIZZES

1. (c)
2. (c)
3. (b)
4. (b), (e)
5. (d)
6. (a)
7. (b)

CONCEPTUAL QUESTIONS

1. (a) higher (b) lower
3. The camera is designed to operate at an assumed speed of sound of 345 m/s, the speed of sound at a room temperature of 23°C. If the temperature should decrease to, say, 0°C, the speed of sound will also decrease, and the camera will respond to the fact that it takes longer for the sound to make its round trip. Thus, it will operate as if the object is farther away than it really is.

It is of interest to note that bats use echo sounding like this to locate insects or to avoid obstacles in front of them, something that they must do because of poor eyesight and the high speeds at which they fly. Likewise, blue whales use this technique to help them avoid objects in their path. The need here is obvious, because a typical whale has a mass of 10^5 kg and travels at a relatively fast speed of 20 mi/h, so it takes a long time for it to stop its motion or to change direction.
5. Sophisticated electronic devices break the frequency range of about 60 to 4 000 Hz used in telephone conversations into several frequency bands and then mix them in a predetermined pattern so that they become unintelligible. The descrambler, of course, moves the bands back into their proper order.
7. A rise in temperature will increase the dimensions of the wind instrument much less than it increases the speed of sound in the enclosed air. This effect will raise the resonant frequencies that are produced by the instrument, which will go sharp as the temperature increases and go flat as the temperature decreases.
9. The echo is Doppler shifted, and the shift is like both a moving source and a moving observer. The sound that leaves your horn in the forward direction is Doppler shifted to a higher frequency, because it is coming from a moving source. As the sound reflects back and comes towards you, you are a moving observer, so there is a second Doppler shift to an even higher frequency. If the sound reflects from the spacecraft coming towards you, there is a different moving-source shift to an even higher frequency. The reflecting surface of the spacecraft acts as a moving source.
11. The center of the string is a node for the second harmonic, as well as for every even-numbered harmonic. By placing the finger at the center and plucking, the guitarist is eliminating any harmonic which does not have a node at that point—that is, all the odd harmonics. The even harmonics can vibrate relatively freely with the finger at the center because they exhibit no displacement at that point. The result is a sound with a mixture of frequencies that are integer multiples of the second harmonic, which is one octave higher than the fundamental.
13. The bowstring is pulled away from equilibrium and released, in a manner similar to the way a guitar string is pulled and released when it is plucked. Thus, standing waves will be excited in the bowstring. If the arrow leaves from the exact center of the string, then a series of odd harmonics will be excited. Even harmonics will not be excited, because they have a node at the point where the string exhibits its maximum displacement.
15. At the instant at which there is no displacement of the string, the string is still moving. Thus, the energy is present at that instant entirely as kinetic energy of the string.
17. As the whistle is approaching you, the sound is Doppler shifted to a frequency higher than the natural frequency of the whistle. You will momentarily hear the natural frequency just as the whistle comes parallel with you and is in the act of passing. As soon as the whistle is past, you will hear sound that is Doppler shifted to a frequency lower than the natural frequency of the whistle. As the frequency of the sound steps down from one constant value to a second one, the intensity of the sound varies continuously. The loudness increases smoothly to a maximum and then decreases.
19. The two engines are running at slightly different frequencies, thus producing a beat frequency between them.

PROBLEMS

1. 5.56 km
3. 32°C
5. 516 m
7. 1.99 km
9. (a) 1.00×10^{-2} W/m^2 (b) 105 dB
11. 37 dB
13. 9 additional machines
15. 66.0 dB
17. (a) 1.3×10^2 W (b) 96 dB
21. (a) 75.2-Hz drop (b) 0.953 m
23. 595 Hz
25. 0.391 m/s
27. 19.3 m
29. 48°
31. 800 m
33. (a) 0.240 m (b) 0.855 m
35. (a) Nodes at 0, 2.67 m, 5.33 m, and 8.00 m; antinodes at 1.33 m, 4.00 m, and 6.67 m (b) 18.6 Hz

37. At 0.0891 m, 0.303 m, 0.518 m, 0.732 m, 0.947 m, and 1.16 m from one speaker.
39. (a) 79 N (b) 2.1×10^2 Hz
41. 19.976 kHz
43. 58 Hz
45. 3.0 kHz
47. (a) 0.552 m (b) 317 Hz
49. 5.26 beats/s
51. 3.79 m/s toward the station, 3.88 m/s away from the station
53. (a) 1.98 beats/s (b) 3.40 m/s
55. 1.76 cm
57. 262 kHz
59. 64 dB
61. 439 Hz and 441 Hz
63. 32.9 m/s
65. 3.97 beats/s
67. 1.34×10^4 N
69. 1 204 Hz
71. (a) 617 m (b) 154 m/s

Credits

Photographs

This page constitutes an extension of the copyright page. We have made every effort to trace the ownership of all copyrighted material and to secure permission from copyright holders. In the event of any question arising as to the use of any material, we will be pleased to make the necessary corrections in future printings. Thanks are due to the following authors, publishers, and agents for permission to use the material indicated.

Chapter 1. **1:** Stone/Getty Images **2**: bottom left, Courtesy of National Institute of Standards and Technology, U.S. Dept. of Commerce; bottom right, Courtesy of Institute of Standards and Technology, U.S. Dept. of Commerce **10:** Billy E. Barnes/Stock Boston **12:** R. Williams (STScI), the HDF-S team, and NASA **19:** top left, Courtesy of George Semple; top right, Courtesy of George Semple **20:** © Sylvain Grandadam/Photo Researchers, Inc. **21:** NASA

Chapter 2. **23:** Caron/Corbis Sygma **39:** Courtesy Amtrak NEC Media Relations **41:** North Wind Archive

Chapter 3. **53:** © Bettmann/Corbis **54:** top, Mack Henley/Visuals Unlimited; bottom, George Semple **62:** HIRB/Index Stock **65:** Mike Powell/Allsport/Getty Images **75:** top right, Joseph Kayne/Dembinsky Photo Associates; bottom left, AP/Wide World Photos

Chapter 4. **81:** top left, © Royalty-Free/Corbis; bottom right, © Lorenzo Ciniglio/Corbis **83:** Roger Viollet, Mill Valley, CA, University Science Books, 1982 **85:** Giraudon/Art Resource **88:** NASA **90:** Jim Gillmoure/corbisstockmarket.com **107:** Guy Sauvage/Photo Researchers, Inc.

Chapter 5. **118:** NASA **121:** top, © Chris Collins/Corbis; bottom, © McCrone Photo/Custom Medical Stock Photo **132:** Wet'n Wild Orlando **141:** top right, Jan Hinsch/Science Photo Library/Photo Researchers, Inc.; bottom right, Gareth Williams, Minor Planet Center **149:** Arthur Tilley/FPG/Getty Images **157:** © Jamie Budge/Corbis **158:** Engraving from *Scientific American,* July 1888.

Chapter 6. **160:** © Reuters/Corbis **162:** © Harold and Esther Edgerton Foundation, 2002, courtesy of Palm Press, Inc. **163:** © Harold and Esther Edgerton Foundation, 2002, Courtesy of Palm Press, Inc. **164:** top right, Tim Wright/Corbis **167:** bottom left, Courtesy of NASA; bottom right, Mike Severns/Stone/Getty Images **171:** Courtesy of Central Scientific Company **181:** Courtesy of Saab **188:** bottom left, Courtesy of CENCO

Chapter 7. **189:** Courtesy NASA **208:** top right, Courtesy of PASCO Scientific

Chapter 8. **226:** © Kevin Fleming/Corbis **230:** David Serway **239:** George Semple **248:** top left, © Benson Krista Hicks/Corbis Sygma; top right, © Benson Krista Hicks/Corbis Sygma; middle left, © Patrick Giardino/Corbis; bottom left, Max Planck Institute for Astronomy and Calar Alto Observatory, K. Meisenheimer and A. Quetz; bottom right, © Smithsonian Institute/Photo Researchers, Inc. **253:** top left, © Eye Ubiquitous/Corbis; top right, Gerard Lacz/NHPA **264:** © Ed Bock/Corbis

Chapter 9. **266:** © Alison Wright/Corbis **267:** Charles D. Winters **271:** NASA **275:** © Royalty-Free/Corbis **276:** Raymond A. Serway **277:** Courtesy of Scientific Company **279:** bottom right, David Frazier **284:** top left, © Royalty-Free/Corbis **285:** top right, © Royalty-Free/Corbis **288:** Andy Sacks/Stone/Getty Images **289:** top right, Kim Vandiver and Harold Edgerton. © Harold and Esther Edgerton Foundation, 2002, courtesy of Palm Press, Inc.; bottom right, George Semple **292:** top left, Corbis-Bettmann; bottom right, Courtesy of Central Scientific Company **299:** Herman Eisenbeiss/Photo Researchers, Inc. **300:** Charles D. Winters **310:** bottom left, Henry Leap and Jim Lehman; bottom right, Pamela Zilly/The Image Bank/Getty Images **318:** The Granger Collection

Chapter 10. **321:** © Jim Sugar/Corbis **328:** top left, George Semple; bottom left, George Semple **329:** AP/Wide World Photos **330:** George Semple **340:** © R. Folwell/Science Photo Library/Photo Researchers, Inc.

Chapter 11. **352:** Steve Bly/Getty Images **353:** By kind permission of the President and Council of the Royal Society **366:** George Semple **371:** Gary Settles/Science Source/Photo Researchers Inc. **373:** bottom left, Daedalus Enterprises, Inc./Peter Arnold, Inc.; bottom right, Photodisc/Getty Images **374:** SPL/Photo Researchers, Inc. **377:** top left, Tom Bean; top right, Tom Bean **380:** Nathan Bilow/Leo de Wys, Inc. **381:** top left, © Jim Bourg/Teuters/Corbis

Chapter 12. **386:** © Tim De Waele/Corbis **404:** Charles D. Winters **405:** J-L. Charmet/SPL/Photo Researchers, Inc. **406:** J-L. Charmet/SPL/Photo Researchers, Inc. **409:** AIP Niels Bohr Library, Lade Collection **413:** top, George Semple; bottom, George Semple **414:** © BSIP/Laurent Science Source/Photo Researchers, Inc.

Chapter 13. **424:** © Rick Doyle/Corbis **429:** Eric Lars Baleke/Black Star **433:** Link/Visuals Unlimited **434:** Telegraph Colour Library/FPG International **448:** Martin

Dohm/SPL/Photo Researchers **451:** Gregg Adams/Stone Getty Images

Chapter 14. **458:** Patrick Ward/CORBIS **460:** Bernard Benoit/Photo Researchers, Inc. **469:** top right, Courtesy Educational Development Center, Newton, Mass. **472:** top right, © 1973 Kim Vandiver & Harold E. Edgerton/Courtesy of Palm Press, Inc. **473:** © 1994, Comstock **476:** all images, © Richard Megna, Fundamental Photographs, NYC **477:** Charles D. Winters **480:** left, top, Courtesy of Professor Thomas D. Rossing, Northern Illinois University; left, bottom, Ben Rose/The IMAGE Bank Getty Images; bottom left, United Press International Photo; bottom right, United Press International Photo **487:** Royalty Free/Corbis **496:** Sidney Harris

Tables and Illustrations

This page constitutes an extension of the copyright page. We have made every effort to trace the ownership of all copyrighted material and to secure permission from copyright holders. In the event of any question arising as to the use of any material, we will be pleased to make the necessary corrections in future printings. Thanks are due to the following authors, publishers, and agents for permission to use the material indicated.

Chapter 2. **24:** Schwadron, Cartoonists & Writers Syndicate/cartoonweb.com **47:** By permission of John L. Hart FLP, and Creators Syndicate, Inc.

Index

Page numbers followed by "f" indicate figures; page numbers followed by "t" indicate tables.

A

Absorber, ideal, 373
Acceleration, 30–33
  angular, 189–192
    average, 192, 217
      of bicycle wheel, 191–192, 192f
    of compact discs, 196–197
    constant
      of rotating wheel, 193–194
      rotational motion under, 193–194
    instantaneous, 192
    tangential acceleration compared with, 195
    torque and, 237–244
      bicycle gear system and, 239, 239f
    in tossing baseball, 242, 242f
  average, 30–31, 46
    in two dimensions, 60, 72
  centripetal, 198–206, 198f, 217
    automobile on curve and, 203–204, 203f
    banked roadway and, 204–205, 204f
    of car on circular road, 198
    forces causing, 201–206
    at race track, 200
    of roller coaster, 205–206, 205f
  constant, one-dimensional motion with, 35–41, 46
    applications of, 37–41
    for car chase, 37–38
    equations for, 36t
    for high-speed train, 38–40
    for landing jetliner, 40–41
    for race car, 37
  free-fall, 41
    at various altitudes, 207t
  as function of time, 435–437
  gravitational, calculation of, 209–210
  at highest point, 61
  instantaneous, 31–33, 46
    of running ball player, 32–33
    in two dimensions, 60, 72
  negative, 31
  net force and, 84–90
  of objects, Newton's second law of motion and, 96–100
  in projectile motion, 60–68, 72
  in simple harmonic motion, 426, 449
  tangential, 195
  total, 199
  in two dimensions, 60
  units of, 85t
Accelerators of automobile, 60
Action force, 90
Adhesive force, 300, 300f, 301f
Adiabatic process, 394–395, 416–417
  adiabatic expansion and, 395
  work and engine cylinder and, 394–395
Air columns, standing waves in, 480–484, 481f
Altitudes, various, free-fall acceleration at, 207t
Amorphous solid, 267, 267f
Amplitude
  of simple harmonic motion, 425
  of wave, 444
Aneurysm, 295
Angle of contact, 301, 301f
Angular acceleration, 189–192. *See also* Acceleration, angular
  average, of bicycle wheel, 191–192, 192f
  of compact discs, 196–197
  constant
    of rotating wheel, 193–194
    rotational motion under, 193–194
  tangential acceleration compared with, 195
  in tossing baseball, 242, 242f
Angular displacement, 190
Angular frequency of periodic motion, 433, 450
Angular momentum, 247–251, 252
  conservation of, 247–249
    in diving, 248, 248f
    in figure skating, 247, 248f
    in merry-go-round, 250–251
    in rotating neutron stars, 248–249, 248f
  definition of, 247
Angular speed, 189–192. *See also* Speed(s), angular
Angular velocity, 201, 201f
Antinode, 475, 475f
Arc length, 190, 190f
Arches, 272f, 273–274
Archimedes, 282, 283f
Archimedes's principle, buoyant forces and, 282–288, 283f, 284f, 285f, 309
Area expansion, coefficients of, 330, 344
Aristarchus, 23
Aristotle, 23, 41
Atmospheric pressure, ballpoint pens and, 282
Atomic clock, 2, 2f
Atom(s), 4
  as solid, 267
  structure of, 4
Automobile, collision of. *See under* Collision(s)
Avogadro's number, 334
  ideal gas law and, 335

B

Barometer, 281, 281f
Basal metabolic rate, 414
Beats, 484–486, 485f, 490
Bell, Alexander Graham, 464
Bernoulli, Daniel, 291, 292f
Bernoulli's equation, 291–294, 309
  aircraft wing lift and, 295–296, 296f
  aneurysms and, 295
  atomizers/paint sprayers and, 295, 295f
  fluid flow in pipe and, 293
  home plumbing vent and, 297, 297f
  phenomena explained by, 294–297, 295f, 296f, 297f
  rocket engine exhaust speed and, 297
  sailing into wind and, 296–297, 296f
  speed of fluid leak and, 293
  spinning golf ball and, 294–295, 295f
  vascular flutter and, 295, 295f
Bioluminescence, 141
Black body, 373
Blood, turbulent flow of, 304–305
Blood pressure measurement, 282
Boiling point, 323
Boltzmann's constant, 339, 345, 411
Boyle's law, 335
Brahe, Tycho, 214
Breaking point, 269
British thermal unit (Btu), 353
Btu (British thermal unit), 353
Bulk modulus, 269t, 270–273
  in compression of lead sphere, 272–273
Buoyant forces
  antifreeze testing and, 284–285, 284f, 285f
  Archimedes's principle and, 282–288, 283f, 284f, 285f, 309
  cerebrospinal fluid and, 284
  crown material and, 285–286, 285f
  floating in two fluids and, 287–288, 287f
  floating object and, 284–285
  hot air balloons and, 284f
  raft in river and, 286–287, 286f
  totally submerged object and, 283–284

C

Calorie, definition of, 353, 377
Calorimeters, 357
Calorimetry, 357–359
Capillary action, 301–302, 301f
  rising water and, 302
Carnot, Sadi, 405, 406f
Carnot cycle, 406–407, 406f, 417
Carnot engine, 405–408
  efficiency of, 407
Carnot's theorem, 406
Cartesian coordinate system, 13, 13f, 15
Cavendish, Henry, 207
Cavendish apparatus, 207, 208f
Cavitron ultrasonic surgical aspirator, 460
Celsius temperature scale, 323, 344
  Fahrenheit and Kelvin scales compared with, 325–327, 326f, 344
  uses of, 358

Centrifugal force, 206
Centrifugation, 308–309, 308f
Centripetal acceleration, 198–206, 198f, 217
  automobile on curve and, 203–204, 203f
  banked roadway and, 204–205, 204f
  of car on circular road, 198
  forces causing, 201–206
  at race track, 200
  of roller coaster, 205–206, 205f
Centripetal force, 201, 202
Cesium atomic clock, 2, 2f
Charles' law, 335
Circle, reference, 435, 435f
Circular motion, uniform, simple harmonic motion and, 432–435
  pistons and drive wheels and, 432, 432f
Clausius, Rudolf, 408, 409f
Coefficient(s)
  of expansion
    area, 330, 344
    linear, 328, 328t, 344
    thermostats and, 329–330, 330f
    volume, 328t, 331–332, 344
  of friction, 100–102, 101t, 108
Cohesive force, 300, 300f, 301f
Collision(s)
  automobile, injury in, impulse/momentum and, 165–166
  elastic, 169, 172–175, 172f, 174f, 180
    billiard ball collision as, 174
    in two blocks and spring system, 174–175
  glancing, 175–177, 175f, 176f, 180
  inelastic, 169, 169f, 180
    glaucoma testing as, 169
    perfectly, 169–172, 180
      of automobiles at intersection, 176–177, 176f
      ballistic pendulum as, 171–172, 171f
      SUV versus compact collision as, 170–171
  momentum and, 168–177
Commutative law of vector addition, 54
Compressibility, 270–271
Compression
  due to tensile strength, of steel beam, 270–271
  sound waves and, 458–459, 459f
Conduction, thermal, energy transfer by, 366–368, 366f, 377
  combining insulation layers and, 369–370
  home insulation and, 368–370
  R value of wall and, 369
  through concrete wall, 367–368
Conductivity, thermal, 367, 367t, 377
Conservation
  of angular momentum, 247–249
    in diving, 248, 248f
    in figure skating, 247, 248f
    in merry-go-round, 250–251
    in rotating neutron stars, 248–249, 248f
  of energy
Conservation *(Continued)*
    asteroid impact and, 141
    first law of thermodynamics and, 389
    in general, 141
    systems and, 140–141, 149
  of mechanical energy, 129–132, 130f, 131f, 132f, 244 (*See also* Mechanical energy, conservation of)
    diving and, 130–131
    grasshopper and, 131–132
  of momentum, 166–168, 166f, 180
    archer and, 168, 168f
    in propulsion systems, 167f
    in rocket propulsion, 177–180, 177f
    in squid propulsion, 167f
    in thrown baseball, 167
Conservative forces, 125–126
Contact forces, 82, 82f, 107
Continuity, equation of, 289–291, 309
  for Niagara Falls, 290
Convection
  energy transfer by, 370–372, 370f, 371f, 377
    in automobile engine, 371
    in lake/pond, 371, 371f
    teakettle and, 371f
  forced, 370
  natural, 370
Coordinate systems, 13–14, 13f, 14f, 18
Copernicus, Nicolaus, 24, 214
Crystalline solid, 267, 267f
Cyclic process, internal energy and, 389

D

Dark matter, 268
Deceleration, 31
Decibels, intensity of sound waves in, 463–465, 464t, 489
Degree of freedom, 391
Degrees, conversion to radians, 190
Density, 274–276, 309
  of common substances, 274t
  definition of, 274
Density wave, 443
Depth, pressure variation with, 277–281
  for oil and water layers, 278
  pyramid building and, 281
  for water in pool, pressure on eardrum and, 278–279
Dewar flask, 375, 375f
Dialysis, osmosis and, 306–307
Diffusion, 305
Diffusion coefficient, 305, 306t
Dimensional analysis, 5–7, 17
Dimensions, as algebraic quantities, 5–7
Displacement, 24–25, 45
  angular, 190
  average velocity and, 26
  components of
    for flying superhero, 57–58
    for hiker, 58–59
  definition of, 24
  force exerted by spring and, 134
  in two dimensions, 59
  in vehicular movement, 24f, 25
Dissipative forces, work and, 121–122
Doppler, Christian, 467
Doppler effect, 467–473, 490
  distance between source and observer and, 468
  siren and, 471
  sound from two speakers and, 470
  train whistle and, 470
Dynamics, 23
  rotational, rotational equilibrium and, 226–265

E

Ear, 487–489
Eddy currents, 288, 289f
Efficiency
  of engine, 400–401
  of human body, physical fitness and, 415–416, 415t
  thermal, 400, 417
Elastic collisions, 169, 172–175, 172f, 174f, 180
  billiard ball collision as, 174
Elastic limit, 269, 269f
Elastic modulus, 268, 269t, 309
Elastic potential energy, 135, 427–431, 428f, 429f, 449
  archery and, 429
  guardrail and, 429–430
Elasticity
  of length, 268–269
  of shape, 269–270
  of solid, 267
  volume, 270–273
Electrical transmission, in energy transfer, 140
Electromagnetic radiation, in energy transfer, 140
Emissivity, 372
Energy, 118–159
  conservation of
    asteroid impact and, 141
    first law of thermodynamics and, 389
    in general, 141
    systems and, 140–141, 149
  degradation of, 413
  elastic potential, 135, 427–431, 428f, 429f, 449
  flow of, direction of, 412
  internal, 140, 377
    definition of, 353
    heat and, 352–355
    of isolated system, 389
  kinetic, 123–126, 123f, 124f, 148–149 (*See also* Kinetic energy)
    average molecular
      in helium tank, 343–344
      temperature and, 341, 345
    collisions and, 169
    in elastic collisions, 172
    rotational, 244–247, 252
    temperature as proportional to, 341
  mechanical (*See also* Mechanical energy)
    conservation of, 129–132, 130f, 131f, 132f
      diving and, 130–131
      grasshopper and, 131–132
    total, 129

Energy *(Continued)*
potential, 126–139
elastic, 135
gravitational, 126–134, 149 (*See also* Gravitational potential energy)
spring, 149
stored in spring, 134–139, 135f
accident reconstruction and, 139
block projected up frictionless incline and, 138–139
circus acrobat and, 138
horizontal spring and, 136–137
of sound waves, 463–465
spring potential, 149
storing, methods of, 140
in thermal processes, 352–385
transfer of, 365–375
by convection, 370–372, 370f, 371f
in automobile engine, 371
in lake/pond, 371
teakettle and, 371f
in cooling hot block, 357–358
exercise and, 354–355
mechanisms for, 140
by radiation, 372–375
in human body, 374–375
night vision and, 374
resisting, 375
sign convention for, 359
by thermal conduction, 366–368, 366f
combining insulation layers and, 369–370
home insulation and, 368–370
R value of wall and, 369
through concrete wall, 337–368
weight loss and, 354
in vertical jump, 145–146
work and, 119–122
zero-point, 325
Engine(s)
Carnot, 405–408
efficiency of, 407
efficiency of, 400–401
heat, 399–408, 417
cycle of, analyzing, 401–403
definition of, 399
second law of thermodynamics and, 399–408
steam, 407–408
Entropy, 408–413, 417
change in, 417
of falling boulder, 410–411
of universe, 410
when condensing steam, 410
when melting ice, 410
when melting lead, 409–410
disorder and, 411–413
of universe, 409
direction of time and, 412
increasing, 413
Equation
Bernoulli's, 291–294, 309
fluid flow in pipe and, 293
phenomena explained by, 294–297, 295f, 296f, 297f
rocket engine exhaust speed and, 297
Equation *(Continued)*
speed of fluid leak and, 293
on continuity, 289–291, 309
continuity, for Niagara Falls, 290
harmonic oscillator, 426
Equilibrium
conditions for, torque and, 230–231, 251–252
objects in, 93, 93f, 108, 234–237
ladder against frictionless wall and, 236–237, 236f
walking horizontal beam and, 235–236, 235f
weighted forearm and, 234–235, 234f
rotational
rotational dynamics and, 226–265
torque and, 227–230 (*See also* Torque)
thermal, 322
Escape speed, 212–213, 213t
Expansion, thermal, 327–334. *See also* Thermal expansion

F

Fahrenheit temperature scale, Celsius and Kelvin scales compared with, 325–327, 326f, 344
Faraday, Michael, 82
Fick's law, 305
Fictitious forces, 20
Field forces, gravitational attraction as, 82, 82f
Fitness, physical, efficiency of human body and, 415–416, 415t, 416t
Flagellar movement, 141
Flow rate, 289
Fluid(s). *See also* Liquid(s)
Bernoulli's equation and, 291–294
buoyant forces and, 282–288
dynamics of, applications of, 294–297, 295f, 296f, 297f
flow of, 288–294
ideal, behavior of, 289
in motion, 288–294
solids and, 266–318
surface of, 300–301
surface tension of, 298–301, 298t
transport phenomena for, 305–309
viscosities of, 303t
viscous, flow of, 302–303, 302f
Foci, of elliptical orbit, 214, 214f
Foot-pound, 119
Force(s), 81–83
action, 90, 90f
adhesive, 300, 300f, 301f
average, impulse-momentum theorem and, 162
buoyant (*See* Buoyant forces)
"centrifugal," 206
centripetal, 201, 202
changes in motion and, 84
cohesive, 300, 300f, 301f
concept of, 81–82, 107
conservative, 125–126
constant, work done by, 119
contact, 82, 82f, 107
Force(s) *(Continued)*
definition of, 81–82
dissipative, work and, 121–122
exerted on object, deformation and, 83
fictitious, 206
field, gravitational attraction as, 82, 82f
of friction, 100–106, 101f
applications of, 108
gravitational, 87–90
on billiard balls, 208–209, 208f
potential energy and, 126–134
motion and, 92
in nature as pairs, 90
net, Newton's second law of motion and, 84–90
nonconservative, 125–126
gravity and, 132–134
skiing and, 133–134, 133f
waterslides and, 132–133, 132f
normal, 91–92, 91f
reaction, 90, 90f
restoring, 134
spring, law for, 425, 425f
units of, 85, 85t
varying, work done by, 146–148, 146f
Frame(s) of reference
for displacement, 24
moving, 68
stationary, 68
Free fall, 41
Free-body diagram, 92, 92f
Freely-falling objects, 41–45, 43f, 46
golf ball as, 42
rocket crashing as, 44–45
thrown stone as, 43
Freezing point, 323
Frequency(ies)
angular, of periodic motion, 433, 450
beat, in tuning musical instruments, 484–485
fundamental, 475f, 476–477
of musical instruments, 482
in harmonic series, 475f, 476–477
of motion, 449
observed, Doppler effect and, 467–470, 468f, 469f
of oscillation of bungee cord, 434
pitch and, 487
resonant, 479
of simple harmonic motion, 425
of simple harmonic oscillator, 434
of string on musical instrument, 477–478
of vibration of automobile shock absorbers, 434–435
of wave, 444–446
Friction
coefficients of, 100–102, 101t, 108
forces of, 100–106, 101f
applications of, 108
kinetic
coefficient of, 100–102
force of, 100
motion of car and, 106
oscillations damped by, 441
static
coefficient of, 100
force of, 100
Frictional work, 121–122

Frictionless surface, simple harmonic motion on, 427
Fundamental frequency(ies), 475f, 476–477
of musical instruments, 482
Fusion, latent heat of, 359–360, 360t

G

Galilei, Galileo, 24, 41, 41f
period of pendulum and, 439
Gas(es)
adiabatic index for, 392t, 394
Bernoulli's principle for, 292
dilute, pressure versus temperature for, 324f
greenhouse, global warming and, 375–377
ideal, 344
definition of, 334
macroscopic description of, 334–339
molecular model for pressure of, 340–341, 340f
kinetic theory of, 339–344
gas in insulated cylinder with movable piston and, 343
work done on, 387–388, 387f, 388f
Gas thermometer, constant-volume, 323–325, 323f
Gauge pressure, 281
Gauss, Karl Friedrich, 207
Gaussian system of units, 3
Gay-Lussac's law, 335
Geocentric model of universe, 23, 214
Geosynchronous orbit, 216–217
Glancing collisions, 175–177, 175f, 176f, 180
Global warming, greenhouse gases and, 375–377
Gravitation, universal
constant of, 207, 217
Newton's law of, 88, 206–214, 217–218
Gravitational constant, measurement of, 207–210
Gravitational force, 87–90
astronaut and, 88f
on billiard balls, 208–209, 208f
Gravitational potential energy, 126–134, 126f, 128f, 149
Earth's mass and, 210
high above Earth's surface, 210–212
of near-earth asteroid, 210–211
reference levels for, 127–129
skier and, 128–129, 128f
Gravity
acceleration of, calculation of, 209–210
artificial, 202
center of, 232–234, 232f
of lab partner, 233–234, 233f
of three-particle system, 233, 233f
of wrench, 232, 232f
conservation of mechanical energy and, 120–132
as conservative force, 125
field forces and, 82
freely falling objects and, 41
nonconservative forces and, 132–134
skiing and, 133–134, 133f
Gravity *(Continued)*
waterslides and, 132–133, 132f
specific, 274
Greenhouse effect, 376
Greenhouse gases, global warming and, 375–377

H

Harmonic motion, simple, 425, 449
on frictionless surface, 427
pendulum motion and, 438, 439f
position of objects with, as function of time, 435–437
uniform circular motion and, 432–435
pistons and drive wheels and, 432, 432f
Harmonic oscillator equation, 426
Harmonic series, 475f, 476
Harmonics
of bugle, 482
of musical instruments, 487f
of pipe, 480–481, 481f, 482–483
of stretched string, 478–479
of waveforms, 486, 486f
Hearing, threshold of, 463
Heat
definition of, 322, 353
in energy transfer, 140, 377
internal energy and, 352–355
latent, 359–365, 377
of fusion, 359–360, 360t
of vaporization, 359–360, 360t
mechanical equivalent of, 353, 377
molar specific, 391
specific, 355–357, 377
of sublimation, 360
units of, 353–355
Heat engine(s), 399–408, 417
cycle of, analyzing, 401–403
definition of, 399
second law of thermodynamics and, 399–408
Heat pumps, 403–404, 403f
Heliocentric model of universe, 24, 214
Hertz, 433
Home insulation, 368–370
Hooke, Sir Robert, 134, 425
Hooke's law, 134, 424–427, 449
Horsepower, 142

I

Ice point, 323
Ideal absorber, 373
Ideal gas law, 335
Avogadro's number and, 335
in gas quantitation, 336
heating bottle of air and, 336–337
helium cylinder and, 343–344
submerging balloon and, 337–338
Ideal gases. *See* Gas(es), ideal
Ideal reflector, 373
Impulse
definition of, 161
momentum and, 160–166, 180
in automobile collisions, injury and, 165–166
Impulse-momentum theorem, 161, 180
apple hit by bullet and, 162f
automobile crash test and, 163–164
boxing gloves and, 162
golf ball struck by club and, 163, 163f
Inertia
definition of, 83, 107
mass and, 83–84, 84f
moment of, 237, 238, 252
of baton, 239–240, 239f, 240f
for composite object, 239–240, 239f, 240f
for extended objects, calculation of, 240–244, 241f
for various rigid objects of uniform composition, 241t
Insulation, home, 368–370
Intensity, of sound waves, 463–465, 489
in decibels, 463–465, 464t, 489
from grinding machine, 464–465
Interference
beats and, 484–486, 485f
constructive, of sound waves, 473–474
destructive, of sound waves, 474
of sound waves, 473–474, 490
stereo speaker connections and, 474, 474f
of waves, 447–448, 448f, 450
constructive, 448, 448f, 450, 490
destructive, 448, 448f, 450, 490
Internal energy, 140
definition of, 353
heat and, 352–355, 377
of isolated system, 389
Inverse-square law, 207
Irreversible process, 405
Isobaric process, 388, 391–393, 416
expanding gas and, 392–393
Isochoric process, 396
Isolated systems, internal energy of, 389
Isothermal process, 396–397, 397f
isothermally expanding balloon and, 397
Isovolumetric process, 396

J

Joule, 119, 353
Joule, James Prescott, 352, 353f

K

Kelvin, Lord, 405f
Kelvin temperature scale, 324–325, 344
Celsius and Fahrenheit scales compared with, 325–327, 326f, 344
uses of, 358
Kepler, Johannes, 214
Kepler's laws, 214–217, 218
first, 214, 214f
second, 215, 215f
third, 215
Kidneys, osmosis and, 306–307
Kilogram, definition of, 2
Kinematics, 23
rotational, slowing propellers and, 193
rotational and translational, 193

Kinetic energy, 123–126, 123f, 124f, 148–149
average molecular
in helium tank, 343–344
temperature and, 341, 345
colliding car and, 124–125, 124f
collisions and, 169
in elastic collisions, 172
moving hammer and, 123–124, 123f
rotational, 244–247, 252
in spinning stool demonstration, 249–250, 249f
skidding car and, 124
temperature as proportional to, 341
Kinetic friction
coefficient of, 100–102, 108
force of, 100
Kinetic theory of gases, 339–344
gas in insulated cylinder with movable piston and, 343

L

Laminar flow, 288
Latent heat, 359–365, 377
of fusion, 359–360, 360t
of vaporization, 359–360, 360t
Laws, of motion, 81–117. *See also* Newton's laws of motion
concept of force and, 82, 82f
Length
approximate values for, 3, 3t
standards of, 2, 17
Lever arm, 228f, 229
Linear expansion
average coefficients of, 344
coefficient of, 328
coefficients of, average, 328t
Linear momentum, 160, 180
Linear speed, of compact discs, angular speed compared with, 197–198
Liquid(s). *See also* Fluid(s)
speed of sound in, 461
thermal expansion of, 327–334
Longitudinal wave, 442–443, 443f, 450
in solid rod, speed of, 461

M

Mach number, 472
Machine(s)
human body as, efficiency of, physical fitness and, 415–416, 415t, 416t
perpetual motion, 409
Manometer, open-tube, 281, 281f
Mass
approximate values for, 3, 3t
center of, 145, 145f, 232
definition of, 83–84, 107
of Earth, 210
inertia and, 83–84, 84f
molar, 334
standards of, 2, 17
units of, 85, 85t
SI, 2, 2f
Matter
building blocks of, 4–5, 17
states of, 266–268
Maxwell velocity distribution, 342
Measurement, uncertainty in, 7–9, 18
Mechanical energy
conservation of, 129–132, 130f, 131f, 132f, 244
diving and, 130–131
grasshopper and, 131–132
total, 129
Mechanical waves, in energy transfer, 140
Membrane, selectively permeable, 306
Mercury thermometer, 323, 323f
Metabolism, human
first law of thermodynamics and, 413–416
rate of
activity and, 414–415
basal, 414
measuring, 414
weight and, 414–415
Meter, definition of, 2
Millikan, Robert, 307
Molar mass, 334
Molar specific heat, at constant volume, 391
Molecule(s)
average kinetic energy per, 341, 345
in helium tank, 34
root-mean-square speed of, 342, 345
Moles, 334–335
Moment of inertia, of baton, 239–240, 239f, 240f
Momentum, 160–188
angular, 247–251, 252
conservation of, 247–249
in diving, 248, 248f
in figure skating, 247, 248f
in merry-go-round, 250–251
in rotating neutron stars, 248–249, 248f
definition of, 247
conservation of, 166–168, 166f, 180
archer and, 168, 168f
in propulsion systems, 167f
in rocket propulsion, 177–180, 177f
in squid propulsion, 167f
in thrown baseball, 167
in elastic collisions, 172, 173
impulse and, 160–166, 180
in automobile collisions, injury and, 165–166
linear, 160, 180
Moon, escape speed for, 213t
Motion(s)
of car, friction and, 106
circular, uniform, simple harmonic motion and, 432–435
pistons and drive wheels and, 432, 432f
fluids in, 288–294
forces and, 92
frequency of, 449
laws of, 81–117 (*See also* Newton's laws of motion)
concept of force and, 82, 82f
linear, 23–52 (*See also* Motion, one-dimensional)
Motion(s) *(Continued)*
one-dimensional, 23–52
acceleration in, 30–33
with constant acceleration, 35–41, 46
applications of, 37–41
for car chase, 37–38
equations for, 36t
for high-speed train, 38–40
for landing jetliner, 40–41
for race car, 37
displacement in, 24–25
of freely falling objects, 41–45, 43f
rotational motion compared with, 193
velocity in, 25–30
average, 26–27
instantaneous, 27–30
of pendulum, 438–440, 438f, 439f
period of, 449
planetary, Kepler's laws of, 214–217, 214f, 215f
projectile, 60–68, 61f
of dropped rocket, 67–68
of emergency rations dropped from plane, 64
of long jumper, 65–66
of thrown stone, 66–67
of water, 290–291
rotational, 189–225
angular acceleration and, 189–192
angular speed and, 189–192
of compact disc, 190, 190f
under constant angular acceleration, 193–194
linear motion compared with, 193
zero torque and, 234
simple harmonic, 425, 449
on frictionless surface, 427
pendulum motion and, 438, 439f
position of object with, as function of time, 435–437
uniform circular motion and, pistons and drive wheels and, 432, 432f
uniform circular motion compared with, 432–435
through viscous medium, 307–308, 307f
two-dimensional, vectors and, 53–80 (*See also* Vector(s))
wave, 442
between wave source and observer, Doppler effect and, 468–470
Motion diagrams, 33–34
Movement, rotational, Kepler's laws and, 214–217

N

Neutron, function of, 4
Neutron star, conservation of angular momentum and, 248–249
Newton, Isaac, 81–100, 85f, 107
Newton (unit), 85
Newton-meter, 119
Newton's law of universal gravitation, 88, 206–214, 214, 217–218
Kepler's first law and, 214

Newton's laws of motion, 81–100, 107–108, 214
applications of, 92f, 108
first, 83–84, 84f, 107
seat belts and, 84, 84f
rotational analog of, 238
second, 84–90, 107–108
accelerating objects and, 96–100
airboat and, 86
applications of, 92–100, 92f
Atwood's machine and, 99–100, 99f
balls connected and, 104–105, 104f
block on ramp and, 102–103, 102f
centripetal acceleration and, 201
crate pulled on frictionless, horizontal surface and, 92–93, 92f
free-body diagram and, 92, 92f
horses pulling barge and, 86–87
momentum and, 161
moving crate and, 96, 96f
rope and, 92
"run-away" car and, 97–98, 97f
sled on frictionless hill and, 95–96, 95f
sliding hockey puck and, 103–104, 103f
traffic light at rest and, 94, 94f
two blocks and cord and, 105–106, 105f
weighing fish in elevator and, 98–99, 98f
third, 90–92, 108
applications of, 91–92, 91f
colliding vehicles and, 90–9192
hammer on nail and, 90, 90f
helicopter flight and, 91
universal gravitation and, 207
Nodes, 475, 475f
Nonconservative forces, 125–126
gravity and, 132–134
skiing and, 133–134, 133f
waterslides and, 132–133, 132f
Normal force, 91–92, 91f
Nucleus, structure of, 4

O

O'Neill, Gerard, 202
Orbit, geosynchronous, 216–217
Order-of-magnitude calculations, 11–13, 18
for gasoline use in U.S., 11–12
for number of galaxies in universe, 12–13
for stacking one-dollar bills to moon, 12
Oscillation(s)
critically damped, 441, 441f
damped, 441, 441f
shock absorbers and, 441, 441f
in harbor, 482
of mass on spring, period of, 434
overdamped, 441, 441f
of standing waves, 475, 475f
underdamped, 441, 441f
OSHA noise level regulations, 465
Osmosis
cell size and, 306–307
definition of, 306
Oxygen consumption
maximum rate of, physical fitness and, 415t
metabolic rates for activities and, 414t

P

Pain, threshold of, 463
Parabola, 61, 61f, 62f
Pascal, 268, 275
Pascal, Blaise, 279
Pascal's principle, 279, 309
car lift and, 280
hydraulic lift/press and, 279, 279f, 280
Pendulum
motion of, 438–440, 438f, 439f, 440f, 450
physical, 440
Period
of motion, 449
of planet, 215
of simple harmonic motion, 425
of object-spring system, 432–435, 450
of simple pendulum, 438–439
Permeable membrane, selectively, 306
Perpetual motion machines, 409
Phase change, latent heat and, 359–365
in boiling liquid helium, 360–362
in comet hitting Earth, 364–365
in melting ice, 362–364
Physical fitness, efficiency of human body and, 415–416, 415t, 416t
Physical pendulum, 440
Physics, thermal, 321–351
Piezoelectric effect, 460, 460f
Plane polar coordinate system, 14, 14f, 15
Plane waves, 466, 466f
Planets
data on, 216t
escape speeds for, 213t
motions of, Kepler's laws on, 214–217, 214f, 215f
period of, 215
Plasma, 267–268
Point source, 466
intensity variations of, 467
Poise, 303
Poiseuille, J. L., 303
Poiseuille's law, 303–304
blood transfusion and, 304
Polar coordinate system, 14, 14f, 15
Position
as function of time, 435–437, 450
velocity as function of, 430–431, 450
Position vs. time graph, 27
Potential energy, 126–139
elastic, 135, 427–431, 428f, 429f, 449
gravitational, 126–134, 149 (*See also* Gravitational potential energy)
stored in spring, 134–139, 135f
accident reconstruction and, 139
block projected up frictionless incline and, 138–139
circus acrobat and, 138
horizontal spring and, 136–137
Pound, 85
Power, 142–146
average, 142, 149
definition of, 142
delivered by elevator motor, 143, 143f
speedboat, 144–145
units for, 141
in vertical jump, 145–146
whale acceleration and, 143–144
Pressure, 274–281, 274f, 309, 345
absolute, 281
atmospheric, ballpoint pens and, 282
average, 275
bed of nails and, 276, 276f
snowshoes and, 275, 275f
waterbed and, 276
gauge, 281
of ideal gas, molecular model for, 340–341, 340f
measurement of, 281–282
variation of, with depth, 277–281
for oil and water layers, 278
pyramid building and, 281
for water in pool, pressure on eardrum and, 278–279
Pressure wave, 443
Problem-solving strategy, 16–17
accelerated motion, 36
for applying conservation of mechanical energy, 130
for calorimetry with phase change, 362
for energy methods and rotation, 244
example of, 17
for forces causing centripetal acceleration, 203
for Newton's second law, 93
for objects in equilibrium, 234
for one-dimensional collisions, 173
for projectile motion, 63
for relative velocity, 69
steps in, 16
for two-dimensional collisions, 176
Projectile motion, 60–68, 61f
of dropped rocket, 67–68
of emergency rations dropped from plane, 64
of long jumper, 65–66
of thrown stone, 66–67
of water, 290–291
Projection angle, 61
Propulsion, rocket, conservation of momentum in, 177–180, 177f
Proton
composition of, 4, 5f
function of, 4
Ptolemy, Claudius, 23, 214
Pulse, of wave traveling on string, speed of, 447, 447f
PV diagram, 388–389, 388f
of Carnot cycle, 407f
of cyclic process, 400, 400f
Pythagorean theorem, 14, 18, 56
total acceleration and, 199

Q

Quark, 4–5, 5f

R

Radians, 190
Radiation
  electromagnetic, in energy transfer, 140
  energy transfer by, 372–375, 377
    in human body, 374–375
    night vision and, 374
Radiation thermometer, 373–374, 373f
Rarefaction, sound waves and, 459, 459f
Rays, 466, 466f
Reaction force, 90
Recoil, 167, 168
Rectangular coordinate system, 13, 15
Reference circle, 435, 435f
Reflection, of waves, 448–449, 449f, 450
Reflector, ideal, 373
Refrigerators, 403–405
Relative velocity, 68–71
  of baseball thrown from train, 69–70
  of boat crossing river, 70
Resonance, 479, 479f, 490
  forced vibrations and, 479–480
  structural, collapses from, 480, 480f
  in tube of variable length, 483–484, 483f
Restoring force, 134
Resultant vector, 54, 55f, 72
  for car trip, 55
Reversible process, 405, 405f
Reynolds number, 304–305
  turbulent flow of blood and, 304–305
Right-hand rule, 201, 201f, 229, 229f
Rocket, escape speed of, 213–214
Rocket propulsion, conservation of momentum in, 177–180, 177f
Root-mean-square speed of molecules, 342, 345
Rotating object, torque on, 237–239
Rotation, axis of, position of, 229
Rotational dynamics, rotational equilibrium and, 226–265
Rotational equilibrium
  rotational dynamics and, 226–265
  torque and, 227–230 (*See also* Torque)
Rotational kinematics, 193
  slowing propellers and, 193
Rotational kinetic energy, 244–247, 252
  of ball rolling down incline, 244, 245, 245f
  of blocks and pulley, 246–247
  in spinning stool demonstration, 249–250, 249f
Rotational motion, 189–225. *See also* Motion(s), rotational
Russell, John Scott, 443

S

Scalar(s)
  addition of, vector addition versus, 54
  multiplication or division of vector by, 55
Scalar quantity, 25, 45
  instantaneous speed as, 29
  work as, 119
Scott, David, 41
Second, definition of, 2
Sedimentation rate, 308–309
Seiche, 482
Shape, elasticity of, 269–270
Shear modulus, 269–270, 269t
Shear strain, 270
Shear stress, 269–270, 269f
  of explosive bolts, 271–272
Shock waves, 472–473, 472f, 473f
SI system, 1–2
SI unit(s), 3, 5
  of angular displacement, 190
  of average acceleration, 30, 60
  of average angular speed, 190
  of average pressure, 275
  of average speed, 25
  of average velocity, 26, 59
  conversion of, to U.S. customary units, 9–10
  of density, 274
  of displacement, 59
  of entropy, 408
  of force, 85, 85t
  of gravitational potential energy, 210
  of heat, 353
  of impulse, 161
  of instantaneous acceleration, 31, 60
  of instantaneous angular speed, 191
  of instantaneous velocity, 28, 60
  of linear momentum, 160
  of mass, 2, 2f
  of moment of inertia, 238
  of power, 141
  of specific heat, 355
  of stress, 268
  of surface tension, 298
  of temperature, 325
  of torque, 227
  of viscosity, 303
  of work, 119
Significant figures, 7, 18
  calculators and, 8–9
  carpet installation and, 8
Slopes of graphs, 27
Solid(s)
  amorphous, 267, 267f
  crystalline, 267, 267f
  deformation of, 268–274
  elastic, stress-versus-strain curve for, 269f
  elasticity of, 267
  fluids and, 266–318
  model of, 267, 267f
  strain of, 268
  stress of, 268
  thermal expansion of, 327–334
Soliton, 443
Sonic booms, 472
Sound, 458–496
  Doppler effect in, 467–473
  quality of, 486–487
  speed of, 461–462
    in various media, 461t
  during storm, 462
Sound waves. *See* Wave(s), sound
Specific gravity, 274
Specific heat, 355–357, 377
Speed(s)
  angular, 189–192
Speed(s) *(Continued)*
    average, 190–191, 217
    of compact disc, linear speed compared with, 197–198
    of helicopter, 191
    instantaneous, 191
    tangential speed compared with, 195
  average, 25, 46
    average velocity versus, 27
  escape, 212–213, 213t
  instantaneous, 29, 46
  linear, of compact discs, angular speed compared with, 197–198
  of longitudinal wave in solid rod, 461
  root-mean-square, of molecules, 342, 345
  of sound, 461–462
    in various media, 461t
  supersonic, shock waves from, 472, 473f
  tangential, 195
    of compact discs, 196–197
    of compact discs versus phonograph records, 197
    satellite launch area and, 196
  terminal, 308
    of object falling through air, 107
  and velocity, difference between, 25
  wave, 444
  of waves on strings, 446–447
Sphygmomanometer, 282, 282f
Spring(s)
  elastic potential energy in, 427–431, 428f
  force law for, 425, 425f
  potential energy stored in, 134–139, 135f
    accident reconstruction and, 139
    block projected up frictionless incline and, 138–139
    circus acrobat and, 138
    horizontal spring and, 136–137
  work required to stretch, 147–148, 147f
Spring constant, 134, 425
  measuring, 425–426, 426f
Spring potential energy, 149
Standing waves, 475–479, 475f
Star, neutron, conservation of angular momentum and, 248–249
Static friction
  coefficient of, 100, 108
  force of, 100
Steam engine, 407–408
Steam point, 323
Stefan's law, 372
Stokes, George, 307
Stoke's law, 307
Strain
  shear, 270
  of solid, 268, 309
  tensile, 268
Streamline flow, 288, 288f, 289f
Strength of materials, ultimate, 269, 271t, 273–274
Stress
  shear, 269–270, 269f
    of explosive bolts, 271–272
  of solid, 268, 309
  tensile, 268
  volume, 270–271

Sublimation, heat of, 360
Superposition principle, 447–448, 450, 475
Supersonic speeds, shock waves from, 472, 473f
Surface tension, 298–301, 298f, 299f
  insect walking on water and, 299–300, 299f
  of various liquids, 298t
System(s)
  definition of, 126
  energy conservation and, 140–141, 149
  isolated, momentum conservation in, 166–167
  nonisolated, and environment, energy transfer between, 140

T

Tangential acceleration, 195
Tangential speed, 195
  of compact discs, 196–197
  of compact discs versus phonograph records, 197
  satellite launch area and, 196
Temperature
  absolute, 324, 325f
  molecular interpretation of, 341–344
Temperature scales
  Celsius, 323
  comparison of, 325–327, 326f
  Kelvin, 324–325
Tensile strain, 268
Tensile stress, 268
Tension, 92
  surface, 298–301, 298f, 299f
    of various liquids, 298t
Terminal speed, 308
  of nonspherical object, 308
  of object falling through air, 107
Thermal conduction, energy transfer by, 366–368, 366f, 377
  combining insulation layers and, 369–370
  home insulation and, 368–370
  R value of wall and, 369
  through concrete wall, 367–368
Thermal conductivity, 367, 367t, 377
Thermal contact, 322
Thermal efficiency, 400, 417
Thermal equilibrium, 322
Thermal expansion, 327–334
  hole in metal and, 330
  metal rings and rods and, 331
  Pyrex glass and, 328
  railroad track and, 329, 329f
  rising sea levels and, 332–333
Thermal physics, 321–351
Thermal processes, energy in, 352–385
Thermals, 355–356
Thermodynamic process, work in, 386–389, 416
Thermodynamics
  laws of, 386–423
    first, 389–399, 416
      adiabatic processes and, 394–395, 399t
      applications of, 391–399
Thermodynamics *(Continued)*
      equation for, 389
      general case and, 398–399, 399t
      heating gas and, 390
      human metabolism and, 413–416
      isobaric processes and, 391–393, 399t
      isothermal processes and, 396–397, 399t
      isovolumetric processes and, 396, 399t
    second, 405, 417
      entropy and, 408–413
      heat engines and, 399–408
      probabilities and, 412, 413f
  zeroth law of, 322, 322f, 344
Thermogram, 373, 373f, 374f
Thermography, 373
Thermometer(s)
  definition of, 322
  gas, constant-volume, 323–325, 323f
  mercury, 323, 323f
  radiation, 373–374, 373f
  temperature scales and, 323–327
Thermos bottle, 375, 375f
Thrust, rocket, 178
Time
  acceleration as function of, 435–437
  intervals of, approximate values for, 3, 3t
  position as function of, 435–437
  standards of, 2, 17
  velocity as function of, 435–437, 450
Tonometer, 169
Torque, 227–230, 251
  angular acceleration and, 237–244
    bicycle gear system and, 239, 239f
  angular momentum and, 247–251
  axis of rotation and, 229
  definition of, 227
  door hinge and, 227, 227f
  net, 227
    angular acceleration and, 237–244
  of revolving door, 228, 228f
  on rotating object, 237–239
  of seesaw, 230–231
  of swinging door, 229–230
  two conditions for equilibrium and, 230–231
  zero, 234
Torricelli's law, 293
Transport phenomena, 305–309
  diffusion as, 305, 306t
  osmosis as, 306–307
Transverse wave, 442, 443f, 450
Traveling wave, 442, 442f, 443f, 444f, 445
  reflection of, 448–449, 449f
Triangle method of vector addition, 54, 54f, 72
Trigonometry, 14–16, 18
  in building height measurement, 15–16
Turbulent flow, 288, 289f
  of blood, 304–305

U

Ultimate strength, 269
Ultrasonic ranging unit, 460
Ultrasonic surgical aspirator, 460
Ultrasonic waves, 459, 489
Ultrasound, 459–460, 460f
Units
  of acceleration, 85t
  conversion of, 9–11, 18
  of force, 85, 85t
  of mass, 85, 85t
  SI, 3, 4 (*See also* SI unit(s))
  systems of, 3
Universal gas constant, 335
Universal gravitation
  constant of, 207, 217
  Newton's law of, 88, 206–214, 217–218
Universe, entropy of, 409
  increasing, 413
U.S. customary system of units, 3
  conversion of, to SI units, 9–10
  power units in, 142

V

Vaporization, latent heat of, 359–360, 360t
Vascular flutter, 295
Vector(s)
  addition of, 54, 54f, 55f, 72
    algebraic, 58
    commutative law of, 54, 54f
    triangle method of, 54, 54f, 72
  angular quantities as, 201, 201f
  components of, 56–59, 56f, 72
  definition of, 25
  displacement, 25
  division of, by scalar, 55
  multiplication of, by scalar, 55
  negative of, 54, 72
  properties of, 53–55
  resultant, 54, 55f, 72
    for car trip, 55
  subtraction of, 55, 55f
  two, equality of, 54
  two-dimensional motion and, 53–80
Vector quantity, definition of, 25, 45
Velocity, 25–30. *See also* Speed(s)
  angular, 201, 201f
  average, 26–27, 46
    average speed versus, 27
    graphical interpretation of, 27, 28f
    of slow train trip, 29–30
    in two dimensions, 59, 72
  escape, 212
  as function of position, 430–431, 449
  as function of time, 435–437
  instantaneous, 27–30, 46
    graphical interpretation of, 29f
    in slow train trip, 29–30
    in two dimensions, 60, 72
  relative, 68–71
  and speed, difference between, 25
  in two dimensions, 59–60
Velocity vs. time graph, 31–32, 31f, 32f
Venturi tube, 292, 292f
Vibration(s), 424–457
  forced, resonance and, 479–480
  in standing-wave pattern, frequency of, 475, 475f

Viscosity
coefficient of, 303
of fluid, 288–289
of liquids, damped motion and, 441
of various fluids, 303t
Viscous fluid flow, 302–303, 302f
Viscous medium, motion through, 307–308, 307f
Volume elasticity, 270–273
Volume expansion, coefficients of, 328t, 331–332, 344
Volume stress, 270–271

W

Water, thermal expansion behavior of, 332–334
Watt, James, 141
Watts, 141, 142
Wave(s), 424–457, 441–443
amplitude of, 444
definition of, 442
density, 443
frequency of, 444–446
interference of, 447–448, 448f
longitudinal, 442–443, 443f, 450
in solid rod, speed of, 461
mechanical, in energy transfer, 140
picture of, 443, 443f
plane, 466, 466f
pressure, 443
reflection of, 448–449, 449f, 450
shock, 472–473, 472f, 473f
sound
audible, 459, 489
categories of, 459
characteristics of, 459–460
Wave(s) *(Continued)*
energy and intensity of, 463–465
infrasonic, 459, 489
intensity of
in decibels, 463–465, 464t
from grinding machine, 464–465
interference of, 473–474
stereo speaker connections and, 474, 474f
plane, 466
producing, 458–459, 459f
from singer, glass shattering by, 480, 480f
in solid bar, 462
speed of, 461–462
spherical, 465–467, 489
ultrasonic, 459, 489
standing, 475–479, 475f, 476f, 490
in air columns, 480–484, 481f
on strings, speed of, 446–447
transverse, 442, 443f, 450
traveling, 442, 442f, 443f, 444f, 445
reflection of, 448–449, 449f
types of, 442–443
wavelength of, 444, 444f
Wave front(s), 466
additional, Doppler effect and, 468, 468f
Wave motion, 442
Waveforms, harmonics of, 486, 486f
Wavelength, 444, 444f
shortened, in Doppler effect, 468
Weight
gravitational force and, 88–90
metabolic rate and, 414–415
Work
definition of, 119, 119f
dissipative forces and, 121–122
Work *(Continued)*
done by constant force, 119
done by varying force, 146–148, 146f
done on gas, 387–388, 387f, 388f
energy and, 119–122
in energy transfer, 140
frictional, 121–122
horizontal movement of bucket and, 120f
lifting and carrying box and, 120, 120f
lifting and lowering box and, 120f
negative, 120
positive, 120
required to stretch spring, 147–148, 147f
as scalar quantity, 119
SI units for, 119
sledding through Yukon and, 120–121
in thermodynamic process, 386–389, 416
Work-energy theorem, 123, 126, 148–149
including rotational energy, 244

Y

Young's modulus, 268–269, 269t, 309

Z

Zero, absolute, 324
Zero torque, rotational motion and, 234
Zero-point energy, 325
Zeroth law of thermodynamics, 322, 322f, 344

PHYSICAL CONSTANTS

Quantity	Symbol	Value	SI unit
Speed of light in vacuum	c	3.00×10^{8}	m/s
Permittivity of free space	ϵ_0	8.85×10^{-12}	$C^2/N \cdot m^2$
Coulomb constant, $1/4\pi\epsilon_0$	k_e	8.99×10^{9}	$N \cdot m^2/C^2$
Permeability of free space	μ_0	1.26×10^{-6} ($4\pi \times 10^{-7}$ exactly)	$T \cdot m/A$
Elementary charge	e	1.60×10^{-19}	C
Planck's constant	h $\hbar = h/2\pi$	6.63×10^{-34} 1.05×10^{-34}	$J \cdot s$ $J \cdot s$
Electron mass	m_e	9.11×10^{-31} 5.49×10^{-4}	kg u
Proton mass	m_p	$1.672\,65 \times 10^{-27}$ 1.007 276	kg u
Neutron mass	m_n	$1.674\,95 \times 10^{-27}$ 1.008 665	kg u
Avogadro's number	N_A	6.02×10^{23}	mol^{-1}
Universal gas constant	R	8.31	$J/mol \cdot K$
Boltzmann's constant	k_B	1.38×10^{-23}	J/K
Stefan-Boltzmann constant	σ	5.67×10^{-8}	$W/m^2 \cdot K^4$
Molar volume of ideal gas at STP	V	22.4 2.24×10^{-2}	L/mol m^3/mol
Rydberg constant	R_H	1.10×10^{7}	m^{-1}
Bohr radius	a_0	5.29×10^{-11}	m
Electron Compton wavelength	$h/m_e c$	2.43×10^{-12}	m
Gravitational constant	G	6.67×10^{-11}	$N \cdot m^2/kg^2$
Standard free-fall acceleration	g	9.80	m/s^2
Radius of Earth (at equator)	R_E	6.38×10^{6}	m
Mass of Earth	M_E	5.98×10^{24}	kg
Radius of Moon	R_M	1.74×10^{6}	m
Mass of Moon	M_M	7.36×10^{22}	kg

The values presented in this table are those used in computations in the text. Generally, the physical constants are known to much better precision.